Airport Systems: Planning, Design, and Management

Richard de Neufville
Amedeo R. Odoni

McGraw-Hill

New York   Chicago   San Francisco   Lisbon   London   Madrid
Mexico City   Milan   New Delhi   San Juan   Seoul
Singapore   Sydney   Toronto

The *McGraw·Hill* Companies

Library of Congress Cataloging-in-Publication Data

De Neufville, Richard, 1939-
 Airport systems planning design and management / Richard
de Neufville, Amedeo R. Odoni.
     p.   cm.
 Includes bibliographical references and index.
 ISBN 0-07-138477-4
 1. Airports—Planning.   2. Airports—Design and construction.
3. Airports—Management.   I. Odoni, Amedeo R.   II. Title.
TL725.3.P5 D4623   2003
387.7'36—dc21   2002028399

6 7 8 9 BKM BKM   0 9 8

387.736
DEN
155359

ISBN-13: 978-0-07-138477-3

ISBN-10: 0-07-138477-4

The sponsoring editor for this book was Shelley Ingram Carr, the editing supervisor was Daina Penikas, and the production supervisor was Pamela A. Pelton. It was set in the Gen1AV1 design in Garamond by Wayne Palmer of McGraw-Hill Professional's composition unit, Hightstown, N.J.

McGraw-Hill books are available at special quantity discounts to use as premiums and sales promotions, or for use in corporate training programs. For more information, please write to the Director of Special Sales, McGraw-Hill Professional, Two Penn Plaza, New York, NY 10121-2298. Or contact your local bookstore.

 This book is printed on recycled, acid-free paper containing a minimum of 50% recycled, de-inked fiber.

*To Ginger and Eleni
in appreciation for their
support and understanding*

Contents

Preface *xix*
Acknowledgments *xxiii*
User's Guide *xxv*

Part One   Introduction *1*

1 The future of the airport and airline industry *3*

1-1 The airport industry at the end of the twentieth century *4*

1-2 Long-term growth *9*

1-3 Commercialization *14*

1-4 Globalization *18*

1-5 Electronic commerce *21*

Electronic ticketing *21*

Electronic commerce *22*

1-6 Implications for airports systems planning and design *25*

Exercises *26*

References *26*

2 International differences *29*

2-1 Introduction *30*

2-2 Some physical differences *31*

Check-in facilities *32*

Aircraft contact stands *33*

2-3 Some useful distinctions *38*

National differences in diversity of decision
making *39*

National differences in performance criteria *43*

2-4 Implications for practice *47*

General implications *47*

Specific implications *49*

Exercises *54*

References *55*

Part Two   System planning   *57*

3 Dynamic strategic planning   *59*

3-1 Forms of planning *60*

Plans *60*

Master plans *62*

Strategic plans *64*

3-2 Airport systems planning *65*

Airport systems *65*

Planning airport systems *67*

3-3 The forecast is "always wrong" *70*

Cost estimation *72*

Aggregate forecasts *74*

Composition of forecasts *76*

Effect of longer planning periods *77*

Effect of economic deregulation *79*

3-4 Implications for planning *80*

3-5 Dynamic strategic planning concepts *81*

3-6 Dynamic strategic planning process and methods *83*

Exercises *88*

References *89*

4 Privatization and deregulation   *93*

4-1 The airport and airline industry before privatization
and deregulation *94*

Airports *94*

Airlines *97*

4-2 Motivations for privatization and deregulation *98*
U.S. airline deregulation *98*
Worldwide airline privatization *99*
Privatization of airports *100*
4-3 The concept of privatization *100*
Rights to residual income *101*
Management control *103*
4-4 Guidelines for airport privatization *106*
4-5 Airline deregulation *110*
4-6 Implications of airline deregulation for airports *111*
Increased volatility *112*
Transfer hubs *118*
Competition between airports *124*
Exercises *125*
References *126*

5 Multi-airport systems *129*

5-1 Introduction *129*
5-2 Basic concepts and issues *132*
Definitions *132*
Prevalence *133*
Unequal size *135*
5-3 Difficulties *138*
Insufficient traffic at new airport *139*
Difficulty in closing old airport *141*
Insufficient traffic overall *142*
Impractical to allocate traffic *144*
Volatility of traffic at secondary airport *146*
Overall perspective *147*
5-4 Market dynamics *149*
Concentration due to sales opportunities *149*
Airlines concentrate on routes *150*
Airlines concentrate at primary airports *152*
Factors favoring multi-airport systems *154*
5-5 Planning and developing multi-airport systems *157*
Landbanking *158*
Incremental development *160*

Flexible facilities   *161*
Careful marketing   *162*
Exercises   *163*
References   *164*

6 Environmental impacts   *167*

6-1 Introduction   *167*
6-2 Fundamentals of noise measurement   *170*
   Measuring aircraft noise   *171*
   Certification of aircraft for noise   *182*
   Practical implications   *185*
6-3 Mitigating airport noise   *186*
   Noise monitoring systems   *187*
   Community relations and public participation
      programs   *190*
   Land-use policies   *191*
   Airport design interventions   *194*
   Surface operations and flight operations   *195*
   Interventions outside airport properties near existing
      sites   *198*
   Access restrictions   *199*
   Economic incentives   *201*
6-4 Air quality and mitigation of air pollution   *202*
6-5 Water quality control   *206*
   Deicing fluids (ADF)   *206*
   Fuel leaks and spills   *208*
   Storm water runoff   *208*
6-6 Control of highway and road access traffic   *209*
6-7 Wildlife management   *210*
Exercises   *211*
References   *212*

7 Organization and financing   *215*

7-1 Introduction   *216*
7-2 Ownership and management of airports   *217*
7-3 Organizational structures   *225*
7-4 Regulatory constraints on airport user charges   *233*
   Price caps   *235*

"Single Till" versus "Dual Till"   *237*
Residual versus compensatory   *241*
7-5 Financing capital investments   *243*
Outright government grants   *243*
Special-purpose user taxes   *244*
Low-cost loans from international or national development banks   *244*
Operating surpluses   *244*
Loans from commercial banks   *245*
General-obligation bonds   *245*
Revenue bonds   *245*
Private financing against specified rights to airport revenues   *246*
Exercises   *249*
References   *249*

8 User charges   *251*

8-1 Introduction   *252*
8-2 Cost and revenue centers   *253*
8-3 Guidelines and background for the setting of user charges   *258*
8-4 The various types of airport user charges   *260*
Landing fee   *261*
Terminal area air navigation fee   *262*
Aircraft parking and hangar charges   *262*
Airport noise charge   *263*
Passenger service charge   *264*
Cargo service charge   *265*
Security charge   *265*
Ground handling charges   *266*
En route air navigation fee   *266*
8-5 Nonaeronautical charges   *268*
Concession fees for aviation fuel and oil   *268*
Concession fees for commercial activities   *269*
Revenues from car parking and car rentals   *269*
Rental of airport land, space in buildings, and assorted equipment   *269*
Fees charged for airport tours, admissions, etc.   *270*

Fees derived from provision of engineering services
and reimbursable utilities by the airport operator to
airport users   *270*

Non-airport revenues   *270*

8-6 Distribution of airport revenues by source   *270*

8-7 Comparing user charges at different airports   *274*

Government funding   *274*

Content and quality of services offered   *275*

Volume of traffic   *275*

Characteristics of traffic   *275*

General cost environment   *276*

Accounting practices   *276*

Treatment of aeronautical users   *276*

8-8 Ground handling services   *277*

8-9 Landing fee computation: average-cost pricing   *283*

8-10 Historical cost versus current cost   *288*

Exercises   *290*

References   *291*

Part Three   The airside   *293*

9 Airfield design   *295*

9-1 Introduction   *296*

9-2 Airport classification codes and design standards   *300*

Practical implications   *302*

Runway designation and classification   *311*

9-3 Wind coverage   *312*

9-4 Airport layouts   *314*

Land area requirements and related observations   *315*

Geometric characteristics   *319*

9-5 Runway length   *328*

Declared distances   *330*

Usability of a runway   *332*

Design length   *334*

9-6 Runway geometry   *336*

Separations from other parts of the airfield   *340*

Vertical profile   *341*

9-7 Taxiways   *343*

Special cases   *347*

9-8 Aprons   *351*
9-9 Physical obstacles   *355*
Exercises   *361*
References   *364*

10 Airfield capacity   *367*

10-1 Introduction   *369*
10-2 Measures of runway capacity   *370*
10-3 Factors that affect the capacity of a runway
         system   *376*
         Number and geometric layout of the runways   *376*
         ATM separation requirements   *377*
         Visibility, ceiling, and precipitation   *388*
         Wind direction and strength   *391*
         Mix of aircraft   *391*
         Mix and sequencing of movements   *394*
         Type and location of runway exits   *396*
         State and performance of the ATM system   *397*
         Noise considerations   *398*
10-4 Range of airfield capacities and capacity
         coverage   *400*
10-5 A model for computing the capacity of a single
         runway   *408*
10-6 Generalizations and extensions of the capacity
         model   *416*
10-7 Capacity of other elements of the airfield   *422*
         Capacity of the taxiway system   *422*
         Capacity of the aprons   *424*
Exercises   *430*
References   *432*

11 Airfield delay   *435*

11-1 Introduction   *436*
11-2 The characteristics of airside delays   *437*
11-3 Policy implications and practical guidelines   *444*
11-4 The annual capacity of a runway system   *450*
11-5 Computing delays in practice   *455*
Exercises   *457*
References   *459*

12 Demand management *461*

12-1 Introduction *462*

12-2 Background and motivation *464*

12-3 Administrative approaches to demand management *467*

Schedule coordination: the IATA approach *469*

Experience in the United States *474*

12-4 Economic approaches to demand management *475*

Congestion pricing in theory *476*

Congestion pricing in practice *480*

12-5 Hybrid approaches to demand management *486*

Slots plus congestion pricing *487*

Buying and selling slots *489*

Slot auctions *491*

12-6 Policy considerations *493*

Exercises *495*

References *496*

13 Air traffic management *499*

13-1 Introduction *500*

13-2 Generations of ATM systems *502*

13-3 Description of ATM system and processes in terminal airspace *504*

Airspace structure *504*

Handling of a typical airline flight *506*

Airport traffic control tower *509*

Terminal airspace control center *510*

Surveillance *514*

Navigation for precision instrument approaches *517*

En-route control center *522*

13-4 Air traffic flow management *525*

Objectives and limitations of ATFM *525*

ATFM operations *527*

Ground delay programs *530*

13-5 Collaborative decision making *534*

Additional technical issues and extensions of CDM *542*

Prospects *544*

13-6 Near- and medium-term enhancements   *545*
  GPS-based navigation   *545*
  Automatic dependent surveillance   *547*
  Digital communications   *548*
  Weather data   *548*
  Automation and decision-support systems   *548*
Exercises   *551*
References   *554*

Part Four   The landside   *557*

14 Configuration of passenger buildings   *559*

14-1 Importance of selection   *560*
14-2 Systems requirements for airport passenger
  Buildings   *563*
  Passenger perspective   *565*
  Airline perspective   *568*
  Owners' perspective   *570*
  Retail perspective   *572*
  Government agencies   *573*
  Balance   *574*
14-3 Five basic configurations   *574*
  Finger piers   *576*
  Satellites   *577*
  Midfield concourses   *580*
  Linear buildings   *582*
  Transporters   *584*
  Centralized and dispersed   *586*
14-4 Evaluation of configurations   *587*
  Walking distances   *588*
  Aircraft delays   *594*
  Transporter economics   *595*
  Flexibility   *599*
14-5 Assessment of configurations   *600*
14-6 Hybrid configurations in practice   *602*
Exercises   *603*
References   *603*

15 Overall design of passenger buildings   *605*

15-1 Specification of traffic loads   *607*
The issue   *607*
Peak-hour basis for design   *608*
Nature of loads   *610*
15-2 Shared use reduces design loads   *611*
Drivers for shared use   *612*
Analysis methods   *616*
Overall implications of sharing   *634*
15-3 Space requirements for waiting areas   *636*
Importance of level of service   *636*
Importance of dwell time   *639*
15-4 Space requirements for passageways   *643*
The formulas   *645*
Effective width   *647*
15-5 Areas for baggage handling and mechanical
systems   *649*
Exercises   *652*
References   *652*

16 Detailed design of passenger buildings   *655*

16-1 Design standards   *656*
16-2 Identification of hot spots   *660*
16-3 Analysis of possible hot spots   *664*
16-4 Simulation of passenger buildings   *669*
16-5 Specific facilities   *673*
Queues   *676*
Check-in areas   *677*
Security and border checkpoints   *678*
Moving walkways   *680*
Waiting lounges   *680*
Concession Space   *682*
Baggage claim areas   *684*
Curbside and equivalent areas   *685*
Exercises   *688*
References   *689*

17 Ground access and distribution   *693*

17-1 Introduction   *694*

17-2 Regional airport access *695*
      Nature of airport access traffic *696*
      Distribution of airport access traffic *699*
      Preferences of the users *701*
      Needs of airport operators *702*
17-3 Cost-effective solutions *703*
      The issue *703*
      Door-to-door analysis *707*
      Rail solutions *710*
      Highway solutions *711*
17-4 Parking *712*
      Hourly parking *714*
      Structured parking *715*
      Long-term parking *716*
      Rental car parking *716*
      Employee parking *717*
17-5 On-airport access *717*
17-6 Within-airport people movers *718*
      Technologies *719*
      Location *722*
      Capacity of network *724*
17-7 Within-airport distribution of checked bags *726*
      Security systems *728*
      Information systems *729*
      Mechanical systems *731*
      Capacity *735*
Exercises *735*
References *736*

Part Five   Reference material   *739*

18 Data validation   *741*
18-1 The issue *741*
      Errors *741*
      Incompleteness *743*
18-2 The resolution *743*
Exercises *745*
References *746*

19 Models of airport operations *747*

19-1 Background   *747*

19-2 Classification of models   *748*
   Level of detail   *748*
   Methodology   *749*
   Coverage   *750*

19-3 Airside models and issues in model selection   *750*
   Principal existing airside models   *750*
   Selection criteria   *753*

19-4 Models of passenger building operations   *756*
   Model availability   *757*
   Data requirements   *758*
   Repeated versus one-time model use   *759*
   Model development process   *760*
   Communicating the results   *761*

Exercises   *762*
References   *763*

20 Forecasting *765*

20-1 Forecasting assumptions   *766*
20-2 Fundamental mathematics   *769*
20-3 Forecasts   *771*
20-4 Scenarios   *775*
20-5 Integrated procedure   *776*
Exercises   *777*
References   *777*

21 Cash flow analysis *779*

21-1 Introduction   *779*
21-2 Discounting and the discount rate   *780*
21-3 Present and annual value of monetary flows   *782*
21-4 Notes on computing   *785*
21-5 Measures of project effectiveness   *788*
Exercises   *797*
References   *801*

22 Decision and options analysis *803*

22-1 The issue   *803*

22-2 Decision analysis concept   *804*

22-3 Decision analysis method   *806*

22-4 Options analysis concept   *812*
Financial options   *814*
"Real" options   *815*

22-5 Options analysis method   *816*

Exercises   *817*

References   *817*

23 Flows and queues at airports   *819*

23-1 Introduction   *820*

23-2 Describing an airport queuing system   *821*
The user generation process   *822*
The service process   *823*
The queuing process   *826*

23-3 Typical measures of performance and level of service (LOS)   *828*
Utilization ratio   *828*
Expected waiting time and expected number in queue   *829*
Variability   *829*
Reliability   *830*
Maximum queue length   *831*
The psychology of queues   *831*

23-4 Short-term behavior of queuing systems   *833*

23-5 Cumulative diagrams   *835*

23-6 Long-term behavior of queuing systems   *842*
Little's law   *843*
Relationship between congestion and utilization   *843*

23-7 Policy implications   *847*

Exercises   *849*

References   *850*

24 Peak-hour analysis   *851*

24-1 Introduction   *851*

24-2 Definition of the design peak hour   *853*

24-3 Conversion of annual forecasts into DPH Forecasts   *854*

24-4 DPH estimates of aircraft movements   *860*

24-5 DPH estimates of flows of arriving passengers and of departing passengers *861*

Exercises *862*

References *862*

Index *865*
About the Authors *884*

Preface

This is a book for all those with a major interest in airport planning, management, and design: owners and operators; architects and engineers; government officials; airlines, concessionaires, and other providers of airport services; travelers and shippers; neighbors and communities, as well as members of the public. Readers need no specific experience or skills to use it. A serious interest in the topic is all that is required to make good use of the text. The authors recognize that most people become involved with airport planning, design, and management later in their careers, and come from a broad range of professional backgrounds.

The book should be useful worldwide. It stresses universally applicable concepts and approaches to airport problems. It refers to several different sets of international and national standards on the airside and the landside and points out both similarities and differences in current airport practices around the globe. The text draws heavily on worldwide experience to bring out the best available approaches to each issue.

The text assumes that readers are professionals who need to deal with current issues in airport planning, management, and design. It focuses on the actual problems that arise, and on practical, effective ways of dealing with them. Theory and methodology appear only to the extent that they are relevant and useful. The authors have tried to illustrate theory and methods with appropriate examples wherever possible.

The text is also suitable for students in planning and design curricula. The authors have used the material which has led up to this book since around 1980, in both their full-term courses at the Massachusetts Institute of Technology (MIT), and professional short courses in North America, Europe, Australia, and Asia.

The book concentrates on medium and large commercial airports, those with more than about 1 million passengers a year. Except when smaller and general aviation airports or military bases provide a region with significant current or prospective capacity to handle airline traffic, the text makes no real attempt to describe them. Likewise, the text does not deal with special facilities such as STOL-ports (for Short Take-Off and Landing aircraft), heliports, or seaplane bases.

The text covers both the development and management aspects of airports. Systems design recognizes that the costs of building and operating a major facility such as an airport are comparable. Good planning and design will thus make sure that the physical configuration of a project facilitates operations, and that the management procedures enable owners to avoid unnecessary capital costs.

The text discusses in detail each of the major development topics:

- Airport site characteristics
- The layout of runways, taxiways, and aircraft aprons
- The design of passenger buildings and their internal systems, including security
- The analysis of environmental impacts
- The planning for ground access to the airport

It also gives equal treatment to the operational and managerial issues of:

- Air traffic control
- Management of congestion and queues
- The determination of peak-hour traffic
- Environmental impacts
- Financing, pricing, and demand management

Competition increasingly provides the context for commercial airports. The success of any airport depends most importantly on its advantages compared to other airports, now and as they may be in the future. The text thus carefully describes competition between airports, both within and between metropolitan areas, as well as in the context of airline networks operating nationally, internationally, and globally. It also discusses how international trends in the industry might change the competitive picture.

Dynamic strategic planning is the approach used to bring these specific topics together. It is the modern method for designing complex systems over time. It builds upon the understanding that all forecasts are unreliable, uses the procedures of decision and options analysis of risky situations, and incorporates the economics of financing. The text covers these topics as needed. The overall object is to plan, manage, and design airports so that they can respond flexibly to the unknown, uncertain future conditions.

The book also describes computer-based approaches and models useful in airport planning and design. The emphasis is on helping the reader understand how to make the best use of these tools, to select the kind that are most appropriate in each case, and to interpret and use the results. A web site for the book provides up-to-date supplements to the text.

The book should be easy to understand. It is free of unnecessary mathematical expressions or technical terms. Those that appear are carefully defined and cross-referenced in the index. The authors have worked hard to make the text easy to use by the many airport professionals who are neither engineers nor native speakers of English.

The book features numerous examples illustrating the application of the concepts and methods. It also cites actual cases drawn from the authors' worldwide experience. The emphasis throughout is on dealing effectively with real issues.

Most of the material is easily accessible to the broad range of persons concerned with airport systems planning, management, and design: engineers and architects as well as managers who do not have a technical background. A separate reference section presents basic theory and, in some cases, background mathematics. Persons who do not need this complementary material can skip any or all of this section. Users can combine this reference material with chapters on specific issues in a variety of ways to meet their personal need for information on a particular topic. In short, as the following User's Guide describes, readers can tailor the material to their own needs and skills.

Richard de Neufville
Amedeo Odoni

Acknowledgments

Many, many people have helped the authors understand the issues of airport planning design and management over the years. Recognizing that they cannot possibly list all those who have been helpful over the years, they particularly would like to thank those who have played important roles in helping us develop material for this text.

From government, our collaborators especially include Larry Kiernan, Ashraf Jan, and Richard Doucette, U.S. Federal Aviation Administration; Zale Anis, Volpe National Transportation Systems Center; Jean-Marie Chevallier, Aéroports de Paris; Dr. Lloyd McCoomb, Greater Toronto Airport Authority; and Flavio Leo, Massport.

Numerous consultants and practicing professionals helped ensure that the work reflected actual conditions in the field. These particularly include: John Heimlich, Katherine Andrus, Tom Browne, Patricia Edwards, Russell Gold, Paul McGraw, and Robert Zoldos, Air Transport Association; Richard Marchi, Airports Council International; Regine Weston and Dr. Alex de Barros, Arup/NAPA, Canada; Cliff King, Louis Berger Associates; William Woodhead, Kinhill Engineers, Australia; Winnie Shi, KPMG, Canada; Harley Moore, Lea + Elliott; Dan Kasper, LECG; Robert Weinberg, Marketplace Development; William Swedish, MITRE; Roxanne Williams, nbbj; President Shota Morita and his Airports Department, Pacific Consultants International, Tokyo; Steve Belin, Simat Helliesen and Eichner; and Thomas Brown, United Airlines.

We also gratefully acknowledge the contributions of our academic colleagues: Prof. Rigas Doganis, Cranfield University, U.K.; Prof. Robert Caves and Dr. Ian Humphreys, Loughborough University, U.K.; Profs. Arnold Barnett, John-Paul Clarke, R. John Hansman, John Miller, and Ian Waitz, and Dr. Husni Idris, MIT; Dr. Steven Bussolari, MIT Lincoln Laboratories; Prof. Tom Symons, Trent University,

Canada; Prof. Chan Wirasinghe, University of Calgary, Canada; and Dr. Gabriel Faburel, Université de Paris XII, France.

We also acknowledge with gratitude our great debt to our administrative and editorial support team at MIT: Paulette Mosley, Sarah Kirshner, and Lauren McCann.

User's Guide

You can create your own book

Readers can tailor the material to their own needs. Persons interested in a specific topic can put together a self-contained set of chapters that will give them what they need to know about that subject. Architects interested in the design of passenger buildings, for example, can assemble an integrated guide to the subject. They can put together the chapter on that topic and the supporting chapters on the analysis of queues, peak-hour analysis, and computer models. Users do not have to get involved in topics of no current concern, and can concentrate on their immediate interests.

Readers can likewise tailor the material to their own skills or depth of interest. Many readers will use the book to get help on a specific project. They will initially want information relevant to only one topic, such as airport financing or airport access, and will be able to get it. The chapters on specific problems, the design of passenger buildings, for instance, are self-contained and provide the necessary guidelines in a way that anyone should be able to understand. Users who do not need the supporting reference material, either because it is not relevant to their job or they know it already, can simply skip it. Thus, both an airport manager and a computer specialist could bypass the chapter on computer models while pursuing an interest in the design of passenger buildings.

The text is modular, in short. Its chapters can be assembled in different ways for a variety of needs. This organization is possible because many of the methods used in airport systems planning are common to several different topics. An understanding of the behavior of flows and queues of traffic, for example, is necessary for the detailed design of both runways and passenger buildings. The reference sections dealing with specific methods fit in with several chapters that deal with specific problems.

How to do it

To appreciate how to tailor the material to your own needs, it is useful to look at the organization of the material. The mode of use then becomes clear.

The text consists of two distinct blocks. As the table of contents indicates, the first block consists of substantive chapters devoted to specific topics in systems planning and management, airside and landside. The second block, Part 5 provides reference on methods of analysis such as forecasting, decision analysis, and queuing theory. These reference materials provide in one place coherent discussions of procedures that apply to several of the substantive chapters.

Table 1 provides a menu that matches the chapters covering specific issues with the most relevant reference materials. To look at a specific issue, a reader will pick a chapter from column A, and appropriate reference material from column B. Readers interested in exploring a larger set of issues in airport planning, such as the problems associated with the airside, will want to select that whole group from column A. They will select from column B as needed.

Recommended combinations

Each of the big blocks on systems planning, airside and landside, is a self-contained unit. Readers can approach them independently of the others. This arrangement should be useful to persons with responsibilities or interests especially in those fields. For example, managers and government officials might focus on systems planning, aviation and air traffic control specialists on the airside, and architects and civil engineers on the landside.

All readers may be interested in Chaps. 1 and 2, which provide context on the future of the airport/airline industry and an international perspective. In addition, the authors suggest these packages for readers with broad interests:

- *Systems planning:* the block in column A plus the block in column B under *Risk*
- *Airside:* the block in column A plus block in column B under *Variable Loads*
- *Landside:* the block in column A plus the block in column B under *Detailed Design*

Table 1.   Menu of Chapters

(A) Issues	(B) Reference
System Planning	**Risk**
Dynamic Strategic Planning	Data Validation
Privatization and Deregulation	Forecasting
Multi-airport Systems	Cash Flow Analysis
Environmental Impacts	Decision and Options
Organization and Financing	Analysis
User Charges	
Airside	**Variable Loads**
Airfield Design	Models of Airport Operations
Airside Capacity	Flows and Queues at Airports
Airside Delay	Peak-Hour Analysis
Demand Management	
Air Traffic Management	
Landside	**Detailed Design**
Configuration of Passenger Buildings	Forecasting
Overall Design of Passenger Buildings	Models of Airport Operations
Detailed Design of Passenger Buildings	Decision and Options Analysis
Ground Access and Distribution	Flows and Queues at Airports
	Peak-Hour Analysis

Aviation Week Books is an imprint of McGraw-Hill Professional Book Group in conjunction with the Aviation Week division of The McGraw-Hill Companies. With nearly 50 products and services and a core audience of some one million professionals and enthusiasts, Aviation Week is the word's largest multimedia information and service provider to the global aviation and aerospace market.

For more information, use www.aviationnow.com

Editorial director: Stanley Kandebo, Assistant Manager Editor, Aviation Week & Space Technology

Part One
Introduction

1

The future of the airport and airline industry

Airport systems exist and must be designed in the context of their major clients, the airlines. To build airport facilities that will perform effectively over the 20 to 50 years of their lifetime, it is necessary both to appreciate the historical context and to understand the current and prospective needs of the users. Understanding the state of the airport industry at the beginning of the twenty-first century gives a perspective on the future of the industry and is thus an important starting point for a forward-looking text on airport systems planning.

Four trends dominate the airport/airline industry at the start of the twenty-first century:

1. *Long-term growth,* on the order of 5 percent a year worldwide, implying a doubling of traffic about every 15 years. This drives the continual demand for expansion and improvement. It also leads to the development of multiple airport systems in metropolitan areas and of niche airports serving leisure traffic or cargo.

2. *Commercialization,* as the rise of business management in a market economy replaces government ownership in a regulated environment. This trend makes economic performance and efficiency salient criteria for good design and radically changes concepts of what should be built.

3. *Globalization,* through the formation of transnational airline alliances and airport companies. This drives the implementation of worldwide best practices in the provision of airport services.

4. *Technical change,* especially electronic commerce, which both propels the rapid rise of integrated cargo carriers and, through electronic ticketing, rearranges passenger handling inside airport buildings.

Taken together, these trends are substantially changing the context, objectives, and criteria of excellence for airport systems planning and design.

1-1   The airport industry at the end of the twentieth century

Airports and air transport at the start of the twenty-first century constitute an exciting long-term growth industry. The industry is large, innovative, and has excellent prospects. This historical base needs to be appreciated before launching into the future. Based on the developments described in the subsequent sections of this chapter, the airport industry is at the beginning of substantial technical and organizational changes that are redefining the practice of airport systems planning and design.

The industry is large. As of 2000, it involved about 1.7 billion airline passengers worldwide plus large amounts of cargo. Its annual revenues are on the order of US $1 trillion (one million million dollars). The world airlines operated approximately 10,000 major jet aircraft, valued in the hundreds of billions of dollars. The annual investments in airport infrastructure ran at about $10 billion a year. To put these figures in perspective, the worldwide air transport industry carries the equivalent of over one-quarter of the world's population each year, and its revenues are about 10 percent of the Gross Domestic Product of the United States. By any measure, this is an important activity.

The industry is actively growing. Over the last third of the twentieth century, the worldwide long-term growth rate in the number of airline passengers was about 6 percent a year—averaging periods of stagnation and boom. During that period the traffic doubled, redoubled, and doubled again, becoming eight times larger in less than 35 years. The rate of growth slowed down in the last decades of the century, to about 4 percent a year, but this still implies between a doubling and tripling of traffic over a 25-year generation. Given that the planning horizon for large-scale infrastructure projects is normally between 10 and 15 years, due to the need to create the designs, assemble financing, and proceed successfully through environmental reviews, the growth in loads on airports means that planners should at any time generally be contemplating 50 to 100 percent increments in capacity.

The growth in air transport translates into major airport projects. About a dozen major programs for airport development, costing over a billion dollars each, have typically been under way at any time in the last decades. Table 1-1 illustrates the situation as of the end of 2001. Naturally, many smaller-scale projects at many other airports accompany these major projects at any time.

The traffic is not evenly distributed, however: it has been concentrated in the United States. During the last half of the twentieth century, about half the air transportation and airport activity took place in the United States. The airports and airlines based in the United States dominate their competitors in size. The largest airlines in the world, in terms of the size of their fleet, have been based in the United States. In 2000, they accounted for 10 of the top 13 airlines (Table 1-2). Likewise, airports in the United States have been the three busiest in the world in terms of the number of passengers. In 2000, they occupied 14 out of the 20 top spots (Table 1-3).

The U.S. share of the traffic fraction decreased in the third quarter of the twentieth century, when Europe and Asia grew rapidly as they recovered from the 1939–1945 war and developed. However, the U.S. share then remained fairly constant at around 40 percent in the last two decades of the twentieth century. It went from about two-thirds of the world traffic in the 1960s, to about one-half in the 1970s, to around 40 percent of the passengers in 1980 and beyond. This continued dominance has almost certainly been due to a steady stream of innovations that has transformed air transportation from an expensive luxury to a relatively cheap and popular mode of travel.

The United States has been a leader in the development of mass use of air transport. As of the year 2000, residents of the United States on average took at least one round trip by air every year. This is about triple the use in Europe and 10 times the use per person in the rest of the world. Much of this has been due to the fact that fares in the United States have tended to be considerably less expensive, on average, than those in Europe.

The air transport industry in the United States has thus faced the challenges of high volumes of traffic well ahead of the rest of the world. It has correspondingly led in the development of major innovations that have and are transforming commercial aviation and airport planning and design worldwide. Table 1-4 indicates some of them. These innovations, together with the trends discussed in the following sections, are radically changing the concept of airport systems planning and design.

Table 1-1.   Billion-dollar-plus airport projects active in the year 2001

City/Airport	Project description
Athens/Venizelos	Totally new airport, opened in 2001
Atlanta/Hartsfield	New runway, new international building
Amsterdam/Schiphol	New runway, passenger building
Bangkok	New Bangkok International Airport under construction
Berlin/Schönefeld	Reconstruction of airport as national gateway
Boston/Logan	New passenger buildings, hotel, garage, and interconnections
Dallas/Fort Worth	New passenger building, new automated people mover
Miami/International	Reconstruction of passenger buildings and a new runway
Nagoya/Chubu	Entirely new airport to be built in the sea
New York/Kennedy	Reconstruction of International, Delta, and American buildings; construction of JFK railroad connection
Madrid/Barajas	New airport passenger buildings and a runway
Paris/de Gaulle	New airport passenger buildings and runways
San Francisco/International	New International passenger building, new rail access
Seoul/Incheon	Totally new airport, opened 2001
Singapore/Changi	New passenger building, new rail access
Toronto/Pearson	Massive reconstruction of passenger buildings and of runways
Washington/Dulles	Development of midfield concourse and people-mover system

Table 1-2. Airlines based in the United States were the largest in the world in 2000, ranked by size of jet fleet

Airline	Aircraft	Airline	Aircraft
American	703	Southwest	332
United	604	Lufthansa	328
Delta	602	British Airways	273
Northwest	424	UPS	234
US Airways	412	Air France	230
FedEx	369	TWA	187
Continental	359	Air Canada	159

Sources: IATA World Air Transport Statistics; Southwest Airlines, www.southwest.com.

Table 1-3. Airports in the United States were the busiest in the world in 2000, ranked by number of passengers

Airport	Passengers, millions	Airport	Passengers, millions
Atlanta	80.2	Denver	38.7
Chicago/O'Hare	72.1	Las Vegas	36.9
Los Angeles/ International	68.5	Seoul/Gimpo	36.7
London/Heathrow	64.6	Minneapolis/St. Paul	36.7
Dallas/Fort Worth	60.7	Phoenix	35.9
Tokyo/Haneda	56.4	Detroit/Metro	35.5
Frankfurt/Main	49.4	Houston/Bush	35.2
Paris/de Gaulle	48.2	New York/Newark	34.2
San Francisco/ International	41.2	Miami/International	33.6
Amsterdam	39.6	New York/Kennedy	32.8

Source: ACI (2001).

The entire context for airport systems planning and design has been significantly different in the United States than in the rest of the world. In the United States, the air transportation and airport businesses have been largely privatized. The airlines (American, Continental, Delta,

Table 1-4.   Late-twentieth-century innovations in air transport from the United States

Decade	Type of innovation		
	Legislative	**Operational**	**Airport design**
1970s	Economic deregulation: U.S. airlines can fly where they want, at any price	Shuttle services Integrated air cargo services: FedEx, UPS	Automated people movers: wide use at transfer hubs
1980s		Transfer hubs: Atlanta, Dallas/Fort Worth, etc. Yield management systems: Sabre, etc. Franchising: Branding of commuter airlines	Midfield concourses: Atlanta, Pittsburgh, Denver, Chicago, Detroit, etc.
1990s	"Open skies" policy: U.S. government promotes free access between countries	Airline alliances: Star, Oneworld Electronic tickets	GPS: Satellite positioning of aircraft for air traffic control

United, etc.) have always been privately owned and financed, in contrast to the standard twentieth-century practice everywhere else. It was only around the 1990s that Britain, the Netherlands, Germany, and Japan began to privatize their airlines, setting off a worldwide trend. Airports in the United States have also been largely designed, built, and operated by private companies. Most important, the money for much of their infrastructure has been paid for by private sources.* Airports in the United States have therefore traditionally had to pay close attention to the returns on investments and ways to make the facilities pay. In this they have contrasted with airports in other countries. Until the privatization trend began in the 1990s, airports outside the United States were virtually all owned, designed, financed, built, and operated by government employees.

The future trends discussed in the next sections continue this past, but in many ways fundamentally change the context, objectives, and cri-

*Major airports in the United States raise capital to build passenger buildings, hangars, garages, and the like through bonds offered to private investors or through fees charged to passengers (the PFC Passenger Facility Charge). The U.S. government, through the Federal Aviation Administration, pays for runways, air traffic control facilities, and safety measures. The government contributions are most significant at smaller airports, but far less important at established major airports—unless they build new runways.

teria of excellence for airport planning, management, and design. Long-term growth in the context of limited sites for airports in urban areas motivates the need for more understanding of how to manage traffic—both overall through economic incentives and in detail through proper control of the varying queues of traffic. The increased commercialization and worldwide privatization of airports calls for an appreciation of the economic and financial aspects of airport operation. New technology, and other changes due to competition between airports, require airport professionals to develop dynamic, strategic plans that incorporate flexible designs and enable airport operators to manage their risks. In short, these factors create the need for airport systems planning and design.

Airport planning previously tended to focus narrowly on technical issues. It did not concern itself intensely with wider issues such as costs and revenues, stochastic traffic and risks, operations and management. Government agencies and international agencies set fixed standards for design that did not admit of trade-offs between cost and service. Textbooks followed the same vein. (See, for example, FAA, 1988; Australia Department of Housing and Construction, 1985; IATA, 1995; ICAO, 1987; ATA, 1977; Horonjeff and McKelvey, 1983; Ashford and Wright, 1991.) Within the traditional context, there was virtually no scope for systems design, and the practice did not consider those issues.

The current environment for airport planning calls for a *systems approach*, a broad consideration of the breadth of factors that shape the performance of the airport and the application of the range of appropriate methods of analysis. In particular, it expands the concept of airport design to include its operational and long-term management through technical and economical measures. Likewise, it uses a wider range tools for analyzing preferable solutions, as indicated in Chaps. 18–24. This approach should be broadly useful to all professionals actively associated with airports.

1-2 Long-term growth

Growth in aviation passenger and cargo traffic is an outstanding characteristic of the last half of the twentieth century. In the third quarter of the twentieth century, over good and bad years, passenger traffic increased an average of about 6 percent worldwide. This meant that the amount of air travel doubled about every 10 to 15 years.

This long-term overall growth rate dropped toward the turn of the century. Worldwide, the rate of growth continued strong at around 6 percent a year, with differential regional rates depending on their relative prosperity and maturity of their markets. In the United States, the more recent long-term growth rate fell to about 4 percent a year, which implied traffic doubling about every 15 to 20 years. The net effect in the United States may be interpreted as almost linear total growth from 1970 to 2000—the annual need for additional capacity stayed the same. See Fig. 1-1.

Cheaper air service augmented by increased globalization has propelled the growth in aviation traffic. Most obviously, the overall real price of air travel has persistently fallen over the last half of the twentieth century. A steady rise in use per person mirrored the long-term drop in prices, as basic economics expounds and Figs. 1-2 and 1-3 confirm. Meanwhile, the safety and smoothness of the flights has increased dramatically. The number of accidents and deaths per trip dropped, for example, by a factor of 3 in the United States over the long term, as Fig. 1-4 shows.* Passengers and cargo now receive far more value for money than they did in the unpressurized small aircraft flying over primitive air traffic control systems in the mid-twentieth century. Meanwhile, increased globalization further increased the demand for long-distance travel. These factors jointly led to the steady growth in aviation.

Future levels of traffic are questionable. Since small differences in assumptions cumulate to enormous differences in consequences 25 years or more from now, airport professionals should be tentative about future levels of traffic. For example, slight deviations of plus or minus 1 percent from a long-term growth rate of 4 percent a year in enplaned passengers lead to substantially different forecasts. A 5 percent annual rate of growth compounded over 25 years gives an end result about 140 percent greater, in terms of the starting amount, than a 3 percent annual rate of growth. Managers should place any assessment of long-range forecasts in a broad range of possibilities. See Chap. 3 for a detailed discussion of this issue.

---

*Note that the apparent rise in accidents after 1997 in Fig. 1-4 is largely a result of a change in definition. Since March 20, 1997, aircraft with 10 or more seats, formerly operated under 14 CFR 135, were included in the records on air carrier aircraft (FAA, 2001). These smaller aircraft have accidents more frequently, and thus increased the overall accident rates reported—although the U.S. aviation system was not less safe after 1997 than before. This is an example of how data are often misleading and need to be validated—see Chap. 18.

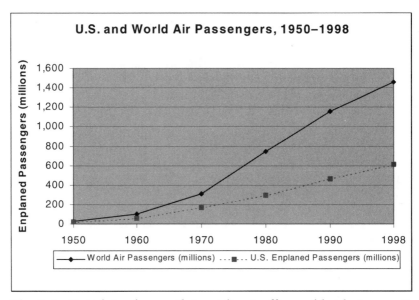

Fig. 1-1. *Rapid steady growth in airline traffic worldwide.* Sources: Air Transport Association, www.air-transport.org; ICAO Bulletin, ICAO Journal [various].

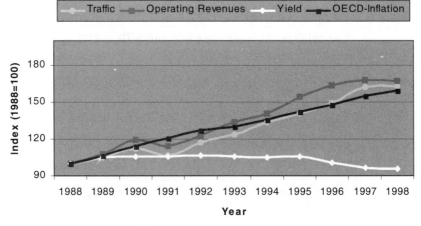

Fig. 1-2. *Worldwide increases in airline traffic have been associated with lower costs of travel in real terms, as inflation decreases the real costs of steady fares per seat-mile.* Source: IATA World Air Transport Statistics, http://www.iata.org/ps/index_products.asp.

Traffic will almost certainly continue to grow substantially. Most of the world rarely flies, and the market is far from saturated. Plausible increases in population, national wealth, the length of paid vacations, and the tendency of members of younger generations to fly, even if only a few percent per year, will lead to more traffic. Increased

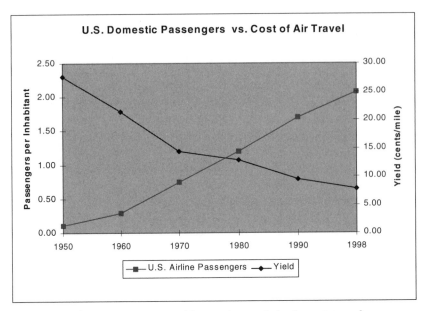

Fig. 1-3. *The rise in air travel has mirrored the long-term decrease in fares per seat-mile.* Source: Air Transport Association, www.air-transport.org.

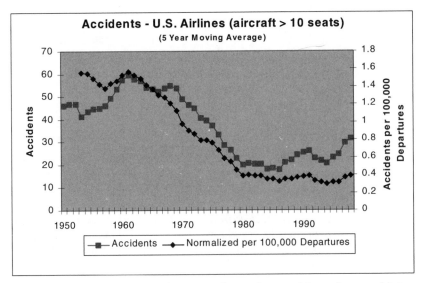

Fig. 1-4. *Long-term, accident rates have dropped by a factor of 3 in the United States.* Source: Air Transport Association, www.air-transport.org.

globalization will impel long-distance travel for business and personal reasons, in general only realistically feasible by air. Even a historically modest 3 percent a year growth rate doubles traffic in 25 years.

No one can count on steady growth, however. The trends may slow down or stop. Historically, a series of major causes steadily reduced costs and drove the historical rise in air traffic. These were:

- Larger, more efficient aircraft, driven by two engines (instead of three or four), with fewer pilots
- Economic deregulation of the airlines, accompanied by competition
- Worldwide privatization of aviation, and increased attention to costs
- The consequent competitive restraint on wages
- Historically low fuel prices (when adjusted for inflation)
- The introduction of yield management systems that raise overall revenues

Some trends may reverse. For example, fuel prices might rise considerably, as they did in the 1970s and 2000/2001. Business travel may give way to inexpensive video conferencing or other communications. Concerns about security may keep people closer to home and limit travel. A worldwide recession might even lower overall traffic. Certainly, the rate of traffic growth is likely to continue its long-term decline. Even so, the likely scenario is that aviation and airport activity will register substantial overall increases. Even lower rates of growth will, when applied to the existing large market, lead to substantial growth.

Major developments with unpredictable consequences reduce the possibility of forecasting the long-term future. What, for example, will be the net effect of affordable video conferencing? Will it substitute for business travel by permitting virtual face-to-face contacts? Will it facilitate globalization and thereby increase overall passenger-miles traveled? What will be the next effect of changing work patterns? Will casual Fridays turn into 3-day weekends and encourage more travel? Will an increasingly wealthy and healthy elder population be inclined to travel more? Alternatively, will long-term concerns about terrorism and security permanently affect worldwide travel patterns? Given all these uncertainties, airport planners should be modest about the possibility of precise forecasts of future traffic.

Overall, it would be reasonable to assume that in 2025 the level of traffic could be two or even three times higher than in 2000. For example, the number of enplaned passengers in the United States in 2025 could be in the range of 1500 million a year, plus or minus 500 million, compared to the 600 million a year flying in 2000. Airport planners should thus prepare for substantial growth (by planning and land acquisition), but not necessarily commit to building facilities for the largest possible levels. In short, they need to manage their risks consciously, as indicated in Chaps. 3, 20, and 22.

The composition of the total traffic may differ significantly from what it was in 2000. Historically, air travel has diffused from the rich to the masses, from the early to the later developed nations. It typically has shifted from being a luxury good for the elite, to a necessary business need, to mass transportation, to international tourism. Airport planners may anticipate an extension of such patterns, both domestically in their own markets and internationally, from North America, Europe, and Japan to the rest of the world.

Cargo traffic may expand dramatically as companies reorganize their distribution systems around electronic commerce. As of 2000, in a development not widely perceived, the integrated package carriers such as UPS and FedEx were among the largest airlines in the world in terms of aircraft operated (see Table 1-2). To the extent that businesses substitute web sites for brick-and-mortar stores, and direct shipments to customers save money by reducing the need for local warehouses and in-store inventories, the integrated cargo carriers may grow rapidly. This traffic may be a driving force for many future airport developments.

1-3 Commercialization

The whole context for airport systems planning and design has been changing fundamentally around the beginning of the twenty-first century. This change can conveniently be labeled as commercialization. It is important to recognize, however, that the evolution goes far beyond an increase in the number of shopping malls and office buildings in and around airports. The entire concept of the purpose and role of airports is evolving. Old ways of thinking are becoming obsolete. New criteria for performance and design are becoming dominant.

Business management in a market economy is replacing government ownership in a regulated environment. An increased orientation to profits and economic efficiency is overtaking political considerations. The standard practice for most of the twentieth century was that government bodies owned, built, and operated the airports. In the United States, local cities or regional airport authorities normally ran the airports. Outside the United States, ministries of transport or aviation or their dependencies typically ran the airports. Their budgets were allocations of the national treasury only loosely connected to revenues. The employees were civil servants. Governments outside the United States meanwhile also owned the major international airlines. The airlines thus benefited from public subsidies and protection. Yet they were also subservient to political dictates concerning routes, discounts for favored groups, and other obligations. Both airlines and airports, furthermore, operated in a highly regulated context, in which government agencies allocated routes, set fares and frequencies, divided revenues between nominally competing airlines through international agreements on the pooling of revenues, and even, in the case of Australia, chose the type of aircraft! This era, and the design and management mentalities that go with it, are fast disappearing.

Airports worldwide are being privatized, as Table 1-5 shows. Governments are transferring their responsibilities to semi-independent airport authorities and private companies. Private companies are taking over day-to-day management, typically hiring new executives recruited from the business world. World-class international consultants rather than ministry officials are designing and building airport facilities. Investment banks are providing the money for major capital projects, instead of the national treasuries. In this regard, airports worldwide are increasingly resembling the major U.S. airports, which obtain large portions of their capital from private sources and whose functions have been among the most privatized in the world.* Revenues are consequently tied much more closely to specific projects, and consumers and market forces thus drive the direction and character of airport development. (See de Neufville, 1999.)

Governments are also privatizing their national airlines, as Table 1-6 shows. The airlines no longer can rely on the support of the national treasury, and must operate profitably to the extent possible. They then find it imperative to drop unprofitable routes, to phase out unnecessary jobs, and eliminate discounts offered to favored clients (such as

---

*At most airports in the United States, a majority of the management functions—such as finance, design, construction, and much of management—are done by private companies, although public agencies own the land. See Chap. 4.

Table 1-5. Examples of major publicly owned airports privatized by 2002

Airport system	Status in 2002
Argentina	All significant airports sold to private consortium
Australia	All significant airports leased for 50 years
Austria—Vienna	A private company
Britain—BAA plc	BAA company runs major London and Scottish airports
Canada	Major airports run as companies by local authorities
Denmark—Copenhagen	A private company
Germany	Several major airports run by private companies
Greece—Athens/Venizelos	A company with government as a minority owner
Italy—Rome	A private company
Malaysia	Airport company shares listed on stock exchange
Mexico	All major airports except Mexico City privatized
South Africa	South African Airports Company runs major airports
Switzerland—Zürich	A private company

government officials). National airlines may even disappear, as Sabena did in 2001. As with airports, the airlines now find that consumers and market forces drive the direction and character of their operations.

The privatization of the air transport industry has been taking place in the context of widespread economic deregulation. Historically, the industry worldwide had been highly constricted. In the United States for example, until 1978 the Civil Aeronautics Board (CAB) had the authority both to license airlines to fly specific routes and to regulate the fares. Moreover, to ensure that airlines did not subtly offer first-class service at cut-rate fares, the CAB would regulate details of the quality of seats and cabin service. Airlines could not start new routes, or cut fares, without lengthy legal proceedings—in which their competitors could strenuously oppose them (Kahn, n.d.; Meyer

Table 1-6. Examples of major publicly owned airlines privatized by 2002

Airline	Status
Aerolineas Argentinas	Completely
Air Canada	Completely
Air France	Partially
Air New Zealand	Completely
Alitalia	Partially
Australian	Completely
British Airways	Completely
Iberia (Spain)	Majority
Japan Airlines	Partially
Lufthansa (Germany)	Completely
Qantas (Australia)	Completely
SAS (Scandinavia)	Majority

and Oster, 1984). Comparable restrictive regulations existed elsewhere, notably in major airline markets such as Australia, Britain, Canada, France, and Germany. International regulations were typically even more restrictive, as each of the two countries at the ends of an airline route would limit the number of flights, destinations, and airlines from the foreign countries. These regulations created local monopolies, and thus restricted competition on prices and service as well as reduced innovation and productivity.

All this began to change when the United States deregulated the airlines economically in 1978. From then on, the U.S. domestic airlines could establish and drop routes as they please, charge whatever prices they wish—and do so at a moment's notice, without having to consult with any government body. This event led to rapid innovation in services, big increases in productivity, and significant fare drops. The example proved contagious, and economic deregulation of air travel has spread to major markets worldwide, notably Australia, Canada, and the entire European Community. More recently, the United States has effectively been promoting international economic deregulation of the airlines through its "open skies" agreements with other countries. These treaties eliminate governmental restrictions on airline destinations, frequencies, and fares, and allow unlimited access to each other's markets (except for domestic flights, known legally as cabotage). The result is that much of the air transport industry now operates in a context completely different from the one that prevailed until the late 1990s.

Worldwide privatization and deregulation are completely changing the criteria for good design in the industry, notably for airports. Economic performance and efficiency are now becoming salient criteria for good design. This is in sharp contrast to traditional design, in which cost was typically not considered a factor in the technical design. In the current environment, cost and economic performance are increasingly crucial criteria, and they are radically changing the timing and nature of what is and should be built.

1-4   Globalization

Globalization of the airport industry is leading toward the adoption of international best practices. This is a remarkable development. Surprisingly for an industry centered on rapid international communication, airport planning and development has typically focused narrowly on local practices. The past organization of the industry kept most airport practitioners in specific cities, and did not give them the opportunity to learn deeply from experience elsewhere. The political bureaucracies controlling airports normally operated within tightly defined regulations established to meet local political purposes. As of the beginning of the twenty-first century, however, salient portions of the industry are beginning to restructure into large international companies. These new groupings are relatively free from local constraints, and highly motivated through the commercialization of the industry to seek out international best practices. This phenomenon is leading to profound changes in airport systems planning and design.

Globalization is setting up international alliances between airports. In the twentieth century the industry was a highly diverse collection of local and national institutions each, with few exceptions, working exclusively in their own territories. Now, however, the industry increasingly involves international organizations working broadly across the world. These groups may be either companies or government-owned authorities. In either case, to compete and succeed internationally they must adopt the best practices they discover among their competitors. This transformation of the industry affects both the airlines and the airports.

Airlines traditionally had specific national characters even though they provided international service. American Airlines, British Airways, and Air France, for example, were each clearly based in their home countries and represented their nationalities proudly. By the end of the twentieth century, however, airlines were developing international characteristics and submerging their nationalities. Throughout the 1990s, for example, the American Northwest and the Dutch KLM airlines worked together to present a single image, so that the U.S. customer could feel comfortable with an American style of service, while the Dutch customer could similarly expect familiar treatment. In this process, each partner adopted better practices of the other: Northwest improved its service inside the aircraft to meet international standards; KLM made its schedules more efficient by adopting the American practice of routing passengers through transfer hubs. Although the airlines did not formally merge, from the customers' perspective they virtually became a single company.

The formation of global airline partnerships embodied this trend toward consolidation. As of 2002, the Star and Oneworld alliances, for example, each linked national airlines into global networks whose stated aim was to provide coherent services to passengers as if they were using one airline. The alliances led to patterns of ownership (for example, American Airlines owned shares in Iberia of Spain, British Airways had major holdings in Qantas of Australia). Table 1-7 illustrates the phenomenon with a snapshot of the situation for the major alliances as of 2002. In the early years of the alliances, the partnerships were unstable and not particularly deep in

Table 1-7. Memberships in Star and Oneworld global airline alliances in 2002

Alliance	Major airline members	Secondary airlines
Star	United, Lufthansa, Air Canada, Air New Zealand, ANA, SAS, Singapore, Thai, Varig	Austrian, British Midland, Lauda, Tyrolean
Oneworld	American, British Airways, Aer Lingus, Cathay Pacific, Iberia, Qantas	Finnair, Lan Chile

terms of their effect on the airlines. Nonetheless, the overall trend toward cooperative integration appears relentless.

Some airports are similarly losing their local character and becoming part of international organizations. International companies have been taking over the operations of all or parts of airports, and delivering their own brand of services. They are developing global "airport chains" similar to "hotel chains." As in the hotel business, the management arrangements are sometimes based on long-term contracts (as when BAA from England agreed to operate Indianapolis airport for 10 years, and Vancouver Airport agreed to operate airports in the Dominican Republic) and sometimes on ownership (as when SEA from Milan bought into the Argentine airports along with Argentine partners). Table 1-8 illustrates the formation of international airport companies.

Large-scale international airport companies can reduce costs and increase performance. Comparably to large international hotel chains that manage properties for local owners, large airport companies should most obviously be able to take advantage of economies of scale, as individual local airports cannot. They should be able to negotiate with suppliers for significant discounts on large orders. More subtly, they should be able to afford to invest in sophisticated operating systems and experiment with and develop new services, since they can spread the costs over many clients. In a commercial environment, their economic efficiency should eventually be a decisive factor in the future salience of large international airport companies.

The large international airport companies will almost certainly redefine airport planning and design. They can be expected to raise

Table 1-8. Examples of global airport companies in 2002

Group	Owns shares in	Operates facilities at
Amsterdam	Brisbane (Australia), Frankfurt, Vienna	New York/ Kennedy, etc.
BAA (UK)	London/Heathrow, London/Gatwick, 5 other UK airports, Naples (Italy), Melbourne and Hobart (Australia)	Indianapolis, Pittsburgh (U.S.), etc.
Frankfurt	Athens, Amsterdam, Brisbane	Athens, Lima (Peru)

individual airports toward international best practices, by improving both the procedures and the skills. As airport activities are increasingly operated as franchises to large international companies, they will be less run as municipal departments or local authorities.

The operators will be pushing toward the best standards available, and ineffective practices will be reduced. The international companies will also be large enough to train personnel efficiently and recruit ambitious managers who wish to succeed in a large international company. They should be able to diffuse best practices across their organization by promoting and transferring their most effective planners, designers, and managers.

1-5   Electronic commerce

The information age will lead to profound changes in airport planning and design. Electronic commerce may be the most important technical innovation to affect airport design in the beginning of the twenty-first century. It will be significant, compared to other developments, because it will lead to some major revisions in the concept of airport facilities. Other developments, such as satellite-based geographic positioning systems (GPS) and new large aircraft (NLA), will certainly affect airports deeply by increasing capacity and demands, but do not imply major conceptual revisions.

Information technology is affecting airports principally in two ways.

- *Electronic ticketing* and waybills enable faster check-in procedures and similar functions. As passengers increasingly check in and print their baggage claim checks electronically, the number of check-in counters required per person will decrease substantially.

- *Electronic commerce* is reorganizing the distribution of goods to customers. It is dramatically increasing the demand for cargo services and cargo airports. By rapidly distributing products directly to customers, manufacturers can reduce the amount of goods in transit, the warehouses needed to stock this inventory, and thus their capital costs.

Electronic ticketing

To appreciate the effect of electronic ticketing on the design of airport passenger buildings, one needs to recognize the extent traditional check-in procedures involve manual processing of information. These functions require staff at desks to read paper tickets, look up passengers' files, verify documents, enter new information, assign seats, and issue boarding passes. Electronic check-in speeds up these operations, doing in microseconds what might otherwise take a minute per person. Electronic ticketing coupled with high-speed communications makes it possible to "check in" remotely throughout the airport, in parking garages or wherever. Passengers can identify themselves with cards or passwords to computers in kiosks, security agents, and wherever needed. Several airlines had already successfully demonstrated this technology in the year 2000 (Fig. 1-5). In short, electronic ticketing speeds up and distributes operations.

Electronic ticketing thus reorganizes the concept of a passenger building. Faster operations increase productivity, which reduces the demand for facilities. The dispersal of check-in points makes it possible for many passengers to avoid detouring through a congested check-in hall and use more direct routes to their aircraft. The net result will be to change the flow of passengers and reduce the number of check-in facilities per passenger.

Electronic commerce

Electronic commerce beyond the airport promotes the demand for rapid transport, for air cargo in particular. Electronic processing of orders gives manufacturers and distributors more timely information

Fig. 1-5. *Electronic check-in devices installed at Seattle-Tacoma Airport.* Source: Zale Anis and Volpe National Transportation Systems Center.

about what their customers want, and allows them to reduce and even eliminate local warehouses and storage rooms in retail shops. These reductions mean that less capital sits idly in inventories, reduces real estate costs, and lessens the probability that the inventories are devalued because they are in the wrong place at the wrong time. To achieve these advantages, manufacturers must have rapid means to get their goods to their customers. Distributors thus substitute relatively high-cost transportation for inventory and warehouse costs. The result is a surge in demand for high-speed distributors of freight, which provide integrated service covering all transactions between the shipper and the customer.

The integrated carriers providing door-to-door service between the supplier and the customer were the fastest-growing major airlines worldwide by the year 2000. Table 1-9 shows that FedEx and UPS grew about three to four times as fast as the conventional airlines (the principal exceptions being airlines that grew by merger: Air France, which took over Air Inter, and Air Canada, which absorbed Canadian). Table 1-10 confirms the domination of these integrated carriers in the freight market. Both FedEx and UPS were three to five times larger than their nearest competitors in terms of the number of tonnes carried in the year 2000.

The surge in cargo traffic due to electronic commerce creates a strong demand for cargo airports. As of 2000, a number of major airports already owed their eminence to cargo traffic (Table 1-11). The

Table 1-9. The integrated cargo carriers FedEx and UPS were among the fastest-growing major airlines in the world between 1995 and 2000, as indicated by growth of their aircraft fleet

Airline	Percent growth	Airline	Percent growth
American	11	Southwest	N.A
United	9	Lufthansa	40
Delta	12	British	23
Northwest	12	**UPS**	**41**
US Airways	5	Air France	47
FedEx	**48**	TWA	22
Continental	14	Air Canada	47

Sources: IATA World Air Transport Statistics, http://www.iata.org/ps/index_products.asp; Southwest Airlines, www.southwest.com.

Table 1-10. The integrated cargo carriers FedEx and UPS dominate their nearest competitors, in terms of freight tonnes carried in the year 2000

Airline	Tonnes, millions
FedEx	5.14
UPS	3.26
Korean	1.28
Lufthansa	1.12
Japan	0.98
Singapore	0.97
Cathay Pacific	0.77
Northwest	0.74
British	0.73

Source: IATA World Air Transport Statistics, http://www.iata.org/ps/index_products.asp.

Table 1-11. Predominantly cargo airports in the United States are a major factor, in terms of landed weight in the year 2000

Airport	Landed weight, pounds, billions
Anchorage	16.2
Memphis	12.6
Louisville	8.0
Miami/International	5.9
Los Angeles/International.	5.8
Indianapolis	5.8
New York/Kennedy	5.6
Dayton	4.5
Chicago/O'Hare	4.1
New York/Newark	3.9
San Francisco/Oakland	3.6
Dallas/Fort Worth	3.4
Philadelphia	2.9
San Francisco/International	2.5

Source: Federal Aviation Administration CY 2000 ACAIS database.

three most important cargo centers in the United States—Anchorage, Memphis, and Louisville—were on average twice as large as traditional cargo gateways such as Los Angeles/International, Miami/ International, and New York/Kennedy. Half the top airports in terms of cargo landed could be considered cargo airports for practical purposes, given their relatively low levels of passenger traffic. In the age of electronic commerce, cargo is no longer a peripheral activity secondary to passenger traffic; it can be a primary driver of airport development.

1-6   Implications for airports systems planning and design

Taken together, the trends in the airport and aviation industry are substantially changing the context, objectives, and criteria of excellence for airport planning and design. A narrow technical focus will no longer be satisfactory; it will simply not be responsive to the range of issues that must now be dealt with by airport professionals.

The context is commercial. Planners and designers are no longer designing primarily for administrators according to standard norms. They must respond to a broad range of business interests, such as the airlines, the airport operators, and concessionaires of all sorts. Through these immediate clients they will have to cater to their customers. This means that airport planners and designers will have to think in terms of profitability, revenues, and service to users.

The objectives consequently focus more on performance than on monuments. Value for money, good service, and functionality will become dominant considerations. Architectural significance and grand visions will be important, but may become secondary considerations. In general, airport planning and design will become more democratic, more in tune with everyday needs, and less directive or technocratic.

The criteria of excellence will correspondingly focus on cost-effectiveness, value for money, efficiency both technical and economic, and profitability. Airport planners and designers will have to factor these considerations into the purely technical analyses of traditional airport engineering. This requires skills not usually part of engineering or architectural training. It calls for a broad range of skills, including economic and financial analyses in particular. It extends beyond construction to operations and the management of risk. In short, it calls for a systems perspective.

A systems approach will be the basis for proper planning and design of airports in the twenty-first century. This will recognize that the technical issues themselves must be considered jointly as part of a larger system evolving over time to meet varying loads and demands. It will also recognize technical issues that should be considered in a wider context of values, objectives, and expectations. This text presents the essential elements of how this can be done.

Exercises

1. Obtain data on national or local growth of airport traffic or airline operations. What are the trends over the last 10 years? How do these compare with international or regional trends? Prepare a discussion of how you think the future traffic might evolve.

2. Investigate the level of privatization of some significant airport. Who owns the airport? Which activities are performed by government officials, by employees of some government-owned corporation, by independent contractors or consultants? How has this picture changed in the last 10 years? According to your sources, how might this picture evolve in the next decade?

3. Using the web and other sources, document the current status of a major international airport group, such as those indicated in Table 1-8. What is the scope of its relationships geographically? Does it specialize in specific kinds of airport activities? How effective does it appear to be?

4. Estimate the growth rate for integrated cargo carriers by comparing current statistics with those in Table 1-9. Use the web to obtain company reports on major carriers to document recent interesting developments. How do you see this activity developing in your region?

References

ACI, Airports Council International (2001) "ACI Traffic Data: World Airports Ranking by Total Passengers—2000," http://www.airports.org/traffic/td_passengers_doc.html.

Ashford, N., and Wright, P. (1991) *Airport Engineering*, 3d ed., Wiley, New York.

ATA, Air Transport Association of America (1977) *Airline Aircraft Gates and Passenger Terminal Space Approximations,* ATA, Washington, DC.

Australia Department of Housing and Construction (1985) *Airport Terminal Manual,* Australia Department of Housing and Construction, Canberra.

de Neufville, R. (1999) "Airport Privatization: Issues for the United States," in *Safety, Economic, Environmental, and Technical Issues in Air Transportation,* Transportation Research Record 1662, Paper 99.0218, pp. 24–31.

FAA, Federal Aviation Administration (1988) *Planning and Design Guidelines for Airport Terminal Facilities,* Advisory Circular 150/5360-13, U.S. Government Printing Office, Washington, DC.

FAA, Federal Aviation Administration, National Transportation Safety Board (2001) "Aviation Accident Statistics Database," http://www.ntsb.gov/aviation/Table2.htm.

Horonjeff, R., and McKelvey, F. X. (1983) *Planning and Design of Airports,* 3d ed., McGraw-Hill, New York.

IATA, International Air Transport Association (1995) *Airport Development Reference Manual,* 8th ed., IATA, Montreal, Canada.

ICAO, International Civil Aviation Organization (1987) *Airport Planning Manual, Part 1: Master Planning,* 2d ed., Doc 9184-AN/902, ICAO, Montreal, Canada.

Kahn, A. E. (n.d.) "Interview with PBS," http://www.pbs.org/fmc/interviews/kahn.htm.

Meyer, J., and Oster, C. (1984) *Deregulation and the New Airline Entrepreneurs,* MIT Press, Cambridge, MA.

2

International differences

Many aspects of airport planning, design, and management practice differ substantially in the various regions of the world. Sometimes these differences represent relative levels of advance in the adoption of innovative procedures or new technologies. In significant instances, however, these differences appear to represent deep-seated cultural perspectives. Various cultural contexts, countries or regions, have evolved different norms about relative competencies and obligations of the several stakeholders in airport operations, specifically concerning

- The role of central political power compared to that of the regions
- The permissible and desirable level of participation of private business
- The relative importance of technical experts and managers
- The criteria for excellent performance
- The rights and capabilities of workers

Practices common in one region may be socially unacceptable in another.

The variety of norms has two immediate consequences:

- There is no single right answer—the concept of excellence depends on the context, and thus
- The "best practice" of one region may not be transferable to another.

Airport planners and operators should recognize how the technical solution depends on the social values, that is, this social construction of technology. Consequently, global organizations should be careful about how they propose to transfer their practices from one region to another. Complementarily, airport operators need to be careful how they import "best practices" from other countries.

In short, airport professionals need to recognize that plans, designs, and operational practices often embody social and cultural assumptions. Therefore, they need to be sure that their proposals suit the local context and future.

2-1 Introduction

Air transport is a global business with remarkably similar international standards. As of the twenty-first century, all significant airlines use aircraft from one of the two dominant manufacturers, Airbus and Boeing. Manufacturers design aircraft to virtually the same standards.* International airlines carry passengers, baggage, and freight that have similar characteristics. In short, the air transport industry places almost identical requirements on airports and airport managers worldwide.

Two regulatory agencies define many of the international requirements for airport design and operation. These are the International Civil Aviation Organization (ICAO) and the United States Federal Aviation Administration (FAA). The ICAO is a United Nations agency that convenes international negotiations about aviation standards and promulgates the results. Their standards for airports are internationally accepted (ICAO, 1983, 1993, 1997a, 1997b, 2000). In parallel, the FAA also sets standards (for example, FAA 1987a, 1987b, 1999) and often establishes the norms that ICAO later follows.[†] See, for example, the discussion of aircraft categories and runway separations in Chap. 10. The FAA has a dominant role because the United States constitutes the largest single market for aviation, and has devoted the most money and research to establishing standards. Moreover, since essentially all aircraft manufacturers want to sell into the big North American market, they make sure their aircraft meet the FAA standards.

The international standards apply most strictly to matters concerned with the safety of aircraft in the air. International practice is thus almost identical for all elements of the airport that concern flight:

---

*This was not the case when the Soviet Union existed and maintained separate standards for its own aircraft such as the Tupolevs.
[†]The URLs for the list of regulatory and other publications from the ICAO and the FAA Advisory Circulars are http://www.icao.org/cgi/goto.pl?icao/en/sales.htm and http://www.faa.gov/circdir.htm.

runway markings and lighting, navigation equipment, and zones to be kept clear of obstructions around the airport. National differences in this regard are small.

National differences are great, however, when it comes to other features of airport planning, design, and management. Although the air transport industry places almost identical requirements on airports and airport managers, many nations develop or adopt their own distinctive solutions to these requirements. The technical requirements are similar, but the technical solutions are not. National differences in social values mediate the translation from the technical specification of the problem to the facilities and services that meet this specification. For example, American and European designers meet the requirement to position aircraft at aircraft gates in strikingly different ways, as Sec. 2-2 indicates. In general, the practice of airport planning, design, and management differs considerably among countries.

National differences in airport practice need special attention. Airport professionals and operators might falsely assume that, since the requirements are similar worldwide, the solutions should be also. The truth is otherwise. What is done in different national contexts, and indeed what should be done to meet local social requirements, often differs greatly from what might appear to be good practice elsewhere. This chapter presents this issue, and offers guidelines for airport professionals on how to cope with this reality.

2-2   Some physical differences

Some obvious physical differences in the design and operation of airports across the world motivate the discussion in this chapter. These illustrate how some seemingly tangential social assumptions and practices can have important consequences in terms of airport design, cost, and efficiency. They provide tangible evidence of the social construction of technology, the way cultural assumptions shape the seemingly technical solutions to design problems.* These examples indicate how this phenomenon occurs with respect to governmental and managerial practices. Readers can easily verify these examples visually by looking at different sites.

---

*Historians of technological development have extensively documented the social construction of technology, that is, the way social patterns and norms shape technological development. See Bijker, Hughes, and Pinch (1987), for example.

Check-in facilities

A passenger approaching a check-in counter in North America will normally encounter an agent standing behind a counter, in an open passageway running between the counter and the parallel baggage conveyor belt. During the check-in process, this agent and others are likely to move up and down the passageway as they sort out issues with colleagues. Finally, somebody will pick up the passenger's bags and place them on the conveyor (see Fig. 2-1). This is the normal practice met by about half the airline travelers in the world.

In much of Western Europe and elsewhere, by contrast, passengers checking in will typically meet an agent sitting down. The agent will not lift the bags. These will move on a small belt between the passenger and the main baggage conveyor belt. These small belts cut across the space behind the check-in counters and effectively prevent the agent from moving directly to other colleagues along the check-in desks. Agents needing to move to the other check-in posts may climb a set of stairs to an elevated passageway that passes over the conveyor belts and permits movement down to other check-in desks (see Fig. 2-2.) This arrangement is the alternative standard for check-in facilities.

Both approaches represent "best practice" in their own context. On the face of it, the North American arrangement is more cost-effective.

Fig. 2-1. *Typical U.S. check-in arrangement: personnel stand and move bags.* Source: Zale Anis and Volpe National Transportation Systems Center.

Fig. 2-2. *Typical European check-in arrangement: personnel sit and do not lift bags.* Source: Munich Airport.

It requires less equipment and permits agents to move around efficiently as needed. The European pattern, however, conforms to their concept of a humane work environment. Clerks and agents are entitled to sit down on the job, whether they work in airports or supermarkets. This social norm is widely established in much of Europe and unlikely to change in the near future. What people think of as the best solution is not a purely technical matter; the judgment rests on social and cultural assumptions.

The issue for designers arises when it comes to designing check-in facilities outside North America or Western Europe. Which tradition should they adopt? This is not merely a technical choice: it is a social judgment.

Aircraft contact stands

Aircraft "contact" stands are those that are in contact with the passenger buildings. They contrast with the "remote" stands away from the buildings. Passengers normally board aircraft at contact stands through some kind of permanent airbridge. To get on aircraft parked remotely, they must either walk or take some kind of bus.

The design of contact stands and the passenger building differs fundamentally between North America and Europe. Specifically, the connections between the passenger building and the aircraft tend to be very different. In North America, the usual arrangement is that aircraft in the contact stands are right next to the passenger building. The aircraft nose may be as close as 10 m, as Fig. 2-3 indicates. In this arrangement, the telescopic, movable airbridges connect directly between the passenger building and the aircraft.*

In Europe, however, aircraft at the "contact" stands typically park relatively far from the passenger building. The nose of the aircraft may easily be 25–40 m away. The system of airbridges can correspondingly be up to 70 m long. Airport operators in Europe, and in many countries worldwide, expect that vehicles operating on the airfield will normally not intersect the paths of aircraft. Ground vehicles will circulate on two-lane roads laid out at the face of each passenger building. Moreover, the design will typically provide parking spaces between this roadway and the building. Consequently, aircraft gates in airport buildings in Europe feature both 15- to 20-m bridges for passengers to cross over the road and substantial piers that support

Fig. 2-3. *Typical U.S. contact position: aircraft connect directly to building through fully mobile airbridge, no intervening road.* Source: Port Authority of New York and New Jersey.

*Exceptions to this rule exist. For example, the International passenger building at Boston/Logan has a road along its face, as do the midfield concourses at Denver/ International. In this case of the design of airport buildings, as for much of airport planning, design, and management in North America, independent local airport operators adopt their own practices.

these bridges and connect with the movable airbridges. The aerial view in Fig. 2-4 illustrates this pattern.

The difference in design of the contact stands can have enormous implications for their cost and the efficient use of extremely valuable airfield space. The European design most obviously requires a much greater investment for each gate, due to the cost of the fixed bridge over the roadway and the pier to support it. Moreover, this additional construction is only a fraction of the added expense. The greater cost comes from the inefficient use of space and the extra cost of making the passenger buildings longer. Example 2.1 illustrates the different implications of the two approaches to designing contact stands for aircraft.

What accounts for this expensive difference in practice? Simply put, Americans expect that the drivers of apron vehicles will drive safely and coordinate their movements with the control tower as necessary. The rate of accidents between aircraft and vehicles driving on the apron appears to be close to zero and is not an issue of real concern. At Washington/Dulles, for example, the airport operators have routinely bused over 10,000 passengers a day between the main passenger building and the midfield concourse. European and other airport operators also routinely transport thousands of passengers across aircraft taxiways to remote stands, as they do at

Fig. 2-4. *Typical European contact position: aircraft connect to building through a fixed bridge crossing an intervening road.*
Source: Munich Airport.

EXAMPLE 2.1: Effect of different standards for aircraft contact stands

Consider a passenger building with eight finger piers, such as has existed at Miami/International and been contemplated for Amsterdam/Schiphol and Taipei/Chiang Kai Shek.

The practice of laying out roadways on both sides widens the effective width of each pier by about 40 m, compared to the North American practice. Applying this standard to the entire eight finger piers lengthens the complex by about 320 m. This arrangement, with roadways in front of the passenger building, makes the central building connecting the piers much longer. This increases walking distances. It also makes the building more expensive. The cost of passenger buildings can be about $1000/m^2$. If the main passenger building is 25 m wide and has two floors, the extra cost of the complex due to the 320-m extension is about $16 million in construction costs—to which must be added operating costs including cleaning, climate control, and maintenance.

An arrangement with roadways in front of the building also limits the capacity of the airfield apron. The width required for each finger pier with roadways is 40 m. Assuming that aircraft are about 70 m long and taxiways are 80 m wide, as for a Boeing 747 or an Airbus 380, the total width for a finger pier with roadways is about 300 m. That is about 15 percent more space per aircraft than for a finger pier without roadways. This corresponding capacity reduction of around 15 percent may be critical at some airports.

Paris/de Gaulle, Zürich, and other airports that use remote parking. Yet European and other operators insist on having separate roadways for apron vehicles. Apparently, they do not have confidence in their employees or control systems. In general terms, this major design difference is another example of the way that social attitudes and assumptions shape technology.

Table 2-1 summarizes the specific differences in design practice these cases demonstrate. As these examples suggest, differences in national or regional practices can have significant consequences on both land-side and air-side operations. The examples are tangible instances of a general phenomenon. As the next sections discuss, the

Table 2-1. Some physical differences in design of airport passenger buildings between Western European and North American traditions

Facility	Design element	Western Europe	North America
Aircraft contact stands at passenger building	Aircraft: Distance from passenger building	25–40 m: Space for a two-way road and parking between aircraft and building	About 10 m: Essentially right at building, with no space for road at face of building
	Apron vehicles: Circulation	On airside road: Between building and aircraft	Across open apron: No dedicated road
	Apron vehicles: Parking	Special parking areas, often at face of building	At face of building, or in space around aircraft stand
Check-in facilities	Workstations	Seats provided: Agents sit, do not lift baggage	No seats: Agents stand, move around, and lift baggage
	Baggage handling	Small belts to main belt: Agents need stairs to move between check-in counters	Agents lift bags: Area between check-in counters is unobstructed

social and cultural differences between regions also lead to significant differences in airport planning objectives, procedures, and criteria. These may fundamentally affect the nature of airport operations in
different contexts.

2-3 Some useful distinctions

Countries and regions differ. Their ingrained habits and concepts are
not the same. Some of these distinctions have significant practical
implications for the planning, design, and operation of airports. This
section presents the dimensions of distinction that seem to drive the
most important consequences.

The important dimensions that characterize national differences are
not necessarily permanent. They reflect patterns of thinking and
behavior that people have learned. They may thus change over time
as the result of either evolutionary or cataclysmic events. In Britain,
for example, government became less centralized in the last quarter
of the twentieth century, as regions such as Scotland and Wales
developed their own legislatures. In some cases, a transition may be
abrupt. The highly centralized Soviet Union, for example, rapidly
dissolved into the Confederation of Independent States and a multi-
tude of autonomous regions. Despite the possibility of change, however, national characteristics are deep-seated. Airport managers can
assume for working purposes that local social assumptions and patterns of behavior are a fact of life.

Experts in comparative government and politics agree that there are
important differences in national attitudes and values. They also know
that it is difficult to describe the full complexity of national patterns
satisfactorily. The literature on the topic is controversial. Specialists
offer alternative, often-conflicting interpretations.* In this context, the
following discussion modestly suggests some of the important considerations. Its purpose is to alert airport practitioners that there are
important national considerations to take into account, and to stimulate them to think about how these considerations affect their practice.

For airport professionals, the important national differences are
those that affect the environment in which they operate. These dif-

*Readers wanting to explore these issues in detail may want to start with some of the
following important works, sorted according to the country they describe: Rose (1969, 1985)
and Hennessy (1989)—Britain; Suleiman (1974) and Cohen (1979)—France; Benedict (1946)
and van Wolferen (1989)—Japan; van der Horst (1996)—Netherlands.

ferences concern who, what, and how things are done. Specifically, airport operators need to understand:

- *Who makes decisions*—central authorities or pluralistic stakeholders?
- *What is the decision-making process*—is it directed autocratically or negotiated among interest groups? Who has the right to participate in this process?
- *Which values and goals are most important*—economic benefits? Social values? Regional prestige?
- *What are the criteria for excellence*—high profits? Good service? Beauty?

Two dimensions of national differences seem most important for airport planning, design, and management. These dimensions usefully define most of the answers to the questions of who, what, and why. These concern the diversity in the decision-making process, and performance criteria.

Diversity in the decision-making process reflects the number and types of stakeholders who strongly influence decisions. In some contexts there is effectively no diversity. This is the case when central authorities or personalities are the final arbiters. In other contexts there is great diversity, as multiple levels of government and numerous stakeholders negotiate resolutions to any issue. Countries also differ in the kinds of goals they promote and the criteria they apply. In some cases, decision makers define goals quite specifically and numerically, either in technical or in economic terms. In other cases there never is any clear definition of goals or objectives.

These two dimensions correlate with each other to some extent. Centralized, directive governments have the ability to impose criteria on the decision-making process. Pluralistic decision-making processes that negotiate developments will not be able to maintain, let alone impose, consistent numerical criteria of performance. Centralized decision-making processes are therefore more likely to be able to impose performance criteria—although they do not have to do so.

National differences in diversity of decision making

Countries that have had salient roles in the development of airport systems differ greatly in the way they expect decisions to be made. Several have strong traditions of central direction and control. Other countries are pluralistic and feature decentralized decision making.

In the United States, decisions about airports are highly decentral-ized. The central national institutions have little influence on specific designs—surprisingly so for persons from outside North America.* Under the Constitution of the United States, the power to make most major decisions—those concerning airports in particular—is in the hands of the states. Moreover, the state constitutions frequently leave decisions about such matters to local communities. Most frequently, local airport authorities and cities are responsible for developing plans and airport proposals. The U.S. Federal Aviation Administra-tion (FAA) can support, encourage, and confirm local decisions, but cannot impose its will.

In the United States, all major stakeholders are entitled and expected to have an active voice in decisions about airports. Airlines, for example, frequently operate their own passenger buildings. Normally, major airlines also participate actively in the design of these facilities. As Chap. 7 indicates, airline tenants at many U.S. airports are guarantors of the revenue bonds and thus effectively have veto power over major investments on airports. Thus, airlines in the United States often can control what is built, how it is designed, and when it is implemented. Additionally, local communities and interest groups expect to be able to participate actively in the decision-making process. Local stakeholders concerned with the airport are likely to have specific rights to intervene. For example, the board of directors for Massport, the independent state agency responsible for operating Boston/Logan airport, by law includes representatives of local com-munities, citizens groups, and labor unions. Decisions about airport planning, design, and management in the United States are negoti-ated among the many stakeholders. These generalizations about the United States have exceptions, in view of the enormous diversity among the 50 states. However, diversity in the decision-making process is a fact in the United States.

To illustrate the decentralization and diversity of authority on airport activities in the United States, consider the proposal to develop a new runway for Boston/Logan airport. Around 1998 Massport, the inde-

---

*To illustrate the force of this pervasive aspect of political life in the United States, con-sider the question of education. Although the United States has a national Department of Education (corresponding to a Ministry of Education in other countries), this institution is a late-twentieth-century innovation and has almost no impact on how schools are run or on curricula. By tradition in the United States, schools are directed by school boards elected by local communities and paid for almost entirely by local taxes.

pendent local authority responsible for the development of this facility, proposed to build a short runway. Airport planners in the FAA widely encouraged this proposal to add capacity. Nominally, the FAA Administrator has the authority to approve this plan. In practice, however, elected officials from the area effectively have the power to block such plans. Local members of Congress have done so for many years, by threatening to block portions of the FAA's budget. In the United States, these kinds of planning issues are resolved through intense negotiations between various local authorities, the airlines, the FAA, as well as numerous advocacy groups represented by lobbyists. The tradition in the United States is that essentially all stakeholders in an issue have the right to participate in their resolution.

France, by contrast, has a history of central direction. The central government announces decisions and implements them. The public elects the government, of course. The government also pays attention to public needs, and has prepared and is beginning to implement mechanisms for compensation and remediation of damages to people and the environment (Faburel, 2001). However, the public has been neither expected nor is entitled to participate in the decision-making process itself (Block, 1975). Thus, in the 1970s, the French government located the Paris/de Gaulle airport, established development zones, built the facility, and directed specific airlines to relocate from the other airport, Paris/Orly. At the end of the century they replicated the process to develop two new parallel runways at Paris/de Gaulle. These developments went forward without significant public hearings or effective protest. French authorities expect to be able to act decisively in the best interests of the public.

Historically, many countries have had traditions of centralized national power, both in general and specifically with regard to airport planning. Typically, countries have had national ministries responsible for the design, construction, and operation of airports. Around the turn of the twenty-first century, however, the degree of centralization of airport planning and management in major aviation markets lessened considerably. Australia and Canada virtually eliminated their federal airport agencies. These countries devolved responsibilities for airports to companies and local authorities in the traditionally autonomous states and provinces. Britain transformed the governmental British Airports Authority into the BAA company, and made local airports—such as Manchester and Birmingham—operate as independent companies. Mexico effectively devolved power from the central government to independent regional companies. This evolution has created more autonomous airport authorities and increased the diversity of airport operators.

Germany, Switzerland, and Italy have traditionally been decentralized. Germany is a federal system whose components (the Länder) have considerable independence. Thus independent groups own and operate the major German airports (Berlin, Frankfurt, Munich, Düsseldorf, Hannover, etc.). Switzerland is also a federation whose parts (the cantons) are notably autonomous. Independent companies thus operate the Zürich and Geneva airports. The situation is comparable in Italy. The Rome, Milan, and Naples airport companies are not only independent but also part of extensive international networks of airport companies.*

Outside the United States, the increased diversity in airport operators in various countries has not generally translated into increased diversity in the local decision-making processes. The power to plan and design airport facilities is still typically in the hands of the airport operator. Airlines typically have little say in the definition of airport investments. For example, British Airways had essentially no part in the design of the proposed billion-dollar Terminal 5 at London/Heathrow, for which it is the presumptive main tenant. Likewise, at Frankfurt/ Main the airport presented Lufthansa in the mid-1990s with an equivalently expensive new passenger building. This building was totally unsuited to Lufthansa's hubbing operations, however. In that case, Lufthansa managed not to occupy the building. Local constituencies likewise generally do not have a deciding role in airport decision making. Environmental and other groups may be heard or consulted, but they do not decide.

Britain has established the rule that an extensive public inquiry must be held for important issues. The investigation into the construction of the second runway for Manchester took about 5 years. The inquiry into the T5 passenger building at London/Heathrow took longer, and reputedly cost over £81 million (about $125 million) (Thorpe, 2001). This practice is totally different from the procedures in France. In Britain, the interest groups have the right to express themselves and delay planning. Legally, however, they have no power. The Minister of State for the central government decides such issues authoritatively.

---

*As of 2002, the Rome airport had a major stake in the South African Airports Company, and the Milan airport had a significant share of the Argentine airports. The British airports company BAA had a major stake in the Naples airport.

A quote from the British political philosopher Edmund Burke illustrates the fundamental differences in perspective on decision-making processes between countries. In this case, he was contrasting the centralized, unitary British view with the pluralistic, negotiated practice of the United States.

> *Parliament is not a* congress *of ambassadors from different and hostile interests....Parliament is a deliberative assembly of one nation, with* one *interest, that of the whole, where not local purposes, not local prejudices ought to guide, but the general good.... (Burke, 1774)*

Overt negotiations among stakeholders to determine airport development are rare outside North America. In many contexts, such negotiations would be taboo. A common sentiment is that the duty of government is to govern, and if they cannot do so, they should resign. The case involving the second parallel runway at Tokyo/Narita illustrates the point. In this situation, several farming families did not wish to sell their land to make way for the construction of this facility. For more than 30 years, a handful of people prevented the completion of a major addition to a significant national asset, yet the national societal and political conventions prevented the authorities from negotiating or adjudicating any kind of compromise that would allow the nation to proceed.

Table 2-2 summarizes these national differences in assumptions about who gets to decide how airports should be planned, designed, and managed. As Sec. 2-4 indicates, these dissimilar perspectives can influence airport development fundamentally.

National Differences in Performance Criteria

Countries that have had salient roles in the development of airport systems also differ greatly in the way they define and state the objectives

Table 2-2. Some national differences in diversity of decision making

		Diversity among airport operators		
		Little	**Some**	**Extensive**
Diversity in decision-making processes	Airports decide	France Japan	Australia Mexico	Germany Italy
	Stakeholders negotiate			United States

for airport planning. In some contexts, the objectives are general and loosely defined. In others, they may be quite specific. Since performance criteria shape the products of design, these differences have significant consequences for how nations develop airports.

The nature of the performance criteria depends on who defines them. It is therefore relevant to look at the kind of people who run airport planning agencies and operators. The differences between countries can be striking. Some countries recruit elite engineers into careers of management of public works and airports in particular. Other countries prefer generalists or economists rather than specialist engineers. Some countries have no particular pattern at all. These national patterns mark the practice of airport planning and management for these countries.

In the United States, there is no visible career pattern for the recruitment of airport executives. Leaders in the field are lawyers, managers, engineers, former military officers, and other professionals. They tend to enter airport planning from some other industry. Typically, they have established themselves in a related field, have been working for one of the stakeholders in the airport business, and have thus become involved in airport planning. Recent leaders of Massport, the operator of Boston/Logan airport, have thus included a lawyer who had been a special assistant to the mayor of Boston, a former local Congressman, an activist for local groups concerned about noise, and an aeronautical engineer who had become a prominent entrepreneur. Such people bring a wide range of perspectives and norms for good performance to airport planning.

Some generalized performance criteria define airport planning in the United States. These emerge from distinct negotiations among interested parties. The Federal Aviation Administration publishes standards for the air-side of the airport, based on their mission to promote safety. These appear in their Advisory Circulars and are readily available on the web and in print. However, the FAA does not establish these norms by itself. It works them out through close discussions with industry groups such as the Air Transport Association (airlines), the Airports Council International (airports), the General Aviation Manufacturers Association, the National Association of State Airport Operators, and so on.

Criteria for economic performance of airports in the United States come from somewhere else entirely. These emerge from the groups that supply the funds, notably the airlines that pay the fees and the investment bankers that loan the money.* These standards are informal and negotiable. Various banking firms distribute their own versions, for example, Moody's (1992, 1997a, 1997b). The consensus is loose but has important implications that imprint a distinctive mark on airport planning and management in the United States. Briefly stated, it is that airports should be:

- Operated as businesses with transparent public accounts (to reassure investors and guarantee repayment of the loans)

- Run as a public service and are not supposed to make profits beyond what is needed to maintain the business

- Charging airlines fees as low as possible consistent with good service and attractive facilities.

France, by contrast, recruits its leaders for airport planning and management from its most talented engineers. Specifically, it has generally obtained the future leaders for its national airports company, the Aéroports de Paris, largely from its most selective national engineering school, the Ecole Polytechnique. It inducts about 30 of its best graduates each year and places them in a quasi-military organization, the Corps des Ponts et Chaussées (the Corps of Bridges and Roads). These persons all share the same background, the same analytic and engineering approach, the same esprit de corps (see Suleiman, 1974). Similarly, it recruits lower-level engineers from less demanding national schools and corps, such as that of the Travaux Publics de l'Etat (State Public Works). As can be expected, these professionals establish analytic and precise performance criteria. Although some say this tradition is in decline, its legacy persists.

Britain has a tradition that prefers generalists to specialists. A common view is that specialists become too involved in their field and cannot be trusted to have a sufficiently broad national perspective. Thus British education at the elite universities such as Oxford and Cambridge stresses education in liberal fields such as political economy, classical literature, and history. The British government has likewise recruited its future leaders from among such people (see

*To put the role of bankers and airlines in perspective, consider the funding for the construction of Denver/International. In this case, bankers raised about $3.2 billion in loans, secured mostly by airline commitments. Thus, United Airlines agreed to pay what amounts to about $200 million/year for its midfield concourse. The Federal Aviation Administration, however, paid only about $800 million of the total capital cost of the new airport.

Hennessy, 1989). These professionals take a broad, pragmatic view of decision making.

An anecdote captures the difference in approach between the technical and generalist approach perspective on airport planning. It involves British and French airport managers in the 1970s. At that time, the British Airport Authority (BAA) had recently introduced peak-hour pricing at their London airports. This practice charges higher prices during peak hours. It thus reduces the peak demands, and the capacity and capital expenditures that the airport operator has to provide. It is an important means of increasing economic efficiency (see Chap. 12). From a theoretical perspective, it is possible to define the best peak-hour prices analytically from the equation of the demand for services. The BAA recognized, however, that the complexity of airline operations made any estimation of the demand speculative. Pragmatically, they thus chose to introduce a flat peak-hour surcharge on each peak-hour operation. They intended to establish the principle of the surcharge, and then to raise or lower the peak-hour price until it achieved the effect they intended. At this point, a French team came to London to learn from the British experience. They asked to see the equations and calculations the BAA used to determine the charge. The BAA said they could not present these, as they did not have any. When the French left the BAA, they exploded in anger at the "uncooperative, untrustworthy British," who refused to share with them. Despite the author's attempts to explain the situation, they would not believe that the British had such a different perspective. Being engineers steeped in equations, they seemingly could not appreciate the possibility of such a different outlook. Yet such differences are real, and do complicate the international understanding of airport planning.

By the beginning of the twenty-first century, British authorities were establishing detailed performance criteria for airports. These were economic, in contrast to those typically prevalent in other countries. The need for these criteria resulted from the British privatization of airports. As Chap. 4 explains in detail, national authorities need to regulate private airport companies to protect the public against monopoly pricing and excessive charges. It is not sufficient to regulate prices, however. Companies can abuse a monopoly position by lowering standards of service. The British regulatory authorities are therefore establishing complex performance criteria in many different areas (see United Kingdom, Civil Aviation Authority, 2000). These criteria will constrain UK airport operators to emphasize the services specified in the regulatory criteria. They will shape the way British airport operators do business.

Overall, the important aspect to retain is that different national traditions emphasize distinct aspects of performance. The criteria prevalent in the United States are generally pragmatic and value the distinct interests of the important consumer groups, such as airlines and passengers. The French and Japanese traditions give more weight to technical factors. The British approach typically favors economy (see Table 2-3). These perspectives strongly influence how airport planners and managers from these traditions build and operate airport systems, as the next section indicates. World travelers will recognize the differences from experience.

2-4 Implications for practice

National differences in concepts of decision making about public projects and of performance criteria lead to significant general and practical implications. In the increasingly global practice of airport planning, these effects are becoming more significant for practitioners. European airport companies operating in the Americas, for example, need to think carefully about how they might modify their practice to suit the local context. North American consultants advising on airport development and operations overseas likewise need to tailor their suggestions to local realities.

General implications

There is no single right answer. This is the most fundamental lesson the reader should appreciate from this chapter. Since the concept of excellence depends on the context, the best practice in one region may be impractical or otherwise unsuitable in another.

Table 2-3. Some national differences in performance criteria

		Diversity among performance criteria		
		Economic	**Mixed**	**Technical**
Diversity among- decision makers	Career specialists			France Japan
	Widely recruited generalists	United Kingdom	United States	

Although analysts may agree on the operational characteristics of different designs, when they disagree on the relative importance of these factors, they may not agree on which is best. For example, it is clear that the North American design for check-in facilities requires less capital investment, and the European practice makes work easier for the check-in agent. Which design is better depends on how the airport operator and the local society value these features.

The related general implication is that "best practices" of one region may not be readily transferable to another. "Best practices" in some places may appear to be poor or unsuitable practices elsewhere. Even when foreign best practices do appear superior to local procedures, they may be sufficiently counter to national norms to make them impractical. For example, it would be difficult to introduce the standard U.S. design for check-in facilities in France or Britain, even if an airline or airport wanted to do so. Such a change would require extensive negotiations with workers and changes in work rules.

Finally, national differences in performance criteria limit the usefulness of international "benchmarking" of airports. "Benchmarking" is the practice of comparing performance at various sites in specific industries. The objective is to identify the sites that perform best in various categories. The best performance in each category then becomes a benchmark, that is, a standard for the rest of the industry. Benchmarking can identify sites that perform better than others overall and that might be taken as models. It can also identify sites that perform poorly overall, and that might need management attention. However, these individual measures of performance are difficult to translate into any internationally meaningful overall measure of performance. Any weighting of the categories represents notions of relative value that will not represent the priorities of all countries (see the discussions in de Neufville and Guzmán, 1998; United Kingdom, Civil Aviation Authority, 2000).

The fact that technical solutions depend on social values means that global organizations should be careful about how they transfer practices from one region to another. The experience of Amsterdam Airport Schiphol (AAS) at New York/Kennedy airport illustrates the issue. In the late 1990s, AAS committed to design, construct, and operate an international passenger building there. They brought with them their excellent reputation and expertise from running an attractive facility at Amsterdam. They proceeded to design the New York facility along the same lines as Amsterdam. In particular, they planned to cater to a variety of foreign airlines, each having a relatively small presence at New York/Kennedy. However, AAS apparently did not understand the power of airlines in the United States to make their own arrangements. In Europe, airlines rarely have much influence on the design of passenger buildings. In the case of New York/Kennedy, airlines disrupted the plans of AAS in two ways. First, significant airlines left the International facility, either to a new building they built themselves (as Air France and its associates did in Terminal One) or to the buildings operated by their alliance partners, such as American Airlines. This phase left AAS with a big investment and insufficient tenants. Secondly, Delta Air Lines then agreed to take over much of the AAS building, provided it were redesigned. Delta wanted a standard U.S. configuration, favoring transfer operations and placing commercial activities near the departure gates, beyond security—exactly opposite to normal practice in Amsterdam. In short, AAS suffered when they tried to transfer excellent Dutch practice to New York. Conversely, the dependence of technical solutions on social values means that airport operators need to be careful how they import "best practices" from other countries.

Specific implications

The differences in decision-making processes and criteria of performance translate into specific differences in how airport operators develop their facilities different countries. These concern

- *artifacts*—what they construct
- *type of service*—the features they stress
- *operations*—how they manage their properties

This section mentions some salient examples. Later chapters discuss these in detail.

In the United States, stakeholders in airport operations participate extensively in the decision-making process. The result is that the design of the airport reflects their concerns. For example, airlines like to minimize the time their aircraft have to taxi. As Chap. 14, on the configuration of airports, explains, efficient designs can save the airlines hundreds of millions of dollars a year. Therefore, when airlines have a strong voice in the design of airports, as they do in the United States, they insist on designs that facilitate easy movement. These stagger the runways, so that landings end and takeoffs start near the

passenger buildings. They also pave over large areas and thus eliminate restrictive taxiways that require aircraft to make many turns.
For example, U.S. airport operators typically pave over the entire
space between finger piers. Elsewhere, at Amsterdam/Schiphol, for
example, large portions of this space may be left unpaved or is set
aside for lights and is otherwise unavailable for aircraft maneuvers.
The comparison of Atlanta and Kuala Lumpur/International illustrates this phenomenon. Both airports feature parallel runways on
either side of passenger buildings. However, the paths the aircraft
follow are much more direct and operationally less expensive at
Atlanta (see Figs. 2-5 and 2-6).

Similarly, airport operators in the United States tend to cater to the
individual desires of passengers. Specifically, they provide extensive
parking facilities at affordable prices and promote easy access for

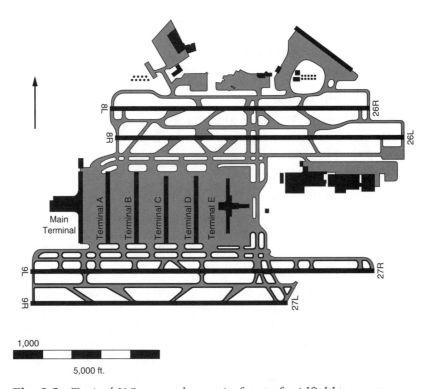

Fig. 2-5. *Typical U.S. apron layout in front of midfield passenger
building: aircraft can access gate with a minimum of turns.*
Source: Atlanta/Hartsfield Airport.

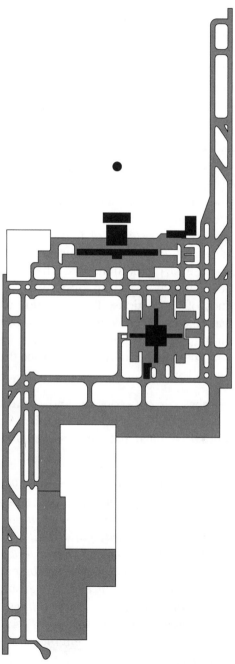

Fig. 2-6. *Taxiway layouts at Kuala Lumpur/International: aircraft require many turns to access gates.*

automobiles. This permits people to proceed from home to airport directly door-to-door and is most convenient for each individual. In countries with centralized decision making, however, airport operators favor collective means of airport access (see Coogan, 1995). They channel travelers into patterns that require combinations of travel by taxis and trains, which are inherently less convenient for individuals although they may be beneficial to the area as a whole. See Chap. 17 on airport access.

National differences in the concepts of excellence also influence the types of service airports offer. The French emphasis on technical excellence, for example, leads them to develop state-of-the-art innovations. The way they have integrated their high-speed rail system, the TGV, into Paris/de Gaulle and Lyons airports illustrates this phenomenon. Moreover, the existence of the TGV itself demonstrates the power of the central government to impose technical excellence for the national cause of public transport. The managing technical elite considers the airports to be an opportunity to develop and showcase all kinds of innovations. These include unique developments such as variable-speed moving sidewalks, baggage belts lifting vertically through several stories, and check-in facilities after passenger control. They deliberately seek to place themselves in the role of technological leaders.

In Britain, the emphasis is on economy and return on investment. Naturally, this leads to less service and elegance. A popular British author thus described Terminal 4 at London/Heathrow in the following terms:

> *Long, slow-moving lines stretch from the check-in desks nearly to the opposite wall of the concourse, crosshatched by two longer lines converging upon the narrow gate that leads to Passport Control, the Security gates, and the Departures Lounge. The queuing passengers shift their weight from one foot to another, or lean on the handles of their heaped baggage trolleys, or squat on the suitcases....[He looks] up at the low, steel-gray ceiling, where all the buildings' ducts and conduits are exposed...which makes [him] feel as if he is working in a hotel basement or the engine-room of a battleship. (Lodge, 1992, p. 3)*

Centralized decision making also leads to operational procedures that are quite different from those prevailing in regions where decision-making power is distributed. In this respect, practice in the North American half of the airports market contrasts with that in the rest of the world. Thus in Europe, Japan, and elsewhere, planning processes are directive and indicate what will happen. In the United States, on the other hand, plans are merely suggestive, as Chap. 3 indicates. As Chap. 12 indicates, busy European and Asian airport operators typically manage their airspace through formal allocations of the "slots" for aircraft arrivals and departures. They also frequently place surcharges on aircraft operations to discourage or effectively ban smaller aircraft from the congested airports. Such procedures are rare in North America. The airlines and operators of small aircraft have rights to operate pretty much when they choose, just as drivers are free to get in their cars and drive.

The pluralistic nature of the United States is evident throughout the operation of the airport itself. In North America, it is usual to have dozens of independent contractors managing various bits of the airport. Airlines, for example, will handle their own baggage and check-in operations, often even their own passenger buildings. Competitive national corporations routinely manage the parking facilities, often several at the same airport. Independent contractors usually do the cleaning and operate security. Architecture and engineering firms will carry out the design and construction management for the airport. As Chap. 4 describes, most U.S. airports are highly privatized in that most of their operations are run by private companies. The situation has been vastly different in the rest of the world. The pattern elsewhere is that the airport operator has provided all services. In Europe, antimonopoly directives now require airports in the European Community to permit competitive services, but the major airports typically offer and provide the whole range of operational services. A comparison between Boston/Logan and Frankfurt/ Main illustrates the difference. Both airports are about the same size and have around 15,000 to 17,000 workers on the airport. At Frankfurt/ Main, most of these employees work for the airport operator, whereas at Boston/Logan, only about 800 work for Massport. Table 2-4 summarizes the range of these particular distinctions.

Table 2-4. Some particular distinctions in airport systems planning and design between North America and the rest of the world

Area of practice	Common practice in	
	North America	**Rest of world**
Facility construction	Generous airfield paving to facility aircraft ground operations	Restrained amount of paving for taxiways and aircraft aprons
	Emphasis on private cars, automobile access, parking	Emphasis on collective transportation, rail access
Planning	Suggestive	Directive
Operations	Airlines schedule freely as they wish (with few exceptions)	Airports allocate landing and takeoff slots to airlines
	No discriminatory pricing; all users have access	Peak-hour pricing common; small aircraft often excluded
	Airport operator has small staff; most services contracted out	Airport operator is a big employer; airport offers most services

Exercises

1. For some airport of interest to you, use the web or other references to identify the managers and their professional backgrounds. What kind of professional formation do they share, if any? How would you characterize this group? What does this imply for their decision making? If time and means allow, repeat this process for a foreign airport and compare the results.

2. Look into the British process for deciding whether the BAA could build Terminal 5 at London/Heathrow airport. The British planning inspectorate has amply documented this exercise (http://www.planninginspectorate.gov.uk/pi_journal/thorpe_art2.htm).

3. Examine the decision-making process concerning the development of a major runway in the United States, for example, at Boston/Logan, Cincinnati, Miami/International, San Francisco/International, or St. Louis/Lambert. Who do you think made the important decisions in this case? What was the power of the

several stakeholders? What do you conclude about airport planning processes in the United States?

4. Repeat the previous exercise for the development of a major new passenger building in the United States, for example, the International Buildings at New York/Kennedy or San Francisco/International, or the American Airlines buildings at Miami/International or Chicago/O'Hare. If time and interest allows, compare your conclusions with the results of the previous exercise.

References

Benedict, R. (1946) *The Chrysanthemum and the Sword*, Charles Tuttle, Rutland, VT, and Tokyo, Japan.

Bijker, W., Hughes, T. P., and Pinch, T. (1987) *The Social Construction of Technological Systems: New Directions in the Sociology and History of Technology*, Papers of a workshop held at the University of Twente, The Netherlands, July 1984, MIT Press, Cambridge, MA.

Block, J. (1975) "Planning the European Environs—A European Viewpoint," *Proceedings ASCE Conference on International Air Transportation*, San Francisco, March, pp. 191–204.

Burke, E. (1774) "Speech to the Electors of Bristol," in *Works of the Right Honorable Edmund Burke*, vol. 1, Henry Bohn, London, 1854, pp. 446–448. Also available in P. Kurland and R. Lerner, eds., *The Founders' Constitution*, 1(13), University of Chicago Press, Chicago, IL. http://press.pubs.uchicago.edu/founders/documents/v1ch13s7.html.

Cohen, S. S. (1979) *Modern Capitalist Planning: The French Model*, University of California Press, Berkeley, CA.

Coogan, M. (1995) "Comparing Airport Ground Access—A Transatlantic Look at an Intermodal Issue," *TR News, 174*, pp. 2–10.

de Neufville, R., and Guzmán, J. R. (1998) "Benchmarking Major Airports Worldwide," *ASCE Journal of Transportation Engineering*, TN 11639, *124*(4), July/August, pp. 391–396.

FAA, Federal Aviation Administration (1987a) *Emergency Evacuation Demonstration*, Advisory Circular 20-118A, U.S. Government Printing Office, Washington, DC.

FAA, Federal Aviation Administration (1987b) *Application for U.S. Airworthiness Certificate*, Advisory Circular 211-12A, U.S. Government Printing Office, Washington, DC.

FAA, Federal Aviation Administration (1999) *Standards for Airport Markings*, Advisory Circular 50/5340-1H, U.S. Government Printing Office, Washington, DC.

Faburel, G. (2001) *Le bruit des avions—evaluation du coût social*, Presses de l'Ecole Nationale des Ponts et Chaussées, Paris.

Hennessy, P. (1989) *Whitehall*, Secker and Warburg, London, UK.

ICAO, International Civil Aviation Organization (1983) *Aerodrome Design Manual, Part 5*, Doc 9157, Montreal, Canada.

ICAO, International Civil Aviation Organization (1993) *Aerodrome Design Manual, Part 4, Visual Aids*, 3d ed., Doc 9157, Montreal, Canada.

ICAO, International Civil Aviation Organization (1997a) *Aerodrome Design Manual, Part 2, Taxiways, Aprons and Holding Bays*, 3d ed. 1991, reprinted July 1997, incorporating Corrigendum 1, Doc 9157, Montreal, Canada.

ICAO, International Civil Aviation Organization (1997b) *Aerodrome Design Manual, Part 3, Pavements*, 2d ed. 1983, reprinted November 1997, incorporating Amendments 1 and 2, Doc 9157, Montreal, Canada.

ICAO, International Civil Aviation Organization (2000) *Aerodrome Design Manual, Part 1, Runways*, 2d ed. 1984, reprinted July 2000 incorporating Amendment 1, Doc 9157, Montreal, Canada.

Lodge, D. (1992) *Paradise Lost*, Penguin Books, New York.

Moody's Investor Services (1992) "Moody's on Airports: The Fundamentals of Airport Debt, Featuring Three Airport Case Studies," Moody's, New York.

Moody's Investor Services (1997a) "Implicit Government Pledges for Airport and Air Navigation System Related Infrastructure Project Debt," Report 25392, August, Moody's, New York.

Moody's Investor Services (1997b) "Project Finance Alternatives to Airport Infrastructure Development—Non Recourse, Non Airline-Secured Special Facilities Bonds," Report 27481, August, Moody's New York.

Rose, R., ed. (1969) *Policy-Making in Britain—A Reader in Government*, Macmillan, London, UK.

Rose, R. (1985) *Politics in England: Persistence and Change*, 4th ed., Faber and Faber, London, UK.

Suleiman, E. (1974) *Politics, Power and Bureaucracy in France: The Administrative Elite*, Princeton University Press, Princeton, NJ.

Thorpe, K. (2001) "The Heathrow Terminal 5 Inquiry: an Inquiry Secretary's Perspective," http://www.planning-inspectorate.gov.uk/pi_journal/thorpe_art2.htm.

United Kingdom Civil Aviation Authority (2000) "Use of Benchmarking in the Airport Reviews," A consultation paper, December, http://www.caaerg.co.uk/.

van der Horst, H. (1996) *The Low Sky: Understanding the Dutch*, Scriptum Books, Schiedan, the Netherlands.

van Wolferen, K. (1989) *The Enigma of Japanese Power: People and Politics in a Stateless Nation*, Knopf, New York.

Part Two
System planning

3

Dynamic strategic planning

Dynamic strategic planning is the approach recommended for future airport development. It is traditional master planning adapted to the realities of the airport and aviation industry of the twenty-first century. It recognizes future uncertainties and leads to a flexible development strategy that positions airports to minimize risks and take advantage of opportunities.

Airport planning continues to be primarily a local concern, done airport by airport. As governments privatize airports, national and regional planning of airports has become rare. Increasingly, however, companies operating many airports in different countries are developing corporate strategies for their collection of physical and operational assets.

The forecast is "always wrong." Modern planners and managers must face this reality in the era of deregulation and competition. Airlines form alliances, merge, and change their routes and services; passengers and shippers reorient their patterns. These variations make forecasts of levels and types of traffic unreliable. Airport professionals must assume that the future reality will easily be different from what seems most likely at present.

Responsible airport planning anticipates the range of possible futures. It then positions the airport to obtain the best performance in the future. It should verify that proposed developments will be able to respond dynamically to these possibilities. It will build in appropriate flexibility throughout the systems to facilitate transitions to new situations. Overall, it will develop a strategy for dealing with future uncertainties.

Dynamic strategic planning augments traditional master planning. It leads planners to consider several possible futures and scenarios of operation, not merely a single forecast. It uses decision and real option analysis to determine the appropriate level of flexibility to incorporate into development plans. It defines organizational relationships that will allow management to develop the right facilities according to future requirements as these emerge. Most important, it provides managers with the support they need to program the implementation of overall strategic direction.

3-1 Forms of planning

The concept of planning needs explanation. It means different things in different contexts, to planning professionals and to airport planners in particular. Specific words and phrases, such as "plan," "master planning," and "strategic planning," have acquired meanings that are not obvious. Persons who have not been intimately involved in these practices or are not aware of local differences may easily get confused. It is therefore useful to identify what the several words for planning can mean in the context of airport systems.

Plans

Professionals from different contexts do not share a common understanding about what the basic concept of planning implies. In a general way, all agree that planning involves the preparation of a response to some possible future events, and that a plan is some conceptual roadmap of what could be done. They disagree, however, between two contrasting perspectives. Is a plan:

- A directive blueprint from top authorities that specifies what is to happen? Or
- A collection of possible local suggestions of what airports might like to do, all of which are debatable or negotiable?

In many contexts, planning is a top-down, directive activity. Elite groups, typically high government officials, prepare plans for important sectors or even the whole economy. They then transmit these plans to powerful subordinates for execution. They allocate resources according to these plans to make sure that they are executed. This approach arguably has some merit, as Chap. 2 discusses. It has been and continues to be the tradition in important capitalist countries, such as France and Japan (see Cohen, 1969; Johnson, 1984).* In those two

---

*France has a centuries-old tradition of control by central government, as Chap. 2 describes. After World War II, France established the Commissariat Général du Plan, an organization that prepared 5-year plans for the economic development of the nation. These exercises

countries and others, the practice has largely been successful in its own terms. In the former Soviet countries, however, directive planning failed miserably and has been discredited. Airport planning in Japan, for example, is directive (de Neufville, 1991). The responsible Japanese national ministry has systematically identified and developed a sequence of major national projects, such as the island airports of Hiroshima, Osaka/Kansai, and Nagoya/Chubu, and of regional airports for each prefecture.*

In the United States and some other countries, planning is a bottom-up, visionary activity. Local authorities prepare their own plans and transmit their efforts, to the extent they want to, to some central office that collates them for presentation. This practice is common in countries that have strong regional governments, such as the provinces in Canada, the Länder in Germany, and the states and cities of the United States. In the United States, for example, every 2 years the Federal Aviation Administration (FAA) has prepared a National Plan of Integrated Airport Systems (NPIAS) that explicitly is an uncoordinated collection of local wishes:

> Because the NPIAS is an aggregation of airport capital projects identified through the local planning process, rather than a spending plan, *no attempt is made to prioritize the projects* that comprise the database *or to evaluate whether the benefits of specific development projects would exceed the costs.* [Italics added] (Secretary of Transportation, 1999, p. vi)

Local airports prepare their lists of possible projects according to what they see as best for them, without consulting other airports and often in direct competition with them.† These "plans" are in no sense guides as to what will happen, and certainly do not dictate any specific allocation of money. These wishful local plans are totally different from directive national plans. Readers should keep this difference in mind whenever they read or listen to international colleagues.

---

are no longer consequential, now that the French government is privatizing many industries that were previously nationally owned, such as banks, telephones, railroads, and so on. However, the tradition of central direction and execution continues.

*As of 2001, the relevant ministries were merged into a new Ministry of Land, Infrastructure and Transportation. The basic planning process did not change, however.

†In the United States, local authorities receiving airport grants from the federal government have been expected to follow some general procedures in preparing state and metropolitan aviation systems (FAA, 1970, 1989). These guidelines do not transform the plans into directives, however.

Master plans

Master plans have a very specific meaning in the context of airport planning. As stated by the International Civil Aviation Organization (ICAO):

> An airport master plan presents the planner's conception of the *ultimate development of a specific airport.* [Italics added] (ICAO, 1987, pp. 1–2)

This definition is widely accepted internationally. Representatives of the states that belong to the United Nations, of which ICAO is a constituent part, developed and agreed to it.

A master plan focuses on an architectural/engineering development at a single airport. Note that the definition of the master plan involves three essential notions. It refers to:

- Ultimate vision, that is, a current view of the possible future a long time in the future, for example 20 years
- Development, that is, the buildings, runways, and other physical facilities—not operational concepts or management issues
- Specific airports, not to a regional or national aviation system

The master plan is thus tightly constricted compared to national plans that governments have prepared and implemented in Australia, Canada, France, Japan, Mexico, and elsewhere.

Traditional practice develops airport master plans in a strict linear process. The ICAO, the International Air Transport Association (IATA, an airline group), and the U.S. Federal Aviation Administration provide the most commonly used guidelines (FAA, 1985; ICAO, 1987; IATA, 1995). These are fundamentally the same, although they differ in detail. The key elements of this process are, as paraphrased from the ICAO:

- Inventory existing conditions
- Forecast future traffic
- Determine facility requirements
- Develop several master plan alternatives for comparative analysis
- Select the most acceptable and appropriate master plan

The master planning process is inherently *reactive.* It represents one prospective response to one specific expectation about what may happen due to forces external to the airport development. As the list indicates, the procedure assumes that the forecast is unalterable, like the weather. The master plan prepares ways to deal with this future. The master planning process does not envisage that airport operators can alter or shape the forecast. This perspective is fundamentally different from an entrepreneurial approach, which might desire to shape the future by the definition of the product.

Proactive planning is the alternative to the conventional master planning process. Although this has not been standard practice in airport planning, it is standard in business and totally possible in airport planning. The TBI airport company demonstrated how this could be done in its development of Orlando/Sanford. Until around 1998, this airport had virtually no traffic and operated in the shadow of Orlando/International, a magnificent first-class facility. A normal forecast would not have projected any significant traffic for the secondary airport in the near future. However, the private owners positioned Orlando/Sanford as an inexpensive base of operations, built appropriate facilities, and teamed up with holiday tours and charter carriers. By 2000, the airport operator had built up the traffic to about 1.2 million passengers, of whom nearly a million were international. Their planning and development shaped the future, rather than responded to it. As private airport companies become more significant in the industry, proactive planning is likely to replace conventional master planning where possible.

The master plan is also inflexible. Its creators focus on a single forecast. They do not consider alternative futures. They therefore have no motivation to include the possibility of alternative sequences or types of development. However, the future usually turns out to be different from what was originally anticipated, as Sec. 3-3 indicates. Consequently, the master plan soon becomes obsolete—airport operators frequently have to junk the ultimate, 20-year vision of the master plan after 3 to 5 years. Sometimes the master plan is "dead on arrival" due to its inflexibility. In the mid-1990s, for example, the board of directors of one of the top airports in the United States voted to "accept" a master plan that had been 5 years in the making. Then, as the next item of business, this same board voted a contract for a new planning process, since they already knew the approved master plan was out of date!

Airport operators will continue for some time to have to prepare master plans, however, despite the deficiencies of these documents. This is because the national and international funding agencies expect to see these kinds of plans. The U.S. FAA, for example, pays for planning processes and results labeled as master plans. In the United States, the general rule is that airports can only get funds from the federal government for projects that are in the National Plan (the NPIAS). Furthermore, projects only get into the NPIAS if they are included in an approved master plan.

The challenge for airport planners is to improve the master planning process, so that it can be both proactive and flexible. In principle, this is not difficult, since it is easy to modify the process from a technical point of view. In practice, old habits die hard, and it may take time for standard processes to evolve. Meanwhile, forward-thinking airport operators should be able to implement better planning procedures. This chapter indicates what these might be.

Strategic plans

This phrase refers to two quite different kinds of activities. In the general field of management, *strategic planning* refers to a disciplined process for analyzing the current situation of a business activity, and identifying the vision of how that entity should position itself with respect to its customers and competitors. This approach comes in several flavors (see, for example, Porter 1980, 1985; and Hax and Majluf, 1996).

The SWOT analysis is a popular generic version of strategic planning. It refers to a process whereby the manager systematically reviews the:

- **S***trengths* of the business group, both internally and with respect to its competition
- **W***eaknesses,* again internally and with respect to the competitors
- **O***pportunities* for the group, in terms of new markets, mergers, technologies, etc.
- **T***hreats* to the group, in terms of the same kinds of events

A SWOT analysis or other form of strategic planning is supposed to guide the managers to an understanding of how they should develop their activity, both physically and organizationally, so that they can shape and benefit from future developments. Physically, they might build new facilities. Organizationally, they might develop relationships with clients, develop a favorable schedule of prices, and change their mix of products. In general, strategic planning in the business sense is a form of proactive, flexible planning.

In airport planning, some professionals have given the phrase "strategic planning" a somewhat different meaning. Caves and Gosling (1999) thus emphasize a "broad multidimensional view of the future." They emphasize the need to define "the role that each airport should play within a group of related airports" and focus on a substantive "understanding of the behavior of the system which is to be developed or improved," as covered in Chap. 5. This view contrasts with the emphasis on procedure and method favored by the managerial view of strategic planning.

In this context, readers should understand that strategic planning as practiced in business has fallen out of favor (see, for example, the discussions by Mintzberg, 1994; Hax, 1997.) In large part, this is because corporate strategic planning in practice evolved into large, expensive, and burdensome processes. In many ways, these efforts were like master planning in that they tried to predict various future states and design corporate responses to these predictions. As the forecasts so often turned out to be wrong, the resulting strategic plans became obsolete, just as the airport master plans do. In any case, these strategic planning efforts did not, or at least did not seem to contribute to improved performance. According to a leading proponent of strategic planning, "The criticism of strategic planning was well deserved. Strategic planning in most companies has not contributed to strategic thinking" (Porter, 1987).

Airport operators need to think strategically. They need to be able to examine the range of future possibilities, position their organizations to respond flexibly to the events that occur, and in fact respond actively when necessary. Metaphorically, they need to think strategically, the way chess players do. Airport operators need to think ahead many moves, establish a position that enables them to respond to threats and opportunities from any direction, and shape the development of their airport properties move by move, year by year, according to actual events as they unfold. Sections 3-5 and 3-6 show how this can be done, using dynamic strategic planning.

3-2   Airport systems planning

This section considers the planning of airport systems. It looks first at the notion of airport systems. It then addresses the operational questions: Who plans the development of airport systems? Who will be planning them in the future?

Airport systems

Airports are integrated into airport systems. Each airport does not operate independently; it is a part of one or more networks connecting other airports. These networks and systems can be defined either geographically or operationally.

Geographically, for example, one can think of:

- *Regional networks* linking smaller airports with a regional center or national center, as commuter aircraft feed traffic from all over the Southeast United States into Atlanta, or Argentine airports connect with Buenos Aires

- *Metropolitan multi-airport systems* serving a single metropolitan area, as Paris/de Gaulle and Paris/Orly do (Chap. 5 discusses multi-airport systems in detail)

- *National networks* linking the major cities of a country, as major airlines do for large countries such as the United States, Germany, and Japan

- *International and intercontinental networks*, connecting countries with each other

Alternatively, one can think of networks and airport systems defined functionally, by the type of traffic or the carrier:

- *Integrated cargo networks*, such as those constituted by major cargo integrators such as UPS or FedEx, which give traffic and meaning to airports such as Louisville, Kentucky, and Ontario, California, which otherwise would have little to do with each other

- *"Cheap fare" networks*, served by no-frills airlines such as Southwest in the United States, or Ryanair in Europe, which serve secondary airports such as Boston/Providence and Miami/Fort Lauderdale, or Oslo/Tore and London/Stansted

In general, an airport is part of several systems of airports simultaneously. Memphis, Tennessee, for example, has been both the major hub for the FedEx system of airports and part of the feeder system for Northwest Airlines. London/Stansted is part of both a "cheap fare" system of airports and the London multi-airport system. As a rule, airport systems cannot be divided into self-contained subsystems or modules, as a car can be divided into the chassis, the engine block, and the drive train. Airport systems overlap. In practice, they do not have a precise definition in terms of the aviation and air transport network.

While national governments do classify airports in a variety of ways, these categories do not necessarily define the systems meaningfully.

For example, in the United States the FAA has organized airports by the relative number of passengers and referred to the largest as major hubs. These definitions, based on current size of a particular type of traffic, have little, if any, consequences for planning and development. In Japan, the governmental distinguishes between "international" airports and others, and this label has had great financial significance. Designated international airports have received far greater support from the central government than the others have. However, numerous Japanese airports that are not officially "international," such as Sendai, do in fact cater to international passengers and cargo. Here again, the governmental label does not identify the functional systems.

The essential point to be retained from this discussion is that governmental jurisdictions do not define airport systems. A single jurisdiction may include two or more reasonably distinct and competitive systems. Thus Germany and the State of California include systems centered, respectively, on Frankfurt, Munich, and Berlin, and on Los Angeles and San Francisco. Conversely, a single system may overlap several jurisdictions. For example, the metropolitan multi-airport system around Boston includes airports in three states (Boston, Massachusetts; Providence, Rhode Island; and Manchester, New Hampshire). Similarly, the feeder system for Amsterdam airport in the Netherlands extends over a large part of Britain.

The consequence of this observation is that governments rarely can plan airport systems effectively. If the government encompasses several airport systems, it will find it politically difficult to choose among the competitive possibilities, to pick "winners" among the competitive systems. Thus, both the California Aviation System Plan and the U.S. National Plan of Integrated Airport Systems (NPIAS) are nonselective assemblies of proposed developments of individual airports. These documents aggregate projects from the "bottom up," as indicated in the previous section. If, on the other hand, the government controls only part of the airport system, it evidently cannot have a forceful impact on it.

Planning airport systems

In the second half of the twentieth century, various governmental powers had a substantial effect on national and metropolitan airport systems. Many governments were able in particular to develop regional airports. Typically, the national ministry in charge of transportation or aviation would use its resources to invest in provincial projects. Thus, for example:

- Australia built excellent facilities in the capital cities of each state and territory.
- Canada invested heavily throughout its provinces and territories, most notably constructing Montreal/Mirabel, the airport with the largest area of property in the world.
- Japan endowed each prefecture with a series of remarkable airports, leveling mountains, filling in valleys, and creating airport islands at Hiroshima, Osaka, and Nagoya.
- Mexico in the 1970s built international airports at coastal resort areas throughout the country, thus strongly promoting the development of tourism in Acapulco, Cancun, and similar sites.
- The United States collected taxes on airline tickets, placed the proceeds in the Airport Trust Fund, and used this money to fund the improvement of runway facilities throughout the United States.
- The European Community largely funded (through a grant and a loan) the construction of Athens/Venizelos airport, opened in 2001, as part of its policy of distributing funds to less developed member states.

Directive national planning of airport systems is generally obsolete, however. Most nations have judged that they can no longer afford the scale of subsidies associated with these programs. Indeed, a significant number of the regional projects sponsored by national planning were plainly not economically efficient, however desirable they might be from a national political perspective. The traffic in many regional airports would never have justified the scale of the investments. At the end of the twentieth century, for example, the newly formed Aéroports de Montréal relocated all scheduled traffic away from Montreal/Mirabel airport, which the Canadian national government had built as the major airport for the region. Moreover, as the size of the airports and the demand for funds increased, taxpayers were reluctant to support the redistribution of resources to peripheral areas. The resulting drive for economic efficiency and the elimination of subsidies has led to the privatization of national airports and their breakup into local companies and authorities, as in Australia, Canada, and Mexico (see Chap. 4). The opportunities for national planning of airport systems are thus largely gone.

Many metropolitan authorities have also developed second airports as part of multi-airport systems.* Their general practice has been to use the revenues provided by major international airports to finance these projects. Typically, these secondary facilities took a long time to build up traffic. They were thus often financially premature and thus economically inefficient, as Chap. 5 discusses in detail. For example:

- The Aéroports de Paris built Paris/de Gaulle to be the premier facility for France, yet this platform took a generation to overtake Paris/Orly as the busiest airport for the region.

- The British Airport Authority (as the government agency now replaced by the privatized BAA) built London/Stansted airport, yet 15 years after its opening it is still largely underutilized compared to its design.

- The Port Authority of New York and New Jersey built major new facilities at New York/Newark, and had a major passenger building built that remained totally unused for over a decade.

- The Federal Aviation Administration built Washington/Dulles and attempted unsuccessfully to build up significant traffic for almost 20 years.

These kinds of long-term subsidized investments in major facilities are likely to be rare in the privatized environment of the early twenty-first century. Airport authorities or companies that have to raise money in the private sector are replacing governmental bodies that could afford to disregard interest payments. The British Airports Authority is now the BAA company. The U.S. government has transferred its responsibilities for the Washington airports to the Metropolitan Washington Airports Authority. This trend limits the opportunities for planning and developing multi-airport systems.

Privatization is thus leading to the end of airport systems planning as it was practiced in the twentieth century. The national governmental bodies that could direct airport development are disappearing, and the local and regional airport authorities will increasingly be required to justify projects for secondary airports to demanding private investors. What will be the airport systems planning of the future?

*This discussion on the development of *second* airports as part of a system deliberately does not mention the many projects planners saw as *replacement* airports, such as those at Dallas/Fort Worth, Denver, Houston, and Kansas City in the United States, or at Athens, Edmonton, Milan, Munich, Osaka, and Kuala Lumpur. In almost every case, the planners intended to close the older airports to commercial traffic, yet many of these facilities stayed active at least for a while. Although the airport developers in effect created multi-airport systems in those cases, that was not their intent.

The inevitable consequence of privatization is that private, local interests prevail. Airport planning in the future is thus likely to focus increasingly on the development of individual airports. Planning efforts will focus on increasing each airport's competitive advantage over other airports. To the extent that a competitive market economy maximizes the public welfare, this is desirable. However, airports do suffer from congestion and create externalities. Basic economics tells us that, under these circumstances, competition is not necessarily in the overall best interests of society, of a nation or a region specifically. As Chap. 4 discusses, this conflict between local efficiency and overall welfare is at the heart of the question of privatization.

Airport planning in the twenty-first century is in practice likely to be narrowly defined around the development of the airport facilities under the control of a single authority or company. The focus will be on the configuration of the airfield; the set of passenger buildings; the supporting people movers, baggage, and communication systems; the complex of cargo and maintenance facilities, and the modes of access. The remainder of this chapter assumes this perspective.

Airport companies may eventually evolve into large international operators of major airports. They may in this case develop strategies for developing airports as part of a coherent global system competing with other chains of airport operators. As of the turn of the century, however, airport companies such as the BAA and Amsterdam Airport Schiphol are holding companies that manage independent airport operations (Moody's Investor Services, 1998). At this time, the large integrated cargo shippers such as UPS and FedEx appear to be closer to planning for their systems of cargo hubs. How this will develop is an open question. Who knows what the future will bring?

3-3   The forecast is "always wrong"

Experience demonstrates that forecasts about airport traffic are "always wrong." Comparisons between what a forecast indicated for a given period and what actually occurred almost invariably show a significant discrepancy. This is especially true when one considers forecasts over 10 to 20 years, that is, over the normal periods for the planning of major airport facilities. The differences between forecast and reality are most apparent when they concern the total level of operations. However, they are equally significant for planning purposes when they concern the composition of the traffic. For example, 10 million passengers at an origin and destination airport require quite different facilities than the same number when half of them are transfers (see Chap. 14). As this section illustrates, the accuracy of forecasts of all types is low.

The fact that forecasts are unreliable has crucial implications for airport planning. Responsible planners consequently must accept that they do not know what levels or types of traffic will use the facilities they design. They need to anticipate that these facilities will have to serve different loads than the ones they now think are most probable. They therefore need to make sure that any design they propose will function well in these different conditions. In practice, they need to check the performance of their designs under different loads and, when they find deficiencies, they need to alter these designs to avoid the potential for future problems. In general terms, the fact that forecasts are unreliable means that designers need to create flexible designs that can adapt easily to the range of future conditions.

The unreliability of forecasts is well documented. Ascher (1978) illustrated the phenomenon through case studies across a variety of issues. Makridakis and colleagues demonstrated the inaccuracy of all kinds of forecasts, even in the short run, through extensive analyses of all the major methods available (Makridakis and Hibon, 1979; Hogarth and Makridakis, 1981; Makridakis et al., 1984; Makridakis and Wheelwright, 1987, 1989; Makridakis, 1990). de Neufville (1976) presents extensive evidence of the poor performance of forecasting for airport systems. The U.S. Office of Technology Assessment (1982) gives an official account of the unreliability of forecasts of airport activities in the United States. This section illustrates this evidence for the benefit of readers who cannot refer to these and other citations.

Forecasts are unreliable because it is basically impossible to get good forecasts. All forecasting is based on some extrapolation of past trends into the future. However, past trends are constantly changing for economic, technological, industrial, and political reasons. The oil crises of the 1970s, the financial crisis in Asia in the 1990s, and terrorism in 2001 caused aviation traffic to subside considerably. New aircraft, larger and quieter, enable new routes, lower fares, and more traffic. Airline mergers and alliances change the services and public consumption of air travel. Political changes, such as the collapse of the Soviet Union and the dismantling of the traffic barriers between Russia, China, and the West, vastly reconfigured traffic patterns. The list of reasons why trends do not continue over a reasonable planning period is practically endless.

Moreover, as Chap. 20 on forecasting indicates, even to the extent that trends do continue, the mathematical methods for determining them are too subjective to permit analysts to determine definitively what that trend might be. In short, better methods or better analysts will not make forecasts more reliable. In fact, in the increasingly deregulated world of air transport, forecasts are likely to become even more unreliable than they have been. (See Armstrong, 1978.)

The presentation of the track record for aviation forecasts first covers three substantive contexts: the estimation of costs, of overall levels of traffic, and of the composition of the traffic. It then indicates how longer planning periods and deregulation of aviation further increase the lack of reliability of forecasts. The object is to provide a sense of the large range of uncertainty that should be attached to any aviation forecast.

Cost estimation

Estimates of construction costs for major projects are notoriously inaccurate. Differences between estimated and actual costs of 30 percent are common on standard projects, due to surprises on site, changed orders from the architects or owner, and the whole litany of things that can go wrong. On innovative, high-technology projects, these differences can be much larger, and analysts of project costs have observed standard deviation equal to the estimate. Benz (1993) provides an account of what has been observed in all kinds of fields of construction and production, and Fig. 3-1 summarizes those find-

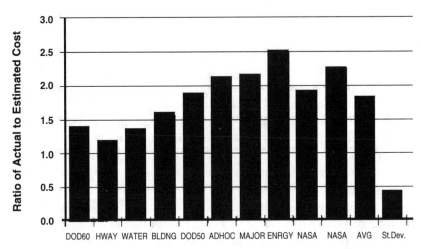

Fig. 3-1. *Average ratios of actual to estimated costs in various areas.*
Source: Benz, 1993.

ings. Notice that he reports an overall standard deviation of about 40 percent. In round numbers this implies that, in one case out of three, actual costs differ from estimated costs by more than plus or minus 40 percent.

An analysis of the cost of resurfacing airport runways illustrates the range of uncertainties in cost estimation. As this particular job is about the simplest to estimate, it provides a conservative indication of the uncertainties to be expected. The process of resurfacing runways uses primitive technology (asphalt is dumped off trucks and rolled to grade) on a clear surface with no hidden surprises. Figure 3-2 illustrates the distribution of the ratio of actual to estimated costs using data from the FAA Western Region of the United States (Knudsen, 1976). The analyst properly adjusted these data for inflation in the cost of construction over the time between the estimate and the execution of the job. Not surprisingly, the average and median actual costs are higher than estimated (about 25 percent in this case). What is remarkable is that the range of costs can be twice or half the average cost!

A general explanation for why actual costs vary from estimates has to do with the fluctuations in the real cost of materials and labor. The production of construction uses commodities, such as steel and cement, whose production consumes a lot of energy. Thus their prices, and those of petroleum derivatives such as asphalt, rise and fall with the cost of petroleum. The price of this commodity is

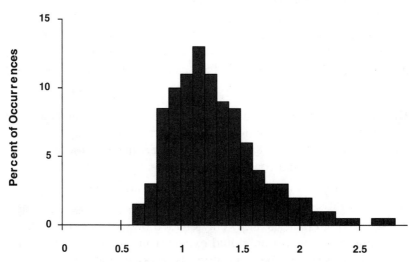

Fig. 3-2. *Distribution of ratio of actual to estimated costs for runway resurfacing projects.* Source: Knudsen, 1976.

highly volatile. Between 1988 and 1994, the price of crude oil in constant dollars doubled and then halved. It repeated this performance between 1994 and 1998 (WTRG Economics, 1998) and again around 2001. The variations in the price of oil, and the effective price of labor, depend largely on the state of the economy. (For empirical evidence on this phenomenon, see de Neufville et al., 1977; de Neufville and King, 1991.) During boom periods, the supplies of oil and labor are tight, so oil prices are high and employers have to pay overtime and premium wages. During recessions, however, supplies are plentiful, oil prices drop, and workers are less demanding.

Aggregate forecasts

The periodic swings in the overall economy naturally affect the overall level of aviation traffic. In boom periods, businesses need and individuals have the money to travel. When there is a global or economic crisis, as during the 1990 Gulf War, the 1998 recession in Southeast Asia, and during the 2001 crisis, the growth in airport traffic correspondingly slows or decreases. The phenomenon makes medium-term forecasts, those covering 5 to 10 years, distinctly unreliable.

As a rule of thumb, half the medium-term forecasts differ from the forecast by more than 20 percent (de Neufville, 1976). This approximation is validated by repeated comparisons between forecasts and actual results over long periods and in different countries. For over 40 years for example, the U.S. Federal Aviation Administration has been publishing national annual forecasts of traffic and airport operations over the following 5 years. It has also reported annually the actual levels observed for all the same categories (FAA, Annual). The comparison of such series, in the United States and elsewhere, demonstrates that 5- and 10-year forecasts are indeed easily wrong by more than plus or minus 20 percent.

The U.S. Office of Technology Assessment (1982) documented the large discrepancies between the 5-year FAA forecasts and reality. Table 3-1 shows their results for a sample of the categories that the FAA forecasted. Readers should consider this document carefully. Notice that:

- The range of discrepancies, even only after a few years, is very large—in one case over 80 percent.
- The discrepancies are evenly divided between being too high and too low, as one would expect from random fluctuations in the economy between periods of boom and crisis.
- Half the errors are over 20 percent, as indicated previously.

Table 3-1. Historical comparison of aviation forecasts with actual results in the United States, in terms of percent differences, negative numbers in parentheses

Forecast year	For year	Enplane- ments	Revenue passenger- miles	GA hours	Ops at FAA towers	
					GA	Total
1959	1964	(1.3)	(6.5)	(0.6)	4.6	9.7
1960	1965	(9.5)	(9.7)	(1.2)	(27.8)	(21.6)
1961	1966	(27.5)	(26.0)	(15.3)	(37.7)	(28.9)
1962	1967	(32.1)	(31.4)	(23.6)	(34.7)	(27.3)
1963	1968	(41.3)	(41.3)	*	(38.4)	(32.5)
1964	1969	(31.4)	(33.6)	(23.5)	(27.3)	(24.9)
1965	1970	(14.1)	(19.8)	(16.3)	(2.6)	(5.2)
1966	1971	9.4	0.5	(1.6)	53.7	42.2
1967	1972	23.6	13.0	9.1	72.5	54.9
1968	1973	23.9	15.9	7.4	78.3	58.4
1969	1974	21.1	21.2	4.6	53.6	42.4
1970	1975	26.3	33.0	(0.6)	80.9	25.9
1971	1976	19.0	28.6	(0.6)	42.9	22.9
1972	1977	22.3	33.7	(6.8)	36.9	4.5
1973	1978	14.0	18.3	(10.4)	14.8	8.8
1974	1979	(9.7)	(7.4)	(13.7)	11.8	9.4
1975	1980	(10.6)	(17.3)	(0.2)	34.6	25.7
1976	1981	4.3	(1.8)	15.7	41.3	32.1

Source: Office of Technology Assessment, 1982.

- The errors are systematically large—less than a third of the time were the errors less than 10 percent.

Simply stated: large forecasting errors are normal.

Anyone can document the phenomenon by making similar comparisons. Table 3-2 shows the same kind of information for the United States, done 20 years later. The results replicate the four conclusions drawn from Table 3-1. Table 3-3 shows similar information from Japan, taken from their periodic forecasts of their national 5-year investment plans and subsequent statistics on passengers. For Japan, the average discrepancy between the forecast and the actual number of international passengers between 1980 and 1995 was 22 percent after 5 years, and 40 percent after 10 years (Nishimura, 1999).

Table 3-2. Recent comparison of aviation forecasts with actual results in the United States, in terms of percent differences, negative numbers in parentheses

Fore-cast year	For year	Enplanements, millions		Revenue passenger-miles, billions		
		Domestic	Internatl.	Domestic	Internatl.	Total
1984	1996	(1.2)	16.5	(1.1)	27.8	6.5
1985	1997	(11.2)	16.2	(8.6)	24.6	0.2
1986	1998	(17.9)	19.8	(15.2)	29.5	(3.3)
1987	1999	(23.9)	5.3	(19.6)	19.0	(9.4)

Source: FAA (Annual).

Table 3-3 Comparison of 10-year forecasts of international passengers to Japan with actual results

Forecast		Passengers (millions)		Percent error
For	Done in	Actual	Forecast	Over actual
1980	1970	12.1	20.0	65
1985	1975	17.6	27.0	53
1990	1980	31.0	39.5	27
1995	1985	43.6	37.9	(13)

Source: Nishimura, 1999, from Japanese Ministry of Justice Embarkation and Disembarkation Statistics, and Ministry of Transportation, Airport Investment 5-Year Plans.

Composition of forecasts

Forecasts of components of the traffic are even less reliable than forecasts of totals. This is because the variations in the levels of components tend to cancel themselves out. For example, if one transatlantic airline suffered a strike and lost traffic, the travelers would use the competing airlines. The result would be large, compensating variations between the airlines, but perhaps not much change in the overall total traffic across the Atlantic. Table 3-2 documents the phenomenon: domestic travel grew much less than anticipated, but international traffic compensated for this deficiency by growing much faster than expected.

Several long-range forecasts made for Sydney, Australia, illustrate the difficulty of forecasting components of an overall total. The team preparing the Environmental Impact Statement for a possible Second

Sydney Airport (Australia Department of Aviation, 1985) used three different authoritative forecasts (Table 3-4). One came from a highly reputed international consultant, a second was the official forecast of the previous planning study, and the third was the forecast of the Australian Ministry of Aviation. None of these forecasts was close to the actual level of traffic some 20 years later.

Table 3-4 provides further lessons about forecasting in practice. Readers should notice the third- and fourth-place accuracy of the forecasts. This kind of precision is wholly unjustified. Most 20-year forecasts for aviation are lucky to get the first two decimal places right. Reporting more decimal places is pretentious. Two of the forecasts for Sydney have the great merit of providing ranges, thus reinforcing the notion that forecasts are not precise. However, these ranges are much too tight. They provide a range of only about plus or minus 20 percent over 20 years. By contrast, as the previous section indicates, about half the forecasts in the United States show as great a discrepancy between the forecast and reality after only 6 years! The lessons are that aviation planners should:

- Focus on the first two decimal points
- Use large ranges, on the order of plus or minus 50 percent over 20 years, especially when dealing with components of the total traffic

Effect of longer planning periods

The discrepancy between forecast and reality increases for longer forecasts. This is entirely to be expected. In the short run, inertia in

Table 3-4 Comparison of actual and forecast international passengers through Sydney

Forecast for year	Source of forecast, in year		
	Consultant, 1974	Regional study, 1978	National ministry, 1983
1980	3.77	2.98–3.46	
1985	7.4	3.87–4.34	2.674–3.047
1990	9.8	4.71–5.51	2.762–3.751
2000 projected	12.0	6.27–8.66	2.938–5.159
2000 actual	10		

Sources: Australia Department of Aviation, 1985; Sydney Airport, 2001.

the system keeps things moving as they were. In the longer run, entirely new trends may set in and make the actual results differ much more. The data on Japan reported in the section on aggregate forecasts indicates the way errors increase for longer-term forecasts.

A comprehensive analysis of airport master plans demonstrated the increase in forecast errors for longer-term predictions (Maldonaldo, 1990). Table 3-5 presents the analyst's results. These clearly show how all measures of the error become larger with longer-term forecasts: the average discrepancy, the absolute range of the error, and the consequence standard deviation of the error.

The discrepancies between the forecasts and actual results apply equally to the content of master plans. By looking at old master plans and comparing them with what actually is built, it is easy to calculate statistics similar to those reported so far. In doing this exercise, the analyst has to consider both the projects in the plan that were not built, and those that were built that were not in the original plan. Maldonaldo (1990) did this as part of his study. Table 3-6 summarizes

Table 3-5 Discrepancies between the forecast and the actual results increase for longer-term forecasts

Forecast years	Error characteristics (%)		
	Average	Range	Standard deviation
Five	23	−36 to +96	23
Ten	41	−22 to +140	34
Fifteen	78	−34 to +210	76

Source: Maldonaldo, 1990.

Table 3-6 Discrepancies between projects forecasted in master plans and actually built increase for longer-term forecasts

Forecast years	Error characteristics (%)	
	Average	Standard deviation
Five	54	27
Ten	58	30
Fifteen	68	21

Source: Maldonaldo, 1990.

his results, showing that the master plan accounted for less than half the projects constructed. As should be expected, the average discrepancy becomes larger for longer planning horizons. In this case, the variations do not increase with time, as Table 3-6 indicates. This result does not mean that the master plans are equally accurate. It merely reflects the statistical fact that when the accuracy is close to zero for all these plans, there is little variation in this conclusion.

Effect of economic deregulation

Economic deregulation increases the volatility of traffic. This is because deregulation removes the barriers to changes in prices, frequency of service, and routes (de Neufville and Barber, 1991). Airlines can and do make sudden major changes in these circumstances, and may radically disturb the patterns and levels of traffic. These moves may have substantial effects on the largest airports. At smaller airports that may cause traffic to double or halve in just a few years. For example, Continental introduced a low-fare service to Greensboro, North Carolina, and doubled traffic from 2 to 4 million total passengers between 1993 and 1995. By 1997 they had terminated this service, and traffic at Greensboro had fallen back to about 2 million passengers a year (Moody's Investor Services, 2000).

"Cheap fare" airlines such as Southwest or Ryanair can suddenly arrive on a market and generate huge increases in traffic. These may persist, or may fall if the airline fails. Southwest, for example, has continued to be successful over a generation, whereas PEOPLExpress rapidly doubled traffic at New York/Newark in the 1980s, from about 10 to about 20 million before it went bankrupt and deflated the traffic by half.

Major airlines can likewise make substantial moves. American Airlines moved a substantial block of its traffic from Chicago/O'Hare to Dallas/Fort Worth at the time deregulation became effective in the United States, thus dropping traffic through Chicago by about 15 percent in one year (see Chap. 4, Fig. 4-1). Similarly, Delta transferred much of its hubbing operations from Dallas/Fort Worth to Cincinnati in the 1990s, thus boosting traffic through that airport substantially in just a few years.

Such radical changes in traffic can obviously affect the performance of an airport drastically. For example, when US Air moved the focus of its international operations from Baltimore to Philadelphia, the Baltimore airport was left with an underutilized international passenger building (Little, 1998). Meanwhile, Southwest Airlines was expanding, so the total traffic at Baltimore stayed steady. However, Baltimore meanwhile had to build new facilities to accommodate this other form of traffic (Baltimore-Washington International Airport, 1998). Chapter 14 describes this case in more detail.

The bottom line for airport planners and operators is that traffic can change rapidly in a deregulated environment. As of the turn of the century, deregulation was already the standard context in the busiest international markets in North America and Europe as well as in Australia. As this pattern spreads to other free-trade areas, and through "open skies" policies allowing airlines to serve destinations in other countries freely, this volatility becomes increasingly important.

3-4 Implications for planning

The fact that the forecast is "always wrong" means that master plans built around specific forecasts will also be "always wrong." This means that to get the planning right, it is necessary to move away from the notion of planning around a fixed forecast.

Good planning needs to deal with reality. For airport systems, the fundamental reality is that future forecasts are highly unreliable. Forecasting errors of 20 percent or more after only 5 years is normal, and errors for longer-term forecasts normally are worse. Good planning therefore needs to deal with a broad range of possibilities.

The range of possibilities to be planned for includes both quantitative and qualitative factors. To demonstrate the inaccuracy of forecasts, the previous section stressed their measurable errors. These stem from changes in economic trends and policies, new technologies, new industrial organizations created by mergers and alliances, as well as from new political possibilities. The same factors clearly also influence the qualitative aspects of the loads on the airports. New technologies impose new requirements on the design of the airport. The introduction of the new large aircraft such as the Airbus A-380 means that many airports have to modify their runways, taxiways, and passenger buildings. New political realities likewise have design implications. The creation of common market areas such as the European Community leads to different, generally easier, requirements for immigration and customs facilities. Meanwhile, a heightened concern for security imposes new requirements for examining and controlling baggage. Good planning needs to recognize the whole range of possibilities, and anticipate solutions for the problems they pose.

Flexibility is essential. It is impossible to build now the facilities that will meet all eventualities. For example, facilities cannot both be large enough to satisfy the highest level of traffic anticipated, yet be small enough to avoid unnecessary expenses if traffic remains steady or drops to a low level. Planners need to establish some middle course, from which they can either grow the facilities as needed, or change them if some newer or lower level of traffic should arise. Consider the case of Baltimore-Washington Airport discussed in the previous section. The situation there was that the level of international traffic dropped suddenly, as their principal international carrier shifted the hub of those operations to Philadelphia. Although events like this are neither usual nor common, they are well within the range of possibilities and have happened elsewhere. Good planning in that case would have anticipated this possibility, and would have designed the international passenger building with the flexibility to accommodate alternative traffic (see Chap. 15). A flexible approach to planning and design would have avoided the difficulties associated with an underutilized building.

3-5   Dynamic strategic planning concepts

Dynamic strategic planning emphasizes flexibility. Its fundamental premise is that airport operators must adjust their plans and designs dynamically over time to accommodate the variety of futures that may occur. This emphasis distinguishes dynamic strategic planning from the traditional master or strategic planning, both of which build upon relatively fixed visions of the future.

Dynamic strategic planning represents a new vision of how airport systems planning should be done. It is particularly suitable for the current situation, in which privatized airlines compete in an increasingly deregulated environment, and increasingly privatized airports respond proactively to the opportunities and threats they perceive.

Although dynamic strategic planning is a new approach, it is entirely compatible with and builds upon the basic elements of traditional airport master planning and with strategic planning in management. It adds to the orderly process of the airport master plans by including the examination of several forecasts rather than one. It also assimilates the proactive approach of strategic planning, by encouraging planners to shape the future loads on the system, rather than reacting passively to whatever loads come to the airport. In short, this approach to planning represents a marriage of the best elements of both master and strategic planning, in a practical form suitable for routine use.

This new approach to planning is an extension of the master planning process outlined in Sec. 3-1 and detailed in standard guidelines (FAA, 1985; ICAO, 1987; IATA, 1995). It is different in two ways. First, it substitutes a range of forecasts for the single forecast that the master planning process normally generates. In this regard, dynamic strategic planning simplifies the process, since it avoids the difficult and unsatisfactory process of trying to choose one forecast from among the many possible candidates. (See the discussion around Table 20-4 in Chap. 20 on forecasting.) In the subsequent phases of the process, dynamic strategic planning directs the planners to consider how each of their plans would

- Perform under the loads implied by the different forecasts
- Adapt to the new conditions these alternative scenarios represent

At this point, the dynamic plan is more complicated than the standard master plan. However, this additional effort can be managed by the appropriate use of computer-based tools such as decision analysis and simulation, as Sec. 3-6 describes.

The new approach is also strategic in that it is proactive. Dynamic strategic planning recognizes that planners can influence the nature of the airport traffic. They may preclude certain types or facilitate others. For example, as Chap. 14 describes, the construction of the passenger buildings at Kansas City made it impractical to service transfer traffic efficiently and thus impelled the locally based airline to establish its hub in another city. On the other hand, the planners went to great lengths to plan Denver/International to service transfer traffic efficiently, and thus maintained that airport as a leading transfer hub in the United States. Likewise, the developers of London/Luton airport consciously targeted the market of price-sensitive travelers, and built their facilities to keep costs low. In a similar vein, Singapore has developed its facilities to offer premium services and thus help establish and maintain that city as a favorite hub for business travelers. In each of these situations, the developments significantly influenced the traffic at the airport. Airport planners need to recognize this potential, and incorporate it into the planning process.

Because it is proactive, dynamic strategic planning recognizes possible relationships between the possible airport designs and the airport loads. That is, it recognizes that the planning process should not apply a single range of loads to all possible plans. Since the plans themselves influence the type of traffic that may use the airport, correspondingly different loads may be applied to different sets of plans being considered. For example, when the planning process examines airport configurations that favor transfer traffic, it should test them against forecasts with higher levels of transfers and of total traffic. Contrarily, when the process looks at plans that favor destination traffic, it should test these against forecasts that have little transfer traffic.

Most important, a dynamic strategic plan is phased. It focuses on finding the most appropriate initial developments. This first phase of development should permit the planners to respond appropriately to the future levels of traffic. For example, they might develop a passenger building that accommodates both international and domestic traffic in a first phase. In a later period, they could expand the capacity to serve either or both activities, or could substitute one capacity for the other, depending on the circumstances. See Example 3-1. The focus is not, as in the master plan, on describing a future long-range vision that in practice never is implemented. The focus of the dynamic strategic plan is on identifying the right initial position that permits effective responses to future opportunities and developments.

Overall, dynamic strategic planning encourages planners to think like players of chess or other strategic board games. Planners should:

- Think many moves ahead
- Choose an immediate development or move that positions them to respond well to whatever develops next
- Rethink the issues after they see what happens in the next phase
- Adjust their subsequent developments or moves correspondingly

Good planners for the uncertain environment of airport systems will, as good chess players do, emphasize good positions and flexibility.

3-6 Dynamic strategic planning process and methods

The process for executing a dynamic strategic plan is a modified form of the master planning procedure described on page 62. In the following list of the essential elements of a dynamic strategic planning

EXAMPLE 3.1

The original master plan for the redevelopment of the passenger buildings for Mombasa airport in Kenya anticipated two distinct buildings, one for domestic and the other for international traffic. Each of these was supposed to be large enough to meet the level of traffic anticipated.

The dynamic strategic plan recognized that one of the major risks was that the proportion of international traffic could shift radically, as passengers might come directly from Europe or transit through Nairobi. If this happened, one or the other of the new buildings might be crowded while the other was underused.

The strategy adopted was to build a single passenger building capable of serving about half the eventual growth. This facility was equipped to serve international traffic on one side, domestic traffic on the other side, and either traffic through shared use in the middle (see Chap. 15 for a discussion of shared-use facilities). This arrangement allowed the building to serve a range of mix of traffic. It also enabled the airport to defer until a future date, when traffic patterns had matured, the decision about how much they should extend the building, for which kind of traffic.

process, these additions appear in italics. The steps for preparing a dynamic strategic plan are thus:

- Inventory existing conditions
- Forecast range of future traffic, *along with possible scenarios for its major components (international, domestic, and transfer traffic, airline routes, etc.)*
- Determine facility requirements *suitable for the several possible levels and types of traffic*
- Develop several alternatives for comparative analysis
- Select the most acceptable *first-phase development, the one that enables subsequent and appropriate responses to the possible future conditions.*

The new elements require different analyses than those involved in master planning. To do a dynamic strategic plan, the analysts need to look at many scenarios, over several periods. In the twenty-first century this wider perspective can be obtained with a reasonable amount of effort, as it was not possible to do when the master planning was first developed around the 1960s (see, for example, Horonjeff, 1962). Planners can do these analyses using computer models and computer-based analyses that speed up the calculations enormously.

Computer models provide the basic tools for investigating the effects of different scenarios. They can analyze hundreds of situations easily. As Chap. 19 on computer models explains, the real effort associated with using computer models is not in the calculations, which are virtually immediate. The tricky part is in finding a model that is easy and cost-effective. Many already exist, however, and consultants are developing many more. While better models will always become available and are to be hoped for, enough exist to make it possible to consider many design alternatives for many scenarios.

Most computer models simulate the performance of facilities under loads. They show how a particular design or configuration—of a runway system, passenger building, or people mover, for example—performs under different conditions. Current practice routinely uses a wide range of models for different situations. These models can be deterministic or probabilistic (see Chap. 19). The former typically use a spreadsheet analysis to specify the result of particular configurations or patterns. Chapter 14, on the configuration of passenger buildings, for example, describes a model that estimates the walking distances that result from various types of traffic on a passenger building. Probabilistic models are most appropriate when there are queues and delays (see Chap. 23). In short, sufficient models exist to enable good analyses of alternative developments under different conditions.

A different class of models facilitates the analysis of risk. Decision and options analysis enable planners to determine which initial developments lead to the best long-term development (see Chap. 22). These tools have the great merit of helping planners estimate the value of the flexibility that they might introduce into the design. As a rule, it costs money to create flexibility, to enable the development of alternative paths of development. Planners must face this issue, and should calculate the value of this flexibility. Chapters 15 and 16 on the design of passenger terminal buildings show how this can be done.

Computer models are not a substitute for strategic thinking, however. They make it possible to carry out the wide consideration of issues under many circumstances. They are necessary for the calculation of the performance of the facilities under different loads. They are not, however, sufficient to develop a good strategic plan.

The ultimate value of a dynamic strategy lies in the way the planners understand the critical issues. They need to overlay the mechanical aspects of the process, as identified at the beginning of this chapter, with strategic thinking, with a strategic framework. This has two elements. The first is a critical assessment of the competitive situation, a step comparable to the first step of the master planning process, that is, the inventory of current conditions. The second is a creative process of identifying the range of good responses to the prospective opportunities and threats. In developing their strategic thinking, planners first need to go through a SWOT analysis, as indicated in Sec. 3-1. Specifically, they need to identify:

- *Strengths* of the existing site or airport, the characteristics that give it advantages over other sites or competitive airports—this may be a site central to an aviation network that favors transfer traffic, for example.

- *Weaknesses* of the same facility, those that may limit its growth or opportunities—this might be the weakness of the major local carrier that might risk being absorbed into some other carrier that focuses traffic elsewhere.

- *Opportunities* for the region, which may favorably enhance its future prospects, such as an increasing economic base that will lead to greater air travel and cargo.

- *Threats* to the airport and region, from competitive airlines, airports, or other factors.

Whereas it is easy to specify a way to inventory the situation, there is no checklist for creative thinking about strategic solutions. Good strategic thinking, like good chess playing, comes from observing examples and practice. Professional, experienced leaders in airport planning and development should be able to think through the possibilities and arrive at good ideas, provided they allow themselves the time and devote careful thought to this issue. By way of illustration, Example 3.2 describes the application of dynamic strategic planning to the development of the passenger buildings for Kuala Lumpur/ International when it was built around 1995.

EXAMPLE 3.2

In the mid-1990s, Malaysia built a superb new airport on a virgin site among rubber plantations. It features enough space for four parallel runways, for access by rail and limited-access highways, and for passenger buildings that can accommodate 100 million passengers. Its original master plan called for a main passenger building to serve domestic traffic and satellite buildings for international traffic. Each was to be designed to fit the traffic forecast for each of these categories.

The development of the dynamic strategic plan for the passenger buildings revisited the master plan and came up with a strategy designed to strengthen the competitive position of the airport and the national airline, Malaysian Air Systems. The SWOT analysis led to these observations:

- *Strengths:* The new airport would have enormous capacity and should thus be attractive for hubbing operations by a major airline alliance.

- *Weaknesses:* The original design of the buildings split the operations of the national carrier between the main building and the satellites, which is terribly inefficient for any airline, and more so for a hubbing operation. Moreover, the division of the buildings into international and domestic facilities lacked flexibility. This was particularly true in this case since the ratio of these operations was likely to be volatile, given the prospects of hubbing, and the possibility that the Southeast Asian countries might form a common market.

- *Opportunities:* To become a leading aviation hub for Southeast Asia, given the capacity and low cost of operation due to inexpensive land.

- *Threats:* Competition from Singapore as the lead hub in the region, and from the prospective Second Bangkok airport, which would be more favorably located for hubbing operations.

Taken together, the SWOT analysis emphasized the opportunities for Kuala Lumpur/International to become a major regional hub— if the design team configured the airport to provide integrated operations for the major carrier and achieve low costs.

The strategic decision was to reconfigure the passenger buildings to make it possible to achieve these objectives, while not compromising the airport's long-term ability to serve different kinds of traffic. The design team achieved this objective by creating operational flexibility in the main facilities. This allowed for efficient joint use of the facilities, and thus made it possible to eliminate one of the satellite buildings from the first phase of construction, which further decreased the cost of the airport.

Specifically, the designers created flexibility by:

- Designing the aircraft gates at the main building for shared use between domestic and international service, thus allowing the national airline to collocate its flights and crews and avoid unnecessary towing of aircraft.

- Reconfiguring the layout of the baggage facilities so that international and domestic flights could share space at different peak periods, and expand independently into reserved vacant space, thus allowing for inexpensive adjustments to capacity as required.

- Laying out the wings of the main passenger building so that these could be easily expanded laterally, as and when needed.

Overall, the dynamic strategic plan for the passenger buildings at Kuala Lumpur/International attempted to position the airport to respond effectively and economically to its range of prospective risks and opportunities.

Exercises

1. Look up previous forecasts for your local (or some other) airport. Compare these with the actual results. Calculate the deviations between forecast and reality, in terms of the percent of what actually occurred. To the extent possible, estimate how this percent error increases for longer-range forecasts.

2. Obtain previous master plans for some airport. Compare these with what actually has been constructed. To what extent has the airport invested in facilities that were not part of the original plan? Not invested in facilities that were in the plan?

3. Start by doing a SWOT analysis for some airport of interest. Which of the issues can the airport operators influence

through their designs and developments? What kind of developments might position this airport to respond most effectively?

4. For the same airport as in Exercise 3, think about what elements of the future traffic might be most uncertain. What could the airport operators do to give themselves flexibility so that they could adjust their future developments to deal effectively and efficiency with these different scenarios?

References

Armstrong, J. (1978) "Forecasting with Econometric Methods: Folklore vs. Fact," *Journal of Business*, *51*(4), pp. 549–564.

Ascher, W. (1978) *Forecasting: An Appraisal for Policy-Makers and Planners*, Johns Hopkins University Press, Baltimore, MD.

Australia Department of Aviation (1985) *Second Sydney Airport: Site Selection Programme*, Australia Department of Aviation, Canberra.

Baltimore-Washington International Airport (1998) "Board of Public Works Gives BWI Airports the Green Light to Begin Design Work on Expansion and Renovation," press release, March 4.

Benz, H. (1993) "Dynamic Strategic Plan for an Embedded Space Computer," S.M. thesis, Management of Technology Program, Massachusetts Institute of Technology, Cambridge, MA.

Caves, R., and Gosling, G. (1999) *Strategic Airport Planning*, Pergamon, Oxford, UK.

Cohen, S. S. (1969) *Modern Capitalist Planning: The French Model*, Weidenfeld and Nicholson, London, UK.

de Neufville, R. (1976) *Airport Systems Planning: A Critical Look at the Methods and Experience*, MIT Press, Cambridge, MA, and Macmillan, London, UK.

de Neufville, R. (1991) "Airport Construction—The Japanese Way," *Civil Engineering*, Geotechnical Issue, December, pp. 71–74.

de Neufville, R., and Barber, J. (1991) "Deregulation Induced Volatility of Airport Traffic," *Transportation Planning and Technology*, *16*(2), pp. 117–128.

de Neufville, R., Hani, H., and Lesage, Y. (1977) "Bidding Models: Effect of Bidder's Risk Aversion," *ASCE Journal of Construction Division*, *103*, March, pp. 57–70.

de Neufville, R., and King, D. (1991) "Risk and Need-for-Work Premiums in Contractor Bidding," *ASCE Journal of Construction Engineering and Management, 117* December, pp. 659–673.

FAA Federal Aviation Administration (1970) *Planning the Metropolitan Airport System*, Advisory Circular 150/5070-5, U.S. Government Printing Office, Washington, DC.

FAA, Federal Aviation Administration (FAA) (1985) *Airport Master Plans*, Advisory Circular 150/5070-6A, U.S. Government Printing Office, Washington, DC.

FAA, Federal Aviation Administration (FAA) (1989) *Planning the State Aviation System*, Advisory Circular 150/5050-3B, U.S. Government Printing Office, Washington, DC.

FAA, Federal Aviation Administration (FAA) (Annual) *Aviation Activity Statistics*, U.S. Government Printing Office, Washington, DC, www.faa.gov/arp/a&d-stat.htm.

Hax, A., ed. (1997) *Planning Strategies That Work*, Oxford University Press, Oxford, UK.

Hax, A., and Majluf, N. (1996) *The Strategy Concept and Process*, 2d ed., Prentice-Hall, Englewood Cliffs, NJ.

Hogarth, R., and Makridakis, S. (1981) "Forecasting and Planning: An Evaluation," *Management Science, 27*(2), February, pp. 115–138.

Horonjeff, R. (1962) *Planning and Design of Airports*, McGraw-Hill, New York.

IATA, International Air Transport Association (1995) *Airport Development Reference Manual*, 8th ed., IATA, Montreal, Canada.

ICAO, International Civil Aviation Organization (1987) *Airport Planning Manual, Part 1, Master Planning*, 2d ed., Doc. 9184-AN/902, ICAO, Montreal, Canada.

Johnson, C. (1984) *MITI and the Japanese Miracle*, Stanford University Press, Stanford, CA.

Knudsen, T. (1976) "Uncertainties in Airport Cost Analysis and Their Effect on Site Selections," Ph.D. thesis, Institute of Transportation and Traffic Engineering, University of California, Berkeley, CA.

Little, R. (1998) "New BWI Pier Is a Dud So Far," *Baltimore Sun*, November 29, p. 1D.

Makridakis, S. (1990) *Forecasting, Planning and Strategy for the 21st Century*, The Free Press, New York.

Makridakis, S., and Hibon, M. (1979) "Accuracy of Forecasting: An Empirical Investigation," *Journal of the Royal Statistical Society*, Series A, *142*(2), pp. 97–145.

Makridakis, S., and Wheelwright, S. (1979) *Forecasting*, Vol. 12, Studies in Management Science, North-Holland, Amsterdam.

Makridakis, S., and Wheelwright, S. (1987) *The Handbook of Forecasting: A Manager's Guide*, Wiley-Interscience, New York.

Makridakis, S., and Wheelwright, S. (1989) *Forecasting Methods for Management*, 5th ed., Wiley, New York.

Makridakis, S., et al. (1984) *The Forecasting Accuracy of Major Time Series Methods*, Wiley, New York.

Maldonaldo, J. (1990) "Strategic Planning: An Approach to Improving Airport Planning under Uncertainty," S.M. thesis, Technology and Policy Program, Massachusetts Institute of Technology, Cambridge, MA.

Mintzberg, H. (1994) *The Rise and Fall of Strategic Planning—Reconceiving Roles for Planning, Plans, Planners*, The Free Press, New York.

Moody's Investor Services (1998) "Airport Credit Goes Global: Privatized Airports Tap the World's Bond Markets," Report 36862, September, Moody's, New York.

Moody's Investor Services (2000) "New Low-Cost Airlines Could Impact the Credit Quality of Canadian and Australian Airports," Report 53600, February, Moody's, New York.

Nishimura, T. (1999) "Dynamic Strategic Planning for Transportation Infrastructure Investment in Japan," S.M. thesis, Technology and Policy Program, Massachusetts Institute of Technology, Cambridge, MA.

Office of Technology Assessment (1982) *Airport and Air Traffic Control Systems*, U.S. Government Printing Office, Washington, DC.

Porter, M. (1980) *Competitive Strategy: Techniques for Analyzing Industries and Competitors*, The Free Press, New York.

Porter, M. (1985) *Competitive Advantage: Creating and Sustaining Superior Performance*, The Free Press, New York.

Porter, M. (1987) "Corporate Strategy: The State of Strategic Thinking," *The Economist*, May 23, pp. 17–22.

Secretary of Transportation (1999) *The National Plan of Integrated Airport Systems (NPIAS) 1998–2002*, Report to the U.S. Congress, U.S. Department of Transportation, Federal Aviation Administration, Washington, DC.

Sydney Airport (2001) "Traffic Statistics," http://www.sydneyairport.com.au/.

WTRG Economics (1998) "Oil Price History and Analysis" www.wtrg.com/prices.htm.

4

Privatization and deregulation

Privatization and deregulation of the airline/airport industry were major trends throughout the end of the twentieth century. These changes in the market for airline and airport services are fundamentally restructuring the industry. This chapter addresses this complex of issues. It discusses the consequences of privatization and deregulation for airports, and provides guidelines on how these processes might best be managed.

Section 4-1 summarizes the situation before the movement for privatization and deregulation. It establishes a benchmark against which to evaluate the consequences of this trend. Section 4-2 discusses the motivations for privatization and deregulation. These aspirations shape the development of what these concepts mean in practice. Section 4-3 follows this thread and discusses the practical concept of privatization. It stresses that privatization is complex and often different from what it seems to be. Paradoxically, privatization can be more than the transfer of ownership of a property from government to private interests, yet it may not even involve this transfer. Section 4-4 builds on these discussions to lay out the major alternatives for managing the privatization of airports, given that most large commercial airports have local monopolies that require regulation. Its conclusion is that major airports are probably best managed through some form of public–private partnership, for which numerous examples exist, particularly in North America, where this is the dominant mode of operation. Section 4-5 shifts attention to the question of economic liberalization of the airlines. In this case, widespread economic deregulation largely eliminates the possibility of local monopolies. There is thus little reason for detailed public participation in the business side of running the airlines. The exceptions to this rule concern nonbusiness or extraordinary issues such as safety and national emergencies. Finally, Sec. 4-6 shows how airline deregulation leads to great volatility in airport traffic and to the development of transfer airports that compete with each other over continental distances. Transfer hub airports and airline deregulation in general thus require planning, management, and design that stress flexibility to adapt to the rapid changes characteristic of commercial competition.

4-1   The airport and airline industry before privatization and deregulation

Privatization and deregulation are two forms of freeing industry from government control, a process also known as liberalization. *Privatization* generally refers to the transfer of ownership from a government agency to private investors, although in practice the idea is more complex, as Sec. 4-3 discusses. In this sense, privatization reverses nationalization, which is the process of government taking over private properties. *Deregulation* is the elimination of government processes that review business decisions. It generally refers to economic deregulation, which removes the need for companies to get permission to raise or lower prices, enter and exit markets, and innovate in the range of services they offer. As Sec. 4-3 indicates, privatization has been the major trend for airports, and deregulation the major trend for airlines. As factors that change the behavior of airlines generally have significant effects on airports, airport operators need to consider both privatization and deregulation carefully.

The patterns of ownership and regulation of airports and airlines have differed considerably between the United States and the rest of the world. Any careful discussion of the liberalization of the industry should recognize this fact. Because their starting points are so different, their evolution and conclusion neither will be nor should be the same.

Airports

Until the British Airports Authority (BAA) became a private company in 1987, governmental agencies owned and operated essentially all significant commercial airports in the world.* With few exceptions,

---

*A few exceptions existed. For example, Lockheed Aircraft Company owned the Hollywood/Burbank airport in Los Angeles for many years.

governmental ownership existed in two different forms, that of the United States and that prevailing in the rest of the world. Whichever form it took, it was understood that airports were governmental institutions.

In the United States, in accordance with the national tradition of local control described in Chap. 2, individual cities, counties, and state agencies own and operate the commercial airports. For example, the cities of Los Angeles and Chicago have traditionally operated their airports through municipal departments, even though Los Angeles/International and Chicago/O'Hare are among the busiest airports in the world. Likewise, Dade County, the City and County of Denver, and the City and County of San Francisco operate Miami/International, Denver/International and San Francisco/International, respectively. Frequently, state governments have established special governmental units called "authorities" to operate airports—and often other infrastructure such as ports and major bridges. State governors and legislatures control these authorities politically. However, the "authorities" typically have distinct "corporate" identities and finance themselves independently with bonds sold to private investors. Thus Massport, the Massachusetts Port Authority, owns and operates Boston/Logan as well as the local toll bridge, major piers, and other properties. The Port Authority of New York and New Jersey answers to the governors of both states, and operates the major airports in the New York metropolitan region. In contrast, the U.S. federal government operated only two major commercial airports—Washington/Reagan and Washington/Dulles—and in 1985 transferred its powers to the Metropolitan Washington Airports Authority.*

The fact that ownership of commercial airports in the United States is distributed among hundreds of organizations has important managerial consequences. There is no coherent institution that trains airport managers, operators, and others concerned with running airports in the United States. In contrast, the Federal Aviation Administration (FAA) develops air traffic controllers and other professionals concerned with runways and other elements crucial to safety. Moreover, since even major airports do not have a steady flow of big projects, they cannot justify having a large range of experienced planners and designers on their staff. Consequently, consultants do most airport planning and design in the United States.

*At various times some cities, such as Albuquerque, New Mexico, operated commercial airports at federally owned military bases under joint-use agreements.

The pattern of independent ownership of airports in the United States has also had important financial implications. Individual airport owners in the United States have had to assemble their own resources. They could neither rely on national budgets, nor were they limited to this source of money. In practice, this has meant that owners of major airports in the United States typically arrange financing from private sources, usually through the sales of bonds to private investors, backed by long-term leases from airlines (see Moody's Investor Services, 1992). This has meant that in the United States both airlines and investment banks have enormous control over airport developments, since they can choose to provide financing—or not!

In the rest of the world, national governments typically controlled the country's major commercial airports. A national department or ministry would design, build, and operate airports with government employees. Thus, Transport Canada ran the Canadian airports, Aeropuertos y Servicios Auxilares ran the Mexican airports, and the Federal Department of Aviation ran the Australian airports.* These governmental airport organizations and their dependent groups enjoyed relatively large staffs of professionals, which could be justified because of their national mission. They represented a central source of expertise that often literally wrote the book on airport planning in their context (for example, Australia Department of Housing and Construction, 1985; Japan International Cooperation Agency, n.d.). Moreover, they often sponsored excellent, innovative research. Transport Canada, for instance, was responsible for the original development of the capacity standards for the design of passenger buildings that are accepted worldwide (see Chaps. 15 and 16). The airports themselves had large staffs of government employees. Amsterdam Airport Schiphol, for example, had about four times as many civil servant employees working on airport matters as Massport—although their traffic was similar. This is because U.S. airport operators hired consultants or subcontractors to do work that would typically be done by airport staff elsewhere.

Airports operated by national departments generally did not have much control over their finances. They typically did not benefit directly from revenues their services provided. Airlines and other users would normally pay fees to the national treasury, which would treat these revenues like other kinds of taxes. Airports would get money for airport projects according to the desires of the national legislature or other powers. Generally, there would be no particular correspondence

*The name of the Australian federal ministry responsible for airports changed in various governments. At one time, it was the Department of Housing and Construction.

between airport revenues and expenses.* The direct drivers of airport developments in this context were the national powers, not the airport passengers, airlines, or other clients.

Airlines

Airlines based in the United States have traditionally been independent businesses. Until 1978, however, they were tightly regulated economically by a national agency, the Civil Aeronautics Board (CAB). Airlines had to have CAB permission to carry passengers between any two cities, and found new licenses difficult to obtain since carriers already in the market would naturally oppose added competition. They could only charge fares according to schedules the CAB promoted. The CAB, moreover, controlled the levels of service in detail, to prevent the airlines from providing insufficient or excessive service for any fare. The regulatory process made it difficult for airlines to offer innovative service. The process also imposed heavy costs on the airlines and their customers, in terms of the efforts to deal with the CAB and its mandates (Jordan, 1970).

National governments typically owned the major airlines based outside the United States. For example, until the last years of the twentieth century, Air Canada, Air France, British Airways, Japan Airlines, and Lufthansa were all national properties. The governments controlled these airlines. They also regulated any other national airlines that competed against them, such as Air Inter in France, British Caledonian in Britain, and ANA in Japan. In effect, national governments organized and maintained cartels that were not supposed to compete. For example, the British allocated routes between its national airlines and the privately held British Caledonian (B-Cal). In general, B-Cal operated from London/Gatwick and could fly overseas to West Africa and Latin America; the national airlines flew from London/Heathrow and served other parts of the world.† Many argued that competition between airlines was neither appropriate nor practical (Payaux, 1984; Ecole Nationale des Ponts et Chaussées, 1987).

In much of the world, airlines simply divided their markets through "pooling" agreements. Governments would agree to set the fares, frequency, and capacity of flights on specific routes, and then

---

*Worldwide, many exceptions to this rule existed. In Britain, for example, many municipalities such as Manchester owned and operated their airports. In Germany, the Länder had dominant stakes in most of their major airports. However, the pattern of national ministries of aviation was most common.
†The British national airlines were BEA and BOAC, which merged into British Airways.

would pool their revenue and divide them by formula. In Australia, the national government regulated nearly everything, even the types of aircraft airlines could use, to the extent that the two domestic airlines offered almost identical schedules. Overall, the general rule was that airlines were not organized to be competitive or economically innovative. They responded slowly to customer pressures for lower fares and more service.

4-2   Motivations for privatization and deregulation

The privatization and deregulation of the airport/airline industry went through three phases at the end of the twentieth century. These were

- Deregulation of airlines in the United States, launched in 1978
- Privatization of airlines worldwide, largely in response to the U.S. deregulation
- Privatization of airports outside the United States

These phases affected airlines and airports in the United States and the rest of the world quite differently.

U.S. airline deregulation

The public desire for effective competition and lower prices was the principal motivation for the economic deregulation of the airlines. People knew that fares could be lower. They had experience with airlines operating charter flights and internally in the states of Texas and California; these were not controlled by the CAB and offered low fares. This factor, together with a variety of politically favorable elements, led to the deregulation of the airlines in 1978, the abolition of the CAB, and the removal of government intervention in the management of the airlines (Kahn, n.d.).

The results were spectacular. After some hesitations, airlines in the United States embraced the new competitive regime aggressively. A burst of economic innovations, price reductions, and new services followed (see Bailey et al., 1985; Meyer and Oster, 1984, 1987). Within only a few years, airlines:

- *Reorganized their networks* around transfer airports, thus providing less expensive, more frequent service
- *Adopted flexible pricing,* to maximize their revenues by discounting fares to fill otherwise empty seats

- *Built up frequent flyer programs,* as a means to build customer loyalty through discounts and free tickets
- *Created "cheap fare" airlines,* of which Southwest was the most successful, emerging as one of the top airlines in the world
- *Established integrated cargo airlines,* such FedEx (see Sigafoos, 1983) and the airline division of UPS
- *Dropped uneconomical services* they had been forced to maintain under regulation—such as regular flights between San Francisco and Sacramento over about 80 miles that could then be covered more quickly by car.

These innovations completely reshaped airport management and planning in the United States, as Sec. 4-5 explains. They also stimulated the privatization of airlines worldwide.

Worldwide airline privatization

The principal driver of the worldwide privatization of airlines has been the need to respond to deregulated airlines or perish. Canada, for example, could not long tolerate the way its traffic moved to airlines in the United States when these could offer low fares and Canadian airlines could not. During this transitional period, many Vancouver passengers would drive to Seattle/Bellingham and Toronto passengers to Buffalo, New York. In the major intercontinental markets, the airlines in the United States could undercut the fares of the foreign airlines by offering inexpensive connections within the United States and free flights through their frequent flyer programs. In order to compete with deregulated airlines and survive, regulated airlines simply had to deregulate.

Privatization has been the means to deregulate the nationalized airline industry. The trend started with Britain and Canada, the closest economic partners of the United States. By the year 2000, about half of the Western European airlines were privatized. This has been a painful process because competition means the elimination of inefficient activities or companies. In the United States, deregulation led to the disappearance of many major airlines such as Pan American, Eastern, and National. Outside the United States, this consolidation has been particularly difficult because it implied the disappearance of airlines such as Swissair and Sabena, that had been symbols of their nations.

Their new status enabled previously national airlines to react more flexibly to the challenge of deregulated airlines. However, possibly because the managers of the newly private airlines had old habits, many of the innovations developed after deregulation in the United

States were slow to implant themselves elsewhere. The system of transfer airports only began to be adopted by Amsterdam/Schiphol and Paris/de Gaulle in the late 1990s, some 20 years after these developments in the United States. Outside the United States, the consequences for airports of airline deregulation have been slow in coming.

Privatization of airports

Disregarding eventual difficulties associated with privatization, governments found that this process gave them an important immediate benefit: it made them a lot of money. By privatizing assets, they got immediate and future payments from the new owners and avoided the need to find money to pay for improvements. (See U.S. Congress, General Accounting Office, 1995). Australia, for example, privatized its airports for over $2 billion, or about $200 for every inhabitant. Events like this appear to be wonderful tax cuts at election time. Greece privatized its new airport at Athens/Venizelos by offering a long-term concession to the company that built the airport. The country got a brand-new airport without having to borrow much money for the project. Similar deals have been organized worldwide.

The desire to emphasize economic performance and responsiveness to consumers has been another rationale for the privatization of airports. The thought is that airports would be better managed if they were run as companies (see Advani, 1998; Kapor, 1995; Vickers and Yarrow, 1988). The idea is to change the management focus from politics to business. However, this line of thinking leaves out of consideration the public concern with equal access to airports and monopoly pricing. The public interest in major airports means that they cannot be run as entirely private companies. Section 4-4 examines this issue and provides guidelines for how airports should be organized as public–private partnerships.

4-3   The concept of privatization

The simple view of privatization is that it involves a change in who owns the property and facilities, specifically the transfer from a government agency to a group of private investors. This dictionary definition is sufficiently misleading in the case of airports to be wrong. Most privatizations of major commercial airports have not involved the actual sale of the property.*

---

*The privatization of the British Airports Authority and its transformation into BAA plc is a notable exception to this statement. Several other British airports have also been sold.

The typical airport privatization involves a long-term lease. For example, when Australia privatized its airports, it actually sold 50-year leases with the option to renew if certain conditions were met. Similarly, the Canadian national government privatized its major airports by leasing them to local public–private agencies operating like companies. Other countries have privatized airports or portions of airports through concessions lasting 30 years or more, for instance Argentina, Greece, Peru, and the Philippines. Privatization thus can and does occur without transfer of ownership.

A useful practical definition is that *privatization* involves the transfer of *some* ownership rights. To understand what this means, we have to look carefully at what rights are associated with ownership. Ownership implies two basic categories of rights. These concern:

- *Rights to residual income,* that is, profits in a general sense, although often not labeled as such
- *Management control,* which covers the range of short-term operational and long-term development issues

Rights to residual income

Residual income is the difference between revenues and costs. The right to this margin is one of the standard benefits of ownership. Readers should note, however, that investors could acquire the rights to residual income from an airport without owning the deed to the property. When governments privatize airports, they offer investors the rights to keep any residual income for a particular period, even when the government keeps title to the land.

The rights to residual income are the prime reason investors are interested in airports. Investors naturally focus on profits. They are willing to pay money now for the opportunity to make profits later. From their prospective, major airports look like good investments because they are local monopolies. For example, if people want to fly to Vienna, Austria, they must go to the Vienna airport. Therefore, the company running the airport has a captive market. Given reasonable conditions, the owners of the rights to the residual income should be able to charge enough both to cover costs and make a good profit, that is, revenues over costs. In everywhere but in the United States, government owners of have been able to set up ways in which they can transfer the rights to residual income to private investors. Outside the United States, governments have thus been able to sell these rights and privatize airports.

In the United States, historical reasons make it almost impossible to sell the rights to residual income for airports (U.S. Congress, General Accounting Office, 1996). The agencies that hold title to airport properties do not have these rights to sell. Paradoxically, moreover, to the extent that the rights to the residual income from an airport define ownership, the actual "owners" of many major commercial airports in the United States are not the local communities but the airlines!

To understand the situation in the United States, it is important to look carefully at the U.S. system of financing airports. Two considerations are fundamental:

1. Airport operators in the United States cannot "divert" funds from the airport to other uses
2. Airlines frequently have "majority in interest" agreements with the airports

Most significantly, the communities owning airports have had to agree not to "divert" money to their own, nonairport purposes as a condition of receiving federal grants for airport projects. The reasoning for this requirement is that federal grants come from airline taxes earmarked for improvements at airports that need support. Thus, local communities should not use airport revenues, derived in part on the basis of federal grants, for other purposes. In short, they should not make profits on public grants. In the late 1990s, the City of Los Angeles tried to use some of the apparent profits from Los Angeles/International for municipal purposes, but a major lawsuit by the airlines led to a judgment against the city. The airport owner in the United States can use the excess of revenues over expenses to improve the airport or to reduce the charges to the airlines. By long-standing national policy, the benefits of investments in airports should go to the air transport industry, that is, the airlines and their customers. The typical airline airport in the United States has no profits for itself, let alone to sell.

Second, many airports have agreements with the airlines that the airline fees cover the difference between the total costs of running the airport and the total revenues from all sources. The practical implication is that if the airport operator is particularly successful in raising revenues, the benefits of these efforts accrue to the airlines. Example 4.1 illustrates the point. Where these arrangements exist, the actual beneficiaries of the rights to residual income are the airlines. To the extent that rights to profits define ownership, airlines in the United States in some sense are "owners" of airports.

EXAMPLE 4.1

Consider an airport operating under the agreement that airline fees should cover the difference between total expenses and revenues from all other sources, such as parking, rental car leases, and other concessions. Suppose that the total expenses in a base year are $300 million, which is about what the Washington Metropolitan Airports Authority spent in 2000. Suppose further that the airport received $250 million from rents, concessions, utility sales, and passenger fees. The airlines would then collectively have to pay the difference:

Airline landing charges in base year = 300 − 250 = $50 million

The airport would prorate this $50 million among the users by adjusting the landing fees.

Suppose that in the following year the airport manages to market its concessions especially well and increases revenues from this source by $25 million. All else being equal, the airline fees would then be reduced:

Airline landing charges in more profitable year
= 300 − 275 = $25 million

Under the assumed conditions, the airlines benefit from the increased profits. They in effect own the rights to residual revenues.

Note that these arrangements mean that airlines share the risks. If revenues go down, the airlines have to pay more through increased landing fees. The airlines paying residual costs thus have another attribute of "ownership." They share risks and the possibility of profits and losses.

Management control

The other important attribute of ownership is control of the property. With respect to airports, this means the ability to run and develop the property. Management control covers many different aspects of the operations, such as

- Planning of new facilities
- Design of these elements
- Financing of capital costs and daily operations
- Operation of activities, including concessions, personnel, and airline activities

- Pricing of the services the airport offers
- Access to the airport services

Private investors want management control so that they have the flexibility to organize facilities and operations in order to maximize profit. Readers should note, however, that typical privatizations of airports have not given private investors complete management control. On the contrary, governments often retain control over at least two important elements: development of new facilities and prices. Private investors thus have limited management control over privatized airports.

Governments generally maintain important controls over the development of major facilities at privatized airports. Insofar as governments privatize through some form of long-term lease or concession, they remain the owners of the property. As landlords, these governments have the right to say how their property will be developed. More specifically, they have the right to prevent their tenants from building runways or buildings that the government does not want. They can use this power to influence other aspects of management, by obtaining concessions in return for their agreement to new projects the private investors desire. However, these powers are negative. The governments can refuse permission, but they cannot force private investors to build facilities that the investors do not think would be profitable for them.

Governments often maintain extensive control over airports through their rights to review substantial projects that will have major impacts on regional development or the environment. Airport operators cannot presume to develop their properties without extensive public review. For example, the operator of London/Heathrow took years to get government permission to develop a major passenger building to the west of its central area. This activity started well before privatization. The latest effort began in 1993 and took 8 years and about $125 million dollars (£81 million) (Thorpe, 2001). This difficulty is not unique to private airport operators: proposals for new runways at San Francisco/International and Boston/Logan have had equally long histories. What is clear is that airport operators do not control their planning and development—these are directed by governmental processes.

Finally, governments control how airports price their services. They recognize that major commercial airports are local monopolies and cannot be allowed to use this fact to extort unreasonable revenues from the airlines, shippers, or passengers. In Britain, for example, the Civil Aviation Authority "has to reset price caps on airport charges generally every five years."* They do this through a lengthy process of inquiry, detailed inspection of accounts, and legalistic hearings. The process scheduled to set the tone for 2003–2008 began in early 2000 and will last three years or more (Civil Aviation Authority, Economic Research Group, 2001). This particular review recognizes that, in order to prevent monopolists from charging excessively, it is necessary to regulate both prices and quality of service, as the U.S. Civil Aeronautics Board used to do for airlines. Consequently, the UK Civil Aviation Authority proposed to set detailed standards of service based on comparisons of about 30 airports worldwide. This is such a large task that it may not be carried out. What is clear, however, is that the privatized UK airports will be continuously controlled in detail (Civil Aviation Authority, Economic Regulation Group, 2000a, 2000b). For practical purposes, the British government is almost constantly investigating the finances of its private airport operators.

The situation in the United States is almost the mirror image of the situation for privatized airports elsewhere. Whereas privatized airports are subject to considerable government regulation and control, government-owned U.S. airports have ceded considerable management control to private interests. Indeed, in the United States private groups lead or play an important role in all aspects of management:

- Consultants do most of the planning—under overall political direction.
- Engineers and architects in private practice do almost all design.
- Major investment banks and the airlines are responsible for organizing the financing of major projects.
- Private companies—airlines, catering companies, baggage handlers, parking specialists, etc.—routinely handle a majority of the operations on the airport.
- Long-term contracts set many prices for airport services, specifically those affecting airlines.
- Federal laws prevent airports from restricting aircraft access to the runways, specifically barring pricing strategies aimed at discouraging small aircraft from major airports.

*This task used to be carried out by the UK Monopolies and Mergers Commission. Parts of the effort may be executed by its replacement, the UK Competitiveness Commission.

The case can be made that the governmentally owned U.S. airports are among the most privatized in the world. Although government agencies own the land, residual revenues may benefit the private airlines and private interests have effective management control over much of the airport operations. This contrasts with privatized airports elsewhere, which must operate with considerable government control.

4-4   Guidelines for airport privatization

Many institutional arrangements for controlling airports are possible and have been used, as the previous discussion shows. The meaningful question is: Which are best? A detailed answer is not appropriate, since local conditions properly influence what can and will be done in each situation. Looking broadly at the worldwide experience, however, it is possible to define overall arrangements that generally work best over the long run.

The root issue concerns how the responsibilities for airport planning, management, and design should be shared between governmental and private groups. There are four basic possibilities for doing this, as Table 4-1 illustrates (de Neufville, 1999). The two elements of ownership, the rights to residual income (profits) and management control, can be shared between public and private groups. Major commercial airports could be run:

- Totally by governments, which for a long time was the standard pattern outside the United States
- Totally by private companies, a logical possibility that does not exist
- Through some kind of shared arrangement

Total control of airports by government or private interests appears to be an inappropriate model for major commercial airports. These facilities are large commercial operations and need to be run with the kind of consumer responsiveness that cannot be provided by government organizations driven ultimately by political considerations. The rationale for total governmental control over commercial airports is most appropriate for developing or remote regions, where local levels of income may not be sufficient to organize airports effectively. Experts now widely accept that, for the kind of mature airport/airline industry that exists at major airports, the participation of private business is essential for achieving economic efficiency and competitive prices and services. On the other hand, the public interest

Table 4-1. Four major possibilities for allocating control of airport planning, management, and design between governmental and private groups

Management control	Rights to residual revenues or profits	
	Government	**Private**
Government	Fully government: complete control by civil service and politics	Regulated control: unilateral, centralized control by government of rates and access
Private	Partnership control: government sets policy as owner, private parties implement	Fully private: complete control by private interests

in preventing monopolistic practices and ensuring coherent regional planning makes it clear that totally private control and management of major commercial airports is unacceptable.

Some form of shared allocation of responsibilities between government and private interests is best. The basic alternatives are either shared or regulated control, as Table 4-1 indicates. The essential difference between them is whether the government is a decision-making partner or a separate entity in a regulatory and, inevitably, an adversarial process. In detail:

- *The Partnership approach* involves government participation as an insider in the policy formation, so that when decisions emerge they already have substantial collective agreement.

- *The Regulated approach* has the government and the private groups acting independently, with their own agendas, and interacting in some strenuous contest to resolve issues.

The current arrangements in Britain exemplify the regulated approach. It is long and argumentative. For instance, in the 1990s the privately owned BAA independently proposed to develop a major project new passenger complex at Heathrow. It advocated its proposal with a full array of legal and other talent. It then took 8 years, $125 million, and considerable anxiety until the company found out what it might do. Similarly with regard to the regulation of prices: the Civil Aviation

Authority investigates the case with its own agenda of how it would like BAA to behave, and BAA defends itself as it can. Conflicting experts argue publicly on both sides, and then the political process tells BAA what it will be allowed to do. This has proved to be an unpleasant process.

Worldwide, privatization of airports along the lines of the BAA has had difficulties. In the short run, these may simply be associated with ineffective private operations. In the case of Argentina and Honduras, for example, it appears that the outside investors paid too much for the airports and were unable to fulfill their contracts to the expectations of the government sellers. In the case of Harrisburg, Pennsylvania, the government agency claimed that the private company simply did not do its job. In any case, private operators that have fixed contracts with governments seem to find it difficult to adjust plans as new regional needs or public pressures arise (Gomez-Ibanez and Meyer, 1993; Armstrong, et al., 1994; Arthur, 1998; Beatty, 2001). In the longer run, the issue is that the regional and private interests diverge. The government and its public may want the airport operator to invest in facilities. However, the private operator may not wish to do so because the regional benefits of jobs or industrial activity do not seem to flow to the private company sufficiently or fast enough to be worthwhile. Under a regulatory regime, the government may be powerful in preventing private operators to do things, but is weak in making them invest when they do not find it attractive to do so. As time proceeds, governments may find that their immediate sense of satisfaction with privatization may fade.

The arrangements in place in North America illustrate the partnership approach. In Canada, for example, major airports are run as "authorities" under the direction of a board of directors chosen from among community leaders who are expected to act in the interests of their region. These groups operate under a common process set up by the national government (Transport Canada, 2001). Their interests may be to:

- Operate as profitably an international airports company as possible and make money for the local community (Vancouver International Airport Authority, 2001)
- Focus on local efficiency (Calgary Airport Authority, 2001)
- Or some other purposes (Greater Toronto Airports Authority, 2001; Ottawa Macdonald-Cartier International Airport Authority, 2001; Aéroports de Montréal, 2001).

The Canadian framework permits and encourages much local enterprise. It appears to work not only for airports but also for many other forms of infrastructure (see, for example, Ontario Department of Finance, 2000). The Canadian approach seems to permit active and effective business operations, encourage local initiatives and direction, and provide generally cooperative arrangements in which to blend public and private concerns.

Airport authorities and other airport operators in the United States operate under a range of similar partnerships between public and private interests. In general, these feature great flexibility in the arrangements, with details dependent on the type of development, the economic capabilities of the developers, and other local conditions. Quite often, the same public airport operator will simultaneously have distinct arrangements with private companies. At Boston/Logan, for instance, both United Airlines and US Airways were developing additions to their passenger facilities around the year 2000. In this case, the stronger airline decided to use its own money to fast-track the construction, whereas the weaker company worked with the public airport operator to finance and build the project. The Port Authority of New York and New Jersey provides an excellent example of the range of possible partnerships possible. For example, at New York/Kennedy it largely let a private power company develop the cogeneration plant—it had little expertise in the field of running power plants or marketing electricity. It also let individual airlines or groups of airlines build their own terminals because they best knew their customers and needs. On the other hand, it took the lead in the planning and design of the rail connection between the airport and the rest of the city, since the public agency may best be suited to deal with the urban political issues (FitzGerald, 1994).

An essential feature of all the partnership arrangements is that potentially contentious issues such as pricing are resolved with a minimum of protracted argument. The public and customers feel that the airport operator is sensitive to its concerns about excessive pricing because people with their collective interests are on the board of directors or equivalent oversight group. However, the airport operators simultaneously run as businesses that have to cover their expenses. The partnership approach appears to represent an effective balance between public oversight and private capabilities.

Definitive answers to the question of what is the best allocation of airport responsibilities between public and private interests may not be available until we have decades of experience. Meanwhile, some conclusions seem appropriate:

- Some form of shared responsibility seems to work best for major commercial airports.
- The exact form depends on local circumstances and capabilities.
- A partnership approach has many advantages over a completely privatized arrangement, and may be better overall.
- Private companies may best focus on the commercial aspects of airport management (such as finance, construction, and operational management), leaving political issues in the domain of government authorities.

In the words of some leaders among the experts arranging privatization processes for airports worldwide, "…a range of issues and complications govern the involvement of the private sector in airport development. Effectively structuring the basic business relationship [between the public and private interests] is the cornerstone of success" (Beatty and Lipson, 1999).

4-5   Airline deregulation

Whereas governments inevitably become closely involved in the central aspects of airport planning and management, the situation is entirely different for airlines. Governments get involved in airports because of their concern about the prices for airport services, the quality of service offered at the going rates, and the quantity of capacity provided—among other factors. Many of these issues are closely linked to the fact that airports are local monopolies to a great extent. If their operators were able to focus on maximizing profits without government supervision, airport prices and services might become intolerable. The situation is different for airlines, since they are much less likely to become monopolies.

The relevant difference between airlines and airports in this regard concerns the nature of their capital assets. Airports are immobile. They expend slowly, over many years. Aircraft, the capital assets of airlines, are highly mobile. In a deregulated environment, owners can deploy them at will. Normally, there is also an extensive international market for new and leased aircraft, so that airlines can shrink or expand their fleets quickly.

The availability and mobility of aircraft mean that, in a deregulated environment, the barriers to entry are low in the airline industry. By contrast, the barriers to entry in the airport business are very high. Given enough time, a prospective competitor may be able establish a new airport to serve a special market, as the developers of London/City airport did for short-range aircraft and as the owners of Orlando/Sanford have done for charter flights. However, it is impractical to establish a significant rival to a major commercial airport in a metropolitan area. (See Chap. 5 for a detailed discussion of competitive airports in a metropolitan region.) In contrast, new airlines such as FedEx, Ryanair, and Southwest were able to develop into very large, valuable airlines within 20 years.*

Because the barriers to entry are low in the airline industry in a deregulated environment, governments can rely largely on competition to keep prices down, to ensure that customers get reasonable service for the money they pay, and to stimulate more capacity in the industry when it is needed. They can do so, that is, provided the governments prevent collusion between the airlines and other forms of cartels. If they do so, they do not need to establish detailed supervision or review of airline management. Thus, in Britain, British Airways operates without detailed governmental supervision of its business, whereas airport operators such as BAA are subject to extensive, detailed governmental review of their activities.

In summary, the issues surrounding the economic deregulation of airlines are significantly different from those associated with the privatization of airports. In thinking about the airport/airline industry, planners, managers, and designers should not assume that concepts that apply to airports are valid for airlines, and vice versa. The reality is that, in a deregulated environment, the pattern of changes for airlines and airports are very different. Changes occur much more rapidly for airlines than for airports. As the next section describes, this fact complicates airport planning, management, and design.

4-6   Implications of airline deregulation for airports

The liberalization of the airline industry through deregulation and privatization affects all those concerned with airports—planners,

---

*As of 2002, the market capitalization of UPS and FedEx were the largest of all airlines, and that of Ryanair was greater than British Airways.

managers, designers, service providers, customers, and neighbors. The effects occur in three major ways:

1. *Increased volatility,* affecting all aspects of the airlines
2. *Development of transfer hubs,* a nontraditional form of airports
3. *Long-distance competition between airports,* in a way that was previously impossible

Increased volatility

Deregulation speeds up change. In a regulated environment, airlines have to go through extensive proceedings in order to establish new routes, offer new services, and change fares. The process of proposing, arguing for, and obtaining approval for change may take years. In a deregulated environment, changes can occur virtually overnight. If the controlling managers wish to do so, they can announce and implement a new direction very quickly. A fare change can be put into the computer reservation systems almost instantaneously. A new route may take longer to implement, but may happen quickly if the airline already operates facilities at the airports at both ends of the route.

An analogy helps visualize the situation. Regulation dampens all changes to a system; it makes all adjustments occur slowly. It is like the shock absorbers on a car. Deregulation removes the impediments to rapid change; it is like removing the shock absorbers. When these are absent, a car feels every bump on the road and, in unfavorable conditions, may resonate with them to amplify changes. Deregulation has a similar effect on airlines.

Deregulation most obviously increases the volatility of overall traffic at airports (de Neufville and Barber, 1991). Airlines cut fares to gain market share to a particular destination, offer new services, or cancel some operations. Overall, the fluctuations of traffic in a deregulated environment are much greater than in a regulated environment. Moreover, airport traffic can rapidly make dramatic shifts. For example, airlines may decide to shift a base of operations, as American Airlines did around 1980 when it moved many of its flights out of Chicago/O'Hare to its new base in Dallas/Fort Worth (which led to a 20 percent drop in traffic), and as Delta did when it reduced its operations into Dallas/Fort Worth and built up Cincinnati as a hub. The record of traffic at Chicago illustrates both effects (see Fig. 4-1).

Increased volatility in overall traffic increases the risk associated with investments in facilities. This is fundamental. The greater the traffic

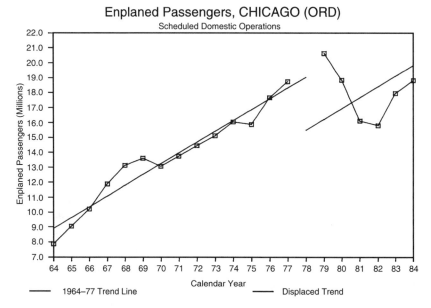

Fig. 4-1.   *Volatility of air traffic at Chicago/O'Hare after deregulation in 1978.*

fluctuations, the greater is the chance that a facility will not be needed at some points in its lifetime. Raleigh-Durham (North Carolina) airport offers an example of this problem. Around 1990, it was a significant hub for American Airlines, offering service to both Paris and Mexico City, for example. Not long thereafter, American decided to reorganize its traffic patterns and withdrew these and many other services. The associated facilities then became redundant. Similarly, when US Airways shifted much of its international traffic from Baltimore/Washington to Philadelphia in the mid-1990s, it left behind an international passenger building that was much too large for the actual international traffic (see Chap. 15). Such shifts of traffic and the associated risks to investments in facilities are common in a deregulated environment.

Deregulation also leads to significant changes in the structure of the airline industry. The competition associated with deregulation normally leads to the disappearance of airlines through mergers or bankruptcies. In the United States, for example, the decades after dereg- ulation witnessed the disappearance of many major airlines, such as Braniff, Eastern, Frontier, National, Northeast, Pan American, PEOPLExpress, TWA, and others. A similar shakeout of airlines began around 2000 in Europe, along with airline deregulation within the Common Market. Competitors took over Air Inter and British Caledonian in the 1990s. In 2001,

Sabena went bankrupt and Olympic and Swissair were undergoing fundamental change.

These changes in the structure of the airline industry can severely impact airports. They worsen the instability of airline traffic changes. The financial collapses of Swissair and TWA in 2001, for example, caused significant changes for the airports that were the bases of these airlines. Once TWA stopped flying, the passengers it carried from the East Coast of the United States to the West Coast no longer went through TWA's hub at St. Louis/Lambert—they went on another airline through another hub.

The formation of airline alliances is another kind of change in the structure of the airline industry that significantly affects the planning, management, and design of airports. Airlines in an alliance want— indeed need—to locate next to each other. They can only fulfill their ambition to provide easy connections and joint routes when they can operate side by side or, better, actually share the same offices, counters, and other facilities. Thus, whenever airlines develop or change an alliance, they want to move their operations at airports around the world so that they can work together with their partners. Their requests place severe demands on the airport operators. Example 4.2 illustrates the changes that can occur.

EXAMPLE 4.2

The cases of Boston/Logan and New York/LaGuardia illustrate the fluctuations in airline traffic and needs at airports in a dereg- ulated context.

The story at New York/LaGuardia concerns the development, during the decade after deregulation, of a new passenger building to serve mostly shuttle traffic to Boston and Washington. In that period the airport operator had to deal with three airlines as prospective clients—first Eastern, which went bankrupt, then Trump Airlines, which took over shuttle service, and finally US Airways. Each of these clients had different objectives, needs, and capabilities. The planners were always struggling.

The story at Boston/Logan begins shortly after deregulation took place in 1978. Two "cheap fare" airlines began service, PEOPLExpress and New York Air. Both offered competitive service to New York. To accommodate these entrants, the airport operator had to create space for them—naturally, none of the established airlines was willing to let them use their gates or facilities. The planners thus found space for these domestic airlines in and around the international passenger building. Around the same time, Northwest contracted with the airport operator for some new cargo buildings on the other side of the airport.

When Northwest officials came to Boston for the opening ceremonies for their new cargo facilities, they informed the airport operator that they would not move into these facilities. Actually, they planned to move their international flights from New York/Kennedy to Boston, and establish many new connecting routes through Boston. This was an offer Boston was delighted to accept, even though it meant that the new cargo building would have to be reconfigured and leased to a new client (eventually, FedEx). As it happened, the airport operator could indeed accommodate Northwest in the international passenger building—because, in the meantime, both PEOPLExpress and New York Air had gone bankrupt. However, the building had to be rearranged to accommodate a large number of international/domestic transfer passengers.

Not long afterwards, Northwest again decided to change its route pattern. This time, it shifted its hub from Boston to Detroit, where it had built up one of its two main bases. Therefore, it pulled much of its traffic out of Boston. The airport operator once more had to reconfigure operations in the international passenger building. Such is the life of airport planners, managers, and designers in the deregulated environment.

Traffic volatility over the short term puts great pressure on airport operators, since they are in the business of developing buildings, runways, and other facilities that have long economic lives. There is a basic incompatibility between the fluctuating level and type of traffic and permanent developments. Airport operators now have to do their best to reconcile these two different patterns of traffic. This new requirement substantially changes traditional views on airport planning, management, and design. Airport operators need to adopt a strategic for minimizing their exposure to the possibility of having inadequate, excessive, or otherwise inappropriate facilities for the actual traffic. Specifically, they need to:

- Develop more flexible designs, so they can adjust their facilities to the changing clientele

- Use short-term leases to limit their financial and operational exposure to changing airline requirements

Flexible Designs   Design flexibility mainly concerns the passenger buildings. Flexibility on the airside of the airport comes mainly from having enough land to be able to extend runways and taxiways. Flexibility in terms of the buildings is achieved by designing them so that individual airlines or other clients can easily expand or contract their activities. As Chaps. 15 and 16 describe, the major design possibilities for developing flexibility are connected buildings, shared-use, and temporary facilities.

Connected buildings allow the airport operator and airline users to shift operations relatively easily. In contrast, it can be difficult if not impractical for airlines to expand their space across separate buildings. At best, this may lead to split operations that confuse both the passengers and airline operations. In other cases, expansion may require the construction of entirely new facilities. This happened at Baltimore/Washington, when Southwest could not move into the unused gates abandoned by US Airways when it shifted many operations to Philadelphia (see Chap. 15). Amsterdam/Schiphol, San Francisco/ International, and Singapore provide good examples of airports with connected passenger buildings.

Facilities that several types of users share provide another way to build flexibility into the design. When many clients share a space, they can vary the proportion of the time they use it, and easily alter the effective space they use. The "international" passenger building at Atlanta/Hartsfield provides a good example of shared use. Its gates can be used by both international and domestic passengers, several times a day if necessary. Enabling airlines to share check-in counters through the use of a common computer system is another good example of shared use, and airport operators are increasingly installing the appropriate Common User Terminals (CUTE). Designs featuring shared use have the further advantage of reducing the total size or amount of facilities that must be provided, since individual users typically have their peak traffic at different times, and can thus share each others' capacity. The design and implementation of shared-use facilities is thus a prime way airport operators are protecting themselves against the risks of volatile traffic. As Chap. 15 explains, shared-use facilities are becoming an integral part of modern airport design, as the new passenger buildings at Boston/Logan (American Airlines), Orlando/Sanford, Singapore, and Toronto/Pearson indicate.

Airport operators can also use temporary facilities to limit their financial exposure to volatile traffic. These can be impermanent structures or other devices for providing capacity. For example, both Los Angeles/International and Boston/Logan have used inflatable structures to provide capacity for passengers or maintenance facilities until they could make more definitive plans. Airport operators can also use buses or other transporters to connect passengers to their aircraft. This approach has the twin advantages of avoiding the significant capital costs of construction of finger piers and other structures, and of permitting the vehicles to be parked or sold if the capacity they provide is no longer needed. Many airports (mostly outside North America) now routinely use transporters to provide for peak and other volatile traffic.

Short-Term Leases Airport operators increasingly use short-term leases to manage their exposure to the risk of volatile traffic. This practice contrasts with the traditional U.S. custom of leases lasting 20 years or more. Long-term leases make it difficult for an airport operator to manage the property when an airline or other client withdraws. For example, when Eastern Airlines went bankrupt and stopped using its passenger building at Atlanta/Hartsfield, the airport was not able for a long time to take control of the property and reassign it to another airline. The airport had to go through lengthy legal procedures and negotiations—meanwhile, the building and its capacity could not be used. Short-term leases permit the airport operator to take control of unused property and reassign it quickly.

Many airports in the United States now use 30-day leases. Subject to restrictions negotiated in advance, these permit the airport to take control of facilities that airlines do not or cannot use. Normally, the lease defines the circumstances that warrant termination of the lease and compensates the airline for the residual value of the improvements it has made to the facility. These short-term leases provide significant management flexibility.

The airport operator might prefer a long-term lease that guarantees that the client will cover all the costs of the relevant facility. Unfortunately, these contracts may not be fully enforceable. Atlanta/Hartsfield, for example, could not enforce its lease with Eastern when it was bankrupt. Long-term contracts may sometimes be enforceable in narrow technical terms, but not meaningfully in terms of airport operations. Thus, when American Airlines abandoned Raleigh/Durham as a hub, it met its legal obligations. However, neither the airport nor the local investors got compensation for the losses incurred from the lack of traffic and of another airline as a hub operator. The airport operator thus requires a combination of long-term and short-term leases to manage the risk associated with the volatility of the airline operations.

Transfer hubs

The development of transfer hubs is a major consequence of the liberalization of the airlines. These facilities have reshaped the pattern of air traffic in the United States and are now taking hold in Europe as the Common Market promotes deregulation. The use of hub airports has also profoundly altered many aspects of airport planning and design, in particular concepts about the configuration of airport passenger buildings, as Chap. 14 details.

A *transfer hub* is an airport that an airline uses to transfer large numbers of passengers between flights.* It is a central element in the operations of most large airlines in a deregulated environment.† It permits the airlines to provide more frequent and less expensive service than they could otherwise do. Transfer hub airports have thus become a major feature of deregulated air transportation networks.

To understand why transfer hub airports provide an attractive configuration for airline services, it is useful to compare a hub-and-spoke network with traditional point-to-point service. When an airline provides point-to-point service, it flies directly between the traveler's origin and destination, for example, between Sacramento and Boston, or Lisbon and Prague (see Fig. 4-2) This service minimizes travel time. However, flights may not be frequent, since the traffic between the two end points may not justify even daily service. It may also be expensive, because the low levels of traffic perhaps only justify smaller aircraft, with higher costs per available seat-mile than larger aircraft.

A hub-and-spoke system flies passengers from end points such as Sacramento or Lisbon into a central airport, then has many of them change planes to other flights to reach their final destinations (see Fig. 4-3). This arrangement has the obvious inconvenience of causing a detour that lengthens the trip. It also requires transferring passengers and bags between aircraft. However, it offers some most attractive

---

*Notice this use of the word "hub" to denote the transfer function going on at an airport. This meaning is different from the way the U.S. FAA has classified airports as "hubs" according to their level of traffic.
†Some airlines, such as Southwest and Ryanair, prefer to provide point-to-point service.

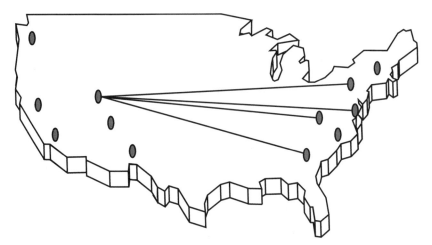

Fig. 4-2. *Traditional point-to-point service provides direct service, even between city pairs with low traffic.*

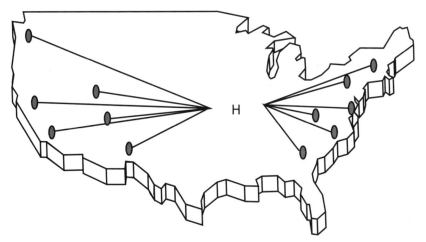

Fig. 4-3. *Airlines route passengers through transfer hubs to achieve economies of scale and high frequency between city pairs.*

compensating features. Hub-and-spoke operations provide service that is:

- More frequent, because flights from the end points (such as Sacramento) to the central hub serve all the end points that together justify many daily flights

- Less expensive, because the airlines can both use large aircraft with lower seat-mile costs and can fill them more consistently because the fluctuations in traffic between the different end points tend to balance out

- More reliable, both for the customers, who can count on back-up flights throughout the day, and for the airlines, who can maintain their supplemental crews and aircraft at the central hub rather than scatter them over the network

On balance, hub-and-spoke networks provide better service to most customers. This is why they have been developing across North America, Europe, and elsewhere. Figures 4-4–4-7 give examples of

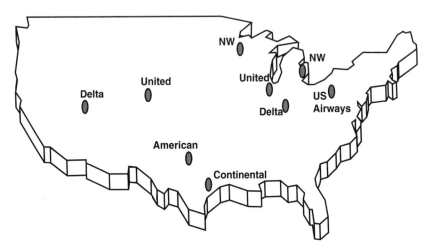

Fig. 4-4. *Transfer hubs in the United States central to the East–West market for air travel.*

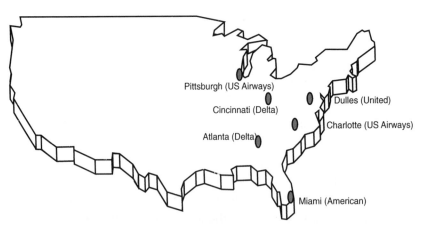

Fig. 4-5. *Transfer hubs in the United States central to the North–South market for air travel.*

hubs serving East–West transcontinental, North–South, and integrated cargo traffic in North America, and of European hubs. The ongoing debate concerns the extent to which hubbing traffic is desirable. Airlines keep experimenting with various combinations, opening up new hubs (as United Airlines did in the 1990s at Washington/Dulles) and

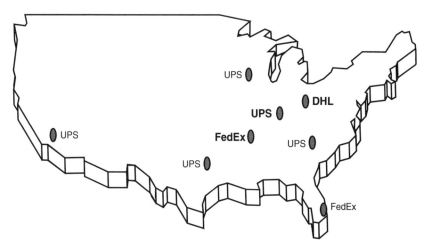

Fig. 4-6. *Transfer hubs in North America for integrated cargo carriers.*

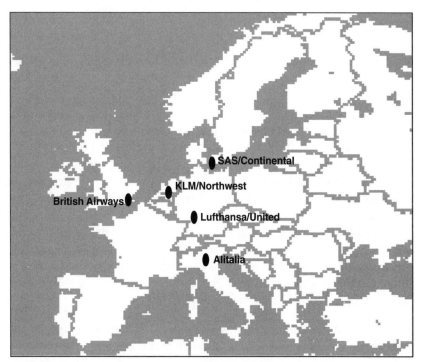

Fig. 4-7. *Transfer hubs in Europe as of 2001.*

closing others (as American Airlines did at Raleigh-Durham). Naturally, these constant changes create difficulties for airport planners, managers, and designers.

Many regions want to have transfer hub airports and compete to get them. They appreciate the jobs associated with the airline operations and the business attracted to the region because of the excellent air service radiating from the hub airport. They believe that the existence of the hub will lead to the economic expansion of the area, just as good ports and rail centers helped establish major cities such as Chicago, San Francisco, Singapore, and others. The issue is: What facilities should an airport have, in order to become and remain a transfer hub?

Criteria for Good Transfer Hubs   To understand what makes an airport a good transfer hub, it is necessary to appreciate how a transfer hub works. A good transfer operation will minimize the delays and unreliability associated with making connections between hundreds of flights a day. On the groundside, transfers will occur efficiently if passengers can move easily between flights. As Chap. 14 describes, some configurations of passenger buildings facilitate transfers and others make them difficult. In general, good transfers require well-connected passenger buildings. Linear midfield buildings in particular serve transfers well.

On the airside, transfers will occur easily if aircraft can land at about the same time and leave shortly, as soon as they can, thereafter. The smoothest transfers occur when arrivals and departures cluster in "waves" or "banks" consisting of about 30 minutes of aircraft arrivals, up to an hour for transfer operations on the ground, and 30 minutes of departures. Watching these operations at Dallas/Fort Worth or Denver/International is an amazing experience. Before the wave starts, the airfield appears empty; few aircraft are on the ground or in the air. Then aircraft start appearing in the sky as steady streams of arrivals, one immediately behind the other, for as many runways as are available. Within half an hour, aircraft are clustered all around the passenger buildings. Then, less than an hour afterwards, steady streams of aircraft start lining up at each active runway until they are all gone and the airport is quiet again. Major transfer hubs will feature many such waves throughout the day (Fig. 4-8).

Airports need enormous runway capacity to handle the peak loads associated with the waves of operations. This accounts for the multiple

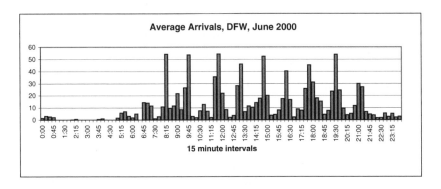

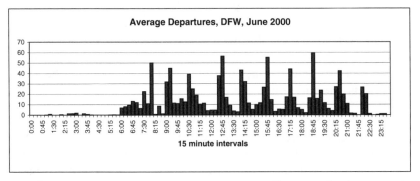

Fig. 4-8. *Waves of arriving and departing traffic at Dallas/Fort Worth.*

parallel runways at transfer hubs such as Atlanta/Hartsfield (four, with a fifth projected), Dallas/Fort Worth (seven, with an eighth in master plan), Denver/International (five, with space for seven more) and Paris/de Gaulle (four). It should be noted that this capacity is idle about half the time during the day, during the lulls in the waves of traffic. In that regard, this capacity is not used efficiently. However, large capacity is essential for effective transfer operations. Good design in this case focuses on the necessary peak periods, just as the number of toilets in a theater should focus on the needs during intermissions. Along with great capacity, airports need good weather to permit them to make good use of this capacity.

An effective transfer hub will be located centrally to the market it serves. This is essential to minimize the costs and delays associated with the detours associated with routes through the transfer hub. Thus the hubs serving East–West transcontinental traffic in the United States are generally located in the Midwest, as Figs. 4-4 and 4-6 show. Likewise, the hubs serving North–South traffic in the United

States are generally in the mid-Atlantic states, as Fig. 4-5 indicates. Miami/International, in Florida, is actually centrally located with respect to its market between the United States, the Caribbean, and South America. Likewise, Washington/Dulles is central to the traffic from Europe to the entire United States.

Airlines also expect close cooperation from the operator of the hub airport. Because transfer hubs are crucial to the operation of the airlines, these companies can operate efficiently only if airport planners and managers cooperate with them to resolve the snags that develop in the operations throughout the day, or over years. Conversely, operators of transfer airports should be motivated to serve their airlines. An airline and its affiliates using an airport as a transfer hub may easily account for 50 to 75 percent of the total airport traffic. This is the case for Atlanta (Delta), Dallas/Fort Worth (American), Denver/International (United), Detroit/Metro (Northwest), and many other major airports. The operation of transfer airports requires close cooperation between the airports and the airlines.

In many respects an operator of a hub airport is "married" to the airline running the transfers. Their economic and financial successes are closely tied. If a hubbing airline has difficulties or fails, its hub airports suffer. When the hub airline is a success, the entire region around its hub benefits. American Airlines, for example, contributed greatly to the development of Dallas/Fort Worth in the last quarter of the twentieth century, which grew from almost nothing to one of the busiest and largest airports in the world.

Competition between airports

The development of transfer airports has brought about long-distance competition between airports, in a way previously impossible. This competition results from the close connection between the airports and the airlines. In a deregulated environment, airlines compete with each other for the same customers. For example, American, Continental, Delta, Northwest, and US Airways all offer service between the East and West Coasts. Depending on which a passenger chooses, that traveler is likely to go through one of the hub airports shown in Fig. 4-3. As one airline succeeds at the expense of the others, so does its hub airport compared with its competing hubs.

The stakes in this competition are very high for the hub airports. As half of its traffic may be going through the airport only because they are transferring to another destination, half of its traffic is volatile and could easily go through another hub. Once TWA ceased operations, for example, the passengers who previously transferred through its St. Louis hub simply reached their destinations by using some other airline and some other hub. Such an event can be financially disastrous for a hub airport.

Airports compete with each other for not only passengers and their business, but also for the hubbing operations themselves. Just as passengers making a transfer do not really care if this happens at Salt Lake City or Denver, so airlines are happy to move hub operations if they find a better deal. Thus SAS moved much of its transfer operations involving North American travelers to Scandinavia from Copenhagen to New York/Newark when it established its alliance with Continental Airlines. Likewise, Cincinnati benefited when Delta selected it as a secondary hub and deemphasized operations at Dallas/Fort Worth. Hub airports need to be competitive in providing services and good prices, both to help their airlines and retain their business.

The competition between hub airports increases the volatility of their traffic. Hub airports cater to a clientele that could and often does move to another airport. Hub airports are vulnerable to changes in traffic patterns, and thus need to do all they can to protect themselves from the associated risks.

Exercises

1. Consider a local or other airport you can examine. Which activities are handled by private companies? Think about and discuss whether this allocation of responsibilities works well for the local situation and context.

2. Review the mission statements of some of the private airport companies, as reported on the web. Contrast these with similar statements presented by airport operators representing public–private partnerships. How do these differ? Which would you prefer to have running your local airport?

3. Pick a transfer airport to examine in detail. Look at appropriate data (FAA, 2001) or an Official Airline Guide (OAG) and determine the pattern of waves or banks of traffic. How many does this airport operate during a typical day?

4. Look up the records on traffic for hub airports that have experienced rapid growth or collapse of traffic, such as Bangkok, Cincinnati, Paris/de Gaulle, Munich, Pittsburgh,

St. Louis/Lambert, or Washington/Dulles. Compare these histories to the traffic records of some airports that are not transfer airports, such as Boston/Logan, Kuala Lumpur, Manchester (UK), New York/LaGuardia, or Paris/Orly. Comment on the relative traffic volatility of hub and nonhub airports.

References

Advani, A. (1998) "Market Orientation: the Case for Privatization," D.Phil. thesis, Oxford University, Oxford, UK.

Aéroports de Montréal (2001) "2000 Annual Report," http://www. admtl.com/index-e.html.

Armstrong, M., Cowan, S., and Vickers, J. (1994) *Regulatory Reform: Economic Analysis and the British Experience*, MIT Press, Cambridge, MA.

Arthur, H. (1998) "Airport Privatization: A Reality Check," *Airport*, September/October, pp. 28ff.

Australia Department of Housing and Construction (1985) *Airport Terminal Planning Manual*, Australia Department of Housing and Construction, Canberra.

Bailey, E., Graham, D., and Kaplan, D. (1985) *Deregulating the Airlines*, MIT Press, Cambridge, MA.

Beatty, S., and Lipson, W. (1999) "Preparation Is the Key," *Airport Finance and Development*, Spring, pp. 24–26.

Calgary Airport Authority (2001) "2000 Calgary Airport Authority Annual Report," http://www.calgaryairport.com/caa/report.html.

Civil Aviation Authority, Economic Regulation Group (2000a) "Quality of Service Issues," Consultation Paper, December, http://www.caaerg.co.uk/.

Civil Aviation Authority, Economic Regulation Group (2000b) "The Use of Benchmarking in the Airport Reviews," December, http://www.caaerg.co.uk/.

Civil Aviation Authority, Economic Regulation Group (2001) "Airport Regulation, Quinquennial Reviews of Designated Airports," http://www.caaerg.co.uk/.

de Neufville, R. (1999) "Airport Privatization—Issues for the United States," in *Safety, Economic, Environmental, and Technical Issues in Air Transportation*, Transportation Research Record 1662, Paper 99.0218, pp. 24–31.

de Neufville, R., and Barber, J. (1991) "Deregulation Induced Volatility of Airport Traffic," *Transportation Planning and Technology*, 16(2), pp. 117–128.

Ecole Nationale des Ponts et Chaussées (1987) *Transport aérien— Libéralisme et déréglementation*, Acte de la journée d'étude 26 mars 1987, Presses de l'ENPC, Paris.

FAA, Federal Aviation Administration, Office of Aviation Policy and Plans, (2001) "Consolidated Operations and Delay Analysis System (CODAS) Database," http://www.apo.data.faa.gov/faacodasall.HTM.

FitzGerald, G. P. (1994) "Public/Private Sector Partnerships: The Port Authority of New York and New Jersey as an Example," AAAE Building Air Service and Airport Privatization Workshop, Boca Raton, FL, November.

Gomez-Ibanez, J., and Meyer, J. (1993) *Going Private: The International Experience with Privatization*, The Brookings Institution, Washington, DC.

Greater Toronto Airports Authority (2001) "Mission and Goals of the GTAA," http://www.lbpia.toronto.on.ca/corporate.

Japan International Cooperation Agency (JICA) and Department of Transport, Civil Aviation Bureau (n.d.) *Textbook for the Group Training Course Seminar on Aerodrome*, Vols. 1 and 2, JICA 68-90, Tokyo, Japan.

Jordan, W. (1970) *Airline Regulation in America—Effects and Imperfections*, The Johns Hopkins Press, Baltimore, MD.

Kahn, A. E. (n.d.) "Interview with PBS," http://www.pbs.org/fmc/interviews/kahn.htm.

Kapur, A. (1995) "Airport Infrastructure: The Emerging Role of the Private Sector," Technical Paper 113, World Bank, Washington, DC.

Meyer, J., and Oster, C. (1984) *Deregulation and the New Airline Entrepreneurs*, MIT Press, Cambridge, MA.

Meyer, J., and Oster, C. (1987) *Deregulation and the Future of Intercity Passenger Travel*, MIT Press, Cambridge, MA.

Moody's Investor Services (1992) *Moody's on Airports: The Fundamentals of Airport Debt, Featuring Three Airport Case Studies*, Moody's, New York.

Ontario Department of Finance, SuperBuild Corporation (2000) "A Guide to Public-Private Partnerships for Infrastructure Projects," Publications Ontario, Toronto, Canada. Also available at http://www.SuperBuild.gov.on.ca/.

Ottawa Macdonald-Cartier International Airport Authority (2001) "Mission Statement and Guiding Principles," http://www.ottawa-airport.ca/airportAuthority/missionStatement-e.php.

Pavaux, J. (1984) *L'Economie du transport Aérien—La Concurrence Impracticable*, Economica, Paris.

Sigafoos, R. (1983) *Absolutely, Positively Overnight!: Wall Street's Darling inside and up Close*, St. Luke's Press, Memphis, TN.

Thorpe, K. (2001) "The Heathrow Terminal 5 Inquiry: An Inquiry Secretary's Perspective," http://www.planning-inspectorate.gov.uk/pi_journal/thorpe_art2.htm.

Transport Canada (2001) "The National Airports Policy and Regional/Local Airports," http://www.tc.gc.ca/airports/nap/english/p7.htm.

U.S. Congress, General Accounting Office (1995) "Budget Issues: Privatization/Divestiture Practices in Other Nations," Report to the Hon. Scott Klug, House of Representatives, GAO/AIMD-96-23, Washington, DC, December.

U.S. Congress, General Accounting Office (1996) "Airport Privatization: Issues Related to the Sale or Lease of U.S. Commercial Airports," Report to the Subcommittee on Aviation, Committee on Transportation and Infrastructure, House of Representatives, GAO/RCED-97-3, Washington, DC, November.

Vancouver International Airport Authority (2001) "Corporate Profile and Mission Statement," http://www.yvr.ca/GeneralInfo/frmGeneralInfo.htm.

Vickers, J., and Yarrow, G. (1988) *Privatization: An Economic Analysis*, MIT Press, Cambridge, MA.

5

Multi-airport systems

Multiple airport systems exist in all the metropolitan areas that generate the largest amount of traffic, as well around many other cities. The general rule is that multi-airport systems perform well only for cities that are the largest generators of originating traffic. Elsewhere, they tend to be economic risks or to disrupt efficient airline operations.

The several airports in each system compete with each other for traffic and services. The dynamics of this competition lead to concentration of traffic at the primary airports and volatile traffic at the secondary facilities. This reality presents special problems and, in many cases, has led to politically and financially embarrassing failures.

Good planning of metropolitan multi-airport systems requires an understanding of the competitive market dynamics that favor their growth and shape their opportunities. It recognizes that planners and operators have limited ability to control the market dynamics and reduce the risks associated with these systems. It also appreciates that the market for airport services has different components with distinct needs. Effective development of multi-airport systems therefore calls for investment in a range of flexible facilities. These will enhance the airport operators' opportunities to respond to the changing market patterns and enable them to cater to the range of services airport users desire.

5-1   Introduction

This chapter shows how to plan, develop, and manage multi-airport systems of metropolitan airports. These systems are sets of two or more commercial airports in a greater urban area, as Sec. 5-2 defines in detail. Many multi-airport systems already exist, for example, around New York and San Francisco, London and Paris, Tokyo and Seoul. Figure 5-1 shows some examples. These systems present

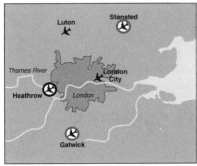

Multi-Airport System: London

Multi-Airport System: Tokyo

Multi-Airport System: San Francisco

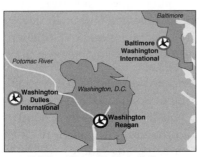

Multi-Airport System: Washington/Baltimore

Fig. 5-1. *Airports in multi-airport systems of London, San Francisco, Tokyo, and Washington.*

unique difficulties for airport planners and operators, because the airports in the system compete with each other for traffic. They thus demand special attention.

The problem with multi-airport systems is that airport planners, operators, and governmental sponsors frequently misjudge the development of individual, constituent airports. Lack of understanding of the way multi-airport systems perform has led to significant expensive and embarrassing failures. Section 5-3 describes some of these cases to motivate concern for the problem and to provide a foundation for the discussion of why and how multi-airport systems tend to evolve. This presentation particularly documents the systematic failure of planning policies that attempt to allocate traffic to the several airports in a multi-airport system. It also illustrates the volatility of customers for the services of secondary airports in a multi-airport system, a factor that has significant implications for how and when airport operators should invest in secondary airports around a city.

Market dynamics provide the basic explanation of the level and distribution of traffic among airports in a multi-airport system. Technical factors, political considerations and chance do modulate the effects of the market. However, the competitive market forces define the underlying structure of the outcomes. In brief, the competition of the providers of airport services for customers concentrates services for any market at specific airports. As Sec. 5.4 indicates, this concentration is a specific example of a wider phenomenon characteristic of the development of regional economies worldwide. Overall, the dynamics of the competition between the markets makes the longer-term outcomes uncertain.

An important practical issue in this context is the question of what services constitute a "market." Indeed, a market may be organized around individual airlines, destinations, types of services, or fare levels. For example, Southwest Airlines has been the prominent carrier at Boston/Providence airport. Paris/Orly has traditionally served North African and Caribbean destinations. Tokyo/Haneda serves internal Japanese destinations; and London/Luton has been a focus for "cheap fare" and charter airlines. Moreover, the situation may change, in some cases rapidly. Washington/Dulles, for instance, evolved from a minor airport serving only a few million annual passengers to a major international/domestic hub once United Airlines decided to use it as a base of operations. Markets can and do evolve in different ways and at different times.

Operators of airports within a multi-airport system may thus face important strategic choices about how they will influence the development of these markets. Airlines also can make similar choices. Jointly, the various airports and airlines are engaged in sequences of actions whose eventual outcome may be difficult to foresee.*

Airport planners and operators dealing with multi-airport systems thus have to deal with uncertain, unstable situations. They should actively seek to influence the situation for the benefit of their airports and the region. At the same time, they should be cautious in their investments, because the volatility in the level of traffic and the nature of their customers may make the facilities at secondary airports obsolete or unnecessary. As Sec. 5-5 concludes, airport planners should

---

*Analysts call these situations "games" and use "game theory" to investigate their properties. These names should not be misinterpreted. These competitive "games" represent deeply serious struggles for economic and other gains.

carefully assess the risks and invest in flexible facilities that give them appropriate options on future developments.

5-2 Basic concepts and issues

Definitions

Market dynamics are a central factor influencing the development of airports around a metropolitan area, as Sec. 5-1 stresses and as Sec. 5-4 demonstrates. It is crucial to the understanding of multi-airport systems to use a concept based on how the customers and users of the system see it. A functional definition that reflects the realities of the market is appropriate.

Thus, for the purposes of airport planners and operators, a *multi-airport system* is the set of significant airports that serve commercial transport in a metropolitan region, without regard to ownership or political control of the individual airports. This definition involves several important points:

1. It focuses on airports serving commercial transport. It leaves out military bases, such as Andrews Air Force Base near Washington, D.C., or Yokota in the Tokyo area. It does not consider airports dedicated to aircraft manufacturing or shows, such as Boeing Field in Seattle and Le Bourget in Paris. It neglects general aviation airfields such as Van Nuys in the Los Angeles area. All these facilities are important from the perspective of air traffic control, but they are not factors in the market for the airline industry.

2. It refers to a metropolitan region rather than a city. As a practical reality, this region may include several distinct cities. The San Francisco metropolitan area from this perspective includes the cities of San Francisco, Oakland, and San Jose, among others. The airports associated with these cities all serve, at different times and in different ways, passengers and cargo associated with the general San Francisco metropolitan area.* To reflect this reality, this book thus precedes references to the individual airports with a

---

*Note that this general definition, useful for market purposes, does not correspond to various governmental definitions that focus on major political divisions. Thus, as of 2001, the U.S. Census Bureau (2001) divides the San Francisco Bay region into several "primary metropolitan areas" and aggregates them into the San Francisco "consolidated metropolitan area." As Chap. 18 notes, these kinds of definitions change, in this case, about every decade.

label indicating the metropolitan region they serve, for example, San Francisco/International, San Francisco/Oakland, and San Francisco/San Jose.*

3. With its focus on the market, the definition does not pay attention to who owns the airport. Thus the London multi-airport system consists of the three airports owned by the BAA (London/Heathrow, London/Gatwick, and London/Stansted) plus the two others that people use to get to and from the region, London/Luton and London/City.† Similarly, the definition does not pay attention to administrative boundaries, unless these segregate the market (as the Berlin Wall did the airports of East and West Berlin). Thus, the multi-airport system for Boston includes the main Boston/Logan, Boston/Providence in the state of Rhode Island, and Boston/Manchester in the state of New Hampshire. The latter two airports are within an hour or less of the Boston suburbs, often closer in terms of travel time than the primary Boston/Logan airport, and provide attractive service to many customers in the greater Boston metropolitan area.

4. Finally, the definition focuses on significant airports, typically those that serve more than a million passengers a year or a comparable amount of freight (about 100,000 tons, in workload units‡). This limitation focuses the discussion on the facilities that contribute meaningfully to the air transport services of a metropolitan region. In the Paris area, for example, it excludes the Beauvais airport, which served less than half a million passengers in 1999, or less than 1 percent of the total regional traffic.

Prevalence

Multi-airport systems constitute a sizable segment of the airport industry. About 30 major, distinct multi-airport systems exist worldwide, involving over 80 airports. They are a feature of all the metropolitan areas with the greatest amount of originating and terminating traffic

---

*Understandably, this is not the way airport operators refer to their airports. The cities of Oakland and Newark do not advertise their roles in the larger San Francisco and New York airport systems.
†The definition thus contrasts with that used by the Airports Council International. This body reports data on their members who own multiple airports. For them, the London multi-airport system includes only the London area airports operated by BAA. This makes sense within their context, but does not reflect how London travelers think about their airports when choosing to fly.
‡A workload unit is either 1 passenger or 100 kg (220lb) of cargo. It is the traditional approximate measure used to compare these two different kinds of traffic.

in the world. As of 2001, they catered to about 1 billion total passengers, well over half of worldwide traffic.

Multi-airport systems have been, without exception over several decades, a feature of all metropolitan areas with the most originating and terminating traffic. This is a remarkable fact. It stresses the strength of market forces to create and maintain multi-airport systems. Table 5-1 demonstrates this phenomenon. It presents the primary (or largest) and secondary airports at all cities that are the largest generators of traffic. Notice that, above a specific level of traffic, all the metropolitan areas feature a multi-airport system.

The level of originating traffic needed to justify and maintain a second airport has not been constant. It keeps rising, as Sec. 5-4 explains. As of 2001, the minimum level is about 14 million annual originating passengers for the entire metropolitan area, as Table 5-1 indicates. This amounts to a minimum of about 30 million total annual passengers, once transfer passengers are included. These thresholds are likely to keep rising over the coming generation.

The emphasis on traffic going to or from the metropolitan region is vital to the understanding of multi-airport systems. The focus on locally generated traffic excludes transfers. The passengers beginning (and ultimately ending) their trips in the metropolitan area create the pressure for multiple airports for their region. The transfers passing through the region want easy connections to their next flights and clearly prefer to be at a single airport. For simplicity, the discussion refers to originating passengers. These equal the number of passengers ending their trips in the region, and both are half the total number of passengers less the transfers. The number of originating passengers for any airport is

Originating passengers $= \frac{1}{2}$ (total passengers $-$ transfers)   (5.1)

This number is not always easy to calculate. Many airports and airlines worldwide do not like to release information about the number of their transfer passengers. Thus, the figures for originating traffic in Table 5-1–5-3 are estimates, valid only to a couple of decimal points. This inevitable lack of precision does not affect the overall association of multi-airport systems with the biggest traffic generators.

A number of cities with lower levels of originating traffic also have multi-airport systems. Some of these are developing multi-airport systems, as Table 5-2 indicates. Other metropolitan areas feature

Table 5-1. Metropolitan multi-airport systems generating more than the 14 million originating passengers threshold of traffic in 2000

Metropolitan region	Multi-airport system, 2001	Traffic, millions of passengers	
		Total	Originating
London	Yes	112	41
Los Angeles	Yes	90	38
Tokyo	Yes	82	35
New York	Yes	95	34
Paris	Yes	74	31
San Francisco	Yes	65	22
Miami	Yes	55	21
Chicago	Yes	88	21
Washington	Yes	55	19
Osaka	Yes	36	17
Boston	Yes	35	16
Hong Kong	Yes	41	16
Seoul	Yes	37	15
Dallas/Fort Worth	Yes	68	14
Atlanta		80	14
Las Vegas		37	14

Source: de Neufville database at http://ardent.mit.edu/airports.

several airports primarily for technical or political reasons (Table 5-3). Technical reasons, for example, led Taiwan to develop a major international airport for its capital, Taipei/Chiang Kai-Shek, equipped with 3600-m runways capable of handling large transoceanic aircraft. The downtown airport, Taipei/Sung Shan, is popular with local traffic but simply cannot handle long-distance aircraft with its 2550-m runways (see database on airport runways in Microsoft, 2000). On the other hand, political and possibly military decisions in the former Soviet Union led to the development of three airports around Moscow.

Unequal size

Airports within a metropolitan multi-airport system characteristically have significantly different levels of traffic. The typical pattern is that a city has a primary airport that has the most traffic, and one or more secondary airports with between 10 and 50 percent of the traffic of

Table 5-2. Metropolitan multi-airport systems almost generating the 14 million originating passengers threshold of traffic in 2000

Metropolitan region	Traffic, millions of passengers	
	Total	Originating
Brussels	26	13
Houston	44	13
Orlando	32	12
Milan	25	11
Toronto	29	11
Shanghai	18	8

Source: de Neufville database at http://ardent.mit.edu/airports.

Table 5-3. Multi-airport systems existing primarily due to political or technical reasons

Metropolitan region	Reason for system	Traffic, millions of passengers in 2000	
		Total	Originating
Düsseldorf/Bonn	Political: former capital	22	11
São Paulo	Technical: runway length	20	10
Taipei	Technical: runway length	21	9
Moscow	Political/military	17	8
St. Louis	Political: mid-America	21	6
Berlin	Political: was divided city	12	6
Buenos Aires	Technical: runway length	13	6
Montreal	Widely viewed as mistake	10	4
Rio de Janeiro	Technical: runway length	9	4
Belfast	Technical: runway length	4	2

Source: de Neufville database at http://ardent.mit.edu/airports.

the primary airport. Table 5-4 indicates relative levels of traffic of the secondary airports, compared to the primary airport in each of the multi-airport systems associated with the cities with the most originating traffic. The secondary airport rarely has as much traffic as the primary airport serving the most passengers.

It should be noted that the primary airports with the most traffic are not necessarily the largest. Montreal/Mirabel, for example, has much

Table 5-4. Traffic at secondary airports is generally significantly less than at primary airports (metropolitan regions ranked by total regional traffic, secondary airports ranked by level of traffic)

Metropolitan region	Traffic at secondary airports (% of primary airport)				
	Second	Third	Fourth	Fifth	Sixth
London	49.5	14.7	6.3	2.2	
New York	95.9	74.3	2.9	1.5	
Los Angeles	11.4	9.9	6.9	1.9	0.9
Chicago	22.1	0.1			
Tokyo	45.6				
Paris	52.8	0.8			
Dallas/Fort Worth	11.7				
San Francisco	31.8	26.7			
Miami	47.3	17.3			
Washington	98.0	78.5			
Houston	25.9				
Hong Kong	15.9	8.0			
Osaka	80.4				
Boston	23.7	4.0			
Orlando	3.9				
Brussels	27.5	1.0			
Milan	38.8	6.5			
Düsseldorf/Bonn	37.7	1.3			
Toronto	NA				
Taipei	30.5				
São Paolo	34.4	1.4			
Shanghai	80.0				
Moscow	43.8	33.3	1.0		
Buenos Aires	79.5				
Berlin	19.8	8.3			
Montreal	16.5				
Rio de Janeiro	41.4	17.2			
Belfast	43.3				
Overall average	41.1	17.8			

Source: Source: de Neufville database at http://ardent.mit.edu/airports.

more land than the downtown primary airport, Montreal/Dorval, and planners destined it to be the main airport for the city. However, it never succeeded in developing as much traffic as Montreal/Dorval—and the airport operator has largely closed it down.* Washington/Dulles has more runways and land than downtown Washington/Reagan, but for about the first 20 years of its existence it developed only about 20 percent of the traffic of Washington/Reagan (about 3 million passengers annually, compared to about 14 million). Similarly, the Aéroports de Paris consistently intended that Paris/de Gaulle would be the major airport for the region, but it took almost a generation before its traffic exceeded that of the smaller Paris/Orly. In fact, airport operators have often built major new airports far away from prospective clients, and these users have preferred to stay with the older primary airport until a major market shift occurs. Washington/Dulles, for example, did not have more than about 3 million passengers a year until United Airlines decided to locate a transfer hub at the airport. This action completely changed the market for local customers.

Exceptionally, secondary airports may have about the same level of traffic as the primary airport. This generally happens when a secondary airport is growing and overtakes a primary airport. Thus, Paris/de Gaulle had about the same number of passengers as Paris/Orly for a few years. However, the Parisian multi-airport system now follows the usual pattern, in which the secondary airport is relatively small compared to the primary airport, as Table 5-4 indicates.

5-3 Difficulties

The development of airports in a multi-airport system has always been problematic. The symptoms of this phenomenon come in several forms. Sometimes these overlap, as the examples indicate. As an overview, the following issues arise.

1. Not enough traffic comes to the new airport, resulting in an expensive and embarrassing "white elephant." Nations and regions all over the world have built major new airports and then found that it was very difficult to attract customers. This has happened at Kuala Lumpur, London, New York, Saint Louis, São Paulo, and Washington.

*The Aéroports de Montréal is the local airport operator. Local government authorities established this corporation to put the management of the airports on a commercial basis. Shortly after the Aéroports de Montréal assumed control of Montreal/Dorval and Montreal/Mirabel, they organized the transfer of all scheduled airline operations to the convenient Montreal/Dorval airport. As of 2001, Montreal/Mirabel serves minimal traffic, almost all cargo and charter flights.

2. It is politically and economically difficult to close an old airport. The traffic therefore divides between the airports, resulting in poor service and insufficient traffic at the new airport until sufficient traffic builds up. Examples of this have occurred at Buenos Aires, Edmonton, Kuala Lumpur, and Osaka.

3. There is not enough traffic to support a multi-airport system. This situation also leads to poor service, low traffic at the new airport, and financial losses. Such situations arose at both Edmonton and Montreal.

4. It is impractical to allocate traffic away from a congested primary airport to alternatives in the region. This frustrates planners, who would like to reduce noise and congestion in one part of the region and provide service to another. Los Angeles, London, Milan, and San Francisco have each experienced this difficulty.

5. Traffic at the secondary airport is volatile, both in level and in type. The consequence is that the operators of these facilities can alternately face underutilization and congestion, and often have inappropriate facilities for the clients they actually have. Operators at London/Stansted, New York/Newark and San Francisco/Oakland have faced these issues.

Difficulties in developing multi-airport systems keep occurring. They have arisen with practically all multi-airport systems. (For details, see de Neufville, 1986, 1994.) As of 2001, the operators of the Milan airports (SEA) were still engaged in a difficult international dispute about which airlines and which flights would be forced to leave the convenient Milan/Linate airport and go to the major new facilities at Milan/Malpensa inaugurated 2 years earlier. Meanwhile, the operators of Kuala Lumpur were trying to attract traffic to their major new facility after years of struggle to close the old airport, located conveniently near the downtown area. Thailand similarly will face immediate issues when it opens its a second major airport for Bangkok in around 2004 or later. Prospectively, the same issues will arise for Lisbon, Madrid, and Mexico City as these cities begin to develop new second airports. To avoid such problems, it is useful to learn from the examples of the past, as discussed below.

Insufficient traffic at new airport

Planners have often designed airport facilities on the mistaken assumption that they would collect traffic from their "catchment areas." Notionally, the catchment area for a facility includes all the places that have easier access to this facility than to its competition. In this case, ease of access can be defined in terms of some expression of overall cost reflecting time, distance, and expense. This concept is a well-established general principle for the siting of facilities (Weber, 1929). It works well for many industries selling undifferentiated products where cost considerations dominate (the shipment of bauxite ore to aluminum smelters, for example). This general approach is not suitable for airport planning, however, as the following examples demonstrate and the theory in Sec. 5-4 explains.

Thus, the Port of New York Authority* in the early 1970s created a major passenger complex at New York/Newark in the state of New Jersey. An important part of the planners' reasoning was that, since this facility would be much more convenient to New Jersey residents and businesses, it would therefore attract about a third of the metropolitan traffic. Indeed, by going to New York/Newark, the travelers from New Jersey avoid the congested crossing of the Hudson River, Manhattan, and the East River to get to the other two major regional airports (New York/LaGuardia and New York/Kennedy). However, for a long time many passengers—and airlines—avoided New York/Newark airport. Travelers would drive by and continue on to New York/LaGuardia, for example. Airlines did not increase their flights or service to make use of the new capacity, since the traffic was not there. The result was that the Port Authority was left with an entire new passenger building literally boarded up and closed, for well over a decade.

Similarly, the British Airports Authority† built London/Stansted airport in the late 1980s to the northeast of London. Their view was that it would serve that sector and relieve the pressure on London/Heathrow and London/Gatwick to the west and south of London. They thus created capacity for between 10 to 15 million annual passengers. However, for most of the next decade London/Stansted had less than 5 million annual passengers. Half its capacity, specifically one of its two midfield concourses, was virtually unoccupied during this time. Travelers from around London/Stansted would systematically bypass this airport to catch flights from London/Heathrow and London/Gatwick. Airlines correspondingly provided service at those airports rather than at London/Stansted. The passenger and airline decisions to avoid London/Stansted reinforced each other.

*Since renamed as the Port Authority of New York and New Jersey to recognize its activities in these two states.
†The British Airports Authority was the governmental organization that preceded the privatized BAA.

A lesson from these examples is that the development of an airport in a multi-airport system requires airlines willing to serve that facility. Passengers do not simply follow the easiest path to an airport. They do not have similar needs. They will not flow, like drops of water, from a "catchment area" to the most accessible exit. Passengers go to an airport to catch flights to specific destinations at an acceptable price. If these services are not available at an airport, they will not go there. The problem for New York/Newark and London/Stansted was that airlines did not want to provide much service to these airports. (As Sec. 5-4 explains, this situation can change. As of 2001, a generation after the opening that left one entire passenger building vacant, New York/Newark was the busiest airport in the region. This change was due to major changes in airline strategy—unanticipated by the original planners.)

Difficulty in closing old airport

Developers of major new airports often assume that they will be able to close the older airport and avoid having two airports active simultaneously. Sometimes this is possible. For example, Denver closed the convenient Stapleton airport when it opened Denver/International. However, it is equally likely that economic and political pressures will intervene to keep the convenient older airport open, despite assurances made by regional authorities and airlines in the original planning process, some 5 to 10 years earlier. Airport planners need to anticipate this possibility.

The case of Osaka illustrates this point. The original plans anticipated that the region's older airport, Osaka/Itami, would close when Osaka/Kansai opened. Situated in the middle of an urban area, Osaka/Itami was crowded and distributed considerable noise and dirt over its neighbors. However, it was also much more convenient, for both passengers and workers, than Osaka/Kansai, located on a man-made island far from the center of Osaka and affordable housing. In any event, Osaka/Itami remains open as one of the busiest airports in the world. The continuing operation of Osaka/Itami has been costly for the new airport. It reduces the traffic and keeps the landing charges at Osaka/Kansai high, at about $10,000 per operation, which further discourages the development of the new airport. Because the runways at the two airports are almost at right angles to each other, their simultaneous operation complicates flight paths. Appropriate contingency planning might have avoided these difficulties.

This kind of difficulty is not unique. Malaysia faced a similar problem when it opened its major new airport, Kuala Lumpur/International. Although the government made its national airline, Malaysian Air System (MAS), transfer operations to the new airport along with foreign carriers, it could not initially close the older, more convenient airport Kuala Lumpur/Subang. Small carriers managed to continue to use the older airport and forced MAS to transfer services back to the older airport in order to compete effectively in the short-range markets. Much the same occurred at Edmonton. In that case, the national government built a new airport for long-distance service associated with the city. However, passengers continued to use the downtown airport for long-distance trips even after the new airport opened. Although there was then no long-distance service available at the downtown airport, travelers simply used the frequent shuttle to Calgary (a city about an hour away by air) to connect to excellent service there.

More recently, when SEA, the operator of the Milan airports, opened their major new facilities at Milan/Malpensa, they attempted to close the older Milan/Linate airport to all international carriers. This policy would have given Alitalia a virtual monopoly on that airport, which is more convenient to downtown Milan. The competitive airlines, backed by their passengers, protested vigorously. In the end, the courts obliged SEA to allow many international flights to stay at Milan/Linate—thus drawing traffic away from Milan/Malpensa.

The lesson from these examples is that markets attempt to find a way to maintain operations at older, more accessible facilities. Often they succeed despite governmental and other commitments. The fact is that the authorities in charge when the new facilities open are generally not those who made commitments during the planning for the new airport, about a decade earlier. Moreover, they inevitably confront new realities—for example, that the access to the new airport is inadequate because a highway or railroad is incomplete or not yet built. Good planning will recognize the likelihood that older facilities often do not close as planned. Good planners will recognize their limited ability to make passengers go where they do not want to go.

Insufficient traffic overall

As Table 5-1 indicates, all metropolitan areas generating more than a threshold of traffic feature a multi-airport system. As Sec. 5-3 explains, these regions have enough traffic to sustain two significant airports at the same time. Conversely, regions with less than the threshold amount of traffic may have difficulty sustaining two airports.

Metropolitan regions with less than the threshold amount of traffic will be able to maintain two airports when there are technical or political reasons that compel these airports to exist. For example, both Taipei and Buenos Aires have multi-airport systems although their current originating traffic is far below the prevailing threshold. This is because their convenient downtown airports (Taipei/Sung Shan and Buenos Aires/Aeroparque) simply cannot handle transoceanic aircraft, so that traffic must go to the alternative airport (Taipei/Chiang Kai-Shek and Buenos Aires/Ezeiza). Theoretically, the airport operators in those cities could close the older, convenient airports. However, such moves would certainly be unpopular with the airport users and difficult to sustain absent compelling reasons. Meanwhile, the split between the airports disrupts international airlines and trade. For example, the most convenient connections between the rest of the world and Mendoza and other Argentine provinces pass through Santiago, Chile, rather than through Buenos Aires.

Metropolitan regions with less than the threshold amount of traffic and no compelling reasons to have two facilities have great difficulty sustaining both airports. The case of Montreal is the prime example. Its convenient older airport, Montreal/Dorval, is fully capable of handling transoceanic aircraft. As of 2000, the region generated about 4 million originating and 10 million total passengers. Neither in 2000 nor a generation before, when Montreal/Mirabel opened, was there enough traffic to sustain two major airports. The authorities originally thought they could force international traffic to go to the distant Montreal/Mirabel by forcing transoceanic flights to land there. Many passengers avoided this displacement by taking flights to Toronto and then proceeding on to Montreal/Dorval. Airlines scheduled flights to serve this alternative pattern, offering fewer flights to Montreal/Mirabel and further weakening its position. This diversion of traffic has been bad for Montreal. As of 2000, the Toronto region had three times as much traffic as Montreal. The development and operation of Montreal/Mirabel has been uneconomical as well. Correspondingly, once the authorities established the Aéroports de Montreal as the commercially-oriented airport operator for the region, this agency moved to bring operations back to Montreal/Dorval and concentrate developments on a single facility.

Looking ahead, this experience raises a warning to airport operators seeking to establish a major second airport before the traffic is sufficiently high, when they otherwise do not need to do so for technical reasons. For example, planners for Mexico City and Lisbon seeking to develop major new airports should be cautious, since their traffic is far below the current threshold of about 14 million originating passengers. In this connection, the Australian planners, who had been seriously contemplating the development of a second airport for Sydney throughout the 1970s and 1980s, wisely decided to defer development once they realized they did not have sufficient traffic (Australia, Department of Aviation, 1985; Australian Federal Airports Corporation, 1990). Moreover, they took the further step of acquiring a site for the second airport without any commitment to develop it at any particular time (de Neufville, 1990, 1991). In effect, they took out a "real option" to protect their future, along the lines indicated in Chap. 22. Their approach has proven to be excellent.

Impractical to allocate traffic

Numerous airport operators have sought to force passengers and traffic to move their activities away from a busy primary airport to a secondary airport with underused capacity. The motivation is straightforward: moving traffic from a crowded busy facility to an uncrowded airport should reduce congestion and delays, make better use of the existing facilities, and perhaps avoid further capital investments. For example, it might seem more reasonable to relocate traffic from congested San Francisco/International to San Francisco/Oakland, which has the capacity to handle more transcontinental flights, rather than build more capacity at San Francisco/International. The city of Oakland has wanted to build up its traffic for over 30 years. However, for this case, and in general with few exceptions, these attempts have been largely futile, as examples show.

The British Airports Authority, as the governmental operator of the London airports in the 1980s and earlier, continuously tried to move traffic out of London/Heathrow and over to London/Gatwick. They attempted to persuade travelers within Britain to shift their travel patterns, but the $20 to $30 discounts (in terms of year 2000 dollars) they built into the regulated fares appeared to have no discernible effect. They pressured foreign countries to have their national airlines fly into the secondary airport. Many, but not all nations successfully resisted second-class assignments to the less popular airport. To this day, London/Gatwick has half the traffic of the primary London/Heathrow.

In the United States, the national government tried unsuccessfully for years to move traffic from Washington/Reagan airport to

Washington/Dulles airport once they opened that airport. To that end, the Federal Aviation Administration (FAA) designated Washington/Dulles as the international airport for the capital, and limited direct flights from Washington/Reagan to airports within 1000 miles (1600 km) (FAA, 1981).* These restrictions did not succeed in forcing either passengers or airlines to move substantially to the distant Washington/Dulles. Airlines scheduled departures from Washington/Reagan to London and Tokyo by the simple device of changing aircraft at intermediate points such as Boston or Chicago. In the early 1990s, for example, almost 20 years after Washington/Dulles opened, the Official Airline Guide showed more international departures to London and Tokyo from Washington/Reagan than from the supposed international airport. Additional flights went overseas from the Baltimore/Washington International airport. As for domestic flights, the airlines and passengers evaded the spirit of the restrictions by scheduling flights from Washington/Reagan to San Francisco, say, via intermediate stops. Moreover, politically influential cities such as Chicago and New Orleans obtained exemptions. In the end, the governmental restrictions did not force traffic to grow at Washington/Dulles. Only when United Airlines decided to make that airport one of its hubs in the 1990s did traffic at Washington/Dulles grow rapidly.

Only the most severe and compelling government pressures can compel the allocation of airlines and traffic between airports. Thus, the Japanese government closed Tokyo/Haneda to international traffic, forcing all service beyond Japan to go to Tokyo/Narita. Likewise, it made Osaka/Kansai an international airport and restricted Osaka/Itami to domestic traffic. The French government largely developed Paris/de Gaulle by compelling Air France (which it owned) to move all its operations to this new airport. This move imposed enormous costs on the airline. The airline immediately had to duplicate facilities that it had at the old airport. For most of the next 20 years, it lost substantial traffic to foreign competitors who continued to operate at the old airport, Paris/Orly, which remained the primary airport and had the best connections throughout France. On some lines, Air France managed to retain only about 40 percent of the French traffic, as domestic passengers connected with Lufthansa and other competitors. Only an airline with generous government financial support could persist in the face of such long-term economic adversity. To the extent

*This regulation was amended many times, to permit specific cities farther than 1000 miles away to have convenient service into Washington/Reagan. As of 2001, the FAA generalized these exceptions with blanket permission for all domestic flights within 1250 miles of the airport (FAA, 2001).

that airlines are owned by private investors, as indicated in Chap. 1, they will not be able to afford to comply with directives that go against the market forces.

The general rule is that market dynamics ultimately prevail. Government efforts to force traffic shifts between airports are impractical, except in limited circumstances. The emphasis is on the dynamics of the market. The outcomes are often unexpected. The process is usually volatile.

Volatility of traffic at secondary airport

Traffic at secondary airports is typically much more volatile than at the primary airports with the most traffic. One explanation for this phenomenon is that secondary airports relieve the congestion at the primary airports. Their traffic thus grows in boom periods, but falls back when traffic returns to the primary airfield during recessions. Another is that a secondary airport often is a base for a new airline. As these ventures often grow rapidly and then collapse, so does the traffic at the secondary airport. For these and other reasons, traffic at secondary airports often grows and falls rapidly. This feature makes planning difficult and investments risky and potentially unprofitable.

The example of Chicago/Midway suggests the point. As of 1987, it was already a sizable airport, with as much traffic as the international airports of Buenos Aires/Ezeiza, Edinburgh, or Rio de Janeiro/Galeao at the turn of the century. As a secondary airport, it was also the home to Midway Airlines, a start-up that chose to operate out of this secondary airport rather than compete with United Airlines at its major hub of Chicago/O'Hare. When Midway Airlines failed in 1992,* traffic at Chicago/Midway dropped over 40 percent, from a high of about 3.2 million enplanements to a low of about 2 million (from about 6.4 to 4 million total annual passengers). Within the following seven years, however, the traffic had tripled to 6.2 million enplanements (12.4 million total passengers). See Table 5-5, noting that two different official reports do not agree on the traffic for 1992; this is an example of the discrepancies that constantly arise in airport statistics, as Chap. 18 notes. These kinds of rapid changes make it very difficult to create the infrastructure needed when the traffic does occur. They also make it difficult to pay for facilities that have become relatively empty when the level of traffic collapses.

---

*Midway Airlines later resumed operations, only to close down again in 2001. At that point, however, its main base was no longer Chicago/Midway but Raleigh-Durham.

Table 5-5. Rapid fluctuations in traffic at secondary airport of Chicago/Midway

Airport/airline	Enplanements in thousands, by year; %					
	1987	1988	1989	1990	1991	1992
Chicago/Midway	2541	3174	3410	3547	2937	1972
Midway Airlines, %	65	65	65	71	69	0

Source: FAA, "Aviation Activity Statistics," www.faa.gov/arp/a&d-stat.htm

Airport	Revenue enplanements in thousands, by year							
	1992	1993	1994	1995	1996	1997	1998	1999
Chicago/ Midway	2029	3051	4213	4267	4492	4426	5059	6219

Source: FAA, 2000.

The experience of Chicago/Midway is not unique, as Table 5-6 demonstrates. Sudden spurts or drops are the common experience of small secondary airports. The traffic at Boston/Manchester, for example, grew gradually for a decade and then at over 50 percent a year for the next 2 years as a new "cheap fare" airline (Southwest) moved in. Traffic at Detroit/Detroit City cycled between zero and about 600,000 total passengers three times within 13 years. Traffic at New York/White Plains and at Los Angeles/Long Beach each dropped within a year, when Continental dropped its service at the former in 1987/1988, and American did the same at the latter in 1991/1992. Boston/Worcester lost 85 percent of its traffic in a decade, and Orlando/Sanford grew from nothing to about 500,000 enplanements or 1 million total annual passengers within 3 years.

Statistically, the traffic at the individual airports in a multi-airport system is much more volatile than it is for the region (see Table 5-7). Moreover, airline traffic worldwide has become more changeable due to deregulation of air transport industry, as Chap. 4 explains. These facts mean that it is more difficult to plan and manage a multi-airport system than a single airport for a city. The traffic varies more; planning has to respond more quickly to these rapid changes, and investments are more risky and difficult to justify.

Overall perspective

Premature ambition to create a major new airport is the cause of many of the difficulties with the development of a second airport.

Table 5-6. Rapid fluctuations in traffic at secondary airports with generally less than 500,000 annual enplanements (1 million annual passengers)

Metropolitan area	Airport	Enplanements in thousands, by year					
		1987	1988	1989	1990	1991	1992
Boston	Manchester	112	169	229	268	293	282
	Worcester	92	142	129	105	74	68
Detroit	Detroit City	0	130	345	363	321	284
Los Angeles	Long Beach	605	579	662	693	650	400
New York	Islip	495	513	427	422	415	375
	Stewart	0	0	0	183	357	325
	White Plains	174	117	145	160	178	203

Source: FAA, "Aviation Activity Statistics," http://www.faa.gov/arp/a&d-stat.htm.

Metropolitan area	Airport	Revenue enplanements in thousands, by year							
		1992	1993	1994	1995	1996	1997	1998	1999
Boston	Manchester	420	395	455	433	486	542	947	1397
	Worcester	109	99	73	33	41	43	38	25
Detroit	Detroit City	289	250	9	0	1	33	138	223
Los Angeles	Long Beach	402	294	247	186	225	325	341	456
New York	Islip	571	566	601	568	545	510	438	942
	Stewart	406	379	398	401	413	419	363	308
	White Plains	375	470	457	471	491	527	478	508
Orlando	Sanford	1	24	10	5	279	501	593	427

Source: FAA, 2000.

Table 5-7. Traffic at individual airports in multi-airport systems is more volatile than for the region

Multi-airport system	Higher traffic volatility at individual airports (%)
New York	+10
San Francisco	+86
Washington/Baltimore	+127

Source: Cohas, 1993.

Worldwide, airport operators have built facilities at second airports that proved to be too large for many years. They counted on being able to move traffic from the crowded primary airports to the new facilities, and were not able to do so. The volatility of the traffic at second airports worsens their difficulties by complicating planning and hurting investments. As Sec. 5-5 indicates, airport planners and operators should develop second airports much more flexibly and incrementally, so that they can avoid these problems. Meanwhile, Sec. 5-4 explains why the difficulties indicated in this Section are inevitable.

5-4   Market dynamics

Concentration due to sales opportunities

Market dynamics in the airline and airport industry lead to concentrations of traffic at specific airports. This concentration is a specific manifestation of a widespread phenomenon. It needs emphasis because this important effect is not generally appreciated. As a leading researcher of international competitive markets wrote:

> ...what is less understood is how prevalent [geographic concentration] is. British auctioneers are all within a few blocks in London. Basel is the home base for all three Swiss pharmaceutical giants. In America, many leading advertising agencies are concentrated on Madison Avenue in New York City.... General aviation aircraft producers are concentrated in Wichita, Kansas.... (Porter, 1998, p. 155)

This concentration effect contrasts sharply with the concept that passenger traffic flows to airports from their catchment areas, defined as those areas that are most accessible to the airport. The model of catchment areas derives from the earliest explorations of location theory (Weber, 1929). It is widely applicable to many situations in which transport costs are most important. However, it does not apply to companies such as airlines, for whom sales, and thus profits, depend on their locations. As a leading later researcher pointed out, "Weber's solution for the problem of location proves to be incorrect as soon as not only cost, but sales possibilities are considered" (Lösch, 1967, p. 25).

The important point is that the "catchment area" notion does not apply to airports. The focus must be on concentration effects.

Airlines concentrate their activities to avoid giving their competitors a decisive advantage in the marketplace. All of these companies make investments and deploy aircraft as strategically as they can. Their goal is to get the largest and most profitable share of the market. Meanwhile, all of their competitors are likewise doing their best to counter these moves. The dynamics of this competition leads to concentration. The parable of the ice cream sellers on the beach, Example 5.1, gives an example of this kind of behavior between economic entities. The situation for the airlines is similar in concept, although the motivation for the concentration is different.

Airlines concentrate on routes

In the case of the airlines, the primary factor that impels the concentration of traffic is the S-shaped relationship between an airline's share of the market on a route and the frequency of service it offers. This kind of relationship applies to the extent that all other factors—such as fares—are equal. Figure 5-2 sketches the essential relationship for the case of two airlines operating in a market. Three points anchor the relationship. First, if the airlines offer identical frequency, then, all else being equal, they will each have half the market. This situation is indicated by the mark in the middle of the sketch. Second, if one of the two airlines withdraws, then it has no frequency, and no market share. At that point, the competing airline will be offering all the frequency of service and have all the market. The latter two points are marked at the end of the dashed line diagonally across the sketch. The crucial factor is what happens between these extreme situations.

Empirical studies have shown that when two airlines compete on a route, the airline with the greater frequency of service gets more than its share of the market (Taneja, 1968; Fruhan, 1972; de Neufville and Gelerman, 1973; de Neufville, 1976). Specifically, for example, this means that if an airline offers 60 percent of the flights along

EXAMPLE 5.1

The parable of the two ice cream sellers on the beach illustrates how sales possibilities can override cost considerations in determining location of economic activities. Suppose that these two operators, A and B, work on a beach 1000 m long. Suppose also that the potential customers are equally dispersed along the waterfront, perhaps because parking is available all along the back of the beach. If we consider only the cost of travel, the optimal location for the sellers is 250 m and 750 m from either end of the beach, as marked on the diagram below. This placement minimizes both the maximum distance customers might have to go (to 250 m) and the average distance (to 125 m). Thus:

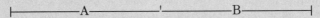

Now let us consider the effect of sales possibilities. Seller A can increase his market by moving toward the center. He will then still be closer to all the people on the left-hand side of the beach. Moreover, he can access some customers who would otherwise belong to B. Suppose then that A moves to the center and becomes the closer seller for over half the beach (625 m, specifically). Thus:

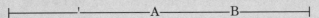

The logical move for competitor B is to move toward the center also. In this way B can recapture the potential customers lost to A. In this situation, both sellers will again have equal access to half the potential market. Thus:

When the sellers concentrate at this central location, neither A nor B can gain any competitive advantage by moving. This location thus represents a stable solution, in contrast to the original location that allows either seller to gain a competitive advantage by moving toward the center. From the perspective of costs alone, however, this is an inferior solution—the maximum and average distances for the customers are 500 m and 250 m, twice the distances associated with the original solution.

The moral of this story is that geographic concentration occurs in a market because participants recognize the importance of sales possibilities. Complementarily, the effect of sales possibilities can override considerations of transportation costs alone.

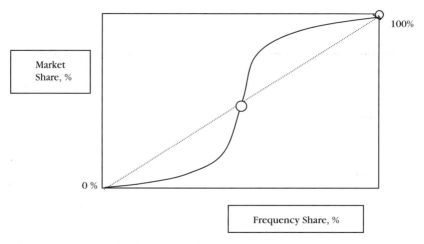

Fig. 5-2. *S-shaped relationship between frequency share and market share for two airlines operating in a market.*

a route, it may get 65 or 70 percent of the traffic—all else such as fares and size of aircraft being equal. Correspondingly, the competitive airline with a frequency share of 40 percent might only get 30 to 35 percent of the traffic. The reason for this phenomenon is simply that passengers prefer to use the airline that offers the better service for the same price. All else being equal, a traveler will go to the airline that has the most departures for his destination, is more likely to provide service when desired, and has more backup in case delays or other setbacks occur. This simple fact has tremendous implications for the profitability of the airlines, as Example 5.2 shows, and thus for their behavior.

The phenomenon of the S-shaped nonlinear relationship between frequency share and market share motivates airlines to match frequencies in a market unless they have some particular competitive advantage. If they cannot match frequency, they may only serve a route occasionally for special reasons—for example, as part of a large route or an extension of a continuing flight—or they may exit a route altogether. Thus in 2001, Delta Air Lines abandoned the shuttle service it had been operating from Washington to Boston. Its competitor, US Airways, had many more flights and was the airline people flocked to when they wanted convenient departures.

Airlines concentrate at primary airports

The matching behavior on routes has important implications for airline and passenger traffic at airports in a multi-airport system. In this

EXAMPLE 5.2

Consider a route that has two airlines operating 100-passenger aircraft. Suppose that there is enough traffic to fill 70 percent of the seats on 20 daily flights, that is, 1400 passengers. Suppose further that the breakeven load factor for the shuttle service is 65 percent. If either airline has a lower load factor, it loses money. If it has a greater load factor, it makes a profit. If both airlines offer the same frequency of service and split the market evenly, they both offer 10 flights daily and carry 700 passengers. They then both make money: each has a profitable load factor of 70 percent.

Suppose that one airline manages to offer 60 percent of the frequency, that is, 12 flights out of the 20. Assume further that, according to Fig. 5-2, it then gets 65 percent of the market or $(0.65)(1400) = 910$ passengers. Its load factor is then:

$$\text{Load factor for more frequent airline} = \frac{910}{12} = 75.8\%$$

Meanwhile, the situation for the competitor is disastrous. With 40 percent of the frequency, it offers 800 seats a day. Yet it carries only 35 percent of the traffic, that is, 490 passengers. Its load factor is ruinous:

$$\text{Load factor for less frequent airline} = \frac{490}{800} = 61.2\%$$

Airlines cannot afford to fall behind on the S-curve. They avoid this situation by matching frequency of service on a route. This effort is conceptually identical to the behavior of the hypothetical ice cream sellers on the beach, presented in Example 5.1.

case, there are two S-shaped relationships at work. One concerns the airlines. The other concerns the airports. Just as the airline with the greater service in a market attracts passengers who appreciate the convenience of more departures and return flights, so the airport with the greater frequency attracts more passengers in a market—all else being equal.

The dynamics of the competition between the airlines serving a multi-airport system then leads them not only to match their flights, but also to place them preferentially into the airports with the greater traffic. Any extra flight that they can allocate to the airport with the most traffic helps their sales. It will either match a flight of their competitors and protect their share of the larger market, or give them an advantage in this larger market (de Neufville and Gelerman, 1973, demonstrate how this works in detail). Although the airlines might provide more convenient service overall at less trouble to themselves if they split their flights proportionally between the airports in a multi-airport system, they will not do this in a competitive economy. In this respect, they behave exactly like the hypothetical ice cream sellers in Example 5.1. The competitors' attempts to gain an edge lead them to a competitively stable position. They tend to concentrate flights at the primary airport in any multi-airport system. This explains the observed pattern of concentration of traffic at primary airports that Table 5-4 presents.

Factors favoring multi-airport systems

The analysis based on frequency share has limits. These define the principal conditions that enable secondary airports to develop. The principal limits concern:

- The assumption that the airlines and airports are operating in the same "market"—however, they may serve distinct markets defined by quality or fare differences
- The idea that "all else is equal," that there is no difference in price, destination, or quality of service for the airport services offered by the airports in the multi-airport system
- The limits to the value of increased frequency of service in terms of attracting passengers, which appears to define the threshold for the meaningful operation of secondary airports
- Geographic advantages associated with the airports in the system
- Technical and other necessities

Secondary airports typically develop around specialized airlines that operate in a different market from that of the airlines at the primary airports, as Table 5-8 indicates. Most frequently, these are "no frills" airlines that appeal to a different range of passengers (and thus a different market) than the "full service" airlines that operate at the primary airports. In their case, the assumption that "all else is equal" does not hold, and the arguments concerning frequency are not decisive. As of 2001, Southwest in the United States and Ryanair in Europe were the prime examples of "cheap fare" airlines. Both have the explicit strategy of implanting themselves at small or secondary airports that are neither congested nor expensive (see Barrett, 2000). In large metropolitan areas, cheap fare airlines have often dominated secondary airports and been responsible for the success of these facilities. A prime example of this phenomenon is the development of Boston/Providence. Southwest implanted itself at this airport in the late 1990s. Its cheap fares attracted passengers, and the traffic tripled to around 6.5 million in just 3 years. Thanks to Southwest, this regional airport, of little consequence for decades, grew to be a major second airport for the greater Boston metropolitan region.

Integrated cargo airlines such as FedEx, UPS, and TNT have also been responsible for the development of secondary airports. These carriers offer special, door-to-door handling of individual shipments and do not compete in the same market as the passenger airlines that also carry belly cargo. Finally, some secondary airports serve

Table 5-8. Secondary airports typically serve special markets

Metropolitan region	Secondary airports	Special roles
London	Gatwick	African, South American destinations
	Stansted	Ryanair, low-fare airlines
	Luton	Easyjet, Holiday charters
	London City	Access to financial district
Tokyo	Narita	International traffic
Paris	Orly	Domestic, African, and Southern Europe
	Beauvais	Ryanair
Dallas/Fort Worth	Love Field	Southwest
San Francisco	San Jose	Hub for American
	Oakland	FedEx, Southwest
Hong Kong	Shenzhen	Cargo hub
	Macao	Macao island
Osaka	Itami	Domestic traffic
Boston	Providence	Southwest
	Manchester, NH	Southwest
Brussels	Charleroi	Ryanair
	Liege	TNT, DHL hubs

special destinations or regions. For example, Tokyo/Narita serves international flights almost exclusively, and Osaka/Itami is a domestic airport. Paris/Orly and London/Gatwick have traditionally served particular regions, such as Africa and South America. None of these airlines competes directly with the airlines at the primary airport. Thus, they are effectively not in the same markets and are not impelled by the market dynamics to concentrate their traffic at the primary airports.

The analysis of the dynamics of competition between the airlines based on frequency presumes that greater frequency is indeed more attractive to their potential customers. At some point, however, additional frequency is no longer attractive. Hourly flights on a shuttle service may be enough, for example. This implies that when the traffic is high enough, airlines will lose interest in further concentration, and will be willing to place additional flights in the secondary airports. Indeed, this seems to be what happens.

Thus, the examination of metropolitan regions with the largest number of originating passengers has consistently shown that, beyond a threshold of traffic, all these areas had a viable multi-airport system. As of 2001, the traffic threshold that seems to justify an effective multi-airport system is around 14 million annual originating passengers for the metropolitan region, as Sec. 5-2 indicates. This threshold has been steadily increasing. In the early 1970s, it was at around 8 million annual originating passengers. In the intervening years, aircraft became larger and airlines could handle more passengers with the same frequency. The interpretation of this evolution is that frequency becomes less important above some level, at which point a second airport can develop more easily. This level translates into a rising number of passengers, as the size of the aircraft increases.

When the importance of frequency diminishes, geographic considerations become more important. At some point, secondary airports receive substantial traffic because they are in fact more convenient. This effect is particularly significant when travel throughout the metropolitan region is inherently difficult. Hong Kong is a prime example of this situation. Although both Hong Kong/Shenzhen and Hong Kong /Macao are geographically close to the primary airport at Hong Kong/Chek Lap Kok, they are actually quite distant in time because of inadequate roads for one and a long water crossing for the other.

Finally, technical factors may impel the development of a multi-airport system. For example, the short runways at Dallas/Love Field led to the development of Dallas/Fort Worth. Likewise, Buenos Aires/Ezeiza has significant traffic because the more convenient downtown airport, Buenos Aires/Aeroparque, simply does not have runways adequate to serve intercontinental aircraft. Similar situations apply to Taipei/ Chiang Kai-Shek, Rio de Janeiro/Galeao, and São Paulo/Garulhos.

5-5   Planning and developing multi-airport systems

As the previous sections discuss, multi-airport systems are both:

- A necessary or inevitable feature of many major metropolitan regions, because of either the high level of locally originating traffic or the technical limitations of an existing airport

and

- A source of problems—for the airport operators due to absence and volatility of the traffic at these airports and the difficulty in paying for them, and for the airlines and the region because of the fragmentation of the traffic and the inefficiency of the operations

Responsible planning agencies and airport operators need to anticipate the development of new airports for many metropolitan areas. At the same time, they should proceed carefully. Specifically, as described below, they should (de Neufville, 1984a, 1984b, 1985, 1995, 1996):

- Secure the possibilities of future developments as necessary, for example, by landbanking sites for new or expanded airports
- Develop new facilities incrementally, in line with demonstrated traffic, rather than speculatively on the hope that traffic will move voluntarily to a second airport
- Build flexible facilities that can serve the several different types of traffic that may develop at the second airports, in light of the experience that the airlines that develop these facilities come and go and each has different requirements
- Work closely with the range of airlines that target markets distinct from those served by the primary airport, and that are consequently most likely to implant themselves at the second airport

These recommendations should also be useful to planners and developers of major new airports that are scheduled to replace older airports. This is because the developers of major replacement airports have been able to close down the existing airports completely in only a few cases. The most evident examples of this are Denver, Munich, and Kansas City. Many of the new large airports intended to replace the older facilities end up being second airports, at least for a while (Table 5-9). Airport developers should anticipate this possibility and plan accordingly.

Landbanking

Landbanking is the practice of securing land for the possible future development of a facility. Properly executed, it represents a major way of implementing long-term plans for the development of new airports at a reasonable cost.

Landbanking is a form of insurance. It protects the region from the risk of needing a site for a future airport, and not being able to find one when the need arises. This risk comes from the fact that significant new airports require at least several square miles (or in the range of 1000 hectares) of vacant, reasonably flat land. However, this is precisely the kind of land that is most attractive for the expansion of the city. If the region waits to acquire land until the city has grown to the point where it needs a new airport, it may well find that no convenient sites are then available. Landbanking gives the region the option (as described in Chap. 22) of building some kind of airport when needed, without requiring the region to do so.

Landbanking is relatively inexpensive. It is obviously much less expensive than buying land and then also building a major airport

Table 5-9. Examples of "replacement" airports that operated as second airports

Metropolitan region	Replacement airport	Actual experience
Dallas/Fort Worth	Dallas/Fort Worth	Dallas/Love became hub for Southwest
Edmonton	International	Edmonton/Municipal kept active for a decade
Kuala Lumpur	International	KL/Subang active for at least 2 years
Montreal	Mirabel	Mirabel virtually closed after 20 years
Osaka	Kansai	Osaka/Itami remains major commercial airport

prematurely—it avoids construction costs that are easily 10 times the price of the land. Moreover, although the initial cost of the land may be expensive in absolute terms, it can be a good long-term investment. As the metropolitan area grows, the value of the land should appreciate. Even if the land is never used for an airport, it will be available for other purposes such as housing and industry. For example, the Australian federal government reportedly paid about US $100 million to acquire land for a possible second Sydney Airport in the late 1980s. This cost only a few percent of the estimated cost for the government's alternative, the construction of a major new international airport. Fifteen years later, this large block of property is worth many times its original price. From an investment perspective, this landbanking was inexpensive. It may even prove to be profitable.

To be effective, landbanking needs to maintain the option the region intends it to serve. Planners setting aside land for a future airport need to make arrangements that will enable them to develop a new airport at the site should they need to do so. They need to control local zoning and development to inhibit obstructions and ensure access. Importantly, they can develop or maintain a small airport at the existing site. This facility and its operations will be useful in maintaining both the principle of the airport at that site and the necessary clear zones for aircraft operations (as Chap. 9 describes). Thus, a great merit of Portugal's decision to locate a prospective second airport for Lisbon at Ota is that a military airbase already exists at that site. In this vein, regional planners and airport operators should try to maintain existing airports in a metropolitan region, as insurance against future needs. For that reason, Massport, the operator of primary Boston/Logan airport, has been subsidizing the continued operation of Boston/Worcester, which otherwise might have closed due to lack of traffic (see Table 5-6).

Conversely, landbanking will fail if planners do not maintain the option to develop the site as an airport. For example, the Toronto region acquired a large convenient site at Pickering for a possible second airport. This area has now been kept as open land for a generation. Meanwhile, the area around it has developed, and enjoys being next to what appears to be a vast nature preserve. For political reasons, the Pickering site may no longer available for a major airport. In practice, the second Toronto airport seems to be developing at Toronto/Hamilton, an airport that has been in continuous operation and that, as of 2001, is a major center for integrated air cargo carriers.

In practice, landbanking often involves the maintenance and eventual recycling of military airports. Examples include London/Stansted and Austin (Texas). Future opportunities in this line exist for Los Angeles (at El Toro), Washington (Andrews Air Force Base), Tokyo (Yokota), Lisbon, and other cities.

Incremental development

In developing new airports, airport operators should anticipate staging development incrementally, along with the actual traffic at the new facility. They will thus save money and be able to build the right facilities for the traffic that eventually occurs. Financial and operational problems arise when the airport operator constructs a first stage of development that is far too big for the traffic that actually occurs.

Airport planners should, and generally do, plan new airports on major sites, typically much larger than the older airports. A large area gives the airport operator room to expand easily when traffic makes this desirable. A large site is a form of landbanking that provides inexpensive insurance for future capacity expansion. Having the option to build in the future is, however, very different from building large at the beginning.

Traffic generally builds up slowly at new airports, unless the airlines are compelled to move. This is primarily because it is advantageous for them to keep flights at the busy airport. So airlines will tend to move away from the established airport slowly. The experience at London/Stansted, Montreal/Mirabel, New York/Newark, and Washington/Dulles document this phenomenon. Both London/Stansted and New York/Newark, for example, had major airport passenger buildings standing empty for a decade or more. Each of these second airports experienced far less traffic than planners expected over a long period, because of the reluctance of the airlines to move to the new facilities.

Moreover, strong financial reasons reinforce the reluctance to move based on market forces. The airlines implanted at the old airport may have major investments in hangars, maintenance facilities, and other properties. They naturally resist abandoning these facilities, and may indeed find it difficult to raise the money to replace them. For example, the national Greek airline Olympic Airlines simply did not have the resources to build an operations base at the new Athens/Venizelos airport as part of its initial development. Therefore, the airline has had to remain at the old Athens/Ellenikon airport, at least for maintenance purposes.

Experience indicates that traffic is likely to develop slowly at a second airport. Airport operators should therefore stage development accordingly. The way the Aéroports de Paris developed Paris/de Gaulle offers a good example of how this can be done. In this case, they built passenger buildings in increments, each capable of handling about 10 million annual passengers. Over a generation, their traffic increased up to almost 50 million annual passengers by the year 2000, and they were then building their sixth major passenger building.* This incremental approach has meant that the airport has been under nearly continuous development for all this time. However, the airport operator anticipated this, and designed the airport so that it could easily accommodate this construction without unduly disrupting ongoing operations.

This incremental approach to the development of a second airport has several advantages. Most obviously, it defers construction of capacity until it is needed. Postponing the construction and maintenance costs for several years may effectively halve the present value of the investments (see Chap. 21). More subtly, but perhaps even more important, incremental development permits the airport operator to design each addition in accord with the then prevailing requirements of the airlines. Each of the increments of passenger buildings at Paris/de Gaulle thus has a different configuration, representing what was needed at the time of construction. The second stage (Terminals 2A and 2B) thus solved a difficulty with baggage handling associated with the first stage (Terminal 1). The stage being built in 2001 was designed to accommodate transfer operations, a form of traffic that had not been significant at the beginning of the development of the airport. Each stage thus represents an important improvement over earlier designs. Overall, the incremental approach has allowed the Aéroports de Paris both to save money and to keep up to date.

Flexible facilities

Since forecasts are uncertain, as Chap. 3 emphasizes, airport operators should always build flexible facilities that can accommodate a range

---

*The interpretation of the number of their passenger buildings is debatable, since many of them are built as part of an extended integrated complex along a spine road. Moreover, they have built a sizeable but inexpensive passenger building that serves charter and cheap fare flights, which is sometimes counted—and sometimes not.

of loads and types of traffic. This recommendation applies especially to the development of second airports, because their traffic is particularly volatile (see Table 5-7).

The traffic at second airports is variable as to both level and type of traffic. Because second airports are often bases for start-up airlines, they go through boom and bust periods. Chicago/Midway went through this as Midway Airlines grew and failed around 1990 (see Table 5-5). New York/Newark had a similar experience when PEOPLExpress grew rapidly and then collapsed in the 1980s. In both cases, passenger traffic rebounded, although with different airlines having different objectives and requirements. In other situations, the type of traffic might change significantly. At Washington/Baltimore, for instance, US Airways pulled out most of its international flights in the 1990s. Although the growth of Southwest maintained the overall level of passengers, the empty international gates were not flexible enough to serve Southwest. The airport thus had to construct a new passenger building, an expense it could have been spared if it had built more flexible facilities in the first place (see Chap. 15 for additional discussion of this point). As another example, the main streams of traffic at London/Stansted have alternatively been long-distance flights with large aircraft, feeder service to Amsterdam with smaller aircraft, and cargo aircraft operated by integrated carriers. Each of these types of traffic would prefer different types of facilities.

Airport operators should therefore configure their facilities so that they can both accommodate different types of traffic and change easily to meet different needs. San Francisco/Oakland offers an example of how this can be done. Their facilities have been inexpensive and designed to cater to domestic, international, and cargo facilities. These developments are perhaps not architecturally impressive, but they have met the varying requirements of the airlines that have come and gone from this airport over the last generation.

Careful marketing

Airport operators should develop a careful strategy for marketing second airports to likely users. These are generally not the carriers that operate at the primary airport. Those airlines will normally be reluctant to withdraw flights and weaken their position at the most important source of traffic.

Airlines or operators serving different markets are the most likely candidates for second airports. These may

- Aim at special market segments such as cheap fares (the approach of Southwest, Ryanair, and Easyjet, which preferentially serve secondary airports)
- Cater to particular clients, for example, holiday tours (such as use London/Luton)
- Orient toward particular destinations (such as business service to Rome out of Milan/Linate) or serve a special business center (as Houston/Hobby does for the NASA Space Center and the refineries, and London/City does for the financial center)
- Provide specialized services, such as integrated cargo (as Los Angeles/Ontario does for UPS and Toronto/Hamilton does for several companies)*

To help entice such clients to secondary airports, operators should develop facilities that particularly serve their needs. The example of Orlando/Sanford shows how this can be done. In the late 1990s, the private operators of this airport aimed to attract "cheap fare" carriers catering to tourists. They therefore made a special effort to reduce the cost of operating at that facility, particularly when compared to the primary airport for the region, Orlando/International. They built inexpensively, for example, managing to construct their parking garages for about half the cost paid by Orlando/International (see Chap. 17). They pioneered the shared use of gates between international and domestic services (see Chap. 15). In short, they had a specific marketing strategy to develop traffic at this secondary airport, and they built their facilities for this market. They were then successful, too. As Table 5-6 shows, the development strategy for Orlando/Sanford built the traffic at this secondary airport from about 10,000 to about 1 million annual arriving and departing passengers in 5 years.

Exercises

1. Select a multi-airport system for which you may be able to obtain data about their operations. Describe how the traffic has developed at each airport over the past 10 years or so. How would you describe the relative size of the primary and

*Toronto/Hamilton is the second airport for Toronto. As of 2001, it primarily caters to cargo traffic such as UPS and FedEx.

the secondary airport(s)? What might account for any changes in this ratio?

2. For the same metropolitan region as in the previous exercise, consider several important destinations for air travelers. What is the distribution of flights from each of the airports in the multi-airport system to these destinations? What do you observe about the concentration of flights at particular airports? If certain flights are spread out among the airports, what factors do you think might account for this distribution: Special markets? Geographic advantage? Frequency saturation at the primary airport? Or some other factor?

3. Consider the capacity of each of the airports in the same system. To what extent does the actual traffic at the airports use this capacity? Would you say that some airports are underutilized? If so, what impact do you think this has had on the financial performance of the investments at these airports?

References

Australia Department of Aviation (1985) *Second Sydney Airport: Site Selection Programme,* Australia Department of Aviation, Canberra.

Australia Federal Airports Corporation (1990) *Proposed Third Runway: Sydney (Kingsford Smith) Airport Draft Environmental Impact Statement,* Federal Airports Corporation, Sydney.

Barrett, S. (2000) "Airport Competition in the Deregulated European Aviation Market," *Journal of Air Transport Management,* 6(1), pp. 13–28.

Cohas, F. (1993) "Market-Share Model for a Multi-airport System," S.M. thesis, Department of Aeronautics and Astronautics and Technology and Policy Program, Massachusetts Institute of Technology, Cambridge, MA.

de Neufville, R. (1976) *Airport Systems Planning: A Critical Look at the Methods and Experience,* MIT Press, Cambridge, MA, and Macmillan, London, UK.

de Neufville, R. (1984a) "Multiairport Systems—How do They Work Best?" *Airport Forum,* June, pp. 55–59.

de Neufville, R. (1984b) "Planning for Multiple Airports in a Metropolitan Region," *Built Environment* (special issue), 10(3), pp. 159–167.

de Neufville, R. (1985) "Systèmes Métropolitains d'Aeroports— comment fonctionnent-ils le mieux?" *Cahiers de Transport, 300,* January, pp. 25–30.

de Neufville, R. (1986) "Multi-airport Systems in Metropolitan Regions: A Guide for Policy," Report to the U.S. Federal Aviation Administration, March, National Technical Information Service, Spring-field, VA.

de Neufville, R. (1990) "Successful Siting of Airports; The Sydney Example," *ASCE Journal of Transportation Engineering, 116,* February, pp. 37–48.

de Neufville, R. (1991) "Strategic Planning for Airport Capacity: An Appreciation of Australia's Process for Sydney," *Australian Planner, 29*(4), December, pp. 174–180.

de Neufville, R. (1994) "Planning Multi-airport Systems in Metropolitan Regions in the 1990's," Final Report for the U.S. Federal Aviation Administration, DTFA01-92-P-012433, May.

de Neufville, R. (1995) "Management of Multi-airport Systems: A Development Strategy," *Journal of Air Transport Management, 2*(2), June, pp. 99–110.

de Neufville, R. (1996) "Policy Guidelines for the Development of Multiple Airports Systems in Major Metropolitan Areas" (in Japanese), *Issues and Direction of Transport Policy for the 21st Century,* Proceedings of the International Symposium for the Commemoration of the Establishment of Institute for Transport Policy Studies, Tokyo, August, pp. 21–27.

de Neufville, R., and Gelerman, W. (1973) "Planning for Satellite Airports", *ASCE Transportation Engineering Journal, 99*(TE3), August, pp. 537–552.

FAA, Federal Aviation Administration (1981) *Federal Aviation Regulations, Part 93-37—Special Air Traffic Rules and Airport Traffic Patterns,* U.S. Government Printing Office, Washington, DC.

FAA, Federal Aviation Administration (2000) "Report 10C—Revenue Enplaned Passenger Activity from CY 1992 to CY 1999," derived from FAA DOT/TSC CY1999 ACAIS database, http://www.faa.gov/arp/a&d- stat.htm.

FAA, Federal Aviation Administration (2001) *Federal Aviation Regulations, Part 93-253—Special Air Traffic Rules and Airport Traffic Patterns,* http://www.faa.gov/avr/AFS/FARS/far_idx.htm.

Fruhan, W. (1972) "The Fight for Competitive Advantage: A Study of the United States Domestic Trunk Air Carriers," Graduate School of Business Administration, Harvard University, Boston, MA.

Lösch, A. (1967) *The Economics of Location* [translated from the German edition of 1939], Science Edition, Wiley, New York.

Microsoft (2000) *Flight Simulator 2000* [software], Professional Edition, Seattle, Washington.

Porter, M. (1998) chap. 4, "The Dynamics of National Advantage," in *The Comparative Advantage of Nations: With a New Introduction,* Macmillan, Basingstroke, UK.

Taneja, N. (1968) "Airline Competition Analysis," FTL Report R-68-2, MIT Flight Transportation Laboratory, Cambridge, MA.

U.S. Census Bureau (2001) http://www.census.gov/population/www/estimates/aboutmetro.html.

Weber, A. (1929) *Theory of the Location of Industries* [translated from the German], University of Chicago Press, Chicago.

6

Environmental impacts

This chapter reviews the fundamental concepts and issues related to the environmental impacts of airports. It provides basic technical background on:

- The physics of noise and the measurement of airport noise
- Airport-related pollutant emissions and their potential impact on global and local environment
- Contamination of water resources from fuel tanks, storm water runoff, and deicing fluids

It identifies potential mitigation actions, ranging from mild to severe intervention, and gives examples of best practices around the world. It emphasizes the great importance of pursuing good community relations and of providing evidence of the airport operator's concerns about environmental issues.

6-1   Introduction

Many aviation experts believe that environmental impacts constitute the single most important impediment to the future growth of the air transport industry. Globally, concerns about noise and air pollution and their effects on health and quality of life have led to increasingly severe environmentally related restrictions on aviation. In the case of airports, these restrictions have greatly affected operating and capital costs, as well as the ability of airport operators to increase capacity in order to meet growing demand. Example 6.1 gives the flavor of the kinds of difficulties airports face on environmental grounds.

Historically, the jet airplane has been responsible not only for a quantum leap in the convenience, affordability, and reliability of air travel, but also for the birth and growth of organized environmental opposition to airports. The first generation of turbojets, which entered service during the late 1950s and early 1960s (Comets, Caravelles,

EXAMPLE 6.1

At Boston/Logan, a heated controversy has been going on since the early 1970s concerning the construction of the proposed 5000-ft. (1524 m) Runway 14/32. Only smaller non-jets and some regional jets would use this short runway. Communities around the airport have strenuously opposed it because of its feared environmental impacts. Its proponents have claimed that it will provide environmental benefits by distributing noise more equitably among the affected communities and by facilitating more over-water approaches to the airport. Practically every local politician and civic organization has taken part in the passionate debate over the years. As of 2002, the issue had yet to be resolved, some 30 years after Massport, the airport's operator, first proposed the new runway.

At San Francisco/International, an offshore runway, badly needed from the capacity viewpoint, is similarly generating fierce opposition. Related environmental impact studies, public hearings, and probable lawsuits will undoubtedly take many years to complete, large amounts of money will be spent on technical studies and legal fees, and the final outcome is highly uncertain.

In Japan, concern about noise impacts, coupled with an often difficult terrain and high population densities, has led to the construction at enormous cost of offshore airports at Hiroshima and Osaka/Kansai (already opened) and at Nagoya/Chubu, Kitakyushu, and Kobe (under development). Moreover, almost 3 mi^2 (1750 acres or 700 ha) or 65 percent of the land area of Tokyo/Haneda, the busiest airport in Asia, is built on reclaimed land at the edge of the Pacific Ocean (see de Neufville, 1991). Nonetheless, Osaka/Kansai has had to pay about $200 million in compensation to local fishermen for interfering with fishing waters and patterns.

Prompted by pressure from local municipalities and from environmental groups, in 2000 the Italian government imposed strict regulations on the flows of arrivals and departures at Milan/Malpensa and a virtual freeze on its traffic growth. This happened only 2 years after the opening of the greatly expanded version of that airport. Interestingly, local opposition to traffic growth at Milan/Malpensa was encouraged by several major European airlines, which wanted to maintain their operations at Milan/Linate, near the city center.

and the far more successful Boeing 707 and Douglas DC-8), were extremely noisy airplanes that generated a very irritating high-pitched whistle. The resulting severe disruption of living patterns in nearby communities unavoidably led to the establishment of formal and informal groups opposing airport expansion, to lots of media attention, and, inevitably, to government intervention (see Nelkin, 1974, for example). Although this type of reaction was initially limited to the world's richest and most developed countries,* airport noise and the other environmental impacts of aviation became a widespread concern by the end of the twentieth century.

Three general trends have unfolded since the introduction of the commercial jet aircraft. First, there has been exceptional progress (see Secs. 6-2 and 6-4) in

- Understanding the physics of noise and, in general, of airport environmental impacts
- Setting, nationally or internationally, ever tighter but technically achievable noise and emissions goals that aircraft and engine manufacturers have to meet
- Developing technologies that have reduced the noise generated by the most recent types of jet engines to a small fraction of that generated by their early predecessors

Second, many national governments and airport operators have developed an array of increasingly sophisticated (and expensive) approaches toward mitigating noise and the other environmental impacts of airports. They have also tried to work closely with local communities and respond to their concerns (see Secs. 6-3 on noise mitigation, 6-4 on air quality, 6-5 on water quality, and 6-6 on motor vehicle emissions).

Third, opposition to airports on environmental grounds is still on the rise worldwide. Although this may sound illogical, in view of the previous two trends, it is not. To begin with, the concerns of airport neighbors and environmentalists are objectively serious. Airports can indeed affect the lives of those living around them and the environment in severely negative ways. In addition, the enormous growth in the volume of air traffic has greatly increased the number and frequency of noise events to which airport neighbors are exposed on a daily basis. Although each event may be far less severe than in the past, the cumulative effect may appear to be just as bad, especially

---

*Airport neighbors in many developing countries are still remarkably passive to levels of airport noise that would be considered intolerable elsewhere.

if one sees an airplane flying above one's head every 2 minutes or so. Furthermore, the growth of cities has increased the number of people in areas likely to be exposed to airport noise. More generally, people, especially in developed countries, became much more sensitive to environmental and quality-of- life issues in the 1980s and 1990s. In democracies, airport neighbors are also far more vocal and politically effective in communicating these sensitivities to local and national leaderships. A more recent development contributing to increased environmental opposition to airports is growing concern about possible harmful effects of aircraft emissions on human health (see Sec. 6-4). In fact, recent public hearings suggest that the principal environmental concerns about airports may be gradually shifting from noise to air quality issues.

It is essential for airport planners and managers to be sensitive to environmental and quality-of-life issues and to remain as current as possible with developments in this area, including familiarity with best practices around the world. At the beginning of the twenty-first century, most developed countries in the world require preparation of a detailed study of the environmental impacts of any major proposed modification, improvement, or expansion of airport airside or landside facilities. The process of preparing these reports and the related public response can add years to the planning and approval of airport projects. For example, the Public Inquiry review of the prospective new passenger building at London/Heathrow took about 8 years.

6-2 Fundamentals of noise measurement

This section provides a brief overview of the salient technical aspects of noise that all airport planners and managers should know. It addresses two general topics:

1. How is aircraft noise quantified and measured?
2. What are the implications of various levels of noise for the living environment near an airport?

Highly detailed answers to these questions are beyond the scope of this book, so this discussion omits many of the relevant subtleties and nuances. The objective instead is to explain in mostly nontechnical terms the intuitive meaning of the principal terminology and noise measures in use, and to discuss the reasons for developing these measures in the first place. It also alerts readers to some common fallacies, and points out some gaps in our understanding of this important subject. This information is essential background for the discussion of noise mitigation measures in Sec. 6-3.

Measuring aircraft noise

From the physical point of view, sound is just energy generated by a source—aircraft, in the case of interest—and propagated in the form of sound waves that travel through the atmosphere. When these waves reach the human ear, they create pressure fluctuations, which are translated into the sounds that human beings hear and process mentally. From this perspective, noise can be defined simply as unwanted sound with little, if any, information content.*

The human ear can receive and process sound in a wide range of pressure intensities. The loudest sound one can hear without pain can exert as much as 1 million times more pressure than a sound one can barely perceive. Moreover, psychometric experts have observed that the human ear perceives changes in loudness of sound nonlinearly. The apparent increase in loudness is approximately proportional to the fractional change from the previous sound pressure, not the absolute value of that change. A change from a sound pressure level of 10,000 on some linear scale to a new sound pressure level of 20,000 will be sensed as a significant increase in loudness. However, a change from 200,000 to 210,000 will be imperceptible, for all practical purposes.

The wide range of pressures to which the human ear responds and the nonlinear response to pressure levels have led to the use of a logarithmic scale for quantifying and measuring the "loudness" of sound or the sound pressure level. The unit of measurement in use internationally is the *decibel* (a tenth of a bell, a unit named after Alexander Graham Bell), denoted as dB. The healthy human ear can hear sound pressure levels in a range of roughly 0 to 120 dB. Sound pressure levels just above 0 dB are barely perceptible by the most sensitive ears in a perfectly quiet environment, while sound pressure levels above 120 dB lie at the threshold of causing pain and physical injury to the ear.

Although the logarithmic scale for measuring the loudness of sound is convenient from the technical point of view, it has also caused immense confusion in informing the public about developments concerning airport noise. When told that measurements at a location

*For detailed discussion, see FAA (1983) or Ciattoni (1997).

near a runway show that the loudness of noise generated on takeoff by the average aircraft has been reduced from a typical value of 100 dB to 92 dB, most people will generally (and not surprisingly) interpret this statement to mean that aircraft noise has been reduced by 8 percent! It is essential to emphasize that studies of human reaction to sound have shown that an increase (decrease) of 6–10 dB in sound pressure level will be perceived as approximately doubling (halving) the loudness of a sound. The logarithmic scale "hides" this fact.*

The A-Weighted Adjustment Measuring the sound pressure level generated by an aircraft movement is not sufficient to characterize the degree of annoyance (or worse) caused by the associated noise. The frequency or pitch of the sound is also important. People may perceive the loudness of two sounds with equal sound pressure but different frequency as significantly different. Although the human ear can hear sounds with frequencies between 16 and 16,000 hertz (Hz) it is most sensitive to sounds in the range of 2000–4000 Hz.

In the case of airport noise, measurements of the loudness of sound thus typically undergo a further calibration intended to capture the sensitivity of the human ear to different sound frequencies. The so-called *A-weighted adjustment* is generally made to measurements of airport noise—and, in fact, to measurements of noise associated with any form of transportation. In practical terms, this adjustment adds roughly 2–3 dB to sounds in the high-sensitivity range of 2000–4000 Hz, and subtracts a few decibels from sounds outside this range. The various noise measurement devices installed around airports are designed to report A-weighted sound levels automatically. To indicate explicitly that the decibel scale has been adjusted to account for the sensitivity characteristics of the human ear, the A-weighted decibel units are denoted as dB(A), or just dBA. Most sounds in a typical living environment are in the range of 30–100 dBA, as Fig. 6-1 shows.

Single-Event Measures of Noise The most commonly used measures of airport noise in practice are subdivided into

- *Single-event measures* associated with a single aircraft movement
- *Cumulative measures,* which seek to capture the cumulative effect of many movements over a specified period.

Audible noise generated by any aircraft movement lasts for some amount of time, T, which may range from about 10 s to minutes,

---

*On the other hand, changes of the order of 1 dB will be virtually imperceptible to most people.

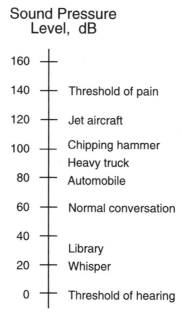

Fig. 6-1. *Examples of noise events associated with a broad range of sound pressure levels.* (**Source:** Based on Ruijgrok, 1993.)

depending on the location of the listener and the aircraft and on the type of movement (approach, departure, overflight, surface movement). Analysts typically use two measures to describe such single events:

- L_{max} measures the *maximum sound level** reached during T. It quite simply is the highest reading, in dBA, recorded by a noise sensor during T.

- SEL, the *sound exposure level,* takes into consideration all the noise readings during the duration T of a noise event—not just the highest reading as in the case of L_{max}. In effect, SEL combines all the readings recorded by a sensor during the interval T to produce a single estimate of the total sound exposure associated with the event. Thus, SEL attempts to measure the "total noise impact" of an event on a listener.

Table 6-1 provides the formal definitions of L_{max} and SEL in quantitative terms. Example 6.2 illustrates their computation.

*Unfortunately, terminology is not standard in this area. One prominent example is the use of the words "sound" and "noise." Many refer to "maximum sound level," others to "maximum noise level." Similarly, for "sound exposure level" versus "noise exposure level," etc. The notation is not standardized, either. For example, sound exposure level is variously denoted as SEL, LAE, and LSE, among others.

Table 6-1 Definitions of L_{max} and SEL

$L(t)$ is the A-weighted sound level (in units of dBA) associated with an aircraft movement at time t. Assume that it takes on nonzero values during the continuous time interval $0 \leq t \leq T$. Then,

$$L_{max} = \max_{0 \leq t \leq T} L(t) \tag{6.1}$$

$$SEL = 10 \cdot \log \left(\int_0^T 10^{L(t)/10} \, dt \right) \tag{6.2}$$

where logarithms ("log") are computed in base 10.

In practice, instead of a continuous function $L(t)$, one must work with readings, L_i, of the sound level taken at N discrete times Δt, $2\Delta t$, $3\Delta t$,..., $N\Delta t \, (=T)$. L_i denotes the reading at instant $i\Delta t$ (i.e., $L_i = L(i\Delta t)$, $1 \leq i \leq N$). Equations (6.1) and (6.2) can then be respectively approximated by

$$L_{max} = \max_{1 \leq i \leq N} L_i \tag{6.3}$$

$$SEL = 10 \cdot \log \left(\sum_{i=1}^{N} 10^{L_i/10} \cdot \Delta t \right) \tag{6.4}$$

Note that Eq. (6.2) defines SEL simply as a measure of the area under the continuous $L(t)$ curve, with appropriate adjustments for the decibel logarithmic scale. Equation (6.4) approximates $L(t)$ through a function that undergoes stepwise changes at intervals spaced by Δt and computes SEL by estimating the area under the step function approximation to $L(t)$.

Example 6.2 illustrates two important practical observations about SEL:

- SEL will always have a higher dBA than L_{max}. This is because the computation of SEL adds the "energy content" associated with noise readings preceding and following the L_{max}.

- For practical purposes, the computation of SEL needs to consider only the sequence of the highest noise readings associated with the relevant aircraft movement. The reason is that, due to the logarithmic scale, the "contribution" of any readings that are more than 10 dBA lower than L_{max} will generally be small. In Example 6.2, for instance, if one disregards the readings at 1, 2, 3, 13, 14, and 15 s (because they are more than 10 dBA smaller than the highest reading of 85.2), the resulting estimate of SEL would be 90.8 dBA, a difference of only 0.4 dBA from the more accurate estimate of 91.2 dBA computed previously.

EXAMPLE 6.2

The readings of a noise sensor near an airport during the 15 "loudest" seconds of a noise event are given below. Readings are in dBA taken at 1-s intervals.

1	2	3	4	5	6	7	8
70.2	72.6	75.1	78.7	77.5	79.0	81.0	83.2

9	10	11	12	13	14	15
85.2	82.7	79.8	76.4	74.2	72.3	70.7

Clearly, $L_{max} = 85.2$ dBA.

From Eq. (6.4) in Table 6.1, SEL is computed as follows:

$$\text{SEL} = 10 \cdot \log[(10^{70.2/10} + 10^{72.6/10} + 10^{75.1/10} + \cdots + 10^{70.7/10})(1)] = 10 \cdot \log(1{,}317{,}172{,}563) \simeq 91.2 \text{ dBA}$$

This second observation facilitates the computation of SEL and leads to a simple rule for selecting the amount of time to use for the duration, T, of a noise event: just choose that (usually continuous) interval during which noise readings are within 10 dBA of L_{max}. The resulting estimate will generally be within about 1 dBA of the more accurate estimate obtained by using all readings associated with the event. (If one chooses the more conservative approach of including all readings within 20 dBA of L_{max}, the approximation will be good to within about 0.1 dBA.)

Cumulative Measures of Noise   *Cumulative measures of noise* estimate the total noise effect of all the aircraft movements taking place over a specified period of time near a particular location. Their definitions attempt to capture the combined impact of the (A-weighted) loudness of the individual noise events and of the frequency of these events. Cumulative measures appropriately "add," on the logarithmic scale, the sound exposure levels (SEL) associated with all movements that take place during the relevant time period.

Two cumulative measures are particularly important:

- *Equivalent sound level* or *equivalent noise level* (L_{eq}), a generic cumulative measure that can be adapted to each specific set of circumstances at hand
- *Day–night average sound level* (L_{dn}), typically expressed as DNL in the United States. It can be viewed as a special case of

L_{eq} that features a significant adjustment for nighttime noise. Importantly, it is the standard metric of the U.S. Federal Aviation Administration (FAA), which should be used before others (FAA, 1983, 1990).

More specifically, L_{eq} can be applied to any time period of interest. It measures noise exposure by computing what is essentially the average dBA of noise per unit of time during the specified period. The unit of time is usually 1 s. For instance, to compute L_{eq} for a 2-h period, the SEL of all the aircraft-generated noise events occurring during that period would be added, on a logarithmic scale, and the resulting total would be averaged (i.e., spread equally) over 7200 s.

In this light, L_{dn} could be viewed simply as L_{eq} computed for an entire day (a period of 86,400 s), were it not for the following refinement: a "penalty" of 10 dBA is added to the SEL of nighttime movements, defined as movements taking place between 10 p.m. and 7 a.m. Table 6-2 gives the formal definitions of L_{eq} and L_{eq}. Example 6.3 illustrates their calculation.

The final observation in Example 6.3 is exactly the reason for the 10-dBA penalty imposed on nighttime aircraft movements by the L_{dn}

Table 6-2 Definitions of L_{eq} and L_{dn}

Let SEL_i denote the sound exposure level associated with the ith aircraft movement occurring during a period of time T near some specified location. Suppose a total of M movements take place during T. The *equivalent sound level* or *equivalent noise level*, L_{eq}, is then

$$L_{eq} = 10 \cdot \log\left(\frac{1}{T} \sum_{j=1}^{M} 10^{\text{SEL}_j/10}\right) \qquad (6.5)$$

T in Eq. (6.5) is typically given in seconds.

Now let T be equal to 1 day (86,400 s) and let M consist of J daytime movements and K nighttime movements. The *day–night average sound level*, L_{dn}, is then given by

$$L_{dn} = 10 \cdot \log\left[\frac{1}{86,400}\left(\sum_{j=1}^{J} 10^{\text{SEL}_j/10} + \sum_{k=1}^{K} 10^{(\text{SEL}_k+10)/10}\right)\right] \qquad (6.6)$$

or, since $10(\log 86,400) \simeq 49.4$,

$$L_{dn} \simeq 10 \cdot \log\left(\sum_{j=1}^{J} 10^{\text{SEL}_j/10} + \sum_{k=1}^{K} 10^{(\text{SEL}_k+10)/10}\right) - 49.4 \qquad (6.7)$$

EXAMPLE 6.3

Consider a situation in which 10 noise events generated by landing and departing aircraft occurred at a particular location, 8 during daytime and 2 during nighttime. The associated SEL values are 91.2, 86.7, 78.8, 82.3, 85.1, 89.8, 83.4, and 80 dBA for the daytime events; and 86.2 and 82.2 dBA for the nighttime events. Assume, moreover, that the first three daytime events took place between 10 and 11 a.m. Then, for this single hour, when 3 of the 10 movements of the entire day occurred, Eq. (6.5) gives

$$L_{eq} = 10 \cdot \log\left[\frac{1}{3600}(10^{91.2/10} + 10^{86.7/10} + 10^{78.8/10})\right] = 57.1 \text{ dBA}$$

Similarly, from Eq. (6.7),

$$L_{dn} = 10 \cdot \log[(10^{91.2/10} + \cdots + 10^{80/10})$$
$$+ (10^{96.2/10} + 10^{92.2/10}) - 49.4 \simeq 50.4 \text{ dBA}]$$

It is easy to verify that the two nighttime movements contributed the "lion's share" to the value of L_{dn} in this case. If there were no daytime movements at all, L_{dn} would have been equal to 48.3 dBA, due to the two nighttime movements alone. If, instead, there were no nighttime movements, the eight daytime movements would have resulted in $L_{dn} \simeq 46.2$ dBA.

measure. L_{dn} emphasizes the point that nighttime movements cause far more disruption of living patterns than daytime movements. Adding 10 dBA to the SEL generated by a particular aircraft's movement during nighttime effectively makes the contribution to L_{dn} of that single movement equal to the contribution of 10 daytime movements by the same aircraft under identical conditions.

Note also that, because L_{eq} and other cumulative measures compute "average noise exposure" over time, they may not be able to distinguish between (1) a case in which only one noise event generates painfully loud noise during some period of interest, and (2) another case in which numerous events each generate moderate noise levels during the same period. The same L_{eq} value may be computed for (1) and (2), but most people would certainly distinguish between these two cases. Public hearings on airport noise often bring up this deficiency of cumulative measures of noise.

L_{dn} and Community Reactions L_{dn} has gradually become the most widely used measure in analyses of airport environmental impacts, worldwide. Table 6-3 summarizes estimated typical effects on airport neighbors of exposure to noise ranging from 55 to over 75 L_{dn}.

In the past, most of the attention in airport environmental assessments and environmental impact studies has concentrated on the people who live in areas that experience L_{dn} values of 65 dBA or higher—i.e., inside the 65-dBA *noise contour* (see discussion of the Integrated Noise Model below). This was based on the premise that this group suffers the most and reacts most strongly to noise, as Table 6-3 suggests. Moreover, buildings and facilities within the 65-dBA contours that are particularly susceptible to disruptions from noise events (homes, schools, churches etc.) qualify for FAA-funded noise mitigation funds for soundproofing or relocation. Through 2000, this involved over $2.7 billion.

Table 6-3 dates to research performed in the early 1990s and may underestimate current reactions to airport noise. It is becoming increasingly evident that people living between the 55-dBA and 65-dBA contours will often be equally vociferous, and more effective politically in objecting to airport operations and projects, as those inside the 65-dBA contour. Part of the reason is that 55–65-dBA zones near airports tend to include far greater populations and wealthier communities than the "65-dBA or greater" zones. Another reason is that people are becoming increasingly sensitive to unwanted disruptions of their living environment of any kind. This trend is therefore likely to turn some of the attention in future environmental impact studies to groups of airport neighbors who were considered in the past to be only marginally affected by noise. Consistently with this trend, the World Health Organization has recommended that an L_{dn} value of 50 dBA in exterior sound levels is necessary to ensure that noise will not have any adverse effects (WHO, 1999).

More generally, there is growing recognition of the fact that human response to noise is highly subjective. It varies widely with sound loudness, frequencies, and tonality; and is affected by many factors, such as background noise, national culture, educational level, age, economic status, etc. Moreover, while scientific and medical knowledge understands the short-term, direct effects of noise reasonably well (e.g., at what point noise disrupts conversation or becomes physically painful), its understanding of the long-term effects of persistent exposure to high levels of noise (such as effects on learning, social behavior, and physical and mental health) is limited.

Other Measures of Noise   The *effective perceived noise level* (EPNL) is an additional measure that deserves mention. Aviation authorities use

Table 6-3. Typical effects of L_{dn} noise levels on people and communities

L_{dn} value (dBA)	Hearing loss (qualitative description)	Percent of population highly annoyed	Typical community reaction	General community attitude
75 or greater	May begin to occur	37%	Very severe	Noise is likely to be the most important of all negative aspects of the community environment
70	Probably will not occur	22%	Severe	Noise is one of the most important of all adverse aspects of the community environment
65	Will not occur	12%	Significant	Noise is one of the important aspects of the community environment
60	Will not occur	7%	Moderate to light	Noise may be considered an adverse aspect of the community environment
55	Will not occur	3%	Moderate to light	Noise considered no more important than various other adverse environmental factors

Source: MIT, 2001.

it in connection with aircraft certification for noise (see subsection on Certification of Aircraft for Noise). It is a single-event measure, analogous to SEL, which computes noise exposure in units of EPNdB (by analogy to dBA). EPNL takes into consideration sound frequencies (as is also the case with A-weighted sound), as well as the tone content of aircraft noise (e.g., by assigning additional weight to certain discrete frequency tones, produced by aircraft engines, that are particularly irritating to the ear). Due to the complexity of the measure's definition, the generation of EPNL estimates for purposes of noise certification requires use of a sophisticated computational procedure. Airport environmental studies typically utilize SEL to measure single-event noise, not EPNL.

Other cumulative noise measures developed over the years are now either obsolete or are used only in specific countries or regions. Examples include the *composite noise rating* (CNR), the *noise exposure forecast* (NEF), and the *community noise equivalent level* (CNEL). These 24-h measures have generally been superseded by L_{dn}*, although as of 2002 the CNEL was still the required metric in California.* The *noise and number index* (NNI) (Ashford and Wright, 1997), used for three decades in the United Kingdom, has been largely replaced by L_{eq}.

Finally, it should be noted that, because human response to noise is so varied and subjective, analysts have devised a number of subsidiary measures to characterize specific aspects of exposure to noise. For example, the *time above* (TA) measure is simply the length of time during a specified period, T, when the A-weighted sound level exceeds a specified threshold (e.g., 75 dBA). The threshold is typically a level at which noise interferes seriously with such activities as conversation, sleep, etc. In the airport context, T is usually taken to be a 24-h period.

INM and Airport Noise Contours   An analysis of noise impacts typically constitutes the principal part of airport-related environmental impact studies. In most cases, the main product of such noise analyses—and certainly the product that undergoes the most intensive public scrutiny—is a set of *noise contours*. These are lines on a map defining the areas around an airport that will be subjected to specific levels of noise after completion of the proposed project (Fig. 6-2 shows an example). These contours are typically drawn for L_{dn} values of 55, 60, 65, 70, and 75 dBA, according to Table 6-3.

*For details about these measures, see, e.g., Horonjeff and McKelvey (1994).

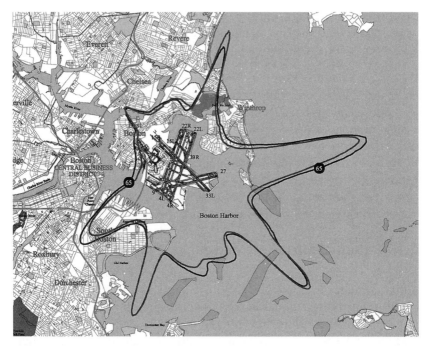

Fig. 6-2. *A set of noise contours for Boston/Logan generated through the Integrated Noise Model.* (**Source:** Massport.)

Computer-based models are needed to estimate single-event and cumulative measures of noise exposure and to prepare the associated noise contours. Academic researchers and others have developed a number of such models. By the mid-1990s, however, the *Integrated Noise Model* (INM), had emerged as a virtual international standard. INM has been developed by the FAA in several successive versions beginning in the late 1970s (FAA, 1982, 2001). It requires two types of input:

1. *Noise and performance data* for all aircraft types operating at the airport(s) or region under study—these data are standard and come with the model.

2. *Location-specific data* on operations at the subject airport(s) or region, such as runway configurations in use and frequency of use, assignment of aircraft types and of arrivals and departures to runways, geometric characteristics of the runways, geometry of flight paths to and from the airport(s), and assignment of aircraft to flight paths.

Users need to spend considerable effort to prepare good-quality, location-specific inputs for INM. They can improve the model's performance through careful calibration, which requires taking appropriate field measurements and using them to adjust the inputs that refer to local conditions.

INM is far from a perfect predictor of noise exposure, as it produces estimates based on specific assumptions about aircraft operating procedures, adherence to prescribed flight paths, etc. Indeed, actual measurements of specific noise events can differ significantly from the values predicted by INM. However, by providing a common noise analysis tool, INM has made it possible to develop comparable estimates of noise performance across airports. As important, it has promoted the standardization of international practices and policies vis-à-vis noise exposure and mitigation.

Certification of aircraft for noise

In the late 1960s, the International Civil Aviation Organization (ICAO) and a number of national civil aviation authorities began the process of adopting noise standards for transport aircraft. They reached an important milestone in this process in 1976, when both the ICAO and the FAA specified very similar standards that subsonic commercial transport airplanes had to meet in order to be designated as Chapter 2 (Stage 2 in U.S. terminology) or Chapter 3 (Stage 3) aircraft. Chapter/Stage 2 aircraft are those complying with Chapter/Stage 2, but not with Chapter/Stage 3 standards. Airplanes not qualifying for either designation are referred to as Chapter 1 (Stage 1) aircraft. These standards were published by the ICAO (1988) in Annex 16 to the Convention on International Aviation* and by the FAA as an amendment to Federal Aviation Regulations (FAR) Part 36 (see CFR, 2001).

Stage 1 aircraft completely disappeared from the fleets of airlines of developed countries during the 1970s and 1980s. However, some airlines in a few developing countries were still operating Stage 1 aircraft at the beginning of the twenty-first century. Similarly, airlines in practically every developed country are required to phase out Stage 2 aircraft by 2003, so that an all-Stage 3 fleet should be operating in these countries by then.

For certification purposes, the noise levels generated by an aircraft cannot exceed certain limits in measurements taken under "standard conditions" (sea level, zero wind, 77°F, and 70% relative humidity, per FAR Part 36) at three points (Fig. 6-3). The regulations specify the

*The standards appear in Chapters 2 and 3 of Annex 16, hence the reference to "Chapter 2 aircraft" and "Chapter 3 aircraft."

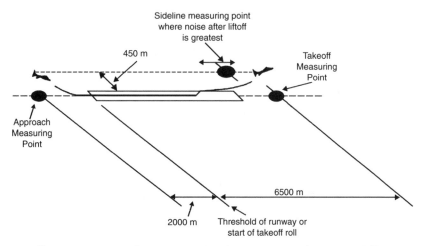

Fig. 6-3. *Locations of measurement for noise certification; different requirements apply at each point.*

limits at each point separately. The FAA and the ICAO standards differ minimally. For example, the distance from the threshold of the runway to Point A should be 1 nautical mile (1852 m) according to the FAA, but 2000 m according to the ICAO. Both standards agree on a most significant feature: the allowable noise (in EPNL) increases linearly as a function of the mean takeoff weight of the aircraft. In short, bigger aircraft generate more noise than smaller aircraft meeting the same standard!

A common misunderstanding about noise certification limits is that *all* aircraft in a particular Stage (2 or 3) satisfy the *same limits*. This is far from true. Two aircraft with very different noise characteristics may both be designated as "Chapter/Stage 3." For instance, a Stage 3, four-engine, very large aircraft (e.g., the A380) can have a EPNL measurement at Point C which is greater by 17 EPNdB (!) or more than the corresponding EPNL measurement for a Stage-3, two-engine aircraft such as the Fokker 100 (106 EPNdB versus 89 or fewer EPNdB). These differences have prompted some airport operators to subdivide Chapter 3 aircraft into more homogeneous (with respect to noise characteristics) groups and to apply different access restrictions, noise-related fees, etc., to aircraft in each of these groups (Sec. 6-3).

Progress in aircraft engine technology coupled with governmental pressure in the form of noise certification standards and phase-out requirements for noisy aircraft have been remarkably successful in bringing about dramatic improvements in aircraft noise characteristics.

Figure 6-4 shows the drop in single-event noise levels, in EPNdB, achieved by different aircraft on takeoff as a function of their date of certification. Note the gradual reduction by about 10–20 EPNdB between the 1960s and the mid-1990s. More strikingly, the number of people subject to high levels of airport noise may have decreased by a factor of 10 over the last generation (Fig. 6-5). The number of people in the United States living within the 55-dBA L_{dn} contour declined from roughly 70 million in 1975 to about 4 million in 2000, and the corresponding numbers for the 65-dBA contour are 7 million and 600,000!

However, it is unlikely that further significant reductions will occur soon. The projected number of persons in the United States affected by noise will remain roughly constant between 2000 and 2020 (Fig. 6-5). The reason is that, after the elimination of Stage 2 aircraft from airline fleets, no further massive phase-outs are currently planned, while much of the existing Stage 3 fleet is still relatively young. Moreover, any further gains achieved during the 2000–2020 period as a result of (limited) advances in engine-noise technologies will be counterbalanced by projected increases in the number of airport movements (MIT, 2001). This, of course, means that, despite the progress made during the last quarter of the twentieth century, emotionally charged controversies regarding airport noise will certainly not disappear

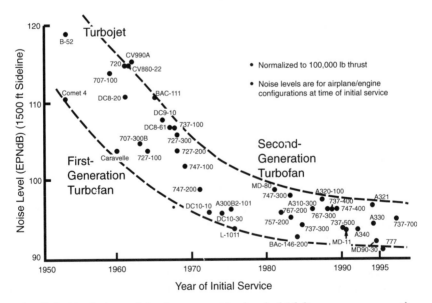

Fig. 6-4. *Evolution of single-event noise levels (sideline measurement) for many well-known commercial jet airplanes* (**Source**: Boeing.)

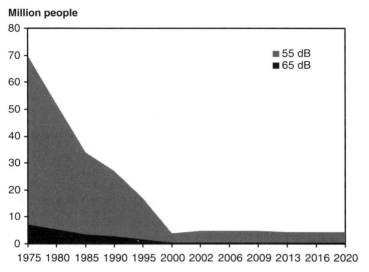

Million people

Fig. 6-5. *Number of people in the United States affected by airport noise: total population within 65 dB and within 55 dBL$_{dn}$ as a function of time.* (***Source:*** MIT, 2001.)

during the twenty-first. Environmental issues may indeed intensify, as quality-of-life and the environment become ever more prominent among citizens' concerns, as noted earlier.

Practical implications

A number of general observations can now be made concerning noise exposure near airports, measurements, and their use in practice:

- The physics of aircraft noise (sources of noise in aircraft engines and hulls, the propagation properties of aircraft noise in the atmosphere, and its frequencies, tonality, and other physical characteristics) is well understood by now, for most practical purposes.

- The state of the art in measuring noise is quite advanced, in terms of both defining appropriate measures and devising measuring scales that take into consideration many of the characteristics of the human ear.

- The aviation community has adequate knowledge about the short-term, direct effects of noise, e.g., at what point noise disrupts conversation or becomes physically painful.

In this context, the maximum noise level and sound exposure level, as single-event measures, and the equivalent sound level and day–night sound level, as cumulative measures, are now the dominant measures of noise in the United States and in much of the world.

6-3 Mitigating airport noise

There are many ways to mitigate the effects of airport noise. The restrictions and/or costs these procedures impose on airport and aircraft operators range from mild to very significant. Table 6-4, prepared by the FAA for the FAR Part 150, provides one set of possibilities. It indicates their applicability to noise-related problems associated with the various activities and aircraft operations at an airport (taxiing, departure, approach, landing roll, training flights, maintenance, and ground vehicle circulation). Note that the FAA indicates that some procedures have important safety implications and therefore belong, in part, to the FAA's sphere of responsibility.

A more complete and concise classification of the ways to mitigate noise follows:

1. Noise monitoring systems
2. Community relations and public participation programs
3. Land-use policies
4. Airport design interventions
5. Surface operations and flight operations
6. Interventions outside airport property near existing airport sites
7. Access restrictions
8. Economic incentives

Conceptually, efforts that involve preventive "land zoning" (i.e., provide in advance for noise-compatible land use near airports, see 3 above) or promote the use of more quiet aircraft (as do some of the alternatives under 7 and 8) may have the greatest potential to produce major benefits. These are "proactive," strategic measures; the others are more tactical and reactive.

The following sections review each category of mitigation briefly, supplemented with some examples of relevant "best practices." Note that international surveys of airport practices on noise mitigation are carried out regularly. For example, as of mid-2001, Boeing (2001) regularly updated information about such practices at 601 airports, 304 of which were outside the United States. All of these airports had in place some measures falling into one or more of categories 1–6 above. About one-third had imposed strict or mild forms of access restrictions (mostly "partial curfews," see Access Restrictions below) and about 20 also claimed to use economic incentives.

Noise monitoring systems

A noise monitoring system has become a virtual requirement for airports in developed countries, as it is essentially a prerequisite to developing dynamic and flexible noise mitigation strategies, both in the short and in the long term. Whereas in the 1970s only a few airports had noise monitoring systems, most of the major commercial airports had installed one by the end of the 1990s and many others were in the process of doing so. The performance and technical sophistication of these noise monitoring systems have improved dramatically over the years. Their fixed cost has increased as well, to as much as several million dollars for a complete system.

The core of a noise monitoring system consists of a number, often 10 or more, remote sensors/microphones located strategically around the airport (with emphasis on points under flight paths or near noise-sensitive facilities or neighborhoods). The sensors collect and transmit data on noise events to a central computing and reporting system. This computer also receives real-time data on aircraft operations from the air traffic management system, primarily the terminal area automation unit (e.g., ARTS or STARS in the United States; see Chap. 13). In this way, the computer can correlate data received from the noise sensors with data on aircraft operations, such as the flight tracks of individual arrivals and departures. The administrators of the system can thus identify the specific aircraft that, for instance, may have deviated from its prescribed flight path or generated an unusually high noise reading. In this manner, they can respond to citizen complaints in an informed way or contact airlines about specific instances that merit investigation. By similarly processing data from individual flights, the system also compiles statistics on each airline's compliance with noise-abatement flight procedures and helps identify persistent violators. Finally, the data obtained by a noise monitoring system are also extremely valuable in calibrating and improving the credibility and reliability of noise estimation and noise forecasting tools, such as the Integrated Noise Model (see subsection on Measuring Aircraft Noise). Citizen groups, for example, will often seek to confirm the validity of L_{dn} noise contours computed by INM through comparison with field data collected by the noise monitoring system.

Table 6-4 Possible actions to mitigate noise impacts of airport operations

	Possible actions	Source of noise	
		Aircraft Operations	Ground Equipment
Airport layout	Changes in runway location, length, or strength	X	
	Displaced thresholds	X	
	High-speed exit taxiways	X	
	Relocated passenger buildings	X	
	Use of test-stand noise suppressors		X
Airport and airspace use	Preferential or rotational runway use*	X	
	Modification of approach and departure procedures*	X	
	Restrictions on ground movement of aircraft*	X	
	Restrictions on engine run-ups and ground equipment	X	X
	Limits on numbers or types of operations and aircraft	X	X
	Use restrictions and rescheduling of operations	X	X
	Raise glide slope angle or intercept*	X	
Aircraft operation	Power and flap management*	X	
	Limited use of reverse thrust*	X	
Land use	Land or easement acquisition	X	X
	Joint development of airport property	X	X

	Compatible use zoning	X	X
	Building code provisions and sound insulation	X	X
	Purchase assurance	X	X
Noise program management	Noise-related landing fees	X	
	Noise monitoring	X	X
	Establish community participation program	X	X

*These restrictions may affect the safe operation of the aircraft. They should not be set by the airport operator, but require approval by aviation authorities (such as the FAA in the United States) and pilots.

Source: Adapted from FAA AC 150/5020-1.

Both the fixed costs and the variable costs of a noise monitoring system may be considerable. In addition to significant maintenance costs associated with the remote sensors, these systems often require expensive technological upgrades. More importantly, a full-time staff of several people is typically required to oversee the operation of the system, analyze its data and prepare relevant reports, respond to inquiries and complaints about noise, and interact with planning staff, citizen groups and airport administrators.

Community relations and public participation programs

Community relations and public participation programs have also become a virtual requirement for airports in many developed countries, especially those operating within pluralistic political systems. Such programs are essential for an airport authority wishing to demonstrate, not just to its neighbors, but also to the public at large, its concern about negative environmental impacts of an airport and its commitment to alleviating problems in concert with the affected communities.

The scope of community relations programs can range widely. At a minimum, it may consist simply of a public information program on noise-related developments, such as on changes in L_{dn} noise contours from year to year or on the projected noise impacts of proposed physical or operational changes at the airport. Increasingly, however, programs of this type include some or all of the following elements:

- *A process of regular consultation* with all key stakeholders (neighboring communities, airlines, air traffic controllers) on environmental issues

- *A citizen advisory group,* a committee consisting of representatives of neighboring communities and other nongovernmental organizations (NGO) that meets regularly with airport officials to advise them on community concerns and discuss planning issues

- *A noise complaint hotline,* a telephone center that people can call to register complaints or ask questions

- *A participatory planning process,* in which neighboring communities and other NGOs have an opportunity to influence and/or review airport plans at an early stage, possibly with assistance from their own independent technical consultants

In recent years, some airport operators have even provided funding to neighboring communities or to citizen groups, so they can hire competent technical consultants to help them address airport-related

EXAMPLE 6.4

Massport, the owner and operator of Boston/Logan, has historically covered a significant part of the costs of consultants to the Community Advisory Committee, a group of community representatives that has generally opposed Massport's initiatives. In 2001, this committee's consultants issued a detailed response to Massport's proposal to construct the new Runway 14/32 (see Example 6.1), strongly opposing the runway.

Palm Beach/International in Florida has created the position of "Noise Officer," a person who is supposed to act as an impartial facilitator of the negotiating process between the airport and community groups. Although this official is on the airport's payroll, he or she does not take sides in disputes and attempts to bring to the fore the concerns of all stakeholders, so they can be properly addressed (Sylvan, 2000).

issues. The rationale for this practice, which is becoming increasingly common in the United States, is that the dialogue between airports and neighboring communities will be facilitated if both sides have access to expert and objective technical information. The validity of this hypothesis is an open question at this point.

Land-use policies

Land-use policies probably have the greatest potential for minimizing the environmental and noise impact of airports on its neighbors. They are "proactive," in the sense that they anticipate future problems and attempt to forestall them through judicious planning. They take two forms that can be used independently or together:

- Zoning restrictions that ensure the compatibility of land-use and noise exposure
- Building code and sound insulation provisions that provide appropriate reductions in noise levels experienced inside structures and buildings

Table 6-5 shows a set of guidelines prepared by the FAA that relate L_{dn} values with potential land use for residential, public service, commercial, manufacturing, and recreational purposes. Note that in a number of instances it indicates the need to reduce the noise levels by up to 35 dBA from outdoors to indoors.

Table 6-5. Guidelines for noise-compatible uses

Use	Category	Yearly day–night average sound level, DNL, dB*					
		<65	65–70	70–75	75–80	80–85	>85
Residential	Dwellings	Y					
Public	Schools	Y					
	Hospitals and nursing homes	Y	25	30			
	Churches and auditoriums	Y	25	30			
	Government services	Y	Y	25	30		
	Transportation and parking	Y	Y	Y	Y	Y	Y
Commercial	Offices	Y	Y	25	30		
	Retail trade	Y	Y	25	30		
	Communication	Y	Y	25	30		
	Wholesale and retail equipment	Y	Y	Y	Y	Y	
	Utilities	Y	Y	Y	Y	Y	
Manufacturing and production	Livestock farming and breeding	Y	Y	Y			
	Photographic and optical	Y	Y	25	30		
	General manufacturing	Y	Y	Y	Y	Y	
	Agriculture and forestry	Y	Y	Y	Y	Y	Y
	Mining and fishing	Y	Y	Y	Y	Y	Y

Recreational					
Outdoor amphitheaters	Y				
Nature exhibits and zoos	Y	Y			
Outdoor sports arenas	Y	Y	Y		
Golf courses, riding stables	Y	Y	Y	25	30
Amusement parks, camps	Y	Y	Y	Y	Y

*Y = fully compatible without restrictions; 25 or 30 means that land use is compatible but that indoor noise reduction of 25 or 30 must be incorporated into related structures; unmarked combinations are not compatible and should be prohibited.

Source: Adapted from FAR 150/5020-1.

Unfortunately, airport operators can only exercise the full force of carefully designed land-use policies at entirely new sites and in sparsely populated areas. There are few such sites left near major cities, and even fewer entirely new airports are being built around the world. In general, airports and planning authorities find it difficult to prevent incompatible developments around airports, given the pressures of urban growth. In many countries, poor people moving to the city simply "invade" protected areas, as has happened at Mexico City, Bangkok, and many other rapidly growing metro-polises. Therefore, guidelines such as those in Table 6.5 serve principally as models for the establishment of land-use policy objectives at existing sites, at new sites that are less than ideal, or for the reuse of property that authorities may acquire around an airport. They suggest the types of land-use interventions governments and airport operators could initiate, as the subsection on Interventions outside the Airport discusses.

Airport design interventions

Airport operators can sometimes reduce noise impacts on airport neighbors by a number of possible modifications, adjustments, or additions to the physical layout and structures on the airport proper. There are numerous examples of potential interventions of this type, such as:

- *Displaced runway thresholds—The declared* threshold of a runway is moved from its physical threshold, with the displacement distance depending on the total length of the runway, the type of operations it is used for and many other operational and physical considerations. Although this action may be taken mostly because obstructed airspace makes it necessary, it does have environmental benefits. It makes arriving aircraft fly over neighboring communities on approach at a somewhat higher altitude than otherwise. It also may have a similarly beneficial effect for takeoffs.

- *Well-placed high-speed runway exits* may reduce the need to apply reverse thrust on landing.

- *Sound barriers*—Since sound propagates along straight lines, the construction of walls, buildings or other structures at strategic points at the airport periphery can help reduce noise exposure levels from sources on the ground such as aircraft on the ground, airport vehicles, and ground support equipment. Numerous airports are creating sound barriers.

Miami/International, for instance, has erected a noise wall 669 m long and 6–9 m high. Civil aviation authorities typically scrutinize measures of this type closely, to ensure that sound barriers do not impinge on flight safety.

- *New runways, taxiways, and buildings*—Airport operators can reduce noise by relocating runways so that the associated approach or departure paths are less noise-sensitive, by placing passenger buildings and other structures where they form a barrier to noise propagation into neighboring communities, and by constructing new taxiways so that aircraft can circulate farther from the airport's periphery.

Only a limited number of possibilities along these lines are typically available at existing airports, especially older ones. With the exception of some types of sound barriers, in most cases there is either not sufficient space for such modifications or their cost is prohibitive.

Surface operations and flight operations

A growing number of airports impose restrictions on aircraft operations, to reduce both noise and engine emissions (see Sec. 6-5).

Surface Operations Some possibilities concerning surface operations include restrictions on:

- *Towing of aircraft* from gates to hangars or from one gate to another when a change of gate is necessary
- *Engine starting, run-ups, and testing.* See Example 6.5.
- *Limits on the number of outbound taxiing aircraft.* At times of air traffic congestion, having too many aircraft taxiing out and queuing at the departure runways is counterproductive (see Chap. 13). By not allowing additional departing aircraft to leave their gates until the number of aircraft already taxiing out has been reduced below some threshold value, the air traffic control tower can reduce noise and engine emissions, without any adverse effects on runway capacity. (However, gate holds may block arriving aircraft, and reduce the operational capacity of the passenger building.)

Flight Operations Noise exposure can be reduced or its effects mitigated through the following types of measures:

1. *Noise abatement procedures on landing and takeoff.* Civil aviation authorities, aircraft manufacturers, and airlines have developed

EXAMPLE 6.5

As of 2001, Amsterdam/Schiphol allowed engine test-running only at special-purpose locations. It is prohibited between 24:00 and 06:00 local time, and requires special permission at all other times. The airport also requires that, safety permitting, engine reverse-thrust should not be applied on the runways between 22:00 and 06:00. While the airport had no formal restrictions on the use of auxiliary power units (APU) in 2001, it encouraged aircraft operators to use the fixed ground power system, when available, for engine start-up instead of running the APU. (Several other airports—e.g., Zurich and Copenhagen—have imposed severe restrictions on APU use.)

noise abatement procedures for landing and takeoff aimed at reducing noise exposure through recommended engine power settings and profile characteristics for aircraft ranging from business jets to large commercial aircraft (FAA, 1993). Further improvements along these lines seem possible (Clarke, 1997, 2001). Another important way to abate noise is to establish, *at individual airports,* arrival and departure flight paths that direct traffic flows over uninhabited or less noise-sensitive areas. In the United States, the FAA works closely with local airport operators and airlines to design such paths. Washington/Reagan, New York/Kennedy, and Boston/Logan, among others, offer well-known examples along these lines. For instance, aircraft departing Boston/Logan toward the southwest (Runway 22R) make a sharp left- hand turn immediately after clearing the runway, to fly over the Atlantic Ocean rather than heavily populated areas under the extended centerline of the runway.

2. *Preferential runway systems.* Preferential use of certain runway configurations, whenever traffic and weather conditions permit, is another important operational tool for noise abatement. A preferential runway system is essentially a set of guidelines for assigning arrivals and departures to runways in a way that minimizes noise impacts. It usually has two objectives. First, it attempts to maximize the time aircraft use runway configurations* that are clearly superior in terms of reducing noise impacts on airport neighbors (e.g., approach and departure paths are over uninhabited areas or over water). Second, it attempts to distribute noise impacts equitably, so that no single area

---

*See Chap. 10 for a detailed discussion of runway configurations.

will be affected excessively. Ideally, this is achieved by agreeing with representatives of the communities involved on a set of targets for the utilization of each of the runway configurations.

Preferential runway systems range widely in terms of level of sophistication. They can consist of just a simple listing of runway priorities (e.g., "for arrivals, use Runway 06R whenever possible, otherwise use Runway 24R"), or they can be computer-based decision support systems that advise air traffic controllers on which runway configurations to use at any particular time. The design and implementation of such "automated" systems requires complex analysis and an advanced ATM environment. They are currently operational at only a small number of multirunway airports around the world. For example, Boston/Logan uses a computer tool, ENPRAS (Enhanced Preferential Runway Assignment System), to help air traffic control select the runway configuration to be used under good weather conditions and to implement the complex, noise-related runway preference rules described in Chap. 10. Also, see Example 6.6.

EXAMPLE 6.6

When visibility is more than 2000 m and the cloud base is above 90 m (300 ft), ATM system operators select the runway configuration in use at Amsterdam/Schiphol according to a preferential runway system, based on the following principles:

- Traffic safety prevails at all times.

- Takeoffs and landings normally take place on separate runways.

- A runway equipped with an instrument landing system (ILS) is preferable for landings.

- The sequence in which runway configurations are used during any time period is determined by a combination of noise impact and traffic handling criteria and must take account of anticipated weather and wind conditions.

- The use of a nonpreferential runway is not permitted, unless specifically requested by the pilot for safety reasons; no deviations from an assigned runway are permitted if the objective is to obtain a shorter taxi distance or shorter departure or approach route.

Interventions outside airport property near existing sites

Increasingly, airport operators have seen fit or have been forced by circumstances to attempt to improve noise-related conditions near existing sites through interventions outside airport property. These can be expensive, time-consuming, and occasionally unpopular. The two most prominent examples are sound insulation and property acquisition.

Sound Insulation   The airport operator and/or a government agency typically pays for sound insulation of noise-affected structures near airports. Sound insulation may be limited to public buildings, such as schools, hospitals, and churches, or extended to private residences. As Table 6-5 suggests, the aim typically is to reduce noise levels by 25–35 dBA from outdoors to indoors. This may also require keeping windows closed and relying on air conditioning during summer months—which will, of course, add to costs (Hye, 2000). The total cost of such sound insulation programs can easily reach into the tens of millions of dollars and, occasionally, exceed $100 million. The average cost of soundproofing each of 600 suburban homes around Chicago/O'Hare was about $27,500 in 1997 (Sylvan, 2000). Through 2001, Boston/Logan had spent around $99 million, mostly on soundproofing; Los Angeles/International had allocated a total of about $119 million for soundproofing and land acquisition; and the total amount spent in the United States for noise mitigation from Airport Improvement and passenger facility charges exceeded $5.2 billion.*

Property Acquisition   Property acquisition is perhaps the most controversial noise mitigation measure. For lack of better alternatives, airport operators and government agencies occasionally acquire (or lease) residential and other property and convert it to uses compatible with prevailing noise levels (see Table 6-5). Acquisition can be either through the real estate market or by invoking *eminent domain* when this is legally feasible. Determining a "fair price" for the property to be acquired is a critical and difficult question in such instances. So are issues related to relocating displaced residents and businesses. Some airports are turning to local communities to implement these programs. For example, Los Angeles/International has regularly granted millions of dollars to the city of Inglewood, to "acquire and recycle incompatible residential property into more compatible use and conduct further residential soundproofing" (Sylvan, 2000; Los Angeles International, 2001).

*Personal communication from Jan Ashraf, U.S. FAA.

Access restrictions

Administrative or legislative restrictions may mitigate noise by inhibiting in some way access to airports. They may apply to specific aircraft types, to specific times of the day, and to complex combinations of the two.

Restrictions on the Operation of Certain Aircraft Types A number of airport operators, especially in Europe, have been imposing restrictions on access that are more elaborate and stricter than those called for by their national policies on Stage 3 aircraft (see Sec. 6-2). In particular, they have drawn distinctions between aircraft meeting Chapter/Stage 3 noise standards, depending on how they meet these standards. Example 6.7 gives a typical instance.

Nighttime Curfews or Restrictions on Airport Access during Night Hours Since nighttime aircraft movements are particularly disruptive of living patterns, nighttime curfews constitute an obvious noise mitigation measure that communities located near airports consistently request. Los Angeles/John Wayne, for example, bans takeoffs between 22:00 and 07:00 (08:00 on Sundays) and landings between 23:00 and 07:00 (08:00 on Sundays). However, only relatively few major airports impose a total ban ("complete curfew") on nighttime movements.* Most national governments and airport operators wish to be able to accommodate some nighttime movements, aware of the fact that certain types of traffic—notably next-day delivery packages and priority cargo—move most naturally during the night. Airport operators are also concerned that any agreement to impose a complete curfew during a number of night hours will almost inevitably be followed a few years later by demands to increase that number.

"Partial curfews" are therefore far more common. Governments and/or airport operators may restrict access during nighttime to only certain types of aircraft. They may also accompany these formal restrictions with informal attempts to dissuade airlines from scheduling flights during late-night hours. Amsterdam/Schiphol offers a typical example of a partial curfew of this type, coupled with restrictions on access by specific types of aircraft. Note that it treats aircraft that meet Chapter 3 noise certification standards a narrow margin—in the

---

*As of mid-2001, London/Heathrow was, by far, the busiest airport in the world with a complete curfew on runway movements that lasts between 23:00 and 07:00 local time. This curfew was brought about as a result of a lawsuit by nongovernmental organizations, which was heard by the European Court of Human Rights. The court ruled that the British Government provided insufficient grounds for overruling this curfew (Sylvan, 2000; http://www.hacan.org.uk/).

EXAMPLE 6.7

As of March 25, 2001 (start of summer season 2001), the following operational restrictions were effective at Amsterdam/Schiphol:

"Chapter 2 Aircraft" (aircraft certified in accordance with the noise standards of ICAO Annex 16 Chapter 2):

- No new operations by aircraft of this type are allowed.

- For aircraft equipped with engines with bypass ratio less than or equal to 2, takeoff and landing are not allowed between 18:00 and 08:00 local time.

- For other aircraft, takeoff and landing are not allowed between 23:00 and 06:00.

"Aircraft narrowly Chapter 3," i.e., aircraft certified in accordance with the noise standards of ICAO Annex 16 Chapter 3, for which the difference between the sum of the three certification noise levels and the sum of the three applicable ICAO Annex 16 Chapter 3 certification noise limits (see Certification of Aircraft for Noise) is less than 5 EPNdB:

- For aircraft equipped with engines with bypass ratio less than or equal to 3, no new operations are allowed.

- For aircraft equipped with engines with bypass ratio less than or equal to 3, takeoff and landing are not allowed between 18:00 and 08:00 local time.

- For other aircraft, takeoffs are not allowed between 23:00 and 06:00.

"Chapter 3 Aircraft": No restrictions on runway movements of aircraft certified in accordance with the noise standards of ICAO Annex 16 Chapter 3, for which the difference between the sum of the three certification noise levels and the sum of the three applicable ICAO Annex 16 Chapter 3 certification noise limits is 5 EPNdB or more.

sense defined in Example 6.7—in the same way as some types of aircraft that only meet Chapter 2 standards.

"Noise Budgets"   A relatively small number of airports in Northern and Western Europe and in North America have adopted "noise budgets." These may result in strict limits on airport operations. They can take one of two forms. In the simpler form, the airport operator (or government) commits to limiting (possibly as a function of time) the annual number of aircraft movements and/or the annual number of passengers—which constitute a rough proxy for the noise generated at the airport. For example, Amsterdam/Schiphol committed to a maximum of 440,000 commercial air traffic movements in 2001 and of 460,000 movements in 2002. Toronto/Pearson does not permit the percent of flights operating during "night restricted hours" (00:30–06:30) to increase in any calendar year more than the percent yearly increase in passenger traffic during the previous year.

It should be noted that setting noise budgets with reference to limits on annual number of passengers has proved to be a risky proposition. The Dutch government committed in the early 1990s to a noise budget for Amsterdam/Schiphol through 2015 based on a limit of 44 million passengers per year, which then was the forecast for 2015. Unfortunately, this forecast proved very wrong—as is so often the case, see Chaps. 3 and 20. Realizing that the limit of 44 million would in fact be reached during the first few years of the twenty-first century, the Dutch rescinded the commitment to this limit in the late 1990s. London/Heathrow had a similar experience in the 1980s.

The second form of noise budgets sets limits with direct reference to noise-related measures. For the period 1997–2004, Copenhagen/Kastrup has agreed to not exceed the annual noise levels reached in 1996 by more than 1 dB. Beginning in 2005 the airport expects to gradually reduce the annual limit by about 5 dB from the 1996 levels. Beginning in 2003, Amsterdam/Schiphol will also transition from a noise budget based on the annual number of movements (maximum of 460,000 in 2002) to one based on a total annual noise volume and a set of maximum noise levels (specified in L_{dn}) for locations around the airport.

Economic incentives

Economic (or market-based) incentives can substitute for or complement administrative, noise-related restrictions on airport access. Incentives can be in one of two forms: noise-related landing fees or penalties for violating noise- related restrictions. Landing fees provide primarily *long-term incentives* to airlines and other aircraft operators to use more quiet aircraft; penalties for violating noise-related restrictions can be effective *in the short run* in ensuring compliance with such restrictions.

Landing fees that include surcharges or rebates to aircraft according to their noise characteristics are becoming increasingly common, especially in Europe. By the beginning of the twenty-first century, practically all the major European airports had adopted differential charging policies in this respect. The surcharges for noisy aircraft can be substantial. For instance, Chapter 2 aircraft operating at night in 2001 in Brussels or Munich had to pay about twice the basic landing charge.

Penalties for violating noise-related restrictions by specific flights are less common, but a few major airports claim to have such systems of fines in place. These typically impose fines on flights that deviate appreciably, without a good reason, from recommended noise abatement departure and arrival routes in an airport's terminal airspace (see Surface Operations and Flight Operations) or operate during curfew hours. Depending on local rules, airlines that seem to be habitual violators of noise-related restrictions may receive substantially greater one-time fines. Noise monitoring systems are obviously essential to an airport operator's ability to implement such a system of fines. Some airport operators publicize habitual violations of noise-related procedures to pressure airlines to abide by such procedures, on the theory that airlines do not wish to be perceived by the public as insensitive to environmental concerns. Whether these economic incentives do change the behavior of an airline depends on its costs of changing equipment or procedures.

6-4   Air quality and mitigation of air pollution

Aircraft engines burn fuel, thus producing emissions with potential effects on global climate change and on local air pollution.* This section focuses on the latter effects, since contributions to local air pollution are another important source of friction between airport operators and neighboring communities. Concerns about the relationship between air quality and health have been growing over the years, thus increasingly making air pollution the focus of environmental controversies involving airports. In public hearings in 2001 about the disputed proposed Runway 14/32 at Boston/Logan (Example 6.1), more comments and complaints were filed about potential effects on air quality than about noise. This was a sharp reversal from what had been the case in earlier years. This may be a problematic development for airport/community relations, as the long-term health effects of air

---

*For a thorough review of aviation's contribution to global climate change, see Penner et al. (1999).

pollutants emanating from aircraft are certainly difficult to measure and so far poorly understood by medical science.

Major airports can indeed be significant sources of air pollutants in metropolitan areas because of emissions from landing, departing, and taxiing aircraft and from airport-related concentrations of automobiles and other ground traffic. About 0.5–1 percent of the exhaust emitted from jet engines consists of some of the air pollutants that are of primary concern in determining local air quality, a combination of nitrogen oxides (NO_x), hydrocarbons (HC), volatile organic compounds (VOC), carbon monoxide (CO), sulfur oxides (SO_x), other trace chemical species, and carbon-based soot particulates. The concentrations of nitrogen oxides, hydrocarbons, and carbon monoxide may be particularly high near busy commercial airports, with the ground operations of aircraft and other vehicles being major contributors. In addition, soot emissions may add to ambient particulate levels, while sulfur oxide emissions from aircraft may be low-level local contributors to acid rain. As an indication of the magnitudes involved, it is estimated that New York/Kennedy and New York/LaGuardia generate about 1900 and 1500 tons per year, respectively, of nitrogen oxide emissions (which are heavy contributors to smog) and volatile organic compounds (VOC). They thus ranked as the first and fourth sources of these pollutants in the Metropolitan New York Area—the others in the top five being two large power plants and the Hempstead incinerator (MIT, 2001).* Figure 6-6 shows a comparison made by the U.S. Environmental Protection Agency between 1990 airport contributions to nitrogen oxide levels in 10 metropolitan areas and those projected for 2010 (EPA, 1999). In every case, the EPA expected the contribution of airports to more than double by 2010.

International and national efforts concerning aviation's role in global climate issues also have important implications for local air quality. The ICAO's Committee on Aviation Environmental Protection (CAEP), which has been charged with studying technology, operational, and market-based opportunities for reducing both the global and local impacts of aircraft engine emissions, has considerable international influence (Crayston and Hupe, 2000). Several national civil aviation organizations are working closely with CAEP on these objectives.

*Strictly speaking, an airport is not a single source of pollution, the way a power plant might be. However, the airport operator is a single entity. As a matter of political reality, many members of the public view airports not only as single sources of pollution, but as ones with access to a lot of money as demonstrated by the millions of dollars they have spent on noise mitigation.

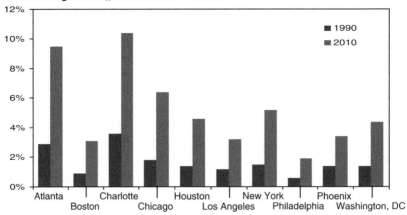

Fig. 6-6. *Airport contributions to nitrogen oxide levels.* (*Source:* EPA, 1999, Table 4-2.)

The ICAO and national standards for emissions certification of aircraft engines currently focus on impacts near airports. They establish limits, expressed in terms of mass of emissions per unit of engine thrust, on nitric oxide, carbon monoxide, unburned hydrocarbons, and smoke for a reference landing and takeoff (LTO) cycle below 915 m (3000 ft) of altitude above the airport. Experts currently believe that aircraft emissions above this altitude do not have any discernible effect at ground level near the airport.

Legislation in several developed countries requires environmental impact statements on proposed airport projects to deal in depth with air quality impacts. Estimation of these impacts typically uses computer databases and models. For example, in the United States, a two-step estimation process is used to compute:

    1. *Total emissions* from all aircraft operations, ground service equipment, motor vehicles, and fuel storage and transfer facilities from appropriate local databases. Estimates must include CO, NO_x, HC, VOC, and particulate total emissions and local concentrations, including odor-causing hydrocarbon emissions.

    2. *Ambient pollutant concentrations* at various locations inside and near the airport, with a computer-based dispersion model. In the United States, since 1998 the FAA mandates the use of its Emissions and Dispersion Modeling System (EDMS) (FAA 1997, 1998). The Industrial Source Complex (ISC) model used by the U.S. Environmental Protection Agency (Gifford, 1976, 1986) is also sometimes applied internationally.

National or local authorities are adopting various mitigation measures in efforts to preserve or improve air quality near airports. As in the case of noise, these measures range from mild to severe interventions. The most important among them include the following.

1. *Extensive air quality monitoring.* For example, Aéroports de Paris (ADP) measures air pollution at 80 locations in the Paris region, analyzes the levels of CO, NO, NO_2, SO_2, O_3, and hydrocarbons and reports results and trends on a quarterly basis. Similarly, Manchester (UK) monitors air quality at and around the airport and publishes the results (Sylvan, 2000).

2. *Aircraft operations.* Possibilities include towing of aircraft between gates and maintenance areas, as is done at San Francisco and at Zürich; limiting the use of auxiliary power units (APU) to certain times of the day (Copenhagen, Zürich); provision of central power via passenger buildings, to reduce need for APU units (many airports in the United States, including most renovated passenger buildings); placing restrictions on the times when engines can be run up (Copenhagen prohibits run-ups between 23:00 and 05:00 local time); and airfield design—or airfield design changes—aimed at reducing taxiing distances (as in the case of many midfield passenger terminals).

3. *Reducing delays while taxiing and idling.* Airport delays contribute to emissions, if these delays are suffered with engines running. Extensive observations have shown that the maximum departure rate at Boston/Logan is reached when nine aircraft are taxiing out for departure. Having more than nine aircraft on the outbound taxiway system does not increase that departure rate (Anagnostakis et al., 1999). Thus, emissions (and noise impacts) can be reduced, without any adverse effects on runway capacity, through the simple expedient of not allowing additional aircraft to begin taxi-out, as long as nine or more aircraft are already on the outbound taxiway system. The critical threshold value (nine in the case of Boston/Logan) will vary, of course, from airport to airport.

4. *Emissions-based landing fees.* A few airports impose an emissions-based surcharge to encourage use of "cleaner" engines by airlines. The first user of this approach has been Zürich. In 1997 it reduced all landing fees by 5 percent and started recovering the lost amount

(5 percent of the total landing fees collected previously) through surcharges based on the emissions characteristics of the different types of aircraft. Specifically, Zürich classified aircraft into five categories, with Category 1 being the worst emission performers, which pay a 40 percent surcharge on their (reduced) landing fee, and Category 5 being the best (Sylvan, 2000). Geneva also adopted the same approach in response to Swiss federal air quality legislation. Several other airports reportedly were considering emissions-based charges at the beginning of the twenty-first century.

6-5   Water quality control

As with any large industrial facility, airport operators need to pay attention to fluid discharges. The activities at the airport produce a wide range of discharges that need to be actively managed. These include discharges associated with industrial activities such as aircraft maintenance and the handling of fuels, storm water runoffs, and ordinary sewage. These should be handled according to locally prevailing regulations. Additionally, airports in snowy climates have to deal with deicing liquids, which need special attention.

Deicing fluids (ADF)

Proper removal and prevention of the formation of ice on aircraft is a critical function for the safe operation of aircraft in areas where snow occurs. Aircraft cannot fly safely when they are covered with ice. Accumulated ice reduces the lift produced by the wings, makes aircraft unstable, and can cause significant accidents. In 1982, an Air Florida flight crashed shortly after takeoff from Washington/Reagan. Airports and airlines in snowy climates thus devote substantial resources to removing ice from aircraft and preventing it from forming before the aircraft has taken off and climbed above the weather. A critical element in this process is the spraying of aircraft with deicing/anti-icing fluids (ADF), which inhibit the formation of ice.

Currently, the most effective ways of deicing and anti-icing aircraft involve the application of heated glycol-based fluids. These chemicals, and the additives necessary to ensure that the ADF meet the safety requirements of the FAA, can pollute the groundwater and thus pose environmental issues. As no effective substitutes have been found so far, the main strategies for managing the environmental impacts of ADF are reduction of the amounts used, collection and disposal of the fluids, and recycling. In general, airports and airlines use a combination of these approaches (see EPA, 2000, for complete, detailed information).

The deicing process consists of spraying quantities of diluted ADF over the aircraft and thus on the ground. Most commonly, hoses are attached to special tanker trucks (see Fig. 6-7). At a number of airports they are attached to a large gantry that passes over aircraft. The pressure of the spray blows the snow and ice away, while the ADF contributes to breaking up the ice and inhibiting its formation for some limited time. The ways to reduce the amount of glycol used are thus to reduce the amount of snow on the aircraft by mechanical means, and to dilute their concentration in the ADF spray according to local conditions.

Some airports and airlines use forced air to remove snow and ice from aircraft. These systems have been available since around 1980, but have not proven to be popular in North America although they appear to be used in Japan. Operators have found that they do not seem to work well with heavy, wet snow, which is common in various areas. The equipment is also expensive. Recently, some carriers and airports have been developing a system of infrared heaters to melt ice and snow. Neither these nor other alternatives eliminate the need for ADF, however. Once the accumulation of snow and ice is off the aircraft, ADF is needed to prevent further ice from forming as the aircraft taxis out to the runway and takes off.

Fig. 6-7. *Deicing operations at Munich.* (**Source:** Munich International Airport.)

Airport operators have focused on collecting the ADF that fall on the ground and preventing them from entering the ground. Some airlines in the United States have built their own deicing facilities. A number of airports, such as Denver/International, Montreal/Dorval, and Toronto/Pearson, have installed common deicing facilities. These are located as conveniently as possible to the ends of the departure runways. They are designed with a collection of drains intended to capture the ADF and channel them to special retaining ponds or tanks. Since they should be able to serve all the aircraft operating in snowy conditions, these facilities can be large. The new facility at Toronto/Pearson will have six bays, capable of accommodating 12 Boeing 747s in all. Each bay will be about 328 ft by 780 ft (100 m by 235 m) and the complex will cover 65 acres (25 ha). These central facilities have proven successful in reducing ADF contamination of the groundwater. Furthermore, in some conditions they make it possible to recycle the wastewater from deicing procedures, and to sell it for other purposes. However, they normally cannot serve all the deicing needs, as some aircraft will have to be serviced at gates to prevent damage during taxiing.

Other airports use a variety of methods to collect wastewater from deicing operations. For example, to prevent the ADF from mixing with the ordinary rainwater, they may install special drainage systems in ramp areas, or use valves or sewer plugs. They may also use special vehicles to vacuum up the glycol. However, neither these procedures nor the central deicing facilities will collect all the ADF. Some of these fluids will normally drip from the aircraft as they taxi around the airport.

Fuel leaks and spills

Airports normally store substantial amounts of aviation and other fuels on or near their property. Under ordinary circumstances, the storage and distribution systems do not pollute the environment, since distributors take care to prevent losses of this expensive resource. However, airport operators need to maintain the distribution systems carefully to prevent leakage and contamination of the groundwater. They also need to protect storage areas against accidental spills or deliberate sabotage. Berms are thus often built around tanks to contain possible massive spills.

Storm water runoff

Because airports consist of large areas of paved surface, rainwater runs off quickly. It cleans the runways, taxiways, aprons, and roadways, and flushes the pollutants that may have accumulated on these surfaces away from the airport. Moreover, the large quantities of water can create flash floods if the drainage system is not designed properly. Airport designs may thus involve a number of settling ponds that will retain water both to prevent floods and to settle the particulate material swept away from the airfield. The issues for the airport in this case are similar to those of any urban area that has large areas of paved and hard surfaces. In the United States, airports may thus have to secure a discharge permit under the National Pollution Discharge Elimination System (see EPA, 2001).

6-6   Control of highway and road access traffic

Airports in North America and Europe are under pressure to reduce automobile traffic to and from airports, and to promote public transport. At the same time, public authorities and airports have invested large amounts of money in the creation of rail links to airports and spent a great deal in subsidizing these operations, as Chap. 17 discusses in detail. A number of airports in the United States have to take special action because they are in metropolitan areas that do not meet the national air quality standards. For many people, the desire to reduce pollution is an important motivation for these efforts.

Although some travelers and airport employees do use public transport, it is often inconvenient or impractical for many potential users. Passengers with bags, and sometimes families with children, may not find it easy to use public transport, especially if they have to make one or more transfers. A trip to a rail station, onto the train, and from its airport stop to a destination at the airport may be difficult. For many people, rail or ordinary public transport is not a possibility because these systems do not serve the right areas at the right times.

Some airports—Boston/Logan, for example—restrict the number of available parking spaces at the airport on the theory that this reduces the number of automobile trips to the airport. Faced with the inability to park at Boston/Logan, however, travelers are likely to take taxis or have someone drive them, and thus cause two airport trips instead of the one that would have occurred if they took their own car and parked it. In short, it is not clear that a push to limit parking has much effect on the amount of automotive pollution associated with the airport trip.

Airport operators are likely to continue to face public pressure to reduce the emissions associated with automotive traffic. The public sees a major airport as a big enterprise that may have one of the single largest concentrations of cars in the region (many airports have over 10,000 parking spaces; see Chap. 17). Some airports also appear to have access to substantial money. Consequently, activists will continue to target airports for measures designed to reduce the use of automobiles and increase the use of public transit (see Coogan, 1995).

This political reality encourages airport operators to adopt a range of ways to limit the way vehicles use roadways near the passenger buildings. They try to constrain the number of vehicles and reduce congestion and thus the amount of idling (which is notorious for producing considerable emissions). In particular, many airports have adopted schemes to restrict courtesy vehicles around the passenger buildings. They reduce the number of these vehicles by limiting their schedules, charging the operators for each trip, consolidating services (as at Sacramento), and replacing them with people movers that carry passengers to remote central rental car facilities (as at New York/Newark and San Francisco/International). Airport operators may also limit curbside stopping and standing, segregate private and commercial vehicles, and provide convenient short-term parking (see Chap. 17).

Many airports, especially in the United States, are furthermore required to use very-low-emission vehicles, such as electric cars. For example, the U.S. Clean Air Act Amendments of 1990 require, in areas of high ozone or carbon monoxide pollution, that all owners of fleets of centrally fueled vehicles acquire low-emissions vehicles. Further, the U.S. Aviation Investment and Reform Act of 2000 mandated a pilot program to promote inherently low-emission vehicles for airport operations (U.S. Congress, 2000). This includes aircraft ground support equipment, service and security vehicles, and parking-lot shuttle buses. This program aims to reduce substantially ozone and carbon monoxide levels at airports that are located in areas where the air quality fails to meet the National Ambient Air Quality Standards enunciated by the Environmental Protection Agency.

6-7   Wildlife management

Airports are home to many kinds of wildlife, including flocks of birds and many small animals. They can become an environmental issue if public groups consider a particular species to be endangered or otherwise of interest. Worldwide, countries that require environmental impact statements typically require information about the population of animals in the area of a proposed airport project. In a number of cases, concerns about wildlife have delayed or altered airport developments.

Wildlife in general, and birds most especially, can pose severe hazards for the safe operation of aircraft. Larger wildlife, such as deer, can create dangerous conditions on the runway, but are more easily controlled by fencing. Birds in flight collide with airplanes and are pulled into engines. A "bird strike" can cause severe damage to the aircraft and endanger the passengers and crew. Expert committees in the United States estimate that wildlife strikes have killed over 400 people worldwide and that this issue costs U.S. civil aviation over $390 million annually (Bird Strike Committee, 2001). Many airports thus have programs to scare off or eliminate birds near runways. Some own impressive aviaries of falcons! For details on handling this problem, see Cleary and Dolbeer (1999).

Exercises

1. Consider a major commercial airport in your region. What is its history of public concern with its noise and other environmental impacts? How has the airport operator responded politically? What has the operator done to mitigate the effects?

2. Find, on the web, some environmental impact statements or noise studies for airport projects. Examine their noise contours. What levels of noise do they reflect? How many people are affected? How does the operator intend to mitigate these effects? Do the noise contours suggest that flight paths could or should deviate from extensions of the runway centerlines to avoid populated areas? Reflect on and discuss your findings.

3. Obtain a recent version of the Integrated Noise Model and exercise it. Evaluate it as a user. Do you feel it provides you with the kind of information you might require as an airport planner? As a resident of the community? As a local political leader?

4. For an airport of interest, explore its air quality issues. As a benchmark, first identify the kinds of air quality controls on automobiles or factories that prevail in its region. Then identify the kinds of controls or mitigations affecting airport sources. Do you think that these are compatible? If not, what might be a more reasonable approach?

5. Examine the water quality issues for some airport of interest. What are the major environmental concerns, if any? How do the airport operator and the local community deal with them?

References

Anagnostakis, I., Clarke, J.-P., Delcaire, B., Feron, E., Hansman, R., Idris, H., Odoni, A., and Pujet, N. (1999) "Observations of Departure Processes at Logan Airport to Support the Development of Departure Planning Tools," *Air Traffic Control Quarterly*, Special Issue on Air Traffic Management, 7(4), pp. 229–257.

Ashford, N., and Wright, P. (1997) *Airport Engineering*, 3d ed., Wiley, New York.

Bird Strike Committee (2001), http://www.birdstrike.org/.

Boeing Aircraft Company (2001) "Airport Information Updates," http://www.boeing.com/assocproducts/noise/airports.html.

CFR, Electronic Code of Federal Regulations (2001) "Chapter 1 (FAA), Part 36, Noise Standards," http://www.access.gpo.gov/nara/cfr/cfrhtml_00/Title_14/14cfr36_00.html.

Ciattoni, J. P. (1987) *Le Bruit,* Collection "Les classiques santé," Privat, Paris.

Clarke, J. P. (1997) "A Systems Analysis Methodology for Developing Single Event Noise Abatement Procedures," Sc.D. thesis, Massachusetts Institute of Technology, Cambridge, MA.

Clarke, J. P. (2001) Class notes, Massachusetts Institute of Technology, Cambridge, MA.

Cleary, E., and Dolbeer, R. (1999) "Wildlife Hazard Management at Airports," FAA, Office of Aviation Safety and Standards, and U.S. Department of Agriculture, Washington, DC, http://wildlife-mitigation.tc.faa.gov/public_html/manuals/title1.html.

Coogan, M. (1995) "Comparing Airport Ground Access—A Transatlantic Look at an Intermodal Issue," *TR News, 174,* pp. 2–10.

Crayston, J., and Hupe, J. (2000) "Civil Aviation and the Environment," in Special Issue on Sustainable Mobility, *UNEP Industry and Environment,* October–December, pp. 31–33.

de Neufville, R. (1991) "Airport Construction—The Japanese Way," *Civil Engineering,* Geotechnical Issue, December, pp. 71–74.

EPA, U.S. Environmental Protection Agency (1999) "Evaluation of Air Pollutant Emissions from Subsonic Commercial Jet Aircraft," Report EPA-420-R-99-013, EPA, Washington, DC.

EPA, U.S. Environmental Protection Agency (2000) "Preliminary Data Summary—Airport Deicing Operations (Revised)," Office of Water, Report EPA-821-R-00-016, EPA, Washington, DC.

EPA, U.S. Environmental Protection Agency (2001) "National Pollution Discharge Elimination System Permit Program," Office of Water, http://cfpub.epa.gov/npdes/.

FAA, Federal Aviation Administration (1982) "INM Integrated Noise Model, Version 3, User's Guide," Report FAA-EE-81-17, Office of Environment and Energy, Washington, DC.

FAA, Federal Aviation Administration (1983) *Noise Control and Compatibility Planning for Airports,* Advisory Circular AC 150/5020-1, FAA, Washington, DC, http://www.faa.gov/arp/150acs.htm.

FAA, Federal Aviation Administration (FAA) (1990) *Day Night Average Sound Level—The Descriptor of Choice for Airport Noise Assessment,* Office of Environment and Energy, Washington, DC.

FAA, Federal Aviation Administration (1993) *Noise Abatement Departure Profiles,* Advisory Circular AC 91-53A, FAA, Washington, DC.

FAA, Federal Aviation Administration (1995) *Noise Standards: Aircraft Type and Airworthiness Certification,* Federal Aviation Regulations, Part 36, FAA, Washington, DC.

FAA, Federal Aviation Administration (1997) "Air Quality Procedures for Civilian Airports and Air Force Bases," Report AEE-AEE-97-03, FAA, Washington, DC.

FAA, Federal Aviation Administration (1998) "Emissions and Dispersion Modeling System Policy for Airport Air Quality Analysis; Interim Guidance to FAA Orders 1050.1D and 5050.4A," http://www.aee.faa.gov/aee-100/aee-120/edms/policy.htm.

FAA, Federal Aviation Administration (2001) "Integrated Noise Model Version 6.0c," Noise Division, AEE-100, http://www.aee.faa.gov/Noise/inm/INM6.0c.htm.

Gifford, F. (1976) "Turbulent Diffusion Typing Schemes: A Review," *Nuclear Safety,* 17, pp. 68–86.

Gifford, F. (1986) "Guideline on Air Quality Models," U.S. Environmental Protection Report EPA-450/2-78-027R, EPA, Washington, DC.

Horonjeff, R., and McKelvey, F. (1994) *Planning and Design of Airports,* 4th ed., McGraw-Hill, New York.

Hye, H. (2000) "Free Air Conditioning Keeps Neighbors Cool about Airport Noise," *Air Conditioning, Heating and Refrigeration News,* 26, June.

ICAO, International Civil Aviation Organization (1988) *Environmental Protection, Annex 16 to the Convention on International Civil Aviation, Volume 1: Aircraft Noise,* ICAO, Montreal, Canada.

Los Angeles International Airport (2001) "Master LAX 2015 Plan," http://www.lax2015.org/frame4.html.

MIT, Massachusetts Institute of Technology (2001) *Mobility 2000,* MIT, Cambridge, MA.

Nelkin, D. (1974) *Jetport: The Boston Controversy,* Transaction Books, New Brunswick, NJ.

Penner, J., Lister, E., Griggs, D., Dokken, D., and McFarland, M. (1999) *Aviation and the Global Atmosphere,* Cambridge University Press, Cambridge, UK.

Ruijgrok, G. J. J. (1993) *Elements of Aviation Acoustics,* Delft University Press, Delft, The Netherlands.

Sylvan, S. (2000) *Best Environmental Practices in Europe and North America,* County Administration of Vastra Gotaland, Sweden.

U.S. Congress (2000) *Aviation Investment and Reform Act,* U.S. Government Printing Office, Washington, DC.

WHO, World Health Organization (1999) "Guidelines for Community Noise," http://www.who.int/peh/noise/ComnoiseExec.htm.

7

Organization and financing

The institutional, organizational, and financial characteristics of airports around the world are changing rapidly, stimulated in large part by airport privatization and airline deregulation. Many different "models" of airport ownership and management exist. The traditional model that places airport management in the hands of a central bureaucracy in the national government does not usually meet the needs of large airports in a fast-changing industry. Several models are centered on the concept of the airport authority, a corporate entity owned by government or private investors or a combination of the two, that acts as an autonomous and flexible airport operator.

Airports must contend with legal, financial, planning, public communication, administration, human resource, environmental, engineering/technical, commercial, and operational issues. Organizational structures are designed to carry out this common set of functions and activities. Structures become increasingly pyramidal as airports grow. They can be complex when the airport operator is responsible for a multi-airport system or when the operator engages in extensive activities outside the core airport business.

Concerns regarding the potential abuse of the quasi-monopolistic position that airports enjoy in serving origin and destination traffic have led to widespread regulation of airport user charges. The focus has been on regulating aeronautical charges through target rates of return, price caps, and restrictions on the annual rate of increase of unit charges. The treatment of nonaeronautical revenues plays a major role in determining the size of aeronautical user charges.

Airport capital investments can be financed in many different ways, ranging from grants from national governments to revenue bonds issued and serviced by airport operators. The available alternatives depend on the size of the airport and on national laws, economic conditions, and practices. The ability of airport operators to obtain

215

favorable terms for the financing of very large projects depends in large part on assessments performed by credit-rating agencies. These agencies have developed a set of common criteria that they use for this purpose.

7-1 Introduction

This chapter presents an overview of the institutional, organizational, and financial characteristics of airports around the world. Familiarity with these characteristics is crucial to understanding many aspects of the diversity in behavior of airports as organizational and economic entities. This material also complements or provides useful background for other chapters in this book, notably the ones on privatization (Chap. 4), user charges (Chap. 8), and demand management (Chap. 12). It is worth noting at the outset that changes in the organizational and economic aspects of airports have been quite dramatic since the late 1980s. This is an area where the "landscape" is evolving rapidly (Deutsche Bank, 1999).

The contents of the chapter are as follows. Section 7-2 presents a brief survey of the arrangements that exist in various parts of the world with respect to the ownership and management of airports. These arrangements increasingly emphasize autonomous management, typically in the form of an airport authority or similar independent corporate entity, as well as participation by private investors in the ownership of airports. These developments, along with the growing complexity of airport operations, underlie the parallel trend toward organizational structures that deviate significantly from traditional forms, as described in Sec. 7-3. As airports move toward operating in many ways like private-sector entities, their revenues and balance sheets are coming under increasingly close scrutiny. Airports are in most cases natural monopolies when it comes to originating or terminating passengers. There is justifiable concern that, in the absence of appropriate economic regulation, pricing practices will take advantage of this monopolistic position. Airport economic and financial practices are therefore being subjected to important regulatory and/or legal constraints in a growing number of countries. Section 7-4 describes briefly some types of constraints, along with relevant examples. Finally, Sec. 7-5 identifies the alternative ways in which airport capital investments can be financed and notes, once again, the considerable diversity of international practices in this respect.

7-2 Ownership and management of airports

Several alternative "models" of airport ownership and management are currently in use around the world, and some are working better than others (Doganis, 1992). Before reviewing these models, it is necessary to establish a clear terminology.

As noted in Chap. 4, the term *privatization* does not reflect accurately the changes that are taking place internationally in airport ownership and management. Most privatizations of major commercial airports have not involved the actual sale of the airport property. The typical airport privatization involves a long-term lease of 20 years or more. What is really transferred through those leases are (1) the rights to residual income, that is, to any profits that may be generated, and (2) management control, that is, the right to operate and develop the airport (see Sec. 4-3 for a detailed discussion). The term *airport operator* will be used to refer to the entity that acquires these rights.

Who, then, should be considered the owner of the airport, or group of airports, assigned to an airport operator? Undoubtedly, the national, regional, or local government granting the license remains, *in principle,* the true owner, even while the license is in force. (In fact, this true owner may retain some critical prerogatives and regulatory controls, as described in Chap. 4.) For most practical purposes, however, it is the licensed airport operator that acts as the property's owner and everyday decision maker during that period. For this reason, the *shareholders of the airport operator* will be treated here as the *owners* of the airport. Note that these shareholders can, in general, be government or private interests or both. Airport ownership, in this sense, can include any combination of

- National government
- Local and/or state/regional governments
- Corporate entities
- Private investors

An important aspect of ownership in the case of government/private partnerships is whether the majority stake belongs to governmental entities or to private interests. Another aspect with major implications for the governance of the airport is the type of private ownership involved. Cases in which private shareholding is limited to a small number of partners (corporate or otherwise) are quite distinct from those where the right to ownership is extended to the general public through a tender of publicly traded shares ("free float").

The airport operator can likewise be anyone of the following types of entities:

- A branch of the national government
- A branch of a local or state/regional government
- An airport authority or other similar corporate entity
- An airport management contractor

As indicated by the last option on this list, the airport operator can be a government-owned or a privately owned company, with expertise in airport management, which provides its services for an agreed fee that may include a percentage of revenues and a number of financial incentives. This option has become quite popular recently, as the management and operation of airports have turned into increasingly specialized and sophisticated activities. Note, as well, that the airport operator has the option to subcontract any set of responsibilities for the entire airport or for parts of the airport to other organizations. As indicated in Chap. 4, outsourcing of many operational and developmental responsibilities is the rule rather than the exception at U.S. airports.

Of the large number of "airport owner"/"airport operator" combinations that can be identified from the two lists, all existing arrangements seem to be consistent with one of eight models, A through H, which are described briefly below.

A. *Owned by a combination of national, regional, and/or local governments and operated by a branch of the national government.* Model A has been by far the most common around the world, especially at secondary airports outside the United States and in developing countries. Typically, a branch of the national civil aviation authority or of the department or ministry of transportation is responsible for the management and operation of all or most airports in a country and appoints civil servants to carry out these functions at each airport. Athens, Bombay, Cairo, Dubai, Helsinki, Moscow, Jeddah, Oslo, Singapore, and Stockholm are all examples of cities whose major airports were managed and operated in this way at least up to 2000. A few countries even have a ministry of airports—or other cabinet-level agency—dedicated to the administration of airports.

B. *Owned by a combination of national, regional, and/or local government and managed and operated by a branch of a local or regional government.*   Model B is common in countries with a strong tradition of decentralized administration and regional autonomy. Many of the most important airports in the United States, including some the busiest airports in the world, are operated by departments in city governments. Chicago/O'Hare, Denver/International, and Miami/International belong to that category. Los Angeles/International, along with the nearby airports of Ontario, Van Nuys, and Palmdale, is operated by the Los Angeles Department of Airports (formally, Los Angeles World Airports, LAWA). The State of Hawaii Department of Transportation operates 16 airports, including the Honolulu International Airport. As noted in Chap. 4, these U.S. airports are also among the most "privatized" in the world, in that they tend to contract out most of the activities and functions of an airport operator.

C. *Owned by a combination of national, regional, and/or local government and, possibly, of private interests and operated under a management contract by a publicly or privately owned company.* The objective here is to obtain expert airport management, responsive to local conditions and cognizant of best practices elsewhere. Several local and/or regional governments, in particular, have contracted with well-known airport operators to implement arrangements consistent with model C. Examples include the contracts that BAA signed with the cities of Indianapolis and of Harrisburg in the United States to manage their airports, and the management (and partial ownership) agreement that Aéroports de Paris has in Beijing.

D. *Owned by a combination of national, regional, and/or local government and managed and operated as an autonomous airport authority.*   Model D applies to some of the busiest airports in the world. It is also the model that has provided extensive experience with the operation of autonomous airport authorities, proving their advantages and effectiveness (Example 7.1). The best-known example is the Port Authority of New York and New Jersey (PANYNJ), which operates, among other facilities, the three main airports in the New York metropolitan area and, as its name indicates, is jointly owned by the States of New York and of New Jersey. Massport, the airport authority that operates Boston's Logan International Airport, is owned by the Commonwealth of Massachusetts, while San Francisco International Airport is operated by an authority jointly owned by the City and County of San Francisco.

EXAMPLE 7.1

Because of the importance of model D and its successful application at several locations in the United States, this example outlines a typical set of the basic terms and provisions for such autonomous airport authorities, fully owned by a state or local government. The provisions are based on those in effect for Massport.

- The authority acts as the owner, manager, and operator of the airport (or group of airports) and all related property, facilities, and services. The authority may also have parallel responsibility for other transportation facilities and services. In the case of Massport, in addition to Boston/Logan and Hanscom Field, a reliever airport near Boston, the authority owns, manages, and operates a major bridge, the seaport of Boston, and several other secondary facilities.

- A board of directors appointed by the shareholders, i.e., the state or local government, acts as the authority's governing body. In the case of Massport, the board consists of seven members, appointed to revolving seven-year terms by the governor of Massachusetts. The governor normally makes one board appointment every year.* Board members receive only nominal compensation.

- The authority operates on a not-for-profit basis and is exempt from taxes. If conditions permit, it makes "voluntary contributions" to its shareholders, in lieu of taxes. In the case of Massport, the airport has been generating an operating surplus for years. Part of this surplus is allocated to providing financial assistance to a number of Boston area cities and townships.

- The authority may acquire land and property at fair value, as necessary for the development and operation of its facilities, and it may undertake construction projects and other enhancements to its property and facilities.

- The authority may issue tax-exempt bonds against future airport revenues[†] to finance necessary capital investments.

*Governors are elected to four-year terms in Massachusetts. Thus, a governor elected to one four-year term will normally obtain a board with a majority of members appointed by him or her only in the last year of his or her tenure. The intent of this appointment system is to insulate, as much as possible, the board of directors of Massport from political changes and pressures and thus to ensure some level of continuity and stability in its policies.
[†]For a brief discussion of revenue bonds, see Sec. 7-5.

- The authority must exercise every effort to achieve economic self-sufficiency, including funding capital investments from its own revenues; the authority may adjust user charges as necessary to reach this objective.

- If the authority is dissolved at some future time, all of its property will be returned to the shareholders, that is, to the Commonwealth of Massachusetts.

Provisions in the same spirit as the ones described in the example also govern the operation of autonomous airport authorities in other regions of the world. There are, however, important differences from country to country with respect to the corporate and legal characteristics of the authority, its responsibilities, and its empowerment.* The makeup of the ownership of airport authorities is also more varied than in the United States. Table 7-1 shows several European examples as of 1999.† Some airport authorities outside the United States are far less local in character. For example, the airport authorities responsible for Dublin and for Madrid also manage and operate other major airports in Ireland and in Spain, respectively: Aer Rianta operates the airports of Cork and of Shannon in addition to Dublin, as well as airport shopping centers and other commercial activities worldwide. AENA's responsibility extends to 40 airports in Spain and to air traffic management.

E. *Owned in the majority by a combination of national, regional, and/or local government with private minority shareholders and with no publicly traded shares; managed and operated as an autonomous airport authority.*   Due to the presence of private interests in the ownership of airports, model E (as well as model F, see below) require particularly thoughtful contractual and regulatory arrangements in order to balance and protect the public interest and the interests of the private investors. The Athens International Airport (AIA) Authority for the new airport at Athens/Venizelos provides an example of model E. The Greek government owns 55 percent of the shares and a consortium of German companies, led by Hochtief, best known as a large construction company, owns the other 45 percent. The creation of

---

*For instance, the right to issue revenue bonds is rarely available to airport authorities outside the United States.

†The ownership status of some of these airports had already changed (e.g., Frankfurt) or was about to change by 2001.

Table 7-1. Makeup of ownership in 1999 of some European airports that were government owned

Airport (operator)	Ownership
Amsterdam (AAS)	75.8% national government; 21.8% City of Amsterdam; 2.4% City of Rotterdam
Berlin	26% national government; 37% State of Berlin; 37% State of Brandenburg
Dublin (Aer Rianta)	100% national government
Frankfurt (Fraport)	26% national government; 45% State of Hesse; 29% City of Frankfurt
Madrid (AENA)	100% national government
Milan (SEA)	85% City of Milan; 14% Region of Lombardy; 1% Chambers of Commerce
Munich	26% national government; 51% State of Bavaria; 23% City of Munich
Paris (ADP)	100% national government

Source: TRL, 1999.

AIA required the Greek Parliament to enact a lengthy special law laying out the terms and conditions of the AIA contract. As a second example, 80 percent of the shares of the Airports Company of South Africa (ACSA) are owned by the national government, while 20 percent have been awarded to Rome's Airport Authority (Aeroporti di Roma) following an international tender.

F. *As in E, but with some portion of the shares traded publicly.* Copenhagen and Vienna Airports are among the examples for model F. Table 7-2 shows the ownership distribution as of mid-2001 for a number of airport operators that had issued publicly traded shares. The percentage of total ownership represented by these shares is indicated in the rightmost column.

G. *Owned fully or in the majority by private investors, with no publicly traded shares, and operated as an autonomous airport authority.* Brussels Airport, in 1998, provided an example of model H. The Belgian government owned 47.5 percent and a group of private and institutional investors owned 52.5 percent. This group previously owned Brussels Airport Terminal, a company responsible for landside operations at Brussels. As mentioned in Chap. 4, under a 30-year contract signed in 1998, Aeropuertos Argentina 2000, whose shares were then divided equally among the Airport Authority of Milan (Italy), Ogden Aviation (USA), and a large Argentine conglomerate, became the owner and operator of 33 airports there.

H. *Owned fully or in the majority by private interests, with some or all shares traded publicly, and operated as an autonomous airport authority.* The British Airports Authority (BAA) is by far the best-known example of model H. The BAA was fully privatized in 1987, with 100 percent of its shares traded publicly.* Other airport operators in this category are TBI, the Mexican Southeast Airports Group, Copenhagen, and Auckland (Table 7-2). In general, interest in offering airport shares to private companies and to the public, at large, has been growing rapidly outside the United States. This is one of the major global trends identified in Chap. 1 and discussed in detail in Chap. 4.

In conclusion, model A can be labeled as the "traditional" model of airport ownership and management. For the most part, it will not work well at the busiest airports for the many reasons discussed in Chap. 4 (large centralized bureaucracies, limited responsiveness to local issues, limited control over finances, few incentives to increase revenues or reduce costs). Model B at first glance seems to have similar disadvantages. However, the model has been mostly successful in the United States because of the way it is practiced. As indicated in Chap. 4, these model B airports, although in theory operated by municipal or regional governments, are in practice among the most "privatized" in the world, due to their extensive reliance on outsourcing for the provision of facilities and services, as well as for planning and development functions. Another critical reason for these success stories is that airport budgets and finances are treated (as legally required) as independent of those of the local and/or regional governments that operate them. Model C can be very appropriate for airports that lack the required expertise in modern airport management. Thus it applies best to either secondary airports in developed countries or to major airports in developing nations. The autonomous airport authority is the common element in models D through H. As airports become busier, more complex, and more important to local and national economies, these organizations have become increasingly commonplace and have been established to

---

*The British government has retained one "golden share" in BAA, which gives it decision-making power in fundamental policy matters.

**Table 7-2. Ownership of some airport
authorities with publicly traded stock in 2001**

Airport	Current ownership	Free float
BAA	100% free float (London Stock Exchange) Source: Company 2000/2001 Annual Report	100%
TBI	100% free float Source: Company 2000–2001 Annual Report	100%
Mexican Southeast Airport Group (ASUR)	15% strategic partner led by Copenhagen Airport 85% free float (NYSE and Mexican Stock Exchange)	85%
Copenhagen (CPH)	33.8% Danish government 66.2% free float (Copenhagen Stock Exchange) Source: Company 2000 Annual Report, April 5, 2001	66.2%
Auckland (AIA)	25.8% Auckland City Council 9.6% Manukau City Council 7.1% Singapore Changi Airport 57.5% free float (New Zealand and Australian Exchanges) Source: Company 2001 Annual Report, June 30, 2001	57.5%
Vienna (VIE)	20% Province of Lower Austria 20% City of Vienna 10% Employee Foundation 50% free float (Austrian Stock Exchange) Source: Company 2000 Annual Report, April 23, 2001	50%
Beijing (BCIA)	65% Beijing Capital Airport Group (government owned) 10% Aéroports de Paris (ADP Management) 25% free float (Hong Kong Stock Exchange) Source: Credit Suisse First Boston, Equity Research, May 8, 2001	35%
Frankfurt/Main (Fraport)	32.1% State of Hesse 20.5% City of Frankfurt 18.4% Federal Republic of Germany	29%

Table 7-2. *(continued)*

Airport	Current ownership	Free float
Frankfurt/Main (Fraport)	29.0% free float (Frankfurt Stock Exchange, June 2001) Source: Company Official Web Page	
Florence Airport	19.3% Florence Chamber of commerce 17.2% City of Florence, 10.5% City of Prato 14.3% other nonprivate investors 28.8% free float (Milan Stock Exchange) Source: Company Report, six months, June 30, 2001	28.8%
Malaysia Airports Holdings Berhad	72% nonprivate investors 28% free float Source: Equity Research. Credit Suisse First Boston, May 3, 2001	28%
Unique Zürich Airport (UZA)	55.7% Canton of Zürich 6.3% City of Zürich 10% other nonprivate investors 28% free float (Zürich Stock Exchange, November 2000) Source: Company 2000 Annual Report	28%
Xiamen Airport	75% nonprivate investors 25% free float Source: Airport Review, Credit Suisse First Boston, September 21, 2000	25%

operate even some airports in developing countries. One of the advantages of the airport authority is that it can accommodate any form of ownership: it can be government owned or privately owned or of mixed ownership. The autonomous airport authority is an institutional device that has proved largely successful in partially insulating airports from political interference and in promoting effective management.

7-3   Organizational structures

Airport organizational structures necessarily reflect the diversity in ownership, management arrangements, and size that characterizes airports worldwide. These can only be reviewed in general terms with the objective of identifying some discernible patterns. A useful approach is to look at how organizational structures tend to evolve as the size of an airport increases.

Operators of modern commercial airports of any reasonable size must contend, to a greater or lesser extent, with a full spectrum of legal, financial, planning, public communication, administration, human resource, environmental, engineering/technical, commercial, and operational issues. The organizational structure of even secondary airports must somehow reflect all of these areas of activity. This gives rise to the most generic of organizational charts of an airport operator shown in Fig. 7-1. The policy source in Fig. 7-1 depends on which of the models A through H identified in Sec. 7-2 applies. In the case where the airport operator is an autonomous airport authority—a common trait of models D–H—the policy source is the authority's board of directors. In the case of models A and B, it is the leadership of the government agency or department to which the airport's management reports. Finally, in the case of model C, policy initiatives may come both from the governmental entities that own the airport and the board of the airport management contractor—with the dominant source of initiatives depending on the level of autonomy granted to the management contractor.

In larger airport organizations a sharp distinction often exists between *staff* (or *supporting*) *units* and *line* (or *operating*) *units*. Staff units pro-

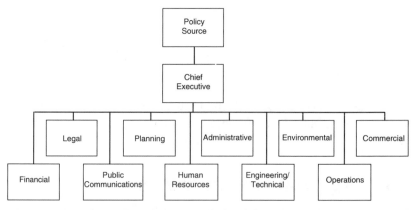

Fig. 7-1. *A generic organizational chart for an airport.*

vide support to the chief executive in managing the airport. Line units, by contrast, are the ones that carry out the day-to-day tasks associated with the operation and serviceability of the airport's facilities. Staff units may often be quite small but, due to their role in decision making, they are often very influential in determining an airport's economic and operational performance and in mapping its future course. Line units, by contrast, may employ hundreds—and, in some instances, literally several thousands—of workers, especially when the airport operator is heavily involved in ground handling. The six units* on the left side of Fig. 7-1 clearly have a primarily staff role, whereas "commercial" and "operations" are primarily line units. "Operations" typically encompasses a wide range of functions that includes public safety (security, fire fighting, emergency medical) as well as passenger handling and ramp handling (Chap. 8). The two other units in Fig. 7-1, "environmental" and "engineering/technical," have both a staff and a line function. They are sometimes also organized along the lines of a department of "infrastructure and environmental affairs," a staff unit responsible for technical support on facility development and environmental issues, and an "engineering and maintenance" department, a line unit responsible for carrying out facility improvements and maintenance.

Depending on the specific circumstances, the relative importance of the organizational units varies, as does the depth and complexity of the organization chart. Secondary airports operated by autonomous or semiautonomous entities are usually structured along the "flat," single-tier lines of the organization chart of Fig. 7-1. In many cases, some of the organizational units shown (e.g., legal and public communications) may be staffed by only one or two persons.

At larger airports the flat organization chart of Fig. 7-1 may take the more pyramidal shape of the chart of Fig. 7-2, which clearly distinguishes between staff and line responsibilities. The position of deputy director is a typical feature of more pyramidal airport organizations. It is designed to relieve the executive director from the task of overseeing daily airport operations and allow him/her to devote more time to strategic management. As noted above, "handling" is typically a part of the operations department, but may appear as a separate

*In practice, the names of these units will change from location to location. Here we have used generic labels to describe their function.

line unit ("Handling Department") in the organization chart of some of the airports that engage heavily in this activity.* A trend has also emerged in recent years toward giving more organizational visibility to public safety activities, due to their growing importance.

In the case of airports operated by national, regional, or local governments (models A and B) the staff functions are typically performed by departments or "offices" located at the administrative headquarters of the governmental agency involved, whereas the line units are positioned at the airports themselves. The staff units may support several airports simultaneously, possibly an entire national system.

Organizational structures analogous to but more elaborate than that of Fig. 7-2 can be found at many of the large airports that are operated by autonomous airport authorities. Wiley (1986) presents a chart (shown in simplified form in Fig. 7-3) of an organizational structure that might be appropriate for an authority that operates several major airports.[†] It can be viewed as an extension of the one in Fig. 7-2. It demonstrates how line units must be replicated at each individual airport operated by the authority. The authority's headquarters provide shared staff support to all the airports.

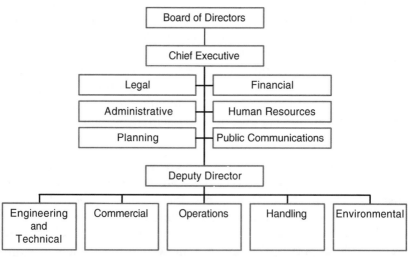

Fig. 7-2. *A two-tier organizational chart showing staff (support) units and line (operating) units.*

---

*Handling refers to the services provided to aircraft on the apron ("ramp"), such as loading and cleaning, and to passengers or cargo in airport buildings. See Chap. 8.
†Wiley's chart is patterned after the organizational structure of the Port Authority of New York and New Jersey at the time.

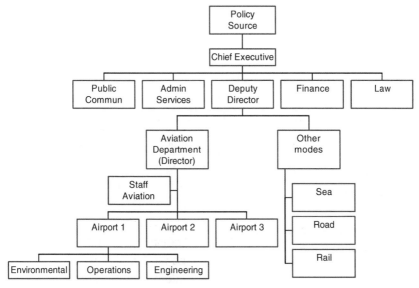

Fig. 7-3. *Organizational structure for a multi-airport Authority*
(*Source:* Wiley, 1986.)

EXAMPLE 7.2

The overall organizational structure of the Massachusetts Port Authority (Massport) is shown in Fig. 7-4. The staff (or "supporting") units and the line (or "operating") units are listed in columns to make the chart more compact.

Fig. 7-5 shows the structure of the Aviation Department at Massport. Note that the Director of Aviation has a staff support unit dedicated to aviation headed by the Director of Aviation Planning and Development. Public safety and security has additional visibility and resources through a distinct line unit that reports directly to the Director of Aviation.

Fig. 7-6 identifies the responsibilities of the Aviation Operations Department and shows how they are distributed among the functions of airside maintenance, airside operations and airside construction.

At the highest level of airport size and organizational complexity, a number of mostly European and Pacific Rim airports are increasingly adopting, explicitly or implicitly, the profile of conglomerates of transportation, commercial, retail, and technical services. This

Fig. 7-4. *Organizational structure of Massport.*

Fig. 7-5. *Aviation Department of Massport.*

shift away from simply serving as operators of transportation hubs implies a marked change in organizational structure as well. The traditional line and staff units of airport operators lose some of their centrality or, at best, are complemented on an equal footing by new units.

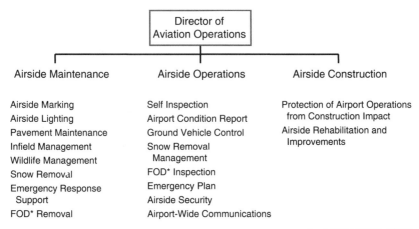

Fig. 7-6. *Aviation Operations Department, Massport.*

EXAMPLE 7.3

Amsterdam/Schiphol provides an interesting example in this respect. The Amsterdam Airport Schiphol (AAS) Authority has evolved into the "Schiphol Group," whose organizational structure is shown in Fig. 7-7. Its stated objective is to go "beyond the development of an efficient transport hub" and work toward realizing the concept of the "Airport City."

> *"The Airport City offers its passengers and visitors, but also the Schiphol based companies (airlines, distribution companies, and logistic and business service providers) 24-hours-a-day services in the field of shops, hotels and restaurants, multimedia services, corporate business and conference facilities, and recreation and relaxation...Schiphol Group is particularly strong in the integration of all these activities into a coherent and durable entity that is closely knit with its geographical surroundings. This knowledge is also internationally in demand. Schiphol Group is therefore increasingly marketing the Airport City concept internationally. (Schiphol Group, 1999)."*

Note that of the nine units shown at the bottom of Fig. 7-7, only the middle three focus directly on the operation and management

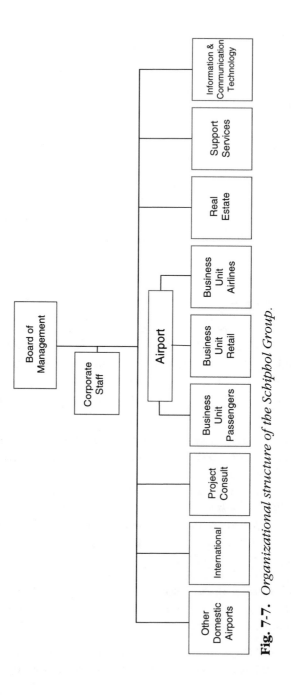

Fig. 7-7. *Organizational structure of the Schipbol Group.*

of Schiphol Airport. Beginning from the left, the "other domestic airports" unit is responsible for the small airports of Eindhoven, Lelystad, and Rotterdam. Schiphol International, BV, is a corporation that oversees large development and management contracts at New York/Kennedy and at Brisbane, Australia. Schiphol Project Consult, BV, is another corporation specializing in project management and consulting at Schiphol itself, elsewhere in the Netherlands, and abroad. Schiphol Real Estate, BV, on the right, is one of the largest real estate agencies in the Netherlands and is active in many aspects of real estate management and development at Schiphol and at other airports internationally.

Of the three units responsible for airport operations at Schiphol, the Business Unit Retail aims at the development and implementation of commercial activities, especially shops, restaurants, and hotels, for passengers and visitors. The organization charts for the other two units, Business Unit Airlines and Business Unit Passengers, appear in Figs. 7-8 and 7-9. These are responsible for all services and facilities offered to airlines and to passengers and visitors. Business Unit Airlines is also involved in marketing and account management vis-à-vis the airlines and third-party ground handling companies (Chap. 8). Note that each of these two units is a large and complex organization in its own right.

Finally, the two rightmost units in Fig. 7-7 acquire and provide products and services in their respective areas for the entire Schiphol Group. Among the responsibilities of Schiphol Support Services, for example, are vehicle acquisition and management, utility services, airport security, access control and identification services, etc., that may be required by facilities managed and operated by the Schiphol Group.

7-4 Regulatory constraints on airport user charges

Concerns regarding the potential abuse of the quasi-monopolistic position that airports enjoy in serving origin/destination traffic have led to widespread regulation of airport user charges. Many airport operators, some fully government owned and others with partial or full private ownership, are subject to regulatory restrictions concerning the pricing of their services and facilities.

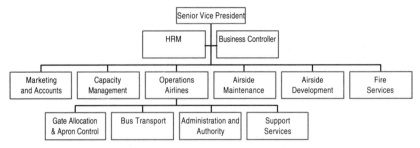

Fig. 7-8. *Business unit airlines, Schiphol.*

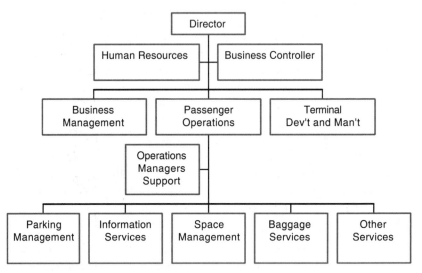

Fig. 7-9. *Business unit passengers, Schiphol.*

A global consensus exists about the need to restrict prices charged for aeronautical facilities and services, that is, for the airside and landside facilities and services essential to the processing of aircraft and their passengers and cargo. On the other hand, nonaeronautical revenues, which are derived from the numerous ancillary commercial services available at an airport, are largely unregulated. (Chapter 8 covers user charges in detail.) The ICAO Council has stated that airport operators may recover the *full cost, but no more* of aeronautical facilities and services (ICAO, 1992). Full cost includes the cost of operations, maintenance, management, and administration, as well as interest on capital investment, depreciation of assets, and, when conditions permit, a fair return on investment.

This still leaves much room for local and national policy makers. Two critical questions that clearly play a major role in determining the airport costs that aeronautical users face are the following:

- What regulatory constraints, if any, are placed on the ability of airport operators to set prices?

- How should nonaeronautical revenues be treated? More specifically, should nonaeronautical revenues affect in any way the prices charged for aeronautical facilities and services?

Price caps

Most countries have responded to the first question by placing explicit or implicit restrictions on the "fair return on investment" that airport operators can derive from aeronautical facilities and services. At U.S. airports, for example, prices are required to be strictly cost-related. This means that they are set at exactly the level required to pay for the variable and capital costs of operating, maintaining, and upgrading the facilities. Other countries specify a "fair target of return" for which airport operators can aim for. For the main private airports in the United Kingdom this target was 7.5 percent per year in 2001; for the Mexican Southeast Airports Group it was 12.5 percent. Note, however, that this is a target of return from *all* the activities of an airport, including nonaeronautical facilities and services. Thus, depending on how the second of the above questions is answered, the rate of return from aeronautical facilities alone can be considerably lower (or, under certain circumstances, higher) than the target.

Some governments or regulatory bodies go further, especially when it comes to regulating "privatized" airports: they place a cap on the maximum amount by which unit charges for aeronautical facilities and services can increase from year to year. Table 7-3 summarizes the restrictions placed on a number of airport operators in this respect. Example 7-4 explains the meaning of the table.

Note that the intent of the regulatory price caps in Table 7-3 is to promote efficiency in airport operations. When x is 0 percent or greater, any improvements in airport financial performance must come from traffic growth and higher productivity, not price increases.*

---

*The Airports Corporation of South Africa is similarly restricted by a "CPI − x" formula (CPI = consumer price index) with x set to 0 percent for the first two years of operation, followed by three years of (CPI = 2%).

Table 7-3. Regulatory price cap formulas for privatized airport operators (RPI = retail price index; CPI = consumer price index)

Airport	Price formula	Review of regulation	Comments
Argentina	RPI−(20% of annual percent traffic growth)	3 years	
Australia	RPI − x	5 years	Aim is to remove regulation eventually
BAA	RPI − x	5 years	"Single till"
Copenhagen	CPI	4 years	Review due 2002
Mexico	RPI − x	5 years	"Dual till"
Vienna	RPI − x (x depends on traffic growth)	5 years	

Source: Enriquez, 2002.

Table 7-4. Productivity changes at the three european airports with publicly traded shares, 1991 versus 2000

	BAA		Copenhagen		Vienna	
	1991	2000	1991	2000	1991	2000
Passengers (m)	72	125	12	18.4	6	11.9
Employees	10,900	10,017	1,300	1,399	1,800	2,644
Passengers/employees	6,600	12,400	9,200	13,200	3,300	4,500
Total annual costs* ($m)	1,100	1,240	89	146	171	224
Staff costs* ($m)	380	441	47	59	91	124
Total cost/passenger ($)	15.3	9.9	7.4	7.9	28.5	18.9
Staff cost/passenger ($)	5.3	3.5	3.9	3.2	15.2	10.4

*All cost figures are in *current* prices.

Table 7-4 provides an indication of productivity improvements at the only three European airport operators with some publicly traded shares that operated throughout the period 1991–2000. BAA, Copenhagen, and Vienna achieved roughly 88, 42, and 35 percent gains, respectively, as measured by annual passengers processed per employee. It is also remarkable that staff cost per passenger declined, even in *current prices,* at all three airports. However, it is not clear—at

EXAMPLE 7.4

The BAA's pricing policies are subject to regulatory review by the UK Civil Aviation Authority (CAA). At the time of BAA's full privatization in 1987, the Monopolies and Mergers Commission specified that the authority's three London airports (Heathrow, Gatwick, and Stansted) would be regulated under the "RPI − x" formula,* which specifies that the annual rate at which an airport can increase its *aeronautical* charges cannot exceed (RPI − x percent). It has been the Civil Aviation Authority's responsibility to review the value of x every 5 years. In 2000, Heathrow and Gatwick were restricted to (RPI − 3 percent) and Stansted to (RPI + 1 percent), i.e., in the latter case $x = −1$ percent. This meant a reduction in aeronautical charges at Heathrow and Gatwick, not only in real-price terms (as the increase was 3 percent below RPI) but in current-price terms, as well, as the RPI itself was equal to only about 2.9 percent in 1999.

In the case of Vienna, a somewhat different regulatory regime is in effect. It ties the allowable annual rate of increase in user charges to the rate of increase in traffic, as shown in Fig. 7-10 for the case in which RPI is assumed to be 3 percent in a particular year. Note that the allowable annual rate of increase in charges is equal to RPI if traffic growth is negative or zero and then declines linearly [with a slope equal to −(3/7) percent] to 0 percent when traffic growth is equal to 7 percent. It remains at 0 percent for growth rates between 7 and 11 percent and then must be negative for growth rates greater than 11 percent, declining again at a −(3/7) percent slope.

least from these very limited data—how much of a role the regulatory restrictions really played. In general, many government-owned or (fully or partially) privatized airports have exercised in recent years considerable self-restraint regarding increases in aeronautical charges.

"Single Till" versus "Dual Till"

The question of how nonaeronautical revenues should be treated has generated a great deal of controversy. It is at the center of the debate outside the United States concerning the relative merits of the

---

*The formula is actually a little more complex, but the "RPI − x" part captures its essence; for details, see Enriquez (2002).

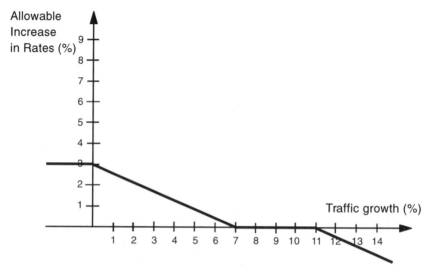

Fig. 7-10. *Vienna—Annual rate of allowable change in user charges as a function of annual traffic growth when RPI is equal to 3 percent.*

single-till and the *dual- till* approaches to setting aeronautical airport user charges. The difference between the amounts airlines pay under the two approaches can be very large. The parallel debate in the United States about the *residual* versus the *compensatory* approach follows along similar lines but has an additional dimension (see below).

As the name suggests ("till" = "a drawer for storing money in a bank"), under the single-till approach any restrictions on the rates of return that an airport operator may earn apply to *total* revenues, that is, to the sum of aeronautical and nonaeronautical revenues. In computing the rate of return on assets, no distinction is made as to the sources of revenue: proceeds from duty-free sales and landing fees are simply added together. This is shown conceptually in Fig. 7-11. The total revenue expected is placed in one "till." The maximum allowable aeronautical charges per unit of traffic can then be computed. This computation takes into consideration the prescribed target rate of return, the asset base, available traffic forecasts, and any regulatory formulae of the "RPI − *x*" type, as shown in Fig. 7-11. In the great majority of cases, nonaeronautical services and facilities, which are usually highly profitable, will help reduce the charges that airport users pay for aeronautical services and facilities (Example 7.5). For this reason, airlines and other aeronautical users strongly support the single-till approach.

The alternative to single till is the dual-till approach shown in Fig. 7-12. Nonaeronautical revenues are now treated separately. The target rate of return for aeronautical services has to be achieved by utilizing aeronautical revenues only. In most cases, this will result in higher aeronautical user charges than under single till. Note that the profits of the airport operator will be higher under dual till. Full cost recovery plus a fair return on investment is achieved on the aeronautical side, while all the profits from the unregulated nonaeronautical facilities and services accrue to the airport operator as well.

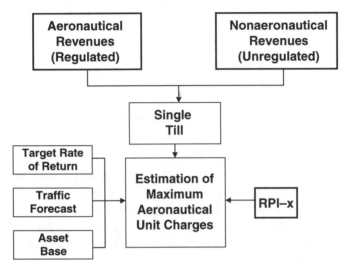

Fig. 7-11. *Schematic explanation of the single-till approach.*

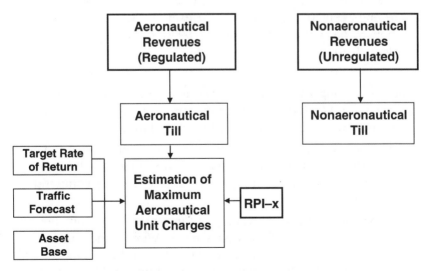

Fig. 7-12. *Schematic explanation of the dual-till approach.*

EXAMPLE 7.5

A stark example of the effects of single till is provided by the history of landing fees at BAA's London/Heathrow Airport. Until the British Airports Authority was privatized in 1987, the landing fees at London/Heathrow were among the highest in the world—and easily the highest in Europe. By 2000 they had declined to rank among the lowest charged by any major airport, including those in developing countries. For example, in 2000 the landing fee paid by a Boeing 747 operating during the peak period of the day in the busy summer season was about $660. This is but a small fraction of the landing fee that the same airplane would pay at practically any other major airport in Europe or in the Pacific/East Asia region (ICAO, annual). Note that London/Heathrow is the world's busiest international airport and one of the most congested. Access to it is highly valued by airlines.

The explanation for this paradox lies in the combination of the single-till approach and the "RPI − x" formula prescribed by the BAA's regulators. With the immensely profitable nonaeronautical activities at London/Heathrow contributing a growing fraction of total revenues, the share of revenue to be raised from aeronautical sources steadily diminished. Landing fees thus reached their low, by any criterion, levels.

It has been estimated that if the BAA operated under a dual-till system and if it sought a 7 percent return on net assets from its revenue centers, landing fees at Heathrow Airport would have been 35 percent higher in the late 1990s (Warburg Dillon Read—UBS, 1999).

There are strong arguments on both sides of the single-till versus dual-till debate (CAA, 2000). Some of the principal arguments of airlines and others who support single till include (Enriquez, 2002):

- The traveling public is the eventual beneficiary of lower airline costs, as some or most of the savings will be passed on to consumers through fare reductions in a competitive system.

- Nonaeronautical activities exist and thrive at airports because of the airlines and their passengers. Without the airlines, no revenue or profit would be derived from these activities. Airlines are therefore entitled to some of the resulting benefits by paying lower user charges. In fact, the more price incentives airlines have at an airport, the more traffic and nonaeronautical business they will generate.

- Airport operators have monopoly power with respect to nonaeronautical services and apply monopoly prices. They should not be allowed to take full advantage by retaining all resulting profits, as they can do under dual till.

Some of the main arguments on the opposite side include:

- Airlines are already protected from monopolistic pricing through regulatory caps on the rate of return that airport operators can seek from aeronautical facilities. There is no reason why airlines, which are for-profit organizations, should not be charged for the fair cost of the facilities and services they require.
- Only aeronautical revenues should be subject to economic regulation; the commercial side of the airport business should not be a factor in such regulation.
- At congested airports and at hubs they dominate, airlines are unlikely to pass on to their passengers a significant part of the cost savings they achieve through single till.
- Single till leads to charges for aeronautical facilities and services below the true costs; this results in economically inefficient use of airport resources, especially at congested airports.
- The profits generated from commercial activities at airports reflect the premium location for such activities that airports provide, not monopolistic pricing.

This debate intensified in the beginning of the twenty-first century. A number of newly privatized airports (Hamburg, Beijing, Mexican airports) operated under a dual-till system. A change to dual till was also under consideration at a number of other airports (Sydney, South Africa) including, notably, London/Heathrow and London/Gatwick (CAA, 2001). ICAO was also undertaking a prioritized study of the matter.

Residual versus compensatory

The approximate counterparts to the single-till and dual-till approaches in the United States are known, respectively, as the *residual* and *compensatory* systems.

The compensatory system is straightforward conceptually: the airlines pay charges that are sufficient to recover the full costs to the airport operator of the facilities and services that the airlines use. The residual system was described in Chap. 4: by analogy to single till, airlines pay only the difference between the total revenue target and the revenues from all nonaeronautical and other sources.*

There is, however, an important difference between single till and the residual system. To benefit from a residual system, airlines in the United States have to take on a financial risk. They sign long-term use agreements with the airport operator under which they underwrite the service of debt issued by the airport. The airlines essentially agree to cover any shortfall that may occur in servicing this debt. Under the compensatory system, by contrast, the airport operator assumes the full financial risk associated with servicing its debt.

Airlines that sign airport use agreements under the residual system are called *signatory airlines*. Among other rights, they can approve or reject plans submitted by the airport operator for new buildings and other capital investments. Signatory airlines may also achieve very significant savings by paying only for residual costs. For example, airlines paid $0.51 per thousand pounds of aircraft landed weight at Los Angeles/International in 1992, the last year in which that airport used the residual system. In 1993, when the airport switched to a compensatory system, the rate tripled to $1.56 per thousand pounds (Enriquez, 2002). At the busiest residual airports, airline fees can be exceptionally low. In 1999, for example, Honolulu charged no landing fee because the residual covered all airfield costs!

The preferred choice between a compensatory and a residual system depends on the circumstances of each case. Generally, secondary airports and hubs dominated by one or two airlines (for example, Cincinnati, Minneapolis/St. Paul, Atlanta, Dallas/Ft. Worth, St. Louis) are better off with a residual system due to their strong dependence on a few airlines to generate traffic. At the opposite end, operators of busy, mostly origin/destination airports with no dominating carriers (New York/Kennedy, New York/LaGuardia, Boston/Logan, Los Angeles/International, Seattle/Tacoma) will clearly prefer a compensatory system. In general, as an airport becomes busier and less dependent on one or two airlines for traffic, it will be more likely to adopt the compensatory system.

*"Other" sources may include aeronautical revenues raised from general aviation and from airlines that have not signed long-term airport use agreements ("nonsignatory" airlines).

7-5   Financing capital investments

Due to the large amounts of required capital, financing of large-scale airport projects is always a central concern of airport owners and operators. It is also a *constant* concern, as these projects are so commonplace at fast-growing airports around the world. Most national governments continue to provide sizable grants and other financing to their airports. However, in the case of major airports, the *relative importance* of government financing of airport capital programs, as measured by the *share* of capital contributed, has been diminishing steadily in recent years. Two factors are responsible for this trend. First, as the number, size, and complexity of major airports grow, government resources are insufficient to meet the need for capital funds. Alternative sources must be sought. Second, air transportation has become a "mature industry" in many regions of the world and governments increasingly expect it to become self-sustaining, including paying—directly or indirectly—for its infrastructure needs, i.e., for airports and air traffic management.

The typical sources of financing available to airports can be classified into the following broad categories.

Outright government grants

Outright government grants are still the most common type of airport financing in most countries. National and, in many cases, regional and local governments worldwide recognize the direct and indirect benefits that economies derive from air transportation. They usually assign high priority to the financing of airport capital investments. In the United States, for instance, the federal government has been providing since 1946 sizable annual grants for airport development under a succession of federal assistance programs, according to an ever-changing set of fund distribution guidelines. The Airport Improvement Program (AIP), for example, distributed amounts in excess of $1.8 billion annually to airports throughout the 1990s—a level that will probably be exceeded considerably in the early part of the twenty-first century, as a result of the AIR 21 legislation. However, most of the AIP funding does not go to the largest airports. For example, the 30 largest airports received only about 15 percent of AIP funds in 2000, roughly $300 million out of a total of $2 billion (FAA, 2001). As these same airports spent about $5 billion on capital investment projects in 2000, the share of AIP funds was only about 6 percent, reflecting the trend mentioned earlier.*

*The ticket taxes paid by passengers at these airports contribute a very large portion of the proceeds of the Airport and Airway Trust Fund, from which AIP is largely funded.

Special-purpose user taxes

Another common way to finance airport projects involves special-purpose user taxes imposed by national, regional, or local governments to finance local airport projects. These are taxes whose proceeds accrue *directly* to *specific* airport operators. A good example is the Passenger Facility Charge (PFC) that the Federal Aviation Authority (FAA) may authorize for certain airport operators in the United States (see also Chap. 8). Revenues from the PFC at each approved site must be spent on projects or programs related to enhancing safety, security, capacity, and noise mitigation at that site. By contrast, another passenger departure tax limited to *international* flights from the United States is intended to raise funds for the entire national airport and air traffic management system and, as such, cannot be considered as targeted to financing any specific local airport projects.*

Low-cost loans from international or national development banks

A number of international and national development banks and funds specialize in financing important infrastructure projects through low-cost loans. This funding is generally intended primarily for airports in developing countries or in economically weaker areas of a region or country. The World Bank (International Bank for Reconstruction and Development), the African Development Bank, the Inter-American Bank, and the European Investment Bank (EIB) are examples of international institutions that have been active in this sector. A number of countries have also established government-owned export credit agencies that may play a similar role. A typical example of the terms offered is a 1997 EIB loan of more than $1 billion for the construction of the new airport at Athens/Venizelos. The loan was for a 20-year term with a 6.05 percent annual interest rate and a 7-year "grace period" on payments. Some loans to developing countries carry even more favorable terms, often requiring little more than repayment of the principal over an extended period of time.

Operating surpluses

A few of the busiest and economically strongest airports in the world have now reached the enviable position of generating sufficient economic surpluses to pay for small-size and medium-size capital improvement projects without needing to seek external financing.

---

*In fact, only a minuscule fraction of the funds raised in this way is allocated back to the international airports where the funds were collected.

This, of course, saves them the interest payments, and the administrative and overhead costs associated with external funding. The number of airports that have reached this level of economic performance keeps increasing.

Loans from commercial banks

A number of large commercial banks, notably several Japanese ones, have been active in providing short- and medium-term (3–10-year) loans for airport capital projects. Such loans are attractive to qualified airport operators because of the flexibility they provide and their ready availability. However, when it comes to large-scale airport projects, only relatively small fractions of the financing requirements are typically covered in this way because of the higher interest costs involved.

General-obligation bonds

General-obligation bonds for financing airport capital improvements may be issued by national, regional, or local governments. As the name suggests, these bonds are secured through the full taxing power of the governmental entity. Should revenues from the airport prove insufficient to cover obligations to bondholders, taxpayers at large must cover the shortfall. Interest paid by these bonds is typically exempt from taxes of the issuing authority.* As a result of being tax exempt and highly secure, general-obligation bonds can usually be sold at very low interest rates. They are thus particularly attractive to airport operators. However, governmental entities are reluctant to finance through general-obligation bonds capital projects at airports with a large user base and revenue base. In the United States, where law strictly limits the total debt that a local or regional government may secure through general-obligation bonds, the financing of airport projects through general-obligation bonds is progressively giving way to airport revenue bonds (see below).

Revenue bonds

Revenue bonds may be issued by airports that are in a position to service debt entirely through their own revenues. The critical difference from general-obligation bonds is that revenue bonds are *not* backed up by the taxpayers. The interest rates that revenue bonds bear may thus be significantly higher than those of general-obligation bonds.

*In the United States, interest from general-obligation bonds issued by local or state authorities is exempt from both federal and state taxes.

The rates depend critically on how secure these bonds are judged to be. An important parameter in this respect is the level of "coverage," essentially the ratio of net airport revenues to debt service requirements in any particular year. The higher the coverage, the lower the interest rate. The portfolio of long-term and short-term airport use agreements that the airport operator has secured with the airlines may also be important in determining the ability of an airport to raise financing through revenue bonds. A summary of the factors used to assess the credit position of an airport is provided at the end of this section. A remarkable fact (Moody's Investor Services, 2000) is that no airport in the United States has defaulted on its debt in the past half-century!

National law in many countries does not permit the issuance of revenue bonds by airports. In others, including several European countries, such bonds constitute a novel and relatively unexplored way to finance airport capital improvements. In the United States, revenue bonds are a particularly popular means of airport financing. Some economically strong airports are, in fact, able to secure revenue bonds solely against their own earning power, without requiring airline guarantees. In a slight variation, the Port Authority of New York and New Jersey and the Massachusetts Port Authority, both of which own and operate seaports and other facilities in addition to airports, issue revenue bonds against earnings of the whole authority (not just the airport). Other airports, however, choose or are forced to secure revenue bonds against long-term airline use agreements under which the carriers commit to cover any shortfall in debt service. In return for backing up the bond issues, the airlines pay only for residual costs and obtain other important rights with respect to airport management, operations and planning (Sec. 7-3).

Private financing against specified rights to airport revenues

Private financing is fast becoming one of the principal means of financing airport capital improvements in both developed and less developed countries. Typically, the airport operator signs a BOT (build, operate, and transfer) contract with a private group that undertakes to finance all or part of a development project against specified rights to its future revenues. This may involve just a single facility (e.g., a multistory automobile parking garage) or a complex (e.g., a new passenger terminal and supporting facilities) or, in a few instances, an entire airport. In this last case, the private group may become the airport operator as well for an agreed period of time. Chapter 4 includes examples of several such arrangements.

Examples 7-6 and 7-7 demonstrate that the financing of very large airport projects typically comes from a combination of sources and that the mix of funding depends strongly on local conditions.

The ability of airport operators to obtain favorable terms for the financing of capital investments depends in large part on the rating of their credit position by specialized companies in this field, such as Moody's Investors Services, Standard and Poor's, and Fitch ICBA. Each of these companies has developed its own rating methodology for airports, but the factors that are taken into consideration are common

EXAMPLE 7.6

The 5-year project to construct the new airport of Athens/Venizelor began in early 1996. The required capital was 658 billion Greek drachmae or, roughly, $2.4 billion at the then-prevailing rate of exchange. The capital was obtained from the following sources:

- 47 percent (roughly $1,128 million) from the European Investment Bank in the form of a low-cost loan under the terms described earlier in this section

- 15 percent ($360 million) from a consortium of commercial banks in loans at market rates

- 12 percent ($288 million) from an airport development fund established by the Greek government in 1993; the fund's revenues came from a special-purpose user tax imposed on all passengers departing from the then-existing airport at Athens/Hellenikon

- 11 percent ($264 million) from grants from the European Union under the Second EU Convergence Program

- 7 percent ($168 million) from grants from the Greek State

- 6 percent ($144 million) from share capital contributed by the shareholders (55 percent Greek State, 45 percent consortium of German private companies)

- 2 percent ($48 million) from secondary debt taken on by the shareholders

Thus about $720 million, or 30 percent of total financing, was essentially grant money (outright government or EU grants and a special-purpose user tax), while another 47 percent was obtained from a low-cost loan.

EXAMPLE 7.7

The financing of the construction of the new Denver International Airport in the late 1980s and of the ongoing airport, seaport, tunnel, and bridge modernization program undertaken by Massport in Boston and centered on Boston/Logan offer an interesting contrast that reflects the different perceptions vis-à-vis the economic prospects of the two airports at the respective times the projects were undertaken.

Most of Denver's financing came from: a federal grant of $500 million, especially authorized by the U.S. Congress; $400 million from PFC revenues at Denver; and contributions from the City of Denver in the form of (1) $900 million in general obligation bonds, (2) $300 million from the sale of the old Denver Stapleton Airport, and (3) a $400 million commercial loan from the Sumitomo Bank.

By contrast, Massport's financing has come from the following sources: $152 million from various federal grants of national (not specially targeted) scope; $420 million from Massport internal funds; $545 million from PFC revenues; $847 million from a series of revenue bond issues; and $939 million from third-party (public/private) development ventures.* Note that neither Massachusetts taxpayers nor any local governments incurred any debt service obligations.

to all (Fitch IBCA, 2000; Standard and Poor's, 2000; Moody's Investor Services, 2000). A listing of these factors (Rivas, 2001) is highly instructive because it summarizes the criteria that define an airport's economic prospects:

- Market strength (geographic location; regional economic characteristics, such as demographics, disposable income, etc.; origin/destination versus hub)
- Air traffic characteristics (air traffic forecast, range and market share of airlines at the airport; strength and commitment of these airlines to the airport)
- Physical infrastructure (utilization of existing facilities; need for new facilities; control of the gates by airport operator)

*Of the total $2.9 billion, approximately $310 million are being invested in seaport, tunnel, and bridge construction and improvements and the remainder in airport and airport-related capital improvement projects.

- Management and operations (cost recovery method and its adequacy to meet the airport's needs; contractual terms in airline agreements, concession contracts, etc.)
- Financing (existing debt burden; share of debt secured by general revenues, PFC, airlines; cash reserves)
- General context (political climate; environmental concerns and disputes).

Overall, airports enjoyed higher credit ratings than airlines in 2000 (Moody's Investor Services, 2000).

Exercises

1. Review the arguments for and against the single-till and the dual-till approaches that are summarized in Sec. 7-4. How valid are these arguments? Can you think of any additional ones, that are not on the list?

2. The six factors used by credit rating agencies to assess airports are listed at the end of Sec. 7-5. The list includes examples of specific criteria considered for each factor. Review these examples and identify the conditions under which each criterion would contribute to the assessment in a positive or a negative way. For example, an unstable political environment is obviously a negative. Several other cases in the list are not as simple.

3. Select an airport with which you are familiar and perform a qualitative assessment of creditworthiness using some of the criteria listed at the end of Sec. 7-5.

References

CAA, Civil Aviation Authority (2000) The "Single Till" and the "Dual Till" Approach to the Price Regulation of Airports, Consultation Paper, London, UK.

CAA, Civil Aviation Authority (2001) "Preliminary Conclusions: Price Caps 2003–2008," Summary Paper, London, UK.

Deutsche Bank (1999) *European Airports: Privatization Ahead,* Deutsche Bank, London, UK.

Doganis, R. (1992) *The Airport Business,* Routledge, London, UK.

Enriquez, R. (2002) "The Mexican Southeast Airport Group Privatization Case," S.M. thesis, Department of Civil and Environmental Engineering, Massachusetts Institute of Technology, Cambridge, MA.

Fitch ICBA (2000) *Airport Revenue Bonds Flying High,* Fitch ICBA, New York.

ICAO, International Civil Aviation Organization (1992) "Statements by the Council to Contracting States on Charges for Airports and Route Air Navigation Facilities," ICAO Doc. 9082/4, ICAO, Montreal, Canada.

ICAO, International Civil Aviation Organization (annual) *Manual of Airport and Air Navigation Facility Tariffs,* ICAO Doc. 7100, ICAO, Montreal, Canada.

Moody's Investor Services (2000) "Worldwide Airport Industry: Rating Methodology," Report 56601, Moody's, New York.

Rivas, V. (2001) "Credit Rating Agencies: Their Role in the Capital Programs of Large US Commercial Airports," term paper, Massachusetts Institute of Technology, Cambridge, MA.

Schiphol Group (1999) *Profile and Organizational Chart,* Schiphol Airport, Amsterdam.

Standard and Poor's (2000), *Finance Criteria: Transportation Bonds* (Airports), New York.

TRL, Transportation Research Laboratory (1999) *Review of Airport Charges 1999,* TRL, Crowthorne, Berkshire, UK.

Warburg Dillon Read—UBS (1999) *Airports Review 1999,* Warburg Dillon Read—UBS, London, UK.

Wiley, J. (1986) *Airport Administration and Management,* Eno Foundation for Transportation, Westport, Connecticut.

8

User charges

Airport operators derive most of their revenues from a wide variety of user charges. Practices differ considerably with respect to what user fees are imposed and what rules are applied to compute these charges. The desirable attributes of any system of charges at an airport include transparency, adequate cost recovery, reasonableness, promotion of efficiency in the use of airport resources, and flexibility. A well-designed system is essential to achieving financial credibility and to being able to obtain funding for capital projects.

The process of developing a system of user charges is quite complex. It requires specification of policy guidelines, definition of revenue centers and cost centers, development of a detailed cost base, allocation of costs to revenue centers, development of a pricing methodology, and consultation with airport users.

Existing policy guidelines at the international level are rather vague and often lead to disputes between airport operators and airport users. In general, airport operators may recover the full cost of facilities and services, including the cost of operations, maintenance, management, and administration, as well as interest on capital investment, depreciation of assets, and, when conditions permit, a fair return on investment. Almost universally, the average-cost pricing method is used to compute unit charges. The method is simple and flexible, but it becomes problematic when it comes to pricing at congested facilities.

User charges are classified into aeronautical and nonaeronautical. The various types of charges in these two categories are reviewed and explained. Major airports currently derive roughly as much revenue from nonaeronautical charges as from aeronautical ones. The latter were far greater than the former until the late 1980s. Comparing in a fair way the size of user charges at different airports is a difficult task because of the numerous factors that influence these charges, many beyond the control of airport operators. Accounting practices may play an important role in this respect. An example is the use of historical cost versus replacement cost in computing charges for depreciation of assets.

An area of significant international controversy concerns handling charges. Airport operators should strive to create a competitive environment for the provision of handling services.

8-1 Introduction

Airport operators derive most of their revenues from a variety of charges they impose on users of airport services and facilities.* Growing pressure to achieve economic self-sufficiency and, when permitted by the regulatory environment, profitability has led to the development of systems of charges that cover every aspect of airport-related activities.

User charges are also part of the direct operating costs that airlines and other aircraft operators face. As such, these charges are carefully reviewed and often criticized by airport users. They are the source of endless controversy, with disputes occasionally taking on an international dimension and involving national governments. An additional benefit of reviewing user charges—as will be done in this chapter—is that through them one can also enumerate and become familiar with all the essential and ancillary services and facilities that are typically provided at an airport.

The contents of the chapter are as follows. Section 8-2 introduces the topic of airports as economic entities through a discussion of revenue centers and cost centers and outlines the methodology typically used to establish systems of user charges. This is an area in which international practices vary widely, partly because there is only limited guidance from such organizations as the International Civil Aviation Organization (ICAO) or from bilateral or multilateral aviation agreements (Sec. 8-3). Sections 8-4 and 8-5 identify and describe in some detail the various types of aeronautical and nonaeronautical charges, respectively, and the rationale behind them. Section 8-6 discusses the size of the revenues that airports derive from these various sources in both absolute and comparative terms. Section 8-7 summarizes some of the difficulties one invariably encounters in attempting to

---

*The term *airport operator* will be used throughout this chapter to refer to any organizational entity responsible for operating an airport; this entity may be an airport authority, a branch of a national government, a department of a local government, etc.

compare charges across airports. It cautions against any facile comparisons. Section 8-8 concentrates on the often-controversial subject of ramp and passenger handling charges and explores the reasons for disputes regarding the provision of these services. Section 8-9 presents a detailed example of the computation of the most common—and still most important—airport charge, the landing fee. This also illustrates the application, advantages, and disadvantages of the average-cost pricing approach, which is widely used by airports throughout the world. A related topic concerns the use of historical cost at some airports and of current cost at others in computing the depreciation costs charged to airport users each year (Sec. 8-10).

8-2   Cost and revenue centers

The development of a system of charges for airport facilities and services is one of the most fundamental tasks facing an airport operator. Growing pressure to achieve economic self-sufficiency and, when permitted by the regulatory environment, profitability, has led to the development of sophisticated systems of charges that cover every aspect of airport activities. This is not simply a matter of good accounting practices. A well-designed system of charges is also essential to achieving financial credibility. It is critical to an airport operator's ability to raise money for capital projects from banks and investors. The desirable attributes and general structure of such charging systems are outlined in this section.

A system of charges should ideally have the following attributes.

1. *Transparency.* Transparency encompasses several characteristics. First, the system should be *simple,* so that prospective airport users can readily understand how much they will be charged and what services and facilities they will be paying for. Moreover, it should be supported by adequate *documentation,* containing the data and explaining the line of reasoning used in developing the price structure. The documentation should also demonstrate a reasonable relationship between airport costs and prices charged. Finally, the charges should be *defensible* in the legal sense. They should not contravene national and international statutes and conventions or violate bilateral or multilateral agreements.

2. *Adequate cost recovery.*   The prices charged for airport facilities and services should meet, with high probability, the airport operator's cost-recovery objective. This objective varies considerably across airport operators. Depending on the regulatory and economic environment in which the airport operates, some will act as profit maximizers, others will seek full recovery of costs—including, possibly, a fair return on investment—whereas a third group may seek only partial cost recovery, in instances when it is unrealistic to expect full recovery.

3. *Reasonableness.* User charges should be "reasonable" with respect to several criteria. In absolute terms, they cannot be prohibitively high for the types of users that an airport wishes to attract. For example, it would be untenable for an airline to pay airport user charges that amount on average to 25 percent of the revenues of its flights. In relative terms as well, the amounts charged should not be out of line with those charged by comparable airports in the same or in neighboring countries. Reasonableness should also extend across different segments of aviation (e.g., foreign airlines should not be charged for a higher fraction of airport costs than their use of the airport warrants).

4. *Promotion of efficiency in airport use.* The fees charged to a user should be closely related to the true costs associated with that user. If a particular class of users is systematically subsidized or undercharged, these users will utilize the facility to a greater extent than economic efficiency would dictate. In developing a system of user charges that promotes economically efficient use, the cost of delays at the airport must definitely be considered. Unfortunately, this is rarely the case in practice, as discussed later in this chapter and especially in Chap. 12.

5. *Flexibility.* Airport pricing systems must be flexible, so that user charges can be easily modified in response to change. Airports operate in a highly dynamic environment. As airline deregulation spreads worldwide, the pace of change, if anything, accelerates (Chap. 4). Regulatory restrictions on how charges can be increased or decreased from year to year, as well as certain long-term use agreements with major airlines, may severely constrain the ability of airport operators to adjust flexibly and dynamically to change. For example, due to regulatory restrictions in several European countries, a number of major airports could not modify their user charges sufficiently following the 1999 abolition of duty-free shopping for intra-European Union passengers and suffered serious economic losses as a result.

The development of a system of charges that satisfies all of the above criteria is obviously not an easy task. To carry out the process successfully a number of steps are necessary, each requiring considerable effort. The principal steps are as follows.

1. *Specification of general policy guidelines.* Top decision makers in the organization must participate in this step. It requires addressing a number of fundamental questions, such as: What are the overall financial objectives of the airport operator? What fraction of total airport revenues will be derived from aeronautical and from non-aeronautical sources? What will be the principal means of financing capital investments? What is the level of cost recovery that will be sought from each of the airport's revenue centers (see step 3)? Will the airport consider congestion costs in developing its pricing structure?

2. *Development of a detailed cost database by item and by cost center.* Many airports still cannot fully account for their costs because they lack an adequate database. The ICAO recommends that airport costs be classified in matrix form by *item* and by *area of service* as shown in Table 8-1. (Areas of service are often referred to as *cost centers.*) In practice, the ICAO classification scheme may often be inadequate. Note, for example, that the areas of service do not explicitly identify "aircraft stands," the area where ramp handling, which is one of the most important and costly airport services, is provided. Thus, numerous variations of the ICAO classification scheme exist. What is crucial

Table 8-1. Classification of airport costs by item and by area of service as recommended by the International Civil Aviation Organization

By item	By area of service (cost center)
Direct personnel costs (wages and benefits)	Aircraft movement areas (runways, taxiways, taxilanes)
Depreciation and amortization	Hangar and maintenance areas
Interest	Air traffic control and communications
Supplies and externally provided services	Meteorological services
Administrative overheads	Fire fighting, ambulance, and security services
Taxes	Passenger terminal facilities
Other expenses	Cargo terminal facilities
	Other facilities and services

Source: ICAO, 1977.

is that item costs be allocated among areas of service (e.g., what wages are paid in connection with aircraft movement areas, hangar and maintenance areas, ramp handling services, etc.). Only in this way can charges be set to reflect the true costs of the services and facilities provided. Note that in some cases (e.g., administrative overheads) the allocation of costs to areas of service may be difficult.

3. *Definition of revenue centers.* Most airports identify a small set of revenue centers, that is, groups of services and/or facilities, which are lumped together for the purpose of setting revenue targets and collecting user charges. One reason for defining such revenue centers is that there are natural groupings of activities and it is convenient to apply a single charge, or a small number of charges, for each such grouping. For example, all aircraft movement areas (runways, taxiways, taxilanes) are typically grouped into a single revenue center, the airfield, and a single charge, the landing fee, is collected to cover all associated facility and service costs. A second, equally important reason is that airport operators often choose to put in effect different pricing policies for different revenue centers. For example, most major airports aim only at recovering full costs from the airfield, but may attempt to extract the maximum possible profit ("charge what the traffic will bear") from the "commercial concessions" revenue center.

4. *Allocation of costs among revenue centers.* The airport operator must allocate the costs associated with each of the areas of service ("cost centers") to the revenue centers. This may be easy to do when a cost center is contained entirely within a single revenue center, so that all the associated costs can be immediately allocated to that revenue center; or it may require considerable work when a cost center overlaps several revenue centers simultaneously.* Table 8-2 lists, as an example, one possible set of cost centers and revenue centers defined at a typical airport in the United States. Note that all the costs associated with the "airfield area" cost center will be recovered through the "landing/traffic area" revenue center. However, costs associated with the "passenger terminal area" must be apportioned among "concessions," "airline leased areas," "other leased areas," "parking and car rentals," and possibly others.

5. *Compute unit charges.* Computation of unit charges is often referred to as the *pricing step*. Steps 2–4 have defined by this point

*One can prepare numerous reasonable variations of this table.

a set of revenue centers and have associated a cost with each of them. Moreover, the airport operator's policy guidelines (see step 1) have specified cost-recovery targets for each revenue center. Given a demand forecast, it is then possible to compute what charges should be imposed on each unit of demand to achieve each revenue center's economic target. At many airports this computation is performed through a simple cost-averaging approach, i.e., essentially by simply dividing the revenue target by the amount of projected demand (see Sec. 8-9). However, more sophisticated approaches, such as marginal-cost pricing, may also be used, especially in the case of congested airports. It is also possible to attribute the costs associated with each revenue center to each of the various categories of users and thus develop different unit charges for each user category.

6. *Establish a framework for interacting with users.* Finally, it is essential to ensure the participation of airport users in the steps just described. Consultation with the users should be an integral part of the process, with regularly scheduled meetings and well-defined procedures for handling user comments and complaints.

This six-step process is dynamic and iterative. Policy guidelines, cost databases, cost assignment to cost centers and revenue centers, and pricing methodologies will necessarily be reviewed and revised over time in response to ever-changing conditions and to criticism from and the requirements of airport users.

Table 8-2. Cost centers and revenue centers at a typical airport in the United States

Cost centers	Revenue centers
Airfield area	Landing/air traffic area
Passenger terminal area (including gates)	Airline leased areas (passenger and cargo)
Hangars	Other leased areas and rental land
Cargo terminals	Concessions
Other buildings and facilities	Parking and car rentals
General and administrative	Utilities
Miscellaneous	Fuel
	Miscellaneous

8-3   Guidelines and background for the setting of user charges

As might be inferred from the previous section, international practices regarding aeronautical and nonaeronautical user charges are enormously diverse, since individual airports or national government agencies must determine their own policies, cost bases, and revenue targets. This diversity has led to widespread conflict between, on the one hand, airlines and other airport users and, on the other, airport operators and civil aviation agencies. At the root of the problem is the fact that the regulatory and statutory framework of international civil aviation lacks specificity when it comes to the subject of user charges and, more generally, of airport economic issues.

The November 1944 treaty of the Chicago Convention on International Civil Aviation, on which much of the international economic and technical regulatory framework of air transportation rests, pays scant attention to the issue of airport user charges.* Only Article 15, Chapter II, of the treaty mentions the subject of user charges. It stipulates that "uniform conditions" must be provided for use of airport facilities by aircraft of all Contracting States and that user charges "against international air services" must be *nondiscriminatory*. The treaty does not suggest any specific methodologies on which the computation of user charges should be based, nor does it refer to the notion of the "reasonableness" of user charges, which is often at the core of current disputes. It does, however, invite states to publish the charges they impose and to communicate them to the International Civil Aviation Organization (ICAO).

The so-called Bermuda 2 Agreement of 1977 between the United Kingdom and the United States was the first to spell out a fundamental principle that is used widely today in computing airport user charges (Doganis, 1992): user charges may reflect, but shall not exceed, the *full cost* to the charging authorities of providing appropriate airport and air navigation facilities and services, and may provide for a *reasonable rate of return* on assets, after depreciation.

This principle was adopted, for all practical purposes, by the ICAO 15 years later. It is echoed in its *Statements by the Council to Contracting States on Charges for Airports and Route Air Navigation Facilities*

---

*This is not surprising, as it was hardly possible at the time to imagine the current volumes of airport activity and the economic stakes involved.

(ICAO, 1992, 2001). This influential document contains the ICAO's guidelines on the subject and stipulates that:

- International users must bear their "full and fair" share of the cost of the airport.
- The "full cost" of the airport and its essential supplementary services should include the cost of operations, maintenance, management, and administration, as well as interest on capital investment and depreciation of assets.
- *"Under the right conditions,"* airports may produce revenues that exceed costs and provide a reasonable return on assets to contribute toward necessary capital improvements.

The ICAO Council Statements offer very useful guidance on developing sound policies vis-à-vis user charges. They invite airports to maintain full financial records that provide a satisfactory basis for determining and allocating the costs to be recovered, as well as to publish their financial statements. Consultation with users should precede any significant change in charges or charging systems. The aim is to seek at an early stage user views on the changes and, if possible, to reach agreement on them. Failing such an agreement, the airport operator should be free to impose the changes. Advance notice of at least two months is recommended before any changes take effect.

Additional aspects of the ICAO Council Statements are worth mentioning because they point to some of the relevant issues of contention on an international scale. The Statements offer valuable perspective by indicating that, while "at an increasing number of airports revenues exceed operating expenses, the large majority of international airports still operate at a loss." It is indeed true that most major and many secondary airports in *developed* countries have reached a point where revenues not only exceed operating costs, but often are sufficient to cover the cost of capital investments as well. However, the overwhelming majority of airports in *developing* countries still operate with substantial economic losses. These airports suffer, in varying degrees, from insufficient traffic, inefficient use of resources, poor labor practices, lagging infrastructure development due to lack of capital financing, and excessive political interference in airport management. In this last respect, the ICAO characteristically states that "airports operated by autonomous authorities seem to outperform financially government-operated ones."

A related issue concerns government subsidies. The ICAO Council does in fact encourage them, when appropriate, by urging that "in shaping policies toward airport finances, States should consider the broader economic implications of airports," such as their contribution to industrial development, cultural exchanges, tourism, etc. Indeed, practically every nation in the world subsidizes to some extent, directly or indirectly, the development of infrastructure at its airports. In a number of instances, the cost of airport *operations* may receive national subsidies as well.

The Statements also discourage the widespread practice, especially in developing countries, of reducing overtly or covertly airport charges to domestic carriers and compensating for this by over-charging foreign carriers. Note that this practice is in direct conflict with the principle of nondiscrimination in the Chicago Convention treaty. The ICAO Council states that a (domestic) airline's capacity to pay "should not be taken into account until all costs are fully assessed and distributed on an objective basis." It is only at that stage that a state may decide to recover less than its full costs from an airline in recognition of other local, regional, or national benefits. In other words, governments may wish to subsidize domestic carriers by assisting them in paying user charges at airports within their own country, as long as this is not done at the expense of foreign carriers.*

The ICAO also expressed concern regarding the "proliferation of charges on air traffic." "Airport users should not be charged for facilities and services they do not use." Indeed, a number of nations around the world have been using charges on civil aviation to pay, in part, for such items as radar installations used primarily for military purposes or capital investments into new airports unrelated to the ones that the airlines paying the charges are using.

8-4   The various types of airport user charges

User charges at airports are classified into two categories: *aeronautical* and *nonaeronautical*. As the names suggest, the former are charges for services and facilities related directly to the processing of aircraft and their passengers and cargo, while the latter refer to charges related to the numerous ancillary commercial services, facilities, and

---

*It is noteworthy that the European Union has explicitly banned subsidies of this kind for EU airlines.

amenities that are often available at an airport. This section identifies and discusses the principal types of aeronautical user charges.

Landing fee

The *landing fee* is the most universal type of aeronautical user charge. It is the fee that aircraft pay for use of the airfield, i.e., of the runway and taxiway systems of an airport. The airfield costs that it covers include capital costs, operations and maintenance costs, and the cost of providing such services as fire fighting, snow plowing, and security.

In the overwhelming number of instances, the landing fee is computed with reference to the weight of the aircraft. Typically the maximum takeoff weight (MTOW) is used for this purpose,* but some airports, especially in the United States, use the maximum landing weight[†] (MLW) instead. The amount to be paid is derived in one of the following ways:

- In direct proportion to the weight (by far the most common in practice)

- As a fixed charge up to a specified weight threshold plus an amount that is directly proportional to any weight above that threshold

- In proportion to the weight of the aircraft but with a changing rate per unit of weight for different ranges of weight (e.g., $x per ton up to 50 tons and $y per ton for any weight above 50 tons)

All these alternatives are consistent with the ICAO Council's Statements on airport charges (ICAO, 1992). There is, however, plenty of room for improvement on current practices. For one, the relationship between the weight of an aircraft and the costs associated with its operation on a runway and taxiway system is not particularly strong. Charging according to weight is, in fact, more related to ability to pay than to the true cost caused by an aircraft's operation on the airfield. Even more important in view of current conditions at many major airports, charges that are solely weight-based do not take congestion-related

---

*In some instances an airport may impose certain limitations on the allowable weight of the aircraft—most often due to inadequate runway length for long-range flights by larger aircraft under certain or all weather conditions. In such cases the maximum allowable weight, given the airport limitations, is used as the basis for computing the landing fee.
†Since practically all airports use the average-cost approach to compute landing fees (see Sec. 8-9), the landing fee paid by any given aircraft at any specific airport will be about the same, no matter whether MTOW or MLW is used as the metric of weight.

costs into consideration. A growing number of important airport operators are therefore currently examining alternative landing fee schemes that encourage efficient use of congested airfields. Such approaches are discussed in detail in Chap. 12.

An ongoing controversy regarding landing fees concerns the practice of charging different landing fees to aircraft depending on flight origin. Sometimes domestic flights or flights to/from certain nations receive preferential treatment by paying lower landing fees. The ICAO Council's Statements are unambiguously negative about this practice, which cannot be justified on technical grounds. Nonetheless, it is used by a number of airports, including some of the busiest in the world— in effect, subsidizing domestic operations or certain airline routes. The European Commission has issued requests to a number of European Union airports to abandon this practice.

Terminal area air navigation fee

Many airports charge a fee for the cost of *terminal area* air traffic management (or "air traffic control" or "air navigation") services and facilities provided to arriving and departing aircraft, including the cost of runway and taxiway lights, airport radar, instrument landing systems, and other landing and traffic control aids (see Chap. 13). A terminal area covers a volume of airspace that typically extends to a radius of 50–80 km around a major airport or a multi-airport system.* The airport operator collects the air navigation fee on behalf of the provider of the air traffic management (ATM) service, which in most cases will be a national civil aviation authority or similar body. In some cases, however, the airport operator may have paid for part or all of the cost of the ATM facilities and equipment. Revenues from the air navigation fee may then be shared proportionately between the airport operator and the civil aviation authority. The terminal area air navigation fee, when one exists, is usually collected as part of the landing fee. An air navigation fee is not imposed at airports in the United States.

Aircraft parking and hangar charges

Aircraft *parking and hangar charges* are for the use of contact and remote apron stands and, if applicable, hangar space. Parking and hangar charges are typically proportional either to the weight of the aircraft or to its dimensions. At many airports, there is no parking charge for "normal" use of a stand, i.e., for occupancies of less than

---

*The shape of this volume as well as the altitude to which it extends varies from country to country (see Chap. 13).

a specified amount of time. This time may vary by type of stand—with a stricter limit sometimes applicable to contact stands. Typical limits for free aircraft parking range from 2 h to as many as 6 h. At contact stands, there may be an additional charge for use of the "aviobridges" (or "jetways") and at remote stands for use of mobile staircases. However, mobile staircases are more often paid for as part of the handling charges.

Airport noise charge

Noise charges have spread rapidly in recent years. The charge often varies by time of day, with considerably higher amounts typically applied to nighttime operations. Noise charges are usually collected as part of the landing fee. However, some airport operators choose to collect them separately, in order to demonstrate both to airport users and, especially, to the airport's neighbors their commitment to addressing noise-related concerns.

The original purpose of noise charges was to cover the costs of mitigation measures that many airport operators have been forced to adopt. These range from installation of noise-monitoring equipment around the airport to noise insulation of public-use buildings and private homes. A more recent parallel use is as a demand control mechanism that penalizes the noisiest aircraft and offers a discount or rebate to the least noisy ones. For purposes of the noise charge, aircraft are typically subdivided into a small number of categories according to their noise characteristics, with a different charge applied to each category. The categories were initially along the lines of the classification of aircraft into Stages 1, 2, and 3 in the United States (or Chapters 1, 2, and 3 of the ICAO). However, more elaborate subdivisions are currently being defined at the national or regional level. Several European airports, for example, assign Chapter 2 and Chapter 3 aircraft into one of four or five categories (see Chap. 6).

The 1992 Statements of the ICAO Council are ambiguous on the subject of noise-related charges, reflecting the tension between less developed countries, many of whose airlines operate older and noisier aircraft, and developed ones, which place a premium on noise mitigation. The ICAO Council stated that such charges "should be levied only at airports experiencing noise problems," that they "should recover only the costs of noise alleviation" and that they "should be non-discriminatory and not prohibitively high for the operation of some aircraft"—obviously meaning Stage/Chapter 1 and 2 aircraft still operated by many airlines in developing nations.

It is safe to expect in the future additional charges related to environmental impacts at airports. For example, few airports (Stockholm, Zürich, Geneva) have already introduced charges related to noxious engine emissions (see Chap. 6).

Passenger service charge

Passenger service charges are also known as *terminal service fees* and are intended to cover costs related directly to the use of passenger buildings. Their application and method of collection vary considerably from country to country. For example, in the United States, no such fee is usually applied to domestic passengers (who account for more than 90 percent of total traffic), as they mostly utilize passenger buildings (or parts thereof) that are operated by the airlines themselves under long- or short-term leases. In such cases the airport operator has no claim to any costs to be recovered. However, U.S. airports apply a charge to all *international* passengers to cover the costs of federal inspection services (FIS), such as immigration, customs, and health. Moreover, in cases where international passengers utilize terminals whose space and gates are shared by several (U.S. and non-U.S.) airlines, a passenger service charge for the general use of the terminal may be applied, in addition to the FIS-related passenger charge.

The collection of the passenger service charge is not a trivial matter. In the great majority of cases today, the airlines, rather than individual passengers, pay this fee to the airport operator. The airlines presumably adjust their ticket prices accordingly, to recover the fee from passengers. The amount paid is computed on the basis either of the actual number of passengers on each flight or of a previously agreed "estimated number" of passengers per flight. However, there have been and still exist cases (e.g., international passengers departing from some Asian airports) in which the airport operator collects the fee directly from the individual passengers. This is sometimes due to the airlines' refusal to have anything to do with fees deemed arbitrary or excessive. This method of collection can be annoying to airport passengers and cause delays and queues.

Some airport-related taxes on passenger tickets constitute another form of passenger service charge when they accrue *directly* to airport operators. A prominent example is the Passenger Facility Charge (PFC) that the Federal Aviation Administration (FAA) *may* authorize for collection by airport operators in the United States, following a complicated application and review process. The PFC applies to both domestic and international passengers. Revenues from the PFC at each approved site must be spent on projects or programs related to enhancing safety, security, capacity, and noise mitigation at that site. By contrast, a number of other taxes on passenger fares are not site-specific and constitute a more general form of taxation. They are intended to support the development of aviation infrastructure at large in a country or simply to contribute to the government's overall tax revenues. An example is the passenger ticket tax* that the United States has used for many years to maintain the Aviation Trust Fund (ATF), which supports improvements in aviation infrastructure in general. In contrast to the PFC, this passenger ticket tax *cannot* be viewed as a passenger service charge.

Cargo service charge

In a manner entirely analogous to the passenger service charge, many airport operators impose a fee per ton of freight (or other agreed measurement unit) to cover the cost of cargo processing facilities and services provided by the airport. This fee is typically collected from the carriers, who, in turn, presumably adjust their cargo rates accordingly. Cargo service charges are not imposed at airports in the United States.

Security charge†

Aviation security services are provided at essentially every commercial airport in the world, and a corresponding charge is almost universally collected. At airports where a security charge is not explicitly identified as a separate fee, it is usually collected as part of the general-purpose passenger service charge. The ICAO Council Statements indicate that security should be a "State responsibility," that "authorities may recover security costs but no more," and that "users requiring additional security services may be charged additionally for them." These have indeed become the guidelines under which security services are generally provided and paid for. The provider of the service is typically the national police or other government security agency. However, in many cases, especially at busy international airports, the national government (or the regional or local government in the United States and a few other nations) may relegate responsibility for the provision of security services either to the airport operator (which hires special personnel for this purpose) or to a specialized third-party contractor. In all circumstances, eventual

---

*This tax has ranged from 8 to 10 percent of the fare price over the years.
†Security arrangements and associated costs were in a state of flux as this text went to print, following the events of September 11, 2001. There is little question that, as a result of the events, security arrangements will become tighter and security-related costs will increase, probably very significantly.

responsibility and oversight rests with national, regional, or local government authorities, as the case may be. The security charge itself is typically imposed on a per-passenger basis, collected by the airport operator from the airlines, and distributed among the ultimate recipients (the service providers) according to the particular circumstances.

Airlines requesting additional security services beyond the standard ones are usually charged an additional amount. An example is departures by U.S. carriers from overseas airports for destinations in the United States. These flights are required by the U.S. government to exercise additional security precautions, such as interviewing of passengers and passport verification, and are typically charged additional security fees at overseas airports. In other cases, some flights may be designated as "high risk," either by the airport operator or by the airport's national government. Such high-risk flights are often charged for the cost of the extra security measures to which they are subjected, even though the airlines performing these flights may not have requested these services in the first place.

Ground handling charges

The diverse and important category of *ground handling charges* is subdivided into *ramp handling charges,* i.e., charges for handling services on the apron ("ramp"), and *traffic handling charges* for services that are provided within the passenger or cargo buildings. Traffic handling services are sometimes further broken down into *passenger handling* and *cargo handling.* The loading and unloading of aircraft, as well as the sorting, bundling, and delivery of baggage to retrieval carousels and other devices, are all considered part of ramp handling. Handling costs to the airlines are often comparable in size to the *total* of all the other aeronautical charges mentioned previously. Due to their importance and the considerable controversy that sometimes surrounds them, ground handling services and charges will be discussed more extensively in Sec. 8-8.

En route air navigation fee

The last type of aeronautical charge is not an airport fee per se, as it pertains to the provision of ATM services in en route airspace, i.e., outside an airport's terminal airspace. Typically, the proceeds from this fee go to national* civil aviation authorities or similar bodies,

---

*In a very few instances, an international body may have responsibility for ATM in some parts of en route airspace. An example is the Eurocontrol organization (see Sec. 8-4), which is responsible for en route ATM over a small part of northwestern Europe.

which are generally responsible for ATM facilities, equipment, and operations in en route airspace.

Charges for en route air navigation have been increasing rapidly. They now constitute an important part of airline costs outside the United States, whereas they were rather insignificant up to the late 1970s. The ICAO Statements stipulate that the costs to be taken into account when determining ATM user charges "should include only those related to services and facilities approved under the relevant Regional Air Navigation Plan* of ICAO." After recognizing that attribution of en route air navigation costs to users is a difficult task, the Statements recommend that en route charges should take into account the distance flown and the aircraft weight, the latter "in less than direct proportion." A growing number of countries around the world are complying with these guidelines. They have adopted approaches for computing air navigation charges that are usually identical to or variations of the "Eurocontrol formula." This formula computes a fee for each flight in direct proportion to the number of *service units* that flight incurs. Specifically, the number of service units, n, incurred by an aircraft with a maximum takeoff weight of T tons flying a great-circle distance of d km[†] in the en route airspace of a Eurocontrol Member State is given by the expression

$$n = \frac{d}{100} \cdot \sqrt{\frac{T}{50}} \qquad (8.1)$$

(As an example, an aircraft with a MTOW of 200 tons flying a great-circle distance of 300 km in French en route airspace is charged for 6 service units in France.) Each Member State imposes its own unit charge per service unit.[‡]

The United States is one of a few countries that do not impose an en route air navigation charge. The cost of ATM facilities and services is recovered through the Aviation Trust Fund (supported primarily by the passenger ticket tax—see above) and from general taxes.

*One of the roles of the ICAO is the preparation of Regional Air Navigation Plans covering the entire planet; these plans specify the number and location of air navigation facilities that are necessary to provide adequate "coverage" and services in each region and are approved by the ICAO Council.

[†]The great-circle distance, d, is reduced by 20 km for every landing and takeoff within the Member State by the aircraft involved.

[‡]Eurocontrol had 27 Member States as of 2000. For the unit air navigation charge for each member state see ICAO's Document 7100 (ICAO, annual).

8-5   Nonaeronautical charges

Airport revenues are also derived from *nonaeronautical charges*. These are also often referred to informally, and somewhat imprecisely, as *commercial revenues*. The importance of nonaeronautical revenues has been growing steadily over the years, as operators of both major and secondary airports have increasingly directed their efforts toward maximizing income from nonaeronautical sources. Many of the busiest airports in the world currently derive more revenue from non-aeronautical than from aeronautical charges. Profit margins from this sector are also typically much larger than those on the aeronautical side. In fact, the ICAO Council's Statements, with unusual candor, encourage airport operators to "develop these revenues to the max-imum extent possible" (ICAO, 1992)—one of the few points on which airports and airlines seem to be in full agreement.

The reasons for the enormous success of commercial activities at airports are several and quite obvious. First, large numbers of travelers and their greeters pass through busy airports each day, creating a huge potential market. Second, air passengers come, on average, from the richer strata of society and include a large fraction of "high-end" business travelers. Third, many air travelers find themselves with lots of free time on their hands in airport terminals—often more than an hour in the case of departing and transfer passengers (see Chaps. 15 and 16). Moreover, the duty-free and tax-free shopping, which are typically available to international passengers, make airports partic-ularly inviting places (to some) to spend money. Passengers of some nationalities are culturally attuned to buying presents for their friends at home when returning from trips abroad—and the final airport stop before they get home is a particularly convenient place for doing so.* The many sources of nonaeronautical revenue can be classified into a few major categories as described briefly below.

Concession fees for aviation fuel and oil

Suppliers of aviation fuel and oil at an airport pay a fee, typically an agreed percentage of gross revenue, to the airport operator. In some cases, the airport operator may itself buy the fuel and resell it to the aircraft operators. The ICAO Council's Statements recommend that fuel concession fees be treated in the same way as "non-discriminatory,

---

*In a fascinating statistical nugget, the BAA found that in 1995 the average Japanese passenger spent approximately $80 at BAA airports, Scandinavians $40, U.K. citizens $16, and U.S. citizens $12.

aeronautical charges," i.e., that only a modest return be sought from such concessions, in view of the importance of fuel and oil costs to the airlines and other aircraft operators.

Concession fees for commercial activities

Concession fees include fees for the operation of duty-free shops, retail shops, bars and restaurants, bank and currency exchange branches, newsstands, game arcades, etc., on airport premises, mostly in passenger buildings. These fees are charged either on a fixed-rent basis for space provided or, very often, on a variable-rent basis. The latter may involve a fee ranging from 20 percent to as high as 60 percent(!) of gross sales, usually supplemented by a guarantee of a minimum annual level of revenue for the airport operator.

Revenues from car parking and car rentals

Car parking and car rentals are fast-growing sources of revenues for airports around the world. They are often the largest generators of non-aeronautical revenues at airports in the United States. Arrangements with regard to automobile parking facilities and services vary considerably across airports. In many cases the airport operator itself will build, manage, and operate the car parking facilities. As an alternative, the airport operator may build the facilities but then contract the management and operation of car parking to a specialized operator. A third, increasingly common arrangement involves a BOT ("build, operate, and transfer") agreement with a contractor who undertakes to finance, construct, manage, and operate car parking over a specified period of time, typically of the order of 10–25 years.

Similarly, a variety of arrangements are in place regarding the provision of car rental services at airports. For example, space for the stationing of rental cars may be collocated with the regular car parking facilities, or provided at remote locations but on the airport's premises, or relegated to locations outside airport property.

Rental of airport land, space in buildings, and assorted equipment

A wide range of possibilities is included under the category of rentals. The most obvious are revenues derived from space rented to airlines for offices and passenger "club" lounges, as well as from facilities and equipment rented to shippers, freight forwarders, etc. Revenues from advertising space may also be significant. Airport property and land may be rented for the development of aircraft maintenance and repair hangars for the airlines and for fixed-base operators. Major airports also make often-complex arrangements for the development of hotels, office buildings, and even shopping centers on their property, thus generating sizable rental and concession revenues.

Fees charged for airport tours, admissions, etc.

Once a significant source of revenue, fees charged for tours and similar activities have now become entirely secondary as revenue sources at all but a few locations.

Fees derived from provision of engineering services and reimbursable utilities by the airport operator to airport users

Revenues from services and utilities are rapidly growing at major airports, in contrast to the previous category.

Non-airport revenues

Non-airport revenues can plausibly be considered as separate from "nonaeronautical revenues" in that the term refers to income from off-airport activities. A number of major airport operators, especially in Europe, are increasingly becoming engaged in providing consulting services and educational and training services to other airports. Other, less common examples include management contracts for hotel and restaurant chains and for duty-free shops; equity investment into various, mostly travel-oriented commercial ventures; and, with increasing frequency, acquisition of shares in other airports, in connection with various privatization schemes.

8-6   Distribution of airport revenues by source

The most striking fact that emerges from any analysis of the sources of revenues at major airports worldwide is that the rate of growth of nonaeronautical revenues over the past 20 years has exceeded by a significant margin growth in aeronautical revenues. The total size of nonaeronautical revenues at the busiest airports in the world was about equal to that of aeronautical revenues in 2000.

The pattern suggested by Example 8.1 is not atypical. Table 8-4 shows the total aeronautical and nonaeronautical revenues at the 30

EXAMPLE 8.1

Table 8-3 provides an approximate breakdown of the revenues of Massport at Boston/Logan in 1998. Note that, in rough terms, 39 percent of the revenues came from aeronautical charges, 48 percent from nonaeronautical ones, and 13 percent from the PFC tax, which supports airport improvements related to capacity, safety, security, and noise. Remarkably, the single largest revenue source was automobile parking and car rentals. Revenue from these sources exceeded the amount collected from landing fees despite the fact that Boston/Logan charges airlines for the full cost of the airfield (compensatory system).

Logan generates about 85 percent of the total revenues of Massport, the remainder coming from Boston's seaport, an important toll bridge, and a number of other transportation facilities.

busiest airports in the United States in 2000, as reported to the FAA through Form 5100-127. Total operating revenues amounted to approximately $7.4 billion, derived in equal amounts from aeronautical and nonaeronautical sources. The three largest single sources of revenue were landing fees, aeronautical uses of passenger and cargo buildings, and automobile parking. As in Example 8-1, the combined revenue from parking and car rentals exceeded the revenue from

Table 8-3. Sources of revenues for
Boston's Logan International Airport, 1998

Source	$ (million)	%
Aeronautical revenues		39
Landing fees	50	18
Passenger and cargo buildings	57	21
Nonaeronautical revenues		48
Concessions (shops, restaurants, duty-free, etc.)	39	14
Automobile parking and rental car fees	64	23
Engineering and utilities (e.g., resale of electricity)	29	11
Tax revenues		13
Passenger facility charge	35	13
Total airport revenue	274	100

Source: Massport.

landing fees. In addition to the amounts shown in Table 8-4, the 30 busiest airports had significant nonoperating revenues in 2000. They raised about $1.1 billion from passenger facility charges (PFC) and received another $350 million in federal grants through the Airport Improvement Program.

Analogous patterns vis-à-vis the relative size of aeronautical and nonaeronautical revenues can be observed outside the United States. Table 8-5 describes the distribution of revenues for many of the largest airport operators in Europe. The classification used is somewhat different, in that a number of items, such as revenues from provision

Table 8-4. Sources of revenues for the 30 airports with the largest number of passengers in the United States in 2000

Source	$ (million)	%	% of total
Aeronautical revenues			
Landing fees	1,465	39.2	19.8
Buildings (passenger and cargo airlines)	1,741	46.5	23.5
Apron	51	1.4	0.7
Fuel flowage	70	1.9	0.9
Utilities	160	4.3	2.2
Cargo	223	5.9	3.0
Miscellaneous	30	0.8	0.4
Total	3,740	100.0	50.5
Nonaeronautical revenues			
Rentals of land and non-terminal facilities	267	7.3	3.6
Concessions (retail, food)	899	24.6	12.2
Automobile parking	1,184	32.4	16.0
Rental cars	577	15.8	7.8
Catering	73	2.0	1.0
Interest	353	9.7	4.8
Miscellaneous	300	8.2	4.1
Total	3,653	100.0	49.5
Total airport revenue	7,393		100.0

Source: FAA Form 5100-127, 2001.

Table 8-5. Revenue distribution in major European airports: 1997 or 1998 results

Airport authority/organization	Aeronautical		Nonaeronautical		
	Aircraft	Handling	Retail*	Other commercial†	Other‡
Aer Rianta (Dublin)	17		39	27	17
AENA (Spain)	60	3	10	21	6
ANA (Portugal)	50		46		4
Amsterdam	40		38	5	17
BAA	29		53	14	4
Copenhagen	46		51		3
Düsseldorf	35	33	24		8
Frankfurt	29	32	21		18
Hamburg	34	37	25		4
Manchester	50	4	23	23	
Milan	25	49	18	3	5
Munich	31	27	30		12
Paris	33	14	41		12
Rome	19	38	23		20
Vienna	33	38	17		12

*"Retail" includes duty-free/tax-free (perfume, gifts, liquor, tobacco), tax-paid shops, shopping centers, currency exchange, restaurants and bars in terminals, bookshops.

†"Other commercial" includes advertising, car parks, car rentals, rentals and other income from property, in-flight catering supplies, and aircraft refueling.

‡"Other" includes engineering services, utilities, consultancy activities, and security services.

Source: Deutsche Bank, *European Airports, Privatization Ahead,* November 1999.

of utilities and from off-airport activities (consultancies, non-airport real estate, management contracts, etc.) are included under the "other" category. Overall, however, aeronautical and nonaeronautical revenues are roughly equal. The possible exception is operators that are heavily involved in ground handling activities (see Sec. 8-8). As a result, in some of these cases (e.g., Düsseldorf, Hamburg, Milan, and Vienna), revenues from aeronautical sources dominate. It is also worth mentioning that, in contrast to North American airports, revenues from retail usually far exceed those from parking and car rentals at European airports.

8-7   Comparing user charges at different airports

It is extremely difficult to compare fairly the magnitude of user charges at different airports. In every case one has to understand well the numerous factors that affect the setting of these charges. One must also look carefully at the detailed description of the charges and at what each charge pays for. This is a perfect example of a case where one must truly read the "fine print." Facile comparisons typically lead to erroneous conclusions. Unfortunately, such facile comparisons occur in practice.

Some of the factors that significantly influence the magnitude of user charges at an airport are summarized next (this is far from an exhaustive list).

Government funding

As noted in Sec. 8-3, practically all governments recognize the national, regional, and local benefits of airports. However, the extent to which they provide direct or indirect funding to airports varies greatly around the world. Obviously, the higher the funding, the lower the user charges will generally be. Government funding can take the form of direct grants-in-aid (mostly for capital improvement projects); special-purpose taxes whose proceeds go directly to airports; general funds benefiting aviation as a whole; and preferential tax treatment. One must also carefully distinguish between funding based on taxes or fees that governments impose specifically on aviation users and funding that comes from general tax revenues. The former essentially redistributes user payments, with governments deciding how to allocate the funds among alternative aviation-related programs. In the United

States, for example, the Airport Improvement Program (AIP), which provides grants to airports for capital investments, is funded from user fees and taxes collected from passengers and aircraft operators through the Airport and Airways Trust Fund. However, funds derived from general tax revenues amount to subsidies of airports and other aviation-related services. For instance, the tax exemptions enjoyed by general-obligation bonds and revenue bonds that finance airport capital projects can be viewed as indirect subsidies. Such direct and indirect subsidies are still commonplace in most countries.

Content and quality of services offered

The services that airport users pay for through any particular type of user charge may vary significantly from airport to airport. An example is the landing fee. At many European and Asian airports this fee includes a substantial charge for terminal airspace ATM services. By contrast, in the United States these services are paid for through a ticket tax and from general tax funds; a charge for ATM services is *not* included in the landing fee. Obviously, the quality of facilities and services offered (as measured by level of comfort, absence of delays, reliability, etc.) also varies immensely from airport to airport.

Volume of traffic

Unit charges (e.g., the landing fee per ton of aircraft weight) are typically computed by averaging costs. This means that the unit charge is derived by essentially dividing the total costs to be recovered by the number of units of demand at each cost center of the airport (Sec. 8-9). Most airport facilities and services are characterized by decreasing marginal costs (at least up to a point) as demand increases.* These economies of scale mean that airports with large volumes of traffic enjoy an advantage with respect to user charges. Airports in the United States have some of the lowest landing fees in the world, partly for this reason.

Characteristics of traffic

Certain types of traffic are inherently less costly to accommodate and process than others. Airport unit costs are therefore strongly influenced

---

*User charges at the overwhelming majority of airports do not take delays into consideration. For this reason, the increasing marginal costs due to delays are only rarely accounted for (see Chap. 12 for an extensive discussion).

by the composition and characteristics of the traffic. For example, passengers on dense domestic "shuttle" routes, e.g., New York–Boston or Frankfurt–Munich or Rome–Milan, can be processed very efficiently with minimal space requirements for passenger buildings, due to short dwell times, simplicity of check-in, few checked bags, etc. (Chaps. 15 and 16). Once again, U.S. airports enjoy a major advantage in this respect due to the large volume of domestic traffic that moves through them.

General cost environment

The general cost environment within which an airport operates is the most obvious and, possibly, the most important factor affecting user costs. The costs of construction, equipment, energy, and technology vary greatly from country to country. The cost of acquiring and maintaining technologically advanced airport equipment is often extremely high in developing nations. National labor regulations and practices deserve special mention, as personnel costs often are the dominant component of airport expenditures. Western European airports, for example, typically operate in environments with high pay scales, generous health and other benefits (vacation days, paid holidays, etc.), strong unions, and limited flexibility in task assignments.

Accounting practices

Accounting practices, in themselves, can also be important in determining the size of user charges. An example, discussed in detail in Sec. 8-10, is depreciation schedules based on historical cost versus those based on current cost. Airports utilizing current-cost accounting charge much higher depreciation costs than historical-cost accounting airports.

Treatment of aeronautical users

Finally, airport policies and/or national policies with respect to the treatment of aeronautical users are also important. In practically every instance, aeronautical users are protected by stipulations requiring aeronautical user charges to be related to costs. However, the details vary widely. In the case of airports utilizing single-till and residual systems, the airlines pay only the difference between airport costs and the revenues that the airport collects from commercial activities and from all other airport users. At the other end of the spectrum, airlines pay for the full cost, including a fair return on investment, of the facilities and services they use at compensatory and dual-till airports. Chapters 4 and 7 discuss these points in more detail.

In conclusion, only some qualitative general statements can be made on the relative magnitude of airport user charges internationally in view of all these difficulties. Examples of such statements, as of 2001, include the following.

- Airport user charges vary widely; excluding ground handling charges, they account for 3–8 percent of total (direct *and* indirect) airline costs, with airports in the United States at the low end of that range and European and Japanese airports at the high end (AEA, 1998).

- The highest landing fees and passenger terminal charges are found in Japan and in many of the European Union countries (Germany, Austria, France, The Netherlands, Scandinavian countries, and, even, Portugal and Greece are among the highest ranking).

- By contrast, these charges are among the lowest in the world at U.S. airports, when it comes to domestic traffic, and, to a lesser extent, at BAA airports. It is noteworthy that, as a result of improved efficiency, strict regulation and the single-till approach (Chap. 7), London/Heathrow and London/Gatwick have moved from being among the airports with the highest landing and passenger service charges in the world in the early 1980s to the ranks of modestly priced airports.

- Many airports in the former Eastern block countries have surprisingly high aeronautical charges, possibly due to low levels of traffic and the large investments of capital needed to upgrade their facilities. High charges, unfortunately, tend to hinder further the growth of traffic. In contrast, many airports in the Middle East tend to have low aeronautical charges, in part due to government policies apparently aimed at stimulating demand.

ICAO Document 7100 (ICAO, annual) and other periodic surveys of user charges at the world's major airports (see, e.g., TRL, 1999) provide some of the data and documentation necessary for updating conclusions of this type. However, readers should be aware that comparisons between airports, no matter how popular in the media and even in regulatory proceedings, are often of dubious value for the many reasons described in this section.

8-8 Ground handling services

Ground handling services are typically subdivided into ramp handling and traffic handling, as noted in Sec. 8-4. The former are essentially airside services provided mostly on the apron ("ramp") and the latter are landside services at passenger buildings and in and around cargo terminals. The providers of handling services are referred to simply as "handlers."

Table 8-6 lists the principal ground handling services at airports. Note that traffic handling services do not typically include passport control, immigration, customs, security control, and health inspection—functions generally performed by government organizations or their contractors. It should also be emphasized that several different handlers may provide the ramp and traffic handling services for any single flight. At many airports, the principal ramp or traffic handlers may not be involved at all in the provision of some of the services listed in Table 8-6. Notable examples are catering transport and aircraft fueling, which may be provided directly by food caterers and by aviation fuel and oil concessionaires, respectively.

At every airport, one or more of the following types of provider may offer ground handling services:

- The airport operator
- The airline itself ("self-handling")
- Another airline ("third-party handling")

Table 8-6. Principal ground handling services at airports

Traffic handling services	Ramp handling services
Passenger handling	Baggage handling and sorting
Ticketing	Loading and unloading of aircraft
Check-in	Interior cleaning of aircraft
Boarding supervision and services	Toilet service
Executive lounge/"club" operation	Water service
Cargo and mail handling	Passenger transport to/from
Some information services	Passenger transport to/from remote stands
Preparation of various handling and load-control documents	Catering transport
Various supervisory or administrative duties	Routine inspection and maintenance of aircraft at the stands
	Aircraft starting, marshalling, and parking
	Aircraft fueling
	Aircraft de-icing

- An independent operator (not an airline) specializing in ground handling; these independent handling companies are sometimes referred to as fixed-base operators (FBOs)

The market share of each of these four types of handlers varies greatly across regions of the world. In the United States, airport operators at major airports do not perform ground handling. By contrast, in Germany and in Italy, not only do many airport operators offer the entire range of ground handling services, but they also enjoyed, until the late 1990s, government-sanctioned monopolies for the provision of all or some of these services. In general, many airport operators in Europe, Asia, and Africa are heavily involved in ground handling.

Similar differences exist with regard to handling performed by the airlines. Whereas self-handling, as well as third-party handling arrangements, are routine at practically every North American airport—and indeed constitute by far the most common mode of ground handling—such arrangements are either prohibited at many airports elsewhere or require stringent and time-consuming review and approval procedures. Moreover, in a number of countries only the national airline has the right to self-handle and to provide third-party services.

Rules and practices can be even more diverse when it comes to independent handlers. Typically, these handlers must receive a concession (often in response to a competitive tender) in order to operate at an airport and must demonstrate that they can satisfy a set of requirements, such as technical know-how, security clearances for their personnel, access to adequate insurance, special vehicle permits, etc. The airport operator will usually also impose a set of service standards that the handler must meet. The independent handler may also sign a concession contract that provides for payment of a fixed or variable fee to the airport operator. Airlines that perform third-party handling are also commonly required to pay similar fees to the airport operator and adhere to similar service standards.

Table 8-7 provides a summary of ground handling arrangements at several major North American airports and illustrates several of the above points. Payment of a fee to the airport operator by independent handlers (where such handlers are allowed) is standard practice outside North America as well.

The issue of competition in the provision of handling services has generated considerable controversy. Several regulatory authorities

Table 8-7. Ground handling practices at North American airports

Location	Number of handlers			Fees	
	Self	Third party	Independent	% to airport operator	Paid by*
Vancouver	Yes	5	1	Fixed fee	Ind.
Montreal/Dorval	Yes	2	2	—	—
Calgary	Yes	1	1	—	—
Atlanta	Yes	Yes	5	—	—
Boston/Logan	Yes	3	4	5%	Ind. + 3P
Dallas/Ft. Worth	Yes	Yes	2	—	—
Denver/International	Yes	3	2	10%	Ind. + NS 3P
Detroit/Metro	Yes, except NS	2	1	—	Ind.
New York/Kennedy	Yes	Yes	3	5%	Ind. + 3P
Los Angeles/International	Yes	Yes	7	10%	NS
Miami/International	Yes	Yes	6	Depends on airline	Ind.
Chicago/O'Hare	Yes	2	5	—	—
Pittsburgh	Yes	Yes	1	—	Ind.
San Francisco/International	Yes	Yes	4	$50	Ind.
Seattle/Tacoma	Yes	Yes	5	—	—

*Ind. = independent handlers; 3P = third-party handlers; NS = airlines that have not signed long-term use agreements with the airport (in Los Angeles a 10% fee is charged on all handling services provided to these airlines, while in Denver third-party handlers are charged a 10% fee only for handling services to nonsignatory airlines).

Source: Toronto/Pearson, Survey, 1996.

have tried to ensure that any airline using an airport—and not self-handling—will have a choice of at least two potential providers of handling services (e.g., the airport operator and an airline providing third-party handling). This would mean competition with respect to both cost and quality of service (Example 8.2). Airports that create a competitive environment will, in the long run, be able to reduce the cost of traffic and ramp handling to their airline customers.

It is difficult to estimate the "typical" charge for handling any particular type of aircraft because reliable data are hard to come by and many service providers treat such data as confidential. It is also standard practice to offer heavy discounts to airlines with large traffic volumes. Undoubtedly, large differences exist in the size of handling charges at low-cost and high-cost airports. As a rough indication, various reported costs for handling services for a routine turnaround of an aircraft ranged in 2000 from $500 to $3000 for narrow-body aircraft, such as a B737 or MD-80, and from $2000 to $10,000 for a B747 on an intercontinental flight.

A related issue concerns the ideal number of competitors for the provision of ground handling services at an airport. Because significant

EXAMPLE 8.2

In October 1996 the European Commission issued a Directive concerning the provision of ground handling services at European Union (EU) airports. The Directive aims at the phased liberalization of the market and requires that: (1) self-handling be permitted and facilitated as of January 1, 1998, at airports with more than 1 million passengers or 25,000 tons of freight per year; (2) third-party and/or independent handling be permitted and facilitated as of January 1, 1999, at airports with more than 3 million passengers or 75,000 tons of freight per year (Deutsche Bank, 1999). "Facilitation" in this case means simplification of approval procedures, as well as the active cooperation of competent authorities with prospective handlers. The Directive had achieved rather modest results by 2001, as some EU Member States were slow to take any action on the matter. Some major airports (e.g., Frankfurt) requested and were granted time extensions in implementing the Directive. In a few cases, however, there has been clear movement toward a more competitive handling environment.

economies of scale are associated with such services, having too many competitors runs the risk of instability, as some of the competitors may be driven out of the market, as well as of deterioration in service quality due to pressure to reduce costs. An anecdotal rule of thumb, whose origin and factual basis are hard to trace, states that a handler providing a full range of services requires a volume of about 4–5 million passengers per year to have a commercially viable operation. Accordingly, an airport with 15 million passengers per year would award a total of three permits to third-party and/or independent handlers to operate on its premises.

The number of companies that specialize in ground handling services has grown quite rapidly in recent years, mirroring the growth of air traffic. Several of these companies have an international presence at many airports. Some are owned in part or fully by a major airline. This reflects the fact that the financial rewards of ground handling concessions have been high. Several tenders for handling concessions at major airports around the world have in fact ended in legal challenges that unsuccessful bidders have mounted to contest the validity of the outcome. Moreover, airport operators and national airlines with monopoly handling rights have fought vigorously to preserve these monopolies.

Another reason for the contentiousness that surrounds ground handling is its labor-intensiveness. Traditional airport operators with heavy involvement in handling employ large numbers of personnel. Downsizing their handling activities can thus be painful in human, economic, and political terms. Table 8-8 shows the total number of employees of some of the most important airport operators in Europe in 2000, along with traffic statistics. The top three in the table had practically no involvement in handling, while Vienna, Frankfurt, Munich, and Rome had a major stake in handling operations.

The rightmost column of Table 8-8 indicates the number of workload units processed in 2000 at each airport divided by the number of employees of the airport operator. Note the difference between the "handlers group" (Vienna, Frankfurt, Munich, and Rome) and the others. Of the 14,300 employees at Frankfurt, about 7400 were in ground handling. The reader is cautioned not to interpret Table 8-8 as suggesting that worker *productivity* at, e.g., Frankfurt is much lower than at Amsterdam. To make such an assessment, one would, at the very least, need to consider the number of *all employees* (whether employed directly by the airport operator or not) who were involved in handling traffic at the two airports. At Amsterdam, for example, four different handlers held concessions in 2000.

Table 8-8. Employees and activity at selected airports (2000)

	Employees	Passengers (million)	Cargo (thousand tons)	Annual WLU*/ employee (thousand)
Amsterdam	1,984	39.6	1,267	26
BAA	10,017	124.7	1,960	14
Copenhagen	1,399	18.4	419	16
Vienna	2,644	11.9	181	5
Frankfurt/ Main	14,271	49.4	1,700	5
Munich	4,059	23.1	133	6
Rome	6,200	27.1	312	5

*WLU = "workload unit" = one passenger or 100 kg of cargo.
Source: Annual reports, 2000, except Munich 1999.

8-9   Landing fee computation: average-cost pricing

The *average-cost method* is by far the most commonly used to compute the unit charges imposed at the various revenue centers of an airport. Economists have widely criticized the use of average-cost pricing at airport facilities that are congested—as a growing number of them currently are (see the end of this section). Despite this major deficiency, its application is almost universal at both noncongested and congested airports worldwide. The method is very simple conceptually and also makes it easy to update the unit charges from year to year—two of the main reasons for its popularity.

To apply average-cost pricing one needs to specify the *revenue centers* of an airport and their associated *cost centers*. Once this has been done, the following three-step procedure is applied to each of the airport's revenue centers separately, or, if desired, even to specific sub-elements of the revenue centers.

Step 1.   At the beginning of any defined time period—typically the beginning of a fiscal year—the airport operator determines a *revenue target, X,* for the revenue center in question. This target is set in accordance with the airport's overall economic policies. For example, *X* may be equal to the *full costs* of the facilities and services, including a fair return on investment, when a dual-till system or compensatory system is in use (Chap. 7). Or it may be less than the full costs, if airport policy calls for preferential treatment of the users of certain types of facilities or if it is decided that full-cost recovery would lead to unreasonably high charges. Or, finally, it may be equal to the *residual cost* to be raised from the revenue center, as is often the case at airports in the United States that use a residual system or those outside the United States using the single-till system* (Chap. 7).

Step 2. For the same time period, typically the new fiscal year, a forecast of demand, Y, is prepared for the same revenue center. Different units of demand would apply to different parts of the airport. For example, "number of tons of aircraft weight" expected to land at the airport might be the natural units in which to forecast demand for the runway and taxiway system, whereas "number of passengers" might be appropriate for computing the passenger service charge for use of passenger buildings.

Step 3. The charge, Z, per unit of demand for the time period of interest is then determined by simply dividing X by Y, that is, $Z = X/Y$.

Example 8.3 illustrates a typical application of the average-cost pricing method to the computation of the landing fee in the United States.

EXAMPLE 8.3

Table 8-9, based on simplified and modified actual figures, shows the computation of the landing fee rate for fiscal year (FY) 2000 at a hypothetical Airport AP, operated by a state-owned airport authority in the United States. Airport AP uses a compensatory system of charges. Thus, revenues from the landing fee are expected to cover the full costs of the airfield of AP, i.e., the part of the system of runways, taxiways, apron taxilanes, and service roads that is accessible to all aircraft and airport surface vehicles. A separate fee is charged for use of aircraft stands, as well as for any parts of the airfield dedicated exclusively to specific airport users.

Item A in Table 8-9 represents the *net value* (or *remaining value*) after depreciation, at the beginning of FY 2000, of the airfield facilities.[†] The capital investments depreciated are only

---

*Note that this requires estimation of the revenues from other revenue centers at the airport (e.g., from terminal buildings) that will be subtracted from the full cost of the revenue center in question (e.g., the airfield).

[†]This is the net value of the *historical cost* of the airfield; see Sec. 8-10.

Table 8-9. Computation of unit rate to be charged for landing fees at Airport AP, fiscal year 2000

	Item	Amount
A.	Capital cost of public part of airfield at beginning of FY 2000	$185,518,266
B.	Depreciation of public aircraft facilities	$7,420,731
C.	Interest on public aircraft facilities	$8,014,837
D.	Depreciation of equipment	$239,483
E.	Interest on equipment	$172,863
F.	Snow removal services	$1,330,000
G.	Maintenance and operations	$14,779,532
H.	Administration	$8,728,346
I.	Allocated portion of estimated tax liability	$1,873,675
J.	Prior year adjustment to projection	($2,390,776)
K.	Annual cost of airfield facilities in FY 2000 (= B through J)	$40,168,691
L.	Projection of scheduled air carrier weight (000 lb)	22,600,000
M.	Landing fee per 1000 lb for FY 2000 (= K/L)	$1.78

those made by *the airport authority itself*. Any grants or other "free" funding received from any other sources do not enter the computation of the landing fee. In this particular case, AP had received by FY 2000 approximately $84 million in federal grants for airfield capital improvements under the Airport Improvement Program. This amount is *not* included in item A.

Item B is the amount of depreciation on item A taken in FY 2000. AP uses a 25-year, straight-line depreciation schedule. The amount charged for item B is then exactly 4 percent of item A. This schedule of depreciation of airfield facilities, which is standard for U.S. airports, is particularly favorable to airport users and results in low depreciation costs over time. Item C covers the amount of interest, primarily on tax-exempt revenue bonds (Chap. 7), that AP must pay in FY 2000 on the outstanding portion of funds borrowed in the past to finance capital investments for airfield improvements. Items D and E are entirely analogous to B and C, respectively, and refer to airfield equipment such as fire trucks and runway inspection vehicles. A 10-year, straight-line depreciation schedule is applied. Item F refers to the amount budgeted for the airfield snow-removal contract that AP signs each year.

Items G and H, which together add up to more than 50 percent of the total cost shown on line K for FY 2000, cover the variable costs of managing, operating, and maintaining the airfield. The cost of personnel, including health and other benefits and employee contributions to retirement funds, is a principal component of both of these items. Because the airport authority owns and operates other facilities in addition to Airport AP, care must be taken so that G and H include only (1) administration, operation, and maintenance costs related *directly* to the airfield of AP, and (2) an allocated portion of the *overhead administrative costs* for the operation of the authority as a whole.

The airport authority that owns and operates AP is exempt from taxes. Instead, each year it makes a voluntary contribution, in lieu of taxes, to various local municipalities. This contribution is, in a sense, partial compensation to the communities for the costs associated with being neighbors of the airport. The authority's board of directors determines the total amount to be contributed each year. Item I is the portion of the authority's total contribution, which is drawn from airfield revenues. Note that the amount involved ($1,873,675) is equal to only about 1 percent of the net assets (item A). This reflects the general philosophy prevailing in the United States: the airfield is considered to be a public-use facility that is made available to airlines and other airport users at the lowest possible cost. Land-side facilities, such as concessions in passenger terminals, generally contribute a far more substantial percentage of revenues to the funds distributed in lieu of taxes.

Item J is an adjustment for the overrecovery or underrecovery of the revenue target during the previous fiscal year. The credit in this case resulted from underestimation of the actual air carrier weight (item L) that landed at AP in the previous fiscal year (FY 1999) and the consequent overrecovery of the revenue target in FY 1999.

The last three lines of Table 8-9 constitute the three-step average-cost pricing procedure outlined earlier in this section, with items K, L, and M corresponding, respectively, to the quantities X, Y and Z. Item L is the amount most subject to uncertainty. Any errors in forecasting the total weight of aircraft landing in FY 2000 will be adjusted for through item J in FY 2001. Note, as well, how easy it is to update Table 8-9 from year to year. All one needs to do is update the database on the various cost items and forecast the level of traffic during the next fiscal year.

A committee representing the principal carriers and regional airlines at Airport AP meets on a regular basis with airport authority officials to review data like those shown in Table 8-9. This is standard practice at all U.S. airports. During these meetings, discussion and any airline criticism typically center on items G, H, and I.

The application of average-cost pricing to other parts of the airport is entirely analogous to the one described in Example 8-3. After the revenue centers and associated cost centers are defined, all that changes from one application to another is the type of costs one is concerned with and the units in which demand is measured. The application can also be easily extended to residual systems (Chap. 7), which will usually reduce the amounts that airlines will pay for landing fees and other aeronautical charges (see also the Exercises at the end of this chapter).

As can be seen from the detailed example, average-cost pricing gives no consideration whatsoever to the possible presence of congestion at an airport. In fact, instead of discouraging potential users from operating at congested airports, average-cost pricing works in exactly the opposite direction. By simply dividing the revenue target, X, by the demand base, Y, to determine the unit cost, Z, average-cost pricing *reduces* the amounts charged for access to a busy airfield—or to any other busy airport facility—as traffic increases, no matter how congested the facility is. Access to the airport becomes less expensive, in terms of user charges, as the airport gets more and more congested. As a result, delay-related costs will grow even more rapidly than otherwise. As discussed extensively in Chaps. 11 and 12, these delay costs can be very substantial and are extremely sensitive to even small changes in traffic volumes.

Average-cost pricing is thus regarded as economically inefficient when it comes to congested facilities. *Marginal-cost pricing* is the alternative approach that economists recommend. The guiding principle of marginal-cost pricing is that "efficient use of a facility is achieved when each user pays a charge exactly equal to the additional (marginal) cost that his/her use of the facility causes to others." This marginal cost has a short-term and a long-term component. In the case of the airfield, for example, the short-term marginal cost associated with an aircraft movement is the cost of (1) the delay to other aircraft caused by the movement and (2) the "wear and tear" to the runways, taxiways, and taxilanes (Carlin and Park, 1970). Long-term marginal costs are those associated with the need to expand the existing infrastructure and facilities to increase airfield capacity (Little and McLeod, 1972).

The application of this principle is far from simple, in practice. Long-term marginal costs cannot be estimated until a plan is developed for exactly what facilities will be built to increase capacity. Short-term, delay-related marginal costs are also difficult to compute. Despite such practical difficulties, a growing number of important airports around the world are adopting modifications to average-cost pricing designed to account for some of these marginal costs. This development and related practices will be discussed in more detail in Chap. 12.

8-10   Historical cost versus current cost

An important source of differences in the magnitude of airport user charges stems from differences in the accounting base used to estimate depreciation and amortization costs. Due to general price inflation, as well as to changes in the relative prices of goods and services, the cost in *current (nominal) prices* of replacing airport facilities and equipment at the end of their lifetime is generally much greater than the amount originally paid for them. A passenger building constructed in 1975 at a major airport of a country might have cost $100 million (in 1975 prices in that country). In 2000 it might easily cost $400 million (in 2000 prices in that same country) to construct the same facility, or one of similar size and quality. Using a depreciation schedule based on *historical cost,* i.e., on the amount originally paid for a facility or a unit of equipment, may lead to underrecovery of the cost of replacing that facility or equipment. *Current-cost accounting* involves the periodic (usually annual) revaluation of *net assets* (i.e., of the remaining value of assets) according to their replacement costs. Under current-cost accounting, the book value of assets is usually increased by the rate of inflation (as measured by some general or specialized

EXAMPLE 8.4

Consider an airport asset and assume for simplicity that: (1) its cost at $t = 0$, the beginning of its economic lifetime, is $100; (2) the asset is depreciated over a 10-year period; (3) the general inflation rate is 8 percent per year; and (4) the replacement cost of this asset increases annually at the same rate as inflation. Table 8-10 compares the depreciation schedule of this asset over its 10-year lifetime on the basis of historical cost (i.e., the cost of the asset at $t = 0$) with the depreciation schedule on the basis of current cost, assuming a straight-line depreciation method when historical cost is used. Note the increasing difference between the amounts charged for depreciation under the two schedules as the asset ages and the gap between its historical cost and its current cost increases.

price index) from year to year (Example 8.4). The book value may also be subjected periodically to more careful revaluation, if it is believed that the replacement cost of the asset is changing at rates substantially different from the general inflation rate.

Table 8-10. Historical-cost and current-cost depreciation schedules

Age of asset (in years)	Historical-cost accounting		Current-cost accounting*	
	Remaining value of asset	Depreciation for the year	Remaining value of asset	Depreciation for the year
0	$100	$0	$100	$0
1	90	10	97.2	10
2	80	10	93.3	10.8
3	70	10	88.2	11.7
4	60	10	81.6	12.6
5	50	10	73.5	13.6
6	40	10	63.5	14.7
7	30	10	51.4	15.9
8	20	10	37.0	17.1
9	10	10	20.0	18.5
10	0	10	0	20.0

*Current-cost depreciation schedule assumes revaluation of remaining value of asset by 8% per year at the end of years 1, 2,...,9.

The use of a current-cost basis may add greatly to the depreciation costs charged at an airport and thus increase user fees significantly. When the BAA switched from historical-cost accounting to current-cost accounting in 1980, the amount charged to users for depreciation jumped from £25 million in FY 1980 to £51 million in fiscal year 1980, while net assets (= remaining value of assets) went from £370 million in FY 1980 to £820 in FY 1981! In the United States, depreciation on the basis of current cost is not an accepted accounting practice and is therefore not used by airports. However, many airports around the world, especially in countries where high rates of inflation are endemic, practice current-cost accounting. *In many cases, current-cost accounting better reflects the realities of the airport environment.* Airport operators should consider its use at locations where this approach is permitted.

Exercises

1. It is often claimed that providers of ground handling services can achieve considerable economies of scale as the volume of traffic they handle increases. Describe some of the conditions that should apply in order to achieve such economies.

2. Consider the situation in which two or more specialized companies have been authorized to provide ground handling services at a major airport. Describe some of the logistical issues that may arise out of the simultaneous presence of several handlers. Think, for example, of the allocation of aircraft stands, the location of ground handling equipment, the redundancy of equipment, etc. Could the handlers benefit from mutual cooperation and coordination under some circumstances?

3. Review the Eurocontrol formula (Eq. 8.1) for computing en route air navigation (air traffic management) charges. What is the reasoning behind it? In what ways does it reflect or not reflect the true cost of air traffic management services in the en route airspace of a country?

4. Consider the computation of the landing fee per unit of aircraft weight at Airport AP in Example 8-3. Airport AP is a compensatory system airport. If, instead, it based charges on residual costs Table 8-9 would have to be modified in a simple way. Indicate what new line item or items would have to be added to Table 8-9 to adjust to a residual system. Would a residual system necessarily reduce the unit charge for the landing fee?

References

AEA, Association of European Airlines (1998), *Benchmarking of Airport Charges: Information Package,* AEA, Brussels, Belgium.

Carlin, A., and Park, R. (1970) "Marginal Cost Pricing of Airport Runway Capacity," *American Economic Review,* 60, pp. 310–318.

Deutsche Bank (1999) *European Airports: Privatization Ahead,* Deutsche Bank, London, UK.

Doganis, R. (1992) *The Airport Business,* Routledge, London, UK.

FAA, Federal Aviation Administration (2001) *Airport Financial Reports, Form 5100-127,* FAA, Washington, DC.

ICAO, International Civil Aviation Organization (1977) *Airport Planning Manual, Part 1: Master Planning,* Doc. 9184-AN/902, ICAO, Montreal, Canada.

ICAO, International Civil Aviation Organization (1992) *Statements by the Council to Contracting States on Charges for Airports and Route Air Navigation Facilities,* ICAO Doc. 9082/4, ICAO, Montreal, Canada.

ICAO, International Civil Aviation Organization (2001) *Policies on Charges for Airports and Air Navigation Services,* ICAO Doc. 9082/6, ICAO, Montreal, Canada.

ICAO, International Civil Aviation Organization (annual) *Manual of Airport and Air Navigation Facility Tariffs,* ICAO Doc. 7100, ICAO, Montreal, Canada.

Little, I. M. D., and McLeod, K. (1972) "New Pricing Policy for British Airports," *Journal of Transport Economics and Policy,* 6(2), May, pp. 101–115.

TRL, Transportation Research Laboratory (1999) *Review of Airport Charges 1999,* TRL, Crowthorne, Berkshire, UK.

Warburg Dillon Read—UBS (1999) *Airports Review 1999,* Warburg Dillon Read–UBS, London, UK.

Part Three
The airside

9

Airfield design

The geometric design of an airfield should provide for efficiency in operations, flexibility, and potential for future growth. It should also comply with an extensive set of design standards and recommended practices developed over the years by international and national civil aviation organizations and intended to promote a maximum level of safety.

The two most influential sets of design standards are those of the International Civil Aviation Organization (ICAO) and the U.S. Federal Aviation Administration (FAA). They are based on similar, but not identical, coding systems that classify airports according to the most demanding type of aircraft they are designed to serve. Once the reference code of an airport has been specified, design standards can be obtained from the relevant manuals and other supporting documents.

Airfields typically account for 80 to 95 percent of the total land area occupied by an airport and affect in critical ways every facet of airport operations. The principal determinants of the size of the airfield are the number of runways, the orientation of the runways, the geometric configuration of the runway system, the dimensional standards to which the airfield has been designed, and the land area set aside to provide for future growth and/or environmental mitigation. These topics are discussed in this chapter to varying levels of detail.

The characteristics and some advantages and disadvantages of a broad set of common airport layouts are reviewed. These range from single runways, to a pair of parallel runways, to intersecting pairs of runways, to systems of three, four, or more runways. Several airports in the United States currently use complex, multirunway layouts to serve very large volumes of nonhomogeneous traffic.

Four common mistakes in planning and designing airfields are failure to provide flexibility for future expansion, overbuilding the airfield in its initial phases, lack of integration and coordination of the planning process, and insufficient appreciation of the economic effects of some decisions involving airport design. The implications of these mistakes for the capital and operating costs of airports and their users can be very serious.

The chapter also reviews many of the FAA and ICAO technical and dimensional standards for the various elements of the airfield. These include coverage for crosswinds, runway length, other runway geometric standards and obstacle clearance requirements, separation of runways from adjacent facilities and static or moving objects, taxiway geometric standards and separation requirements, apron layouts and separation requirements, and obstacle limitation surfaces (or "imaginary surfaces") in the airspace in the vicinity of airports. The objective is not to reproduce the requirements and extensive and detailed manuals, but to summarize the standards, outline their rationale, and indicate where further information can be found.

9-1   Introduction

The geometric design of an airfield affects in critical ways every aspect of airport operations. This includes land-side facilities and services, as the layout of the runway system largely dictates the general placement of the passenger, cargo, and other buildings, as well as the interfacing of airside and landside operations.

Because of the overwhelming importance of safety for aviation operations, airfield design must comply with a very detailed and, by now, voluminous set of standards and recommended practices developed over the years by national and international civil aviation authorities and organizations. Two organizations in particular, the International Civil Aviation Organization (ICAO) and the Federal Aviation Administration (FAA) in the United States play a central role, in this respect. The design standards and practices adopted by national civil aviation authorities are largely based on or, in most cases, are identical to those specified by the ICAO in Annex 14 to the International Convention on Civil Aviation and in related documents. Any national civil aviation authority that adopts regulations and practices that differ in any way from the international standards set in Annex 14 is required to notify the ICAO ("file a difference"). These differences are then published for the information of all the other member states of the ICAO.

The largest number of differences has historically been filed by the United States. The FAA has developed a set of airport design standards and recommended practices, which are very similar to those of the ICAO but also differ in several details. Several examples will be given later in this chapter. In practice, the FAA plays as important a role as the ICAO in setting airport design standards. One reason is that the United States is far and away the largest market in the world for air travel and commercial aircraft and has most of the busiest airports (Chap. 1 and Table 9-2 below). Second, the FAA and the U.S. government have traditionally invested heavily into research on aviation, including airports and air traffic management. As a result, the design standards that the FAA has adopted or updated have often preceded the adoption of identical or very similar standards by ICAO.

For these reasons, airport professionals should be cognizant of both of these sets of standards. Annex 14 (ICAO, 1999) was first published in 1951 and, by 2001, had been amended more than 40 times by the ICAO Council, usually following reports and studies by committees and panels of experts. Annex 14 is supplemented by the multi-volume Aerodrome Design Manual (ICAO, 1983a, 1983b, 1984, 1991, 1993) and many other documents, e.g., ICAO (1987), that amplify on aspects of the Annex and provide more detailed guidance.[*] The principal document on the FAA side is Advisory Circular 150/5300-13 (FAA, 1989, 2000) on Airport Design, which was amended six times between 1989 and 2000. This also references numerous other related advisory circulars and federal aviation regulations (FAR).[†]

These extensive sets of standards and guidelines still give airport planners a great deal of discretion. Subject to the environmental, political, and economic constraints at each site, they must address such fundamental issues as:

- How much land should be acquired or reserved for a new airport?
- What should be the overall geometric layout of runways, taxiways, and aprons?
- What size of aircraft should the airfield be designed for?
- How should the construction of airside facilities be phased?

Variants of these questions also come up all the time, when it comes to modifying or expanding the airfields of existing airports. Indeed, this is by far the more common context for airport planning and design, in view of the very small number of entirely new major airports

---

[*]The site http://www.icao.org/cgi/goto.pl?icao/en/sales.htm provides the list of publications and on-line order forms.
[†]The list of FAA publications can be found at http://www.faa.gov/circdir.htm. Many of these are available on-line at no cost. FARs are published in the Code of Federal Regulations.

currently being built or planned anywhere in the world. The trend is clearly toward expanding or reconfiguring existing airports, a task that is often equally complex as designing new ones—and sometimes more so. One of the most difficult challenges in this respect is developing a design and a schedule of construction activities that will allow the airport to continue operations during the expansion and/or reconfiguration project.

Four generic types of mistakes are common in planning and designing airfields:

- Failure to provide flexibility for responding to future developments
- Overbuilding in the initial stages of an airport's operation
- Adopting a hierarchical, nonintegrated approach to design that does not consider adequately the interactions among the various elements of the airport
- Insufficient appreciation of the economic implications of design choices

The first of these applies to *long-range planning*. One example is the severe, often insuperable, constraints that too many airports currently face due to the failure of designers and planners to anticipate the eventual land area requirements of the airfield (Sec. 9-4). A second is the design and construction of runways, taxiways, and aprons in a way that makes it impossible to accommodate new, larger types of aircraft in the future without either making very expensive changes to existing facilities or having to rebuild them from scratch (Secs. 9.5–9.8).

The converse of the above is the second type of mistake, the tendency to overbuild the airfield in the initial stages of airport operations. For example, an airport in the early phases of its development may not need the full runway system it has been planned for, or it may need only part of the eventual full length of one or more of its runways. Similarly, constructing only a part of the planned taxiway system may be more than sufficient for a period of 10 years or more. This may mean, for instance, building only one full-length taxiway running parallel to a main runway, instead of the planned two parallel taxiways in the full development phase.

The failure to adopt an integrated approach to planning for the various parts of the airfield is the third weakness encountered in practice. Airports tend to be planned and designed in a hierarchical fashion without full consideration of the interactions among the various "subsystems" (runways, taxiways, aprons, passenger and cargo buildings, service areas, etc.). For example, on the airside, the typical focus at the highest level of the design process is on settling the configuration of the runway system, with limited analysis of what this implies for the other components of the airfield. Similarly, on the landside, passenger buildings are often designed with inadequate understanding of how they interface with the apron areas, taxilanes, and taxiways. Because of this absence of a "systems" viewpoint, taxiway and apron systems, in particular, are often inefficient and sometimes include parts that are obvious candidates to become congestion points ("hot spots") for air traffic (see also Chaps. 15 and 16). Airfield design also needs to consider safety-related criteria such as minimizing the number of runway crossings and reducing the likelihood of runway incursions. This can be achieved only through an integrated approach to planning and design.

Finally, the economic implications of some design choices are often not fully appreciated and analyzed. Design choices are sometimes made that save some capital costs but greatly increase the operating costs of airport users—for example, by increasing taxiing times on the airfield. This is because the planners and designers do not have a good grasp of the cumulative economic value of saving, e.g., an average of 2 min of taxiing time for each of hundreds of thousands of aircraft movements per year (Sec. 9-4).

This chapter has the dual objective of reviewing some of the most important airfield design standards and recommended practices and of providing a perspective on how these are, or should be, applied. The ICAO and the FAA use simple classification schemes to develop two-element reference codes for each airport. The design standards to be used at each airport depend on these reference codes. Section 9-2 explains the airport reference codes and discusses their application to major airports. The section also provides relevant background and terminology for the chapter. Section 9-3 reviews "wind coverage" requirements that determine whether there is a need for runways in more than one orientation at an airport site. Section 9-4 offers a brief tour through progressively more complex airfield configurations, using several important airports as examples. It shows how the requirements for separations between runways largely dictate the overall layout of the airport as well. It also points to some systemic differences between traffic characteristics in different regions of the world. Section 9-5 provides an overview of the complex topic of runway length. The emphasis is on explaining the fundamental concepts and the meaning of various important technical terms without going into much technical detail. Section 9-6 summarizes some of the most important design standards for runways, as they apply to major airports. Sections 9-7 and 9-8 do the same for taxiways, elements of taxiway systems such as high-speed exits and taxilanes, and apron stands. In all cases, the principal concern centers on the practical implications for busy airports serving large commercial airplanes. Section 9-9 goes beyond the airport's boundaries to compare the standards that the ICAO and the FAA have developed for protecting the airspace in the immediate vicinity of airports from natural or man-made obstructions that may pause a threat to the safety of runway operations. The section describes the various obstacle limitation surfaces (or "imaginary surfaces") that form the basis for these standards.

Airfield design is also greatly affected by the desire to maximize airside capacity, a subject covered in detail in Chap. 10. Airfield design issues will therefore be discussed further in that chapter. The discussion of the design of passenger buildings in Chaps. 14 and 15 also bears directly on the design of aprons for contact and remote aircraft stands.

Finally, it should be noted that the review of design standards in Secs. 9.6–9.8 is far from exhaustive, as it omits several topics altogether—such as taxiway curves and fillets, visual aids and marks, and emergency and rescue services—and leaves out numerous details on others. Covering all these topics and details is beyond the scope of this text. For those engaged in the detailed design of airfield facilities, there is really no alternative to consulting the voluminous materials referenced earlier and other related documents. These professionals also invariably consult with the competent government organizations and regulators and typically work with specialized engineering consulting firms.* The FAA also provides a periodically updated, free computer program that incorporates many of its standards and performs several types of analyses in support of airfield design projects.†

9-2 Airport classification codes and design standards

Both the ICAO and the FAA use simple classification schemes to develop a two-element *reference code* for each airport. These schemes are

---

*In fact, both the FAA and the ICAO offer guidance on obtaining consultation services, for example in ICAO (1983c).
†The program is available for downloading at http://www.faa.gov/arp/software.htm (Office of Airports).

Table 9-1. ICAO airport reference code

ICAO code element 1		ICAO code element 2		
Code number	Aeroplane reference field length (RFL)	Code letter	Wing span (WS)	Outer main gear wheel span (OMG)
1	RFL < 800 m	A	WS < 15 m	OMG < 4.5 m
2	800 m ≤ RFL < 1200 m	B	15 m ≤ WS < 24 m	4.5 m ≤ OMG < 6 m
3	1200 m ≤ RFL < 1800 m	C	24 m ≤ WS < 36 m	4.5 m ≤ OMG < 9 m
4	1800 m ≤ RFL	D	36 m ≤ WS < 52 m	9 m ≤ OMG < 14 m
		E	52 m ≤ WS < 65 m	9 m ≤ OMG < 14 m
		F	65 m ≤ WS < 80 m	14 m ≤ OMG < 16 m

Source: ICAO, 1999.

Table 9-2. FAA airport reference code

FAA reference code element 1		FAA reference code element 1	
Aircraft approach category	Aircraft approach speed (AS) in knots	Airplane design group	Aircraft wing span (WS)
A	AS < 91	I	WS < 49 ft (15 m)
B	91 ≤ AS < 121	II	49 ft (15 m) ≤ WS < 79 ft (24 m)
C	15 m ≤ AS < 141	III	79 ft (24 m) ≤ WS < 118 ft (36 m)
D	141 ≤ AS < 166	IV	118 ft (36 m) ≤ WS < 171 ft (52 m)
E	166 ≤ AS	V	171 ft (52 m) ≤ WS < 214 ft (65 m)
		VI	214 ft (65 m) ≤ WS < 262 ft (80 m)

Source: FAA, 1989.

shown in Tables 9-1 and 9-2. For any type of airplane, the first element of the ICAO code is determined by the *airplane reference field length,* the minimum field length required by that aircraft for takeoff at maximum certificated takeoff weight (MTOW), sea level, standard atmospheric conditions,* no wind, and level runway (see Sec. 9-5). The second element is determined by the most demanding of two physical characteristics of the airplane: its wingspan and the distance between the outside edges of the wheels of the main gear.[†] The reference code of an airport thus corresponds to the code for the most demanding type of aircraft ("critical aeroplane") served by the airport in each element.

In an analogous way, the FAA uses aircraft approach speed to determine the first element of its reference code and wingspan to determine the second. The aircraft approach speed is defined as 1.3 times the stall speed in the aircraft's landing configuration at maximum landing weight.

Practical implications

When it comes to the first element of the ICAO reference code, virtually all the major commercial airports have airport code number 4. The most demanding airplane types using these airports almost always have a reference field length greater than 1800 m. The second element of the ICAO reference code, for all practical purposes, is determined by the wingspan of the most demanding aircraft at major airports. This is because, for the existing types of important commercial jet airplanes, the distance between the outside edges of the wheels never places these airplanes in a code letter category higher than the one to which they would be assigned based on their wingspan. For example, no airplane assigned code letter D on the basis of its wingspan would be assigned code letter E or F on the basis of its outer main gear wheel span. It follows from these two observations that the ICAO reference code for major airports can only be one of 4-C (in the rather unusual case where the largest aircraft that the airport is designed to serve is of a similar size to the A-320 or the B-737, see Table 9-3) or, far more often, 4-D, 4-E, or 4-F.

Turning to the FAA airport reference code, note that its second element is determined solely by wing span (as is also the case in practice for the ICAO code) and that the limits that separate airplane Groups I

---

*The standard atmosphere is defined as temperature of 15°C and pressure of 76 cmHg at sea level, with a temperature gradient of −0.0065°C/m from sea level to an altitude of 11,000 m.
†Note that the outer main gear wheel span limits for ICAO code letters D and E are identical.

through VI from one another are *exactly* the same as the ICAO limits. This means that the second elements of the FAA and the ICAO reference codes correspond perfectly. The only difference is that the FAA uses Roman numerals and the ICAO uses capital letters. An airport in Group V per the FAA will also be in Group E per the ICAO and vice versa—certainly a fortunate circumstance for airfield designers. This is pointed out in Fig. 9-1, which plots the length and wingspan of many of the most important types of recent commercial jet airplanes and identifies the second code element to which they belong.

The second code element is the one that largely determines the geometric design standards at airports, because it reflects the physical characteristics of aircraft, using wingspan as the primary indicator of airplane size. Since (1) any airport will be classified in the same way by the ICAO and the FAA, on the basis of wing span, and (2) the dimensional standards used by the FAA and ICAO for each reference code are usually either identical or very similar, it makes little difference in most instances whether an airport is designed to FAA or ICAO standards. There are, however, a few significant exceptions to this (Secs. 9.6–9.9).

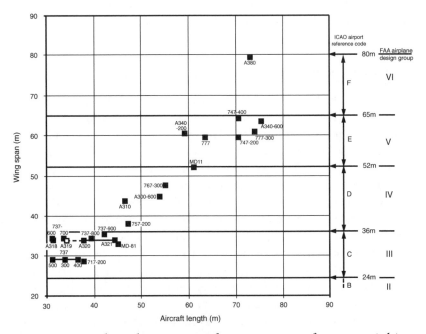

Fig. 9-1. *Length and wingspan of current types of commercial jet transport airplanes. The corresponding ICAO and FAA reference codes are indicated along the vertical axis on the right.*

Planners and designers of new airports or of major improvements to existing ones have to make the critical decision of what airport reference code they should design for. Building for a more demanding aircraft than necessary means incurring unnecessary capital and maintenance costs: the dimensions of runways, taxiways, and aprons and the separations between them will be larger than required. On the other hand, it may be even costlier to "underdesign" the airport. If airlines try to initiate service to/from an airport with a type of aircraft that the airport is not designed to handle, then this service must either be denied, or arrangements must be made to accept that aircraft under some special handling provisions, or the airport's facilities must be modified to make them compatible with the aircraft in question. The first two choices are unattractive in the long run, especially if the popularity of the aircraft in question increases over time. The third choice can be very expensive and disruptive if adequate provisions were not made at the outset for the possibility of redimensioning airfield facilities in the future (see also Examples 9.1 and 9.2).

The message is clear: *design for flexibility and build in stages*. The geometric design should provide, whenever the land area available makes this feasible, for future adjustments of the geometric characteristics of the airfield to accept larger aircraft than an airport was

EXAMPLE 9.1

By the end of 2001, Airbus Industrie had received about 100 orders for the A380, the new large airplane scheduled at that time to enter service by 2006. This airplane, with a wingspan of 79.8 m, could then become the first nonmilitary airplane in FAA Group VI or with ICAO reference letter F that is widely introduced at commercial airports around the world. Airbus carried out a survey of 81 leading airports around the world, considered the top candidates to receive A380 service, in order to identify potential airport/aircraft compatibility problems. The survey found that the three principal problems were runway and taxiway dimensions and separations, weight effect on taxiway bridges, and the effects of aircraft size and capacity on passenger buildings. An Airports Council International survey of 30 of these airports found an average cost of about $100 million per airport for the adjustments needed to accept the A380 at these airports, or a total of $3 billion for these 30 airports (Airbus Industrie, 2001).

EXAMPLE 9.2

The FAA, anticipating the future development of aircraft larger than the Boeing 747, published as early as 1983 design standards for aircraft in Group VI with a wingspan of up to 80 m (FAA, 1983). The ICAO did not officially publish its corresponding standards (code letter F) until 1999. Prudent planners at airports that may be used in the future by new large airplanes in Group VI or Code Letter F, such as the Airbus A380, consulted the FAA standards between 1983 and 1999 in designing new airports or planning for the expansion of existing ones.

originally built for. To follow such a strategy, airport designers and planners should be aware of the full foreseeable range of potential aircraft sizes and associated design standards (Example 9.2).

Table 9-3 lists some of the principal characteristics of many common types of commercial jet airplanes. It is subdivided into parts for wide-body aircraft (Table 9-3a), narrow-body aircraft (Table 9-3b), and, due to their growing importance, regional jets (Table 9-3c). The information provided concerns:

- Wing span and length
- Maximum certificated structural takeoff weight (MTOW), maximum landing weight, and operating empty weight
- Number of passenger seats
- Range of the aircraft
- FAA takeoff field length at maximum certificated structural takeoff weight, sea level, standard atmospheric conditions, no wind, and level runway

The weights shown are only indicative and have been rounded off to the nearest 100 kg. They can vary for the same aircraft model as a result of modifications and improvements over time and of new options becoming available. The last three columns in Table 9-3 should be treated in the same way. In fact, variability within each aircraft type is even greater for these three characteristics. The number of seats depends on such factors as the intended use of the aircraft (e.g., short hops versus longer-range flights), the seat pitch desired by the airline, and the class configuration of the airplane cabin. The numbers indicated in Table 9-3 are mostly for three-class (first, business, and economy) or two-class configurations and are therefore on the conservative side.

The flight ranges shown are for a full complement of passengers, but

Table 9-3a. Characteristics of common wide-body turbofans

	Wing span (m)	Length (m)	Max takeoff weight (tons)	Max landing weight (tons)	Empty weight (tons)	Passenger seats	Range (km)	FAA takeoff field length (m)
Airbus								
A300-600R	44.8	54.1	171.7	140.0	90.1	266	7,700	2,320
A310-200	43.9	46.7	150.0	123.0	80.8	210	8,100	2,200
A310-300	43.9	46.7	164.0	124.0	82.6	220	9,600	2,260
A330-200	60.3	59.0	230.0	180.0	120.5	253	12,300	2,550
A340-200	60.3	59.4	275.0	185.0	129.0	239	14,800	3,060
A340-600	63.5	75.3	365.0	254.0	177.0	380	13,900	3,300
A380*	79.8	73	560	386	277	555	14,800	n.a.

Boeing

747-200B	59.6	70.6	377.8	255.8	169.7	366	12,700	3,320
747-400	64.4	70.6	396.9	295.7	183.6	416	13,600	2,990
767-200ER	47.6	48.5	179.2	129.3	86.2	181	12,200	2,530
767-300ER	47.6	54.9	186.9	145.2	92.3	218	11,300	2,830
777-200	60.9	63.7	247.2	201.9	140.7	305	9,600	2,580
777-200ER	60.9	63.7	297.6	213.2	142.6	301	14,300	3,370
777-300	60.9	73.9	299.4	237.7	159.8	368	10,400	3,690
MD-11†	51.8	61.2	286.0	199.6	134.1	285	13,200	3,050

*Preliminary data.
†Boeing/Douglas airplane.
Sources: Manufacturers' data; FAA takeoff field length from *Aviation Week and Space Technology*, 2000.

Table 9-3b. Characteristics of common narrow-body turbofans

	Wing span (m)	Length (m)	Max takeoff weight (tons)	Max landing weight (tons)	Empty weight (tons)	Passenger seats	Range (km)	FAA takeoff field length (m)
Airbus								
A318	34.1	31.4	68.0	57.5	38.4	107	3,300	1,310
A319-100	34.1	33.8	75.5	62.5	40.1	124	6,800	1,920
A320-200	34.1	37.6	77.0	66.0	41.0	150	5,700	2,160
A321-100	34.1	44.5	93.0	77.8	47.7	185	5,600	2,680
Boeing								
717-200	28.5	37.8	54.9	49.9	32.1	106	3,800	1,800
727-200Adv	32.9	46.7	86.6	70.1	44.6	145	2,800	3,050
737-200Adv	28.4	30.5	58.3	48.5	28.3	120	4,900	2,090
737-300	28.9	33.4	56.5	51.7	32.8	126	4,200	2,030
737-400	28.9	36.6	68.0	56.2	34.6	147	3,800	2,360
737-500	28.9	31	60.6	49.9	32.0	110	4,400	1,860
737-600	34.3	31.2	65.1	55.1	36.5	110	5,600	1,880
737-700	34.3	33.6	70.1	58.6	38.0	126	6,000	2,040
737-800	34.3	39.5	78.2	66.4	41.1	162	5,400	2,320

737-900	35.8	42.1	78.2	66.4	42.5	177	5,100	2,500
757-200	38.1	47.3	115.7	95.3	58.4	201	7,200	2,380
MD-81*	32.9	45.1	63.5	58.1	36.6	143	2,800	1,870
MD-87*	32.9	39.7	63.5	58.1	34.5	117	4,400	1,860

*Boeing/Douglas airplane.

Sources: Manufacturers' data; FAA takeoff field length from *Aviation Week and Space Technology*, 2000.

Table 9-3c. Characteristics of common "regional jets" (narrow-body turbofans)

	Wing span (m)	Length (m)	Max takeoff weight (tons)	Max landing weight (tons)	Empty weight (tons)	Passenger seats	Range (km)	FAA takeoff field length (m)
BAE Systems								
RJX-70	26.3	26.2	38.1	37.9	24.1	70	4,000	1,050
RJX-85	26.3	28.6	42.2	38.6	24.8	85	3,900	1,020
Bombardier/Canadair								
CRJ200ER	21.2	26.8	23.1	21.3	13.7	50	3,000	1,770
CRJ700ER	23.3	32.5	34.0	30.4	19.7	66	3,700	1,680
Embraer								
ERJ135ER	20.1	26.4	19.0	18.5	11.1	37	2,300	1,610
ERJ145ER	20.1	29.9	20.6	19.3	11.8	50	1,900	1,970
Fairchild/Dornier								
328JET-310	21.0	21.2	15.7	14.4	9.4	32	1,400	1,380

Sources: Manufacturers' data; FAA takeoff field length from *Aviation Week and Space Technology*, 2000.

may vary greatly depending on model option. On a day-to-day basis, range also depends on the selected trade-off between payload and the amount of fuel carried and on weather conditions such as winds aloft. Finally, the actual takeoff field length required by any given flight depends on numerous factors, such as the takeoff weight of the aircraft, the range of the flight, the elevation of the airport, the external temperature, the wind conditions, and the physical characteristics of the runway (Sec. 9-5).

Runway designation and classification

Every runway is identified by a two-digit number, which indicates the magnetic azimuth* of the runway in the direction of operations to the nearest† 10°. For example, a runway with a magnetic azimuth of 224° is designated and marked as "Runway 22" (for 220°). Obviously, the identification numbers at the two ends of any given runway will differ by 18. For instance, the opposite end of Runway 22 is designated as Runway 04, and the runway may be referred to as "Runway 04/22." In the case of two parallel runways, the letters R, for right, and L, for left, are added to distinguish between the runways. Thus, Boston/Logan has a pair of close parallel runways designated as 04R and 04L when the runways are operated to the northeast and, respectively, as 22L and 22R when operated to the southwest. If three parallel runways exist, the letters R, C, for center, and L are used. If four parallel runways are present, then one pair is marked to the nearest 10°, with the additional indications R and L, and the other to the next nearest 10°, with the additional indications R and L. For example, Atlanta (see also Fig. 9-7) has four parallel runways arranged in two close pairs, the pair 8L/26R and 8R/26L and the pair 09L/27R and 09R/27L. This also indicates that, with operations to the east, the magnetic azimuth of the four Atlanta runways is between 80° and 90°.

For the purpose of specifying design standards, runways are also classified as noninstrument and instrument. A *noninstrument* (or *visual*) *runway* is intended for the operation of aircraft using visual approach procedures. An *instrument runway* permits the operation of aircraft using instrument approach procedures. Instrument runways are further subdivided into nonprecision and precision approach. A *nonprecision*

---

*An angular difference may exist between the magnetic azimuth and the true azimuth. In the United States this difference may range from 0° to a little more than 20°.
†This is true as long as there are three or fewer parallel runways at the airport; see the case of four or more parallel runways later in the text.

runway has visual aids and, at a minimum, a navigation aid that provides at least directional guidance adequate for a straight-in approach (Chap. 13). A *precision approach runway* allows operations with a decision height and visibility corresponding to Category I, or II, or III limits (Chap. 13).

9-3   Wind coverage

The construction of systems of intersecting runways is usually motivated by the requirement to provide adequate coverage for crosswinds at an airport. Landings and takeoffs are typically conducted *into* the wind.* For instance, during times when the wind is from the north, runways with a northerly orientation, if available, will be preferred over others, and landings and takeoffs will be performed in a generally south-to-north direction. When any given runway is in use, the *crosswind component* is the component of the surface wind velocity vector that is perpendicular to the runway centerline. Its magnitude can be computed easily by multiplying the speed of the prevailing wind by the sine of the angle between the wind direction and the runway centerline. The ICAO specifies that a runway should not be used if the crosswind component exceeds (ICAO, 1999):

- 19 km/h (10.5 knots) for airplanes whose reference field length is less than 1200 m
- 24 km/h (13 knots) for airplanes whose reference field length is 1200–1499 m
- 37 km/h (20 knots) for airplanes whose reference field length is 1500 m or greater, except that with poor braking action (for instance, when the runway surface is wet) the limit is 24 km/h (13 knots)

FAA requirements call for crosswinds not exceeding:

- 10.5 knots (19 km/h) for airport reference codes A-I and B-I
- 13 knots (24 km/h) for A-II and B-II
- 16 knots (30 km/h) for A-III, B-III, C-I through C-III, and D-I through D-III
- 20 knots (37 km/h) for all other airport reference codes (A-IV through D-VI)

Naturally, these limits are somewhat conservative. For example, aircraft in FAA Groups IV through VI (or with ICAO code letter D, E, or F)

---

*Operations on a runway may sometimes be conducted with a slight tailwind, instead of a headwind. The maximum tailwind permitted is typically about 5 or 6 knots (about 9–11 km/h).

can maneuver with crosswinds as high as 25–30 knots (46–55 km/h). The actual selection of the runway(s) to be used at any given time at an airport is made by the provider of air traffic management (ATM) services (the FAA in the United States), *not* the airport operator, taking into consideration prevailing winds (see Chap. 10). A pilot may request reassignment to a different runway on account of crosswinds.

Both the ICAO and the FAA recommend that the number and orientation of runways should be such that *crosswind coverage* (or the *airport usability factor* in ICAO terminology) is at least 95 percent. In other words, the percentage of time during which the use of a runway system is restricted because of crosswinds should be less than 5 percent. Note, however, that the 95 percent target may still leave approximately 18 days per year without crosswind coverage. For many major airports this may not be acceptable. In practice, these airports usually provide runways in a sufficient number of orientations, when needed, to ensure usability factors higher than 95 percent. National civil aviation authorities, in fact, may impose more stringent crosswind coverage requirements than 95 percent at the principal airports of their countries.

Airport designers use historical wind statistics collected at an airport's site to determine the orientation of the runways that should be provided to achieve adequate crosswind coverage. *Wind roses* provide a convenient way for summarizing these statistics graphically. A wind rose (Fig. 9-2) consists of a series of concentric circles, representing wind speed groupings from 0 to 10 knots, 10 to 16 knots, 16 to 21 knots, etc., and a set of radial lines, usually drawn at intervals of 10°, that cut through the circles. The figure within each resulting "box" indicates the percentage of time during which observed winds at the site are within the corresponding orientation limits and speed limits.

In a wind analysis, two parallel lines are drawn for each runway orientation examined. These lines are tangent to the circle corresponding to the allowable crosswind limit. For example, one of the runway orientations of interest in Fig. 9-2 is 105°–285° (true) and the allowable crosswind limit is 10 knots. All winds within the rectangle drawn in this way are "covered" for crosswinds of 10 knots or less by a runway with that orientation; those outside the rectangle are not. In the example of Fig. 9-2, two bi-directional runways, one with a 105°–285° orientation and the other with 15°–195°, provide together 98.84 percent crosswind coverage when the crosswind limit is 10 knots.

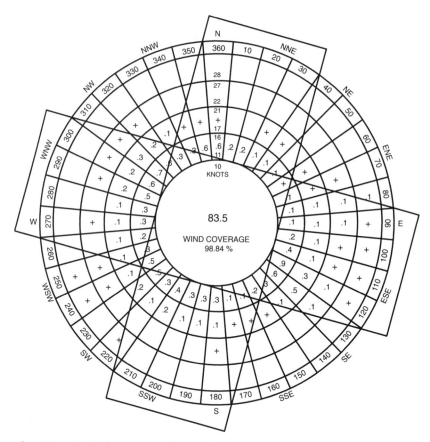

Fig. 9-2. *A wind rose and the wind coverage of two runways. Numbers indicate percentage of time and a "+" indicates "less than 0.1 percent." The total coverage achieved by the two runways is 98.84 percent.* (**Source**: FAA, 1989.)

Wind analyses are important both in the design of new airports and in any analysis of the effects of winds and weather on existing airports. For this reason, the FAA and the ICAO recommend that extensive historical wind records be analyzed, preferably covering up to 10 consecutive years.

9-4   Airport layouts

This section provides a brief descriptive review of airport layouts, with emphasis on the configuration of the airfield and of systems of runways. Several general observations are made in the beginning concerning the range of land areas that airports occupy and the principal factors that influence this parameter. A survey of some commonly used generic types of airport layouts and a summary of some of their properties and characteristics follow these observations.

Land area requirements and related observations

Airfields largely determine the total land area occupied by an airport and play a critical role in determining the airport's functionality and capacity. How much land area is actually occupied by an airfield depends on many factors. Principal among them are:

- The number, orientation, and geometry of the runways, including length, separations between parallel runways, airport reference codes selected for the purposes of airfield design, etc.
- The location of the landside facilities relative to the airside facilities
- The additional land area held in reserve for future expansion or to provide a "buffer" area for mitigation of noise and other environmental effects

Some relevant relationships are now examined in qualitative terms. The land areas that airports occupy span an enormous range, as suggested by Table 9-4. Note that the land area of New York/LaGuardia is equal to only about 2 percent of Denver/International's! The correlation between land area and the amount of traffic *actually processed* at the airports shown in Table 9-4 is rather weak. New York/LaGuardia had 25.4 million passengers in 2000, equal to roughly 65 percent of Denver/International's 38.8 million; Atlanta at 80.2 million had more than twice Denver's traffic. There is, however, a strong correlation between an airport's land area and its *potential capacity* to handle passengers and aircraft.

The airfield takes up most of the land area occupied by an airport. Depending on the size of the airport, the landside facilities (passenger buildings, cargo areas, on-site access roads, car parking, etc.) typically take up only between 5 and 20 percent of the total land area, with the larger percentages applying to airports with small land areas, such as New York/LaGuardia and Washington/Reagan. The other 80–95 percent is dedicated to the complex of runways, taxiways and aprons.

The number of runways needed to serve air traffic demand is a critical factor in determining land area requirements for airfields. In this respect, there exists an interesting systemic difference among regions of the world. This is suggested by Table 9-5, which lists the 30 busiest

Table 9-4. Land areas occupied by airports

In the United States			Outside the United States	
Airport	Land area (acres)	Land area (million m²)	Airport	Land area (million m²)
Denver/International	34,000	136	Buenos Aires/Ezeiza	34
Dallas/Ft. Worth	18,000	72	Paris/de Gaulle	31
Orlando/International	10,000	40	Amsterdam/Schiphol	22
Kansas City	8,200	33	Frankfurt/Main	19
Chicago/O'Hare	6,500	26	Athens/Venizelos	16
New York/Kennedy	4,950	20	Munich	15
Atlanta	3,750	15	Singapore	13
Los Angeles/International	3,600	14	Brussels	12
Miami/International	3,250	13	Milan/Malpensa	12
New York/Newark	2,300	9	London/Heathrow	12
Boston/Logan	2,250	9	Tokyo/Haneda	11
Washington/Reagan	960	3.8	Sydney	9
New York/LaGuardia	650	2.6	Zürich	8
			London/Gatwick	8
			Osaka/Kansai	5

Sources: ACI, 1998; airport web sites.

airports in the world in 2000, ranked by number of passengers. Of those, 19 are in North America, 5 in Asia, and 6 in Europe. The table shows the number of aircraft movements at these airports and the number of passengers per movement. One distinct characteristic of North American airports is that they process very large numbers of aircraft movements every year. London/Heathrow, Dallas/Ft. Worth, and Tokyo/Haneda had similar numbers of passengers in 2000, but Tokyo/Haneda served only 55 percent of the number of movements at London/Heathrow and 30 percent at Dallas/Ft. Worth. This also means a small number of passengers per movement at North American airports. Asian airports are at the opposite end and European airports in the middle between these two extremes. For North American airports in Table 9-5 the number of passengers per movement ranges from 96 for San Francisco/International to 56 for Phoenix. The five Asian airports have an average of 178 passengers per movement and the six European airports an average of 105.

The explanation for these large differences between regions lies in the fact that the aircraft mix at the busiest airports in Asia is heavily tilted in favor of large, wide-body jets. For example, wide-body aircraft performed 96 percent of the aircraft movements at Tokyo/Narita in 1999. By contrast, the aircraft mix at North American airports is dominated by a combination of smaller narrow-body jets, regional jets, and nonjets flown by regional airlines. A large fraction of the passengers on the latter two types of aircraft connect to flights on the larger airplanes. Europe is again in the middle. Following the 1993 deregulation of the airline industry in the European Union, there has been a sharp increase in regional flights at European airports, including "feeder" flights with smaller aircraft. Thus, Europe is getting closer to the North American model rather than to the Asian one, in terms of the aircraft mix at some of the busiest airports.

These observations have major implications for runway requirements at each region's principal airports. The busiest North American and, to a lesser extent, European airports need more runways than their Asian counterparts to serve the same number of passengers. This explains why practically all airports that operated with three or more runways in 2001 were in North America (primarily) and in Europe.

There is no indication that the average number of seats per aircraft will increase substantially in the immediate future in any region of the world. Since 1983 the average number of seats per departure on jet

Table 9-5. Traffic at the world's 30 busiest airports in 2000

	Passengers (million)	Movements (thousand)	Passengers/ Movement
Atlanta	80.2	915	88
Chicago/O'Hare	72.1	909	79
Los Angeles/International	66.4	783	85
London/Heathrow	64.6	467	138
Dallas/Ft. Worth	60.7	838	72
Tokyo/Haneda	56.4	256	220
Frankfurt/Main	49.4	459	108
Paris/de Gaulle	48.2	518	93
San Francisco/International	41.0	429	96
Amsterdam/Schiphol	39.6	432	92
Denver/International	38.8	520	75
Las Vegas	36.9	521	71
Minneapolis/St. Paul	36.8	523	70
Seoul/Gimpo	36.7	236	156
Phoenix	36.0	638	56
Detroit/Metro	35.5	555	64
Houston/Bush	35.3	484	73
New York/Newark	34.2	450	76
Miami/International	33.6	517	65
Madrid	32.9	358	92
New York/Kennedy	32.9	345	95
Hong Kong/Chep Lap Kok	32.8	194	169
London/Gatwick	32.1	261	123
Orlando/International	30.8	359	86
St. Louis/Lambert	30.6	481	64
Bangkok	29.6	195	152
Toronto/Pearson	28.9	427	68
Singapore	28.6	184	155
Seattle/Tacoma	28.4	446	64
Boston/Logan	27.4	479	57

Sources: IATA web site; ACI web site.

aircraft flown by scheduled carriers around the world has remained relatively constant around 190 seats and has actually declined slightly from a peak of 197 in 1989 to 189 in 2000. In Europe and in the United States, the average number of seats per departure was 168 and 157, respectively, in 1990 and 168 and 145 in 2000 (Swan, 2001).

The pressure to add new runways to existing airports will therefore continue for as long as traffic growth is outpacing any gains in runway capacity that are due to improvements in the ATM system alone. When feasible, new runway projects at sites that are already developed come at a very large cost for acquiring property, mitigating environmental impacts, and building the runways and associated taxiway system. A survey of five projects aimed at adding one runway each at Atlanta, Miami/International, Minneapolis/St. Paul, Seattle/Tacoma, and St. Louis found that their estimated cost in 2000 ranged from $490 million at Minneapolis/St. Paul to about $1 billion at Atlanta, Seattle/Tacoma, and St. Louis (Fortner, 2000).

Geometric characteristics

The geometric characteristics of the runway system and of the airfield are also important in determining land area requirements. Length of the runways is a simple example. Runways at major airports can be as short as 2000 m (roughly 6500 ft) or even less and sometimes as long as 4000 m (13,000 ft)—or even longer at high elevations. Obviously, the longer the runways, the more land area needed, especially if the airport property is to have a regular and compact shape, such as a rectangle. When parallel runways are present, the separation between the center-lines of the runways is another parameter critical to determining the total space the airport occupies, as well as the placement of land-side facilities. Finally, the physical dimensions of runways, taxiways, and aprons (width, separations between runways and neighboring taxiways, taxilanes, aprons and buildings, separations between parallel taxiways, etc.) depend on the airport reference code selected for design purposes. The following brief survey of some important types of airport layouts illustrates these points.

Several major airports and a large number of secondary ones have only a single runway. Due to limitations on land availability, it is also unlikely that most of these airports will ever add second runways. Their geometric layout is quite simple. Figure 9-3 sketches the layout of London/Gatwick. Although the airport seems to have a pair of close parallel runways, it actually operates as a single-runway airport: runway 08L/26R is normally used as a taxiway for 08R/26L, except at times when 08R/26L is closed down for maintenance and repairs. The landside facilities are to the side of the runway at these airports, with passenger buildings and cargo buildings possibly on opposite sides. Due to site limitations, the buildings may sometimes be inconveniently located relative to the runway. Note that at London/Gatwick the location of the main passenger building necessitates taxiing distances of about 3500 m (2.2 mi) for aircraft taking off from 08R, which is the case most of the time. The nature of the traffic that single-runway airports handle is dictated in large part by the length of the runway. Milan/Linate, with a 2700-m (8860-ft) runway, is limited to short- and medium-range flights. Tokyo/Narita, Osaka/Kansai, London/Gatwick, and Geneva all have long runways and can accommodate intercontinental, long-range flights. Single-runway airports may be able to handle surprisingly large numbers of passengers, especially if the mix of aircraft includes a high fraction of wide-body aircraft. London/Gatwick had 32.1 million passengers in 2000 and ranked 23rd in the world, and Tokyo/Narita had 26 million. San Diego, heavily utilized by medium-size and smaller aircraft, processed 207,000 aircraft movements and 15.8 million passengers.

The runway systems of many major airports consist of two parallel runways. Depending on the separation between the centerlines, the airport is said to have "close," "medium-spaced," or "independent" parallel runways. Although standards may differ somewhat from country to country, *close parallel runways* are generally those with

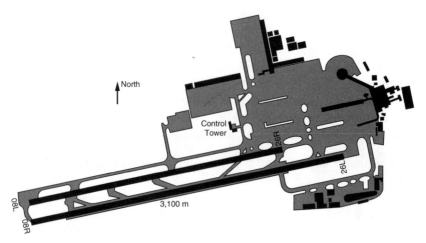

Fig. 9-3. *Layout sketch of London/Gatwick.*

centerline separations of less than 2500 ft (762 m). Under instrument flight rules (IFR), movements of aircraft on the two close parallel runways* must be carefully coordinated (Chap. 10). Independent parallel runways, on the other hand, are those whose centerlines are separated by distances greater than 3400 ft (1035 m) or 4300 ft (1310 m) or 5000 ft (1525 m), depending on airport and country (see also Chap. 10). As the name suggests, any pair of aircraft movements on the two independent parallel runways need not be coordinated, as long as a number of air traffic management and terminal airspace conditions are satisfied. Independence allows simultaneous parallel approaches to the two runways. Between these two extremes, medium-spaced parallel runways make possible independent departures from the two runways or independent "segregated" parallel operations, meaning that one runway can be used for arrivals and the other, independently, for departures. However, arrivals on two medium-spaced runways are not independent.

Obviously, the capacity of the runway system typically increases as one moves from a close to a medium-spaced to an independent pair of parallel runways. However, close and medium-spaced runways may be able to generate capacities equal to those of an independent pair when operated under visual flight rules (VFR) in good weather, as they often are in the United States. Parallel runway operations under VFR can be conducted on pairs separated by as little as 700 ft (214 m) according to the FAA and the ICAO, although the FAA recommends 1200 ft (366 m) for airplane Group V and VI runways. Chapter 10 provides details on all these points.

Close parallel and medium-spaced parallel runways do not provide sufficient space for the development of a landside complex between them. Thus, landside facilities at these airports are generally located to one or both sides of the runway pair. This arrangement is necessitated by some combination of limited land availability, environmental restrictions, and irregular shape of the site. Important examples include Philadelphia, New York/Newark, Frankfurt/Main, Seattle-Tacoma, and Milan/Malpensa.† Figure 9-4 shows a sketch of Frankfurt/Main, where the close pair of parallel runways is supplemented by a third runway. This third runway usually serves only takeoffs (and only to

*The ICAO specifies this distance as 760 m (ICAO, 1999), while the FAA converts the distance of 2500 ft to 762 m (FAA, 1989). Many minor discrepancies of this type exist, due to different practices in converting between systems of units.
†The first three of these airports actually have a third runway with a different orientation from the close or medium-spaced parallel pair. However, the great majority of movements take place on the parallel pair.

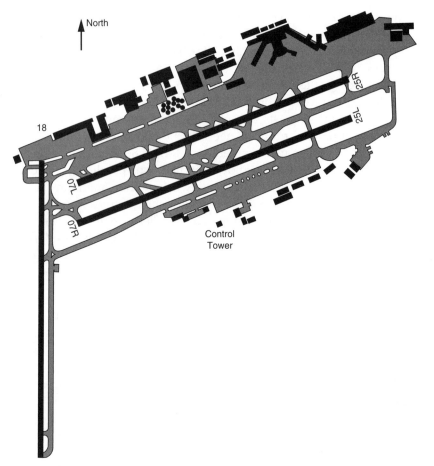

Fig. 9-4. *Layout sketch of Frankfurt/Main.*

the south), primarily due to environmental restrictions. Note that all these airports are on older sites. Due to their local and regional importance, they have all undergone (or are undergoing) major infrastructure improvements and, in some cases, limited expansion of the available land area. One important disadvantage these airports share is that aircraft operating on the runway farther from the passenger building and main apron area must usually cross an active runway or its extension. This increases surface traffic delays and taxi times, as well as air traffic controller workload (Chap. 10).

When the runway system occupies the central part of an airport's site, as in the cases of single and close or medium-spaced parallel runways, it is very important to have all passenger buildings on the same side of the runways. A few airports, such as Moscow/Sheremetyevo, Sydney, and Athens/Hellenikon (replaced in 2001 by Athens/Venizelos), have buildings on both sides of the runway system. Transfers of passengers and bags between buildings are then very difficult, expensive, and time-consuming. There is also wasteful duplication of services and of facilities on the two sides of the airport and limited opportunity to achieve economies of scale through sharing of common areas (see Chap. 15). For these and related reasons, airlines dislike operating at airports with these types of landside arrangements.

Independent parallel runways offer sufficient space between them to accommodate the bulk of an airport's landside facilities, especially when spaced by 5000 ft (1525 m) or more, as is usually the case. The landside facilities are built mostly along the central axis of the airport. Some of the busiest airports in the world and most new airports that have started operations since 1990 are in this category. Examples include Singapore, Beijing, Kuala Lumpur/International, Munich, Hong Kong, Seoul/Incheon, and Athens/Venizelos. Figure 9-5 sketches the airport of Munich as an illustration. Some of the main advantages of this family of layouts are:

- *Efficient utilization* of the vast area between the independent runways, which would otherwise be greatly underutilized

- *Reasonable proximity of passenger and cargo buildings* to both runways, assuming that the land-side configuration and apron and taxiway systems are well designed (Chap. 15)

- *Better airfield traffic circulation,* as aircraft can reach either runway without having to cross another active runway

- *Ability to isolate the airport's landside from the surroundings* of the airport and thus better control the landside's development, as well as ground access to the airport

There are disadvantages, as well. One stems from the fact that these layouts typically feature multi-lane access roads that provide ground connection to the local highway system, as well as possible tracks for rail access. Such transportation links may cut across the entire length of the airport or at least a major part of it. To ensure good circulation of aircraft on the airport's surface, this necessitates the construction of an extensive taxiway system that includes expensive taxiway bridges passing over the access roads. Munich (Fig. 9-5) has eight such bridges. A second disadvantage is that the placement of landside facilities along the central axis of the airport restricts somewhat the flexibility for expansion of these facilities when traffic grows. London/Heathrow is an extreme case in point.

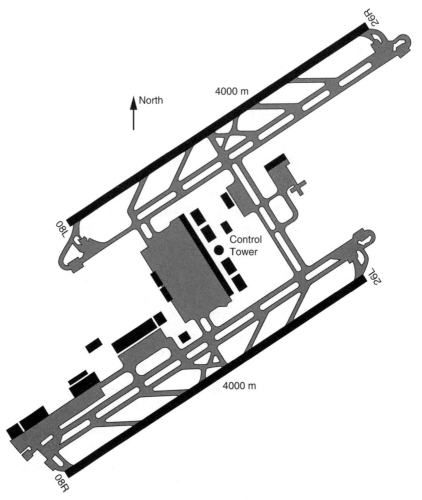

Fig. 9-5. *Layout sketch of Munich Airport.*

As Table 9-4 suggests, a site of at least 11 million m², or roughly 5 km by 2.2 km, is needed to accommodate an airport with two long parallel runways of 3.5–4 km length, separated by at least 1525 m between their centerlines.

Figure 9-5 shows that the independent parallel runways at Munich are "staggered": the threshold of runway 08L is farther along the central axis of the runway than the threshold of runway 08R, and the same is true for the thresholds of runways 26R and 26L. One of the benefits of staggered runways is that they provide additional vertical separation between aircraft operating on the two runways. For example, when two aircraft are performing simultaneous parallel approaches to the two runways, the aircraft landing on the "farthest" runway is at a higher altitude than the aircraft aiming for the closer runway.

Another advantage is reduced taxiing distances when one of the two runways is used for arrivals only and the other for departures only, as is often done in practice. For instance, when Munich is operating in a northeastern orientation, the use of runway 08R for arrivals and 08L for departures reduces the taxiing distances for both landing and departing aircraft from/to the arrival/departure runway to/from the apron area and passenger terminal. Conversely, when operations are to the southwest, assigning arrivals to 26R and departures to 26L will accomplish the same objective. Munich indeed obtains tangible taxiing distance benefits in this way. However, as traffic increases, airports are often forced to mix arrivals and departures on both runways in order to increase runway capacity (see Chap. 10). In such cases, the reduced taxiing advantage of staggered runways is mostly lost.

A possible disadvantage of staggered runways is that they increase the land area required if the airport property is to retain a rectangular or nearly rectangular shape. Land acquisition is almost always a problem in airport development.

Airports often have runways whose orientations differ. The runways intersect, either physically or along their projected centerlines.* New York/LaGuardia (Fig. 9-6) is an example. Airfield geometries with intersecting runways may be necessary at sites that often experience strong winds from several different directions. The different orientations of the runways make it possible to operate the airport under most weather conditions and provide the 95 percent or greater crosswind coverage recommended by the ICAO and the FAA (Sec. 9-3).

Airports with intersecting runways, such as New York/LaGuardia, are often difficult to operate from the air traffic management (ATM) point of view. From Fig. 9-6 it is clear that when both runways are active, aircraft movements on each must be carefully coordinated with those on the other runway. Moreover, the capacity of the runway pair will vary depending on the direction in which the operations take place and the location of the intersection point (Chap. 10). When strong winds in one direction force one of the two runways to close down, the airfield capacity is also affected in a major way. Thus, airports with intersecting runways often present difficult operational challenges.

*Two runways, which intersect only at a point along their projected centerlines, are often called "converging" or "diverging," depending on the direction of operations.

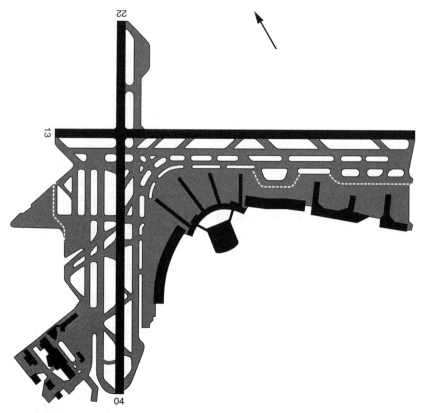

Fig. 9-6. *Layout sketch of New York/LaGuardia.*

A number of airports provide two independent parallel runways for operations in a primary orientation, but also offer reduced capacity in a secondary orientation through one intersecting ("crosswind") runway. London/Heathrow, Miami/International,* Brussels, and Tampa are examples. When winds are calm, such airports may operate with three active runways, if needed.

Atlanta (Fig. 9-7) provides an example of a runway layout that is well suited for airports operating at the next level of airfield capacity and processing 50 million or more passengers per year. It involves two pairs of close parallel runways, one pair on each side of the land-side complex. Each close pair of runways is typically operated by having departures on the inner runway (08R/26L and 09L/27R in Atlanta) and arrivals on the outer (08L/26R and 09R/27L). The distance between the two runways used for arrivals is sufficient to operate the two

*Miami's crosswind runway may be decommissioned after a new, third parallel runway goes into operation in 2003.

runways independently. Los Angeles/International and Paris/de Gaulle have a similar four-runway layout. The complex of the four runways can provide a total capacity of 140 or more movements per hour, even under IFR. Airports with a land area of 15 million m^2 or more may be able to accommodate this type of layout. For airports, such as Munich, that are currently operating with two parallel independent runways or with three parallels (e.g., Orlando/International), this layout constitutes the obvious path for future expansion.

Airports that occupy 30 million m^2 or more of land have enormous capacity potential, as they can accommodate six or more parallel (or nearly parallel) runways in the primary orientation of operations, if necessary.* Moreover, at least three of these runways can be separated

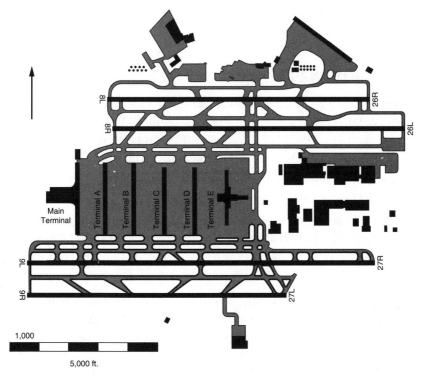

Fig. 9-7. *Layout sketch of Atlanta Airport.*

*Only one of the six airports with a land area of more than 30 million m^2 in Table 9-1 had six or more runways in 2001. This was Dallas/Ft. Worth, with seven runways and an eighth in the master plan (see Fig. 9-8). The master plan of Denver/International calls for a total of 12 runways, eight in the primary direction and four in the secondary. The airport had five runways in 2001.

by more than approximately 5000 ft (1525 m) from one another, as required by the FAA in order for all three to be used for approaches simultaneously. This means a capacity of 100 or more arrivals per hour or 200 or more movements per hour for all six runways (see Chap. 10). The United States has at least three potential such "mega-airports" in terms of land area, at Denver/International, Dallas/Ft. Worth (Fig. 9-8), and Orlando/International.

Allowing for local variations, the airfield layouts discussed are typical of those encountered at the vast majority of major airports around the world. However, they certainly do not exhaust the range of possibilities. Several multirunway airports have layouts that do not fit any of the models described. San Francisco/International, Amsterdam, and Zürich are examples of airports with four, five, and three runways, respectively, that fall in this category. Boston/Logan has five runways with three different orientations, as will be shown in Fig. 10-5. Chicago/O'Hare has seven runways with four different orientations!

9-5   Runway length

Many factors affect the runway length required by any airplane movement on any given day. The most important among those are:

- Weight of the aircraft on takeoff or on landing and the settings of its lift- or drag-increasing devices (e.g., wing flaps)
- Stage length (or nonstop distance) to be flown
- Weather, particularly temperature and surface wind
- Airport location, notably airport elevation and the presence of any physical obstacles in the general vicinity of the runway
- Runway characteristics, such as slope and runway surface condition (wet or dry pavement, surface texture)

The qualitative relationships between runway length required and these factors are quite obvious (ICAO, 1984). The greater the total weight (operating weight empty plus payload plus fuel) of an airplane, the longer are the takeoff or landing distances. Longer stage lengths mean more fuel and thus increased weight and longer takeoff distances. The greater the headwind, the shorter the required length; conversely, a tailwind increases the length of runway required. High temperatures create lower air densities, resulting in lower output of thrust and reduced lift, thus increasing runway length required. An airplane taking off on an uphill gradient requires more distance than one on a level or downhill gradient. Similarly, a wet

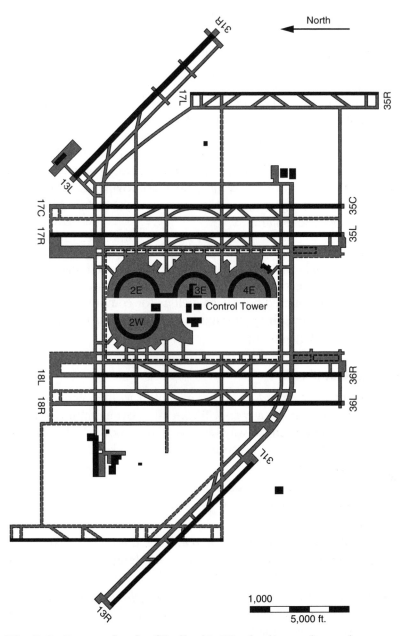

Fig. 9-8. *Layout sketch of Dallas/Ft. Worth; the north–south runway at the lower left (without a runway number) had not been constructed as of 2001.*

runway will increase the runway length required, especially on landing. The higher the elevation of an airport, the longer will be the runway required, everything else being equal. Finally, the presence of hills or other obstacles near an airport will increase the runway length requirements or may reduce the allowable takeoff weight of the aircraft. Approximate quantitative relationships are given in the ICAO (1984) that provide corrections for elevation, temperature and runway slope, to "basic length," i.e., runway length computed at sea level, zero wind, zero runway slope, and standard atmosphere. For example, runway length should be increased by 7 percent for each 300 m (~1000 ft) of airport elevation. Such approximations, however, are valid only as long as the total correction does not exceed about 35 percent of basic length.

These relationships are presented in graphical or tabular form in aircraft flight manuals and can also be approximated using mathematical formulas and correction factors. These tools form the basis for the type of analysis outlined below, as well as for computing on a daily basis the runway length requirements and constraints for the takeoff or landing of any given airplane.

Declared distances

The concept of *declared distances* is central to understanding the usability of a runway for any specific aircraft movement. It also helps explain how the design runway lengths at an airport are determined in the first place. For any given runway, four declared distances are defined (ICAO, 1984):

- TORA, the *takeoff run available:* the length of runway declared available and suitable for the ground run of an airplane taking off

- TODA, the *takeoff distance available:* the length of the takeoff run available (TORA) plus the length of the clearway, if provided

- ASDA, the *accelerate–stop distance available:* the length of the takeoff run available (TORA) plus the length of the stopway, if provided

- LDA, the *landing distance available:* the length of the runway declared available and suitable for the ground run of an airplane landing

These definitions are now explained through Fig. 9-9. When there is no clearway, stopway, or displaced threshold (see below), the four declared distances are all equal to the length of the runway (Fig. 9-9A). However, some runways may have one, two, or all three of these features.

A *clearway* (CWY in Fig. 9-9), when available, is a rectangular area, beginning at the end of the runway and centered on the runway's extended centerline, over which an airplane can make the initial portion of its flight on takeoff. The clearway can be on ground or on water. It must, however, be under the control of the airport operator or other appropriate organization, and it must be clear of any obstacles or terrain at an upward slope of 1.25 percent. Its width must be at

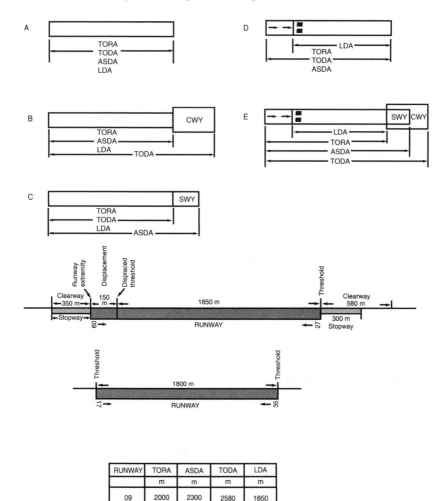

RUNWAY	TORA	ASDA	TODA	LDA
	m	m	m	m
09	2000	2300	2580	1850
27	2000	2350	2350	2000
17	NU	NU	NU	1800
35	1800	1800	1800	NU

Fig. 9-9. *Declared distances for a runway.* (*Source:* ICAO, 1999.)

least 500 ft (150 m per ICAO recommendations) and its length cannot exceed 1000 ft, according to the FAA's regulations.* When a clearway is available (Fig. 9-9B), TODA is equal to the sum of the lengths of the takeoff runway and of the clearway.

A *stopway* (SWY), when available, is a rectangular area, beginning at the end of the runway and centered on the runway's extended centerline, which has been prepared as a suitable area where an aircraft can be stopped in the case of an aborted takeoff without suffering structural damage. The stopway must be at least as wide as the runway. When a stopway is available (Fig. 9-9C), ASDA is equal to the sum of the lengths of the takeoff runway and of the stopway.

The *runway threshold* is the beginning of that portion of the runway which is usable for landing. Due to the presence of obstacles on the approach path or for environmental or other local reasons, the threshold of a runway is sometimes *displaced* and does not coincide with the physical beginning of the runway. When this is the case (Fig. 9-9D), LDA is reduced by the distance the threshold is displaced. Note that this does not affect LDA in the opposite direction of operations.

Figure 9-9E shows the four declared distances when a clearway, stopway, and displaced threshold are all present. Note that the stopway and the clearway will necessarily overlap partially, if both exist. The lower part of Fig. 9-9 shows a way to display declared distances at an airport. The values of TORA, TODA, ASDA, and LDA for each of the four runway orientations available are given in the table. Runway 09/27 has asymmetrical clearways and stopways on its two sides, as well as a displaced threshold for landings on 09. Runway 17/35 is assumed to be unidirectional: it can be used for arrivals only on runway 17 and for departures only on 35. The notation "NU" means "not usable."

Usability of a runway

Consider now the usability of a runway by a particular type of turbine-powered aircraft[†] (turbojet or turbofan). For the runway to be usable for landing, the aircraft *must be able to come to a full stop within a distance of at most 60 percent of the landing distance available*, LDA, assuming the aircraft makes a normal approach to the runway

---

*For additional details on design standards for clearways, see FAA (1989) and ICAO (1999).
†Clearways and stopways, even if present, are not taken into consideration when computing the declared distances in the case of piston aircraft. Thus, TORA = TODA = ASDA in this case. Balanced field length (see farther down in the text) always applies to takeoffs of piston aircraft.

and flies over the threshold of the runway at a height of 50 ft (15 m). Note that this leaves a large margin of safety to account for deviations, such as coming over the threshold at a higher altitude or landing at a higher than normal speed.

The requirements are more complex when it comes to takeoffs. Both the takeoff distance available, TODA, and the accelerate–stop distance available, ASDA, must be considered. Two cases must now be analyzed, a normal takeoff and a takeoff during which the failure of an engine occurs. In the first case, a distance, TOD1, is computed that is equal to 115 percent of the distance needed by the aircraft to reach a height of 35 ft (10.7 m) with all engines assumed available throughout. This provides for a margin of 15 percent to allow for variability in performance and pilot technique, much greater than the variability normally expected.

The second case assumes that the failure of one engine occurs during takeoff and the following strategy is used by the pilot. A *decision speed* V_1 (also known as *critical engine-failure speed*) is defined. If the failure occurs before the aircraft has reached V_1, then the takeoff is aborted and the aircraft is brought to a stop. If, on the other hand, the failure occurs at a speed greater than V_1, the takeoff must continue as there is not enough distance left on the runway and, possibly, the stopway to brake to a stop. Two distances are then computed for the situation when the engine fails at *exactly* the decision speed V_1. One is the distance, TOD2, which is equal to the total distance (from start of the takeoff run) needed for the aircraft to attain an altitude of 35 ft (10.7 m) if it continues the takeoff under the conditions described. (Note that no margin—like the earlier 15 percent—is applied, as this is presumably a low-probability case, given the reliability of turbine engines.) The second quantity, ASD, is the total distance (from start of the takeoff run) needed to bring the aircraft to a full stop if the takeoff is aborted.

For the runway to qualify for takeoffs by the aircraft in question, the following two conditions must both be satisfied:

- TODA must be greater than or equal to the greater of TOD1 and TOD2; *and*
- ASDA must be greater than or equal to ASD.

In words, this requires that TODA should be sufficient to accommodate a normal takeoff to 35 ft with a 15 percent margin, as well as an engine failure at a speed of V_1 or greater; and, at the same time, ASDA should provide sufficient distance for the aircraft to stop if an engine failure occurs at a speed of V_1 or less.

Obviously, both TODA (through TOD2) and ASDA depend on the decision speed V_1. What should this speed be? Note that as V_1 increases, the distance needed to come to a stop will also increase, but the distance needed to attain the height of 35 ft, if the takeoff continues with one failed engine, will decrease. The converse is also true.

It follows that, *when there is no clearway and no stopway, V_1 should be such that the distance between the point where V_1 occurs and the point where the aircraft comes to a stop should be exactly equal to the distance between the same point and the point where the aircraft reaches the required altitude of 35 ft with one engine out.* This means that TODA must be equal to ASDA. Moreover, both of these quantities should be equal to the length of the runway (or TORA); otherwise, the airport could not take advantage of the full length of the available runway. This simple relationship among these three quantities is referred to as the concept of *balanced field length*.

In the presence of a clearway and/or stopway, a somewhat different value of V_1 may be adopted, depending on the relative cost of preparing a clearway and or constructing a stopway. However, at major airports, where the length of the runway (TORA) is typically much greater than the length of any stopway or clearway that can be provided, the balanced field length approach provides optimal or near-optimal values for TODA and ASDA.

Design length

The approach for deciding on the appropriate length of the runway (or runways) to be provided at an airport follows directly from the preceding discussion. The airport designer must select the most demanding aircraft type (*critical aircraft*) to be accommodated by a runway and the *conditions of use* by that critical aircraft. The runway length that will accommodate that aircraft under these conditions is then computed. "Conditions of use" essentially mean:

- The longest nonstop distance (stage length) to be flown by the critical aircraft from/to the runway; and
- The most demanding environmental conditions during runway use, such as the mean daily temperature for the hottest month of the year at the airport.

The FAA suggests (FAA, 1990) that the critical airplane/flight combination should be a service operating for at least 250 days in a year—for instance, a flight scheduled on a Monday-through-Friday basis every week. A better approach is to identify target markets that an airport should be designed to serve as part of its long-term strategy and design, but not necessarily build, accordingly. It should be emphasized, however, that such target markets should be chosen realistically. Several secondary airports in Europe built runways long enough to serve scheduled intercontinental flights to the United States with Boeing 747 aircraft. Such flights never materialized.

Another common mistake in selecting runway lengths is building too many very long runways. Consider an airport with two parallel runways in its primary direction of runway operations. At least in the early phases of the airport's development, the airport operator may not wish to construct both runways to equal length, if the number of flights requiring a long runway is not large. For example, building one of the runways to a length of 3600 m (11,800 ft) and the other to 2700 m (8900 ft) may be perfectly adequate for an airport at sea level in an area with nonextreme summer temperatures. The long runway is sufficient for the takeoff of practically any long-range flight, while the shorter one is adequate for nearly any short- and medium-range flight. The two runways then can share all operations, except for long-range flights, which must be assigned to the long runway. If long-range traffic grows to the point where two long runways are necessary, then the shorter runway can be lengthened—assuming adequate planning exists for this eventuality.

The following should be interpreted as *only rough* (and somewhat conservative) *indications* of the typical length of runway needed to serve various types of airline flights with jet aircraft at sea level and nonextreme climates. They should *not* be used as substitutes for detailed computation of runway length requirements:

- 2000 m (6600 ft) will accommodate regional jets and many short-range flights (up to roughly 2000 km or 1200 mi) by narrow-body conventional transport jets.
- 2300 m (7500 ft) will accommodate practically all short-to-medium-range flights (3000 km or 1900 mi).
- 2700 m (8900 ft) will accommodate practically all medium-range flights (4500 km or 2500 mi).
- 3200 m (10,500 ft) will accommodate many longer range flights (9000 km or 5600 mi).
- 3500–4000 m (11,500–13,100 ft) will accommodate all feasible stage lengths in other than extremely high temperatures.

9-6   Runway geometry

The ICAO and the FAA specify on the basis of the applicable airport reference code the design standards for all the geometric characteristics of runways, other than runway length. The required or recommended dimensions and separations ensure the safe operation of the critical aircraft type for which the runway is designed. The FAA provides dimensional standards for the runway itself and for runway-associated elements that include:

- Runway shoulders
- Runway blast pad
- Runway safety area
- Obstacle-free zone
- Runway object-free area
- Clearway
- Stopway

Section 9-5 described the dimensional standards for any clearways or stopways at an airport. The text below briefly describes the other elements. Table 9-6 summarizes the associated FAA dimensional standards. Additional details can be found in FAA (1989), its amendments, appendices, and supporting documents. The ICAO design standards are quite similar, although the terminology may be somewhat different.

The *runway shoulders* (Fig. 9-10) are adjacent to the structural pavement of the runway. They provide resistance to jet blast erosion and accommodate maintenance and emergency equipment. A natural surface with dense, well-rooted turf may suffice for the runway shoulders, but paved shoulders are recommended for runways (as well as taxiways and aprons) that accommodate Group III or higher aircraft. Groups V and VI normally require paved shoulder surfaces.

Runway blast pads provide blast erosion protection beyond runway ends. They should extend across the full length of the runway plus its shoulders.

The *runway safety area* (RSA) includes, but is not limited to, the structural pavement, runway shoulders, runway blast pads, and stopways (Fig. 9-10). It was formerly known as the "landing strip." It must be: cleared, graded, and free of hazardous surface variations; drained through sufficient grading or storm sewers; capable of supporting snow-removal equipment, rescue, and firefighting equipment and

Table 9-6. Dimensional standards for runways (FAA, 1989)

	Airplane design group					
	I	**II**	**III**	**IV**	**V**	**VI**
Runway width	100 ft	100 ft	100 ft	150 ft	150 ft	200 ft
	30 m	30 m	30 m*	45 m	45 m	60 m
Runway shoulder width[†]	10 ft	10 ft	20 ft	25 ft	35 ft	40 ft
	3 m	3 m	6 m*	7.5 m	10.5 m	12 m
Runway blast pad width	120 ft	120 ft	140 ft	200 ft	220 ft	280 ft
	36 m	36 m	42 m*	60 m	66 m	84 m
Runway blast pad length	100 ft	150 ft	200 ft	200 ft	400 ft	400 ft
	30 m	45 m	60 m	60 m	120 m	120 m
Runway safety area width			500 ft			
			150 m			
Runway safety area width beyond RW end[†]			1000 ft			
			300 m			
Obstacle-free zone width			400 ft			
			120 m			
Obstacle-free zone length beyond RW end			200 ft			
			60 m			
Runway object-free area width			800 ft			
			240 m			
Runway object-free area length beyond RW end[‡]			1000 ft			
			300 m			

*For runways in Group III serving airplanes with MTOW greater than 150,000 lb (68.1 tons), the standard runway width, runway shoulder width, and runway blast pad width are the same as for Group IV runways.
†See FAA (1989) for some special cases.
‡The runway safety area and runway object-free area lengths begin at each runway end, when a stopway is not provided; otherwise, they begin at the stopway end.

the occasional passage of aircraft without causing structural damage to the aircraft; and free of objects, except those that need to be located in the RSA because of their function (such as runway lights). Objects higher than 3 in (7.6 cm) above grade should be constructed on frangible mounted structures with the frangible point no higher than 3 in above grade.*

*Frangible objects are those with low mass designed to break, distort, or yield on impact, so as to present the minimum hazard to aircraft (ICAO, 1999).

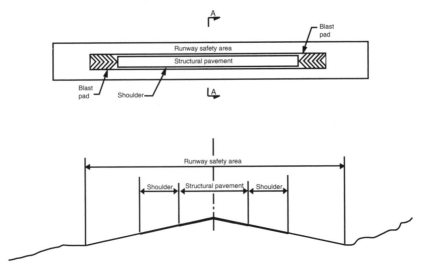

Fig. 9-10. *The runway safety area.* (**Source:** FAA, 1989.)

The *obstacle-free zone* (OFZ) defines a volume of protected airspace below 150 ft (45 m) above the established airport elevation (Fig. 9-11). It is centered above the runway and the extended runway centerline and is intended to provide clearance protection for aircraft landing or taking off from the runway and for missed approaches ("balked landings" in ICAO terms). It must be clear of all objects, except for frangible visual navigational aids that need to be located within the OFZ. The OFZ is subdivided into:

- *Runway OFZ,* the airspace above a surface centered on the runway centerline
- *Inner-approach OFZ,* centered on the extended runway center-line and applicable only to runways with an approach lighting system
- *Inner-transitional OFZ,* the airspace above the surfaces located on the outer edges of the runway OFZ and the inner-approach OFZ and applicable only to runways with an approach visibility minimum lower than ³/₄ of a statute mile

The *runway object-free area* (OFA) occupies ground centered on the centerline of a runway, taxiway, or taxilane. It is an area kept free of all objects, except those needed for air navigation or aircraft maneuvering purposes. It is acceptable to taxi or temporarily hold aircraft within the OFA, when necessary. Note (Fig. 9-12) that the OFA surrounds a runway and extends beyond its ends. Beginning in 2000,

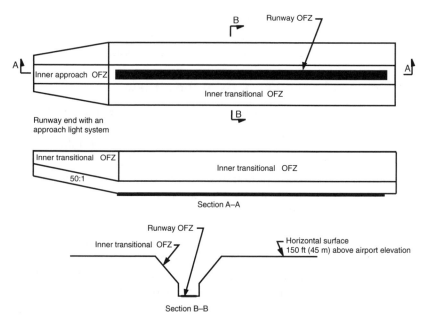

Fig. 9-11. *Obstacle-free zone for runways serving large aircraft with lower than ³/4-statute mile (1200 m) approach visibility minimums.* (*Source:* FAA, 1989.)

the FAA requires that any new authorized precision approaches with less than ³/4-mi visibility also provide a *precision object-free area* (POFA). The POFA is centered on the runway centerline, begins at the runway threshold, extends to 200 ft along the runway's centerline, and is 800 ft wide (FAA, 2000). In Fig. 9-12, the POFA would occupy the gap between the runway threshold and the runway protection zone and extend 400 ft to each side of the extended runway centerline.

In addition to the elements identified above, the *runway protection zone* (RPZ) is an area off the end of the runway intended to enhance the protection of people and property on the ground. As Fig. 9-12 shows, the RPZ contains a part of the OFA. The portion of the RPZ beyond and to the sides of the OFA, the *controlled activity area,* should be under the control of the airport operator, as much as possible. The controlled activity area should be reserved for uses and activities that do not interfere with airport operations and with navigational aids. Some agricultural activities that do not attract wildlife are expressly permitted, but residences, places of public assembly, and fuel storage facilities are prohibited. For precision approach runways, the length, L, of the RPZ (see Fig. 9-12) should be 2500 ft (750 m) and the widths, W_1 and W_2, 1000 and 1750 ft, respectively. This results in a total land area for the RPZ of about 80 acres (32 ha).

Separations from other parts of the airfield

The required distances between runways and other parts of the airfield are an important aspect of geometric design. Table 9-7 summarizes some of the FAA standards in this respect. Holdlines keep aircraft waiting to use a runway at a distance sufficient to ensure that no part of the aircraft penetrates any obstacle limitation surfaces. They should also be located so that aircraft do not interfere with the operation of navigation aids.

The most significant differences between Table 9-7 and the corresponding ICAO standards concern the distances between runway centerlines and taxiway centerlines for instrument runways.* For ICAO code letters C, D, E, and F, the ICAO standards call for 168, 176, 182.5, and 190 m of separation, respectively—compared to the 120, 120, 120, and 180 m shown for Groups III through VI in Table 9-7. For the other separation distances listed in Table 9-7, the ICAO standards are also more conservative, generally.† For example, for the distance between runway centerlines and holdlines (as well as holding bays, see

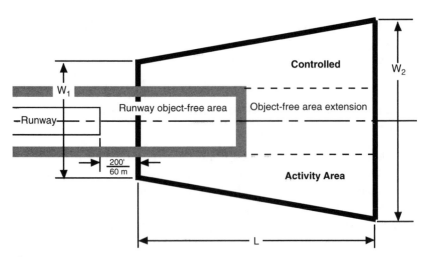

Fig. 9-12. *Runway protection zone.* (***Source:*** FAA, 1989.)

*For detailed descriptions and special cases, see Tables 3-1 and 3-2 in ICAO,1999.
†Comparisons cannot be exact because of the somewhat different ways the ICAO and the FAA specify these standards in terms of their respective airport reference codes and runway instrumentation classes.

Table 9-7. Separation standards between runways and other facilities or parts of the airfield for approach categories C and D (FAA, 1989, 1999)

	Airplane design group					
	I	II	III	IV	V	VI

Visual runways and runways with not lower than ¾ statute mile (1200 m) approach visibility minimum

Runway centerline to:

	I	II	III	IV	V	VI
Holdline†			250 ft			
			75 m			
Taxiway/taxilane centerline	300 ft 90 m	300 ft 90 m	400 ft 120 m	400 ft 120 m	400 ft* 120 m	600 ft 180 m
Aircraft parking area	400 ft 120 m	400 ft 120 m	500 ft 150 m	500 ft 150 m	500 ft 150 m	500 ft 150 m

Runways with lower than 3/4 statute mile (1200 m) approach visibility minimum

Runway centerline to:

	I	II	III	IV	V	VI
Holdline†	250 ft 75 m	250 ft 75 m	250 ft 75 m	250 ft 75 m	280 ft 85 m	280 ft 85 m
Taxiway/taxilane centerline	400 ft 120 m	400 ft 120 m	400 ft 120 m	400 ft 120 m	400 ft* 120 m	600 ft 180 m
Aircraft parking , area			500 ft 150 m			

*The distance increases to 450 ft (135 m) for airports with elevations between 1345 and 6560 ft (410–2000 m) and to 500 ft (150 m) for airports with elevations above 6560 ft (2000 m)
†For additional separations to holdlines at higher elevations for Design Groups III–VI, see FAA (1999.)

Sec. 9-7) at airports designed to ICAO code number 4, the ICAO calls for distances of 75 m for noninstrument, nonprecision approach and takeoff runways, of 90 m for precision approach runways and of 107.5 m for precision approach runways serving ICAO code letter F (or FAA Group VI) aircraft. The FAA counterparts of these three values in Table 9-7 are 75 m, 75 m (or 85 m), and 85 m, respectively.*

Vertical profile

The FAA and the ICAO also have strict standards for runway and taxiway surface gradients and lines of sight. This brief review is limited to

*The FAA provides for the possibility of increased distances to holdlines if aircraft intrude on areas critical to the operation of instrument landing systems.

FAA standards for runways designed to serve aircraft in approach categories C, D, and E of the FAA's airport reference code. The longitudinal and transverse gradient standards are shown in Figs. 9-13 and 9-14, respectively. Analogous but different standards apply to approach categories A and B.

The *longitudinal* standards apply restrictions on the runway grades allowed, on the changes in grades, and on the distances between changes. In general, it is desirable to keep the longitudinal grades, as well as the number and size of changes in grades, to a minimum. The maximum longitudinal grade allowed is ±1.5 percent. However, in the first and last quarter of the runway length, the grade may not exceed ±0.8 percent. At the same time, the maximum allowable grade change is ±1.5 percent. Parabolic vertical curves should be used to effect longitudinal changes in grade. To ensure smooth transition between grades, the length of the vertical curve should be at least 1000 ft (300 m) for each 1 percent of change in grade. Moreover, to avoid frequent changes in grade, the minimum allowable distance between successive changes in grade is 1000 ft (300 m) multiplied by the sum, in percent, of the grade changes, associated with the two vertical curves. Thus, the distance between a +1 percent grade change and a −0.5 percent grade change must be at least 1500 ft. The longitudinal grades applied to a runway should also be applied to the entire runway safety area (RSA). Figure 9-13 also shows the longitudinal grade standards for the parts of the RSA immediately beyond the runway ends.

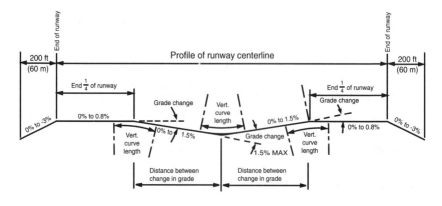

Minimum distance between change in grade = 1000 ft (300 m) x sum of grade changes (in percent)
Minimum length of vertical curves = 1000 ft (300 m) x grade change (in percent)

Fig. 9-13. *Longitudinal grade limitations for FAA aircraft approach categories C and D.* (**Source:** FAA, 1989.)

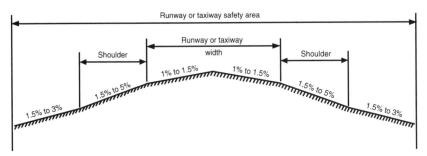

Fig. 9-14. *Transverse grade limitations for FAA aircraft approach categories C and D.* (***Source:*** FAA, 1989.)

Figure 9-14 shows the maximum and minimum transverse grades for runways, taxiways, and stopways, including the runway shoulders and the runway safety area. In general, the smallest transverse grades that satisfy local drainage requirements should be used.

To enhance safety, line-of-sight standards have also been set for runways. An acceptable runway profile permits any two points 5 ft (1.5 m) above the runway centerline to be mutually visible for the entire runway length. However, if the runway has a full-length parallel taxiway, this requirement is reduced to the two points being mutually visible for *one-half* the runway length. A clear line of sight between the ends of intersecting runways is recommended, as well. For detailed guidance on this point, see FAA (1989). There are no line-of-sight requirements for taxiways.* However, the sight distance along a runway from an intersecting taxiway must be sufficient to allow a taxiing aircraft to enter safely or cross the runway.

9-7 Taxiways

Inspection of Figs. 9-5, 9-6, and 9-7 shows that taxiway systems at major airports can be extensive, complex in configuration, and costly to construct and maintain. The taxiway system at Munich, for example, has a length of approximately 30 km (about 19 mi), not including the taxilanes in apron areas. In contrast, the combined length of the two runways is 8 km (8 mi).

All too often in airfield design, the taxiway system is almost an afterthought. Typically the positioning and configuration of the runways

---

*The ICAO does have a sight distance requirement for taxiways; for example, where the code letter is C, D, E, or F, an observer 3 m above the surface of the taxiway should be able to see the whole surface of the taxiway to a distance of at least 300 m (ICAO, 1999).

and of the landside facilities, including the ground access roadways and other guideways, are fixed first. The taxiway system is then designed to provide connections between the runways and the apron areas near and around passenger and cargo buildings, maintenance areas, etc. This can be a costly approach in terms of both fixed and operating costs. It may, for example, lead to the construction of an unnecessarily large number of expensive (segments of) taxiways on bridges, so that road traffic can access the passenger buildings. It may also require aircraft to take circuitous routes between the runways and the apron stands, increasing airline operating costs and wasting time. Airports with landside facilities located at midfield are particularly prone to such problems, if not designed in a manner that integrates the planning of the taxiway system into the overall process.

To appreciate the magnitude of the economic quantities involved, consider the taxiing time that aircraft experience in traveling between aprons and runways at a busy airport handling, for example, 300,000 movements and 27 million passengers in a year. A design improvement that reduces taxiing time by 1 min, on average, translates to savings of 5000 aircraft hours and 450,000 passenger hours per year. If the average direct operating cost for the mix of aircraft at this airport is $2400 per aircraft hour,* this means savings of about $12 million per year in direct airline costs alone. Assuming a 25-year economic lifetime for airfield facilities and a 7 percent discount rate in constant prices, $12 million in annual savings would justify a capital investment of up to $140 million to obtain the reduction of 1 min in taxiing times! This amount would, in fact, double to a total of approximately $300 million, if savings in passenger time at $30 per hour saved (typical for such analyses in the United States) were also taken into consideration.†

Another common mistake is overbuilding in the early stages. Runways that are not used intensively during the early years of an airport's operations can usually be adequately supported, depending on the airport's geometry, by a single taxiway running parallel to the entire length of the runway ("full-length taxiway"). Even when a second parallel taxiway may be needed to ensure smooth circulation of airport

---

*An average direct operating cost of $2400 per hour is reasonable for an airport with an average of 90 passengers per movement (= 27,000,000/300,000), which suggests a limited presence of nonjets and regional jets.

†One problem with this type of reasoning is that the direct recipients of the benefits are the users (airlines and passengers), while the capital costs are incurred by the airport operator. If, however, benefits exceed costs, then the airport operator can, at least in theory, recover its costs through user charges and still leave the users better off than they would have been without the capital investment that provides the savings in taxiing time.

surface traffic under all circumstances, that second taxiway need not be full-length. It may be sufficient in most cases to build that second parallel taxiway for only part of the length of the runway. However, airport designers tend to emphasize symmetry. Almost reflexively sometimes, they design taxiway systems with two full-length parallel taxiways per runway and then build the entire taxiway system in a single step. The key to economic efficiency is to design carefully a flexible taxiway system and then build it up in phases over the years as traffic grows.

Design standards for taxiways include recommendations for taxiway width; taxiway curves; minimum separation distances between taxiways and parallel taxiways, taxiways, and objects; longitudinal slope changes, sight distances, and transverse slopes (FAA, 1989; ICAO, 1991, 1999). Both the FAA and the ICAO make special reference to taxiways located around and in apron areas. These taxiways are divided into two types: *apron taxiways,* which may surround aprons or may provide a route across them; and *taxilanes* (or *aircraft stand taxilanes*), which are areas in an apron designated as taxiways and providing access to the aircraft stands only. Standards and recommendations for apron taxiways are essentially the same as for regular taxiways. However, some separation requirements for taxilanes may be less conservative than for taxiways. This is because aircraft move more slowly in apron areas and thus their paths typically adhere more closely to the centerlines of taxilanes than to those of regular taxiways.

Table 9-8 lists the principal FAA dimensional standards for taxiways (upper part of table) and the minimum separations between the centerline of a taxiway and either a parallel taxiway/taxilane or a fixed or movable object (lower part). The *edge safety margin* is defined as the minimum acceptable distance between the outside of the airplane wheels and the pavement edge. The other dimensions in the upper part of Table 9-8 are analogous to the dimensions defined earlier for runways. For instance, the *taxiway safety area,* by analogy to the runway safety area, includes, but is not limited to, the taxiway and taxiway shoulders and must be cleared and graded, drained, and free of objects.

The ICAO has similar dimensional standards for taxiways [see Chap. 3 of ICAO(1999) and, especially, ICAO (1991)]. For airports serving large aircraft, the principal difference with the FAA standards is that for its code letter F, the ICAO recommends a taxiway width of 25 m and an edge safety margin (*clearance distance*) of 4.5 m, by comparison to the FAA's 30 m and 6 m, respectively, for Group VI.

Table 9-8. Taxiway design standards (FAA, 1989)

	Airplane design group					
	I	**II**	**III**	**IV**	**V**	**VI**
Dimensional standards						
Width	25 ft	35 ft	50 ft*	75 ft	75 ft	100 ft
	7.5 m	10.5 m	15 m	23 m	23 m	30 m
Edge safety	5 ft	7.5 ft	10 ft*	15 ft	15 ft	20 ft
margin	1.5 m	2.25 m	3 m	4.5 m	4.5 m	6 m
Shoulder width	10 ft	10 ft	20 ft	25 ft	35 ft	40 ft
	3 m	3 m	6 m	7.5 m	10.5 m	12 m
Safety area width	49 ft	79 ft	118 ft	171 ft	214 ft	262 ft
	15 m	24 m	36 m	52 m	65 m	80 m
Taxiway object-	89 ft	131 ft	186 ft	259 ft	320 ft	386 ft
free area width	27 m	40 m	56 m	79 m	97 m	118 m
Taxilane object-	79 ft	115 ft	162 ft	225 ft	276 ft	334 ft
free area width	24 m	35 m	49 m	68 m	84 m	102 m
Separation distances						
Taxiway centerline to:						
Parallel taxiway	69 ft	105 ft	152 ft	215 ft	267 ft	324 ft
or taxilane	21 m	32 m	46.5 m	65.5 m	81 m	99 m
centerline						
Fixed or movable	44.5 ft	65.5 ft	93 ft	129.5 ft	160 ft	193 ft
object	13.5 m	20 m	28.5 m	39.5 m	48.5 m	59 m
Taxilane centerline to:						
Parallel taxilane	64 ft	97 ft	140 ft	198 ft	245 ft	298 ft
centerline	19.5 m	29.5 m	42.5 m	60 m	74.5 m	91 m
Fixed or	39.5 ft	57.5 ft	81 ft	112.5 ft	138 ft	167 ft
movable	12 m	17.5 m	24.5 m	34 m	42 m	51 m
object						

*For airplanes in Group III with a wheelbase equal to or greater than 60 ft (18 m), the standard taxiway width is 60 ft (18 m) and the taxiway edge safety margin is 15 ft (4.5 m). *Source:* FAA, 1989.

Note that the FAA separation distances in the lower half of Table 9-8 are different for taxiways and for taxilanes. These separation distances are based on a set of simple formulas that may be used to take into consideration any special local conditions. The separation between a taxiway centerline and a parallel taxiway or taxilane centerline is given by 1.2 times the maximum wingspan of the airplane design group plus 10 ft or 3 m. For Group VI, with a 262-ft (80-m) maximum wingspan (Table 9-3b), this gives 324 ft (99 m) of separation, as shown in Table 9-8. Note that this leaves a margin of 62 ft or 19 m for the total possible deviation from the centerlines of the parallel taxiways when two Group VI airplanes are moving (usually in opposite directions) on the taxiways. The margin is, of course, larger when one or both of the aircraft are smaller than Group VI.

For the separation between the centerlines of a taxilane and a parallel taxilane, 1.1 times the maximum wingspan (instead of 1.2) plus 10 ft or 3 m is used. Similarly, a taxiway centerline should be separated by 0.7 times the maximum wingspan plus 10 ft (3 m) from a fixed or movable object, while 0.6, instead of 0.7, is used in the case of a taxilane. The ICAO uses somewhat different formulas than the FAA to determine a set of separation standards [see Table 3-1 in ICAO (1999)], which are very similar to the ones shown in the lower part of Table 9-8—in most cases within 1.5 m of the corresponding values. ICAO, however, does not make a distinction between taxiways and taxilanes, when it comes to separation from a parallel taxilane.

Special cases

A taxiway system includes segments and special-purpose elements that require more detailed design considerations, due to their particular characteristics. Included in this category are

- Curved segments of taxiways
- Taxiway intersections or junctions
- Taxiways on bridges
- Exit taxiways, including high-speed (or rapid or acute-angle) exit taxiways
- Holding bays and bypass taxiways

For *curved segments of taxiways* and for *intersections and junctions* with runways, aprons, and other taxiways, dimensional requirements are driven by the FAA's edge safety margins (Table 9-8) or the ICAO's clearance distances and the *wheelbase* of the design aircraft. The wheelbase is the distance from the nose gear of the aircraft to the geometric center of the main gear. Additional taxiway width or fillets* are provided to ensure that the applicable edge safety margins

---

*The term *extra taxiway width* is often used for curved segments of taxiways, whereas *fillet* is used to refer to additional width provided at junctions and intersections; in either case, the strength of the additional pavement provided must be the same as that of the taxiway.

are maintained, assuming the aircraft's nose gear stays on centerline markings (Fig. 9-15). Detailed guidance on the design of these curved segments can be found in ICAO (1991), FAA (1989), and Horonjeff and McKelvey (1994).

Taxiways on bridges are becoming increasingly common, especially at airports where the landside facilities occupy the central portion of the airport* (Sec. 9-3). Due to the high cost of these structures, airport designers try to minimize their number. This often leads to difficult trade-offs between construction (fixed) costs, on the one hand, and taxiing and operations (variable) costs, on the other, as noted earlier. The ICAO and the FAA guidelines regarding taxiway bridges are very similar. The width of the part of the taxiway that lies on the bridge should be at least as great as the taxiway safety area (FAA) or the graded part of the taxiway strip (ICAO). Where this may not be possible, edge protection should be provided on the bridge, as well as engine blast protection for vehicles or people crossing under the bridge. There should also be adequate space on the bridge for access by firefighting and rescue equipment from both sides and for aircraft evacuation slides. The bridges should be on straight segments of taxiways and away from high-speed exit taxiways or other special-purpose parts of the airfield (FAA, 1989; ICAO, 1991, 1999).

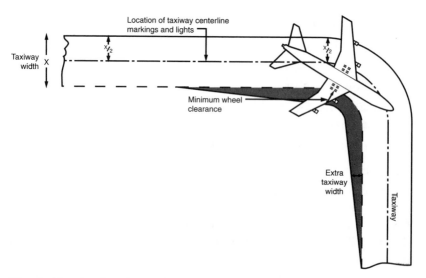

Fig. 9-15. *A curved segment of a taxiway; additional taxiway width may be needed to ensure that taxiway edge safety margins are preserved. (**Source:** ICAO, 1999)*

---

*A small number of airports also have parts of runways on bridges over highways.

Exit taxiways provide egress paths from runways for arriving aircraft. The centerlines of conventional exits form a 90° angle with the centerlines of the runway (Fig. 9-16). *High-speed exits* (or *rapid exit taxiways* or *acute-angle exit taxiways* in ICAO and FAA terms, respectively) are those whose centerlines form an angle significantly less than 90° with the runway centerline. With a 30° high-speed exit, aircraft can theoretically initiate a turn while traveling as fast as 90 km/h (56 mi/h) or more. However, in practice, most pilots are more conservative. Figure 9-17 shows a typical geometric design of a high-speed exit. An angle of 30° between the centerlines of the high-speed exit taxiway and the runway is common. The ICAO specifies that the radius of the turn-off curve of a high-speed exit (Fig. 9-17) should be at least 550 m to enable exit speeds of 93 km/h (about 58 mi/h) under wet runway surface conditions for ICAO code numbers 3 or 4 aircraft (ICAO, 1999). For a high-speed exit of this type the FAA requires at

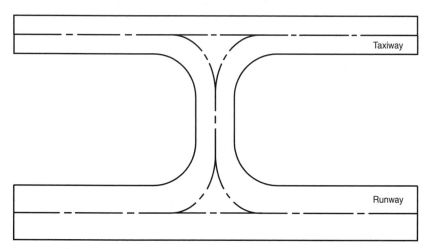

Fig. 9-16. *Conventional, right-angle exit taxiway.* (*Source:* FAA, 1989.)

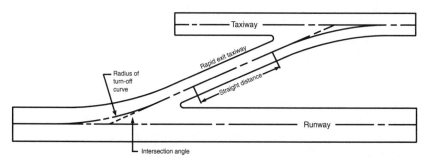

Fig. 9-17. *Acute-angled exit taxiway.* (*Source:* FAA, 1989)

least 180 m (600 ft) of separation between the centerline of the runway and the centerline of the parallel taxiway (FAA, 1989). Note that the FAA's runway-to-taxiway separation distances shown in Table 9-7 apply to the case where conventional, right-angle exit taxiways are used. These distances may have to be increased in order to accommodate high-speed exits.

The location of exit taxiways plays a significant role in determining runway occupancy times and, under certain conditions (Chap. 10), runway capacity. To take an extreme example, average runway occupancy times will obviously be much longer on a long runway which has a single exit at its far end than on an equally long runway that has several well-located exits. There is, however, a point of diminishing returns, after which little is gained by constructing additional exits. High-speed exits contribute to reducing runway occupancy times. Given, however, their higher cost and diminishing returns as their number increases, it is difficult to make a case for constructing more than, at most, three high-speed exits for each direction of operation of a runway, if other conventional exits at 90° are provided. Moreover, this can be justified only for intensively utilized runways, with more than 30 peak period movements per hour.

Several studies have been performed on the optimal location of exit taxiways, beginning in the 1950s (Horonjeff, 1959; Daellenbach, 1974; ICAO, 1991). However, few general statements can be made in this regard, due to the many local factors that play a major role in the exit selection process. Included among those are the mix of aircraft types using the runway, pilot technique, the condition of the runway surface (wet or dry), and the location of aircraft stands relative to the runway. On this last point, it has been observed that pilots often aim for the exit(s) that will most facilitate access to the stands and adjust the deceleration of the aircraft accordingly. As an approximate rule, a long runway that may be used in either direction at a busy airport, in addition to its two ends, should have exit taxiways at approximately 450-m (~1500 ft) intervals toward its middle, up to a distance of about 600 m (~2000 ft) from its ends. Thus, a long 3500-m (~11,500-ft) runway should have seven or eight exit taxiways, including the two at its ends.

One of the most common mistakes in airport design is the placement of high-speed exits on runways (or in runway directions) that will seldom, if ever, be used for arrivals. It is often entirely appropriate to equip one direction of operation of a given runway with high-speed

Fig. 9-18. *Runway 01L-19R and exit taxiways at Amsterdam/Schiphol.*

exits and the other direction with only conventional exits. The example of Runway 01L/19R at Amsterdam/Schiphol is shown in Fig. 9-18. The asymmetry in the types of exits is explained by the fact that Runway 01L is almost always used for departures only, while 19R for arrivals only. In general, it is useful to remember that high-speed exits offer essentially no capacity benefits at runways used only for departures, some capacity benefits at runways used only for arrivals, and significant capacity benefits, under some movement-sequencing strategies, at runways used in a mixed operations mode (Chap. 10).

Holding bays are areas adjacent to taxiways where aircraft may be held temporarily without impeding the circulation of other taxiing aircraft. Holding bays are usually placed close to runways, so they can provide a waiting area for aircraft that are not yet ready for takeoff, as well as allow air traffic controllers to sequence departing aircraft in a particular way, if desired. Figure 9-5 shows the holding bays next to the four runway ends at Munich. Holding bays take several different geometric shapes (ICAO, 1991). They should be located so they do not interfere with instrument landing systems and other navigational aids and keep aircraft out of obstacle-free zones and runway safety areas. *Dual taxiways* provide a second taxiway segment that makes it possible to bypass aircraft near critical points of the airfield, typically runway ends. Figure 9-5 shows four dual taxiway pairs next to the four runway ends at Munich.

9-8 Aprons

Aprons provide the interface between airside and landside facilities at airports. They can be classified, according to the type of aircraft stands they hold, into passenger building aprons, cargo building aprons, long-term parking aprons, service and hangar aprons, and general aviation aprons. Long-term parking aprons, if any, are located at remote areas of the airfield and can be used for aircraft parked for periods ranging from overnight to months (due to temporary grounding). General aviation aprons may be separated into areas for "itinerant" aircraft passing through and areas for aircraft based at the airport in question. These latter aprons are also known as "tiedowns."

Passenger building stands, on which this section focuses, are subdivided into *contact* and *remote,* depending on their location relative to the buildings. Figure 9-19 shows schematically generic configurations of passenger building aprons. The configuration of the aprons clearly depends on the passenger building concept adopted. For this reason, a more detailed discussion of the advantages and disadvantages of the different apron configurations is provided in Chaps. 14 and 15. Only brief related observations are made here.

The objective of apron design is to develop a configuration that respects all safety-related requirements, while maximizing efficiency

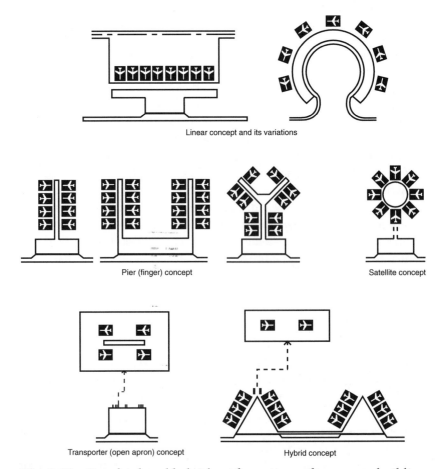

Fig. 9-19. *Standard and hybrid configurations of passenger building and aprons.* (***Source:*** ICAO, 1991.)

for aircraft moving in and out of the apron and providing flexibility. On the safety side, ICAO requires (ICAO, 1991, 1999) that the following minimum clearances be provided at an aircraft stand between any part of the aircraft and any adjacent building, aircraft on another stand, or other object, except for vehicles and equipment servicing the aircraft:

- 3 m for code letters A and B
- 4.5 m for code letter C
- 7.5 m for code letters D, E, and F

However, these may be reduced in the specific case of the clearance between the nose of the aircraft and the passenger building (including passenger loading bridges).

Of special importance are the expandability of the apron area and its ability to accommodate the full range of aircraft using the airport. The latter poses a particularly difficult challenge. At one extreme, the dimensions of the stands in an apron could be large enough to accommodate all potential aircraft sizes (e.g., Group V or smaller, for airports not expected to serve Group VI aircraft). This, however, would be highly inefficient in most cases. At the opposite end, the mix of stand sizes could be identical to the current mix of aircraft parked at the airport during peak demand periods. The obvious disadvantage is that this offers little flexibility if the mix changes in the direction of a higher fraction of larger aircraft. The proper compromise is to provide a mix of stand sizes that is biased toward having a higher fraction of larger stands than is warranted by the current aircraft mix. However, developing the specifics of this approach is a complex task that depends on much more than apron-related considerations alone. For example, if the airport relies on contact stands, the choice of the mix of stand sizes clearly affects directly the dimensions of the passenger building. In this case, any bias toward a high percentage of larger stands must be tempered by the associated very large capital costs. Chapters 14 and 15 address this question further.

The efficiency of operations associated with different apron designs also involves complex questions. For example, trade-offs must be made between ease of movement of aircraft versus passenger convenience and passenger building operating costs. To take an obvious example, the transporter (or "open apron") concept shown at the lower left of Fig. 9-19 greatly facilitates the movement of aircraft—as they can park in less space-constrained areas and do not usually have to be pushed back from their stands. However, this requires transporting passengers to/from the aircraft, with buses or special-purpose vehicles, implying higher variable costs and necessitating "closing" of acceptance of passengers for departing flights 30 min or more before departure time. As a second example, the linear concept (top of Fig. 9-19) also facilitates aircraft movements, but makes inefficient use of the frontage of the passenger building, as aircraft can park on only one side of the building. It also requires considerable duplication of landside services and, thus, leads to higher operating costs (Chaps. 14 and 15). For high-volume connecting airports, a concept that has emerged as a good solution to this difficult design optimization problem is the midfield satellite terminal (Chap. 14). The best-known—and original—example of this particular design is Atlanta (Fig. 9-7).

An important problem that occurs frequently at aprons of major airports is caused by the blocking of aircraft movements in the taxilanes that serve sets of stands. This problem occurs primarily around pier buildings and may reduce significantly the capacity of the apron. A typical example is the one involving parallel piers, as shown in Fig. 9-19. Ideally, the distance between the piers should be sufficient to allow for two parallel taxilanes, one serving incoming and the other outgoing traffic (lower part, Fig. 9-20). Note that the distances shown in Fig. 9-20 are based on the FAA's formulas for separations between parallel taxilane centerlines and between a taxilane centerline and an object described in Sec. 9-7. If the distance between the piers is not sufficient, a single taxilane must serve all the stands in the apron. Whenever an aircraft is either pushing back from a stand or moving on the taxiway in either direction, it blocks circulation in and out of the stands.* Serious delays to traffic may result. A simple rule of thumb is that such delays will indeed occur if there are *more than 4–6 stands* on each side of the single taxilane. The extent of the delays and the number of stands at which the problem will set in depends on the size of aircraft served by the stands and the nature of the flights affected (short-range versus long-range, etc.). The same problem arises with numerous variants of the single-taxilane geometric configuration. As an example, Fig. 9-21 shows the narrow entrance and exit to an apron between passenger buildings B and C at Boston/Logan. This apron is shaped like a horseshoe. Any aircraft occupying the narrow space between the two passenger buildings effectively blocks circulation into or out of that apron.

---

*An important factor in apron operations is the turning radius of aircraft. The turning radius is the distance between the pivot around which the aircraft turns and the part of the aircraft farthest from the pivot. The pivot is located along the centerline of the main undercarriage of the aircraft, at a distance from the center of the aircraft's fuselage, typically under the inner part of the wing. The point farthest from the pivot is usually the wing tip, but, for some aircraft it can be the nose or the tips of the horizontal stabilizers at the tail of the aircraft. The turning radius of the common types of commercial jets is typically equal to between 65 and 100 percent of the wing span of the aircraft.

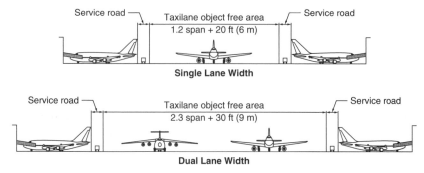

Fig. 9-20. *Single versus dual taxilanes and wingtip clearances at an apron.* (***Source:*** FAA, 1989.)

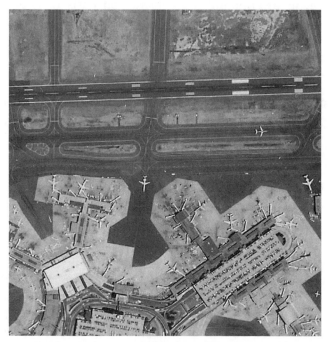

Fig. 9-21. *The "horseshoe" apron between passenger buildings B and C at Boston/Logan and the entrance to that area.* (***Source:*** Massport.)

9-9 Physical obstacles

To ensure and preserve the safety of operations in the airspace in the immediate vicinity of airports, both the ICAO and the FAA have established a series of obstacle limitation surfaces that define the limits to which objects may project into airspace. These surfaces protect approaches to runways, takeoffs, and missed approaches ("balked landings") from obstructions. Objects that penetrate these surfaces are considered obstacles to air navigation and should be removed when possible. An obstacle can be any fixed or mobile object, including terrain, natural objects such as trees, and man-made ones such as antennas or buildings. Several types of airport charts display these obstacles for each airport. The International Air Transport Association, among others, maintains an on-line, current Airport and Obstacle Database with information on more than 2700 airports worldwide (IATA, no date).

Obstacle limitation surfaces are very important to aviation safety. They provide guidance for zoning restrictions on the height of buildings, antennas, and other structures near airports. They may also play a major role in determining the construction costs of new airports when these are located in difficult terrain. More than 50 million m^3 of soil, rocks, and other materials had to be removed from hills around the site of the new airport at Athens/Venizelos in order to remove obstacles to air navigation per Annex 14 of ICAO.

The obstacle limitation surfaces defined by the ICAO are the following:

- Conical surface
- Inner horizontal surface
- Approach surface
- Transitional surfaces
- Inner approach surface and inner transitional surfaces
- Balked landing surface
- Takeoff climb surface

These surfaces are shown schematically in planar view at the top part of Fig. 9-22 and in side view for two different sections at the lower part (ICAO, 1999). They are described briefly below. The relevant dimensions are given in Table 9-9 for approach runways and in Table 9-10 for departure runways. The FAA has established a very similar set of surfaces, called *imaginary surfaces*, which are defined in Part 77, paragraph 25, of the Federal Aviation Regulations (FAA, 2001). Some differences between the ICAO and FAA specifications are identified at the end of this section. FAR Part 77 also describes the requirements for adequate advance notice to the FAA of any proposed construction that may affect navigable airspace (Subpart B), as well as the procedure for evaluating, reviewing, and determining the need for remedial action (Subparts D and E).

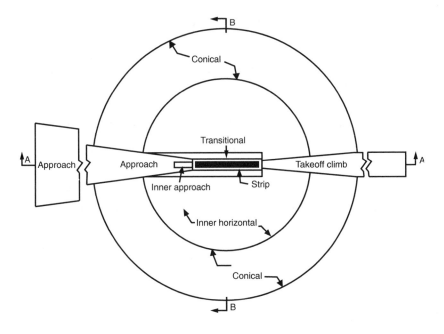

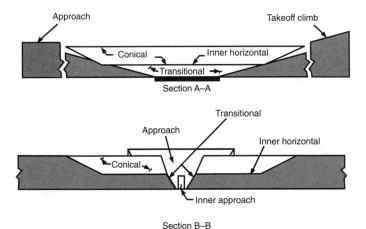

Fig. 9-22. *Obstacle limitation surfaces as defined by the ICAO.*
(***Source:*** ICAO, 1999.)

The *inner horizontal surface* is a horizontal plane above an airport and
its environs. It should normally be a circle whose radius depends on
the type of runway(s) available (noninstrument approach, nonprecision approach, or precision approach) as Table 9-9 shows. The circle
is centered at the airport's reference point, the designated geographic

Table 9-9. Obstacle limitation surfaces, approaches (ICAO, 1999)

	Nonprecision approach runway			Precision approach runway		
				Cat. I		Cat. II or III
Surface and dimensions	Code number			Code number		
	1, 2	**3**	**4**	**1, 2**	**3, 4**	**3, 4**
Conical						
Slope	5%	5%	5%	5%	5%	5%
Height	60 m	75 m	100 m	60 m	100 m	100 m
Inner horizontal						
Height	45 m	45 m	45 m	45 m	45 m	45 m
Radius	3.5 km	4 km	4 km	3.5 km	4 km	4 km
Inner approach						
Width	—	—	—	90 m	120 m	120 m
Distance from threshold	—	—	—	60 m	60 m	60 m
Length	—	—	—	900 m	900 m	900 m
Slope	—	—	—	2.5%	2%	2%
Approach						
Length of inner edge	150 m	300 m	300 m	150 m	300 m	300 m
Distance from threshold	60 m	60 m	60 m	60 m	60 m	60 m
Divergence (each side)	15%	15%	15%	15%	15%	15%
First section						
Length	2.5 km	3 km	3 km	3 km	3 km	3 km
Slope	3.33%	2%	2%	2.5%	2%	2%
Second section						
Length	—	3.6 km	3.6 km	12 km	3.6 km	3.6 km
Slope	—	2.5%	2.5%	3%	2.5%	2.5%
Horizontal section						
Length	—	8.4 km	8.4 km	—	8.4 km	8.4 km
Total length	—	15 km	15 km	15 km	15 km	15 km
Transitional						
Slope	20%	14.3%	14.3%	14.3%	14.3%	14.3%

Table 9-9. (*continued*)

Inner transitional						
Slope	—	—	—	40%	33.3%	33.3%
Balked landing surface						
Length of inner edge	—	—	—	90 m	120 m	120 m
Distance from threshold	—	—	—	see text	1.8 km	1.8 km
Divergence (each side)	—	—	—	10%	10%	10%
Slope	—	—	—	4%	3.33%	3.33%

Table 9-10. Obstacle limitation surfaces, takeoff climb (ICAO, 1999)

	Code number		
	1	**2**	**3 or 4**
Length of inner edge	60 m	80 m	180 m
Distance from runway end	30 m	60 m	60 m
Divergence (each side)	10%	10%	12.5%
Final width	380 m	580 m	1,200 m
			1,800 m
Length	1,600 m	2,500 m	15,000 m
Slope	5%	4%	2%

location of the airport.* The height of the inner horizontal surface is 45 m (150 ft in FAR part 77) above the established elevation of the airport.

The *conical surface* projects upward at a slope of 5 percent from the periphery of the inner horizontal surface to a specified height above that surface. That height depends on the type of runway(s) available.

The *approach surface,* as the name suggests, protects the approach to the runway from obstructions. In planar view, it looks like a trapezoid inclined relative to the horizontal plane and may have a first section and a second section of different slopes. For nonprecision or precision approaches with code number 3 or 4, the approach surface has a horizontal section, as well (see Section A-A in Fig. 9-22). In these

---

*The airport reference point is located "near the initial or planned geometric center of the airport" and is reported in degrees, minutes, and seconds (ICAO, 1999).

cases, a 3-km first section begins 60 m from the runway threshold with a slope of 2 percent. This is followed by a 3.6-km second section with an upward slope of 2.5 percent and finally by a horizontal section (at an elevation of 150 m) with a length of 8.4 km. The total length of these three sections is 15 km (~≈ 9 mi), although in some cases it may be even longer (see ICAO, 1999).

The *transitional surfaces* on either side of the runway and of the approach surface slope upward and outwards to the height of the inner horizontal surface (see planar and side views in Fig. 9-23). The elevation of any point at the lower edge of the approach surfaces is given by either the elevation of the approach surface at that point (when the point is along the side of the approach surface) or by the elevation of the runway strip at that point (when the point is along the strip).

For precision approach runways, an *inner approach surface* and associated *inner transitional surfaces* are also defined. The inner approach surface protects the part of the approach closest to the runway threshold, while the inner transitional surfaces are the controlling obstacle limitation surfaces for navigation aids, aircraft, and other vehicles that must be near the runway. The inner transitional surfaces are not to be penetrated except for frangible objects.

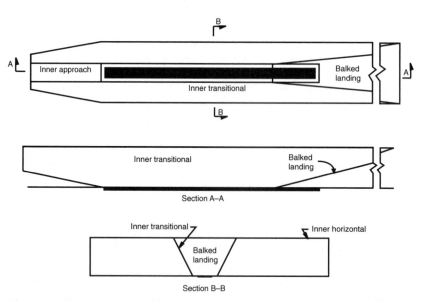

Fig. 9-23. *Inner approach, inner transitional, and balked landing surfaces.* (**Source:** ICAO, 1999.)

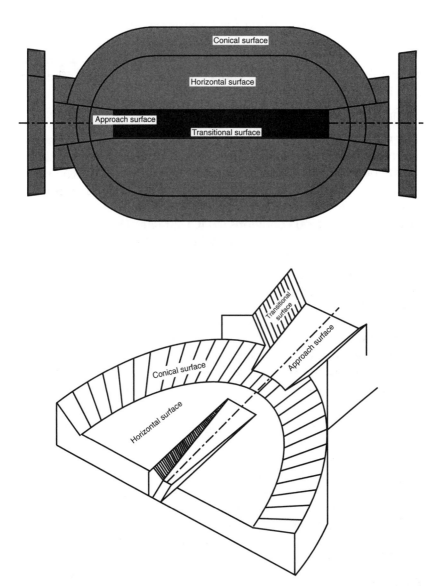

Fig. 9-24. *Imaginary surfaces as defined by the FAA.* (*Source:* FAA, 2001.)

The *balked landing surface* is also defined only for precision approach runways. It provides an obstacle-free volume of airspace at the back end of the approach runway. Note in Table 9-9 that the balked landing surface begins at the end of the runway strip or at 1800 m from the approach threshold of the runway, whichever is less. When a runway can be used for precision approaches from both directions, the approach surface provided at the opposite end of the runway, with a first-section slope of 2.5 or 2 percent, is more restrictive than the balked landing surface that requires a 4 or 3.33 percent slope. This means that a balked landing surface at both ends of the runway is provided by default in such cases.

Finally, the *takeoff climb surface* is an inclined plane intended to prevent obstructions to the paths of departing aircraft near a runway. Note that for code numbers 3 and 4, the final width of the surface is shown as 1200 or 1800 m. The larger number applies to cases where the flight track includes changes of heading greater than 15° for operations conducted in instrument meteorological conditions (IMC) or in visual meteorological conditions (VMC) at night. The 2 percent slope for code letters 3 and 4 may be reduced, if local conditions make this desirable. The ICAO recommends that if no object reaches the 2 percent takeoff climb surface, new objects should be limited to a slope of 1.6 percent to preserve future options.

As already mentioned, the standards for the imaginary surfaces established under FAR part 77 for the FAA (Fig. 9-24) are very similar in concept and in specified parameters to those of the ICAO. One notable difference is that the inner horizontal surface under FAR Part 77 is not a circle, but an oval (FAA, 2001). The oval is determined by drawing two semicircles at the two ends of the runway and connecting their corresponding end points with straight lines. Each semicircle is centered at one end of the runway and has a radius of 10,000 ft. Another significant difference is that the approach surface defined in FAR Part 77 is the same for landings and takeoffs. The inner edge of the approach surface for precision instrument runways is 1000 ft wide and its outer edge 16,000 ft. Its total length is 50,000 ft (roughly the same as the ICAO-specified 15 km), with a slope of 2 percent for the first 10,000 ft and 2.5 percent for the next 40,000 ft.

Exercises

1. Consider a single runway airport where the most demanding aircraft served are in FAA approach category D and airplane design Group V. This airport has a long linear passenger building running parallel to the runway, with contact stands next to the building (on the side of the building that faces the runway). There are 14 contact stands, 5 of which can accommodate the Boeing 747-400, 4 the MD11 and other Group IV aircraft, and 5 the 757-200 or smaller aircraft. The 9 stands for the larger aircraft are at the central part of the building and the other 5 at the two end parts. Arriving aircraft park nose-in and are pushed back on departure. A vehicle road that is 13 ft (4 m) wide lies behind the aircraft stands followed by a taxilane that runs in parallel to the full length of the face of the passenger building in a manner analogous to that shown in the upper part of Fig. 9-20. The taxilane provides sufficient space for the aircraft push-back maneuver. Beyond this taxilane, two full-length taxiways are provided running parallel to the entire length of the runway. They are located between the apron and the runway. Assume the runway is 11,200 ft (or 3400 m) long and that it is used in either direction, depending on prevailing winds, for both arrivals and departures.

Aircraft not served at the contact stands are parked at remote stands in an apron area that does not affect the operation of the main apron next to the passenger building or the required distance between the main apron and the taxiways.

Provide an approximate layout plan for this airfield showing its key dimensions. Make sure to indicate the minimum linear frontage of the passenger terminal, the approximate dimensions of the main apron, the width of runways and taxiways, and the separations between the centerlines of the taxiways and adjacent areas, taxiways, or runways. Do not go into details such as the design of curved segments and fillets, but indicate high-speed versus conventional exit taxiways and their approximate location. Feel free to work in the set of units you are most comfortable with.

2. Consider again Exercise 1, but now assume that the runway system consists of two medium-spaced parallel runways, with a distance of 2500 ft (760 m) between their centerlines. The runway farthest from the passenger building is used for arrivals and the other for departures. Modify the layout plan of Exercise 1 to account for the new runway. Would you construct a new parallel taxiway between the two runways? What are the advantages and disadvantages of doing so?

3. Return to Exercise 1 and assume now that the most demanding aircraft is in FAA approach category D and airplane design Group VI. For simplicity assume that, as far as contact stands are concerned, the airport operator will want to convert the 5 stands that can now accommodate the Boeing 747-400 into stands that can accommodate Group VI aircraft, while still maintaining the 4 stands for the MD11 and other Group IV aircraft, and the 5 for the 757-200 or smaller aircraft. Indicate what other changes will have to be made to the approximate airport design that you developed in Exercise 1.

References

Airbus Industrie (2001) "A380 Airport Compatibility," http://www.airbus.com/A380/.

ACI, Airports Council International (no date), "Airport Traffic Statistics," http://www.airports.org/.

ACI, Airports Council International (1998), *Airport Capacity/Demand Profiles,* Geneva, Switzerland.

Aviation Week and Space Technology (2000) "Commercial Jet Transports," January 17.

Daellenbach, H. G. (1974) "Dynamic Programming Model for Optimal Location of Runway Exits," *Transportation Research, 8.*

FAA, Federal Aviation Administration (1983) *Airport Design,* Advisory Circular 150/5300-12, U.S. Government Printing Office, Washington, DC.

FAA, Federal Aviation Administration (1990) *Runway Length Requirements for Airport Design,* Advisory Circular 150/5325-4A, incorporates Change 1, U.S. Government Printing Office, Washington, DC.

FAA, Federal Aviation Administration (1999) *Standards for Airport Markings,* Advisory Circular 150/5340-1H, U.S. Government Printing Office, Washington, DC.

FAA, Federal Aviation Administration (2000) *Airport Design, Change 6,* Advisory Circular 150/5300-13, U.S. Government Printing Office, Washington, DC.

FAA, Federal Aviation Administration (2001) "FAR Part 77—Objects Affecting Navigable Airspace," in *Code of Federal Regulations (CFR) 14,* Office of Federal Register, National Archives and Records Administration, Washington, DC.

Fortner, B. (2000) "New Runway Projects," *Civil Engineering,* August, pp. 40–45.

Horonjeff, R. (1959) *Exit Taxiway Location and Design,* Institute of Transportation and Traffic Engineering, University of California, Berkeley, CA.

Horonjeff, R., and McKelvey, F. (1994) *Planning and Design of Airports,* 4th ed., McGraw-Hill, New York.

IATA, International Air Transport Association (no date) "Airport and Obstacle Database," http://www.aodb.iata.org/.

IATA, International Air Transport Association (no date) http://www.iata.org/index.asp.

ICAO, International Civil Aviation Organization (1983a) *Aerodrome Design Manual, Part 5—Electrical Systems,* 1st ed., reprinted April 1998, Doc. 9157, ICAO, Montreal, Canada.

ICAO, International Civil Aviation Organization (1983b) *Aerodrome Design Manual, Part 3—Pavements,* 2nd ed., reprinted November

1997, incorporating Amendments 1 and 2, Doc. 9157, ICAO, Montreal, Canada.

ICAO, International Civil Aviation Organization (1983c) *Airport Planning Manual, Part 3, Guidelines for Consultant/Construction Services,* 1st ed., reprinted December 1996, Doc. 9184, ICAO, Montreal, Canada.

ICAO, International Civil Aviation Organization (1984) *Aerodrome Design Manual, Part 1, Runways,* 2nd ed., reprinted July 2000 incorporating Amendment 1, Doc. 9157, ICAO, Montreal, Canada.

ICAO, International Civil Aviation Organization (1987) *Airport Planning Manual, Part 1, Master Planning,* 2nd ed., reprinted April 1999, Doc. 9184, ICAO, Montreal, Canada.

ICAO, International Civil Aviation Organization (1991) *Aerodrome Design Manual, Part 2—Taxiways, Aprons and Holding Bays,* 3rd ed., reprinted November 2001, Doc. 9157, ICAO, Montreal, Canada.

ICAO, International Civil Aviation Organization (1993) *Aerodrome Design Manual, Part 4—Visual Aids,* 3rd ed., Doc. 9157, reprinted July 1997, ICAO, Montreal, Canada.

ICAO, International Civil Aviation Organization (1999) *Aerodromes, Annex 14 to the Convention on International Civil Aviation, Volume I: Aerodrome Design and Operations,* 3rd ed., ICAO, Montreal, Canada.

Swan, W. (2001) "Airline Route Development: A Review of History," Seminar Presentation, November 30, Massachusetts Institute of Technology, Cambridge, MA.

10

Airfield capacity

This chapter reviews the subject of airfield capacity, a topic fundamental to modern airport planning and design. The capacity of the airfield and especially of runway systems typically determines the ultimate capacity of an airport.

Maximum throughput capacity is the principal and most fundamental measure of the capacity of a runway system. It indicates the average number of movements that can be performed on the runway system in 1 h in the presence of continuous demand, while adhering to all the separation requirements imposed by the air traffic management (ATM) system. Practical hourly capacity, declared capacity, and sustained capacity are all measures designed to estimate the number of hourly movements at which operations can be performed over an extended period of time at acceptable levels of delay. These are measures derived from and based on the maximum throughput capacity. They are typically equal to 80 to 90 percent of the maximum throughput capacity.

The (maximum throughput) capacity of a runway system depends on many parameters and factors. The most important of these are the number and geometric layout of the runways, the ATM separation requirements, weather conditions (visibility, precipitation, wind direction and strength), mix of aircraft types, mix and sequencing of runway movements, type and location of runway exits, performance of the ATM system, and noise restrictions on operations. The runway system capacities that one encounters at major airports in various parts of the world span a wide range. In developed countries, capacity per runway at major airports ranges from 20 to 60+ movements per hour. A few airports in the United States operate with as many as four to seven simultaneously active runways and serve more than 200 movements in 1 h.

The range of capacities available at an airport over a long period of time, such as a year, and the frequency with which these capacities are available can be summarized through the capacity coverage chart (CCC). The CCC makes the assumptions that (1) the operations mix consists of 50 percent arrivals and 50 percent departures and (2) the runway configuration in use at any given time is the one that provides the highest capacity under the prevailing conditions. An "uneven" CCC indicates an airport where the supply of runway capacity is not reliable. In practice, this may create serious operational problems.

Simple mathematical models can be used to obtain good approximations to the capacity of simple runway systems. Moreover, these models provide insight into the sensitivity of capacity to changes in such parameters as separation requirements, traffic mix and characteristics, etc. Runway capacity envelopes, which can also be computed approximately from such models for simple runway systems, indicate the capacity that can be achieved for all possible mixes of arrivals and departures. A number of computer-based simulation and mathematical models are available to assist in investigating issues related to airfield capacity and delay at airports with complex airfield layouts.

The capacity of taxiway systems depends greatly on local conditions and the geometric configuration at hand at each airport. The capacity of the taxiway system of major airports almost always exceeds the capacity of the runway system by a considerable margin. Delays sustained at specific "choke points" are typically much smaller than the delays experienced due to the capacity limitations of the runway system, but some exceptions may exist at older, space-constrained airports. Taxiway capacity problems are airport-specific and must be resolved in the context of local conditions.

It is important to distinguish between the static capacity of an apron, i.e., the number of aircraft that can be stationed there at any particular instant, and the dynamic capacity, which indicates the number of aircraft that can be served at the apron per unit of time. The dynamic capacity depends strongly on the stand blocking time. Its determination is often difficult, due to the differences in the sizes of stands and the large number of constraints and conditions on stand assignments. These constraints and conditions also vary greatly among airports. For the same reasons, it is also difficult sometimes to compare apron capacity with runway system capacity. In the long run, all but the most space-constrained airports should be able to increase their apron capacity to a level greater than the capacity of the runway system.

10-1   Introduction

This chapter reviews the subject of airfield capacity. The emphasis is on the capacity of runway systems of major commercial airports. This is a topic fundamental to modern airport planning and design, because it is the capacity of the airfield and especially of runway systems that typically determines the ultimate capacity of an airport. The runway complex is usually the principal "bottleneck" of the air traffic management system because, quite simply, it is at the runway and its immediate vicinity that air traffic transitions from three-dimensional flows in airspace to the "single-file" regime that must be followed for runway operations. Moreover, it is usually extremely difficult and time-consuming to increase substantially the capacity of the runway system of a major airport. New runways, along with associated protection zones, noise buffer space, etc., typically require acquisition of a large amount of additional land area. Equally important, they have significant environmental and other external impacts that necessitate long and complicated review-and-approval processes with uncertain outcomes. By contrast, the capacity of landside facilities (passenger and cargo terminals, road access, etc.) and of other airfield facilities (taxiways, apron stands) can usually be increased, in one way or another, to equal or exceed the capacity of the runway system.

The subject of airport capacity and delay has received a great amount of attention, not only by airport professionals but also by the public at large, as air traffic delays have increased and spread geographically. The problem is particularly acute in North America, Western Europe, and the Pacific Rim. Many airline executives and aviation officials believe that the principal threat to the long-term future of the global air transportation system is the apparent inability of available runway capacity to keep up with growing air traffic demand at many of the world's most important airports.

The chapter begins with a review of the several definitions of airfield capacity that are in use and have been the cause of much confusion among airport planners (Sec. 10-2). The various factors that determine the capacity of a runway system are then discussed in some detail in Sec. 10-3. The objective is to give the reader an appreciation of the complex relationships that play a role in determining runway capacity, as well as the reasons why it may be very difficult to increase this capacity beyond a certain level at any particular location. It is also noted that the capacity of a runway system is not a constant, but a highly variable quantity, as it depends on several parameters that vary probabilistically and dynamically. Section 10-4 presents a brief survey of the range of runway system capacities at major airports around the world. It also introduces the notion of the capacity coverage chart as a means of capturing the range of capacity values associated with a runway system over time and the relative frequency with which these values occur. Section 10-5 turns to the issue of computing the capacity of an airport. The standard approach for computing the capacity of a single runway is outlined and one of the best-known mathematical models available for this purpose is presented in some detail. Section 10-6 discusses generalizations of the single-runway model of Sec. 10-5 and describes briefly more detailed versions of the model. It also introduces the important concept of the capacity envelope and suggests ways to extend the single-runway methodology to more complex runway systems. Finally, Sec. 10-7 is concerned with the capacity of the taxiway and aircraft stand system of an airport, concentrating primarily on the estimation of the capacity of the apron area, a quite complex problem.

10-2   Measures of runway capacity

Several alternative measures of runway capacity are in use, all of them intended to provide an estimate of how many aircraft movements (arrivals and/or departures) can be performed on the runway system of an airport during some specified unit of time. To utilize them properly and to avoid confusion, one must understand clearly the definitions of these alternative measures.

It is essential to realize at the outset that, from a long-term perspective, *runway capacity* is a probabilistic quantity, a random variable, which can take on different values at different times, depending on the circumstances involved. For a simple example, note that the number of arrivals and departures that can be performed on a runway during any particular hour at a busy airport will depend on the "mix" of aircraft that will be using the runway during that hour. If, for instance, the mix happens to include a high percentage of wide-body aircraft (B747, MD-11, A340, etc.), the capacity will generally be lower than at times when the mix consists, for the most part, of smaller aircraft (regional jets, turboprops, B737, etc.). The reason is that bigger airplanes generate wake vortices that may pose a threat to aircraft flying immediately behind them. To ensure safety, providers of air traffic management (ATM) services (e.g., the Federal Aviation Administration, FAA, in the United States) require longer separations (in terms of time or of distance) between pairs of successive aircraft whenever the first aircraft in the pair is a heavy one. Even with identical mixes of aircraft, the number of movements that can be performed may vary depending on winds, visibility, proficiency of air traffic controllers working at the time in question, and many other factors. Thus, the numbers that are cited for the runway system capacity of any airport, typically refer to the "average number" or, more formally, the *expected number* of movements that can be performed per unit of time.

The first and, as will be seen later, most fundamental measure of runway capacity can now be introduced. The *maximum throughput capacity* (or *saturation capacity*) is defined as the expected number of movements that can be performed in one hour on a runway system without violating ATM rules, assuming continuous aircraft demand.

Two points should be noted about this definition. First, in order to compute the maximum throughput capacity, one needs to know the specific conditions under which runway operations are conducted. This means specifying the ATM separation requirements in force, the mix of aircraft, the mix of movements (arrivals and departures), the allocation of movements among the runways (if the runway system consists of more than one runway), and several other factors that will be described in Sec. 10-3.

Second, the definition of maximum throughput capacity makes no reference to any level-of-service (LOS) requirements. In other words, all one cares to know is how many aircraft movements can be processed on average per hour, if the runway system is utilized to its maximum potential in the presence of "continuous aircraft demand." Whether this means a delay per movement of a few minutes or of several hours is immaterial, as far as this measure of capacity is concerned.

It is the absence of any reference to LOS that has motivated the occasional use in practice of three other measures of hourly capacity. The common characteristic of all three is that they define capacity indirectly, through the explicit or implicit specification of an acceptable threshold of LOS or of air traffic controller workload. The three measures are the practical hourly capacity, the sustained capacity, and the declared capacity.

The *practical hourly capacity* (PHCAP) is the oldest of these measures, having originally been proposed by the FAA in the early 1960s. It is defined as the expected number of movements that can be performed in 1 h on a runway system with an average delay per movement of 4 min.

Note that this definition specifies a threshold value for acceptable LOS ("average delay of 4 min per movement") and states that the runway system "reaches its capacity" when that threshold is exceeded. As a rule of thumb, the PHCAP of a runway system is approximately equal to 80–90 percent of its maximum throughput capacity, depending on the specific conditions at hand. Note that today the average delay per movement is considerably higher than 4 min at practically every major airport, especially during peak traffic hours. This does not invalidate the notion of a "practical" hourly capacity tied to a threshold of acceptable LOS.* Instead, it simply indicates that the failure of runway capacity to keep up with demand at many airports has forced these airports to operate routinely at a LOS much lower than what was considered acceptable in 1960. In fact, the selection of the particular threshold value of 4 min is not unreasonable, as it is based on a quite solid rationale that draws on queuing theory, the mathematical theory of waiting lines, as will be seen in Chap. 11.

The *sustained capacity* of a runway system is a measure defined, rather ambiguously, as the number of movements per hour that can be reasonably sustained over a period of several hours. "Reasonably sustained" refers primarily to the workload of the ATM system and of air traffic controllers. The rationale is that, to achieve maximum throughput capacity, the ATM system should work to its full potential all the time. However, operations at such a level of full efficiency and maximum performance often cannot be sustained in practice for periods of more than one or two consecutive hours. It is thus argued that one should specify a more realistic target than "maximum throughput capacity" when it comes to operations over a period of several hours or an entire day of air traffic activity.

A good example of the application of the notion of sustained capacity is the setting of performance targets† at many major airports in the United States. These targets are determined after discussions between

---

*For many airport facilities and services, the idea of associating capacity with some acceptable LOS has great merit. As will be seen in Chaps. 14–16, this is a key concept when it comes to defining the capacity of airport passenger buildings. While, in principle, one could, for example, jam four people in an area of 1 m^2 (about 2.5 ft^2 per person), no one would dare claim that the capacity of a 2000-m^2 lobby in a terminal is 8000 passengers, as the crowding would be intolerable under such conditions. The capacity, in this case, would undoubtedly have to be determined with reference to an acceptable level of personal comfort (i.e., to a LOS).

†These performance targets were called "engineered performance standards" (EPS) when they were first used.

FAA specialists and the local air traffic controller teams and specify desirable levels of runway system capacity to be achieved at each participating airport over periods of several hours. For example, the sustained capacity for Boston/Logan in good weather conditions and operations to the northeast was set in 2000 to approximately 110 movements per hour. This capacity is usually further subdivided into a sustained arrival capacity, the airport acceptance rate (AAR), and sustained departure capacity, the airport departure rate (ADR)—see also Chap. 13. Typically, the sustained capacity is set to approximately 90 percent of maximum throughput capacity when runway configurations with high maximum throughput capacity are in use (e.g., in good weather conditions) and to almost 100 percent of maximum throughput capacity with configurations with low maximum throughput capacity.* The reasoning is that low-capacity conditions, usually associated with poor weather, prevail for only a few consecutive hours at a time and it is critical to operate as close as possible to the maximum available capacity during those periods.

Declared capacity is another measure based on the same general notion as sustained capacity. It is defined, again somewhat ambiguously, as the number of aircraft movements per hour that an airport can accommodate at a reasonable LOS. Delay is used as the principal indicator of LOS. Declared capacity is widely used outside the United States, especially in connection with "schedule coordination" and the allocation of "slots" at congested airports (Chap. 12). Under this practice, each airport that experiences congestion "declares" a capacity, which is then used to set a limit on the number of movements per hour that can be scheduled at this airport. For example, in the summer of 1998, the declared capacity of London/Heathrow was about 80 movements per hour, on average. As a result, an average of 78.8 movements per hour was allocated that year among the airlines wishing to operate at London/Heathrow over the most active 17-h period (06:00–23:00) of a typical summer weekday.†

Unfortunately, there is no generally accepted definition of declared capacity and no standard methodology for setting it. It is essentially left up to local or national airport and civil aviation organizations, in

---

*Sections 10-3 and 10-4 will provide more details about high- and low-capacity runway configurations.
†It should be noted, however, that the actual number of movements scheduled at London/Heathrow ranged from 96 per hour between 10:00 and 11:00 to 32 between 06:00 and 07:00. Thus, in this case, declared capacity is viewed as a target for the average number of movements to be scheduled per hour. At other airports, however, the declared capacity is often interpreted as a limit on the maximum number of movements that can be scheduled per hour.

cooperation with other interested parties, to compute and set the declared capacity (Chap. 12). The approaches used for this purpose vary from country to country and even from airport to airport. In fact, there are instances of airports where the declared capacity is dictated by the capacities of the passenger terminal or of the apron area, which are believed to be more constraining than the capacity of the runway system. In most instances, however, the declared capacity seems to be set close to roughly 85–90 percent of the maximum throughput capacity of the runway system. As in the case of sustained capacity, the reasoning is that this choice will ensure reliability of airport operations over extended periods of a day, as well as a reasonable LOS.

The advantages and disadvantages of the measures of runway capacity introduced in this section can now be summarized. Among these measures, the maximum throughput capacity is clearly the most fundamental and least subjective. It provides an estimate of capacity in its truest sense: how many operations can be performed per unit of time, on average, when the runway system is pushed to its limits. Indeed, it is possible to obtain rough estimates of maximum throughput capacity by collecting data in the field. All one needs to do is observe the runways and count the number of movements taking place during a continuously busy period, i.e., when all movements experience some delay. Note this means that it is much easier to measure maximum throughput capacity at a very congested airport than at a mildly busy one. Moreover, the data should be collected during peak traffic hours rather than at off-peak. Equally important, maximum throughput capacity can be computed quite accurately through a number of existing analytical and simulation models—see Secs. 10-5 and 10.6. These models make it possible to estimate capacity under hypothetical future conditions, in addition to existing ones.

The principal disadvantage of maximum throughput capacity is that it does not consider LOS in any way. In fact, as will be seen in Chap. 11, extremely long delays will be experienced whenever the average number of movements scheduled at an airport is very close to the runway system's maximum throughput capacity for several hours in a row. By contrast, when demand remains close, on average, to the practical hourly capacity or to sustained capacity,* the LOS, as measured by the amount of delay per flight, will usually remain at acceptable

---

*This is also true, in most cases, for declared capacity. However, there exist a few instances of airports that have "declared" a capacity roughly equal to their maximum throughput capacity; if the number of scheduled movements is approximately equal to declared capacity, very long delays will be suffered routinely at these airports.

levels. Thus, measures such as PHCAP, sustained capacity, and, in most cases, declared capacity are good indicators of how much demand can be accommodated at a reasonable LOS. PHCAP, sustained capacity, and declared capacity are also useful measures for planning purposes. When the average demand per hour, over a period of several hours of a day, grows over the years to a level close to the PHCAP or the sustained capacity or the declared capacity, this is a clear signal that an increase of the airport's capacity is highly desirable. Even relatively small increases in demand beyond that critical level will probably lead to unacceptable delays and airfield congestion.

In conclusion, PHCAP, sustained capacity, and declared capacity are rather subjective measures of capacity that can be highly useful in some instances. They are also "derivative" measures, in the sense that one needs to compute the maximum throughput capacity before one can estimate these other capacity measures. This will be discussed further in Chap. 11.

Henceforth in this and subsequent chapters, the term *runway capacity* will be used to refer to the *maximum throughput capacity* of a runway system. Whenever reference is made to some other measure of capacity (e.g., the "declared capacity"), this will be stated explicitly.

The reader should also bear in mind the following convention: the often-heard statement, "The capacity of Airport A is X movements per hour," typically makes the implicit assumption that X consists of approximately 50 percent arrivals and 50 percent departures. When this is not the case, the statement is usually more detailed (e.g., "The arrival capacity of the runway is Y," or "When two runways are used for arrivals and one for departures, the arrival capacity is Z and the departure capacity is W").

Finally, note that all the measures of capacity mentioned so far use the hour as their unit of time. Another natural measure of great practical interest is the *annual* capacity of an airfield. This is a number that can be compared readily with airport demand forecasts that are typically given in terms of annual estimates ("500,000 aircraft movements expected by 2010"). In fact, the FAA has been using for some years the measure of *practical annual capacity* (PANCAP) for this purpose. As will be discussed in Chap. 11, PANCAP and other similar estimates of annual capacity can be derived from the fundamental measure of maximum (hourly) throughput capacity. These annual measures of capacity must necessarily be tied to a LOS and should take into consideration the daily and seasonal patterns of demand at an airport.

10-3   Factors that affect the capacity of a runway system

The dependence of the capacity of any runway system on many different factors was emphasized in the previous section. This section provides an overview of the most important of these factors and of the ways in which each affects runway capacity. These are:

- Number and geometric layout of the runways
- Separation requirements between aircraft imposed by the ATM system
- Visibility, cloud ceiling, and precipitation
- Wind direction and strength
- Mix of aircraft using the airport
- Mix of movements on each runway (arrivals only, departures only, or mixed) and sequencing of movements
- Type and location of taxiway exits from the runway(s)
- State and performance of the ATM system
- Noise-related and other environmental considerations and constraints

One of the objectives is to make the reader aware of the complex relationships that are often at play.

Number and geometric layout of the runways

The most obvious, and probably single most important, factor influencing a runway system's capacity is the number of runways at the airport and their geometric layout. From a practical point of view, the surest way to achieve a "quantum increase" in the capacity of an airport is by constructing a well-located (relative to the other existing runways) and well-designed runway.* Unfortunately, as noted at the beginning of this chapter, adding a new runway is a task that today ranges from "very difficult" to "nearly impossible" at most of the world's busiest and most congested airports. The following are some introductory observations on how the number and geometric layout of runways affect capacity.

First, it is important to distinguish between the number of runways at an airport and the number that are *active* at any given time. For example, Boston/Logan and Amsterdam/Schiphol have five runways each, but

---

*The meaning of "well-located" and "well-designed" in this context is discussed in several parts of this book, including a number of relevant points later in this section.

no more than three of these runways are ever active simultaneously, due to the geometric layouts of the runway systems and to noise restrictions. By contrast, Atlanta/Hartsfield has four runways* and uses all four simultaneously during most of the busy hours of the day. Similarly, Dallas/Fort Worth has seven runways and typically uses six or all seven during busy hours. It is the number of *simultaneously active runways* that is a primary factor in determining airfield capacity.

Second, the number and identity of runways in use at any given time, as well as the allocation of types of aircraft and movements to them, may change several times a day at many airports. The selection of the specific set of runways to be operated at any one time depends on many of the factors to be discussed further in this section: demand (e.g., during periods of low demand an airport may accommodate all its traffic on a single runway, even though more than one runway may actually be available); weather conditions, including visibility, precipitation, and wind speed and direction; mix of movements (e.g., during peak periods for flight arrivals, one or more runways may be dedicated to serving arrivals exclusively—and conversely for peak departure periods); and noise restrictions, which, for example, may prohibit or discourage the use of certain runways during the night or during certain parts of the year. For an airport with several runways, there can be a large number of combinations of simultaneously active runways, weather conditions, and assignments of aircraft types and movements (arrivals and/or departures) to the active runways. Each of these combinations is called a *runway configuration*. For example, Boston/Logan, with five runways, can operate in about 40 different configurations!

Third, the precise geometric layout of any set of runways is extremely important, as it determines the degree of *dependence* among them. Section 9-4 has already provided an overview of this topic, and more details will be given below in connection with the description of the relevant ATM separation requirements.

ATM separation requirements

Every ATM system, no matter how advanced or how primitive, specifies a set of required minimum separations between aircraft flying under instrument flight rules (IFR). Obviously, the purpose of these requirements is to ensure safety. In turn, the separation requirements determine the maximum number of aircraft that can traverse each part of the airspace or can use a runway system per unit of time.

---

*A fifth runway has been approved for Atlanta, parallel and to the south of the existing complex of four runways.

In most instances, required separation distances between aircraft operating under IFR at major airports in the United States are the smallest (or "least conservative") anywhere, reflecting in part the need to maximize airport capacity, as well as the outstanding proficiency and training of the air traffic controllers. Several major European airports, such as London/Heathrow and London/Gatwick, Frankfurt, and Amsterdam/Schiphol, have also come to be operated in recent years with separation requirements that are essentially identical to those used at the busiest airports in the United States. Such "tight" separation requirements recognize the need for more capacity at these airports and have been made possible by the outstanding ATM capabilities that have been developed there.

Separation requirements for aircraft operating to/from the same runway.
The longitudinal separation requirements for aircraft landing on or departing from the same runway are of particular importance in determining runway capacity. Typically, each type of aircraft is assigned to one of a small number (usually, three or four) of classes according to the aircraft's size and/or weight. The separation requirements are then specified in units of distance or of time. Each set of requirements gives *the minimum separation that must be maintained at all times* between two aircraft operating successively on the runway. The requirements are specified for every possible pair of classes and every possible sequence of movements: "arrival followed by arrival," A-A; "departure followed by departure," D-D; "arrival followed by departure," A-D; and "departure followed by arrival," D-A (Example 10.1).

EXAMPLE 10.1

In the United States, the FAA assigns all aircraft to three classes, according to their maximum certified takeoff weight (MTOW): heavy (H), large (L), and small (S). Aircraft with

- MTOW greater than 255,000 lb (~116 tons) are in class H

- MTOW between 41,000 lb (~19 tons) and 255,000 lb (~116 tons) are in L

- MTOW less than 41,000 lb (~19 tons) are in S

- The FAA also identifies the Boeing 757, whose MTOW places it at the borderline between the L and H classes, as an aircraft class by itself for some terminal airspace separation purposes, due to its strong wake-vortex effects.

Wide-body commercial jets generally belong to the H class (see also Table 9-3). Class L includes practically all types of narrow-body commercial jets and regional jets and some types of turbo-props used by regional (or "commuter") air carriers. Finally, Class S includes most general aviation aircraft, including many general aviation jets, as well as many types of nonjet aircraft used by regional air carriers and by air taxi operators.

Table 10-1 summarizes the FAA separation requirements for movements on the same runway under IFR (FAA, 2001b). Note that requirements are specified for all four possible combinations of movements (A-A, A-D, D-D, D-A) and for all possible pairs of aircraft classes. For example, it can be seen (A-A separations) that when the landing of a large (L) aircraft is followed immediately by the landing of a small (S) aircraft, the minimum separation allowed between the two aircraft when the leading one is at the threshold of the runway is 4 nautical miles (nmi). If a departure is to be followed immediately by an arrival (and regardless of the classes of aircraft involved) the arriving aircraft must be at least 2 nmi away from the runway at the time when the departure run begins* *and* cannot touch down on the runway before the preceding departing aircraft has lifted off (D-A separations). In the reverse situation, i.e., when an arrival is followed immediately by a departure, the arriving aircraft must be safely out of the runway before the takeoff run can begin, again regardless of the classes of the two aircraft involved. Note that some separations are specified in terms of time or of occurrence of an event ("clear of runway"), while others are specified in units of distance.

Aircraft pairs in which the first aircraft is in class H or B757 generally require greater separations than other pairs, both in the A-A and in the D-D cases. The reason is that these aircraft classes generate severe *wake turbulence (wake vortices)* behind them. A wake vortex poses the threat of destabilizing a trailing aircraft that runs into it, especially if the trailing aircraft belongs to Class S.

---

*The 2-nmi requirement is usually more restrictive than the requirement that the departing aircraft be clear of the runway before the arriving aircraft touches down. In visual meteorological conditions (VMC) only the latter requirement applies. A more precise statement of the former requirement is that "a departure should be separated from a trailing arrival on final approach by 2 nmi, if separation will increase to 3 nmi within 1 min after takeoff." Controllers refer to this as "2 increasing to 3."

Table 10-1. Single-runway IFR separation requirements in the United States in 2000

Arrival followed by arrival (A-A)

A. Throughout final approach, successive aircraft must be separated by at least the distance (in nautical miles) indicated by the table below:

		Trailing aircraft		
		H	L + B757	S
Leading	H	4	5	5/6*
aircraft	B757	4	4	5
	L	2.5 (or 3)	2.5 (or 3)	3/4*
	S	2.5 (or 3)	2.5 (or 3)	2.5 (or 3)

*Separations indicated with an asterisk are distances required at the time when the leading aircraft is at the threshold of the runway.

B. The trailing aircraft cannot touch down on the runway before the leading aircraft is clear of the runway.

Arrival followed by departure (A-D)

Clearance for takeoff run of the trailing departure is granted after the preceding landing is clear of the runway.

Departure followed by departure (D-D)[†]

Clearances for takeoff run of successive aircraft must be separated by at least the amount of time (in seconds) indicated by the table below.

		Trailing aircraft		
		H	L + B757	S
Leading	H	90	120	120
aircraft	B757	90	90	120
	L	60	60	60
	S	45	45	45

[†]D-D separations shown here are simplified and approximate—see details in the text.

Departure followed by arrival (D-A)

The trailing arrival on final approach must be at least 2 nmi away from the runway at the time when the departing aircraft begins its takeoff run. The departing aircraft must also be clear of the runway before the trailing arrival can touch down on it.

As can be seen in Table 10-1, the A-A case requires that two conditions be satisfied: (1) the two landing aircraft should not be on the runway at the same time*; *and* (2) while airborne on final approach, the two aircraft must be separated by a minimum distance specified in units of nautical miles. The separations of 4, 5, and 6 nmi in the A-A case are all related to the potential presence of wake turbulence due to the leading aircraft, while the 2.5-nmi separations apply in pairings where it is believed that wake vortices are not a factor. The two A-A separation requirements denoted with an asterisk in Table 10-1 apply only at the time when the leading aircraft (H or L, as the case may be) is at the runway threshold, i.e., about to touch down on the runway. All the other A-A separation requirements apply at all points of the final approach path to the runway. (In the specific case of a H-S pair, for example, the trailing Class S aircraft is required to be at least 5 nmi behind the leading Class H aircraft *at all points* on final approach and at least 6 nmi behind *at the instant when H is at the runway threshold*—with the more restrictive of the two requirements dictating the actual separation.) It should be mentioned, as well, that a 2.5-nmi separation between the indicated pairs of aircraft classes is used *at the busiest airports* in the United States. At other airports 3 nmi is used, as indicated in parentheses in Table 10-1.

Turning next to the D-D case, the requirements shown in Table 10-1 give *approximate* (and somewhat conservative) estimates of the time separations that result in practice[†] from the following more complicated set of rules:

1. In the case where the leading departing aircraft belongs to Class L (Class S) and thus wake turbulence is not a factor, the takeoff run of the trailing aircraft can start after the leading aircraft is airborne *and* (*a*) is at a distance of more than 6000 ft (4500 ft) from the trailing aircraft *or* (*b*) has either cleared the runway end or has turned out of conflict.

---

*A somewhat less strict requirement may apply at a number of airports, where the trailing aircraft cannot touch down on the runway unless the leading aircraft is more than a specified long distance (e.g., 8000 ft) down the runway and heading toward a runway exit.

[†]An important consideration to achieve these separations is that the departure courses diverge by 15° within 1 mi of the runway end (7110.65 para. 5-8-3). Controllers refer to this as "fanning" the departures.

2. In the case where the leading departing aircraft belongs to Class H or to Class B757 (and thus wake turbulence is a factor), the takeoff run of the trailing aircraft can start as soon as *one* of the following two conditions has been satisfied: (*a*) 2 min have elapsed since the start of the takeoff run of the leading aircraft *or* (*b*) the following separations, in nautical miles, have been assured when the trailing aircraft becomes airborne:

		Trailing aircraft		
		H	**L or B757**	**S**
Leading	**H**	4	5	5
aircraft	**B757**	4	4	5

The separations of 90 and 120 s behind H-class and B757 aircraft shown in Table 10-1 are again approximate estimates of the earliest time it takes to satisfy, in each case, the less constraining of the above two conditions.

Finally, for the case of A-D pairs, an approximate estimate of the minimum time separation required is 60 s, corresponding to a typical runway occupancy time of an arriving aircraft. A similar amount of time (roughly 60 s) provides an approximate estimate of the time required for a D-A pair, the 60 s now corresponding roughly to the time it takes a landing aircraft to fly the last 2 nmi of the final approach. The true amount of time, however, depends on the actual approach speed, which, in turn, depends on the type of aircraft involved. The estimate of 60 s is rather conservative in VMC when the only applicable requirement is that the departing aircraft be clear of the runway before the arriving aircraft touches down (see footnote on p. 381). This is especially true for the case when the departing aircraft is in Class S, which typically take about 40–45 s or even less for their takeoff run.

Obviously, the larger (or more "conservative") the separations required by the ATM system, the lower the runway capacity. To emphasize this point, Table 10-2 lists the separation requirements used until 1998 at the busiest airports in Italy, Rome/Fiumicino, Milan/Malpensa, and Milan/Linate.* A comparison with Table 10-1 shows that the separations in Italy were considerably more conservative than those in the United States. It stands to reason that the runway system capacity of an airport in the United States was typically much higher than the capacity of a similar airport in Italy, assuming the same mix of aircraft and of movements. For instance, the *declared* capacity of the

Table 10-2. Simplified single-runway IFR separation requirements in effect in Rome's and Milan's airports until 1998

Arrival followed by arrival (A-A)

A. Throughout final approach, successive aircraft must be separated by at least the distance indicated by the table below (in nautical miles):

		Trailing aircraft		
		H	**L**	**S**
Leading	**H**	5	5	7
aircraft	**L**	5	5	5
	S	5	5	5

B. The trailing aircraft cannot touch down on the runway before the leading aircraft is clear of the runway.

Arrival followed by departure (A-D)

Clearance for takeoff run of the trailing departure is granted after the preceding landing is clear of the runway.

Departure followed by departure (D-D)

Clearances for the takeoff run of successive departures must be separated by at least 120 s.

Departure-followed by arrival (D-A)

At the start of the takeoff run of the leading (departing) aircraft the trailing landing aircraft must be at least 5 nmi away from the end of the runway. (Typically this translates to about 2–2.5 minutes between the beginning of the takeoff run of the leading departure and the subsequent touchdown on the runway of the trailing arrival.)

These separation requirements have now been tightened considerably at both Rome and Milan.

single-runway airport at Milan/Linate in 1998 was 32 movements per hour. By comparison, the single-runway San Diego airport often handles as many as 60 movements per hour.

Separation requirements for aircraft operating to/from parallel runways. The separation requirements for aircraft landing on or departing from a pair of *parallel* runways play a critical role at those major airports that often operate with more than one active runway. Most of these multirunway airports rely largely on operations to/from parallel runways.

Table 10-3 summarizes the FAA separation requirements for operations on parallel runways under IFR. The "arrival/arrival" column refers to the required separation between a pair of arriving aircraft, the first of which is landing on one of the parallel runways and the second on the other. Similarly, "departure/arrival" refers to the situation in which the first aircraft in the pair will depart from one of the parallel runways and the second will land on the other. The "departure/departure" and "arrival/departure" columns should be interpreted in a similar way.

The critical parameter is now the distance between the centerlines of the runways. For runway centerlines that are separated by less than 2500 ft (762 m), the separation requirements in the "arrival/arrival"

Table 10-3. IFR separation requirements between aircraft movements on parallel runways in the United States (* = arriving aircraft must be over runway threshold and committed to land)

Separation between runway centerlines	Arrival/ arrival	Departure/ departure	Arrival/ departure	Departure/ arrival
Up to 2500 ft (up to 762 m)	As in single runway	As in single runway	Arrival touches down	Departure is clear of runway
2500–4300 ft (762–1310 m)	1.5 nmi	Independent	Independent	Independent
4300 ft or more (1310 m or more)	Independent	Independent	Independent	Independent

Source: FAA, 1989.

case are the same as those in Table 10-1, which apply to the case when the two aircraft are landing on the same runway.* In other words, the second aircraft should follow the first by 2.5 (or 3), 4, 5, or 6 nmi, depending on the classes of the two aircraft. Similarly, the separations in Table 10-1 apply to the "departure/arrival" case (the landing aircraft must be at least 2 nmi from the parallel runway when the departure run begins and should not touch down on its runway before the preceding departure on the other parallel runway has lifted off) and to the "departure/departure" case. The only change from the separation requirements of Table 10-1 occurs in the "arrival/departure" case: the departing aircraft does not have to wait for the landing aircraft to exit the parallel (arrival) runway—as was the case with a single runway—but can begin its takeoff run as soon as the landing aircraft touches down on the parallel runway (or, in a less conservative interpretation of the rules, crosses the threshold of that runway).

The situation changes considerably when the separation between the centerlines of the two parallel runways exceeds 2500 ft (762 m). Now, the two parallel runways may operate independently when both are used for departures or when one is used for arrivals and the other for departures. "Independently" means that, absent airspace constraints, the movement on one runway does not have an impact on the movement on the other. When both runways are used for arrivals, then the trailing aircraft has to be at least 1.5 nmi behind the leading one when the two centerlines are between 2500 ft (762 m) and 4300 ft (1310 m). The 1.5-nmi distance is measured diagonally, i.e., represents the direct distance between the two aircraft (Fig. 10-1). Finally, when the centerlines are separated by more than 4300 ft (1310 m), the two parallel runways may be operated independently, even if both are used for arrivals. At a small number of airports, the FAA has also authorized simultaneous approaches to parallel runways separated by as little as 3400 ft (1035 m) when a precision runway monitor system (PRM) is available. The FAA will, in fact, consider (FAA, 1989) on a case-by-case basis, authorizing simultaneous approaches to parallel runways with centerline separations down to 3000 ft (915 m).

Two additional points should be mentioned. First, when both runways are used for departures, independent movements are allowed only if the aircraft departing from each of the parallel runways will follow

---

*The FAA recommends a separation of at least 1200 ft (366 m) between the centerlines of runways that are used for IFR operations by airplane design groups V and VI. In other words, two parallel runways used by wide-body jets cannot be simultaneously active unless separated by 1200 ft or more.

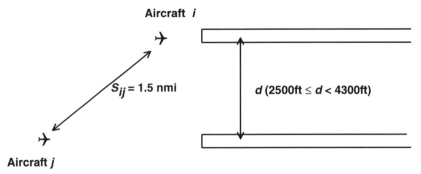

Fig. 10-1. *The diagonal separation between two aircraft approaching medium-spaced parallel runways.*

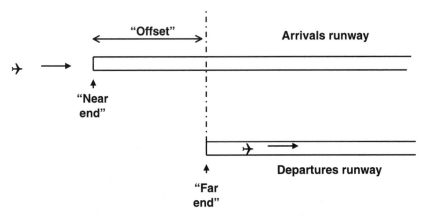

Fig. 10-2. *Staggered parallel runways; the "near" runway is used for arrivals and the other for departures.*

diverging climb paths after takeoff—as is most often the case in such circumstances. If not, one must apply the same separation requirements as for departures from a single runway just as in the case of close-spaced (under 2500 ft) parallel runways. Second, Table 10-3 assumes that the parallel runways are not "staggered"—i.e., their thresholds are not offset (see Fig. 10-2). If they are staggered, then an "effective separation distance" between the centerlines of the two runways should be computed. Specifically, when arrivals are to the "near end" in the direction of operations (Fig. 10-2), the 2500-ft (762-m) separation requirement between runway centerlines is reduced by 100 ft (30 m) for each 500 ft (150 m) of threshold offset, down to a minimum of 1200 ft (366 m). For example, when the offset is 1000 ft (300 m), a separation of 2300 ft (690 m) between the runway centerlines is equivalent to 2500 ft (762 m) when there is no offset. In other words, arrivals on one runway and departures on the other can be performed independently on a pair of parallel runways whose centerlines are 2300 ft apart *and* whose thresholds are staggered by 1000 ft, as long as the arrivals are assigned to the runway with the near threshold, as shown in Fig. 10-2. The reverse applies when arrivals are to the far threshold: the 2500-ft separation between runway centerlines must be increased using the same method.

While Table 10-3 shows the separation requirements that apply to operations on parallel runways in the United States, the pattern it presents is also typical of separation requirements for parallel runways elsewhere. Generally speaking, the greater the distance between runway centerlines (and the greater the offset of the runway thresholds), the "less dependent" operations on the two runways are. However, differences with the specific values used in the United States abound. Following the FAA's lead, the International Civil Aviation Organization (ICAO) recommends that a minimum distance between centerlines of 1035 m (3400 ft) be authorized for simultaneous instrument approaches, provided appropriate instrumentation and procedures are in place (ICAO, 1999). Most countries, however, require at least 5000 ft (1525 m) of separation between runway centerlines for independent simultaneous approaches to a pair of parallel runways.

For independent approaches to three parallel runways under IFR, the FAA requires a 5000-ft (1525-m) separation between the centerline of the middle runway and the centerlines of *each* of the outer runways. An approved FAA aeronautical study is also required for triple approaches at airports above 1000 ft (305 m) mean sea level.

Separation requirements for aircraft operating on intersecting, converging, or diverging runways. When it comes to runways that either intersect physically or converge/diverge (so that the projections of their centerlines intersect), the applicable operating procedures and separation requirements vary from airport to airport and from country to country. Examples of the considerations involved include the location of the intersection of the runways, the angle between their centerlines, the mix of aircraft and of movements on each runway, and the local missed-approach procedures. It is therefore impossible to provide a general summary, analogous to Table 10-3, for the separation requirements that apply to such cases. However, these requirements can be specified for any set of local conditions of use.

Clearly, the combined capacity of runways that intersect or converge
/diverge will vary significantly, depending on all the factors mentioned.
The highest capacities for pairs of runways that intersect physically
are usually achieved when the intersection is at the very beginning
of both runways in the direction of operations.* A pair of intersecting
runways can then provide as much capacity as a pair of close-spaced
parallels or even medium-spaced parallels, under some modes of use.
At the opposite extreme, when crosswinds preclude operations on
one of the two runways, their capacity will be the same as that of a
single runway.

Visibility, ceiling, and precipitation

Airport capacity is affected in important ways by weather conditions.
Cloud ceiling and visibility are the two parameters that determine
the weather category in which an airport operates at any given time.
Figure 10-3 shows a classification of weather conditions according to
these two parameters at a typical airport in the United States. The

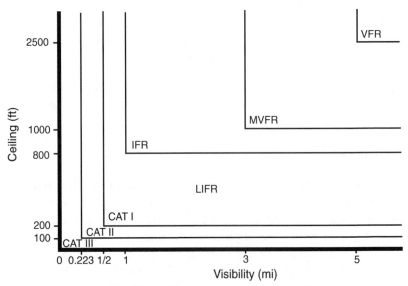

Fig. 10-3. *A typical classification of weather conditions (ceiling and
visibility) at an airport in the United States. For CAT II operations the
minimum visibility should be 1200 ft runway visual range (RVR),
approximately equal to 0.223 statute miles.*

*High capacities are also attained in the United States when the intersection is at the far
end of two long runways, if both runways are used for arrivals; this is achieved through
use of the "land and hold short" of the intersection (LAHSO) procedure.

designations "VFR," "MVFR," "IFR," and "LIFR" for the various regions shown are informal, but widely used in practice. MVFR stands for "marginal VFR" and LIFR for "low IFR."

The regions denoted as VFR, with a cloud ceiling of 2500 ft (762 m) or higher *and* visibility of 5 mi or more, and as MVFR are associated with visual meteorological conditions (VMC). The other two correspond to instrument meteorological conditions (IMC) of increasing severity. Note that Category I, II, and III conditions (see Chap. 13) are all part of LIFR. Depending on the instrumentation of the runways and on local topography, different approach, spacing, and sequencing procedures may be used under ceiling/visibility combinations associated with MVFR, IFR, and LIFR. This means that airport capacity may also change considerably.

An important example of these effects on runway capacity is the occasional use of visual separations on final approach at major airports in the United States. Almost as a rule, IFR separations, such as those in Tables 10-1–10-3, are maintained between landing and/or departing aircraft at major commercial airports outside the United States, regardless of prevailing weather conditions. However, in the United States, under VMC, pilots are often requested by air traffic controllers to maintain visual separations from preceding aircraft during the final spacing and final approach phases of flight. This practice results in higher capacities per runway than can be achieved with strict adherence to IFR. Equally important, this allows for more efficient use of parallel runways than suggested by Table 10-3.

Boston/Logan illustrates this last point well (right side of Fig. 10-4). Procedures have been established that allow simultaneous, parallel landings *in VFR weather* on runways 04L and 04R, which are separated by only about 1600 ft (490 m). Nonjets land on 04L and practically all jets on 04R. These procedures have been extended for use in MVFR weather as well.

More generally, FAA procedures in VMC allow for parallel landings and takeoffs on pairs of parallel runways whose centerlines are separated by as little as 700 ft (214 m) when the runways are used by aircraft in airplane design Groups I through IV, and by 1200 ft (366 m) when aircraft in airplane design groups V and VI are involved.* It is for such reasons that the capacity of several major airports in the United States

---

*Even in VMC, however, operations on parallel runways with a centerline spacing under 2500 ft (762 m) are treated as in Table 10-3 whenever wake turbulence may be a factor.

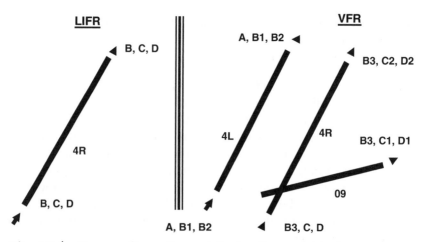

Fig. 10-4. *Two configurations at Boston/Logan with the same orientation, but under different weather conditions. The notation A, B, B1, etc., indicates classes of aircraft assigned to each runway. For example, C1 are narrow-body jets on short- and medium-range flights.*

under VMC is, in practice, considerably higher than would have been predicted if one applied strictly the IFR separation requirements of Tables 10-1 and 10-3. A few busy European airports have moved toward application of similar procedures in VMC and have gained capacity.

A second example of the effect of weather is provided by operations under low-ceiling and low-visibility conditions (LIFR). A first and obvious effect is that only certain runways, those equipped with a qualified Instrument Landing Systems (ILS), can be used under such conditions. For example, Runway 04L at Boston/Logan cannot be used for arrivals in LIFR conditions, as it is not equipped with an ILS, in part due to its proximity to 04R. This means that, when operations are to the northeast in LIFR, Boston/Logan may operate with only one arrival runway, 04R, which must accommodate all aircraft (left side of Fig. 10-4). A second effect is that, to minimize interference with the ILS signal that each aircraft receives (Chap. 13), separations between aircraft landing consecutively on the same runway in Category II and Category III weather are typically increased to as much as 9 nmi or to several minutes. This, of course, reduces dramatically the arrival capacity under these conditions (cf. Table 10-1).

Finally, precipitation and icing may severely affect runway capacity due to poor visibility, poor braking action, and the need for aircraft deicing. For example, when braking action is poor, the crosswind limits for approaches to a runway may be reduced. More extreme weather events, such as snowstorms and thunderstorms, often lead to the temporary closing of airport runway systems.

Wind direction and strength

Winds may also affect airport capacity in crucial ways. As explained in Sec. 9-3, a runway can be used only when crosswinds are within pre-scribed limits and tailwinds do not exceed 5 or 6 knots (9–11 km/h). This means that the orientation of runway operations and, more gen-erally, the combination of the active runways largely depend on the direction and strength of the prevailing winds at any given time. At locations that may experience strong winds from several different directions at different times, this can be the cause of considerable variability in the available capacity of the runway system. Boston/Logan again provides a good example. With strong winds from the north-east or the southwest, the airport usually operates in VMC with two arrival runways—04L and 04R when operations are to the northeast, 22L and 27 when operations are to the southwest (Fig. 10-5). How-ever, with strong winds from the northwest, only one runway, 33L, is truly available for arrivals, because runway 33R is very short (2300 ft, approximately 700 m) and can be used by only some nonjet aircraft (Fig. 10-6). This means that with strong northwest winds, aircraft arrivals at Logan International, even in VMC, experience severe delays. These may exceed 2 h on some days.

When wind speed is less than 5 knots ("calm"), air traffic controllers have considerable latitude as to which runways will be used (if more than one exist) and in which direction. Such decisions must be made often, as calms prevail a large percentage of the time at most airports. To select the active runway(s) and the direction(s) of operations on these occasions, a combination of such criteria as maximizing runway capacity and minimizing noise impacts may be used (see also below and Chap. 6).

Mix of aircraft

Tables 10-1–10-3 suggest why the *mix of aircraft* is another important factor in determining runway capacity. Consider, for example, a run-way used only for arrivals and assume that the mix of aircraft consists of 50 percent "heavy" (H) and 50 percent "small" (S). Everywhere in the world, arriving aircraft are currently sequenced for access to a run-

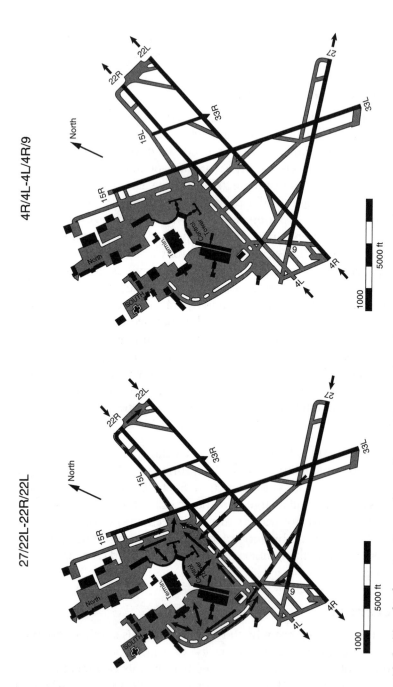

Fig. 10-5. *Two high-capacity configurations with opposite orientations at Boston/Logan. The configuration on the left is Configuration 9; the one on the right is Configuration 1.*

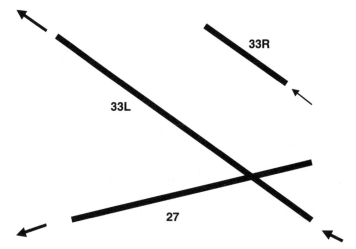

Fig. 10-6. *A low-capacity configuration in VMC at Boston Logan.*

way according to a first-come, first-served (FCFS) queue discipline.* With FCFS sequencing, about 25 percent of aircraft pairs $[(0.5)(0.5) = 0.25]$ will then be "H followed by S" (H-S) pairs and, according to Table 10-1, will be separated by 6 nmi at the runway threshold. Another 25 percent of the pairs will be separated by 4 nmi (H-H), and 50 percent will be separated by 2.5 nmi (S-H and S-S). By contrast, if the traffic consisted of 80 percent L-class aircraft and 20 percent S-class, then 16 percent $[(0.8)(0.2) = 0.16]$ of *all* possible pairs of aircraft would require a separation of 4 nmi and the other 84 percent would require 2.5 nmi. The runway capacity, as measured by the expected number of landings performed per hour, will clearly be considerably greater in the second case than in the first, as can be confirmed by using the simple mathematical model of Sec. 10-5.

In general, a relatively homogeneous mix of aircraft (i.e., a mix consisting of one or two dominant classes) is preferable to a nonhomogeneous mix from the point of view of runway capacity. Moreover, a homogeneous mix also offers advantages for ATM purposes, as it simplifies the work of air traffic controllers, who have to make fewer adjustments for wake vortex separations of varying magnitude, for

*Departures are also typically sequenced according to FCFS. It should be noted, however, that advanced ATM systems occasionally deviate from the FCFS order so as to avoid particularly wasteful sequences of aircraft, such as a stream of consecutive H-S arriving pairs. In fact, some of the new ATM automation systems that are currently being installed at busy terminal areas in the United States and in Europe include software that assist controllers in performing a limited amount of aircraft resequencing to increase capacity and efficiency, as described in Chap. 13.

different approach speeds, and for other aircraft characteristics. In fact, when the mix of aircraft is very nonhomogeneous, air traffic controllers at multirunway airports often attempt to "segregate traffic" by assigning different aircraft classes to different runways, as suggested by the jet runway versus nonjet runway examples already mentioned for Boston/Logan.

This also explains why the combined capacity of two independent parallel runways, if operated well by the ATM system, can provide *more than twice* the capacity of a single runway: the two runways provide an opportunity to optimize the assignment of aircraft types to each runway, as well as the mix and sequencing of movements (landings and/or departures) on each runway—see below.

Mix and sequencing of movements

Another factor that influences runway capacity is the mix of movements (arrivals versus departures) at the airport as a whole, and on each runway separately. For most, but not all, ATM systems, separation requirements are such that the capacity of a runway that is used only for departures is higher than the capacity of a runway that is used only for arrivals, given the same mix of aircraft. At some major airports in the United States more than 60 departures may be performed in 1 h from a single runway when the traffic mix includes only a small percentage of aircraft in class H. By contrast, it is difficult to perform more than 45 arrivals per hour per runway with a similar aircraft mix.

At busy airports there are typically some periods of the day when arrivals dominate and vice versa.* Hub airports, in particular, experience surges of arrival activity several times a day, followed about 1 h or less later by surges of departures.[†] The capacity of the airport may vary correspondingly. For instance, the number of runway movements per hour that can be performed at New York/Kennedy during the early afternoon hours, when many flights from Europe arrive, is significantly smaller than can be performed late in the evening, when a similar number of flights depart for Europe.

A related issue is the assignment of arrivals and departures to runways at airports operating with more than one active runway. When

---

*Such surges of arrivals or of departures are often referred to as "an arrival push" or "a departure push."

[†]Europeans use the term "wave" and Americans the term "bank" to refer to such surges of connecting arrivals and departures by any particular airline at a hub airport (e.g., "American Airlines schedules 8 banks a day at Dallas/Ft. Worth" or "KLM schedules 5 waves a day at Amsterdam").

given the opportunity, air traffic controllers often prefer to use separate runways for arrivals and for departures. This is especially common at European and Asian airports that operate with two parallel runways, as several do. This practice may simplify ATM operations, but is not necessarily optimal as far as overall airport capacity is concerned. It may overload one runway and underutilize another at times when the number of arrivals differs significantly from the number of departures. This may also create a serious imbalance between the delays experienced by arrivals versus those experienced by departures. In fact, a better way to operate an airport with two parallel runways, when feasible, is to assign some arrivals to a runway used primarily for departures, whenever arrivals "overflow" their primary runway, and do the reverse whenever there is an excess of departures in the mix. It may be even more efficient to mix arrivals and departures on two or more runways at airports where the ATM system is sufficiently advanced to sustain this mode of operation well. Frankfurt Airport, on its two closely spaced parallel runways, as well as several airports in the United States, achieve high processing rates through such a mixed runway-use strategy. The mathematical models discussed in Secs. 10-5 and 10-6 and in Chap. 19 can be helpful in assessing what benefits can be obtained from alternative assignments of operations to runways for each particular set of local conditions.

The *sequencing* of movements on a runway also influences runway capacity, especially whenever a runway is used for mixed operations (arrivals and departures). As noted earlier, arriving aircraft are generally sequenced in roughly FCFS order for access to a runway, and so are departing aircraft. Air traffic controllers, however, have considerable latitude regarding the sequencing of arrivals versus departures on the runway. It is possible, of course, to maintain an approximate FCFS discipline and sequence arrivals and departures roughly according to the time when they can first make use of the runway—the earlier the time, the higher the priority.* More typically, though, arrivals are given priority over departures for reasons of safety, controller workload, and aircraft operating cost. However, the strictness with which this practice is applied in practice varies considerably from one ATM system to another and from airport to airport. Quite often, for example, air traffic controllers will process a string of several consecutive landings until the queue of arriving aircraft is practically exhausted and will then process a string of several consecutive departures. Air traffic controllers will also look for some "free departures," i.e., they will try

---

*This, in fact, is approximately the case in practice at times when the airport is not heavily utilized.

to insert one or more departures between two arrivals without seriously disturbing the arrival stream and, thus, without reducing the arrival processing rate. This can often be done when there is a long gap between two landing aircraft, e.g., due to a 6-nmi separation between a leading aircraft of type H and a trailing one of type S. There are also occasions when a long queue of departures may form on the ground because the runway is continually busy with arrivals. In such instances, ATC may decide to interrupt the arrival stream for a while, assigning temporary priority to takeoffs until the departure queue returns to a reasonable length.

Finally, alternating arrivals and departures on the runway can be a very effective strategy for maximizing overall runway capacity, as measured by the *total number of movements* per unit of time. This sequencing strategy can be implemented by "stretching," as necessary, the separation between a pair of consecutive arriving aircraft, in order to create a gap that is just sufficiently long to allow insertion of a departure between the two arrivals. In a number of countries, ATM separation requirements make it possible to achieve such insertions with only a relatively modest amount of stretching of the required A-A separations. Indeed, this happens to be the case with the separation requirements shown in Table 10-1 for the United States. Thus, by "sacrificing" only a modest amount of arrival capacity per unit of time, the number of departures served by the runway per hour becomes roughly equal to the number of arrivals. However, this type of separation-stretching procedure is somewhat more demanding from the ATM point of view and requires skilled air traffic controller teams. Thus, its application is still limited primarily to some of the busiest airports of the United States and of Europe.

Type and location of runway exits

The *runway occupancy time* of an arriving aircraft is defined as the time between the instant the aircraft touches down on the runway and the instant it is on a runway exit, with all parts of the aircraft clear of the runway. Since the location of runway exits ("exit taxiways") plays a significant role in determining runway occupancy times, it may also have an impact on runway capacity. In particular, it can be seen from Table 10-1 that reducing runway occupancy times will contribute to increasing runway capacity in:

- The A-D case, where the earlier the arriving aircraft leaves the runway, the earlier the trailing departure's takeoff run can begin, provided the departing aircraft is set to go

- The A-A case, but only if the requirement that two arriving aircraft should not occupy the same runway simultaneously is the more restrictive of the two requirements listed in Table 10-1—the other requirement being the longitudinal separation of 2.5 or 4 or 5 or 6 n mi on final approach.*

Well-placed high-speed exits can be helpful in reducing runway occupancy times and increasing capacity. However, as noted in Sec. 9-7, the cost of constructing a high-speed exit may be considerably higher than that of a conventional exit forming a 90° angle with the runway centerline. When the benefits obtained from a high-speed exit, in terms of additional runway capacity, are compared with this additional cost, it is difficult to justify the construction of more than two or three high-speed exits for any single direction of runway operation.

In general, it is useful to remember that high-speed exits offer essentially no capacity benefits at runways used only for departures, some capacity benefits at runways used only for arrivals (primarily at those airports where visual separations are in use under VMC), and significant capacity benefits, under some movement-sequencing strategies, at runways used in a mixed operations mode.

State and performance of the ATM system

A high-quality ATM system with well-trained and motivated personnel is a fundamental prerequisite (but not a sufficient condition by itself) for achieving high runway capacities. To use a simple example, tight separations between successive aircraft on final approach (i.e., separations that are as close as possible to the minimum required in each case) cannot be achieved unless (1) accurate and well-displayed information on the positions of the leading and trailing aircraft is available to air traffic controllers through the ATM system, and (2) the controllers themselves are skilled in the task of spacing aircraft accurately during final approach. ATM systems for airports and for terminal airspace are reviewed in Chap. 13.

Air traffic controllers are the core element of ATM systems and will continue to be so in the foreseeable future. Human factors and ergonomics therefore play a central role in determining airport capacity. Air traffic controllers at most of the busiest airports in the world are highly qualified and, typically, well-paid professionals. The synergy

---

*As will be seen in Sec. 10-5, the longitudinal separation on final approach is, with few exceptions, the more restrictive of the two requirements. This means that the principal benefits from high-speed exits usually come from the A-D case, not the A-A.

between air traffic controllers and aircraft pilots is also very important. If air traffic controllers perceive that a pilot is inexperienced or has difficulty understanding instructions, they will slow down operations considerably to allow for additional margins of safety, thus reducing airport capacity.

Noise considerations

Last, but certainly not least, environmental considerations, especially noise impacts, exert an important influence in determining runway system capacity at an ever-growing number of airports. In the daily course of airport operations, noise is one of the principal criteria used by air traffic controllers to decide which one among several usable alternative runway configurations to activate. As indicated earlier, a choice among two or more alternative configurations exists whenever weather and wind conditions are sufficiently favorable. As a simple example, at a single-runway airport, air traffic controllers can choose to operate in either of the two directions of the runway when the weather is fair and there is little wind. The noise impacts associated with each direction of operation will then often be the principal criterion that will determine the choice between the two options.

Noise-related considerations work, in general, as a constraint on airport capacity, since they tend to reduce the frequency with which certain high-capacity configurations may be used. Example 10.2 illustrates the types of noise-related restrictions and configuration-selection practices that one increasingly encounters at major airports worldwide.

EXAMPLE 10.2

At Boston/Logan, no turbojet departures are permitted on Runway 04L, except in special cases, despite the fact that this runway is sufficiently long to accommodate the landing and takeoff requirements of most jet flights. The reason for this policy is noise mitigation for densely populated areas under the takeoff paths from Runway 04L. Few jet arrivals are also assigned to 04L, again in order to avoid noise-related complaints from airport neighbors living under the 04L approach paths. Similarly, jet landings are generally not permitted on Runway 22R, due to noise considerations.

Another type of noise-related constraint with an impact on capacity at Boston/Logan takes the form of a set of long-term and short-term goals for runway utilization and related restrictions. Specifically, the Massachusetts Port Authority ("Massport"), owner and operator of Boston/Logan, has agreed with representatives of the communities surrounding the airport on the following noise-related operating guidelines:

- *Annual goals* have been set for the utilization of every runway end. The goals are stated in terms of the desired percentage of *effective jet operations* that should be performed annually over each runway end (nonjets are not considered). The number of effective jet operations is obtained by multiplying the number of nighttime (22:00–07:00) operations by 10 and then adding this product to the number of operations during the rest of the day (see discussion of day–night noise levels in Chap. 6). For instance, the goal for Runway 33L is that it be used for 42 percent of effective arrivals and 12 percent of effective departures in a year. The reason for the high goal of 42 percent is that aircraft landing on 33L approach the airport over the Atlantic Ocean and thus have little noise impact on neighboring communities. The overall objective of the annual goals is to "distribute" noise among neighboring communities in a way that is considered fair by the parties involved.

- No runway can be used continually for more than 4 h in any single direction. This so-called "persistence" restriction is aimed at preventing the continuous exposure of any single community to noise on any particular day.

- No runway can be used for more than 24 h in any 72-h period. This restriction, too, is intended to prevent excessive, even if intermittent, exposure of a community to noise within a relatively short time span of 3 consecutive days.

It should be noted that these noise-related restrictions are applied only if weather conditions permit. When weather conditions are unfavorable, air traffic controllers have little or no choice as to the runway configuration to be used.

The overall effect of these restrictions is to inject noise as the second criterion (in addition to making optimal use of runway capacity) in the selection and use of active runways. For example, during periods when weather permits a choice among a number of alternative runway configurations, air traffic controllers may elect to use that configuration which will bring Logan closer to meeting the annual goals for the use of the runway ends, rather than the configuration that will offer the highest capacity. Indeed, this is very often the case, especially at times when demand is relatively low. While adherence to the three restrictions mentioned above is currently voluntary, community and Massport representatives meet regularly to review how well the airport meets each.

10-4   Range of airfield capacities and capacity coverage

The capacities of runway systems of major airports around the world span a wide range. Some single-runway airports have a capacity as low as 12 (!) movements per hour, due to inadequate air traffic control systems or other local factors. At the opposite end, a few airports in the United States operate with as many as 4 to 7 simultaneously active runways and accommodate more than 200 (and in the case of Dallas/Ft. Worth up to 300) movements in 1 h. At locations with reasonably advanced ATM systems—and absent noise-related or other restrictions—capacities range from about 24 per hour *per runway* to as many as 60 *per runway*, depending on the many factors that were reviewed in the previous section. Airports in the United States are typically at the high end of the capacity-per-runway range. At many airports, the capacity of the runway system may also be highly variable over time, primarily due to sensitivity to weather and wind conditions.

To illustrate these points, Table 10-4 shows the FAA's 2001 estimates of the maximum throughput capacities of 31 of the busiest airports in the United States under optimum weather conditions and under weather conditions that lead to reduced capacity.* In the former case, unlimited ceiling and visibility permit visual approaches and visual separations between landing aircraft. In the latter, reduced visibility conditions necessitate IFR separations. The capacities shown are for the most commonly used runway configurations under these

---

*The "benchmark" capacities shown in Table 10-4 should be interpreted as only partial indications of the overall capacities at these airports.

Table 10-4. Approximate capacities of 31 of the busiest airports in the United States

Airport	Optimum conditions	Reduced conditions
Atlanta	185–200	167–174
Baltimore/Washington	111–120	72–75
Boston/Logan	118–126	78–88
Charlotte	130–140	108–116
Chicago/O'Hare	200–202	157–160
Cincinnati	123–125	121–125
Dallas/Ft. Worth	261–270	183–185
Denver/International	204–218	160–196
Detroit/Metro	143–146	136–138
Honolulu	120–126	60–60
Houston/Bush	120–123	112–113
Las Vegas	84–85	52–57
Los Angeles/International	148–150	127–128
Memphis	150–152	112–120
Miami/International	124–134	95–108
Minneapolis/St. Paul	115–120	112–112
New York/Kennedy	88–98	71–71
New York/LaGuardia	80–81	62–64
New York/Newark	92–108	74–78
Orlando/International	144–145	104–112
Philadelphia	100–110	91–96
Phoenix	101–110	60–65
Pittsburgh	140–160	110–131
Salt Lake City	130–132	95–105
San Diego	43–57	38–49
San Francisco/International	95–99	67–72
Seattle/Tacoma	90–91	78–81
St. Louis	104–112	64–65
Tampa	110–119	80–87
Washington/Dulles	120–121	105–117
Washington/Reagan	76–80	62–66

Source: FAA, 2001a.

conditions. Note that 24 of the 31 airports had a capacity under optimum conditions that exceeded 100 movements per hour. In contrast, only three airports outside the United States, Amsterdam/Schiphol, Paris/de Gaulle, and Toronto/Pearson, could regularly handle more than 100 movements per hour in 2001! These are also among the very few airports outside the United States that operate with three simultaneously active runways.

The difference between the optimum and the reduced capacities for some of the airports in Table 10-4 is also remarkable. In several cases, this difference is of the order of 30 percent or more. As explained in the previous section, these differences usually stem primarily from the geometric configuration of the runways that forces a reduction in the number of runways that can be operated simultaneously for approaches in IMC, and the use of IFR instead of visual separations in IMC. Note that certain airports, such as Cincinnati and Minneapolis/St. Paul, have runway configurations that are little affected by reduced visibility.

A particularly convenient way to summarize the range of capacities at an airport and the frequency with which various levels of capacity are available is the *capacity coverage chart* (CCC). The CCC shows how much runway capacity is available for what percentage of time at a given airport under the assumptions that (1) the operations mix is 50 percent arrivals and 50 percent departures and (2) the runway configuration in use at any given time is the one that provides the highest capacity under the prevailing weather conditions.

EXAMPLE 10.3

Figure 10-7 shows the CCC of Boston/Logan Airport. A capacity of 132 movements per hour is available for approximately 60 percent of the time, a capacity of 120 movements per hour for approximately 18 percent of the time, and so on. The airport's capacity declines to the range of 60 movements per hour—i.e., to less than half of the peak capacities of 132 and 120 movements—for about 14 percent of the time. For about 1.5 percent of the time, when the airport is closed due to snowstorms or severe thunderstorms, the capacity is zero. This CCC is obtained by looking at historical statistics regarding the frequency with which each of the possible combinations of visibility, ceiling, and wind conditions at Boston/Logan occur during the course of a year and identifying the runway configuration that provides the highest capacity for each set of weather conditions. For instance, Configuration 1, which consists of arrivals on runways 4R and 4L and of departures on runways 4R, 4L, and 09 in visual meteorological conditions (on the right in Fig. 10-5), is the Boston/Logan configuration with the highest capacity. When the mix of movements is 50 percent arrivals and 50 percent departures, this capacity has been estimated at 132 movements per hour, a number that can be obtained either on the basis of empirical data or from a mathematical or simulation model like the ones described in Secs. 10-5 and 10-6 and in Chap. 19. Weather records indicate that Configuration 1 can be used about 60 percent of the time. From the assumption that the available configuration with the highest capacity will be selected at all times [assumption (b) in the definition of the CCC], it follows that Configuration 1, the highest capacity configuration at Boston/Logan, will be used whenever possible, i.e., for about 60 percent of the time. This is shown in Fig. 10-7, where a capacity of 132 movements per hour for 60 percent of the time is indicated at the left-hand part of the CCC, along with the indication that this capacity is associated with Configuration 1—denoted by the number 1 above the relevant part of the CCC. Proceeding now toward the right in Fig. 10-7, it can be seen that, when weather conditions do not permit use of Configuration 1, the configuration with the next highest capacity, 120 movements per hour, is Configuration 9 (shown in Fig. 10-5 on the left), which consists of arrivals on runways 22L and 27 and of departures on runways 22R and 22L in VMC. From weather statistics at Boston/Logan, the percentage of time when Configuration 9 is available *and Configuration 1 is not*, is equal to 18 percent. This is once again shown in Fig. 10-7. Continuing in the same way toward the right part of the CCC, one encounters ever-smaller capacities, as the airport "runs out" of high-capacity configurations, until "100 percent of the time" is accounted for.

The CCC obviously provides highly useful information for airport planners and managers. However, it is also important to keep in mind its underlying assumptions. By assumption (a) in the definition of the CCC, the capacities of 132, 120, etc., for each configuration in Example 10.3 are computed for an operations mix of 50 percent arrivals and 50 percent departures. By assumption (b), the operator of the runway

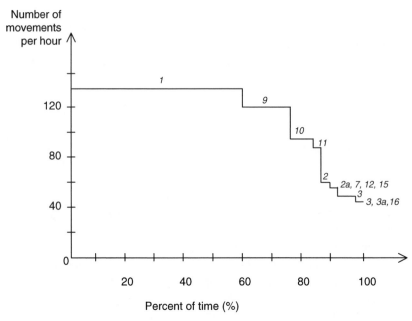

Fig. 10-7. *The capacity coverage chart for Boston/Logan.*

system (i.e., the FAA or other ATM service provider) *will choose at all times the available runway configuration with the highest capacity.* For instance, as noted above, in VMC and with calm winds, both Configuration 9 and Configuration 1 (Fig. 10-5) are available for use at Boston/Logan. According to assumption (*b*), Configuration 1 will always be selected in such cases, because it has the higher capacity.

In practice, assumptions (*a*) and (*b*) are only rough approximations to reality. The mix of movements is rarely exactly 50 percent arrivals and 50 percent departures. When the mix is significantly different from that (e.g., 65 percent arrivals and 35 percent departures), the capacity of the runway system may also differ significantly from the number indicated on the CCC. Fortunately, the operations mix at busy nonhub airports during peak hours typically falls in the range between 40 percent arrivals, 60 percent departures and 60 percent arrivals, 40 percent departures. Therefore, the capacities indicated under the 50 percent–50 percent assumption are usually fairly representative of the capacities available during peak hours. At hub airports, where waves of arrivals are followed by waves of departures, the CCC must be supplemented by an analysis of the capacity to handle these surges of arrival and departures. Mathematical or simulation capacity models can be used for this purpose.

Regarding assumption (*b*), noise considerations may dictate use of a configuration other than the one with the highest capacity, especially during hours when demand is not at its peak, as already seen in the previous section. In the Boston/Logan case, Configuration 9 is often chosen over Configuration 1 during periods when they are both available, in order to "distribute the noise more equitably" among the airport's neighboring communities and meet annual noise-related goals (see Example 10-2). In this light, the CCC can better be viewed as showing *the upper limit* of how much runway capacity is available at an airport over time.

This last point is underscored by Fig. 10-8 (Idris, 2001), which summarizes the usage of runway configurations at Boston/Logan during January 1999. A very low-capacity configuration* that uses Runway 33L for arrivals and Runway 15R for departures (i.e., the same runway in opposite directions) is utilized heavily during the six hours of 00:00 to 05:59. This is because the configuration in question has the least noise impact of any at Boston/Logan, as both arrival and departure paths are over the sea and avoid populated areas. That the capacity of this configuration is low does not really matter, because traffic is also very low during the period when it is used. An intermediate-capacity configuration with arrivals on runways 33L and 33R and departures from Runway 27 is used quite intensively during the early morning to noon hours, when traffic demand is not very heavy (Fig. 10-8). Finally, the two highest-capacity configurations that figure so prominently in the CCC, Configurations 9 and 1, are utilized very heavily during the peak traffic hours between 14:00 and 21:00, when their high capacity is truly needed. Figure 10-8 confirms that the noise impact of a runway configuration is often the dominant selection criterion during periods of low demand, while the CCC is a good indicator of what configurations will be used during peak demand periods.

Thus, the CCC essentially provides a summary description of the relative frequency with which different values of capacity are available at an airport during periods of high demand. For complex, multirunway airports, the computation of the CCC may require considerable effort. For example, in the case of Boston/Logan one needs to compute the capacity of each of the 39 different runway configurations, along with the percentage of time when each is available on the basis of historical weather/wind data.

An "uneven" CCC, like the one of Fig. 10-7, indicates an airport

---

*The capacity is so low that the configuration does not even appear in the Boston/Logan CCC of Fig. 10-7.

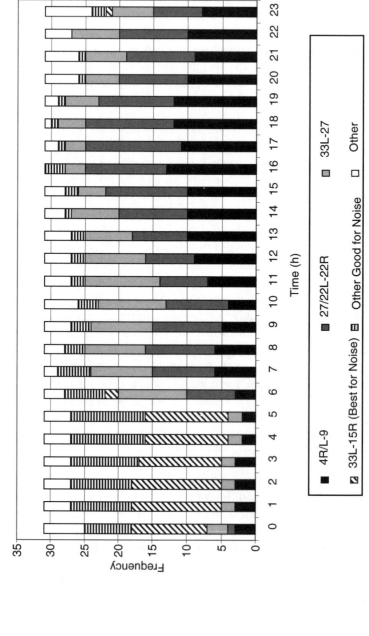

Fig. 10-8. *Runway configuration usage at Boston/Logan, January 1999 (from Logan FAA tower logs).*

where the supply of runway capacity is not reliable. This may result in long delays and serious operational problems when the typical demand levels during peak periods are close to the higher capacity values of the CCC. Consider again Boston/Logan. Because the airport's capacity is 120 or more for about 78 percent of the time, airlines are consistently scheduling close to 110–115 movements per hour for several hours each day during the peak summer season.* This means that for about 22 percent of the time (Fig. 10-7), or on one out of every 5 days on average, the available capacity may fall considerably short of demand during peak periods, resulting in serious delays. On truly bad days, when the capacity is 60 or lower for several hours in a row, very long delays and many flight cancellations occur. An extreme alternative to this scheduling practice would be to restrict airport demand to a low level, for example, to a maximum of 60 movements per hour. This would guarantee that demand is always (or almost always) exceeded by available capacity. While this ensures the virtual absence of delays and a high level of service, it also means wasting a great amount of available capacity for 80–90 percent of the time. This type of dilemma will be discussed further in connection with the subject of airport demand management (Chap. 12).

A "flat" (or "even") CCC, on the other hand, is characteristic of airports where the runway capacity stays relatively constant over time. For example, the single-runway Athens/Hellenikon Airport, which almost always operated under good weather conditions, had an almost completely even CCC, with a declared capacity of 32 movements per hour (and a maximum throughput capacity of about 40) for essentially 100 percent of the time. A flat CCC means more predictable airside performance and more effective utilization of airport resources and facilities, as the number of operations at the airport can be scheduled with reference to a stable level of runway capacity.

Recall now that runway capacity is defined as the *expected* ("average") number of movements that can be handled per hour. This is what the CCC shows. In practice, the actual number of movements that can be performed during each hour may be greater or less than the expected number shown. For example, instead of the capacity of 132 movements per hour shown in Fig. 10-7, the actual number performed during a particular hour when Configuration 1 is in use may be 127 or 140, depending on the exact traffic mix during that hour, the performance of the team of air traffic controllers on duty at the time, the strength and variability of the prevailing winds, etc.

*This describes the situation during the late 1990s.

Finally, the reader who is familiar with probability theory will recognize that the CCC is essentially a graphical representation of the probability distribution of an airport's (maximum throughput) runway capacity. Figure 10-7 indicates that, at any randomly chosen instant, Boston/Logan's maximum available runway capacity will be equal to 132 movements per hour with probability 0.6, to 120 movements per hour with probability 0.18, etc., and to 0 with probability 0.015.

10-5   A model for computing the capacity of a single runway

In addition to understanding qualitatively the definitions and complex relationships that determine capacity, it is essential in practice to have access to computational tools that provide reasonably accurate estimates of the capacity of runway systems under any set of specified conditions. Fortunately, a number of mathematical and simulation models have been developed over the years that make this possible. In this section one such mathematical model will be described in some detail because, despite its many simplifying assumptions, it yields good approximations to the capacities observed in practice. The model is also particularly convenient for sensitivity analyses that explore the effects of many of the factors reviewed in Sec. 10-3 that affect airport capacity. Finally, this model illustrates well the conceptual approach taken by virtually all the computer-based mathematical models that are now used widely to estimate capacity and delays at runway complexes (Chapter 19).

The simple mathematical model in question is originally due to Blumstein (1959). It estimates the capacity of a single runway used solely for arrivals. The same approach, however, can be readily extended to runways used solely for departures or runways used for mixed movements.

Consider a single runway, shown schematically in Fig. 10-9, which is used for landings only. Aircraft descend in single file along the final approach path until touching down on the runway, whereupon they decelerate and exit onto the taxiway system. The paths of arriving aircraft merge in the vicinity of the "gate" to the final approach, typically 5–8 nmi away from the runway threshold (Fig. 10-9). Throughout the final approach, aircraft must maintain a safe longitudinal distance from each other, in compliance with the ATM system's separation

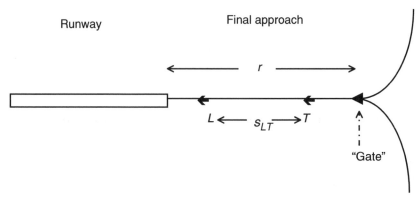

Fig. 10-9. *A simple representation of a runway used for arrivals only under IFR.*

requirements, as explained in Sec. 10-3. Moreover, single occupancy of runways is required (Table 10-1): each aircraft must be safely out of the runway before the next landing can touch down. These safety rules impose limits on the maximum acceptance rate of the runway, i.e., on its maximum throughput capacity.

Define now the following quantities for an aircraft of type i:

r = the length of the common final approach path

v_i = speed on final approach assuming, as a reasonable approximation, that aircraft i maintains a constant speed throughout the approach

o_i = runway occupancy time, i.e., the time that elapses from the instant when the aircraft touches down on the runway to the instant when it leaves the runway at one of the runway exits

Consider the case in which an aircraft of type i is landing, followed immediately by another aircraft of type j. Denote by s_{ij} the minimum separation required by ATC between the two aircraft while they are both airborne. For example, in Fig. 10-9, s_{LT} indicates the minimum separation required between a leading aircraft of type L and a trailing aircraft of type T. Let T_{ij} denote the minimum time interval (in the sense of not violating any ATM separation requirements) between the successive arrivals at the runway of the type i and type j aircraft. The two fundamental equations that determine T_{ij} can be written as follows:

$$T_{ij} = \max\left[\frac{r + s_{ij}}{v_j} - \frac{r}{v_i} , o_i \right] \qquad \text{when } v_i > v_j \qquad (10.1a)$$

$$T_{ij} = \max \left[\frac{s_{ij}}{v_j}, o_i \right] \qquad \text{when } v_i \leq v_j \qquad (10.1b)$$

The situation in which $v_i > v_j$ is known as the "opening case" because the distance between the two aircraft keeps increasing as they fly in single file along the final approach path on the way to the runway. In this case, the two aircraft are closest to each other at the instant when the first of the two, of type i, is at the gate of the final approach path, a distance r from the threshold of the runway (Fig. 10-9). If at that instant the two aircraft are separated by the minimum allowable separation s_{ij}, then the type j aircraft will be a distance $r + s_{ij}$ from the runway. The difference between the times when the leading aircraft (type i) and the trailing aircraft (type j) will touch down on the runway is then equal to

$$\frac{r + s_{ij}}{v_j} - \frac{r}{v_i}$$

However, the time interval between the successive arrivals at the runway must also be at least o_i long, to allow enough time for the leading aircraft (type i) to exit the runway before the trailing aircraft touches down. The minimum time interval, T_{ij}, between the successive arrivals at the runway is then equal to the larger (maximum) of the quantities

$$\frac{r + s_{ij}}{v_j} - \frac{r}{v_i} \qquad \text{and} \qquad o_i$$

and is thus given by Eq. (10.1a). By contrast, in the closing case ($v_i \leq v_j$) the two aircraft are closest to each other at the instant when the first aircraft is at the runway threshold. The minimum interval, T_{ij}, is then given by Eq. (10.1b).

Suppose now that the probability of the event "a type i aircraft is followed by a type j aircraft" is p_{ij}. Then

$$E[T_{ij}] = \sum_{i=1}^{K} \sum_{j=1}^{K} p_{ij} \cdot T_{ij} \qquad (10.2)$$

where $E[T_{ij}]$ denotes the expected value of T_{ij}, i.e., the value of T_{ij} "on average," and K is the number of distinct aircraft classes ($K = 4$ in Example 10.4). A numerical example illustrates the application of the model.

Table 10-5. Data for the example
(1 knot = 1 nmi/h = 1.15 statute miles/h = 1.852 km/h)

i (a/c type)	p_i (probability)	v_i (knots)	o_i (s)
1 (H)	0.2	150	70
2 (L)	0.35	130	60
3 (S1)	0.35	110	55
4 (S2)	0.1	90	50

Table 10-6. Separation requirements
(in nautical miles) on final approach

	Trailing aircraft		
Leading aircraft	**H**	**L**	**S1 or S2**
H	4	5	6*
L	2.5	2.5	4*
S1 or S2	2.5	2.5	2.5

*Indicates that the separation applies when the leading aircraft is at the runway threshold.

EXAMPLE 10.4

As noted in Sec. 10-3, the FAA subdivides aircraft into three classes with respect to the separations s_{ij} required on final approach: "heavy" (H); "large" (L); and "small" (S). (The special case of the B757 will not be considered here.) Because different types of aircraft in class S have quite different approach speeds, this class will be subdivided in this example into two more homogeneous subclasses, S1 and S2, as is often done in airport capacity analyses. Denote the classes H, L, S1, and S2 with the indices 1 through 4, respectively.

Assume now that, at a major airport, a runway, which is used for long periods of time for arrivals only, serves an aircraft population with the characteristics given in Table 10-5. Note that the probabilities, p_i, indicate the traffic mix at this runway (e.g., 20 percent of the aircraft are of type H). Assume, as well, that the IFR separation requirements, s_{ij}, in use are as shown in Table 10-6. Note these are the same as in Table 10-1 with some simplifications.

Let now the length, r, of the final approach path be equal to 5 nmi. Applying Eqs. (10.1a) and (10.1b), one can compute the matrix, **T**, of minimum time separations, T_{ij}, in seconds, at the runway shown as Table 10-7. To obtain Table 10-7, Eq. (10.1a) has been used to compute T_{12} and T_{34} and Eq. (10.1b) to compute all the other elements of the matrix. Note that the two equations give the same result when it comes to the diagonal elements T_{ii} of the matrix and that Eq. (10.1b) has been used in the cases of T_{13}, T_{14}, T_{23}, and T_{24} because the separation requirement in these cases applies when the leading aircraft is at the threshold of the runway.

As mentioned in Sec. 10-3, air traffic controllers use first-come, first-served (FCFS) sequencing of aircraft wishing to land at an airport. This makes it reasonable to assume that, for any pair of aircraft, the probability that the leading aircraft will be of type i is simply equal to p_i, the proportion of aircraft of type i in the mix, and the probability that the trailing aircraft is of type j is equal to p_j. This means that the probability of an i-followed-by-j pair is given by

$$p_{ij} = p_i p_j \tag{10.3}$$

The matrix, **P**, of aircraft-pair probabilities, p_{ij}, can be computed in this way, as shown in Table 10-8.

Multiplying the corresponding elements of the matrices **T** and **P** to apply Eq. (10.2) yields an expected value $E[T_{ij}] \approx 103$ s. In other words, if the ATC system could somehow always achieve the minimum allowable separations between landing aircraft, the runway of this example could serve one arrival every 103 s, on average, or up to about 35 arrivals per hour.

Table 10-7. The matrix T of minimum time separations, T_{ij} (s) for our example

	Trailing aircraft			
Leading aircraft	1 (H)	2 (L)	3 (S1)	4 (S2)
1 (H)	96	157	196	240
2 (L)	60	69	131	160
3 (S1)	60	69	82	136
4 (S2)	60	69	82	100

Table 10-8. Matrix P of pair probabilities p_{ij}

	Trailing aircraft			
Leading aircraft	**1 (H)**	**2 (L)**	**3 (S1)**	**4 (S2)**
1 (H)	0.04	0.07	0.07	0.02
2 (L)	0.07	0.1225	0.1225	0.035
3 (S1)	0.07	0.1225	0.1225	0.035
4 (S2)	0.02	0.035	0.035	0.01

In practice, it is extremely difficult to achieve the perfect level of precision in spacing successive landing aircraft on final approach implied by the matrix **T**. With human factors playing a key role in the spacing between aircraft, it is reasonable to expect some deviations from the separations suggested by the elements T_{ij} of **T**. In fact, in view of the natural tendency of both pilots and air traffic controllers to "err on the conservative side," one would expect the separations between given pairs of aircraft types to be, on average, *larger* than the corresponding values of T_{ij}. This is indeed the case: for example, in the United States, average spacing under instrument meteorological conditions exceeds the minimum required separations by about 10–25 s. The model presented here can capture this effect, if the matrix **T** is modified in a simple way. A "buffer time" can be added to every element T_{ij}, with the value of the buffer time chosen to account for the spacing added, intentionally or unintentionally, in practice to each *i*-followed-by-*j* pair of aircraft. For instance, under a particularly simple but reasonable approximation, one could just add the same constant buffer time, *b*, to all the elements T_{ij}, obtaining a new matrix **T′** whose elements t_{ij} give the *average* (*not* the minimum possible) separation achieved for an *i*-followed-by-*j* pair of aircraft. In this case,

$$t_{ij} = T_{ij} + b \qquad (10.4)$$

The expected value of t_{ij} gives the average time interval between successive landings on the runway. By analogy to Eq. (10.2), one can now write this expected value as

$$E[t_{ij}] = \sum_{i=1}^{K} \sum_{j=1}^{K} p_{ij} \cdot t_{ij} \qquad (10.5)$$

Finally, if the expected amount of time between successive landings has been computed, the maximum throughput capacity is simply given by

$$\text{Maximum throughput capacity} = \mu = \frac{1}{E[t_{ij}]} \qquad (10.6)$$

where $E[t_{ij}]$ is measured in hours.*

EXAMPLE 10.4 (continued)

Suppose that $b = 10$ s in Eq. (10.4). This means that all intervals between successive landings are 10 s longer than the minimum, due to inaccuracies in spacing of aircraft, conservatism on the part of pilots and controllers, etc. Obviously, the expected amount of time between successive landings is then $E[t_{ij}] \approx 113$ s $= 0.03139$ h. This leads to a capacity estimate of $\mu \approx 32$ aircraft per hour, a number typical of the service rates that would be observed at an airport in the United States with a traffic mix similar to this example's operating with IFR separations.

It is easy to use this model to assess the sensitivity of airport capacity to changes in various input parameters that may result from changes in the ATM system, airline fleet composition, terminal area procedures, etc. Consider a few instances.

First, a comparison of Table 10-7 with the o_i column of Table 10-5 indicates that the runway occupancy time of the leading aircraft is not the constraining factor for *any* of the 16 possible pairs of consecutive landing aircraft. All 16 values of the T_{ij} in Table 10-7 are greater than—and, in a single case, equal to—the value of the corresponding o_i. (The one case in which equality applies is T_{21}, for which the minimum separation dictated by the final approach spacing requirement of 2.5 nmi for the "type-2-followed-by-type 1" pair is equal to $(2.5)(3600)/(150) = 60$ s, the same as the 60-s runway occupancy time of the class 2 (L) aircraft that leads the pair.) This means that any reductions of the runway occupancy times, given as inputs in Table 10-5, will *not* increase arrival capacity. (Such reductions in o_i could, for instance, be obtained through the construction of high-speed runway exits.) In practice, it is indeed true that, for practically all ATM systems in the world, the final approach IFR spacing

---

*The Greek letter μ has been used to denote capacity. This is in line with standard notation in queuing theory (see Chap. 23).

requirements, such as those shown in Tables 10-1 and 10-2, are significantly more restrictive than the runway occupancy times.*

Second, suppose this airport was still operating with a 3-nmi separation requirement (instead of 2.5 nmi) for the L-H, L-L, S-H, S-L, and S-S aircraft pairs as is the case at less busy airports (cf. Table 10-1). The reader can verify that this would reduce capacity by approximately 2.5 arrivals per hour (from 32 to 29.5), or by about 8 percent. $E[t_{ij}]$ would be equal to approximately 122 s.

Similarly, air traffic controllers at the busiest airports in the United States often attempt to achieve more uniform final approach speeds, typically by recommending that pilots fly the smaller and slower aircraft at speeds more similar to those of some of the commercial jets on final approach. For instance, if, in this example, the ATM system could achieve $v_3 = 130$ knots and $v_4 = 110$ knots through higher final approach speeds of S1 and S2 aircraft, then $E[t_{ij}] \approx 103$ s, or approximately 35 arrivals per hour, an increase of about 9 percent over the 32 arrivals computed with the original approach speeds.

Note that a combination of (1) reducing the 3-nmi separations to 2.5 nmi, (2) increasing the final approach speeds of S1 and S2 aircraft to 130 and 110 knots, respectively, and (3) a potential reduction of the safety buffer to $b = 5$ s, instead of $b = 10$ s, has the overall effect of reducing $E[t_{ij}]$ from 122 to 98 s and increasing capacity from 29.5 to about 37 arrivals per hour, a 25 percent increase! It is the cumulative effect of relatively small changes such as these that has prevented airport capacity from falling hopelessly behind growing demand over the past 20 years.

Other possibilities for increasing runway capacity can be assessed by exploiting simple mathematical models. For example, inspection of the matrix **T** (Table 10-7) indicates that certain aircraft sequences are more desirable than others. For example, the sequence 1-4, or H-S2, requires at least 4 min of separation between successive landings, while the sequence 4-1, or S2-H, requires only a 1-min separation. This suggests the possibility of computer-aided sequencing of aircraft waiting to land at an airport, an idea that has been investigated in detail by several researchers (Dear, 1976; Psaraftis, 1980; Venkatakrishnan

*On the other hand, it is possible that if runway occupancy times were reduced considerably from their current values, the longitudinal separations required on final approach might be reduced as well.

et al., 1992) and is now being partially implemented through advanced ATM decision-support systems (Chap. 13). Note that, when sequences not based on a FCFS discipline are in use, Eq. (10.3) is no longer necessarily valid and must be replaced by an expression—or an algorithm—for computing probabilities p_{ij} that reflect the sequencing scheme actually in use.

10-6 Generalizations and extensions of the capacity model

The capacity model presented in the last section can be extended in a number of ways. Its accuracy can be improved as well. Before discussing some of these extensions and improvements, it is important to summarize the basic approach that the model follows. For all practical purposes, all mathematical models of runway capacity follow essentially this same approach, consisting of three basic steps.

Step 1. For all possible pairs, i and j, of aircraft classes and for all permissible combinations of movements ("arrival followed by arrival," "arrival followed by departure," etc.) involving a type i aircraft followed immediately by a type j aircraft, compute the *expected* time interval t_{ij} between successive movements, such that no ATM separation requirements are violated. Note that the average time interval t_{ij} is greater than or, at best, equal to T_{ij} the minimum interval between successive movements for that aircraft pair.

Step 2. Compute p_{ij} the probability of occurrence of each of the expected time intervals, t_{ij} obtained in step 1.

Step 3. Compute the overall expected time of the interval between any two successive movements,

$$E[t_{ij}] = \sum_{i=1}^{K} \sum_{j=1}^{K} p_{ij} \cdot t_{ij} \qquad (10.7)$$

and from that the (maximum throughput) capacity

$$\mu = \frac{1}{E[t_{ij}]} \qquad (10.8)$$

Example 10.5 illustrates the application of the three-step approach, previously used in the "all arrivals" model, to the case where a runway is used only for departures.

EXAMPLE 10.5

Consider a runway with the same aircraft mix as in Table 10-8, but with all aircraft now performing takeoffs from the runway. Assume that the separation requirements, s_{ij}, that apply in this case are as follows, in units of seconds:

	Trailing aircraft		
Leading aircraft	**H**	**L**	**S1 or S2**
H	90	120	120
L	60	60	60
S1 or S2	60	60	60

In the case of departures, it is reasonable to assume that, because of the simplicity of the control process, the average interval between the beginning of the takeoff run of two aircraft of types i and j is roughly equal to the minimum separation required between these aircraft. Thus, $t_{ij} \approx s_{ij}$ for all pairs i and j in this case. Assuming, as before, FCFS sequencing of departures on the runway, the pair probabilities, p_{ij}, are the same as in Table 10-8, since the aircraft mix is the same. Applying Eq. (10.7) (i.e., multiplying each of the probabilities in Table 10-8 by the appropriate 60-, 90-, or 120-s separation requirement) one obtains $E[t_{ij}] \approx 71$ s and $\mu \approx 51$ departures per hour.

An analogous three-step approach can also be used to estimate the capacity of a runway used for both landings and takeoffs. As already noted in Sec. 10-3, it is now important to identify the strategy employed by ATC controllers to sequence landings and takeoffs on the runway. Under the strategy most commonly used, controllers during peak demand periods may serve a string of successive arrivals (e.g., 5–10 arrivals in a row), then a string of successive departures, then another string of arrivals, and so on. The runway capacity can then be approximated as a simple weighted average of μ_a, the runway capacity when the runway is used only for arrivals, and of μ_d, the runway capacity when the runway is used only for departures, the weights being equal to the fractions of time spent in serving arrivals and departures, respectively (Odoni, 1972). An alternative strategy used occasionally by controllers at busy airports in the United States has arrivals alternating with departures; i.e., the separations on final approach between successive arriving aircraft are "stretched" so that a departure can take off during the time interval between the two arrivals. This is a procedure that requires considerable skill but, if performed accurately, can increase significantly the capacity of the runway, as measured by the total number of movements (landings *and* takeoffs) performed. A model of this operating strategy was developed by Hockaday and Kanafani (1974) and was subsequently generalized by several researchers (Swedish, 1981). In Exercise 3 at the end of this chapter, the reader is guided through the application of the three-step approach to this case.

Some improvements and extensions to the generalized three-step approach can now be reviewed briefly. To begin, it is obvious that some of the parameters which are treated as constants in the examples presented so far, such as the approach speeds, v_i, and the runway occupancy times, o_i, for each class of aircraft, can be treated more realistically as random variables with associated probability distributions. Most important, the distances between successive aircraft on final approach are random variables whose probability distribution depends on the ATM system's separation requirements and on the characteristics and performance of the terminal area ATM system, including the controllers and pilots (Harris, 1972; Odoni, 1972).

Computer-based mathematical models developed recently address all these possibilities (Lee et al, 1997; Stamatopoulos, 2000; Andreatta et al, 1999). These are generalized probabilistic models for computing capacity, when a runway is used for arrivals only or for departures only or for mixed operations. For instance, much as a controller would do, the models compute the spacing required between landing aircraft as they enter the common approach path so that, with reasonable confidence, no violations will occur later on as the aircraft fly toward the runway.

The principal output of these models is the *runway capacity envelope* (Gilbo, 1993), i.e., a boundary that defines the envelope of the maximum throughput capacities that can be achieved at the runway under the entire range of possible arrival and departure mixes (Fig. 10-10). Any point inside the envelope is *feasible,* and any point outside is *infeasible.* The runway has sufficient capacity to serve x arrivals per hour *and* y departures per hour, as long as the point (x, y) is within the runway capacity envelope.

The models actually compute the coordinates of only the four points on the envelope, which are denoted as points 1, 2, 3, and 4, and then approximate the entire envelope by interpolating between

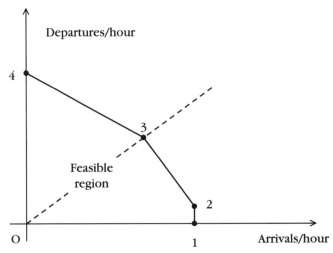

Fig. 10-10. *A typical capacity envelope for a single runway.*

them with straight-line segments in the manner shown in Fig. 10-10. The four points are the following.

Point 1. This is the "all arrivals" point, i.e., it indicates the capacity of the runway when it is used for arrivals only.

Point 2. This is known as the "free departures" (or "arrival priority") point, because it has the same capacity for arrivals as point 1 and a departures capacity equal to the number of departures that can be inserted into the arrivals stream without increasing the separations between successive arrivals (and, thus, without reducing the number of arrivals from what can be achieved in the all-arrivals case). Thus, the "free departures" are obtained by exploiting large interarrival gaps.

Point 3. The "alternating arrivals and departures" point, i.e., the point at which an equal number of departures and arrivals is performed through an A-D-A-D-A...sequence. As indicated, such a strategy can be implemented by "stretching," when necessary, interarrival (and interdeparture) gaps by an amount of time just sufficient to insert a departure (arrival) between two successive arrivals (departures).

Point 4. The "all departures" point, i.e., the capacity of the runway when it is used only for departures.

Capacity envelopes such as the one shown in Fig. 10-10 provide a complete description of the capacity made available by a runway under any specific set of conditions. Note that different capacity envelopes may (and, most probably, will) apply to VFR or IFR or LIFR operating conditions (Sec. 10-3). Instead of computing a capacity envelope for 1-h periods, one may also compute the envelope for 15-min or 30-min periods. This is sometimes useful for airports that practice slot coordination (Chap. 12).

The same modeling approach can be extended quite readily to airport configurations with two simultaneously active runways. The simplest possible case involves two parallel runways, of which one is used solely for arrivals and the other solely for departures, independently. As noted in Chap. 9, many airports outside the United States typically operate in this way. In such instances, the capacity envelope can be obtained simply by computing the "all arrivals" (point 1) and "all departures" (point 4) capacity of a single runway and combining the results as shown in Fig. 10-11. Note that point 2 lies above the 45° line in Fig. 10-11, indicating that the departure capacity is higher than the arrival capacity in most cases.

More generally, the case of two parallel runways is quite tractable, no matter what types of movements (arrivals, departures, or mixed) each of the runways is used for or the degree of interaction between the two runways* (Swedish, 1981; Stamatopoulos, 2000). The case of intersecting pairs of runways is also quite tractable, as long as local procedures for operating the runways are well understood. Given a set of priority rules for sequencing operations on the two active runways,

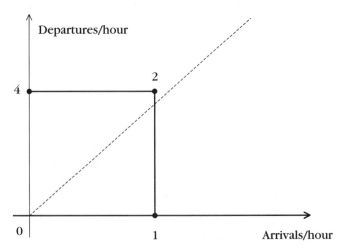

Fig. 10-11. *A capacity envelope for two parallel runways, one used for arrivals only and the other for departures.*

---

*As seen in Sec. 10-3, the capacity of a pair of parallel runways depends on the distance between runway centerlines, cf. Table 10-3.

one can compute the elements, t_{ij}, of the time separation matrix for successive operations and approximate quite accurately the available capacity. For example, in the case of New York/LaGuardia (see Fig. 9-6) a configuration commonly used has departures on Runway 13 and arrivals on Runway 04. Air traffic controllers will typically alternate arrivals and departures in this case (a departure from Runway 13, then an arrival on Runway 04, then a departure from Runway 31, etc.). Given the location of the runway intersection, one can then compute the airport's capacity. Note that the assignment of landings and takeoffs to runways may change, depending on wind direction, thus giving rise to additional configurations. Because of the change of the location of the runway intersection relative to the points where takeoffs are initiated or where landing aircraft touch down, the capacities of these various configurations at New York/LaGuardia (all involving two active runways) may be far from equal.

With configurations involving three or more active runways, the three-step approach described at the beginning of this section becomes very cumbersome. The interactions among the runways are usually too numerous and complicated to permit development of matrices of separations between all possible pairs of movements on all active runways, as called for under step 1. Instead, one of two approaches should be used: "decomposition" of the configuration or simulation. The former involves decomposing the configuration in use into parts, each of which consists of either a single runway or a pair of runways. This is followed by estimation of the capacity of each of the parts, using the one- and two-runway models just described. For example, Atlanta (Fig. 9-7) can be viewed as consisting of two independent pairs of close-spaced parallel runways. The capacity of the full runway system can be approximated by computing, first, the capacity of each of these two pairs of runways separately and then adding the results. The use of simulation models to compute airfield capacity is described in Chap. 19. Approximate estimates of capacity can also be obtained through the FAA's "Airfield and Airspace Capacity and Delay Analysis" (FAA, 1981), but this is not recommended because the document is obsolete.

Through these approaches it is also possible to estimate approximately the capacity envelopes of airports with complex runway configurations. Figure 10-12 sketches what a capacity envelope for a multirunway airport might look like. Approximate capacity envelopes for the 31 airports listed in Table 10-4 are given in (FAA, 2001a).

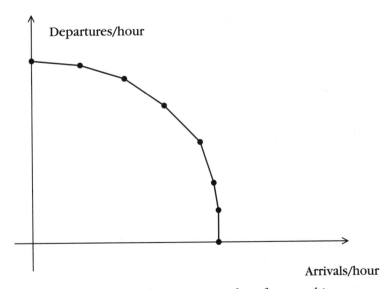

Fig. 10-12. *A hypothetical capacity envelope for a multirunway airport with mixed use of the runways.*

10-7   Capacity of other elements of the airfield

This section reviews briefly the capacity of the other elements of the airfield, namely, the taxiway system and the apron. Some general statements can be made about the capacity of these elements. However, airport-specific factors also play a primary role in determining taxiway and apron capacity.

Capacity of the taxiway system

The overall capacity of the taxiway system can be determined, in theory, by the number of aircraft per hour that the taxiway system can deliver from the apron areas to the runway system and vice versa. From a practical viewpoint, however, it suffices to know that a fully developed and reasonably well-designed taxiway system, like those that one is likely to encounter at major airports, will not, in general, be a factor limiting airport capacity. This can be seen by considering the most fundamental component of a taxiway system, a full-length taxiway, i.e., a taxiway that runs parallel to the entire length of a runway and serves aircraft moving to/from the apron areas from/to the runway (Chap. 9). These long taxiways are typically used as one-way traffic lanes for any given runway configuration. The flow capacity of a full-length taxiway typically exceeds by a considerable margin the capacity of the associated runway. For example, if aircraft travel on the taxiway at a speed of 36 km/h (roughly 22 mi/h) and the separation between (noses of) successive aircraft on the taxiway is a conservative 400 m, the flow capacity of the taxiway is 90 aircraft per hour, far more than a runway can typically handle. The flow capacity will, of course, be higher if taxiing speeds are higher or if headways between successive airplanes on the taxiway are smaller.

This does not mean that a taxiway system may not have local bottle-necks—points where aircraft may sustain some taxiing delays which are additional to the delays suffered while waiting to use the runway system. Taxiway intersections, short taxiway segments between two intersections, points where taxiing aircraft must cross an active runway, and locations where high-speed runway exits merge with taxiways can all be potential local "choke points" on a taxiway system. Figure 10-13, for instance, identifies pictorially a set of points where arriving air-craft must cross the departure runway 22R, when 22L and 27 are used for arrivals and 22R and 22L for departures at Boston/Logan (Idris, 2001). During periods when 22R is busy with departures, delays at these crossing points can be significant. Equally important, air traffic controllers may occasionally have to interrupt the flow of departures on

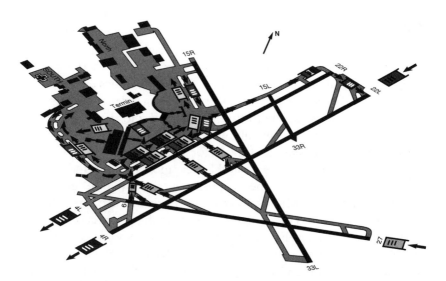

Fig. 10-13. *Potential congestion points at Boston/Logan when runways 22L and 27 are used for arrivals and runways 22R and 22L are used for departures; locations of queues on the airport surface are noted.*

Runway 22R in order to give waiting arriving aircraft an opportunity to cross 22R and reach the apron areas. Such interruptions reduce the departure capacity of the runway system at Boston/Logan when the configuration of Fig. 10-13 is in use.

Several such flow-constraining points typically exist on the taxiway systems of older, space-constrained airports. These points can usually be identified easily, essentially by inspection of the layout of the airfield. Problems are specific to the local airfield geometry and can be solved only through location-specific measures. Ground controllers, that is, the air traffic controllers responsible for directing traffic on the airport's surface, are generally well aware of the presence of these potential bottlenecks on taxiway systems and try to anticipate and prevent localized delay problems from spreading throughout the airfield.

In conclusion, the overall capacity of the taxiway system of major airports almost always exceeds the capacity of the runway system and does not constitute a significant constraint on runway capacity. Delays sustained at specific "hot points" are typically much smaller than the delays experienced due to the capacity limitations of the runway system. Some exceptions may exist at older, space-constrained airports. The general rule is that taxiway capacity problems are airport-specific and must be resolved in the context of local conditions.

Capacity of the aprons

In contrast to the taxiway system, the capacity of aprons *can* occasionally be a constraining factor on the overall airside capacity of airports with small land area. Aprons consist of areas reserved for *remote* and *contact* aircraft stands and for the taxilanes that traverse these areas. Stands can be further subdivided into those designated for *exclusive use* by a single airline (or possibly a small group of affiliated or allied airlines) and those for *shared* (or *common*) use. At many major airports in the United States, most of the stands are for exclusive use, whereas the opposite is usually true elsewhere (Chaps. 14 and 15). When stands are for exclusive use, the scheduling of occupancy and the assignment of aircraft to stands is managed by the airlines themselves or by a contractor responsible for ramp handling in that part of the apron area (Chap. 8). When stands are shared, it is either the airport operator or a handling contractor who performs these tasks. Each stand is also characterized by its *size*— the dimensions of the largest aircraft it can accommodate.

Some general statements can be made about the capacity of aprons, but airport-specific conditions usually dominate. At the most obvious, a good indication of the available apron capacity is given by the number of stands at hand. This is sometimes referred to as the *static capacity* of the apron, because it indicates the maximum number of aircraft that can be occupying simultaneously the apron at any given instant.

Static capacity, while informative, provides only a "snapshot" of the instantaneous capacity of the apron and cannot be readily compared to the runway capacity of the airport or to the capacity of other parts of the airport, which are specified in terms of number of units (such as aircraft movements) processed *per unit of time*. For this reason, the *dynamic capacity of aprons* is also a widely used measure. Dynamic capacity is defined as the *number of aircraft per hour* that can be accommodated at the stands and is more consistent with the notion of runway capacity.

To compute dynamic capacity, it is necessary to consider the time interval between successive occupancies of a stand by two different aircraft. By analogy to the approach used to compute runway capacity, a minimum interval and an average interval should be determined. The minimum interval consists of two components:

1. The amount of time that an aircraft is scheduled to spend at a stand; this will be referred to as the scheduled occupancy time (SOT) and is also known as the scheduled "turnaround" time of an aircraft.

2. The time needed to position the aircraft into and out of the stand; during that *positioning time* (PT) the stand is unavailable to other aircraft.

Typical values of SOT range from 20 min for flights of regional airlines, making a stop at an airport to unload some passengers and take new ones on board without aircraft servicing, to 4 h for wide-body aircraft turning around on an intercontinental route. Note this does not cover occupancy times as long as 10–12 h for "overnighting" aircraft. Typical values for PT are of the order of only 2–4 min for remote stands,* to as much as 10 min or more for contact stands (due primarily to the time-consuming pushback maneuver).

---

*The short positioning times are for remote stands with taxilanes behind and in front of the stand.

To determine an average interval between successive stand occupancies, one must again consider "buffer times" built into the schedules of stands at all major airports. Managers of stands, whether airport operators or airlines, are forced to provide buffer times, especially at contact stands, because air traffic is always subject to delays and short-term schedule changes. If a given stand is to be assigned successively to two aircraft, there should be sufficient time between the *scheduled departure time* of the first and the *scheduled arrival time* of the second to ensure that, with high probability, deviations from schedule will not necessitate a change in stand assignments. Airport operators and airlines prefer to avoid last-minute changes in stand assignments, because they are disruptive and costly. They inconvenience departing passengers in the case of contact stands and they require reassignment or repositioning of ramp equipment and of aircraft handling and passenger handling personnel. The buffer times (BT) actually used will depend greatly on local circumstances, such as the length of typical flight delays, airport policies vis-à-vis stand assignments (preference for given flights being assigned to the same stand each day versus variable assignments from day to day), stand type (remote or contact), apron geometry, passenger terminal configuration (for contact stands), aircraft handling agreements, and exclusive or shared use of stands. Typical buffer times can range from a few minutes for stands serving remotely parked regional aircraft to 1 h or more for contact stands for intercontinental flights.

Example 10.6 illustrates how dynamic capacity is affected by these parameters.

EXAMPLE 10.6

Consider a simplified case in which all stands at an airport are of the same size and can accommodate all aircraft using the airport. Assume there are 60 stands and that the average scheduled occupancy time (SOT) for all aircraft is 50 min. The naive approach in this case is to estimate that each stand can serve an average of 1.2 aircraft per hour, so that the dynamic capacity of the apron is $60/(50/60) = 72$ aircraft per hour.

If the positioning time, PT, is also taken into account—by adding, for example, 8 min to the 50 min of SOT—the apron capacity is substantially reduced to roughly 62 aircraft per hour

[= 60/(58/60)]. This is the maximum achievable dynamic capacity, assuming the schedule of flights is executed perfectly every day.

For purposes of this example, a buffer time, BT, of 30 min is now added to the 58 min previously allowed for SOT and PT, for a total of 88 min during which a stand is "blocked" from access by other aircraft. The dynamic capacity of the 60 stands is then further reduced to roughly 41 aircraft per hour, 43 percent less than the naive estimate of 72.

The procedure outlined above can now be summarized and somewhat generalized. Assume that a set of n stands exists at an airport, with each stand capable of accommodating all types of aircraft.

Step 1. Subdivide arriving aircraft into a small number K of classes according to an appropriate combination of criteria such as aircraft size and/or type of flight and/or airline. For instance, class i might consist of wide-body aircraft on longer range (5 h or more) international flights. Note that the classes specified for the purpose of computing apron capacity are not necessarily the same as the classes (e.g., "heavy," "medium," "small") specified for the purpose of computing runway capacity.

Step 2. For each class i estimate the typical ("average") time between occupancies of the stand as the sum of SOT, PT, and BT for that class. For class i, call this sum the "stand blocking time," SBT_i.

Step 3. Compute the expected ("average") stand blocking time for the airport

$$E[\mathrm{SBT}] = \sum_{i=1}^{K} p_i \cdot \mathrm{SBT}_i$$

where p_i is the fraction of arriving aircraft that belong to class i.

Step 4. The dynamic capacity of the apron is then approximately equal to $n/E[\mathrm{SBT}]$ aircraft per hour.

It should be emphasized that this procedure is very approximate and will yield only a rough estimate of dynamic capacity. This is especially true when, as happens at practically every airport, all stands cannot accommodate all aircraft, either because of physical limitations (e.g., size of the stands) or because of operational constraints (stands reserved for international flights versus stands for domestic flights, etc.). For a more accurate estimate of apron capacity under these more complicated conditions, the best approach is to subdivide the stands into groups to which reasonably homogeneous conditions of use apply. One can then perform a separate capacity analysis for each such identifiable group of stands using the procedure outlined above and then combine the results. However, this analysis can be quite tedious, even when the number of stands is relatively small. It can be facilitated through use of several available computer-based tools that can assign aircraft to stands by taking into consideration many of the constraints, operational rules, and airline priorities and preferences that are typically encountered in practice. It should be noted, however, that these tools are intended primarily for the task of fitting *a specified daily schedule of arrivals and departures* into the available set of stands at an airport. Thus, they can assist only indirectly in estimating the apron's dynamic capacity by indicating whether *a given hypothetical daily schedule* of flights can be accommodated by an airport's set of stands. By preparing many variations of daily schedules with increasing numbers of flights, one can use these tools to help determine the apron's dynamic capacity.

A last related question concerns the comparison of the dynamic capacity of the apron with the capacity of the runway system. This question often arises in the context of determining the number of slots that many airports around the world use for "schedule coordination" purposes (Chap. 12). Note that apron capacity is measured in terms of number of *aircraft* per hour and runway capacity in terms of *movements* per hour. Obviously, as a quick approximation, one can simply multiply the dynamic capacity of the apron by 2 to convert it to a number that can be compared to runway capacity, as the occupancy of a stand is associated with two movements on the runways, an arrival and a departure. A more prudent approach, however, takes into consideration the fact that the daily flight schedule at any airport contains periods during which there are considerably more arrivals than departures, and vice versa. Surges in arrivals may "flood" the apron with aircraft. This approach consists of two steps. First, the schedule of runway movements during the busy hours of the day is scanned to identify the largest fraction of arrivals in the traffic mix during any time interval of length comparable to $E[SBT]$, as defined in step 3 of the procedure described above. The apron's dynamic capacity is then divided by this fraction to obtain the equivalent capacity expressed in terms of runway movements per hour. This is illustrated in the next example.

The advantage of the more conservative approach is that it protects the airport from overestimating the apron's ability to cope with the

EXAMPLE 10.6 (Continued)

Consider again the situation described earlier, in which $E[SBT]$ was set equal to 88 min, about 1.5 h, and the dynamic capacity of the 60 stands was consequently estimated as 41 aircraft per hour. Multiplying by 2 gives an estimate of 82 as the number of movements per hour on the runway system that can be accommodated in the apron area. Stated differently, if the runway system has a (maximum throughput) capacity of 82, then the runway system is able to "feed" the apron about 41 arriving aircraft per hour, a number equal to the rate at which the apron can serve aircraft.

The better approach, however, calls for scanning the schedule of arrivals and departures during the busy part of a typical day (e.g., from 07:00 to 21:00 local time) to identify the most "arrival-intensive" 1.5-h interval of the day. Suppose this interval occurs between 08:10 and 09:40 local time and that, during the interval, arrivals constitute 62 percent of the scheduled runway movements. One then obtains an estimate of 66 $[\approx 41/(0.62)]$ as the apron's capacity, expressed in terms of runway movements per hour. The statement here is that, if a runway system with a maximum throughput capacity of 66 movements per hour is available and if the airport's strongest surge of arrivals over a 1.5-h period results in 62 percent arrivals and 38 percent departures during that period, the runway system will send about 41 aircraft per hour to the apron—if working at full capacity—a number equal to the capacity of the apron. Note that the new equivalent capacity of 66 movements per hour is significantly lower than the 82 obtained through the "simple method."

unavoidable fluctuations in arrivals and departures during the day. It is particularly useful for hub airports that experience several major surges in arrivals and departures in the course of a day. Note, however, that when applying this approach to project apron stand needs for a future time, one requires both good historical data on the dynamic mix of arrivals and departures over the course of a day and a reasonable guess as to what this dynamic mix will look like in the future.

The following practical rule of thumb can also be stated: to convert the dynamic capacity of the apron to an equivalent number of runway movements per hour, multiply the dynamic capacity by 1.67. Note that the coefficient 1.67 [≈ 1/(0.60)] implies a roughly 60–40 percent mix of arrivals and departures during the peak arrivals surge of the day. This is reasonable and works quite well for busy, nonhub airports.

Exercises

1. The following information is given about air traffic at a particular runway of an airport:

a. Aircraft can be classified into three types: heavy (H), large/medium (L), and small (S).

b. Some relevant aircraft characteristics are as follows:

Aircraft type	Approach speed (knots)	Mix (%)	Runway occupancy time on landing (s)
H	150	20	70
L	135	40	60
S	105	40	50

c. The length of the final approach to the runway is 6 nmi.

d. The minimum separation requirements (in nautical miles) between successive landing aircraft on final approach are given by the matrix below (rows indicate the leading aircraft and columns the following aircraft):

	S	L	H
S	2.5	2.5	2.5
L	4*	2.5	2.5
H	6*	5	4

[*These separations apply only when the leading aircraft is at the runway threshold; all the other separations apply throughout the final approach.]

e. A "buffer time" of 15 s (see Sec. 10-5) is added to all the minimum separation times between successive landings to account for uncertainties.

f. The minimum separation requirements (in seconds) between successive departing aircraft are given by the matrix below (rows indicate the leading aircraft and columns the following aircraft):

	S	L	H
S	45	45	45
L	60	60	60
H	120	120	90

Part 1. Suppose this runway is used for departures only. Find its (maximum throughput) capacity for departures. [No buffer times are added for departures.]

Part 2. Suppose this runway is used for arrivals only. Find its (maximum throughput) capacity for arrivals.

2. Consider the model for the capacity of a single runway with arrivals only presented in Sec. 10-5. Let W be a constant greater than 1. Assume that the runway occupancy times are negligible compared to the time intervals between arrivals dictated by the longitudinal separation requirements on final approach. Assume you have been assigned the task to assess two alternative proposals to improve the capacity of a runway: (*a*) multiply the final approach speeds of all aircraft types by W; and (*b*) divide the length of the final approach path by the same constant W. With the exception of the proposed changes, everything else in the model remains the same.

   Would the two proposals lead to the same improved runway capacity? If not, which of the two proposals would lead to the higher runway capacity? Justify your answer using the analytical model of the runway capacity.

3. The airport of Exercise 1 is sometimes forced to use only a single runway during IFR weather periods. Thus, the runway must accommodate both landings and takeoffs during these periods. The data for arriving and departing aircraft given in Exercise 1 also apply here, unless noted otherwise. The following rules/assumptions apply.

   a. The local air traffic controllers use an operations sequencing strategy of alternating landings and takeoffs on the runway; i.e., during periods of continuous demand, a landing is always followed by a takeoff, which is then followed by a landing, etc. Thus, when the minimum required time gap between two landing aircraft, i and j, is not sufficient to insert a takeoff, the time gap will be increased by ATC appropriately.

   b. There is no uncertainty about the position of aircraft on final approach. Thus, this is an entirely "deterministic" problem and no buffers are added to minimum aircraft spacing; ignore assumption *e* of Exercise 1.

c. Takeoffs wait next to the threshold of the runway. As soon as a landing aircraft crosses the runway threshold, the next departing aircraft enters the runway and prepares for the takeoff run. It takes 30 s for a departing aircraft to enter the runway and set up for takeoff. (Note that, in the meanwhile, the arriving aircraft that just landed is moving down the runway toward a runway exit.)

d. A takeoff run cannot begin until the preceding landing aircraft has cleared the runway.

e. Once a takeoff run begins, the runway occupancy time for all departing aircraft (time from the beginning of the takeoff run to clearing the runway) is equal to 60 s.

f. The takeoffs of successive aircraft must be separated by at least 90 s in this case (i.e., disregard the separation matrix shown in *f* of Exercise 1).

g. A landing aircraft is not allowed to cross the runway threshold unless the runway is clear of all landing or departing aircraft. (Note that this is assumed to be the *only* "departure followed by arrival" separation requirement.)

Part 1. Find the capacity of this runway (total number of landings and takeoffs per hour) when it is used for both arrivals and departures in the manner described.

Part 2. In *e* above, it was stated that all aircraft, independent of type, occupy the runway for 60 s on departure. In practice, the three different types of aircraft [H, L, or S] have different (typical) runway occupancy times on takeoff. *Without doing any calculations,* write a couple of paragraphs, explaining how your work in Part 1 would have to be modified to take this into consideration. Please make sure to indicate what your "time separation" matrix and your matrix of probabilities would look like.

References

Andreatta, G., Brunetta, L., Odoni, A., Righi, L., Stamatopoulos, M., and Zografos, K. (1999) "A Set of Approximate and Compatible Models for Airport Strategic Planning on Airside and on Landside," *Air Traffic Control Quarterly,* 7, pp. 291–317.

Blumstein, A. (1959) "The Landing Capacity of a Runway," *Operations Research,* 7, pp. 752–763.

Dear, R. (1976) "The Dynamic Scheduling of Aircraft in the Near-Terminal Area," Technical Report R76-9, Flight Transportation Laboratory, Massachusetts Institute of Technology, Cambridge, MA.

FAA, Federal Aviation Administration (1981) "Airfield and Airspace Capacity/Delay Analysis," Report FAA-APO-81-14, Office of Aviation Policy and Plans, Washington, DC.

FAA, Federal Aviation Administration (1989) *Airport Design,* Advisory Circular 150/5300-13, incorporates subsequent Changes 1 thru 5 (Change 5 in 1997), U.S. Government Printing Office, Washington, DC.

FAA, Federal Aviation Administration (2001a) *Airport Capacity Benchmark Report,* U.S. Government Printing Office, Washington, DC (http://www.faa.gov/events/benchmarks/).

FAA, Federal Aviation Administration (2001b) "Air Traffic Control: Order 7110.65M," includes Change 3, July 12, Washington, DC (http://www.faa.gov/ATpubs/ATC/index.htm).

Gilbo, E. (1993) "Airport Capacity: Representation, Estimation, Optimization," *IEEE Transactions on Control Systems Technology, 1,* pp. 144–154.

Harris, R. M. (1972) "Models for Runway Capacity Analysis," Report FAA-EM-73-5, The MITRE Corporation, McLean, VA.

Hockaday, S., and Kanafani, A. (1974) "Developments in Airport Capacity Analysis," *Transportation Research, 8,* pp. 171–179.

Idris, Husni (2001) "Observation and Analysis of Departure Operations at Boston Logan International Airport," Ph.D. thesis, Massachusetts Institute of Technology, Cambridge, MA.

ICAO, International Civil Aviation Organization (1999) *Aerodromes, Annex 14 to the Convention on International Civil Aviation, Volume I: Aerodrome Design and Operations,* 3rd ed., ICAO, Montreal, Canada.

Lee, D., Kostivk, P., Hemm, R., Wingrove, W. and Shapiro, G. (1997) *Estimating the Effects of the Terminal Area Procuctivity Program,* Report NS301R3, Logistics Management Institute, McLean, VA.

Odoni, A. (1972) "Efficient Operation of Runways," in *Analysis of Public Systems,* A. W. Drake, R. L. Keeney and P. M. Morse (eds.), MIT Press, Cambridge, MA.

Psaraftis, H. (1980) "A Dynamic Programming Approach for Sequencing Groups of Identical Jobs," *Operations Research, 28,* pp. 1347–1359.

Stamatopoulos, M. A. (2000) "A Decision Support System for Airport Planning," doctoral dissertation, Athens University of Economics and Business, Athens, Greece.

Swedish, W. (1981) "Upgraded FAA Airfield Capacity Model," Report MTR-81W16 and FAA-EM-81-1, The MITRE Corporation, McLean, VA.

Venkatakrishnan, C., Barnett, A., and Odoni, A. (1993) "Landings at Logan Airport: Describing and Increasing Airport Capacity," *Transportation Science, 27*(3), pp. 211–227.

11

Airfield delay

Airport delays and congestion constitute a major threat to the future of air transportation. The dynamic characteristics of airport delays are difficult to predict accurately. Advanced computer-based tools are typically needed to obtain good estimates of delay-related measures at a busy airport over time. In general, delays:

- May be present even during periods when the demand rate is lower than capacity
- Depend in a nonlinear way on changes in demand and/or capacity, becoming very sensitive to even small changes when demand is close to or greater than capacity
- Exhibit a complex dynamic behavior over any time span (e.g., a day of operations) when the runway system is utilized heavily

In the long run, both the expected length and the variance (a measure of variability) of airport delays increase nonlinearly with increases in the utilization ratio of the runway system, i.e., the ratio of the demand rate divided by the (maximum throughput) capacity. Delays will be very long and highly variable from day to day at runway systems operated with utilization ratios in excess of the 0.85–0.9 range during the most active 15–18 traffic hours of the day.

A quantity of great practical interest is the approximate annual capacity of a runway system. To estimate this number it is necessary not only to compute the capacity coverage chart of an airport, but also to make projections or assumptions regarding: daily demand patterns, day-of-the-week demand patterns, seasonal demand patterns, and acceptable levels of delay. Historical evidence provides some useful guidelines with regard to these parameters.

A number of simulation models and mathematical, computer-based models are available to assist in investigating issues related to airfield capacity and delay. A crucial decision on the part of the prospective user concerns the level of modeling detail needed for the analysis. The more detailed (and more costly and complex) models are not necessarily best suited for many of the questions that come up in practice.

11-1   Introduction

The principal consequence of the lack of adequate airside capacity at an airport is delays to landings and takeoffs, with their attendant economic and other costs. When delays become large, other undesirable consequences such as missed flight connections, flight cancellations, and flight diversions to other airports can also become commonplace. Airport and air traffic congestion is a growing problem on an international scale and is widely considered one of the principal constraints to the future growth of the global air transportation industry. In the United States alone, delays at each of 29 of the busiest commercial airports* were estimated in 1999 to be in excess of 20,000 aircraft hours per year†, the threshold at which an airport is considered "congested" by the Federal Aviation Administration (FAA). More than 65 percent of all air travelers in the United States were enplaned at these 29 airports. The major U.S. airlines have been required by law since 1987 to report statistics on all flight delays to the U.S. Department of Transportation (DOT). This requirement reflects the great interest that the subject of air traffic congestion has attracted from travelers and the general public. The government processes and publishes these statistics monthly (U.S. Department of Transportation, monthly) and the media review them carefully. The statistics also provide a rich data source for researchers and for developers of computer-based models of airport delays.

The situation is almost as bad on the other side of the North Atlantic. The Performance Review Commission of EUROCONTROL has stated

---

*The airports are Atlanta, Baltimore, Boston/Logan, Charlotte, Chicago/O'Hare, Cincinnati, Dallas/Ft. Worth, Denver/International, Detroit/Metro, Houston/Bush, Indianapolis, Las Vegas, Los Angeles/International, Memphis, Miami/International, Minneapolis/St. Paul, New York/Newark, New York/Kennedy, New York/LaGuardia, Orlando/International, Philadelphia, Phoenix, Pittsburgh, Salt Lake City, San Francisco/International, Seattle/Tacoma, St. Louis, Washington/Dulles, and Washington/Reagan.
†To appreciate the severity of this delay, consider that a typical jet aircraft of a major airline in the United States typically performs about 2500–3000 h of commercial service per year. Thus, 20,000 h per year is roughly equivalent to wasting the potential services of the equivalent of 7–8 jet aircraft per year. Several of the airports listed in the previous footnote had air-side delays of more than or near 100,000 aircraft hours per year in 1999. Overall, the total number of potential jet aircraft years wasted annually in the United States in recent years due to air-side delays is of the order of 400—about the same as the size of the fleet of the fourth or fifth largest airline in the United States and almost twice the size of the fleets of the largest European carriers.

(Fron, 1998) that, in the absence of capacity improvements, airport and air traffic management (ATM) delays in European airspace will increase by a factor of 10–20 between 1995 and 2015! This is, of course, an untenable prospect, if one considers that congestion had already reached virtually unacceptable levels by the beginning of that period.

This chapter reviews briefly and descriptively some fundamental points concerning the estimation, characteristics, and measurement of airside delays at airports. The characteristics of air-side delay and congestion in a short-term, dynamic sense are reviewed in a qualitative fashion in Sec. 11-2. Section 11-3 discusses long-term characteristics and some of their policy implications. It provides several important practical guidelines that airport operators should follow for the purpose of maintaining an adequate level of service on airside. Section 11-4 addresses the question of the annual capacity of a runway system. Estimating this annual capacity requires consideration of local demand patterns and of level-of-service issues. Finally, Sec. 11-5 covers briefly some issues related to the computation of delays and to the availability of delay data. A more quantitative discussion of certain aspects of airport congestion can be found in Chap. 23, which also provides a short introduction to queuing theory—the mathematical theory of waiting lines. Approaches for reducing the magnitude and/or cost of airside delays through demand management and through better air traffic management are described, respectively, in Chaps. 12 and 13.

11-2 The characteristics of airside delays

It is useful to begin with a qualitative look at the relationship between air-side demand and capacity, on the one hand, and delays, on the other. Figure 11-1 shows schematically the typical weekday demand profile at Boston/Logan airport during the late 1980s and compares it with three different levels of (maximum throughput) capacity, labeled "high," "medium," and "low." As noted in Chap. 10, weather is the principal factor that determines the level of capacity at which the airport operates at any given time.* The demand profile, which has

---

*As explained in Sec. 10-3, the capacity of the runway system actually varies with the mix of arrivals and departures. The capacity is typically higher during periods when the demand consists mostly of departures and lower when the reverse is true. However, for the purposes of this example, it is assumed that the arrival–departure mix is about 50–50 percent during all hours. In this way the hourly capacity can be approximated by a single value that may vary by time of day, depending only on weather conditions.

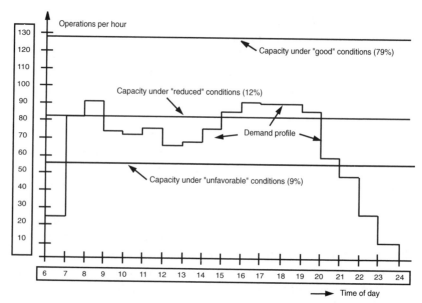

Fig. 11-1. *Weekday demand profile at Boston/Logan airport compared with three different levels of the capacity of the runway system.*

two peaks, as is typical of many busy airports with large volumes of business traffic, shows the total number of aircraft movements *scheduled* per hour. Note that the number of movements scheduled in an hour is not necessarily the same as the number of movements that will *actually* be requested on a day-to-day basis during that hour. Because of mechanical or logistical problems with aircraft, flight cancellations, late-boarding passengers, late-arriving crews, delays at other airports, etc., the number of movements actually requested at a given airport during any particular period of time will fluctuate around the number scheduled. In this sense, just like capacity, the number of movements scheduled for an hour can be viewed as only an *expected* (or "average") *value.* This expected value will be referred to, henceforth, as the *demand rate* for that hour.

A few observations can now be made about the delays associated with each of the three levels of capacity. First, and obviously, queues of landing and departing aircraft will almost certainly form and delays (often called *overload delays*) will occur during those parts of a day when the demand rate exceeds the capacity for any significant time interval. This is because aircraft will seek to use the runway system at a rate greater than the system's capacity. Colloquially, "demand exceeds capacity" during such periods. In general, if there is an interval of length T during which the demand rate continually exceeds the service rate, then both the expected length of the aircraft queue and the expected waiting time per aircraft during that interval will grow in direct proportion* to T and to the difference between the demand rate and capacity during T. In Figure 11-1, the period during which demand exceeds the "low," level of capacity runs from 07:00 to 21:00 and the "medium" level from 08:00 to 09:00 and from 15:00 to 20:00. As one can infer, when the capacity is "low," delays keep building up throughout a day. Aircraft scheduled to arrive or depart during the afternoon and evening hours may be subjected to horrendous delays. Indeed, the airlines simply cancel numerous flights to Boston/Logan during such days due to the size of the expected delays.

Less obviously, significant delays may also be observed during time periods when the demand rate is *less than but reasonably close* to the service rate—a notion that is sometimes confusing to airport operators. This may include days during which the demand rate is less than the capacity for the *entire day,* as is the case when the capacity is at a "high" level in Fig. 11-1. Such delays are due primarily to the variability of the time intervals between successive requests for use of the runways, as well as to the variability of the time it takes to process ("serve") each landing and takeoff. The sources of this variability are several:

- The time instants at which demands (arrivals and, especially, departures) are *scheduled* to take place are generally not evenly spaced, but are often "bunched together" around certain times, which aircraft schedulers prefer (e.g., "on-the-hour" or "on-the-half-hour" departure peaks).

- The time instants at which demands *actually* occur on a day-to-day basis are "randomized" as a result of the inevitable deviations from schedule due to the many reasons already mentioned (mechanical problems, delays at other airports, etc.).

- The amount of time it takes to serve departures and arrivals on the runway system is not constant, but varies with the many factors discussed in Chap. 10 (type of aircraft, separation requirements from preceding aircraft, runway exit used, etc.).

*The reader will find more details in Sec. 23-3, which derives this result through an example based on the notion of *cumulative diagrams.* Cumulative diagrams can be very useful for approximate analyses of queuing phenomena during periods when demand rates exceed capacity. This is often the case, for instance, when a passing weather front reduces considerably the capacity of a runway system for several consecutive hours.

The net effect is the presence of time intervals during which "clusters" of several closely spaced demands and/or of longer-than-usual service times occur. Queues of airplanes will then form on the ground and/or in the air.* When the demand rate is smaller than the capacity but close to it, a long time may pass before such queues dissipate. In fact, new clusters of demands or of long service times may come along before the previously formed queue has dissipated and the waiting line(s) may get longer for a while, not shorter. The resulting delays are often called *stochastic,* to distinguish them from overload delays. In summary, long queues may form even if the demand rate is smaller than capacity in cases where (1) there is considerable variability in the times between successive demands on the runway systems and/or in the service times at the runway system and (2) the demand rate is close to the runway system's capacity (Example 11.1). What "close" means in this context is discussed shortly.

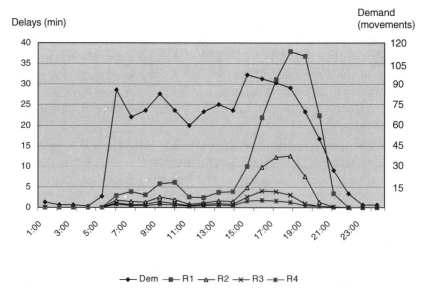

Delays (min)

Demand (movements)

—◆— Dem —■— R1 —▲— R2 —✕— R3 —✳— R4

(R1= capacity is 80 movements per hour; R2 = 90; R3 = 100; R4 = 110)

Fig. 11-2. *The dynamic behavior of aircraft delay for four different levels of capacity ranging from 80 to 110 movements per hour; the scale for the demand profile is on the right and for the expected delay on the left.*

*The exact physical location of these queues depends on ATM policies and on the length of the queues themselves. For example, in the case of departing aircraft, the queue(s) in most instances will form on the taxiway(s) leading up to the departure runway(s); however, when the taxiway queue becomes very long, airplanes will often be held at their departure stands, so that a second queue of airplanes waiting to enter the taxiway system is created.

EXAMPLE 11.1

Figure 11-2, based on a study of delays at a major airport, illustrates all these points. The curve Dem shows the demand profile* during a day when the total number of scheduled runway movements is 1200 and the peak-hour demand rate is 94 movements during the hour between 15:00 and 16:00. (The scale for the demand profile is the vertical axis on the right-hand side of Fig. 11-2.) Practically all the demand, with the exception of 32 nighttime movements, is concentrated in the 16 "busy" hours of the day between 06:00 and 22:00 local time.

Figure 11-2 shows the estimated expected waiting time in queue ("average delay") when the maximum throughput capacity at the airport is 110, 100, 90, or 80 movements per hour throughout the day. Specifically, for each time t on the horizontal axis, each of the four graphs shows how much delay a landing or takeoff requesting use of the runway system at time t would be expected to suffer for the corresponding capacity level. The delay estimates have been obtained through a computer-based queuing model called DELAYS (see Chap. 19), which approximates a runway system as a dynamic queuing system with demands that are generated through a time-varying Poisson process[†] and service times whose expected value and variability the user of the model specifies.[‡]

Figure 11-2 makes several noteworthy points. First, as already mentioned, delays occur not only under overload conditions, when the demand rate exceeds capacity (as happens for short or for longer parts of the day when the capacity is equal to 90 and 80 movements per hour), but also when the demand rate is less

---

*The demand profile graph, Dem, connects with straight-line segments the 24 points that indicate the demand rate during each of the 24 h of the day. For example, the demand rates between 11:00 and 12:00 and between 12:00 and 13:00 are equal to 60 and 71 movements per hour, respectively. For the purpose of computing delays, these values have been assigned to the mid-hour points (11:30 and 12:30, respectively). The instantaneous demand rate between 11:30 and 12:30 is then read from the Dem graph by interpolating between the values of the demand rate at the two mid-hour points.

[†]The Poisson process is used in probability theory to model events that occur "perfectly randomly" over time; for a more complete discussion of demand generation and of service times at a queuing system, see Chap. 23.

[‡]Service times in this model are selected from the family of Erlang probability distributions. Depending on which member of the Erlang family is chosen, the user can model service times that range from "perfectly random" (i.e., with a negative exponential distribution) to constant. See also Chap. 23.

than capacity throughout the day (as happens when the capacity is equal to 100 or 110). Note, in fact, that the overall shapes (i.e., the peaks and valleys) of the four graphs for the expected waiting time during the day are not fundamentally different in the former and in the latter cases. It is only that the overall magnitude of the expected waiting time and its peaks increase greatly as one transitions from a capacity of 110 to a capacity of 80.

Second, note that, as capacity decreases, the magnitude of the delays grows dramatically, in a *nonlinear* way. Table 11-1 highlights this aspect of the behavior of this queuing system (ignore for now the two rightmost columns). The second column from the left shows the maximum value, to the nearest minute, that the expected waiting time takes during the 24-h day. For example, with a capacity of 80 movements per hour, an aircraft requesting access to the runway system shortly after 18:00 can expect a delay of roughly 39 min. The third column indicates the estimated overall expected waiting time per movement over the entire day. Note that, when the hourly capacity is reduced from 110 to 100 movements per hour (a 9.1 percent reduction), the overall expected delay per movement doubles (from 0.8 to 1.6 min), while an 11.1 percent reduction, from 90 to 80, leads to a *tripling* of overall expected delay per movement, from 4.3 to 12.8 min. In this sense, Fig. 11-2 and Table 11-1 illustrate one of the principal results of queuing theory: expected delay changes in a nonlinear way as the demand rate and/or the capacity change; the closer the demand rate is to capacity, the more sensitive delay is to even small changes in demand and/or capacity. On days with a capacity of 80, aircraft using the airport incur a total of about 15,000 min [$\approx$(12.8)(1200)] or 250 h of delay. At a

Table 11-1. Some queuing statistics for Example 11-1

Capacity (movements per hour)	Expected waiting time (min)		Utilization ratio	
	Maximum	Per movement	24 h	6:00–21:59
110	2	0.8	0.455	0.664
100	4	1.6	0.500	0.731
90	13	4.3	0.556	0.812
80	39	12.8	0.625	0.913

current direct operating cost of about $2000 per aircraft hour, this is equivalent to a daily cost of $500,000!

It should be emphasized that the waiting times shown in Fig. 11-2 and in the second and third columns of Table 11-1 are *expected values*. At each capacity level and at each time of the day, the delay that aircraft will actually suffer on a given day is a random variable. Thus, some aircraft may suffer, on a day-to-day basis, considerably longer or shorter delays than indicated by the expected values in Fig. 11-2. In fact, another fundamental result of queuing theory states that the *variability of delay*—as measured by the *variance* of the delay or its *standard deviation*—also increases nonlinearly as the demand rate gets closer to capacity. One might therefore expect that, on certain days, especially when the capacity is low (e.g., 80 movements per hour), the delays experienced will be considerably longer than are shown* in the figure.

Figure 11-2 also illustrates some of the complex dynamic characteristics of queues and delays. The delay (and queue length) aircraft experience during any particular time interval depends strongly on the waiting times and queue lengths during previous intervals. For example, it can be seen in Fig. 11-2 that, for each of the four levels of capacity, the expected delay per aircraft during the morning hour of 06:00–07:00 is very small compared to that during the three afternoon hours that begin at 15:00. This despite the fact that the demand rate between 06:00 and 07:00 is only about 10 percent smaller than the average demand rate during the three hours that begin at 15:00 (86 per hour versus about 94 per hour). Part of the explanation is that the morning peak hour of 06:00–07:00 is preceded by a period of practically no demand (and no delays), whereas a queue has already started building up well before the beginning of the afternoon peak period. Moreover, the morning peak lasts for only about 1 h, whereas the one in the afternoon persists for several hours in a row.

Another interesting aspect of the dynamic behavior of airside queues is that a considerable lag often exists between the time

---

*In fact, the delays may become so bad on some days that the airlines may decide to cancel some flights. Indeed, cancellations often act as a safety valve at congested airports. In effect, they reduce demand during days when delays are at their worst and consequently also reduce the delays suffered by those flights that are actually performed.

when the demand rate peaks and the time when delays reach their peak. This time lag may be a long one on days when the demand rate exceeds the capacity continuously for a significant period.* For example, in the case when the capacity is 80 movements per hour in Fig. 11-2, one can see that the peak of the expected delay occurs between 18:00 and 19:00, while the peak demand hour is between 15:00 and 16:00. The reason for this time lag is that queues "build up" during periods of high traffic demand. Thus, those flights that request access to the runways near the end of these periods must necessarily join the end of the queues that have already formed and, as a result, experience the worst delays. This is a phenomenon that is often observed at busy airports. It is also one that motorists often experience when driving *after* the peak of the morning or afternoon "rush hour" on congested highways.

11-3　Policy implications and practical guidelines

Of the many characteristics and properties of air-side delays that have been discussed in connection with Example 11-1, the one that has the most important implications for the long-term growth of congestion at busy airports, is the nonlinear relationship between delays, on the one hand, and demand and capacity on the other. The key parameter in this respect is the *utilization ratio,* typically denoted in queuing theory by the Greek letter ρ ("rho"), which is defined as the average demand rate over a specified period of time divided by the average capacity over that time. For instance, suppose that, in Example 11.1, the specified "period of time" is a 24-h day. Since the demand rate is 1200 movements per day, one obtains $\rho = 0.625$ (= 1200/1920) for the 24-h day in the case where the hourly capacity is 80—or, the daily capacity is 1920 (= 24 × 80). This value of ρ is shown in the last row of column 4 of Table 11-1.

An untenable situation would result if anyone ever attempted to operate an airport with a daily utilization ratio, ρ, which is greater than 1. For this would mean that the number of movements requested per day would be greater than the daily capacity of the airport. Thus, on

---

*However, a noticeable time lag of the delay peak may exist even when the demand rate never exceeds (but is close to) capacity.

average, some movements would be "left over" each day in a queue, to be "processed" on the next day.* The leftover demand would then be added to the already scheduled demand for the next day—which, by assumption, was greater than the daily capacity in the first place. The queue of movements accumulated at the end of the day would thus grow *ad infinitum* from day to day in the long run.

Clearly, the situation just described is so extreme that it makes little practical sense. But it points to a general condition that must be satisfied if a queuing system is to be operating in a stable way over a long period of time. This condition states that a queuing system *cannot* be operated *in the long run* with a utilization ratio greater than 1, because delays for its use will never reach equilibrium and will grow without limit.

Having established that ρ, in the long run, must be less than 1, the following fundamental property of queuing systems can now be stated informally as follows: in the long run, both the expected waiting time and the expected queue length at any queuing system that reaches equilibrium ($\rho < 1$) increase nonlinearly with ρ, in proportion to the quantity $1/(1 - \rho)$.

This very important relationship is shown schematically in Fig. 11-3 for a typical queuing system. Suppose a queuing system has been operating for a sufficiently long time that its long-term equilibrium characteristics can be observed and measured statistically. Let W_q denote the time that a random user spends waiting in queue before being served by the system. The quantity plotted in Fig. 11-3 is the *expected value $E[W_q]$* of W_q. For example, in the context of a runway used for arrivals only, $E[W_q]$ would be the expected (or "average") time an aircraft requesting to land would spend waiting for its turn to approach the runway.[†] Note that as ρ approaches the value of 1 in Fig. 11-3—or, as the demand rate approaches the service rate or, more colloquially, as "demand approaches capacity"—$E[W_q]$ grows rapidly. $E[W_q]$ finally becomes infinite for $\rho \geq 1$, as suggested by the necessary condition for equilibrium described above. A plot of the quantity $E[N_q]$ versus ρ would have a similar shape. $E[N_q]$ is defined as the expected value of N_q, the number of users waiting in queue. The

---

*In practice, of course, the "leftover movements" would probably be cancelled each day, indicating again that, over a long period, any demand in excess of capacity is unsustainable.
[†]In most instances the wait would take place while the aircraft is airborne. However, in cases of extremely long delays, the aircraft may be "held" on the ground at its airport of origin to avoid excessive waiting times in the air.

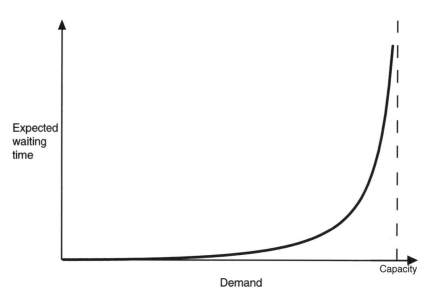

Fig. 11-3. *Typical relationship between the expected waiting time and the demand at a queuing system.*

exact mathematical expressions for $E[W_q]$ and $E[N_q]$ depend on several parameters. In general, the higher the variability of user service times (i.e., of the processing time per movement in the case of runways) and of the intervals between successive demands by users (i.e., between successive aircraft requesting use of the runways), the faster $E[W_q]$ and $E[N_q]$ will increase as ρ increases.

From the practical point of view, this observation and the generic "shape" of the function $E[W_q]$ shown in Fig. 11-3 (as well as of $E[N_q]$) have important implications for airports at the policy level.

First, they provide a warning to airport and ATM policy makers and managers not to operate runway systems at levels of utilization that are very close to 1 over an extended period of time. If they do so, long aircraft delays and queues will occur on a routine basis. Moreover, queuing theory has shown that the variability of W_q and N_q, as measured by their standard deviations, $\sigma(W_q)$ and $\sigma(N_q)$, also increases in proportion to $1/(1 - \rho)$. This means that, when ρ is close to 1, a queuing system not only experiences serious congestion on average, but is also subject to great fluctuations of behavior over time. Under the same set of *a priori* conditions (similar demand rates, weather conditions, etc.), delays to landings and takeoffs may be modest and tolerable on a particular day and, on the following day, they may be extremely long and unacceptable. This is a phenomenon observed very often at the busiest airports throughout the world. A good rule of thumb is that runway systems should not be operated at more than 85–90 percent of their capacity for the duration of the consecutive busy traffic hours of the day. The number of consecutive busy traffic hours in a day is considerably less than 24 at the great majority of airports: typically even the busiest of them have at most 16–18 h per day of high traffic activity, while little happens during the remaining 6–8 h, usually the nighttime hours.

Table 11-1 illustrates these ideas clearly. In col. 5, note that, with a capacity of 80 movements per hour, the utilization ratio during the 16 busiest hours of the day is equal to 91.3 percent. The associated delays are essentially unacceptable, with the average waiting time for all movements during the day and during the peak hour equal to about 13 min and 39 min, respectively. Column 4 shows that the utilization ratio for the entire day is quite modest (0.625) in this case, but this is deceptive and simply reflects the fact that there is little demand between 22:00 and 6:00, or for one-third of the day.

A second major point at the policy level is that, when a runway system operates at high levels of utilization, small changes in demand or in capacity can cause large changes in delays and queue lengths. This is a direct consequence of the fact that both the expected value and the standard deviation of W_q and N_q are proportional to $1/(1 - \rho)$: when ρ is close to 1, $1/(1 - \rho)$ is large and its value is highly sensitive to even small changes in ρ. This point is illustrated in Fig. 11-4. Note that the same amount of change in demand has very different consequences in terms of increases in expected delay, depending on whether the initial demand was high (relative to capacity) or low. This motivates much that is being done today at major airports around the world to contain delays. Many initiatives are currently underway aimed at either managing demand (Chap. 12) or at increasing the airside capacity of busy airports through improvements of the air traffic management system (Chap. 13). Airport operators generally recognize that most of these initiatives will produce only small changes in demand or in capacity. However, these small changes are still expected to produce significant reductions in delays, because many of the facilities and services at these airports operate at very high utilization ratios. It is hoped that these reductions in delay will make it possible to maintain acceptable levels of service at these airports until more dramatic improvements in capacity can be achieved. Experience as well as queuing models and simulations show that, at

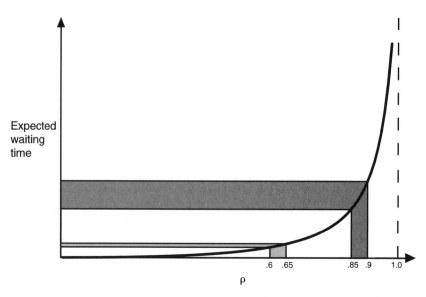

Fig. 11-4. *Nonlinear response of delay to demand changes; when an airport operates close to its capacity, delays are very sensitive to changes in demand or in capacity.*

very congested airports, a 1 percent reduction in daily demand for airside operations or a 1 percent increase in runway system capacity may result in a 5 percent or more reduction in delay (Fan and Odoni, 2001).

One can also revisit some of the definitions of capacity in Chap. 10 in light of this discussion. Remember that practical hourly capacity (PHCAP) is defined as the number of movements at which the average delay for use of a runway is equal to 4 min. The 4-min criterion was derived from graphs like those of Figs. 11-3 and 11-4, which the FAA prepared in the early 1960s. These suggested that expected waiting time at a typical airport would start to increase rapidly at levels of airport utilization corresponding to about 4 min of delay per aircraft (see Fig. 11-5). It was therefore decided to use 4 min as the threshold value at which a runway would be said to have reached its "practical capacity."

Similarly, the notion of *declared capacity*—which is used widely outside the United States, as noted in Chap. 10—is tied directly to the observation that delays will reach unacceptable levels at airports operated at their maximum throughput capacity for long periods. By "declaring" capacities that are equal to about 85–90 percent of the maximum throughput capacity, airport operators seek to maintain an adequate level of service as well as utilize their runway systems intensively.

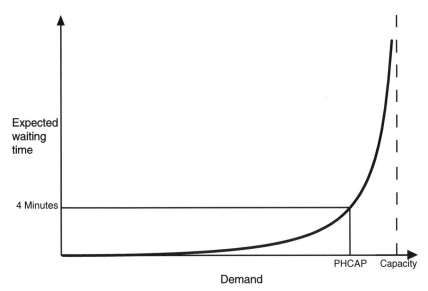

Fig. 11-5. *Determination of the practical hourly capacity (PHCAP) for a runway.*

Finally, the fact that airside delay is highly variable at airports that are intensely utilized has important implications for performance measurement and assessing level of service. Specifically, airport managers should utilize for such assessment measures that describe not only the average magnitude of delays, but also their dispersion around these average values. All of the following are examples of the types of performance metrics that should be of interest in an analysis of airside delays:

- Expected delay per movement during a typical day
- Variance (or standard deviation) of the delay per movement during a typical day
- Probability that delay will exceed some specified high value (e.g., 15, 30, or 45 min)
- Expected delay during the peak hour of the average day of the busiest month of the year

Expected delay per movement is undoubtedly the principal measure of interest. However, airlines are almost equally concerned about the predictability of delays, i.e., about how "tightly" delays are distributed around that average. If delay at an airport has a large variance (or standard deviation), the implied high variability in operating conditions means low reliability in carrying out daily airline schedules. Delays that exceed certain large values are also particularly damaging, as they result in missed passenger connections and in flight delays that propagate throughout an airline's network. Thus, the probability of such delays is another important measure of performance. Lastly, some specially "targeted" measures, such as the expected delay during the peak demand hours of the year, can be very helpful in assessing performance at times that are particularly critical to an airport's operations.

11-4   The annual capacity of a runway system

The reviews of capacity coverage charts in Chap. 10 and of the characteristics of airside delays in the last two sections provide the necessary background for a discussion of the *annual capacity* of the runway system of an airport. The annual capacity is a quantity of great practical interest. Airport demand forecasts are typically given in terms of annual figures ("X million passengers and Y thousand aircraft movements are forecast for 2015"). By comparing such demand forecasts with estimates of annual capacity, airport operators can roughly determine the timing of major capital investments and plan accordingly.

Annual capacity can be estimated only in very approximate terms, due to two principal difficulties. First, to compute annual capacity, planners and managers must make a number of rather subjective choices concerning the minimum acceptable level of service. There are no international standards at this time to assist in making these choices. Second, annual capacity depends in large part on daily and seasonal demand patterns that are difficult to predict far into the future. These points are best explained through a detailed example.

EXAMPLE 11.2

Consider again the case of Boston/Logan airport, whose capacity coverage chart (CCC) was described in Chap. 10. From the CCC, one can compute the average long-run hourly capacity of the airport's runway system. This capacity for Boston/Logan (see Example 10.3) is equal to approximately 115 movements per hour $[(132)(0.6) + (120)(0.18) + \cdots + (0)(0.015) \approx 115]$. One could thus infer that the airport can accommodate up to roughly 1,000,000 movements over an entire year $[(115) \times (24 \text{ h/day}) \times (365 \text{ days/year}) = 1,007,400]$.

In truth, however, this is an entirely theoretical number, very far from being attainable in practice. This is because the estimate of 1,000,000 movements per year implies that the airport will be utilized at 100 percent of its maximum capacity for 100 percent of the time. It is necessary to adjust this estimate downward following the type of reasoning indicated below.

Adjustment 1. Very limited or no runway activity takes place for between 6 and 8 h in a day (or for one-quarter to one-third of the time) at almost every major airport in the world. Passengers generally prefer not to arrive at or depart from airports between midnight and 6 a.m., and the marketplace usually accommodates this preference. In addition, to alleviate environmental impacts, most major airports either strongly discourage all but a small number of flights during nighttime or, in a few cases, ban all nighttime flights outright through the imposition of curfews. In the case of most airports in the United States, for example, passenger flights essentially cease from roughly 23:30 to about 06:00 local time, every day. Moreover, traffic is usually at a low level between 22:00 and 23:30 and between 06:00 and 06:30, when the morning flight activity truly picks up. Because the issue here is the *ultimate* annual runway capacity of the airport, it may be reasonable to assume that the airport will eventually "stretch" intensely utilized hours to the equivalent of roughly 16–17 h per day. It is highly doubtful that a runway system could be utilized heavily for much longer than that every day. This assumption will reduce the initial estimate of annual capacity at Boston/Logan to the range of 670,000–710,000 movements [(115) × (16) × (365) = 671,600; (115) × (17) × (365) = 713,500].

Adjustment 2. Even this range, however, is unrealistic. It posits a demand level of about 115 movements per hour for 16 or 17 consecutive hours every day, 365 days a year. This would certainly lead to intolerable delays on the many days when the runway system operates for several hours in a row at capacities below 115 per hour. But airside operations would also experience serious problems on those days when average hourly capacity is somewhat higher than 115. The utilization ratio, ρ, will be very close to 1 for extended periods of the day, and as indicated in Secs. 11-2 and 11-3, this would result in long expected delays and high delay variability. It is therefore necessary to assume that demand will not exceed a certain percentage of the average hourly capacity of 115 during the 16–17 "useful" hours of the day, if delays are to be kept at an acceptable level. As discussed in Sec. 11-3, this means a roughly 85–90 percent utilization of the average maximum throughput capacity of 115. The approximate estimate of the annual runway capacity of Boston/Logan would then be reduced further (and the range of the estimate broadened) to 570,000–640,000 movements per year [(671,600) × (0.85) = 570,860 at the low end; (713,500) × (0.90) = 642,150 at the high end].

Adjustment 3. However, seasonal variations in demand have not yet been considered! At most airports in the Northern Hemisphere, for instance, demand during the summer season exceeds demand during the winter season, often by a considerable margin. If during the summer, Boston/Logan is utilized at 85–90 percent of its capacity during the 16–17 useful hours of the day, the utilization during the winter season will necessarily be less because both regular and charter airlines will reduce their schedules, especially to international destinations and to summer resorts. (The terms "summer season" and "winter season" are used here to denote 6-month periods from May to October and from November to April, respectively.) In 1999, the number of movements at Boston/Logan on an average day during the summer season was approximately 18 percent higher than during the winter season. (This is a rather small difference in comparison to most other locations, as the number of summer season movements per day often exceeds winter season movements by 25 percent or more at many major airports.) Thus, the estimates of the range of annual capacity must be updated once more to account for seasonal peaking. It may be reasonable to assume that the intensity of summer peaking will decrease further as traffic grows, although by not much below its already low levels. Using a 15 percent seasonal peaking percentage, the range of the annual capacity estimate then becomes 530,000–600,000 movements per year [(570,860/2) + (570,860/2) × (1/1.15) = 533,630; (642,150/2) + (642,150/2) × (1/1.15) = 600,270]. The lower end of this wide range is the one that can be considered the more reasonable, as the high end is based on the rather extreme assumptions of 17 heavy traffic hours during a typical summer season day with a 90 percent utilization ratio over the entire day.

It can be noted parenthetically that the number of movements at Boston/Logan in 1998–2000 was about 500,000, i.e., the airport was operating within about 6–25 percent of its estimated ultimate annual capacity.* On the other hand, the estimated range does not take into consideration the possibility of capacity increases due to air traffic management developments or to expansion/improvements of the system of runways.

Several comments can now be made on the subject of the annual capacity of a runway system and the procedure for its estimation as Example 11.2 illustrates.

1. Just like "practical hourly capacity," "sustained capacity" and "declared capacity," *annual capacity* is a derivative measure. One must first compute the maximum throughput capacity of an airport— and, indeed, its entire capacity coverage chart, indicating how much capacity is available for what percent of time—to be able to obtain an estimate of annual capacity for an airport.

2. The estimation of annual capacity requires a number of implicit or explicit assumptions concerning, at the very least: future daily demand patterns, such as the assumption of 7 or 8 essentially idle hours in the Boston/Logan example; acceptable levels of delay, e.g., limiting operations to 85–90 percent of full capacity during the 16–17 useful hours of the day; and future seasonal demand patterns, e.g., 15 percent more operations per day in the summer season, on average. As the appropriate assumptions are likely to vary from airport to airport, so will the relationship between the maximum throughput capacity per hour of the airport and its annual capacity. In other words, the estimation of annual capacity is very much dependent on local factors and considerations.

3. The straightforward approach of multiplying the average hourly maximum throughput capacity of an airport by the number of hours in the year [(24)(365) = 8760] to compute the annual capacity of an airport greatly overestimates the true annual capacity. After all the necessary adjustments are made for level of service and for daily and seasonal demand patterns, the true annual capacity will be much smaller than the initial estimate (e.g., 530,000–600,000 versus 1,000,000 in the Boston/Logan example).

*At a moderate 2 percent annual growth rate in the number of operations, 6–25 percent represents the equivalent of 3–11 years of growth. A further concern is that the accelerating replacement of nonjet regional aircraft by regional jets at Boston/Logan will reduce the average hourly capacity of the airport to a value less than 115.

4. Industry practices, as well as public perceptions of what is "accept-able" or "reasonable," may change over time. This would also mean a change in estimated annual capacities. In the Boston/Logan example, if nighttime air travel increased or if winter became a major travel season for leisure passengers—perhaps stimulated by airline price incentives—the annual capacity of the airport would also increase! Indeed, experience suggests that annual airside capacities may have been seriously underestimated in the past. For example, as shown in Table 11-2, several airports in the United States were handling 25-75% more aircraft operations in 2000 than was thought possible 20 years earlier when the FAA had estimated their annual capacities. The reason was not so much an increase in the hourly capacities of these airports—capacities per hour in-creased only modestly over these 20 years. What happened instead was that airport operations were spread into longer and longer parts of the day, airlines increased the number of mid-day operations (formerly a slack period at many airports) as they increased their flight frequencies on many routes, and both airlines and passengers were forced to tolerate delays that would have been considered unacceptable in the early years of aviation.

Table 11-2. Actual number of movements often exceeds early annual capacity estimates

Airport	Annual operations (thousands)			Excess (%) of 2000 actual over PANCAP
	PANCAP (1980)*	Actual (1990)	Actual (2000)	
Los Angeles/International	448	680	781	74
St. Louis/Lambert	280	392	484	73
Boston/Logan	303	425	508	68
Philadelphia	295	407	484	64
New York/Newark	280	379	457	63
Seattle/Tacoma	280	355	446	59
Minneapolis/St.Paul	360	322	522	45
New York/Kennedy	272	302	359	32
New York/LaGuardia	247	354	323	31
Chicago/O'Hare	616	811	770	25

*PANCAP = "practical annual capacity" as estimated by the FAA.
Sources: FAA, 1981 and 2001.

These observations can now be generalized. Let A denote the number of annual movements obtained by multiplying the number of hours in the year (8760) by the average value of the maximum throughput capacity available per hour at an airport. The annual capacity of the airport will then be equal to kA, with the coefficient k usually in the range 0.50–0.60. The appropriate value of k, in each instance, will depend on local demand characteristics and willingness to accept delays. Values of k in the low end of the range (0.5–0.55) will apply to airports with relatively sharp daily and seasonal peaking, little or no (nighttime) activity for 7–8 h per day, and limited tolerance for long delays. Values of k in the high end of the range (0.55–0.60) will apply to airports with moderate daily and seasonal peaking, intensive utilization during all but a few nighttime hours of the day, and high tolerance for delays.

Many airports outside the United States use *declared capacity,* instead of maximum throughput capacity, as their measure of hourly capacity. If A has been computed by multiplying the declared capacity—instead of the maximum throughput capacity—by the number of hours in the year (8760), then the annual capacity should be estimated by using a coefficient k in the range 0.60–0.70 (instead of 0.50–0.60). The reason is that, as explained in Chap. 10, declared capacity is typically set to about 85–90 percent of maximum throughput capacity in order to ensure that delays will be reasonable, if the number of movements per hour scheduled at an airport is set equal to the declared capacity. Thus, Adjustment 2 of Example 11-2 is performed implicitly when using declared capacity as the starting point.

11-5   Computing delays in practice

The estimation of the delays to be expected at a heavily utilized runway system is usually a difficult task because of the complex dynamics of queuing systems and the nonlinear relationships that drive the behavior of queues. However, this is a task that airport planners and managers must deal with repeatedly. A complete analysis of airside capacity and delay requires a multistep procedure that can be summarized as follows.

Step 1.   Identify all possible runway configurations and the weather conditions in which they are used.

Step 2.   Compute the maximum throughput hourly capacity of each of these configurations.

Step 3. From historical records of weather conditions—and after taking into consideration local policies* regarding selection among alternative runway configurations when more than one configuration is available—estimate the annual utilization of each runway configuration, i.e., the approximate percent of time in a typical year during which each configuration is in use. At airports where the policy is to choose at all times the available configuration with the highest capacity, this step is equivalent to determining the capacity coverage chart.

Step 4. Prepare typical daily profiles of demand on the runway system (hourly number of arrivals and departures, mix of aircraft types, seasonal variations in the profiles).

Step 5. Estimate the delays associated with appropriately selected combinations of demand profiles and runway configurations in use.

Step 6. Estimate overall delay statistics on the basis of the results of step 5 and of the frequency with which each runway configuration is used as determined in step 3.

This procedure is obviously not simple. It requires access to extensive data on traffic to prepare demand profiles and on historical weather patterns to estimate the utilization of the various runway configurations. Computer-based capacity models are usually necessary to carry out step 2 for all but the simplest runway configurations. Reasonable approximate estimates of delays for step 5 can be obtained in some simple cases through mathematical formulae available from queuing theory. (Chap. 23 reviews this topic and provides a number of good references.) However, these formulas assume conditions, such as constant demand and capacity for extended parts of the day and a utilization ratio of less than 1 throughout the day ($\rho < 1$), which are often not met at busy airports. Therefore, such formulas should be used with caution. In particular, one should avoid the use of simplistic graphs provided in handbooks (e.g., FAA, 1981) for general use. Planners are strongly advised to use either computer-based mathematical queuing models or simulation models to estimate delays in all cases that involve time-varying demand and capacity at a runway system. Both of these types of models, as well as capacity models for step 2, are reviewed in Chap. 19 and in great detail by Odoni et al. (1997).

The type of delay model that will be most appropriate in each case depends on the requirements of the analysis. Issues related to model selection are also discussed in Chap. 19. If a very high level of detail

---

*As explained in Chap. 10, such local policies involve consideration of air traffic management procedures as well as of environmental impacts.

is desired, such as computing delays at every point of the airfield including taxiways, apron areas, stands, etc., then one of the standard large-scale simulation models (e.g., SIMMOD or TAAM) should be used. In most instances, however, the questions of interest will center on delays associated with the runway system, typically by far the most important bottleneck at an airport. In such cases, computer-based mathematical queuing models, which are far more flexible and easier to use than the available simulation models, should be quite adequate and often more informative.

Exercises

[Additional exercises on delays and congestion can be found in Chapter 23.]

1. Suppose that an airport has a maximum throughput capacity of 100 movements per hour in good weather, which prevails about 80 percent of the time, and of 60 movements per hour in poor weather (about 20 percent of the time). To estimate delays at this airport, Consultant A has computed an expected capacity of 92 per hour [= (0.8) × (100) + (0.2) × (60)] for the airport. He has then obtained delay estimates through a computer-based queuing model that uses as inputs the daily demand profile at the airport and an airport capacity of 92 per hour.

   Consultant B has used the same computer model as A with the same daily demand profile as A. However, she has run the model twice, once for a capacity of 100 per hour and then for a capacity of 60 per hour. She then took the weighted average of the delays computed through the two runs by multiplying the delays obtained from the first run by 0.8 and those from the second by 0.2.

   a. Which consultant's approach is more correct and why? Explain with reference to Fig. 11-3 or 11-4.

   b. Which consultant's delay estimates will be higher?

   c. Would you use the same daily demand profile for good-weather days and poor-weather days?

2. In this assignment you are asked to assess the impact, if any, that constraints in the airport system have on a major airline with which you are familiar. You should consider the general question of whether the infrastructure of the principal airports that your airline uses is adequate to support your airline's operations. You should also consider what the future portends at these principal

airports. Make recommendations on what you think your airline should do regarding the key airports and their constraints. You do not have to address all the issues raised below, but the list may be helpful in structuring your report.

a. For your airline, review briefly the situation at its top two airports. Make an assessment about how congested these airports are and their potential for capacity increases.

b. How sensitive is the runway capacity of these airports to weather conditions?

c. Do the airside constraints appear mostly on the surface of the airport (aprons and taxiways, including crossing of runways), at the runway system, or in the airspace?

d. How "delay-prone" are the airports you have examined and your airline? Examine some delay statistics for these airports.

e. To what extent does airline scheduling ("banks" or ""waves" of connecting flights) contribute to the problems you have identified?

3. Consider Exercise 1 of Chapter 10, in which you computed the maximum throughput capacity of a single runway, which is used for arrivals only. Based on your work in that exercise, you can easily compute, not only the expected value of the service time at this runway, but also the variance of the service time.

Assume below that the instants when the demands for landing at this runway occur can be approximated as being generated according to a Poisson process.

a. Estimate the practical hourly capacity of the runway, using the approximate queuing expression given in Eq. (23.10) in Chap.23.

b. Suppose that the cost of 1 min of delay in the air is $80, $30, and $15 for H, L, and S aircraft, respectively. Suppose, as well, that this runway has a demand of 27 arrivals per hour for about 6 consecutive peak hours during the day, a demand of 20 per hour for about 10 h, and a demand of 10 per hour for 8 h. Estimate approximately the annual delay costs incurred at this runway due to traffic delays. [This part, of course, oversimplifies what happens in practice: in truth, capacity changes over time with weather and runway configuration changes and demand will typically be more variable than is indicated here.]

References

FAA, Federal Aviation Administration (1981) "Airfield and Airspace Capacity/Delay Analysis," Report FAA-APO-81-14, Office of Aviation Policy and Plans, Washington, DC.

FAA, Federal Aviation Administration (2001) "Air Traffic Activity Data System (ATADS) Database," http://www.apo.data.faa.gov/faaatadsall.htm.

Fan, T., and Odoni, A. (2001) "The Potential of Demand Management as a Short-Term Means of Relieving Airport Congestion," EUROCONTROL-FAA Air Traffic Management R&D Review Seminar, Santa Fe, NM, http://atm2001.eurocontrol.fr.

Fron, X. (1998) "ATM Performance Review in Europe," EUROCONTROL-FAA Air Traffic Management R&D Review Seminar, Orlando, FL, http://atm-seminar-98.eurocontol.fr.

Odoni, A., Deyst, J., Feron, E., Hansman, R., Khan, K., Kuchar, J., and Simpson, R. (1997) "Existing and Required Modeling Capabilities for Evaluating ATM Systems and Concepts," International Center for Air Transportation, Massachusetts Institute of Technology, Cambridge, MA (prepared for NASA Advanced Air Transportation Technologies Program), http://web.mit.edu/aeroastro/www/labs/AATT/aatt.html.

U.S. Department of Transportation (monthly) "Air Travel Consumer Report," http://www.dot.gov/airconsumer.

12

Demand management

Demand management refers to any set of administrative or economic measures and regulations aimed at constraining the demand for access to a busy airfield and/or at modifying the temporal characteristics of such demand. It is widely practiced at major airports throughout the world. Despite a number of valid concerns about its use, there is little doubt that airport demand management is here to stay, as it is unlikely that the supply of airport capacity will be able to accommodate growing demand in the foreseeable future. The issue, therefore, is not whether to apply demand management, but how to apply it best.

The available approaches can be subdivided into three categories: purely administrative, purely economic, and hybrids, i.e., combinations of the former two. The fundamental rationale for the last two categories is that they force airport users to consider the full costs of access, including the delay costs they impose on other airport users—costs that users can largely ignore under most circumstances at present.

Schedule coordination, conducted under the aegis of the International Air Transport Association (IATA), is currently used at the great majority of the busiest airports outside the United States. This is an administrative demand management procedure that uses historical precedent as the primary criterion for allocating airport slots. It can be effective at mildly congested airports, but will cause serious market distortion and affect competition adversely at airports where unconstrained demand exceeds available capacity by a significant margin.

All purely economic approaches to demand management involve some form of congestion pricing; i.e., they are based on the principle that, to optimize use of a congested facility, users should be forced to internalize, preferably fully, the external costs imposed by their use of the facility. Congestion pricing is an economically efficient approach, but it may be difficult to apply in practice for both technical and political reasons.

Hybrid demand management systems combine elements of administrative and economic approaches. Their common characteristic is the use of administrative procedures to specify the number of slots available at an airport. Hybrid systems rely on such economic devices as congestion pricing, slot markets, and auctions to arrive at the final allocation of slots. The worldwide use of hybrid demand management systems will probably increase very significantly in the future.

12-1   Introduction

The ever-tighter relationship between demand and capacity at the world's major commercial airports and the worsening air traffic congestion have led to increased interest in airport demand management. Demand management refers to any set of regulations or other measures aimed at constraining the demand for access to a busy airfield and/or at modifying the temporal characteristics of such demand. Examples are slot restrictions and airport pricing schemes aimed at discouraging the scheduling of flights during peak traffic hours and inducing airlines to shift some operations to off-peak hours.

Until the early 1980s, demand management was practiced at only a small number of airports around the world. Primarily because of its potential, under certain conditions, for adversely affecting competition, it was considered to be a method of last resort for reducing airport congestion and congestion costs. However, with few exceptions, notably the United States, the debate on demand management today has shifted from whether it should be used at all to how it can be applied most effectively. The overall premise is as follows: Capacity expansion should generally be the principal means for accommodating growth in airport demand, but it may require a long time or may even be entirely infeasible. In such circumstances, some form of demand management may be the only available alternative, at least in the short and medium terms, for keeping delays within reasonable bounds.

Demand management is currently practiced, in one way or another, at virtually all the busiest airports outside the United States, as well as at many secondary ones. It is viewed as an essential complement, on the demand side, to "supply-side" efforts to increase capacity. As this chapter shows, all demand management approaches have some weaknesses, which are recognized even by most proponents. It is typically argued, however, that the overall effects on competition and on access to airports and markets are mild compared to the significant benefits that stem from the resulting reduction in air traffic congestion. Regardless of their own views on demand management, airport professionals need to be familiar with this subject, as its implications for current and future airport operations, and for the air transportation system in general, are potentially very important.

This chapter reviews the principal approaches to demand management, as currently practiced or proposed, and discusses some of the advantages and disadvantages of each. The approaches can be subdivided into three categories: purely administrative, purely economic, and hybrids, i.e., combinations of the former two. The fundamental objective of the last two categories is to force airport users to consider the full costs of their usage, including the delay costs they impose on others Airport users can largely ignore these costs in most circumstances.

By definition, purely administrative approaches do not use economic incentives, such as landing fees that may vary by time of day, to induce prospective airport users to request one particular time over another for the operation of a flight. If such economic incentives are used, along with administrative measures, the approach is classified as a hybrid. In purely economic approaches, on the other hand, there is no administrative interference with airline decisions on selecting the times for the operation of flights. These three types of demand management approaches will be covered in Secs. 12-3–12-5, while Sec. 12-2 briefly recaps the rationale for applying any form of demand management at congested airports. Section 12-6 presents some policy-related conclusions.

It is important to clarify at the outset that this chapter is concerned with *strategic* demand management approaches, i.e., measures that become part of the institutional and regulatory framework within which an airport is operated. Some air traffic management organizations, notably EUROCONTROL and the U.S. Federal Aviation Administration (FAA), also apply a *tactical* form of demand management to relieve airport congestion on a day-to-day basis: *air traffic flow management* (ATFM) controls the flow of traffic into congested airports and congested parts of the airspace on a dynamic, "real-time" basis. A typical tactic is to postpone for some time the departure of an airplane if it is expected that, once airborne, it will be subjected to a long delay. ATFM is described in Chap. 13.

The chapter will also refer primarily to approaches for managing access to an airport's *system of runways*. This is by far the most common context in which demand management is applied. Entirely analogous ideas can be applied in the context of access to passenger terminals, aprons, or other parts of the airport. In these latter cases, passenger service charges or aircraft parking charges may sometimes be used as the instrument for managing demand (see Sec. 12-5 and, especially, Chap. 8). It should be noted, however, that the coordination and harmonization of landside demand management measures with those aimed at the airside can be a complex task and may involve difficult economic and regulatory issues. Very few airports practice simultaneous airside *and* landside demand management at this time.

12-2   Background and motivation

The principal objective of demand management is to assist in maintaining efficient operations at airports that are threatened by congestion. This is not done through capital investments or changes in traffic handling procedures aimed at increasing capacity, but through regulations or other measures that aim at some combination of (1) reducing overall demand for airfield operations, (2) limiting demand during certain hours of the day, and (3) shifting demand from certain critical time periods to other, less critical ones. The means used to accomplish this objective is what differentiates one demand management approach from another. In all cases, the net effect is to restrain access to airports in some way, either at all times or at selected times. For this reason demand management is also often referred to as *access control.*

In the absence of a demand management program, access to commercial airports throughout the world is governed by essentially one set of rules. Any aircraft technically qualified to operate at a particular airport, i.e., that fulfills air traffic management and airworthiness requirements, can utilize the airport by paying a landing fee proportional to the weight of the aircraft.[*] In this environment, any airline may schedule a flight at the airport for any time it wishes, outside any curfew hours that may exist.[†] A demand management program modifies these conditions. For example, an upper limit may be placed on the number

---

[*]The maximum take-off weight (MTOW) of the aircraft is typically used as the basis for computing the landing fee. Other types of user charges (passenger service fees, aircraft parking fees, hangar fees, etc.) may also be imposed, depending on the type of flight involved. Chapter 8 discusses airport user charges in detail.

[†]International flights must be authorized in every case under the applicable bilateral or multilateral agreements.

of operations that can be scheduled during a particular period of the day, or a surcharge may be imposed on the landing fee during peak hours.

The motivation for demand management comes directly from a fundamental observation in Chap. 11, namely, that when the utilization of a service facility is high (or when demand approaches the capacity of a system), the relationship between delay, on the one hand, and capacity or demand, on the other, becomes *very nonlinear:* a small increase in capacity or a small reduction of the demand rate results in a proportionally much larger reduction in delay (with the reverse also being true). Demand management aims at achieving those small reductions in demand (or the shifts in demand from peak to off-peak periods) that will bring about these large delay benefits. In the process, additional benefits may be achieved, such as reduced operating costs through a more efficient utilization of available personnel, equipment, and resources.

It is sometimes argued that demand management of any form is unnecessary, even at the busiest airports, because delay will act by itself as a "natural" access-control mechanism. According to this argument, as delays at an airport grow, more and more prospective aircraft operators will deem the situation unacceptable and will choose not to use the airport. At some point, the costs associated with delay, as perceived by individual users, will become so high that demand will cease to grow and "equilibrium" will be reached.

This line of reasoning misses a critical point. The equilibrium reached in this way will, in general, be inefficient economically, as (1) the associated level of delays to aircraft and passengers will be excessive and (2) the resulting mix of airport users may include a large fraction who have a low value of time and whose use of the airport cannot be justified on economic grounds.* This will be explained in detail in Sec. 12-4. Suffice to note here that, in recent years, a growing number of aviation experts, managers, and operators have come to realize that the "do nothing" alternative (i.e., allowing demand to grow unabated until the users themselves become discouraged by the high cost of delays) is wasteful and inefficient. This has motivated the extensive ongoing examination of the relative merits and effectiveness of the various demand management approaches that will be reviewed next.

*This statement is not entirely accurate at airports where traffic is dominated by a single airline—see Sec. 12.3.

Example 12.1 illustrates clearly what may happen if delay is allowed to serve as the only access-control mechanism. It also indicates the large benefits, in terms of delay reductions, that may be obtained from even primitive demand management measures under the right conditions.

EXAMPLE 12.1

For more than 30 years before 2000, the number of aircraft operations at New York/LaGuardia (LGA) was constrained by the number of slots authorized under the FAA's High Density Rule (HDR)—see Sec. 12-3. Early in the spring of 2000, an average of about 1050 aircraft movements (arrivals and departures) took place at LGA on a typical weekday. However, the Wendell H. Ford Aviation Investment and Reform Act for the 21st Century (AIR-21), enacted in April 2000, exempted from HDR slot limitations aircraft with a capacity of 70 seats or fewer performing scheduled flights between LGA and small airports. In the first 7 months after AIR-21 was enacted, airlines sought to schedule more than 600 new movements a day at LGA. As of November 2000, about 300 of those new movements had begun operations, bringing the average number of movements on a typical weekday to 1350. The result was unprecedented levels of delay and numerous flight cancellations on a daily basis. LGA alone accounted for more than 25 percent of the serious delays (more than 15 min) experienced at *all* commercial U.S. airports in the fall of 2000. Yet airlines kept announcing the scheduling of additional regional flights at the airport.

As an interim solution, the FAA, with strong support from LGA's operator, the Port Authority of New York and New Jersey, imposed a limit on the number of slot exemptions granted under AIR-21. It allocated the pool of "AIR-21 slots" among eligible flights through a lottery that took effect on January 31, 2001. The lottery was designed to impose an hourly cap of approximately 75 scheduled movements per hour, a number that the airport was deemed adequate to accommodate at reasonable levels of delay in good weather conditions (FAA, 2000).

Indeed, the severity of delays and the number of cancellations at LGA declined enormously after January 2001 from the fall of 2000 levels. The slot lottery reduced the number of weekday scheduled flight movements (take-offs and landings) from about 1350 in November 2000 to 1205 in August 2001, a roughly 10 percent reduction in frequency. Figure 12-1 compares the profile of hourly flight operations before and after the lottery. It is noteworthy that the level of flight operations prior to the slot lottery exceeded the sustainable capacity of 75 for most of the day, with virtually no time for schedule recovery. Prior to the slot lottery, delay rose continuously from the early morning till 8 p.m., reflecting the fact that scheduled demand exceeded (good-weather) capacity throughout that period. At its daily peak, expected delay was more than 80 min per movement in November 2000; it declined to a peak value of about 15 min per movement in August 2001 (Fan and Odoni, 2001). Total expected delay for a typical weekday was about 900 aircraft-hours prior to the slot lottery, but only about 150 aircraft-hours after it. The 10 percent reduction in the number of movements imposed through the lottery thus led to a reduction of 80 percent in total aircraft delays! At an average of about $1600 per hour in direct operating costs, this translates to savings of about $1.2 million in direct operating costs per weekday for the airlines, without including savings associated with less passenger waiting time and reduced schedule recovery costs.

The FAA also stated its intention to replace the lottery with a more appropriate long-term demand management mechanism for preventing excessive congestion at LGA. In June 2001 it published a Notice for Public Comment (FAA, 2001), which described several potential approaches for managing demand at LGA in the future. The alternatives included schemes variously based on purely administrative allocation, on congestion pricing, or on auctioning of slots (see Secs. 12-3–12-5). This process may lead eventually to adoption of demand management measures at LGA that have little resemblance to the slot lottery.

12-3   Administrative approaches to demand management

A fundamental element of all administrative approaches to demand management is the concept of a slot. A *slot* is an interval of time reserved for the arrival or the departure of a flight and is allocated to

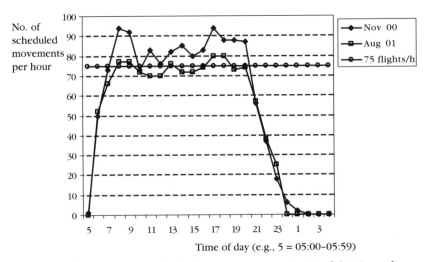

Fig. 12-1. *Profile of scheduled movements at New York/LaGuardia before and after the slot lottery.*

an airline or other aircraft operator for a specified set of dates. Thus, the statement, "Airport X can offer up to 60 slots between 09:00 and 10:00 local time for the summer season of 2001," means that the administrative entity responsible for Airport X is willing to allocate among prospective aircraft operators up to 60 time intervals for scheduling arrivals and departures during the hour in question. Once a slot is allocated, it typically becomes associated with a specific flight operation, e.g., "Airline A has been given the 09:10 slot for the arrival of its Flight 124 at Airport X during the summer season of 2001." This does not mean that Flight 124 necessarily *has* to land at 09:10, but that Flight 124 has been *scheduled* to arrive at 09:10 and will be expected to arrive at about this time throughout the summer season of 2001.

Administrative approaches to demand management require the selection of a set of criteria for allocating slots among prospective users. Some examples of potentially reasonable criteria and of related considerations may include:

- The length of time for which a flight has been operating already (flights that have been operating for a long time may be deemed to deserve priority for continuation of service)
- The regularity of the flight (scheduled flights operated on a daily or weekly basis may be given preference over occasional charter flights)

- The origin or destination of the flight (service to/from certain locations or to new markets may be deemed particularly important)
- The characteristics of the airline requesting the slot (in the interest of more competitive service, for example, "new entrant" airlines that have not previously served a particular route or a particular airport may be given priority for slots)

Schedule coordination: the IATA approach

All the criteria mentioned above are used to some degree in connection with the most widely used approach to administrative slot allocation, the *schedule coordination approach* of the IATA. The following description omits many important details that can be found in *Worldwide Scheduling Guidelines* (IATA, 2000).

For purposes of schedule coordination, airports are classified into three categories. *Level 1* (or *"noncoordinated"*) airports are those whose capacities are adequate to meet the demands of users. *Level 2* (or *"coordinated"* or *"schedules facilitated"*) airports are those where demand is approaching capacity and "some cooperation among potential users is required to avoid reaching an over-capacity situation" (IATA, 2000). To this purpose, a "schedules facilitator" is appointed who seeks voluntary cooperation on schedule changes by the airlines to avoid congestion. *Level 3* (or *"fully coordinated"*) airports are those deemed sufficiently congested to require the appointment of a schedule coordinator, usually supported by a committee of experts and stakeholders, whose task is to resolve schedule conflicts and allocate available slots. *All* requests for slots at Level 3 airports must be reviewed and cleared by the schedule coordinator. As of 2001, approximately 140 airports worldwide were designated as Level 3 and used the IATA's schedule coordination approach. Their list includes practically every one of the busiest airports of Europe and of Asia and the Pacific Rim, as well as many secondary ones.

Schedule coordination is carried out at Schedule Coordination Conferences (SCCs) organized by the IATA every November and June* and attended by numerous representatives of airports, airlines and civil aviation organizations from around the world. Each fully coordinated airport must first specify a *declared capacity* (Chap. 10), which indicates the number of aircraft movements per hour (or per other

*The November SCC is concerned with schedule coordination for the upcoming summer season (in the Northern Hemisphere) and the June SCC with the upcoming winter season.

unit of time) that the airport can accommodate. Under the IATA system, responsibility for determining the declared capacity of each airport rests with local and national authorities. Declared capacity need not be determined solely by the capacity of the runway system. Constraints due to the availability of aircraft stands, passenger terminal processing capacity, and even aircraft ramp servicing capacity can be taken into consideration. This is one of the reasons there are many secondary airports in the Level 3 category. Some airports with severe landside constraints accept only a very limited number of movements per hour and are therefore classified as fully coordinated.

Prospective users of Level 3 airports must submit a formal request for each and every desired slot. The declared capacity is rationed according to a set of criteria, among which the principal and overriding one is *historical precedent:* an aircraft operator who was assigned a slot in the same previous season ("summer" or "winter") and utilized that slot for at least 80 percent of the time, is entitled to continued use of that "historical slot." Second priority is assigned to requests for changing the time of historical slots. In addition, a slot awarded on the basis of historical precedent may be used in the new season to serve a different destination from the one served in the previous season. Slot exchanges between airlines, on a one-for-one, nonmonetary basis, are also allowed, as are short-term leases to code-sharing partners.

Slots for services that have been discontinued, or not used at least 80 percent of the time,* are returned to a "slot pool" for reallocation. Any new slots made available through increased airport capacity are also placed in the slot pool. All requests for new slots are served from the slot pool. In order to encourage competition, to establish new markets or to strengthen previously underserved ones, at least 50 percent of the slots in the pool within each coordinated time interval are assigned to airlines designated as *new entrants*. However, the definition of a "new entrant" is very restrictive: an aircraft operator qualifies for this designation as long as it does not hold more than four slots in a day, *after* receiving any new slots from the slot pool (IATA, 2000). Thus, a new entrant can be awarded at most two flights (or four runway operations) per day, hardly sufficient to establish a significant foothold at a major airport.

After new entrants, priority is given to requests for extending seasonal scheduled service (previous winter or previous summer) to year-round

*Some exceptions may be made for chartered services provided by tour operators.

scheduled service (to the next summer or the next winter, respectively). Any further remaining slots after this step are distributed according to a number of additional criteria, such as the size and type of market involved, contribution to competition on routes, the existence of any curfews at the airports of flight origin or flight destination, etc.

The schedule coordinator obviously plays a central role in this process. The manner in which the coordinator is selected varies from country to country and even from one airport to another in some countries. In most cases, the national airline or a major airline of the country is asked to designate an experienced employee or a team of employees to serve in this capacity.

There are also significant variations in the level of sophistication with which these demand management procedures are applied at different airports. For example, some airports utilize a simple limit on the number of movements that can be scheduled in any single hour of the day, while others employ combinations of limits that may restrict the number of movements for intervals smaller than an hour. This is shown in Table 12-1 for a number of major international airports. Note, for example, that Sydney controls the number of movements down to 5-min intervals and utilizes a staggered enforcement period in order to smooth any intrahour peaks in the traffic schedule. At

Table 12-1. Declared capacity at selected international airports

Airport	Limit on scheduled aircraft movements per interval (2001)						
	1 day	3 h	1 h	30 min	15 min	10 min	5 min
London/Heathrow			79–85*				
Tokyo/Narita	367[†]		26–32*				
Frankfurt/Main			78	43		16	
Seoul/Inchon			37[‡]				
Sydney			80[§]		21		8
Osaka/Kansai		81	30				

*Depending on departure and arrival mix.
[†]This is a noise-related limit and is subdivided into a limit of 349 international flights and 18 domestic flights.
[‡]Enforced in 1-h periods staggered by 30 min, e.g., 1000–1059, 1030–1129, etc.
[§]Enforced in 1-h periods staggered by 15 min, e.g., 1000–1059, 1015–1114, etc.

Tokyo/Narita, the hourly limit in 2001 was set between 26 and 32 movements per hour, depending on the anticipated mix of departing and arriving aircraft, whereas at Frankfurt/Main there were also separate quotas for the maximum number of departures and of arrivals that could be scheduled in any hour (48 and 43 per hour, respectively). An even more flexible approach is in use at London/Heathrow. First, as in Tokyo/Narita, the declared hourly capacity may change by hour of the day and depends on the mix of departures and arrivals in each hour. Moreover, a marked difference exists between the number of slots available in the summer and winter seasons (Fig. 12-2), in order to take into consideration the impact of unfavorable weather conditions in the winter. The number of slots at London/Heathrow is adjusted from year to year with the objective of maintaining the level of airborne holding to an average of 10 min or less per arriving flight.

The declared capacity at an airport may increase gradually over the years as a result of airport infrastructure expansion, air traffic control improvements, and airport operator experience with the slot coordination system. The number of daily slots made available at London/ Heathrow increased by 8 percent between the summer seasons of 1991 and 2001, from 1246 to 1347, and at Frankfurt/Main by 18 percent, from 1056 to 1248 (Fan and Odoni, 2001).

The main appeal of the IATA's purely administrative schedule coordination approach is that it "has been singularly successful in maintaining

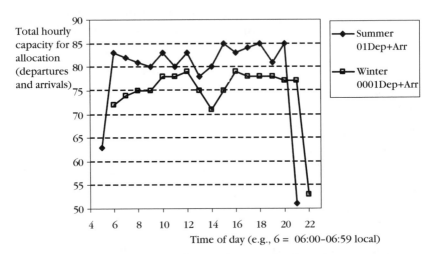

Fig. 12-2. *Number of slots allocated at London/Heathrow in winter and summer of 2001.*

a high degree of coherence and stability in the international air transport system" (IATA, 2000). Indeed, this demand management mechanism has worked well in practice in instances where demand exceeds the supply of airport capacity by a relatively few operations and for only a small number of hours in a day. However, when a significant excess of demand over capacity exists, there is a clear risk that an approach entirely detached from economic considerations and incentives may lead to serious distortions of the marketplace. Indeed, it can be argued that, at some of the most congested airports in the world, the schedule coordination process currently serves as a means for preserving the status quo, effectively acting as a regulatory device at the airport level. New competitors may be prevented from entering markets in an effective way: they may be denied slots altogether or be relegated to slots at inconvenient times of the day or awarded far fewer slots than necessary to establish a truly competitive schedule of flights. Some airports have in fact gone as far as segmenting airport capacity to serve perceived public policy goals. Examples include the designation of blocks of slots reserved for international, domestic, and general aviation traffic at Tokyo/Narita, and setting aside slots to accommodate regional services within the State of New South Wales at Sydney Airport.

In response to such concerns, a number of governments around the world have been examining closely the IATA's schedule coordination procedures. This has been particularly true in the European Union (EU), whose air transport system was largely deregulated in 1993. Regulation 95/93, which became effective in January 1993 and applies to all EU Member States, attempts to modify the IATA schedule coordination approach and make it more impartial and more accommodating of change. This regulation also provides a good illustration of the many complications that must be dealt with. Noteworthy provisions concern schedule coordinators, new entrants, and an appeals process. Regulation 95/93 explicitly assigns to national governments responsibility for appointing schedule coordinators, thus attempting to take slot allocation out of the hands of the airlines. As a practical matter, however, several EU members have continued to give in various ways to their flag carriers a leadership role in coordinating congested airports in their national territory, reasoning that the airlines possess the necessary expertise. The definition of new entrant airlines is more "liberal" under Regulation 95/93. An airline qualifies if it holds fewer than 3 percent of the slots* at the airport on the day for which it requests a new

---

*Note that at a large airport with, for instance, 1000 movements per day, 3 percent amounts to 30 movements, a large increase over the limit of 4 that IATA sets.

slot, as well as fewer than four slots on that day for the specific market for which it requests a slot. Finally, a process is established under which aircraft operators may appeal slot allocation decisions, first, to the schedule coordinator, then to the Member State where the airport is located, and eventually to the European Commission itself. The Commission, in fact, retains the right to cancel some or all of the slot allocations made by a schedule coordinator.

Recently, the European Commission has repeatedly considered significant changes to Regulation 95/93, such as allowing some trading of slots via an auction and a secondary market, and easing further market entry into congested airports by new carriers (Pagliari, 2001). It is probable that the existing purely administrative slot allocation systems at most EU airports will be replaced in the future by hybrids combining economic and administrative mechanisms.

Experience in the United States

As demand for slots at a congested airport increases, so clearly does the complexity of the task of the schedule coordinator. The situation may eventually become untenable: no matter how slots are allocated administratively, the interests of several carriers will be materially damaged. The rejection of many slot requests will certainly lead to distortion of market forces and dilution of competition.

A case in point is the experience with schedule coordination in the United States during the years immediately following airline deregulation in 1978. (For a detailed description see U.S. Department of Transportation (1995).) In November 1968, the United States initiated the airport High Density Rule (HDR). Five busy airports, New York/Kennedy, New York/LaGuardia, New York/Newark, Washington/Reagan, and Chicago/O'Hare, were designated HDR airports and hourly limits were placed on the number of operations that could be scheduled at each of them.[*] Four of these airports[†] continue to operate today with such limits.[‡]

Schedule Coordination Committees were established at each HDR airport, one to administer and coordinate slots designated for use by

[*]In fact, two limits were specified for each airport: one on the number of air carrier operations allowed per hour, and a second on the number of hourly commuter/regional airline operations.
[†]New York/Newark was exempted from slot limits in October 1970.
[‡]The limits are specified for IFR weather conditions. In VFR weather, additional flights can be scheduled on an ad hoc basis, in response to requests for permission to use these airports. Naturally, most of these are general aviation flights, since they can be scheduled on a short-term basis.

air carriers and the other to do the same for slots designated for commuter airlines. After a difficult first meeting in 1969, which required a full month of intense negotiations before consensus was reached, subsequent HDR meetings generally went smoothly for more than a decade, with participants making minor adjustments from year to year to the schedules and slot assignments. This was the preregulation era, when there were few, if any, new entrants from season to season into routes served by the airports involved. In fact, HDR made no special provisions for transferring slots to new entrants.

The situation changed completely soon after the 1978 deregulation of the airline industry. In the new environment, prospective new entrants pressed for the acquisition of large numbers of slots at the four HDR airports. The Schedule Coordination Committees were unable to satisfy these requests, despite some changes that the Department of Transportation made to the HDR.* After a period of several years during which the Schedule Coordination Committees were deadlocked on a virtually continuous basis, the Department of Transportation finally abolished these committees on December 16, 1985. The slots at the HDR airports were declared by the Department to be available for use by the airlines that held them as of that date, and a "buy-and-sell" environment was established at the four HDR airports (Sec. 12-5). No Schedule Coordination Committees have operated in the United States since that time. With the exception of the four HDR airports, no other airport in the United States currently uses the device of slots and of slot allocation procedures in the demand management sense.†

The experience with schedule coordination in the United States may be a precursor of future developments in other parts of the world, especially in Europe, which is now seeing the appearance of new low-cost carriers, just as happened in the United States in the early 1980s.

12-4   Economic approaches to demand management

Economic approaches to airport demand management utilize various forms of *congestion pricing* to exercise some control over airport access. A system of fees based on congestion pricing principles takes into consideration the pattern of delay at an airport over time and

---

*One of these changes was a requirement for a minimum 70 percent utilization of slots by current slot holders.
†However, as noted earlier, the notion of a slot is used in a tactical context by the Air Traffic Management system in connection with dynamic, day-to-day air traffic flow management, as described in Chap. 13.

attempts to reduce these delays to an economically efficient level. The access fees typically vary with time of day, as well as possibly by season and even by day of the week, with higher fees during peak demand periods and lower fees during off-peak periods. For this reason, congestion pricing is sometimes called *peak period pricing*.

Congestion pricing can serve as either an alternative or a complement to the purely administrative slot allocation procedures described in the previous section. When used as a complement (as is currently done at a few of the busiest airports in Europe), the pricing scheme is designed to discourage users from requesting slots during the most "popular" hours of the day by specifying higher charges for those hours. This kind of system is a "hybrid" and will be discussed in Sec. 12-5. This section concentrates on purely economic demand management approaches.

Congestion pricing in theory

As noted in Sec. 12-1 and in Chap. 8, access to airports today is typically paid for through a landing fee, which is proportional to the weight of the aircraft. The landing fee per unit of weight is determined through the average-cost pricing method, described in detail in Chapter 8.* Two aspects of this virtually universal practice are truly striking. First, as the amount of traffic at a congested airport increases, the landing fees will decrease, because the cost of the airfield is divided among more users (and more aircraft weight). Thus, with average-cost pricing, access to the airport becomes cheaper as congestion worsens! Second, there is a tenuous, at best, relationship between the landing fee an aircraft pays and the true costs imposed by that aircraft's operation. A fee based solely on the landing weight of the air-craft essentially charges aircraft according to their "ability to pay," rather than in proportion to the costs they cause to others by operating at the airport. A 360,000-kg wide-body commercial jet will pay a landing fee 60 and 18 times greater, respectively, than a 6000-kg general aviation aircraft and a 20,000-kg aircraft of a regional airline, using the same runway. Yet, as noted in Chap. 10, all three aircraft will occupy the runway and associated final approach path for roughly similar amounts of time—and occupancy time is what really counts in the case of congested facilities.

*Briefly, average-cost pricing consists of three basic steps: (1) a target amount of revenue, X, to be collected from landing fees, is specified at the beginning of the airport's fiscal year (typically, X is equal to the annual cost of the airfield, including a reasonable return on the airport's investment); (2) a forecast is made of the total number of units of weight, Y, of all the aircraft that will utilize the airport during that year; and (3) the landing fee per unit of weight, Z, is set equal to the ratio X/Y.

This fundamental inconsistency between the price charged and the true cost of using congested airport facilities has long been pointed out by economists [see, e.g., Levine (1969), Carlin and Park (1970), de Neufville and Mira (1974), Morrison (1987)] and is being increasingly recognized by airport and civil aviation experts and administrators. With growing congestion, the use of airport landing fees as an instrument to reduce delays and maximize efficiency at these very expensive and scarce facilities may be equally important as the use of these fees to generate airport revenues. This is especially true at a time when landing fees, which have traditionally been one of the principal sources of revenue for airports, are becoming far less dominant in this respect* (Chap. 8). Thus, many of the strongest economically and most congested airports around the world can afford to experiment with alternative structures of the landing fees that draw on the principles of congestion pricing.

The theory of congestion pricing is well established by now in the literature of economics [see, e.g., Hotteling (1938, 1939), Vickrey (1969), Carlin and Park (1970), Daniel (1995)]. Only its main points are summarized here. Consider any facility that experiences congestion, all or part of the time. Every user who obtains access to the facility during periods when delays exist generates a congestion cost that consists of two components: (1) an "internal delay cost," or "private delay cost," i.e., the cost that this particular user will incur due to the delay that user suffers; and (2) an "external delay cost," i.e., the cost of the additional delay to all other prospective facility users which is caused by this particular user. For example, if airplane A, which uses a runway during a peak period, will delay 30 other aircraft by 2 min each—a very common occurrence at congested airports today—then the external cost generated by airplane A is the cost of the 60 min of delay to the other aircraft. (The 2 min correspond roughly to the "service time" of aircraft A, i.e., to the time during which A occupies the runway/final approach, excluding all other aircraft from it—see Chap. 10.) At a cost of $40 per minute of delay to an airborne aircraft—a cost typical of airports with a significant fraction of wide-body airplanes in the traffic mix—this comes to $2400. This is an amount much greater than either the internal delay cost experienced by most individual aircraft using the runway system or the weight-based landing fee that a small airplane would pay at most of the world's busy commercial airports.

When an aircraft pays only the traditional, weight-based landing fee to operate at an airport (no matter how congested that airport might

*For example, revenues from automobile parking now exceed revenues from landing fees at several major airports in the United States!

be), the only cost, in addition to the landing fee, that the aircraft's operator will perceive is the internal delay cost, the cost due to the delay that airplane suffers. Those airport users with the highest tolerance for internal delay costs, i.e., those with a low cost of delay time, will be the ones who will persist the longest in using an airport as congestion and delays grow. By contrast, high-value-of-time operations, such as airline flights with large numbers of passengers, tight connections, and short turnaround times on the ground, are the ones that will be the most sensitive to worsening congestion.

The fundamental principle that the theory of congestion pricing applies in such cases is that, to achieve an economically efficient use of the facility, one must impose a congestion toll on each user which is equal to the external cost associated with that user's access to the facility. This is what economists refer to as forcing users to "internalize external costs." The underlying rationale is clear. Those who can pay the congestion toll, i.e., can compensate "society" for the external costs they impose, must be deriving an economic value from the use of the facility that exceeds the external costs. In other words, their use of the facility increases total economic welfare. Conversely, a user who is not able to pay the congestion toll must be deriving a net economic benefit from the use of the facility that is less than the cost imposed on others. (Otherwise, the user would be willing to pay the toll.) Prospective users in this category are denied access to the facility through the device of the congestion toll, because such access reduces total economic welfare.

The congestion toll thus serves to optimize use of the facility: absent any constraints, optimal use is achieved through a toll equal to the external cost associated with an additional ("marginal") user. Such a congestion toll not only contributes to a socially desirable result, but also is necessary to reach such a result. In the case of airport runways, the congestion toll is the airport landing fee.

Figure 12-3 illustrates the situation. Curve D is the demand "curve" at the runway system of an airport facing capacity constraints. Curve I shows how the expected cost of delay suffered by each individual aircraft movement ("internal delay cost") increases as a function of demand.* Curve T shows the sum of the internal delay cost and the external delay cost generated by each additional aircraft movement at every level of demand. The difference between curve T and Curve

---

*Note that curve I for the cost of delay has the shape of the nonlinearly increasing expected delay functions seen in Chap. 11.

Delay cost per aircraft

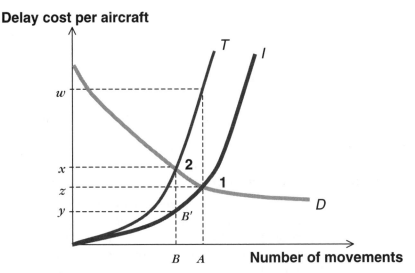

Fig. 12-3. *The effect of charging for external delay costs.*

I at each level of demand is equal to the external delay cost generated by an additional (or "marginal") aircraft movement at that level
of demand.

When there is no congestion toll, i.e., when aircraft do not pay for
any part of the delays they cause to other aircraft, the equilibrium
point is at 1, where the demand curve *D* intersects the internal delay
cost curve *I*. The equilibrium level of demand is equal to *A,* the projec-
tion of point 1 on the demand axis. When a congestion toll is imposed
with a value equal to the external delay cost caused by the marginal
runway system user, the new equilibrium is at 2, the point where the
demand function *D* intersects the total cost curve *T.* Thus, the new
equilibrium level of demand is equal to *B.* The demand has been
reduced by an amount equal to $(A - B)$ because of the congestion
toll. In turn, the congestion toll is equal to the external cost corresponding to a demand equal to *B,* i.e., to $(x - y)$ in Fig. 12-3. The
total amount collected from congestion tolls is $B(x - y)$, the area
corresponding to the rectangle *yx2B′* in Fig. 12-3.

Figure 12-3 and this discussion explain the statement in Sec. 12-2
regarding the economic inefficiency of the "do nothing" alternative. As
mentioned then, the use of delay as a "natural" mechanism for access
control results in an equilibrium (point 1 in Fig. 12-3) at which
(1) the delay level will be excessive and (2) the resulting mix of airport
users may include a large fraction who have a high tolerance for delays

because of a low value of time lost. Access to the airport by these low-value-of-time users cannot be justified on economic grounds, as they cannot compensate others for the external delay costs they cause.

A very important side benefit of congestion pricing is that it provides information to decision makers about the need for investing in additional capacity and the value that users attach to such capacity. Equilibrium ("market-clearing") congestion tolls help establish a market price for airport capacity (Morrison, 1983; Oum and Zhang, 1990).

Congestion pricing in practice

The application of the theory of congestion pricing to airports is far from simple. At the technical level, it is not easy to estimate accurately the marginal external costs for any given level of demand, although considerable progress has been made in this direction in recent years (Jansson, 1998). It is even more difficult to predict the exact effects of any proposed system of congestion tolls on demand, because existing information about the elasticity of airport demand with respect to the landing fee is limited (see, however, Cao and Kanafani, 2000). Consequently, it is also difficult to determine the size of the landing fees that will lead to a stable situation ("equilibrium"), i.e., that will not drive away too many or too few users (Odoni, 2001).

The principal practical problem, however, is more often a political one. The impact of congestion pricing is most severe on general aviation and on regional airlines. These two classes of users are the ones that can least afford to compensate others for external costs and oppose congestion pricing as being discriminatory against them. Such opposition, when politically strong, as is the case in the United States, can arrest or slow down attempts at implementation. Smaller and remote communities, which depend on regional airlines for access to major airports and to the national and international aviation systems, typically join in opposition.*

The major airlines often find themselves in an ambivalent position in this respect. In principle, these carriers stand to benefit the most from congestion pricing. When the traffic mix includes flight operations by general aviation, regional airlines, and the major carriers, the application of congestion pricing is likely to reduce significantly delays to

---

*On the opposite side, airport congestion pricing is generally supported politically by environmentalists, as well as by the neighbors of major airports, who see in this approach a means of postponing airport expansion through access control.

major carriers, by "driving away" from the peak traffic hours many general aviation and regional airline operations. As a result, major carrier operations face reduced costs, even after paying the higher landing fee. Yet many airlines, especially in the United States, have to date assumed a stance on congestion pricing that ranges from guarded to adversarial. The reasons are several. For one, many major carriers have alliances with regional and commuter airlines (or even own such airlines as subsidiaries) and are reluctant to support measures that are perceived as detrimental to them. Major carriers at busy airports also benefit from "feeder" traffic carried by smaller aircraft to/from smaller communities. Such flights may be affected or inconvenienced by congestion pricing. It is also probable that major carriers are, in general, uneasy about congestion pricing because they perceive it as a significant change to the *status quo*, with conceivably broader implications for the existing relationship between airports and airlines. The principal concern in this respect is the possibility that airports will abuse their inherent monopoly power or significant market power (Chap. 4).

As a consequence of such practical and political considerations, the congestion pricing mechanisms that have been proposed to date or have actually made their way to implementation are far less sophisticated than the theory suggests. They also generally impose or propose congestion tolls that are much lower than the true marginal external costs at congested airports suggest. Typically, these congestion pricing schemes involve one of the following approaches.

1. A *surcharge* is applied to the weight-based landing fee paid by aircraft operating during the airport's peak period(s). For example, all aircraft landing at an airport between 07:00 and 10:00 and between 16:00 and 19:00 local time might be required to pay a surcharge of $250, in addition to the weight-based landing fee to which they are subject.

2. A *flat fee,* entirely or partly independent of the aircraft's weight, is imposed on all aircraft operating during the peak period. For example, all aircraft, no matter what their weight, may be required to pay a landing fee of $500 if operating during peak hours. Or, aircraft under 50 tons may be required to pay $300 and aircraft over 50 tons $600.

3. A *multiplier* is applied to the weight-based landing fee charged to aircraft operating during the peak period. For example, if the multiplier is 1.25, an aircraft that would be subject to a $400 landing fee during off-peak hours would pay $500 if operating during the specified peak period.

4. A *minimum landing fee* is specified for aircraft operating during peak hours, to be applied only to aircraft that would otherwise have paid less than that amount. For example, if the minimum landing fee is specified as $150, an aircraft that would have paid $80 during off-peak will be charged $150 if operating during the peak period; however, an aircraft that would have paid $250 during off-peak will still be charged the same $250 for a peak-period operation.

Clearly each of the approaches outlined above has different impacts on different categories of airport users. For instance, a minimum landing fee affects only light aircraft that would have otherwise paid a fee smaller than the minimum landing fee specified. By contrast, a multiplier fee increases the fee paid in direct proportion to an aircraft's weight, thus "penalizing," in absolute terms, heavier aircraft more than light aircraft for operating during peak periods. One can, in fact, question whether an approach based on multipliers bears any relationship to the principle of charging in proportion to external costs.

A number of additional practical problems must be resolved in developing congestion pricing schemes. A particularly thorny one concerns the use to which an airport puts the funds collected through the congestion tolls. Presumably these funds should be used locally to support projects, such as construction of new runways, aimed at relieving congestion by increasing airport capacity. However, at many congested airports around the world, there is little that can be done to increase capacity, other perhaps than waiting for improvements in the traffic handling efficiency of the ATM system (see Chap. 13). Airports such as New York/LaGuardia, Washington/Reagan, Boston/Logan, London/Heathrow, Frankfurt/Main, etc., already fully utilize their "real estate" and are not able to expand substantially their airside facilities without acquiring adjacent parcels of land—a step both politically difficult and time-consuming. In such cases, the collection of congestion tolls may be viewed as just a punitive measure or, worse, as a way for the airport authority to increase its revenues by taking advantage of the presence of congestion. It has been suggested that one possible approach to countermanding this criticism is to reduce off-peak period charges in such a way that the total revenue the airport derives from landing fees remains the same as before the imposition of a congestion toll.

Example 12-2 describes a demand management system that illustrates many of the points discussed in this section, as well as additional practical issues that may come up in practice.

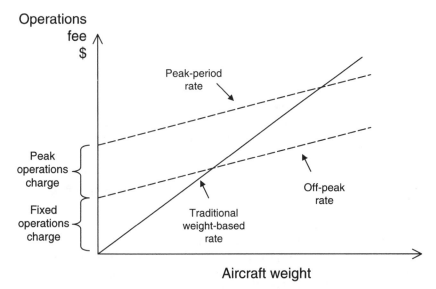

Fig. 12-4. *New landing fee structure versus traditional weight-based fee.*

EXAMPLE 12.2

Massport has examined over the years a broad range of congestion pricing options for Boston/Logan. Peak-period pricing was one of the principal alternatives explored in the Draft Environmental Impact Statement submitted by the agency in 1999, which proposes a number of improvements to the airport. Figure 12-4 summarizes the structure of one of the pricing schemes examined in detail at Massport (see also Barrett et al., 1994). Its main elements are outlined below, along with the underlying rationale.

The landing fee consists of a fixed charge per flight, a variable charge proportional to the weight of the aircraft, and a surcharge that applies only to aircraft movements (landings and take-offs) taking place during peak periods. The peak-period surcharge would be modest—of the order of $100–$200. This range was selected so that the surcharge would have some noticeable impact on the traffic, while not making access to Boston/Logan prohibitively expensive for smaller aircraft. The actual marginal external delay costs per aircraft movement at Boston/Logan are many times higher during peak traffic periods.

A peak period is defined as any period of three or more consecutive hours when the moving 3-h average of expected demand exceeds the range of 100–110 movements per hour throughout the period. This threshold was selected because it is close to the average of the maximum throughput capacity of Logan Airport (about 115) when this average is computed over the entire year.[*] There are two reasons for requiring peak periods to last for at least three consecutive hours. First, serious delays occur only if high demand levels persist for an extended period of time. When a brief period of high demand is followed by a period of low activity, the airport has a chance to recover before delays become disruptive. Second, if peak periods are too short, many users may be able to avoid paying the surcharge by making only small changes to their schedules to "shift out" of the peak period. For example, if a peak period happened to be only 1 h long, then about half of the affected aircraft operations could avoid the surcharge by shifting their scheduled times, forward or backward, by at most 15 min.

The landing fees are computed so that the pricing structure is "revenue neutral." This means that total revenue from runway fees will be the same under the new, congestion-pricing system as under the existing system that determines landing fees solely on the basis of aircraft weight. This avoids any appearance that the proposed system may be only a pretext for increasing airport revenues. The net result (see Fig. 12-4) is that smaller aircraft will end up paying a significantly higher landing fee than under the existing system, especially if they operate during peak traffic periods. However, some of the largest aircraft will generally be charged a *lower* landing fee than under the existing system, even during peak periods when the surcharge is in effect.

A potential variation would exempt from the peak-period surcharge two movements per day to or from the 12 smallest markets served by Boston/Logan. This provision is designed to cushion small local communities from the impact of the proposed system and partly defuse political pressure and criticism.

---

[*]The average of the maximum throughput capacity is computed from the capacity coverage curve (see Chap. 10) for Boston/Logan.

Example 12.3 illustrates the use of congestion pricing in passenger terminal buildings and on aprons.

EXAMPLE 12.3

The British Airports Authority has pioneered the use of marginal cost pricing and congestion pricing at airports (Little and McLeod, 1972). It has been using congestion-pricing schemes at the passenger terminals and the aprons of London/Heathrow since the mid-1970s. This practice was subsequently extended to London/Gatwick and to London/Stansted. In 1998, for instance, the schedule of passenger service charges (see Chap. 8) at the three airports was as shown in Table 12-2. Note that a Boeing 747 with 400 passengers on board would be charged the substantial *additional* amount of £1240 (approximately $2000 in 1998) in passenger service charges for a departure scheduled during a peak period at London/Heathrow.

On aprons, the standard charge in 1998 for parking an aircraft at London/Heathrow was based on the duration of the stand's occupancy and on the MTOW of the aircraft and was equal to £3.30 per quarter-hour or part thereof plus 5.4 p per ton. However, each minute's occupancy of a contact stand counted as 3 min for the period between 07:00 and 12:29 GMT. A similar arrangement was in effect at London/Gatwick, with triple charges for contact stands during the defined peak periods.

Table 12-2. Passenger terminal service charge payable per departing passenger at Heathrow, Gatwick, and Stansted airports, 1998 (a £1.50 rebate per passenger is provided in all cases for flights departing from remote stands)

	Heathrow		Gatwick		Stansted	
Flight type	Peak*	Off-peak	Peak[†]	Off-peak	Peak[†]	Off-peak
International[‡]	£8.50	£5.40	£6.95	£4.00	£6.75	£2.90
Domestic	£3.40	£3.40	£3.30	£3.30	£3.20	£3.20

*Aircraft departing between 09:00 and 15:29 GMT, April 1–October 31.
[†]Aircraft departing between 06:00 and 15:59 GMT, April 1–October 31.
[‡]Passengers from Gatwick and Stansted to Ireland pay a slightly different charge from the ones shown.

Finally, two practical observations are particularly relevant to congestion pricing as it applies to airports. The first is that congestion pricing will be most effective when applied to airports with non-homogeneous traffic. This is the case, for example, when, in addition to major carrier traffic, a significant fraction of the traffic mix consists of general aviation and/or regional airline flights; or when there is a reasonable mix of short-range flights by small aircraft and long-range flights by large ones. If such conditions do not exist, the ability of any landing fee to achieve price differentiation among users will be limited and a demand management system based solely on congestion pricing is likely to prove ineffective. Hybrid demand management systems, such as those to be discussed in the next section, would probably work better in such cases.

The second observation pertains to hub airports, where traffic may be dominated by a single airline. (An extreme example may be Memphis at nighttime, when essentially all the traffic consists of FedEx cargo aircraft.) Note that, in such cases, aircraft belonging to the same airline will absorb nearly all the external delay costs generated by any aircraft movement. Thus, the dominant airline internalizes external delay costs and will endeavor to operate exactly the number of flights that maximizes its total economic welfare. It follows that the application of congestion pricing is more appropriate at congested airports with a large number of competing carriers and with no dominant operator or operators. New York/LaGuardia (Example 12.1) is an excellent case in point.

12-5   Hybrid approaches to demand management

Hybrid demand management systems combine administrative and economic mechanisms. The starting point for all hybrid systems is the determination by some administrative authority of the number of slots to be made available at an airport. However, instead of (or, in addition to) schedule coordinators, hybrid systems rely on such economic devices as congestion pricing, slot markets, and slot auctions to arrive at the final allocation of slots among potential airport users. These three approaches will be reviewed in this section. Additional combinations of administrative and economic approaches are possible, as suggested at the end of the section.

Slots plus congestion pricing

The *slots plus congestion pricing* approach will quite possibly be widely employed in the future. Its application involves the following steps.

Step 1. "Declare" the airport capacity, i.e., specify the number of slots available in each time period.

Step 2. Develop and announce a schedule of landing fees (and possibly other airport charges) that vary by time of day and/or day of week and/or season.

Step 3. Invite requests for slots from prospective users.

Step 4. Use a schedule coordinator (or other administrative mechanism) to allocate slots, whenever the number of requests for a time period exceeds the number of available slots.

Note that the main difference between this and the purely administrative schedule coordination approach is that prospective airport users must now also consider the cost of access to the airport at different times when preparing their requests for slots. The higher cost of access during peak periods may dissuade some prospective users from requesting slots for these times. A few European airports have already adopted this approach, and several more are seriously considering it. Two examples are given below.

EXAMPLE 12.4

In 1998, Brussels Airport, which is fully coordinated, used the following formula for computing landing fees:

$$\text{Landing fee} = T \cdot P \cdot K \cdot W$$

where T = a unit rate specified in Belgian francs (Bf)
P = a peak-period multiplier
K = a noise-related multiplier
W = the MTOW of the aircraft in metric tons

For 1998, T was set to 139 Belgian francs (BEF) or, roughly, $4 at the time per metric ton; P was set equal to 1.5 for operations taking place during the periods 08:00–11:00, 17:00–20:00, and 01:00–05:00 local time and to 1.0 for all other times*; and K had

---

*Note that the use of $P = 1.5$ for the period 01:00–05:00 has little to do with peak-period pricing, as there is hardly any traffic during that time; instead, this is clearly intended to discourage use of the airport during nighttime for noise-mitigation reasons (and thus to enhance the effectiveness of the noise-level coefficient K).

10 possible values ranging from 2.0 to 0.85, depending on which one of five different "noise-level" categories an aircraft belonged to and whether or not that aircraft operated during nighttime, defined as 23:00–06:00 local time.* For example, an Airbus 310-300, with W equal to 150 tons, landing at 09:00 local time, would have $P = 1.5$ and $K = 0.9$ (the A310 is classified as belonging to noise-level Category 4, which has the second lowest noise impacts). It would thus pay a landing fee of about 28,200 BEF, or approximately $810, in 1998.

EXAMPLE 12.5

The landing fee schedule imposed by the British Airports Authority at its three airports in London in 2001 is shown in Table 12-3. All three airports are fully coordinated, so the schedule shown, coupled with the slot allocation procedure, creates a hybrid slots-plus-congestion-pricing system.

Note that during peak periods the landing fee is essentially independent of aircraft weight, whereas off-peak the fee is directly proportional to the weight of the aircraft. Comparison of Tables 12-2 and 12-3 also reveals that different peak periods are defined for aircraft traffic and for passenger traffic.

Table 12-3. Landing fees for domestic and international flights at Heathrow, Gatwick, and Stansted airports, 2001 (a 30% surcharge is applied to ICAO Annex 16 Chapter 2 jet aircraft)

Fee per landing	Heathrow		Gatwick		Stansted	
Aircraft weight	Peak*	Off-peak	Peak†	Off-peak	Peak‡	Off-peak
MTOW ≤ 16 tons	£418	£130	£310	£75	£80	£70
16 < MTOW ≤ 50	£465	£195	£345	£85	£120	£85
50 < MTOW	£465	£335	£345	£115	£195	£95
Special charge for MTOW > 250	—	—	—	—	£335	£335

Additional noise surcharges apply.
*07:00–09:59 and 17:00–18:59 GMT, April 1–October 31.
†06:00–11:59 and 17:00–18:59 GMT, April 1–October 31.
‡April 1–October 31.

*The use of values of K such as 0.9 and 0.85 for the two "best" noise-level categories essentially offers a discount or rebate to the least noisy aircraft.

Buying and selling slots

Another hybrid approach to demand management treats airport slots as a commodity that can be bought and sold. Once the number of slots available has been specified and the slots have somehow been allocated to prospective users, they become the property of their current holders, who may continue utilizing them for the operation of their own flights, lease them to another user, or sell them, just like any other asset.* An essential part of this hybrid system is a clear explication of the extent and duration of the rights inherent in a slot. The buy-and-sell approach, as well as the slot auctions described below, enjoys an advantage over congestion pricing in at least one respect. Whereas congestion pricing has difficulty in determining, a priori, equilibrium (or "market clearing") congestion fees, buy-and-sell and slot auctions permit these prices to be determined directly by the market.

By far the most important existing example of this approach can be found at the four High Density Rule (HDR) airports in the United States. As already noted in Sec. 12-3, the U.S. Department of Transportation abolished schedule coordination at these airports in 1985 and authorized the airlines that held them as of that date to continue utilizing their slots. Since that time, the HDR airports have operated under the following set of rules.†

1. Slots are subdivided into three categories: air carrier, commuter/regional, and other (general aviation, military, and charter). Air carrier and commuter slots can be further subdivided into domestic, Essential Air Service‡ (EAS), and international.

2. Slots are authorized by the Department of Transportation and are not technically the property of their current holders. However, domestic slots can be bought, sold, leased, and used as collateral. They may therefore be of considerable economic value to their holders. Major carrier domestic slots can be utilized by aircraft of any size, but regional domestic slots can be used only by aircraft with a maximum seating capacity of less than 75 and turbojet aircraft with seating capacity less than 56. (Note that this restricts the market for regional slots.)

3. Any slots that are not utilized for at least 80 percent of the time during any 2-month period are taken away from

---

*Under the High Density Rule in the United States, slots have even been used as collateral for securing loans.
†The rules have undergone several changes since 1985.
‡EAS refers to flights serving small remote communities in the United States; such flights are subsidized by the U.S. government.

their current holders and transferred to the pool of "available" domestic slots.

4. All domestic slots are assigned a *priority withdrawal number*. This number indicates the order in which slots can be claimed back by the U.S. Department of Transportation for designation as EAS slots or as international slots. The Department thus retains the right to withdraw, as necessary, some of the domestic slots. The priority withdrawal number indicates to a holder or to a potential buyer the likelihood that a slot may be claimed back. However, slots held by users who possess eight or fewer slots at a HDR airport are exempt from slot withdrawals at that airport.

5. New slots, i.e., those made available either through an increase in airport capacity or because of underutilization of an existing slot, are distributed periodically, with the first 15 percent of the slots reserved exclusively for prospective new entrants.

The HDR environment has given rise to an active market for slots at the four airports. Major carrier peak-time slots at these airports have been traded over the years at prices which are consistently well in excess of $1 million per slot. However, the entire HDR program has also been the subject of considerable controversy over the years. The most persistent issue is whether it shields the holders of slots at the four airports from competition. In 2000, the AIR-21 legislation (see Example 12-1) mandated the phasing out of HDR by 2007.

Of the many interesting questions raised by the buy-and-sell approach to demand management, one is particularly worth mentioning. It concerns the proper method for starting the process, i.e., for making the initial allocation of slots to the "first generation" of slot holders. For example, the method used by the U.S. Department of Transportation in the case of the HDR airports (assigning, by Executive Order, ownership to those who held the slots on December 16, 1985) has been criticized as amounting to a "windfall" worth several billion dollars to the beneficiary airlines. It can be argued that an alternative method, such as an auction (see below) that would involve some form of payment by slot recipients, would have been preferable. It is plausible that the airport operators involved should have received at least a part of these payments, in view of the capital investments they have made into the development of the four HDR airports. The federal government might also justifiably lay claim to a considerable part of the proceeds, as (1) it has funded part of the development of the airports through federal grants and other subsidies and (2) the slots would not have existed in the first place were it not for the air traffic management (ATM) infrastructure and services provided by the FAA. In general, the issue of who the "original" owners of airport slots are (before the slots become salable commodities) is bound to be contentious, as these slots become increasingly valuable to air carriers.

Slot auctions

Another hybrid approach to airport demand management would use auctions to allocate slots. A number of economists have advocated this approach over the years, and several airport operators have considered its adoption at airports such as Sydney, New York/LaGuardia, and a number of European sites. A rough description of the main steps involved is as follows.

Step 1. Provide a clear explication of the extent and duration of the rights inherent in a slot.

Step 2. The capacity of the subject airport, i.e., the number of slots available in each time period, is specified. Let the capacity per time period be equal to C slots.

Step 3. Airlines and other prospective airport users submit sealed bids for the slots they wish to obtain. A bidder can request more than one slot in any time period and can offer a different amount for different slots (e.g., "we wish to obtain 3 slots between 14:00 and 15:00 and we bid \$X for the first of these slots, \$Y for the second, and \$Z for the third").

Step 4. After all the bids are received, the slots in each time period are awarded to the C highest bidders. (If there are fewer than C bids for a particular time period, then all slot requests for that time period are accepted.)

Step 5. The price that a user actually pays for an awarded slot is set equal to the lowest successful bid in each time period, i.e., to the Cth highest amount offered for a slot in that period.* The rationale is that two successful bidders should not end up paying different amounts for slots that fall within the same time period.

No precedent exists to date for the allocation of airport slots through auctions, so there are many issues concerning steps 1–5 that have yet

*The obvious alternative is that the price actually paid for an awarded slot will be equal to the amount originally bid for that slot.

to be resolved. It is more important, however, to discuss briefly here the complexity of auctions in the airport context. This complexity stems from the strong interdependence of slots, both at the local level and across airports. Consider, for example, an airline that wishes to obtain an arrival slot for some particular flight at Airport A. Assume that the preferred time for the arrival of that flight is between 09:00 and 09:30 local time, but that a slot between 08:30 and 09:00 may be acceptable as well. The airline may then decide to bid a certain amount for a slot between 09:00 and 09:30 and a smaller amount for a slot between 08:30 and 09:00, for "insurance." Note now that, if the airline wins both slots, the 08:30–09:00 slot is essentially of no value to that airline. At the same time, if the turnaround time for the aircraft involved is about 60 min, the airline must make sure to obtain a departure slot between 10:00 and 10:30, so that the arriving aircraft can operate efficiently. Otherwise, the value of the 09:00–09:30 slot will be greatly diminished. Moreover, once the airline obtains a 09:00–09:30 arrival slot at Airport A, it must obtain a corresponding earlier departure slot at the airport from which the flight will originate.

Because of these strong interdependencies, the true value of the slots acquired will not be clear to an airline until all the slots are allocated. At that point, the airline will probably wish to dispose of some of the slots it has been awarded, revise the price it has offered to pay for others, and possibly acquire some additional slots. To make such postauction adjustments possible, a follow-up slot market is needed. This follow-up market is, in fact, an indispensable part of any demand management system based on auctions. Thus, a more viable hybrid system may be one in which the slots at an airport are auctioned off to the highest bidders by the airport operator and/or by a civil aviation organization and then become commodities that can be bought and sold. This is essentially a combination of slot auctions and the buy-and-sell system.

Finally, note that a major airline, which will bid for many slots at a particular airport, probably enjoys an inherent advantage over a smaller airline, because the major airline can choose among many possible combinations of slot usage once it obtains its slots. It is the realization of such subtleties and complications, in addition to uncertainties about legal authority, that has so far dissuaded airport officials from using slot auctions as a demand management tool.

12-6 Policy considerations

It is now possible to summarize the attributes of an ideal demand management system. Such a system would:

- Promote economically efficient use of scarce airport capacity by discouraging access by aircraft operators whose use of the airport generates more costs to others than the benefits they derive for themselves.

- Maintain access to a congested airport for all users willing to pay the full economic cost.

- Not be perceived as unfair or discriminatory by any class of prospective airport users.

- Not serve as another way to regulate air transport.

- Not provide opportunities for collusion among airlines and for anticompetitive practices.

- Not result in any "windfall" revenues to airport operators or airport users; any additional revenues accruing through demand management would be used to increase airport capacity and relieve congestion.

- Be transparent in its methodology and easy to administer and modify, if necessary.

None of the approaches described in this chapter fully satisfies all these criteria. However, given any specific set of circumstances, certain approaches are undoubtedly superior to others. For policy-setting purposes, a number of important points to bear in mind are summarized below.

The traditional weight-based landing fees do not take into consideration the costs associated with airport congestion. In fact, the weight-based fees, if anything, contribute to congestion, by lowering the cost of airport access as demand grows and by encouraging users with low direct operating costs and low value of time to use busy airports.

Schedule coordination, if relying solely on administrative procedures, can be effective at mildly congested airports and in regulatory environments where change is slow and gradual. However, in a dynamic, deregulated environment and at airports facing severe congestion, purely administrative procedures almost unavoidably will cause significant market distortion in the long run and will inhibit competition.

All purely economic approaches to demand management involve some form of congestion pricing, i.e., are based on the principle that, to optimize use of a congested facility, users should be forced to internalize, preferably fully, the external costs imposed by their use of the facility. Congestion pricing can be particularly effective at airports where traffic is nonhomogeneous and is not dominated by one or two carriers. Although in theory the principles of congestion pricing are well understood, it is difficult to apply them in practice, both for technical and for political reasons. As a result, the congestion-related landing fee schedules that have been implemented to date impose relatively low tolls on peak-period operations and are greatly simplified in structure.

Hybrid demand management systems combine elements of administrative and economic approaches. Their common characteristic is the use of administrative procedures to specify the number of slots available at an airport. Hybrid approaches rely on such economic devices as congestion pricing, slot markets, and slot auctions to arrive at the final allocation of slots. Several major airport authorities already use a combination of schedule coordination and congestion pricing to effect demand management. The establishment of slot markets ("buy-and-sell") is another interesting concept. However, it requires resolution of the difficult issue of who the original owner or "provider" of airport slots is. Auctioning of airport slots is an idea that is largely unexplored in practice and whose application entails many complications.

The implementation of any demand management system requires attention to the details of the proposed approach, a significant amount of effort and analysis, consultation with all parties involved, and resolution of numerous difficult issues. Examples of issues raised implicitly or explicitly in this chapter include the number of slots to be offered for distribution, the duration of peak and off-peak periods, the use of scheduled or actual time of operation for application of the charge, possible exemptions from the demand management system, rights and obligations of slot holders, and duration of slot ownership.

Interest in airport demand management is growing quickly throughout the world. In general, access to many of the busiest and most congested airports seems to be seriously underpriced at present. The use of demand management measures will probably expand in the future, with hybrid systems likely to be widely adopted. Movement in this direction will be gradual, however, as the strengthening of the economic components of hybrid systems will undoubtedly encounter serious opposition from several segments of the air transportation community.

Exercises

1. Example 12.1 indicates that the limited numbers of AIR-21 slots at New York/LaGuardia were allocated among eligible users through a lottery that was conducted in December 2000 and whose results took effect at the end of January 2001. This method was selected because of the pressure to deal quickly with delays that had reached crisis levels. It was acknowledged to be only a temporary solution to the problem. What is wrong with using a lottery to allocate scarce airport capacity? What are the advantages, if any, of using this approach?

2. Go to FAA (2001)—see reference list below—in the U.S. *Federal Register* to check the Notice concerning alternative demand management options for New York/LaGuardia. Specifically, you will find two alternatives based on congestion pricing and two based on slot auctions as proposed by the Port Authority of New York and New Jersey. Select any one of these four alternatives and describe clearly how it would work. What kinds of aircraft and flights would be most affected? What are the strengths and weaknesses of the option? [The site of the *Federal Register* is http://www.access.gpo.gov/su_docs/aces/aces140.html].

3. Consider an airport experiencing serious congestion during peak traffic hours. Under such peak conditions, the arrivals of airplanes at the vicinity of the airport can be assumed to be approximately Poisson with a rate $\lambda = 55$ aircraft per hour. Of these airplanes, 40 on average are commercial jets and 15 are small general aviation and commuter airplanes. The probability density function for the duration of the service time, t, to a random landing aircraft is uniformly distributed between 48 and 72 s.

   a. Peak traffic conditions occur during 1000 h per year, and the average cost of 1 min of airborne waiting time (i.e., of time spent in the air while waiting to land) is $40 for commercial jets. (This accounts for additional fuel burn, extra flight crew time, and other variable operating costs.) Estimate the yearly costs to the airlines of peak traffic conditions. Assume the model described by Eq. (23.10) (Chap. 23) for estimating waiting time is valid for this case.

b. In order to alleviate congestion under peak traffic conditions, the airport's managers are considering an increase in the landing fees at the airport. They have concluded that demand by commercial jets is completely insensitive to moderate increases in the landing fee (i.e., demand will continue at the level of 40 per hour). However, demand by smaller aircraft is expected to drop drastically as the landing fee increases. (There are several good small airports near the city in question that offer alternatives to the main airport.) A study of the small aircraft segment at that airport has shown that the relationship between demand by small aircraft and the increase in the landing fee is given by the relationship

$$Y = 15 - \frac{X}{16} \qquad \text{for } 0 \leq X \leq 240$$

where X is the amount added to the landing fee and Y is the number of small aircraft per hour demanding access to the airport. (Note that when $X = \$0$, $Y = 15$, and when $X = \$240$, $Y = 0$.) What is the most desirable amount of increase in the landing fee from the point of view of the airlines? (Remember that the airlines will also be paying the higher fees.) [*Note:* The variance of the service times in this problem is equal to $(72 - 48)^2/12 = 48 \text{ s}^2$.]

References

Barrett, C., Drazen, M., Hoffman, W., Lewis, S., Murphy, R., Odoni, A., and Pearson, L. (1994) "Peak Period Airport Pricing as It Might Apply to Boston-Logan International Airport," *Transportation Research Record, 1461,* pp. 15–23.

Cao, J-M., and Kanafani, A. (2000) "The Value of Runway Slots for Airlines," *European Journal of Operational Research*, 126, pp. 291-300.

Carlin, A., and Park, R. (1970) "Marginal Cost Pricing of Airport Runway Capacity," *American Economic Review, 60,* pp. 310–318.

Daniel, J. (1995) "Congestion Pricing and Capacity at Large Hub Airports: A Bottleneck Model with Stochastic Queues," *Econometrica, 63*(2), pp. 327–370.

de Neufville, R., and Mira, L. (1974) "Optimal Pricing Policies for Air Transport Networks," *Transportation Research, 8,* pp. 181–192.

FAA, Federal Aviation Administration (2000) "High Density Airports—Notice of Lottery of Slot Exemptions at LaGuardia Airport," *Federal Register, 65*(233), December, pp. 75765–75771.

FAA, Federal Aviation Administration (2001) "Notice of Alternative Policy Options for Managing Capacity at LaGuardia Airport and Proposed Extension of the Lottery Allocation," *Federal Register,* *66* (113), June 12, pp. 31731–31748.

Fan, T., and Odoni, A. (2001) "Demand Management as a Means of Relieving Airport Congestion," Working Paper, Department of Aeronautics and Astronautics, Massachusetts Institute of Technology, Cambridge, MA.

Hotelling, H. (1938) "The General Welfare in Relation to Problems of Taxation and of Railway and Utility Rates," *Econometrica, 6,* pp. 242–269.

Hotelling, H. (1939) "The Relation of Prices to Marginal Costs in an Optimum System," *Econometrica, 7,* pp. 151–155.

IATA, International Air Transport Association (2000), *Worldwide Scheduling Guidelines,* IATA, Montreal, Canada.

Jansson, M. (1998) "Marginal Cost Congestion Pricing under Approximate Equilibrium Conditions," S.M. thesis, Department of Electrical Engineering and Computer Science, Massachusetts Institute of Technology, Cambridge, MA.

Levine, M. (1969) "Landing Fees and the Airport Congestion Problem," *Journal of Law and Economics, 12,* pp. 79–108.

Little, I. M. D., and McLeod, K. (1972) "New Pricing Policy for British Airports," *Journal of Transport Economics and Policy, 6*(2), May, pp. 101–115.

Morrison, S. (1983) "Estimation of Long-Run Prices and Investment Levels for Airport Runways," *Research in Transportation Economics, 1,* pp. 103–130.

Morrison, S. (1987) "The Equity and Efficiency of Runway Pricing," *Journal of Public Economics, 34,* pp. 45–60.

Odoni, A. (2001) "Congestion Pricing for Airports and for En Route Airspace," in *New Concepts and Methods in Air Traffic Management,* Lucio Bianco, P. Dell'Olmo, and A. R. Odoni (eds.), pp. 31–44, Springer-Verlag, Berlin.

Oum, T., and Zhang, Y. (1990) "Airport Pricing: Congestion Tolls, Lumpy Investment and Cost Recovery," *Journal of Public Economics, 43,* December, pp. 353–374.

Pagliari, R. (2001) "Selling Grandfather: An Analysis of the Latest EU Proposals on Slot Trading," *Air and Space Europe, 3,*(1/2), pp. 33–35.

U.S. Department of Transportation (DOT) (1995) "Report to Congress: A Study of the High Density Rule," DOT, Washington, DC.

Vickrey, W. (1969) "Congestion Theory and Transport Investment," *American Economic Review, 59,* pp. 251–260.

13

Air traffic management

This chapter provides an introduction to air traffic management (ATM), with emphasis on terminal airspace and airport operations. ATM in developed countries has evolved into a complex, large-scale system that depends heavily on the smooth interaction of a highly skilled labor force with increasingly advanced technologies and on close cooperation with a diverse community of users. ATM planning, investments, and operations must constantly make trade-offs among objectives involving safety, efficiency, and cost.

ATM systems can be classified into four generations depending on their technological characteristics. Several developed countries are beginning to operate fourth-generation systems that take advantage of satellite-based technologies, collaborative decision making, and advanced automation and decision-support tools. Many less developed countries, however, still operate first-generation systems, little different, in some ways, from the earliest air traffic control systems.

Aspects of third- and fourth-generation systems that are relevant to terminal airspace and airport operations are described briefly. These include the classes of airspace, the operation of airport traffic control towers and of terminal airspace control centers, principal types of surveillance equipment, and instrument landing systems.

Air traffic flow management (ATFM) is playing an increasingly important role in airport operations, as a way of avoiding traffic overloads and excessive congestion and delay costs. ATFM systems in Europe and the United States evolved greatly in complexity and sophistication during the last part of the twentieth century. In particular, the adoption by the U.S. Federal Aviation Administration (FAA) of a collaborative decision-making (CDM) approach to ATFM is one of the most significant events in the history of ATM, in general. It marks a major change in the philosophy under which air traffic control has traditionally been operated. A description of some early steps in the application of CDM to ATFM is provided.

The chapter concludes with a brief survey of expected near-term and medium-term innovations in surveillance, navigation, and automation as they apply to terminal airspace and airports. It is difficult to forecast precisely the impact that these innovations will have on airport capacity, but an estimate of a 10–20 percent increase over a period of 10–15 years seems reasonable.

13-1   Introduction

Air traffic management is essential to the operation of airports. Any airport planner or manager must therefore have some familiarity with at least its most fundamental aspects. It is the objective of this chapter to provide the requisite basic background, with special emphasis on ATM in terminal area airspace. Selected references dedicated exclusively to air traffic management in general [see, e.g., Nolan (1999)], to terminal area operations (Mundra, 1989), to en-route operations (MIT Lincoln Laboratory, 1998), and to future plans (FAA, 1999, 2001) offer far more detailed treatment.

The ATM system provides a set of services aimed at ensuring the safety and efficiency of air traffic flows. Advanced ATM systems are becoming increasingly complex. They must:

- Accommodate growing numbers of users with different capabilities and requirements
- Achieve exceptional levels of safety under close scrutiny from the public and the mass media
- Mesh seamlessly a large labor pool of skilled human operators (the air traffic controllers and other technical staff) with a network of computers and other sophisticated communications, surveillance, and navigation equipment
- Take advantage of technological developments, while evolving gradually to allow users to keep pace with the rate of change
- Accomplish all this at reasonable cost to service providers and users

Viewed from a long-term perspective, ATM systems have been reasonably successful on all these scores. Undoubtedly, their greatest achievement is the extraordinary level of safety attained by commercial jet travel in developed countries throughout the world and the resultant benefits to nearly 2 billion passengers per year (Barnett and Wang, 2000; Barnett, 2001).

Any ATM system is comprised of the six components listed below (Braff et al., 1994). In fact, ATM systems can best be compared with one another—and their evolution over time can be traced—by making reference to the state of advancement of each of these critical components:

- The *procedures* and *regulations* according to which the ATM system operates and the *organization of airspace* around airports and en route
- The *human air traffic controllers,* who are responsible for providing ATM services
- The *automation systems* (e.g., computers, displays, and special-purpose software) that provide information to the controllers on the status, location, and separation of aircraft in the system and assist them ("decision support") in processing safely and expediting the flow of traffic
- The *communications systems* that enable air–ground, ground–ground, and air–air voice communications and data exchange and sharing
- The *surveillance systems* (e.g., radar) that provide real-time positional information to air traffic controllers—and, possibly, to the cockpit—for tracking aircraft and hazardous weather
- The *navigation systems* that provide real-time information to individual aircraft on their own position so they can navigate through airspace and on the airport surface

Section 13-2 presents a brief description of the types of ATM systems that exist around the world, ranging from first-generation systems still operating in many developing countries to advanced-technology, fourth-generation ones currently being implemented in a number of developed nations and regions. Section 13-3 covers aspects of third- and fourth-generation systems that are particularly relevant to ATM in terminal airspace and airports. The three types of control centers and the various air traffic control positions that monitor and serve a typical IFR (instrument flight rules) flight are described. In addition, the section discusses briefly the principal types of surveillance equipment in terminal areas, along with the instrument landing system (ILS), currently which provides the navigation required for precision approaches to runways. Air traffic flow management

(ATFM) is the subject of Secs. 13-4 and 13-5. ATFM systems in Europe and the United States have become essential to airport operations. In particular, Sec. 13-5 provides a description of ATFM as conducted under collaborative decision making (CDM) in the United States, using a simple example to explain how CDM motivates the timely exchange of information between the FAA and the airlines. Finally, Sec. 13-6 offers a brief survey of near-term and medium-term innovations in surveillance, navigation, and automation in terminal airspace and airports and identifies the types of impacts they may have.

13-2   Generations of ATM systems

Enormous differences exist among ATM systems around the world with respect to technology and level of sophistication. It is therefore useful to classify ATM systems conceptually into "generations," according to their principal characteristics.* It should be emphasized, however, that ATM systems evolve slowly and do not change overnight. Thus, the boundaries between successive generations are blurred, as a transition from one to the next may take 10 years or more. Four generations of ATM systems can be identified.

Many countries are still operating *first-generation* ATM systems, similar in many respects to the ones that existed in North America and Western Europe before and during World War II. First-generation systems are characterized by the definition of a system of airways and the absence of radar coverage. Air traffic controllers cannot observe airborne aircraft on radar screens, but are kept informed of the current positions and altitudes of aircraft along the airways through voice communications from pilots. Controllers keep track of and update this information manually, sometimes by moving plastic strips ("shrimp boats") representing each aircraft on a map that depicts the geographic area for which they are responsible (Gilbert, 1973). Air traffic in vast parts of the en-route airspace over Africa, Asia, and South America is still controlled through essentially first-generation systems.

The transition from the first to the *second generation* of ATM systems was marked by the introduction of radar after World War II. *Primary radar systems* were developed, consisting of medium-range *airport surveillance radar* (ASR) and longer-range *air route surveillance radar* (ARSR), for terminal airspace and en-route airspace, respec-

---

*This classification, although commonly used, is by no means standard.

tively. These provided surveillance by relying on the "skin effect," i.e., the reflection of the transmitted radar signal from the aircraft's metallic skin. Air traffic controllers were thus able to observe the horizontal position of aircraft, although the quality of the information was not particularly good. This generation evolved over a period of about 25 years, from the late 1940s to the early 1970s, in the United States and in Western Europe. A large number of terminal areas in less developed countries and parts of the airspace in some developed countries are still controlled by essentially second-generation ATM systems.

The adoption of digital technology as a means of acquiring, processing, and distributing information is the main distinguishing feature of third-generation systems,* which were introduced during the late 1960s and the 1970s in a number of countries in North America, Western Europe, East Asia, and the South Pacific. Important components of third-generation systems include *secondary surveillance radar* (or ATC Radar Beacon System, ATCRBS, in the United States), which interrogates aircraft every few seconds and receives back digitized messages that report each aircraft's identification and altitude, among other data; automatic aircraft tracking and alphanumeric displays using digitized radar data at both terminal airspace and en-route airspace control centers; extensive data processing by networks of computers; and automation of some routine ATM tasks, such as the automatic distribution and updating of information about a flight at all controller positions that will handle that flight. Another characteristic of third-generation systems is upgraded voice communication systems, primary radar systems (for tracking en-route, terminal airspace, and airport surface traffic) and airborne navigation systems.

Finally, some of the most developed countries have entered the era of *fourth-generation* ATM systems, characterized by progress in three general areas: development of automation tools, both on the ground and in the aircraft, that aid controllers and pilots in ways that go far beyond the routine processing and updating of data; use of advanced technologies such as satellite-based communications, navigation, and surveillance (CNS), Global Positioning System (GPS)-based precision instrument approaches, digital data links, and advanced weather systems; and partial decentralization of ATM decision making, primarily through real-time collaboration and exchange of information among ATM service providers, airlines, and flight crews.

---

*A distinction is sometimes made between third-generation and "upgraded third-generation systems," depending on specific automation and equipment features.

This chapter will concentrate on describing some of these developments in connection with third and fourth generation systems, primarily as they pertain to ATM for terminal airspace and airports. The reader, however, should not lose track of the fact that first- and second-generation systems are still very much in evidence in a great number of countries and geographic regions around the world. It is remarkable, for example, that until the late 1990s, those taking a 3.5-h flight from London to Athens also traversed the technological history of ATM, starting with some of the most advanced ATM systems in the world (in the United Kingdom and elsewhere in Northwestern Europe) and ending with a first-generation en-route system in Greece* and an early third-generation system in the terminal airspace of Athens.

13-3   Description of ATM system and processes in terminal airspace

Flights are conducted under either *visual flight rules* (VFR) or *instrument flight rules* (IFR). When flying under VFR, pilots must try to stay outside of clouds at all times and are responsible for maintaining safe separation from all other aircraft by visually scanning their surroundings (Braff et al., 1994). By contrast, air traffic controllers on the ground are responsible for maintaining safe separation between any two aircraft flying under IFR in controlled airspace (see below). While IFR flight was initially associated with *instrument meteorological conditions* (IMC), essentially all airline jet flights, as well as many other commercial and general aviation flights, now operate under IFR even in good weather (*visual meteorological conditions,* VMC). In the presence of increased traffic density, this ensures the availability of surveillance and of separation assistance by air traffic controllers during all phases of flight. Although, technically, IFR pilots are responsible for separation between their aircraft and VFR traffic when outside clouds, in practice they receive a lot of help from air traffic controllers in this respect. Flights operating under IFR must file a flight plan with the ATM service provider (the FAA in the United States) and must receive clearance of the flight plan from an ATM facility.

Airspace structure

To facilitate the air traffic control process, airspace is subdivided into three types: positive controlled airspace, controlled airspace, and

*In early 1999, Greece inaugurated use of a state-of-the-art en-route and Athens terminal area ATM system.

uncontrolled airspace. In *uncontrolled airspace,* the ATM system does not provide any aircraft separation services. Responsibility for maintaining safe separation rests with pilots and, consequently, uncontrolled airspace is normally populated solely by VFR flights. The volume of uncontrolled airspace is gradually diminishing internationally. In the United States, for example, the only uncontrolled airspace left is essentially Class G airspace* (FAA, 1992a), i.e., the airspace below 1200 ft above ground level and away from any busy airports.

Certain parts of the airspace, which are heavily populated by IFR flights, are designated as *positive controlled airspace.* Access to such airspace by VFR flights is either prohibited altogether or is limited by a number of restrictions. For example, in the United States, only IFR flights can operate at an altitude between 18,000 ft mean sea level and FL600.[†] This is called *Class A* airspace, and it is the part of the airspace that airline jet flights almost always utilize during their en-route phase (Fig. 13-1). The second class of positive controlled airspace, *Class B,* is of particular relevance to this text, as it comprises the terminal area airspace around the busiest airports. Access to Class B airspace is not limited to IFR flights, but it is restricted: to be admitted to this airspace, aircraft operating under VFR must be equipped with appropriate communications and navigation equipment, must obtain clearance from air traffic control to enter the airspace, and must be operated by pilots holding a private certificate or a student certificate with appropriate instructor endorsements (FAA, 1992b). Class B airspace (also referred to as *terminal control airspace,* TCA) is shaped like an inverse wedding cake and is usually centered on a major airport (Fig. 13-1). VFR aircraft which are accepted in Class B airspace must comply with instructions issued by air traffic controllers, who are responsible for ensuring standard separation *between every pair of aircraft* in that airspace, whether operating under visual or instrument flight rules. As a rule, Class B airspace is fully contained within the jurisdiction of a Terminal Radar Control (TRACON) facility (see below) in the United States.

In *controlled airspace,* air traffic controller responsibility for maintaining standard separation is limited to *pairs of IFR aircraft only,* as well as to runway operations, when applicable. IFR aircraft are still responsible for separating themselves from VFR aircraft, while VFR

---

*Airspace in the United States is classified into Classes A, B, C, D, E, and G. This is consistent with the ICAO classification, which, however, also includes Class F airspace, which does not exist in the United States.

[†]"FL" refers to flight level as measured according to constant atmospheric pressure related to a reference datum of 29.92 in.Hg. Flight levels are expressed in units of hundred of feet, so FL600 corresponds to 60,000 ft.

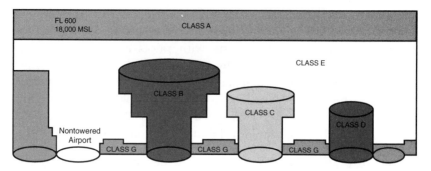

Fig. 13-1. *Classes of airspace, United States.* (*Source:* Modified from Nolan, 1999.)

aircraft are responsible for separating themselves from all other aircraft. Controlled airspace includes Classes C, D, and E. Airspace around medium-sized airports which are not close to major airports is usually in *Class C*. In Class C airspace, air traffic controllers will issue traffic advisories and conflict resolution for IFR/VFR aircraft pairs and traffic advisories for VFR/VFR pairs. *Class D* and *Class E* refer to airspace around smaller airports with air traffic control towers, and controlled airspace and control zones around airports without air traffic control towers, respectively.

Handling of a typical airline flight

In countries and regions where the ATM system is reasonably well developed, the successive phases of a typical airline IFR flight between two sizable airports are controlled by three types of facilities that play a dominant role in ATM. In generic terms, these are the *airport traffic control tower,* the *terminal airspace control center,* and the *en-route control center.* Different names and acronyms may be used in different countries for these facilities. In the United States, the major terminal airspace control centers are called *approach control facilities* or *Terminal Radar Approach Control* (TRACON) facilities—names that do not fully reflect their function, as these facilities handle departures in addition to arrivals. Similarly, en-route control centers are known as *Air Route Traffic Control Centers* (ARTCC). Figure 13-2 summarizes the role that each of these three types of facilities plays in controlling a typical flight. While the figure is drawn for the ATM system in the United States, an analogous allocation of responsibilities among ATM facilities exists elsewhere. The same is the case for the typical flight described briefly below.

Type of Facility	Terminal Area Facilities			En-Route Facilities
Controlling Facility	Airport Traffic Control Towers	Approach Control Facilities		Air Route Traffic Control Centers (ARTCCs)
Type of Control	Ground Traffic Control Takeoff and Landing Control	Approach and Departure Control		ATC during Transition and Cruise
Airspace	Airport Traffic Area Typically 5 nmi and 3000 ft AGL	Approach Control (Tracon Area) Typically extending up to 40 nmi + 10000 ft from the airport	En-Route Airspace	
			Transitional Phase Typically 50–150 nmi from airport	Cruise Phase Up to 60000 ft
Typical Flight Time	Typical Ground Time 5–10 min	Typical Flight Time 10–20 min	Typical Flight Time 10–20 min	Typical Flight Time 20 min to several hours
Flight Profile				

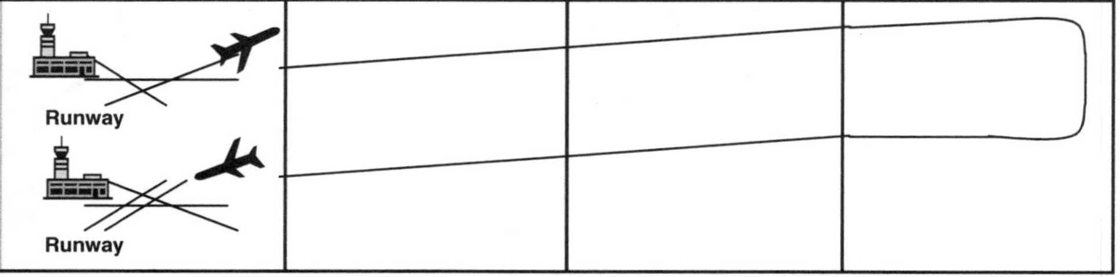

Fig. 13-2. *Role of the three principal types of ATM facilities in a typical flight.* (*Source:* Mundra, 1989.)

The operator (airline or other) of an IFR flight is required to file an IFR *flight plan* with the ATM service provider (the FAA, in this case). This is usually done one or more hours before the expected time of departure. Flight plans for regularly scheduled airline flights are typically stored in a computer and activated automatically some time before a flight. The flight plan contains a detailed description of the route to be flown and must receive clearance by the ATM service provider responsible for the airport of departure. In the United States, clearance is typically given by the ARTCC with jurisdiction over the departure airport, possibly after suggested modifications have been made to the flight plan. The operator of the flight is expected to adhere as precisely as possible to the flight plan and to notify the FAA of any significant changes before or during the flight. The approved flight plan is also entered into the FAA's computer system and may be updated several times prior to and during the actual flight. The flight plan is used to notify the various air traffic control jurisdictions and positions that will handle a particular aircraft of the imminent entry of that aircraft in the airspace or airport area of their responsibility. These notices give air traffic controllers advance information about near-future traffic loads and typically appear at each position 20–30 min before the aircraft's arrival in the form of either a printed message ("flight strip") or a message on an electronic display.

Figure 13-3 summarizes, in a general way, the flows of data and other information and commands among the principal "agents" involved in the execution of a flight plan: the pilot and the flight management

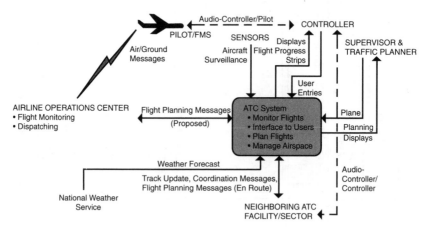

Fig. 13-3. *The flow of data among the major participants in the ATM system.* (***Source:*** MIT Lincoln Laboratory, 1998.)

system (FMS) of the aircraft; the air traffic controller, supervisors and air traffic flow planners (see below); the airline operations centers (AOC); neighboring ATM facilities and sectors of responsibility; and weather information providers. Voice communications are indicated with dashed lines.

The facility-specific aspects of the handling of flights by the airport traffic control tower, the terminal airspace control center, and the en-route control center will be described briefly next.

Airport traffic control tower

The airport traffic control tower provides a good vantage point for observing most of the airfield under good-visibility conditions. The following traffic control positions are located in the tower: clearance delivery; gate hold (only at some of the busiest airports); ground control; and local control. Depending on the size and complexity of the airport, these positions may be staffed by more than one controller. For example, two local controllers are active most of the time at airports that operate with two independent runways. In addition to these positions, each of which can make voice contact with pilots, a tower supervisor oversees operations, while one or more other controllers provide support services, primarily in the form of entering, updating, processing, or distributing flight data.

The first contact between an aircraft and the traffic control tower prior to a flight's departure takes place when the pilot requests pre-departure clearance for the flight from the *clearance delivery* position. If a *gate-hold* position exists in the tower, the pilot is then transferred to the frequency of that position. The pilot may be informed at that time of a "gate holding" (or "ground holding") delay due to traffic flow management restrictions or other reasons (Sec. 13-4). When the aircraft is finally ready to leave its apron stand, the clearance delivery position—or the gate-hold position, if one exists—is contacted again for permission to push back from its gate or move out of a noncontact stand, as the case may be. Once permission to leave the apron area is given, the aircraft is handed off to a *ground controller,* under whose instructions and supervision it proceeds through the taxiway system to its assigned departure runway.* Just before it reaches the departure runway (or the queue of aircraft awaiting takeoff), the aircraft is handed off to a *local controller,* who supervises the aircraft's takeoff. Soon after

---

*At some airports, a taxiing aircraft may come under the control of more than one ground controller on its way from/to the apron to/from the runway.

the aircraft is clear of the runway, the local controller hands it off to the appropriate *departure control* position in the TRACON (see below).

Conversely, during the arrival phase of a flight, responsibility for a landing aircraft is handed off by one of the *final control* positions in the TRACON to a *local controller* in the tower. This happens when the aircraft is on its final approach segment to the arrival runway. After the aircraft is safely off the runway, usually on an exit taxiway, it is handed off to a ground controller, who guides the aircraft to its apron area.

The airport tower is equipped with several visual displays, the most important of which are the radar display for the terminal airspace, weather-related displays, and the radar display for the airport surface (see below). The latter is particularly important when visibility from the tower is poor—for example, when it is surrounded by fog—and visual contact with the airfield is fully or partly lost. Electronic displays of conditions in apron areas (e.g., showing occupied stands) may also be important, especially at airports where physical obstructions may impede visual contact between the tower and such areas. Other controller aids may include an air traffic situation display tied to the air traffic flow management (ATFM) system (Sec. 13-4) showing traffic headed toward and away from the airport in question; flight-strip trays or other equipment showing imminent aircraft movements; and displays indicating equipment outages, pending runway closures, etc. Under various modernization programs, airport towers are increasingly being equipped with various decision-support ("automation") tools intended to assist controllers in increasing the efficiency of airport operations, as well as safety-related monitoring and alert systems for preventing the occurrence of hazardous events, such as runway and taxiway incursions by taxiing aircraft or ground vehicles.

Terminal airspace control center

The physical location of the terminal airspace control center varies. In many cases it is in the same building as the airport control tower. However, it can also be at a different location, especially in the case of multi-airport systems controlled by a single terminal airspace control center (New York, London, Paris, Milan, etc.). Terminal airspace control centers near major metropolitan regions are also responsible for coordinating traffic headed to/from that region's major airport(s) with traffic to/from secondary, mostly general aviation airports in these regions.

Figure 13-4 shows an idealized two-dimensional depiction of arrival patterns in a terminal airspace "feeding" two parallel arrival runways, 27R and 27L. It helps explain the way operations are carried out in major terminal areas in the United States and the allocation of responsibilities in a TRACON (Mundra, 1989). Despite local differences, this description is also typical of advanced ATM systems elsewhere. As can be seen from Fig. 13-4, aircraft enter the airspace that the TRACON controls over one of four navigational *arrival fixes* or *approach feeder fixes,* A through D, located at a distance of 30–40 mi (50–65 km) from the airport. At these fixes arriving aircraft are handed off, usually by an en-route control center, to one of the *arrival control* positions in the TRACON. The particular symmetric configuration of fixes

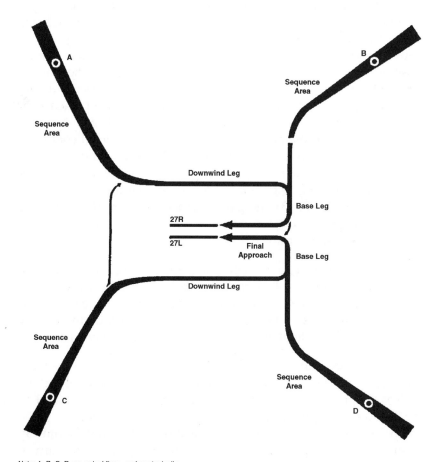

Note: A, B, C, D are arrival fixes, and are typically
    located 30–40 mi from the runway.

Fig. 13-4. *Idealized planar view of terminal area arrivals pattern.*
(***Source:*** Mundra, 1989.)

shown in Fig. 13-4 is called a *four-corner* or *four-post arrangement* and is encountered at locations where airspace is unrestricted by major physical obstacles or other constraints. Two arrival control positions will typically be active with an arrival configuration such as this. One arrival position ("Arrival North") will be responsible for aircraft entering the TRACON's airspace from the north (fixes A and B) and the other ("Arrival South") for those entering from the south (fixes C and D). It is the arrival position's responsibility to develop a desirable *sequence* of landing aircraft while funneling them through the *sequencing area* and onto either the *downwind leg* (e.g., for aircraft entering from C) or the *base leg* (e.g., for aircraft entering from D), where they are handed off to a *final control* position in the TRACON. For the configuration shown, there will typically be two final control positions ("Final North" and "Final South"), with each position being handed aircraft primarily by the corresponding arrival control positions ("Arrival North" and "Arrival South," respectively). The final control positions are responsible for directing aircraft through the *pattern area,* which consists of the *downwind leg* and the *base leg* and onto the final approach. At or near the beginning of the final approach, control over landing aircraft is transferred from the TRACON's final control position to the airport traffic control tower's local controller. The final control positions in the TRACON have a critical role in *spacing* aircraft for the final approach and in *merging* streams of aircraft at the base leg (e.g., aircraft entering from arrival fix A with those from arrival fix B). Merging may also occur at the beginning of the final approach, especially when only one exists, as in the case of airports with a single runway.

It is also possible that aircraft entering the terminal airspace from the south (fixes C and D) will be routed onto the north downwind leg or final approach (or vice versa) in order to better balance the traffic load on final approaches and runways.* This is indicated by the narrow arrows on the left side of Fig. 13-4 and at the end of the two base legs.

The geometric configuration shown in Fig. 13-4 is by no means a standard one. The actual configuration depends on several factors, such as the precise location of the primary airport or airports served by the TRACON, the presence of important physical obstacles, the traffic composition (high-altitude versus low-altitude traffic), and the runway configurations available. Three-corner arrangements of the arrival fixes are used at several locations. The number of controller positions

---

*This type of operation requires good coordination between the relevant control positions in the TRACON.

also varies. The Boston TRACON, for instance, has two arrival positions for traffic into Boston/Logan, but only a single final position.

Departures must also be handled by the TRACON following takeoff. The local controller in the tower hands off aircraft to one of the *departure control* positions in the TRACON. For a configuration such as the one shown in Fig. 13-4 there will usually be two such positions. Departure controllers oversee the ascent of an aircraft through the TRACON's airspace and eventually hand it off to another facility, usually an en-route center, for the transitional and cruise phases of the flight.

The volume of airspace associated with a terminal control facility, which in the United States may extend as much as 17,000 ft above ground level, is structured so that departing traffic is separated in altitude from arriving traffic. In every horizontal section of the airspace, certain altitude bands are reserved for arrivals and others for departures. In those cases where a facility serves more than one major airport, the overall three-dimensional structure of the airspace may be complex. For example, the structure of the New York airspace changes dynamically, with certain portions of it allocated to either LaGuardia or Kennedy, depending on the runway configuration in use at each airport.

In addition to the arrival, final, and departure control positions in the TRACON, which are dedicated to controlling traffic to and from the primary airport(s), a flexible number of other arrival and departure positions may be dedicated to the control of secondary airport arrivals and departures. Terminal airspace control centers in the United States and elsewhere are also staffed by a *supervisor,* a varying number of personnel performing *flight data processing* and other support functions, and possibly by *air traffic flow management specialists,* who provide coordination with the national (or international, in the case of European countries) ATFM system (Sec. 13-4).

In general, centers controlling the terminal airspace of hubs and of major international airports often perform the most complex tasks in the ATM system. These centers are critical to the efficient operation of the entire air transportation system. For this reason they are staffed by some of the most experienced and skilled air traffic controllers. Not surprisingly, these centers are also the focus of many programs aimed at introducing advanced ATM automation aids and decision-support tools.

Central to the operation of a terminal airspace control center are the information processing systems and associated displays that serve as the air traffic controllers' primary source of information for managing and controlling traffic. At the beginning of the twenty-first century, many countries in Europe, North America, and Asia and the Pacific Rim are in the process of installing much improved systems and displays in terminal airspace control centers, often as replacements of antiquated ones. For example, the FAA is replacing ARTS, the Automated Radar Terminal System,* first installed in the 1970s, with STARS, the Standard Terminal Automation Replacement System.[†] STARS receives and processes traffic and weather data from the primary and secondary traffic and weather radars and presents this information to air traffic controllers in high-quality, color displays. By displaying six distinct levels of weather "intensity" (identified by different colors), as defined by the National Weather Service, and by superimposing traffic and weather data, STARS assists controllers in directing air traffic around bad weather. It can track up to 1350 airborne aircraft simultaneously within a terminal area and can interface with up to 16 short- and long-range radars, 128 controller positions, and 20 remote airport towers in a 400-by-400 mi region. Equally important, it has been designed with an open architecture that facilitates integration with advanced decision-support tools, such as CTAS (Sec. 13-7). STARS also has a built-in backup in case of failure of the primary system.

Surveillance and navigation are two of the fundamental functions of ATM systems. Some of the most important types of navigation and surveillance equipment used in terminal airspace operations are briefly described next.

Surveillance

Surveillance is the function that provides the current location of the aircraft to air traffic controllers. This can be accomplished in three different ways: *pilot reporting,* via voice communications, of the aircraft's position and altitude; returns from *primary surveillance radar,* and automatic responses to *secondary surveillance radar.* Modern terminal airspace control centers—and, more generally, all advanced ATM systems—rely heavily on the second and third approaches,

---

*Of the several versions of the ARTS system, the ARTS III, used at the busiest terminal areas, is the most complex.
[†]The ARTS replacement program was delayed considerably by the need to make adjustments to STARS displays and other features in response to extensive controller comments and requests. This is a typical problem with all ATM systems in which human factors and ergonomic considerations play a critical role.

which are described briefly below. Interestingly, an automated form of the first approach, *automatic dependent surveillance* (ADS), is likely to become increasingly important as a means of surveillance in the near future (Sec. 13-6).

Primary surveillance radar relies on "skin effect" or "skin tracking" to obtain information about aircraft position. A rotating antenna on the ground emits pulses, which are reflected by the metallic exterior of aircraft and returned to the antenna. This process generates the information needed to determine the polar coordinates—distance and angle (or *azimuth*)—of each aircraft relative to the antenna. By measuring the time it takes for a round trip of the pulse, the distance of the aircraft from the antenna is computed, while the azimuth is derived from the corresponding angular position of the antenna. Note that no information is obtained about the altitude of the aircraft in this way. This must be determined either through pilot reporting or, now routinely, through the secondary surveillance radar (see below). This type of surveillance is also referred to as *independent,* because it requires no avionics equipment on the aircraft.

The primary radar used to track traffic in terminal airspace is known as *airport surveillance radar* (ASR). ASR technology has advanced greatly in recent years, providing good-quality target resolution and high levels of reliability. Its most modern versions are digital—the ASR-9, installed at the busiest airports, and the ASR-11. ASR is sometimes referred to as "short range," to distinguish it from long-range radar, which is used for en-route airspace (*air route surveillance radar,* ARSR). ASR typically performs 10–15 revolutions per minute, i.e., updates information about each aircraft's position every 4–6 s. It has a principal range of 30–60 nmi, which may extend as far as 120 nmi in some of the most recent versions.*

Another type of primary radar, known as *airport surface detection equipment* (ASDE) or just *surface radar,* is used to track aircraft and other vehicles on the airport's surface. ASDE plays an important role during periods of poor visibility, including nighttime. Despite considerable progress, ASDE still presents difficult technical problems, as it is highly sensitive, by design, and must cope with the many reflecting surfaces and physical obstacles that may provide false or distorted signal returns. ASDE-3 is the most recent version of surface radar. Due to its cost, it is installed only at the busiest airports.

---

*ARSR have a slower revolution rate (about 6 per minute, or 10-s updates) and a range of the order of 250 nmi.

However, ASDE may be instrumental in helping prevent runway and taxiway incursions during periods of low visibility, and its use may be expanded in the future (Barnett, 2000 et al), as long as it is not overtaken by alternative technologies (Sec. 13-7).

The secondary surveillance radar (SSR) is a rotating antenna on the ground, which emits interrogation messages at a frequency of 1030 MHz that trigger automatic responses from a transponder on the aircraft in the form of a digitized message on a different frequency* (1090 MHz). SSR-based surveillance is called *cooperative* because it requires that aircraft carry a transponder. The transponders are distinguished into "modes," depending on the format of their response messages. For practical purposes, two such modes are currently in use,[†] Mode C and Mode S. Mode C transponders automatically report aircraft identification (four-digit code) and altitude at 100-ft increments in 13-bit responses. They cause significant congestion of the 1090-MHz frequency because they generate multiple responses to a single interrogation. As well, Mode C responses from different aircraft on the same angle from the radar beacon often interfere with one another ("garbling"), resulting in loss of information. In the United States, a Mode C transponder is the minimum requirement for all aircraft flying above 10,000 ft or within 30 mi of a major airport.

The more recent Mode S transponders offer improved capability for aircraft-specific interrogation and provide for response messages with greatly increased information content. Each aircraft is assigned a 24-bit identification number, and the ground antenna can selectively interrogate any specific aircraft through that number. Moreover, the interrogation message can request a reply in any one of 256 message formats. Reply messages are 56 bits long. Far more information than just aircraft identification and altitude can be transmitted in these 256 formats. Mode S transponders respond only once to each interrogation. This reduces frequency congestion and the possibility of message garbling. Finally, in addition to responses to interrogation messages, Mode S transponders automatically broadcast about every second a short message (called a "squitter") with their identification. This message is broadcast throughout a flight (whether or not the aircraft is within range of a SSR) and provides part of the basis for the operation of *traffic alert and collision avoidance systems* (TCAS). Due to this essential role in TCAS, Mode S transponders are required

---

*In the United States, SSR systems are often referred to as *air traffic control radar beacon systems* (ATCRBS).

[†]The original Mode 3/A transponders are essentially obsolete in advanced ATM environments.

on all jet aircraft with more than 10 seats and on all commercial aircraft with more than 30 seats in the United States.

Navigation for precision instrument approaches

The ability to continue operating at near-normal levels in instrument meteorological conditions is essential to major airports. The *instrument landing system* (ILS) is by far the most widely used navigation aid for conducting precision approaches to an airport under low-visibility conditions. Practically every major airport in the world is equipped with at least one ILS. The ILS provides landing aircraft with a *single, straight-line path* that they can follow along their final approach to the runway. The straight-line path may extend as far as 25–30 km (~15–20 mi) from the near end of the arrival runway. This straight line is the intersection of two planes that are defined by two different types of transmitters (Fig. 13-5):

- *The localizer*, which defines a plane (sometimes referred to as the *runway centerline plane*) that extends vertically up from the runway centerline and the extension of that centerline

- *The glide slope*, which defines a "ramp," typically inclined at an angle of 2.5–3° to the horizontal plane*

The localizer operates at a VHF frequency in the range of 108.10–111.95 MHz and provides lateral guidance by indicating whether the aircraft is to the left or to the right of the extended runway centerline. It is a

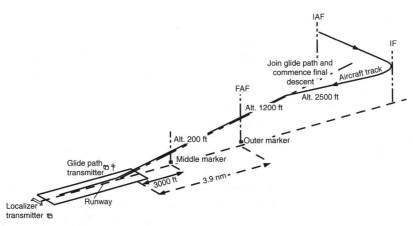

Fig. 13-5. *Schematic configuration of an Instrument Landing System.*
(***Source:*** Mundra, 1989.)

*The angle defined by the glide slope may range from a minimum of 2° to a maximum of 7°.

system consisting of a localizer transmitter, located roughly 1000 ft from the far end (or "departure end") of the runway and a localizer transmitter building, located about 300 ft to one side of the localizer antenna.

The glide slope operates at a UHF frequency of 329.3–335.0 MHz and provides vertical guidance by indicating whether the aircraft is above or below the correct line of descent to the runway. The *glide slope transmitter building and antenna* are located to one side of the runway centerline at 250–600 ft from the centerline and at a typical distance of 1000 ft from the near end (or "approach end") of the runway.

Two or three *marker beacons,* transmitters that emit a cone-shaped local signal, supplement the localizer and glide slope. When this signal is received by aircraft flying over the beacons, it "marks" the position of that aircraft along the final approach course. The two standard marker beacons on all ILS are the *outer marker,* at roughly 5 mi from the near end of the arrival runway on the extension of the runway centerline, which marks the *final approach fix* (FAF) to the runway, and the *middle marker* at about 3000 ft. A Category II ILS (see below) is also equipped with an *inner marker* located approximately 1000 ft from the end of the runway.

The sequence of events that take place in connection with an ILS approach is indicated in Fig. 13-6. The aircraft "captures" the ILS-defined final approach path, typically at a distance of 5–10 nmi from the runway threshold, and follows that path until it reaches the appropriate decision height (see below). A missed approach is executed if visual contact with the runway has not been made by that point.

Instrument landing systems are classified into three categories, depending on decision height and on visibility or runway visual range (RVR), as described in Table 13-1. Runway visual range is measured by RVR equipment along the runway being utilized. It is defined as the range over which the pilot of an aircraft on the centerline of a runway can see the runway surface markings or the lights delineating the runway or identifying its centerline (ICAO, 1999).

The standard ILS is a Category I system. Category II systems vary in relatively minor ways from Category I, e.g., requiring an inner marker and some additional approach lighting. However, Category III ILS are significantly different (requiring specially designed localizer and glide slope systems), can be deployed only at sites satisfying stringent landscaping requirements, and are far more expensive. Only specifically certified pilots and aircraft may perform Category II

A: ATC Events in Final Approach – Vertical View

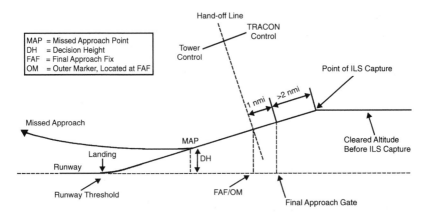

B: ATC Events in Final Approach – Horizontal View

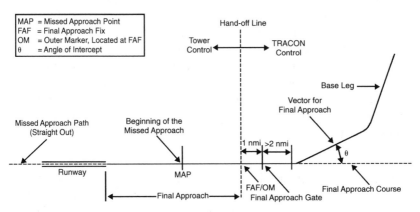

Fig. 13-6. *Side and planar view of a precision approach.*
(***Source:*** Mundra, 1989.)

and Category III approaches. It should also be emphasized that runways approved for Category I, II, and III approaches should satisfy various requirements with regard to obstacle clearances and parallel taxiways (see also Chap. 9), minimum paved runway length (4200 ft or 1280 m), runway markings, holding position signs and markings, runway threshold location, runway edge lights, and approach lighting systems. These requirements (e.g., for approach lighting systems) may be different for each of the three categories. Details and additional guidance can be found in FAA (1989) and FAA (2000).

Table 13-1. ILS categories (ICAO, 1999)

Instrument landing system	Decision height	Visibility or runway visual range (RVR)
Category I	60 m (200 ft)	Visibility: 800 m (0.5 mile) *or* RVR: 550 m (1800 ft)
Category II	30 m (100 ft)	RVR: 350 m (1200 ft)
Category III-A	0 m	RVR: 200 m (700 ft)
Category III-B	0 m	RVR: 50 m (150 ft)
Category III-C	0 m	RVR: 0 m

Instrument landing systems have a number of disadvantages. First, the quality of their signals can be affected seriously by distortions and reflections caused by both stationary and moving objects and vehicles. This is particularly true of the localizer signal, but the glide slope is also susceptible to similar problems. For example, the presence of several inches of snow or of high sea waves, in the case of coastal airports, may affect significantly the accuracy of the glide slope transmission. These systems therefore require careful calibration, potentially difficult adjustments to adapt to local topography and stringent monitoring of performance. Other precautions include the designation of a *localizer critical area* and of a *glide slope critical area,* where the presence of aircraft and vehicle traffic or of objects and obstacles (e.g., snow) when aircraft are performing ILS approaches must be either completely prohibited or strictly controlled.

A second family of problems is caused by the fact that the ILS provides only a single, straight-line path to the runway for instrument approaches. Aircraft may be forced to fly persistently over areas that are sensitive environmentally. In addition, having aircraft fly in a single file toward the runway will cause a loss of operating efficiency when aircraft with different approach speeds use the same runway. In such cases, the spacing between a "fast" arriving aircraft and a "slow" one that trails immediately behind will increase as the two aircraft fly down the final approach path,* thus reducing airport capacity (Chap. 10).

Finally, there is a scarcity of frequencies for installation of new ILS in major terminal areas where several of them are already in operation.

---

*By contrast, the longitudinal distance along the final approach path between a slow leading aircraft and a fast trailing one cannot be reduced below a specified separation minimum.

The resulting inability to develop additional ILS approaches to existing runways may restrict airport capacity.

Microwave landing systems (MLS) address many of the problems associated with ILS. As the name suggests, MLS operate in the UHF frequency band, thus being less susceptible to signal distortion and reflections than ILS. Rather than a single, straight-line final approach path, a MLS provides guidance (glide slope, azimuth, and continuous updates on distance to the runway) over an entire volume of airspace leading down to the arrival runway, as shown in Fig. 13-7. The volume extends from a minimum of 40° to a maximum of 60° to each side of the runway centerline (total of 80–120°), from glide angles between 0° and 15–30°, and up to a range of 20 nmi horizontally from the runway and 20,000 ft in altitude. The resulting availability of multiple final approach paths may make it possible to segregate traffic according to approach speed in the early segment of the final approach, thus providing airport capacity benefits. It is only at the very last segment, when all final approach paths essentially merge, that differences in approach speeds become a problem. Moreover, standardized final approach paths that minimize environmental

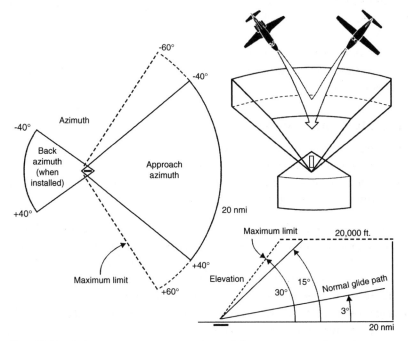

Fig. 13-7. *Schematic representation of the coverage of a microwave landing system.*

impacts can be established within the volume of airspace in which MLS guidance is available. Finally, more than 200 MLS channels are available, as opposed to 40 for ILS, thus eliminating the problem of frequency availability in major terminal areas.

In the 1980s, the International Civil Aviation Organization (ICAO), with agreement from the FAA, decided to adopt MLS as the eventual replacement for ILS worldwide. The FAA, in fact, made plans for the installation of more than 1200 MLS in United States airports by roughly 2005, when a phased decommissioning of ILS would begin. This was met with strong opposition from many aircraft operators, including the major airlines in the United States, due to the significant equipment and training costs associated with replacing ILS with MLS. The airlines felt that adoption of MLS could not be justified on economic grounds, given the wide availability of ILS in developed countries and the high level of operational reliability already achieved in LIFR ("low-IFR") conditions. When it became clear that GPS-based precision approaches (Sec. 13-5) might eventually replace ILS-based approaches (at the very least for Category I and possibly down to Category III conditions), the FAA withdrew its commitment to a near-term massive adoption of MLS in the United States. As of 2001, a number of European nations were still confirming their intention to provide concurrent ILS and MLS capabilities at their major airports and eventually to replace ILS with MLS. However, a massive replacement of ILS with MLS on a global scale seems unlikely.

En-route control center

En-route control centers handle IFR traffic outside terminal airspace and thus handle practically all airline en-route traffic. This description of en-route control centers here will be very brief. For a thorough description, see MIT Lincoln Laboratory (1998).

Each en-route center is given jurisdiction for a part of the airspace of a country or a region, the number of centers varying with the size of the area to be covered. En-route airspace over the continental United States is controlled by 20 such centers (ARTCC), as shown in Fig. 13-8, while the en-route airspace of a small country may be controlled by a single center. The airspace controlled by an en-route center is subdivided, in turn, into *sectors* that constitute the fundamental "unit" of airspace volume from the ATM point of view. Figure 13-9 shows

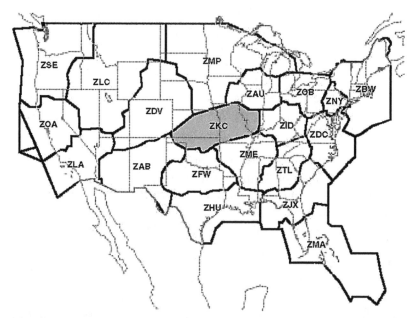

Fig. 13-8. *En-Route Air Traffic Control Centers, United States.*
(*Source:* MIT Lincoln Laboratory, 1998.)

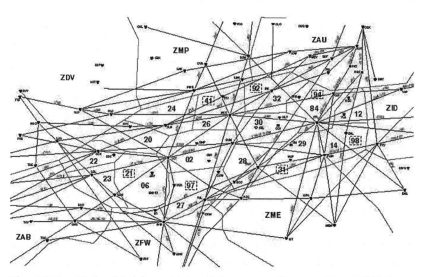

Fig. 13-9. *High-altitude sector and jet routes, Kansas City ARTCC.*
(*Source:* MIT Lincoln Laboratory, 1998.)

some of the sectors of the Kansas City ARTCC. In the United States, en-route sectors are further subdivided into super-low, low, high, and super-high, according to altitude, with varying floors and/or ceilings for each one of these types. For example Fig. 13-9 shows sectors whose floor is at FL240 but whose ceiling varies from FL310 to FL370. These are traversed by jet routes.The general locations of the super-high sectors 21, 31, 41, 92, 94, 97 and 98, with floors at FL350 or higher, are also indicated.

En-route navigation still depends primarily on systems of airways, essentially networks of "highways in the sky." A set of ground-based navigation aids, the *VHF omnidirectional range* (VOR) *finders,* generally define these airways. In the United States, the network of *Victor airways* extends up to (but not including) 18,000 ft above mean sea level, and the network of *jet routes* from 18,000 to 45,000 ft. The centerline of each airway coincides with a radial (i.e., a straight line at a given angle) projecting out of the antenna of a VOR station. Each VOR station has its own frequency. By tuning a navigational radio to the proper frequency, a pilot can fly along straight radial paths from one VOR station to another. To fly between two points A and B on the earth's surface, an aircraft may, for instance, fly from A to X, from X to Y, and from Y to B, where X and Y are the locations of the antennas of VOR stations. This accounts for the "dogleg" paths that most aircraft still fly in traveling between two points on the earth's surface.

Practically all VOR stations are also equipped with some type of *distance-measuring equipment* (DME), which an aircraft can use to determine the distance between itself and the VOR station. Stations that combine VOR and distance-measuring capabilities (known either as VOR/DME or as VORTAC facilities) thus provide the means of navigation by making it possible for each aircraft to determine its polar coordinates (distance and angle) relative to the (known) locations of these facilities.

Area navigation (RNAV) refers to the capability to navigate directly between two defined points without having to adhere to the system of airways. RNAV allows airspace users to specify, through a set of waypoints, an optimal origin-to-destination path—a *user-preferred route*—according to some performance criteria such as minimum time or fuel burn. RNAV can be performed by utilizing various navigation aids, including the network of VOR/DME or VORTAC facilities; the long-range (civilian) navigation system (LORAN-C); inertial navigation systems (INS); microwave landing systems (MLS) near to or in terminal airspace; and, most important, the Global Positioning System (GPS) or other global navigation satellite system (Sec. 13-7). While all commercial jet aircraft now have RNAV capabilities, the ATM system was unable, until recently, to accommodate requests for user-preferred routes routinely, due to the additional complexity that such routes imply for the prediction and resolution of conflicts. The National Route Program, underway in the United States since the early 1990s, authorizes user-preferred routes in phases, beginning at the highest flight levels and lowering progressively over time the altitude above which NRP flights are routinely authorized.

13-4   Air traffic flow management

The development of advanced *air traffic flow management* (ATFM) systems in the United States and in Europe during the 1980s and 1990s has already had an enormous impact on air traffic management (ATM) and airport operations. ATFM has become an essential instrument for keeping delays to airborne aircraft within manageable levels, for reducing the cost of delays to airlines and other airspace users, and for achieving better utilization of airport and ATM resources. ATFM undoubtedly produces important safety benefits as well, by controlling the flows of aircraft into crowded portions of airspace and by reducing the probability that aircraft will be subjected to excessive airborne delays. On the other hand, ATFM has also been criticized at times for deficiencies that occasionally contribute to slowing down air traffic operations and exacerbating, rather than reducing, delays. These flaws, however, are gradually being corrected as expertise, technological support, and decision-making processes improve. The adoption of a *collaborative decision-making* (CDM) approach to ATFM, possibly the single most important current development in the entire ATM system, has greatly accelerated progress in this respect. While ATFM and the CDM approach to it affect every part of the airspace, their effects on airport and terminal airspace operations are particularly critical.

Objectives and limitations of ATFM

The objectives of ATFM can be summarized as (1) preventing any overloading of airports and ATM facilities and services that might affect safety and (2) minimizing the economic and other penalties imposed on aircraft operators by air traffic congestion. This is accomplished by adjusting dynamically the flows of aircraft so that demand matches as well as possible available capacity at airports, in terminal airspace, and in en-route airspace.

The extensive use of ATFM is a relatively recent development. While essentially ad hoc ATFM systems operated both in the United States and in Europe during the 1970s, the event most responsible for stimulating interest in advanced ATFM was the 1981 strike of air traffic controllers in the United States. To reduce pressures on the ATM system, it was decided at the time that airborne delays should be avoided as much as possible. This was accomplished by holding aircraft on the ground prior to takeoff. An aircraft was not allowed to take off unless there was reasonable assurance that, after departure, it would be able to proceed to its destination with a minimum amount of delay in the air. This marked the first extensive use in ATM of the strategy of *ground holding* and provides a good illustration of the meaning of "matching demand to available capacity." After the ATM system returned to normalcy, use of ground holding was maintained as an option for dealing with the most serious instances of air traffic congestion. Equally important, the 1981 experience resulted in better appreciation of the great potential of ATFM.

It is useful to note that, while ATFM can ease the effects of congestion and overloading, the extent of its contributions has some inherent limitations. A helpful distinction in this respect is between ATFM interventions caused by "bottlenecks" in en-route sectors and those due to inadequate airport capacity (runway system and terminal airspace). In the former instance, the bottlenecks can often be bypassed at modest cost to aircraft operators, primarily through countermeasures of a local nature, such as rerouting of aircraft and restructuring the flows of traffic in the airspace. In such cases, ATFM essentially generates some capacity that would not have been available in ATFM's absence and can thus reduce the total delay that aircraft operators would otherwise experience collectively. However, when the bottleneck is at the flight's airport of destination, delay becomes unavoidable and ATFM cannot reduce the total delay to all aircraft operators. ATFM can, nonetheless, accomplish two things in such situations:

- Reduce the cost of unavoidable delay—for instance, by delaying aircraft on the ground and thus saving on fuel consumption
- Modify, if desirable, the way in which unavoidable delay is distributed among aircraft operators.

On the other hand, if not performed properly or if applied excessively, flow management may, in practice, *increase* total delay. For example, ATFM actions may sometimes lead to underutilization of the available airport capacity, as will be explained below.

ATFM operations

The ATFM systems of Europe and of the United States are operated in a hierarchical fashion, as might be expected of systems that attempt to coordinate traffic flows over vast geographic regions, with local units acting under the direction and control of a centralized unit. In the United States, the Air Traffic Control System Command Center (ATCSCC), an impressive FAA facility located near Washington, DC, has the role of national coordinator, while Traffic Management Units (TMU) operate at each of the regional en-route control centers (ARTCC) and at the major terminal airspace (TRACON) facilities. In addition to implementing directives of national scope issued by the ATCSCC, the regional and local TMU may also take actions of more limited scope on their own, in order to relieve problems of a local nature.

The ATFM system is more centralized in Europe, where local actions at all times must be coordinated with and, in principle, approved by the Central Flow Management Unit (CFMU), another impressive facility, which is operated by EUROCONTROL in Brussels.* Beginning in March 1996, the CFMU has been charged with providing flow management services for all country members of the European Civil Aviation Conference (ECAC). As of 2000, a total of 37 states participated, spanning practically the entire airspace of Europe, including Turkey's, with the notable exception of Russia. Flow Management Positions (FMP) operating at each of Europe's Area Control Centers (ACC) and major terminal airspace control facilities implement the directives ("regulations") of the CFMU.

The principal functions that ATFM performs are three:
- Prediction of the location of potential overloads
- Development of strategies for relieving these overloads
- Overseeing implementation of these strategies

Implementation often includes revising strategies in "real time."

---

*The CFMU replaced five regional flow management centers previously operating in Europe.

The prediction of overloads in Europe tends to be more "proactive" and more focused on en-route airspace compared to the United States. EUROCONTROL's CFMU has a strategic planning phase that begins with the submission of airline flight schedules for the following (winter or summer) season, i.e., about 6 months in advance. The CFMU reviews these schedules and associated probable flight routings and projects typical air traffic flows through en-route sectors and at airports. It then consults with the airlines, trying to relieve anticipated habitual overloading and to balance en-route sector workloads by suggesting alternative flight routings and even some modifications to flight schedules. No similar process takes place in the United States, where advance flow planning is limited to special events that may attract heavy volumes of air traffic (e.g., the Super Bowl football game) or may significantly affect air route capacity or airport capacity at some locations (e.g., the closing of a runway for repairs).

Once airline and other flight schedules are in place, ATFM must predict overloads on a daily and indeed hourly basis. In Europe, the CFMU routinely performs most such predictions 24–48 h in advance.* In the United States, this is done on a shorter-term basis, at the beginning of each day (typically around 6 a.m., Eastern time). In both cases, the initial predictions are constantly updated. In the United States, the principal concern is with weather conditions that may affect the capacity of *key airports*. These are the primary and most common cause of serious delays, although portions of en-route airspace may also become problematic in the presence of weather fronts. In contrast, for European ATFM, *en-route sector capacity*† is as much or more of a concern as airport capacity. This is partly due to the use of schedule coordination, which limits *a priori* the potential for demand overloading at most of the major European airports (Chap. 12).

The strategies that ATFM deploys to deal with overloads employ three principal types of interventions: *ground holding,* i.e., intentionally delaying an aircraft's takeoff for a specified amount of time; *rerouting,* i.e., changing or restructuring some flight routes to

---

*Naturally, these predictions may be revised later, if necessary.
†En-route sector capacity in Europe is determined primarily by the day-to-day availability of personnel to staff sector positions, as well as by the directionality and overall configuration of traffic flows. Personnel shortages or other events may at times require merging or partial reconfiguration of sectors.

modify the distribution of traffic flows; and *metering,* i.e., controlling the rate at which traffic crosses some specified spatial boundaries by adjusting the spacing between aircraft. Of these, the first is the most drastic, as it controls the number of aircraft moving through the ATM system, while metering is the most tactical in nature.

Whereas the ATFM strategies employed on the two sides of the Atlantic are similar, utilizing various combinations of these three types of interventions, philosophies, once again, differ on how active ATFM should be. The role of ATFM in the United States is viewed as mostly reactive: ATFM intervenes only as called for by weather conditions or other circumstances. In Europe, by contrast, all aircraft for which a flight plan is received by the CFMU must, in turn, be cleared by the CFMU, i.e., receive a "departure slot," before they can leave their parking stand. As is the case with overload predictions, the CFMU typically develops its overload-resolution strategies earlier in advance than the FAA (24–48 h ahead of the event, versus 4–6 h, respectively). In both instances, strategies are revised, if necessary, in response to developments in the field.

The acquisition, processing, and display of accurate and timely information is probably the single most important prerequisite for a successful ATFM system. A noteworthy technical achievement of the FAA's ATFM system has been the development of the Enhanced Traffic Management System (ETMS) and the associated Aircraft Situation Display (ASD). ETMS has amassed an enormous and constantly expanding information base, of both a historical and a current nature, comprising geographic, air traffic, weather, and traffic management data.* Most of this information can be readily displayed through a menu-driven interface at any location[†] equipped with the ASD.

Finally, it should be emphasized that ATFM operates in an extremely difficult decision-making environment, which can be described as *information intensive,* inherently *stochastic* (i.e., subject to uncertainty),

---

*Additional types of information are constantly being added to the ETMS database; in fact, one of the principal consequences and benefits of the application of CDM to ATFM (Sec. 13-5) has been the inclusion in the ETMS database of large amounts of information provided by the airlines and other CDM participants on a dynamic basis.

[†]Access to ETMS or to certain parts of it has been made available not only to airlines, but to many other non-FAA organizations, as well. Several of those use ETMS-derived information to support commercial activities. For example, some operators of limousine services at airports use ETMS and ASD to obtain accurate, real-time information on the time of arrival of the flights of their customers.

and highly *dynamic*. As indicated above, ATFM receives and processes large amounts of information every day and must decide what part of that information is relevant to support flow management and what specific data are needed at each level of decision making. Key parameters describing future operating conditions, such as the available runway capacity at an airport, are often subject to a high level of uncertainty, even on a time horizon of less than 1 h, due to their dependence on unpredictable or partly predictable variables, such as the incidence and intensity of fog or of thunderstorms. Moreover, operating conditions change constantly during the course of each day. To operate well in this challenging environment, ATFM must adopt strategies that take into consideration the level of uncertainty associated with key parameters and are flexible so they may be easily revised as new information becomes available. Even the best-trained human operators must be assisted in their tasks by well-designed, computer-based *decision-support systems* (DSS).

Ground delay programs

Ground delay programs (GDPs) illustrate how ATFM works, as well as the complexity of the problems that must be addressed. A GDP is initiated in the United States whenever a serious and possibly persistent overload is predicted for the next several hours at an airport. Each GDP is specific to an airport. Thus, several such programs may be run on any single day, often simultaneously.* Boston/Logan, Chicago/O'Hare, New York/LaGuardia, New York/Newark, and San Francisco/International were the airports with the highest incidence of GDPs. These are also airports where the difference between the VMC and IMC capacities of the runway system is large (cf. Table 10.4). The corrective action taken by ATFM during a GDP requires delaying on the ground, prior to takeoff, aircraft bound for the airport in question. As noted already, the rationale is that it is both less expensive and safer to absorb unavoidable delays on the ground rather than in the air. The duration of a GDP, i.e., the length of time during which restrictions on access to the airport are in force, may be as long as 12 h or more, generally as a result of persistent adverse weather conditions. However, a more typical duration is a few hours. Example 13.1 provides a brief and simplified description of how a GDP worked in the United States prior to the initiation of collaborative decision making in 1998.

---

*In 2000 about 2.5 GDPs were initiated, on average, per day in the United States.

Table 13-2. The original time schedule and the initial GDP

Airline (1)	ETA (2)	CTA (3)	Delay minutes (4)
A1	0700	0700	0
A2	0700	0710	10
B1	0705	0720	15
B2	0705	0730	25
B3	0710	0740	30
B4	0710	0750	40
A3	0720	0800	40
C1	0720	0810	50
B5	0740	0820	40
C2	0740	0830	50
A4	0820	0840	20
B6	0840	0850	10
Total A			70
Total B			160
Total C			100
Total			330

EXAMPLE 13.1

Consider airport XYZ, where 12 flights* have been scheduled to arrive between 0700 and 0900, local time, of a particular day. A forecast of heavy fog that will begin at 0700 and end at 0900 has been issued for XYZ. It is estimated that, as a result, the arrival capacity of XYZ will be reduced to 6 per hour for the duration of this event. The capacity estimate of 6 per hour is called the *airport acceptance rate* (AAR) and plays a crucial role in GDPs. The original schedule of arriving flights at XYZ, beginning with 0700, is indicated in the first two columns of Table 13-2. The first column identifies the airline (A, B, and C) and flight number, and

---

*The number of flights and airport capacity are both unrealistically small in this example, to facilitate the presentation. The example is otherwise typical of what may occur in practice.

the second indicates the "estimated time of arrival" (ETA) of the flight to the nearest 5 min, absent any GDP restrictions or other disturbances.

Given this situation, the FAA "will run a GDP" by assigning a new time of arrival, the *controlled time of arrival* (CTA), to each flight, as shown in column 3. The basic strategy is very simple. When airport capacity is limited, that capacity will be rationed among airlines in accordance with the original schedule of flights. Since in this example the capacity is down to 6 per hour, the arrival of flights will be set for the "slots" of 0700, 0710, 0720, etc., that is, they will be evenly spaced during the hour according to the rate indicated by the AAR. Column 4 indicates the resulting delay to each flight, the difference between the flight's CTA and ETA. The bottom four rows of Table 13-2 show the total delay suffered by each airline and by all the flights.

The FAA will next delay the departure for XYZ of each of the flights A1–B6 in Table 13-2 by the amount of time shown in column 4. A *controlled time of departure* (CTD)—also known as *expected departure clearance time* (EDCT)—is assigned to each flight at its airport of origin. If Flight C1 were originally scheduled to leave BOS at 0540 to make the 0720 ETA at XYZ, it will now be assigned a CDT of 0630, for a ground delay ("ground hold" or "gate hold") of 50 min.

Note that one of the objectives of the GDP is to follow a procedure and develop an arrival sequence that is considered fair to all users. This is accomplished in two ways: by allocating slots according to the original schedule of flights, as described (this is called *rationing by schedule,* RBS), and by scheduling the CTA's in the same order as the ETA's, i.e., by implementing a "first scheduled, first served" policy.

The revised schedule of arrivals, as indicated by column 3 in Table 13-2, usually turns out to be only a preliminary indication of what will eventually happen. In practice, some flights may be cancelled, either for reasons independent of the GDP or because the delay assigned to them is so long that it makes little sense to perform the flight. It is also possible that a flight may be unable to meet the CTA assigned to it due to delays caused by mechanical problems or other reasons. The CTAs could be revised dynamically, if information about such flight cancellations or delays was made available to the ATFM system in a timely manner.

Consider, for example, the case in which Airline A decides to cancel flight A2 for some reason. If the airline makes this known to ATFM soon enough, Table 13-2 would be revised as shown in Table 13-3. The CTA of every flight from B1 through B6 has been changed to 10 min earlier, taking advantage of the gap created by the cancellation of A2.

On second thought, however, the GDP shown in Table 13-3 might never come to pass. By informing ATFM of the cancellation of A2, Airline A has reduced its own total delay by 30 min, but the delay of its competitors by a total of 80, compared to Table 13-2. Moreover, A has disclosed to B and C the fact that A2 has been cancelled, information that might be valuable to them in

Table 13-3. The GDP obtained if Airline A cancels Flight A2 and so informs ATFM in a timely fashion

Airline (1)	ETA (2)	CTA (3)	Delay minutes (4)
A1	0700	0700	0
B1	0705	0710	5
B2	0705	0720	15
B3	0710	0730	20
B4	0710	0740	30
A3	0720	0750	30
C1	0720	0800	40
B5	0740	0810	30
C2	0740	0820	40
A4	0820	0830	10
B6	0840	0840	0
Total A			40
Total B			100
Total C			80
Total			220

other ways, in addition to saving delay minutes.* It is therefore entirely possible that Airline A will simply not inform the ATFM system of the cancellation of A2 until it is too late to take advantage of the gap in the schedule. Thus, the final GDP schedule might end up being the one shown in Table 13-4, instead of Table 13-3. The 10-min slot between 0710 and 0720 has now been wasted. This suggests that airlines may need incentives in order to share information with the FAA and their competitors during a GDP. They need to know that sharing will work to their benefit or, at least, will not put them at a disadvantage.

Another important practical consideration in planning a GDP is that airlines may desire to change the order in which CTAs are assigned to their own flights. Consider, for instance, Flights B1, B2, B3, and B4 in Table 13-2, all scheduled originally to arrive within the 5-min interval from 0705 to 0710. Because of the respective ETA, ATFM assigned to Flight B1 a delay of 15 min and to Flight B4 a delay of 40 min. To Airline B, however, it may be far more important to have Flight B4 arrive at XYZ as close as possible to schedule than Flight B1. For example, Flight B4 may be bringing to XYZ many passengers who will be connecting to other flights of Airline B or pilot crews who will operate subsequent flights from XYZ. Note that it is perfectly feasible to exchange the assigned CTAs of Flights B1 and B4 in Table 13-2, so that B4 will suffer a delay of only 10 min and B1 of 45. This may be much more palatable to Airline B, under the circumstances described, and results in the same total number of delay minutes as before for Airline B and for all the other airlines.

13-5   Collaborative decision making

Example 13.1 points to the need for a number of ways to improve the GDP planning process. At the most obvious, a very fast, two-way communications environment is required so that the ATFM system and aircraft operators can exchange information efficiently. Second, and more subtly, there is a need for partial decentralization of decision making. The FAA (or, more generally, any provider of ATFM and

*Information about the cancellation of A2 is obviously valuable to A's competitors on A2's market, as they may be able to attract some passengers originally booked on A2. If these competitors were also considering canceling their own flights on that market that day, knowledge of the cancellation of A2 might persuade them not to do so.

Table 13-4. Revision of the GDP of Table 13-2, if Airline A fails to inform the ATFM system of the cancellation of Flight A2 in a timely fashion

Airline (1)	ETA (2)	CTA (3)	Delay minutes (4)
A1	0700	0700	0
Void	—	0710	—
B1	0705	0720	15
B2	0705	0730	25
B3	0710	0740	30
B4	0710	0750	40
A3	0720	0800	40
C1	0720	0810	50
B5	0740	0820	40
C2	0740	0830	50
A4	0820	0840	20
B6	0840	0850	10
Total A			60
Total B			160
Total C			100
Total			320

ATM services) does not have the information needed to make certain decisions on behalf of the users. The example of exchanging the CTAs of Flights B1 and B4 is a case in point. Airline B, not the FAA, is the only one qualified to decide whether this feasible exchange is worth making. Third, if decision making is to be partially decentralized in this way, it is essential that all participants in the process operate from a *shared knowledge base,* i.e., that all have a "common picture" of the current situation at all times, so they can take into consideration everyone else's actions. Fourth, the process of revising the schedule must be concluded quickly, despite some of the complicated decisions and extensive information exchanges that must take place. This calls for the availability of (preferably common) *computer-based decision-support tools* to facilitate the work of participants. Finally, there are many instances in GDPs where, for competitive reasons, an airline may prefer to withhold certain information from the FAA and other airlines for as long as possible. The GDP planning process should therefore offer aircraft operators incentives for sharing information.

The *Collaborative Decision Making* (CDM) approach aims at addressing all these requirements and marks a fundamental change in ATFM's operating philosophy. The basic premise is that "shared information and collaboration in planning and executing ATFM initiatives benefits all ATM users as well as the ATM service provider" (Metron, 2000). CDM's specific stated goals (Metron, 2000) are to:

- Provide the FAA and the airlines with a common picture of current and predicted air traffic conditions by having them look at the same data

- Allow each decision to be made by the person or organization in the best position to make it

- Make these decisions in an open manner so that all know what is happening and can contribute as necessary or desired

"Prototype" GDPs using CDM began on an experimental basis at two airports, San Francisco/International and New York/Newark, in January 1998. The experiments were judged so successful that, beginning in September 1998, all GDPs at all airports in the United States are conducted via CDM. A unified environment for conducting collaborative ATFM with the active participation of every airline of significant size in the United States is now in place. Its scope goes well beyond GDPs. The communications infrastructure for this collaborative system is an Internet-like network called the *CDMnet,* which gives CDM participants the capability for two-way exchange of real-time information. Airline Operations Centers (AOC) make available via CDMnet information regarding changes to flight arrival and departure schedules, assignments of flights to airport arrival slots and delays, cancellations, and newly created flights. This information is used by the ATCSCC to revise ongoing GDPs and to determine whether any capacity/demand imbalances exist that warrant additional ATFM intervention. The demand information is consolidated approximately every 5 min by the U.S. Department of Transportation's Volpe National Transportation Systems Center and returned to aircraft operators in an "aggregate demand list" (ADL). The ADL allows airlines to see where their flights fit in the traffic flow and to plan accordingly.

The *Flight Schedule Monitor* (FSM) is the software that provides the shared knowledge base of current and predicted conditions for all

CDM participants and makes possible user collaboration in GDP decision making. The FAA and the AOCs use it to implement and manage all GDPs. FSM provides a graphical presentation of airport demand and capacity information, as received through CDMnet, displaying flight-specific information, airport arrival and departure rates, open arrival slots, and other traffic flow information. FSM also provides many of the requisite decision-support tools. It contains a set of computer algorithms and utility programs to support GDP management and analysis so users can react quickly to airport and airspace capacity constraints. FSM users can also test alternative ATFM scenarios involving flight cancellations, delays, or substitutions, and observe the results before taking any action on their flights.

EXAMPLE 13.1 (Continued)

Table 13-2 can be used to illustrate some features of GDP in the CDM environment. It will be assumed again that Airline A has decided to cancel Flight A2 and that Airline B has internally assigned priorities among Flights B1, B2, B3, and B4 in the order of B4, B2, B3, B1, top to bottom.

Under CDM, the proposed CTA schedule shown in Table 13-2 is sent to the airlines via the CDMnet in the form of a "GDP Advisory." Airlines and other ATM users have until the cutoff time to reschedule, substitute, cancel, or delay flights and send back to ATFM their proposed revisions. It is possible that demand will be reduced sufficiently in this way to eliminate the need for a GDP or to make it possible to delay the start of the GDP. This happens quite often and is one of the benefits that CDM offers.

With regard to substitutions and cancellations, the airlines and the FAA have adopted, in connection to CDM, two GDP operating rules that are particularly relevant to this example*:

- Each airline may freely substitute flights within the set of its own flight slots and may move any flight to a slot which is not earlier than that flight's ETA.

- An airline that cancels a flight has the right to advance its later flights to the first feasible slot which becomes available as a result of the cancellation.

---

*The two rules are described here in a somewhat simplified form, which, however, captures their essence.

Under the first of the rules, Airline B is free to use its six slots (at 0720, 0730, 0740, 0750, 0820, and 0850) to assign its Flights B1–B6 in any way it wishes, as long as any flight is not assigned to a slot earlier than that flight's ETA. This means, that Airline B can now assign Flight B4 to its first slot at 0720 (as B4's ETA is at 0710) according to its preferences. Similarly, Airline B will assign flights B2, B3, and B1 to the 0730, 0740, and 0750 slots, respectively, reflecting its preferences.

Consider now the second of the operating rules above. If Airline A cancels Flight A2 and if B rearranges the CTA of B1–B4, the situation shown in Table 13-5 is obtained. Under the second rule Airline A has priority for utilizing the 0710 slot vacated by

Table 13-5. Revision of the GDP of Table 13-2, if Airline A cancels Flight A2 and B rearranges the order of B1, B2, B3, and B4

Airline (1)	ETA (2)	CTA (3)	Delay minutes (4)
A1	0700	0700	0
Void	—	0710	—
B4	0710	0720	10
B2	0705	0730	25
B3	0710	0740	30
B1	0705	0750	45
A3	0720	0800	40
C1	0720	0810	50
B5	0740	0820	40
C2	0740	0830	50
A4	0820	0840	20
B6	0840	0850	10
Total A			60
Total B			160
Total C			100
Total			320

the cancellation of A2. However, the ETA of A3, the first flight of Airline A after 0710, is 0720. This means that A3 cannot be moved to the 0710 slot. Under the CDM procedures, Flight B4, the next eligible flight, will then be moved to the 0710 slot. (Note that B4's ETA is 0710.) This vacates the 0720 slot, which is a feasible one as far as A3 is concerned, because A3's ETA is also 0720. Thus, A3 will now "leapfrog" over B2, B3, and B1 to occupy the 0720 slot. Essentially, because Airline A gave back to the pool a slot through the cancellation of A2, it is rewarded with use of the first slot that becomes available at or after A3's ETA. These rearrangements will lead to the situation shown in Table 13-6. Note that the empty slot has now moved to 0800, the spot vacated by Flight A3.

Table 13-6. Revision of the GDP of Table 13-2, after Airline A cancels Flight A2 and some flights are moved to fill the vacant slot

Airline (1)	ETA (2)	CTA (3)	Delay minutes (4)
A1	0700	0700	0
B4	0710	0710	0
A3	0720	0720	0
B2	0705	0730	25
B3	0710	0740	30
B1	0705	0750	45
Void	—	0800	—
C1	0720	0810	50
B5	0740	0820	40
C2	0740	0830	50
A4	0820	0840	20
B6	0840	0850	10
Total A			20
Total B			150
Total C			100
Total			270

Following a similar line of reasoning, the final GDP shown in Table 13-7 is obtained. C1 will now be moved up to occupy the 0800 slot, B5 will occupy the 0810 slot, and A4 will leapfrog over C2 to occupy the 0820 slot. This is the earliest slot that A4 can occupy without violating its ETA. Airline A (and Flight A4) became eligible for preferential treatment when A3 vacated the 0800 slot and moved up to occupy the one at 0720. Compare Table 13-7 with Table 13-3, obtained under the assumption that Airline A informs ATFM of the cancellation of Flight A2 in a timely fashion. Both these GDPs result in exactly the same total number of delay minutes, 220 (saving 110 min from the original GDP of Table 13-1), but with a very different distribution of the resulting benefits. The main beneficiary, by far, in Table 13-7, is Airline A, whose three remaining flights now suffer no delay. It can be seen that Airline A now is motivated to report early the cancellation of A2. Ball et al. (1998) have reported that before CDM, airlines informed the ATCSCC of the cancellation of flights, on average, about 50 min *after* the scheduled time of departure of the cancelled flights! Under CDM, this time became 45 min *before* the scheduled departure time, a difference of more than 1.5 h.

Note also that Airline B receives benefits, which cannot be quantified in terms of delay savings, by having its Flights B4 and B1 exchange positions, per its preference. This points out the fact that some of the (real) economic benefits of CDM can be quantified only by the airlines themselves.

The two CDM operating rules just described in connection with Example 13.1 are referred to as a *substitution rule* and a *compression rule,* respectively. The latter refers to the process used to fill in gaps in the GDP created by the cancellation of flights. The steps involved in a GDP under CDM can now be summarized as follows.

Step 1. The ATCSCC obtains on a daily basis an estimate of the airport acceptance rate (AAR) for each airport where capacity may be reduced due to unfavorable weather conditions or other reasons.

Step 2. If delays are projected to be severe at an airport, the ATCSCC prepares to run a GDP by assigning slots to airlines on a first-scheduled, first-served basis ("ration by schedule," RBS) using the predicted AAR.

Table 13-7. The final GDP

Airline (1)	ETA (2)	CTA (3)	Delay minutes (4)
A1	0700	0700	0
B4	0710	0710	0
A3	0720	0720	0
B2	0705	0730	25
B3	0710	0740	30
B1	0705	0750	45
C1	0720	0800	40
B5	0740	0810	30
A4	0820	0820	0
C2	0740	0830	50
B6	0840	0840	0
Total A			0
Total B			130
Total C			90
Total			220

Step 3. A GDP Advisory is sent to the airlines and other users via CDMnet that includes the planned controlled time of arrival (CTA) for all arrivals that will be affected by the planned GDP.

Step 4. Each airline informs the ATCSCC by a cutoff time on how it plans to use its slots, including any substitutions and flight cancellations.

Step 5. After receiving user responses, the ATCSCC performs compression to take advantage of any empty slots and finalize slot assignment to flights. (If flight delays have been reduced sufficiently due to cancellations, the GDP may be cancelled altogether at this point.) The ATCSCC thus computes the final CTA assigned to each flight.

Step 6. By working backward from the CTA, the ATCSCC estimates the controlled time of departure (CTD) at which each flight affected by the GDP will depart from its airport of origin for the GDP airport. The CTA and CTD for each flight is then communicated to the airlines.

This process is typically repeated in intervals of a few hours, so that the GDP can be revised in light of the latest information about the expected AAR, flight cancellations, deviations from the GDP plan, etc.

Additional technical issues and extensions of CDM

Numerous other issues arise in practice and are gradually being dealt with by expanding the procedures and capabilities of CDM and associated tools. As an example of the kind of detailed question one must contend with, consider the example of the "double-penalty problem." Suppose an airline is forced to delay the departure of a flight due to a mechanical problem. Under earlier GDP rules, if the airline informed the FAA of the new departure time, the FAA would compute a new ETA for that flight. If a GDP were then initiated for the airport of destination, that flight would receive a CTA, which would force it to suffer *additional* delay on top of the delay due to the mechanical problem. Clearly, airlines had no incentive to report such mechanical delays to the FAA. GDP slots were wasted as a result, when aircraft failed to meet the CTA assigned to them. Under CDM, the rationing of slots is based on the original ETA, not the revised one. Thus, an airline can truthfully report a mechanical delay, knowing that its flight will be assigned the slot it was originally entitled to and, through compression, will be able to use the earliest slot that the flight can make after departing late due to the mechanical delay. Similar special-purpose policies have been developed for several other problems, such as accommodating, without further delay, flights that were previously diverted to other airports due to weather.

However, some ATFM problems of a more technical nature are particularly difficult to solve. Certainly the most fundamental is the setting of the AAR, which is typically estimated several hours in advance of the starting time of the GDP, based primarily on weather forecasts that are highly uncertain in many cases. Think, for example, of the difficulty in predicting when exactly heavy fog will roll in or burn off at an airport, or when a line of thunderstorms will arrive near an airport, when it will move away, and how severely it will affect the airport's operations. Since CTA schedules and ground delay assignments are all based on the AAR, a wrong prediction of the AAR will lead to one of two types of errors. If the estimate is too high, i.e., if it turns out that the airport is not able to accept as many arriving aircraft per hour as predicted, then these aircraft will suffer additional, possibly long, *airborne* delays, on top of the ground delays already assigned to them. If, on the other hand, the estimate is too low, aircraft will be held for an excessive amount of time on the ground and suffer unnecessary delay. Note that in this second case valuable airport capacity will be wasted, as the rate of arrivals at the airport (restricted to 6 per hour in Example 10-1) will be lower than what the airport could accommodate. This is sometimes referred to as "starving the runways." Airlines often complain that this type of error is all too common. They argue that ATFM tends to adopt worst-case scenarios regarding the AAR, i.e., is by nature biased toward low-side estimates of capacity, because its primary concern is avoiding overloads.

One possible response to this type of problem is the *managed arrival reservoir* (MAR). Under this approach, the ATFM system plans for some amount of airborne delay when determining the controlled time of departure (CTD) of an aircraft during a GDP. For example, with reference to Table 13-7, Flight C1 could be given a CTD from its airport of origin, which is only 25 min later than its scheduled departure time, not 40 min. If everything goes according to schedule, 25 of the planned 40 min of delay will then be taken on the ground prior to takeoff and 15 min in the air.* The intent is to create an airborne queue near the GDP airport with arriving aircraft having to wait roughly 15 min before landing, if the predicted AAR proves to be correct. If, however, the true AAR turns out to be higher than the predicted one, the 15-min queue will provide a "reservoir" of aircraft, which will be available to utilize the additional available slots and avoid wasting airport capacity. Note that if the true AAR is *lower* than predicted, the MAR will add further to the resulting unplanned airborne delays. A more genuine and better-performing solution than MAR to the critical problem of the uncertainty associated with the AAR can come only from a combination of improved weather forecasting technology and more advanced methodologies for setting the AAR based on stochastic optimization (Richetta and Odoni, 1993; Ball et al., 2002).

An approach under consideration for improving compliance with CTA under CDM is referred to as *control by CTA*. The notion here is that the ATFM system will specify only the CTA, not the CTD, for each flight. It will then be up to each airline or aircraft operator to determine, according to its own economic and other criteria, the best time for a flight's departure from the airport of origin, taking advantage of the shared information and data exchanges afforded by the CDM environment. This will essentially do away with Step 6 of the GDP planning process outlined above. Note that control by CTA allows users to determine for themselves what fraction of an expected delay they wish to take on the ground before takeoff and what fraction while airborne. Based on experience with airline attitudes gained

---

*Note that, under this scheme, a flight that has been assigned a delay of less than 15 min by the GDP will not be required to take any ground delay at all.

through CDM, it may be true that airlines are less concerned than previously thought about reducing fuel costs through ground holding. They seem to attach greater importance to adhering as closely as possible to their published schedules by taking maximum advantage of all available airport capacity each day. Several difficult issues must still be resolved, however, before control by CTA can be fully implemented. In the meanwhile, interim versions are being used that, subject to certain restrictions, give airlines more latitude in specifying the CTD of flights.

The CDM environment offers the potential to develop other important extensions of GDP or new applications of ATFM. Two examples can be described briefly. The first is motivated by the observation that GDP's currently are limited to allocating arrival capacity only.* If the allocation also included departures, while taking into consideration the trade-off between the arrival capacity and the departure capacity of an airport (Chap. 10) this would provide a great deal more flexibility and could be most helpful in increasing the reliability of flight connections at airline hubs (Gilbo, 1993; Hall, 1999).

A second important example, but less relevant to airports, is *collaborative routing* (CR). As the name suggests, this involves coordinating through CDM the rerouting of aircraft in *en-route airspace,* whenever it appears likely that a sector will be overloaded with traffic or whenever a weather front might necessitate such action. CR requires (1) availability of timely data and of reliable methods for predicting delays en route, so an airline can decide which alternative routes between two points are, *a priori,* the most attractive, as well as (2) a procedure and set of rules for allocating in a fair and efficient manner the alternative routes requested by the airlines. A major concern is the balancing of resulting flows, so that rerouting does not create overloads and congestion on the alternate routes.

Prospects

ATFM systems now play a central role in air traffic management in both the United States and Europe. Similar systems may eventually be developed in other regions of the world. The most important weakness of ATFM at this time is the limited availability of decision-support tools that are sufficiently advanced to cope with the complexity of

*EUROCONTROL allocates slots to both arrivals and departures; however, it works with a fixed arrival capacity and a fixed departure capacity for each airport and does not consider potential trade-offs between the two.

the ATFM environment. The CDM program has made significant progress in this respect through the development of the FSM software, which incorporates several powerful decision-support algorithms and utilities. Still, much remains to be done.

The CDM approach marks a turning point both for ATFM and, quite possibly, the entire ATM system as well. It takes advantage of the opportunities afforded by information processing and communications technology and introduces an entirely new philosophy to the allocation of decision-making responsibilities between service providers and users. The notion of collaboration to make decisions in real time with a common data basis is a fundamental one, so that applications to other domains of ATM are only a matter of time. In the sense that it contributes to decentralization of decision making, with allocation of additional responsibilities to ATM system users, and that it removes operating restrictions, CDM can also be considered a contribution to progress toward the concept of "free flight."

13-6   Near- and medium-term enhancements

Many near- and medium-term enhancements are planned for the ATM system. This section identifies and describes briefly a subset of those that are expected to bring about improvements in the safety and efficiency of operations in terminal airspace.

GPS-based navigation

Because of its accuracy and global coverage, the Global Positioning System* (GPS)—and possibly other existing or future global navigation satellite systems† (GNSS)—has enormous potential for ATM applications. With limited investment in equipment, it gives aircraft the capability to fly user-preferred routes anywhere in the world. It also can reduce greatly the need for governments to invest in ground-based navigation facilities and equipment with high fixed and maintenance

---

*The GPS consists of 24 orbiting satellites, 21 primary and 3 backup, arranged spatially so that any point on the earth's surface is within line-of-sight of at least four satellites. For reasons of security, the GPS provides a lower level of service, the Standard Positioning Service (SPS), to nonauthorized users. The horizontal positioning information provided by GPS in the SPS mode is within approximately 100 m of the exact position with probability of 95 percent (Nolan, 1999).

†As of 2001, the Global Navigation Satellite System (GLONASS), deployed by the former Soviet Union, was the only other existing GNSS (but had become problematic due to lack of funding); the European Union was planning deployment of its own GNSS, called Galileo.

costs (Braff et al. 1994). GPS has already had a great impact on oceanic air traffic, where it has been approved as the primary means of navigation. It is also used extensively as a supplemental aid to navigate in en-route and terminal area airspace and for nonprecision approaches. The horizontal accuracy provided by the GPS service available to all users is sufficient for these applications, which the FAA has authorized since the early 1990s.

The positioning accuracy of the GPS (as well as its availability and integrity) does not, however, meet requirements for precision approaches. For this reason, "augmentation" of the GPS positional information is needed to attain the required performance. A *local area augmentation system* (LAAS) essentially consists of a reference station located at or near an airport and a monitor station that make it possible to measure precisely any GPS errors *at that airport* and transmit corrective information ("differential data") to aircraft. A LAAS can support Category I, II, and III approaches by providing sufficient navigation accuracy to distances of 20 mi or more from an airport. According to the FAA's Operational Evolution Plan, Category I approaches based on LAAS will become available in the United States by 2004, and Category II and III by 2007 (FAA, 2001). LAAS may also provide the means to navigate on the airport's surface in low-visibility conditions. Eventual installation at 150 airports in the United States is expected.

A *wide area augmentation system* (WAAS) is designed to provide corrections to the GPS signal on a regional or national basis, for example, over an area comparable to that of the United States. It consists of a network of ground reference stations, master stations, and a geosynchronous communications satellite, which broadcasts corrections to the GPS signal to aircraft. A WAAS can support Category I approaches. A WAAS capability in the United States is slated to become available by 2003.

Because both LAAS and WAAS provide corrective information ("differential data") for GPS signals, they are examples of *differential GPS* (DGPS) systems. Details and references on DGPS are provided by Braff et al. (1994) and Nolan (1999). The advantages that DGPS-based precision approaches offer over ILS, include all those associated with MLS (Sec. 13-3) plus availability of a signal whose accuracy does not decline with distance from the runway (within bounds in the case of LAAS). On the other hand, as in the case of MLS, airlines operating in developed countries do not seem to attach as high a priority to a DGPS capability for precision approaches as one might expect. This can be attributed to the fact that instrument landing systems are now available at practically all the major and secondary airports in these countries.

Automatic dependent surveillance

As noted in Sec. 13-2, pilot reports of their aircraft's position and altitude via voice communications constituted the earliest form of surveillance in air traffic control. *Automatic Dependent Surveillance* (ADS) is also based on the notion of relying on aircraft to report their own position. In ADS, the aircraft determines its position, primarily through the GPS (or other GNSS), and automatically reports that position through data-link technologies. Depending on who receives the position reports, a distinction is made between two types of ADS systems. Under ADS-A (for "addressable"), the aircraft exchanges information with the ATM system provider upon request. This is designed to support oceanic operations. Under ADS-B (for "broadcast"), the aircraft broadcasts periodically information about its position to ATM facilities and to all other aircraft in its vicinity. That information can then be processed as needed by the ATM system, as well as displayed directly in the cockpit of other aircraft which are equipped with a CDTI (Cockpit Display of Traffic Information).

The ADS-B system, when used in terminal airspace, may provide safety benefits by increasing the awareness of pilots equipped with a CDTI of the position of other (ADS-B-capable) aircraft in their vicinity. Another potential benefit is increasing airport capacity. For example, if, in IMC, pilots can obtain visual information through the CDTI about the position of other aircraft in their vicinity, more accurate spacing between aircraft near the airport and on final approach might be achieved, in much the same way as when pilots separate themselves visually in VMC. A third potential application concerns surveillance of traffic on the airport's surface in conditions of poor visibility. By making it possible to monitor the positions of aircraft and ground vehicles, ADS could reduce the likelihood of runway and taxiway incursions, as well as assist in reducing taxiing times. This would lessen dependence on the ASDE radar—and possibly eventually replace it altogether. As noted in Sec. 13-3, ASDE is an expensive system to acquire and maintain.

Digital communications

The ATM system relies on VHF and UHF voice radio for communications in nonoceanic airspace. With growing traffic, this has led to saturation of the frequency spectrum at many locations, as well as added to controller and pilot workload. The use of data-link technology to transmit routine and repetitive messages between the ground and the aircraft is planned in a number of countries. In the United States, the FAA will make available *controller-pilot data link communications* (CPDLC) for this purpose. It has also proposed the longer-term *next-generation air/ground communications* (NEXCOM) program, which would replace existing ground radios and would provide significant efficiency in the use of VHF frequencies.

Weather data

Important steps are also expected in the area of acquiring as well as integrating and presenting weather-related information for terminal airspace and airport traffic control. In the United States, *terminal Doppler weather radar* (TDWR) is being enhanced to improve detection of windshear and prediction of gusts. The *integrated terminal weather system* (ITWS) is an automated weather system that provides short-term (0–60 min) predictions of significant terminal-area weather at major airports (FAA, 1999). ITWS provides predictions or information about windshear and microbursts, storm-cell hazards, lightning, and strong winds. In addition to safety, ITWS may provide very significant benefits by helping reduce delays related to thunderstorms or other weather events. Allan et al. (2001) present an interesting case study in this respect, based on a ITWS demonstration program at the New York City airports.

Automation and decision-support systems

Researchers have developed many automation and decision-support tools over the years for use in terminal airspace and airports. While only a small number of these have been implemented in the field, their usability and viability have been improving rapidly. Several more such tools will therefore become part of terminal area ATM systems in the near and medium term.

The earliest advanced tools of this type concentrated on the spacing and sequencing of arrivals at runway systems. The first large-scale integrated system to be successfully implemented, COMPAS, became operational at Frankfurt/Main in the early 1990s (Volckers et al., 1993). COMPAS, developed by the German Aerospace Research Establishment (DLR), assists air traffic controllers in transitioning arriving airplanes from en-route airspace to the beginning of final approach, allocating arrivals between the two close parallel runways used for landing in Frankfurt (see Fig. 9.4), and spacing these arrivals longitudinally as well as diagonally for approaches to the two runways.

A more comprehensive system along these lines—and the most advanced to date—is the *Center TRACON Automation System* (CTAS), which has been developed by NASA and is operating at Denver/International and at Dallas/Ft. Worth (NASA, no date). CTAS consists of three key modules that assist controllers in performing most of the tasks essential to roughly the last 40 min of a flight [see Erzberger (1995) for detailed descriptions]. The *Traffic Management Advisor* (TMA) facilitates the planning of the sequence of arrivals and their allocation to active runways while aircraft are still in the en-route phase of flight. The *Descent Advisor* (DA) is used in developing near-optimal four-dimensional paths from start of descent in an en-route sector to delivery at the approach feeder fixes (Sec. 13-3) in the terminal area. The DA provides recommendations concerning the point where descent should be initiated and the descent rate, path, and speed. Its objective is to deliver each aircraft at the feeder fix at a specified time, while minimizing fuel consumption. Finally, the *Final Approach Spacing Tool* (FAST) aims at maximizing runway acceptance rates and balancing traffic on multiple runways. It assists arrival controllers in the TRACON (Sec. 13-3) in determining loads on the runways, generating a desirable sequence of landings on each runway, and achieving tight spacing between them. The latter is done through speed, heading, and turn advisories to aircraft. An upgraded version* of FAST will provide greater precision by using "real-time" aircraft performance data and aircraft wake vortex information to refine the recommended spacing and sequence (FAA, 1999). Preliminary indications are that FAST produces an increase in capacity at Dallas/Ft. Worth ranging from almost 20 percent in IMC with two arrival runways to about 10 percent in VMC with three arrival runways.

One of the features of CTAS (as well as of COMPAS) is the ability to deviate from first come, first served (FCFS) sequencing of arriving air-

---

*The initial version of FAST is referred to as p-FAST (for "passive") and the later version as a-FAST (for "active").

craft in order to increase capacity and reduce delay. This is achieved by moving an aircraft up to a specified maximum number of positions from its FCFS order. For example, if a particular airplane is 12th in line for landing according to FCFS and if the maximum number of position shifts is 2, then that airplane can be the 10th, 11th, 12th, 13th, or 14th to land. This is called *constrained position shifting* (CPS). It has been shown that, with a maximum position shift of 2 or 3, CPS can reduce significantly delays by avoiding the most undesirable landing sequences, such as a "Heavy" aircraft followed by a "Small" and requiring 6 nmi of longitudinal separation. This issue was also discussed in Sec. 10-5. Note that when the maximum position shift is as small as 1, 2, or 3, CPS also maintains a sense of fairness by guaranteeing that no aircraft will be given a position in the landing sequence that differs significantly from the FCFS order (Dear, 1976; Neuman and Erzberger, 1991).

Automation and decision-support systems also exist or are being developed for handling departures and airport surface traffic. With regard to the former, several ongoing efforts aim at optimizing flows of departing traffic on the airport's surface, as well as in terminal airspace, immediately following takeoff (Idris et al., 1999; Idris, 2001). An important observation is that, for each airport and for each runway configuration, one can determine the optimal number of departing aircraft to be allowed onto the taxiway system, such that the runways can operate at their full capacity, while the amount of fuel burn and engine emissions from taxiing aircraft is minimized (Pujet, 1999). NASA is developing a set of tools, under the generic name Expedite Departure Path (EDP), that assists controllers in managing the load of aircraft on departing runways and in sequencing, spacing, and merging departing traffic into en-route traffic streams (NASA, no date).

The most ambitious effort with regard to airport surface traffic is the *Advanced Surface Movement Guidance and Control System* (A-SMGCS), currently under development in Europe with funding from the European Commission. The A-SMGCS provides a real-time capability for planning and implementing traffic flows on every part of an airport's surface. Demonstration projects of prototype A-SMGCS were under way at a small number of European airports as of 2001. A parallel program of the FAA is the *Surface Management System* (SMS), which was to be tested in Memphis in 2003. Several related research projects in this area are also underway in North America and in Europe. Surface traffic guidance and control tools have the potential to generate both safety-related and efficiency-related benefits. It is difficult to quantify these benefits, however, as they depend on airport-specific characteristics and on how one defines, in the first place, the "baseline" against which the benefits are assessed.

Eventually, automation aids for controlling arrivals, surface traffic, and departures will be merged into a single integrated system. However, achieving such integration is far from simple, because of:

- The interactions among the various flows involved (optimizing one type of flow usually has negative consequences for the other flows)
- The difficulty of predicting accurately (even a few minutes into the future) the time and location where critical events (such as the exit of an arriving aircraft from the runway) will occur
- The complexity of realistic objective functions

Thus, the development and implementation of an integrated system that performs to the satisfaction of all stakeholders may be a long time off.

Exercises

1. One of the most interesting ideas under consideration in terminal ATC involves the "sequencing" of aircraft on arrival to an airport in order to achieve certain benefits. In this problem we examine the effects of various sequencing schemes in a terminal area. We consider for this purpose the terminal airspace around an airport at which a single runway is used exclusively for landings. Aircraft arrive at this terminal area at random times, are sequenced by air traffic controllers, and land at the runway. We assume the following somewhat simplified conditions:

   (i) All aircraft fly a 5-nmi final approach.

   (ii) The minimum airborne longitudinal separation between successive landing aircraft is 3 nmi for *all* possible pairs of aircraft, *except* behind aircraft approaching at 150 knots. In this latter case, 4 nmi are required, regardless of the type of the second aircraft in the sequence.

   (iii) No buffers are added to the minimum separations between aircraft. (Note that in computing the minimum separations between successive landing aircraft, the runway capacity model presented in Sec. 10-5 is used.)

Table 13-8 Data for Exercise 1

Aircraft identification number	Terminal area entrance time* (s)	Approach speed (knots)	Terminal area transit time† (s)	Nominal arrival time at the runway (s)
1	0	120	990	990
2	20	135	930	950
3	55	150	900	955
4	110	120	990	1100
5	180	135	930	1110
6	350	150	900	1250

*Time when the aircraft enters the terminal area.
†Time it would take an aircraft to travel through the terminal area and reach the runway threshold in the absence of any other traffic.

(iv) Assume now that, just before t = 0, there are no aircraft in the terminal area or on the runway and that, beginning at t = 0, a sequence of 6 aircraft enter the terminal area according to the data shown in Table 13-8.

(Note that the "nominal arrival time at the runway" indicates when an aircraft would reach the runway in the absence of any other traffic.)

Consider now the following four different sequencing strategies:

Strategy 1.   Aircraft are sequenced according to the time of entrance into the terminal area, i.e., first come, first served according to entrance time. (This means that aircraft would land at the runway in the sequence 1–2–3–4–5–6, spaced apart by (at least) the required ATC separations between them.)

Strategy 2.   Aircraft are sequenced according to their nominal arrival time at the runway, i.e., first-come, first-served according to nominal arrival time at the runway. (This means a 2–3–1–4–5–6 order.)

Strategy 3.   Aircraft are sequenced in order to minimize total delay, i.e., the sum of the differences between the actual time when each aircraft is assigned to reach the runway and the "nominal arrival time at the runway" for that aircraft.

Strategy 4.   Aircraft are sequenced so as to maximize "throughput," i.e., so as to land the last aircraft to reach the runway as soon as possible (this is equivalent to maximizing the "flow rate" of aircraft onto the runway).

a. For each of the four strategies described above, indicate:

    (i)   The sequence in which the 6 aircraft in our example should land. (You have already been given the answer for Strategies 1 and 2.)

    (ii) The total delay corresponding to this sequence.

    (iii) The time when the last of the 6 aircraft in the sequence would reach the runway.

b. Suppose now that an aircraft cannot be re-sequenced by more than one position from the order in which it entered the terminal area. For example, aircraft no. 3 can land second, third, or fourth but not first, fifth, or sixth. Repeat part *a* under this restriction for strategies 3 and 4. [The restriction in this part is often referred to as constrained position shifting (CPS), see Sec. 13-6.]

c. Comment briefly on the advantages and disadvantages of each of the four strategies described above, as well as on CPS. Why might CPS (with the objectives of strategy 3 or 4) be an attractive idea? Please write a thoughtful few sentences. Think of such items as delay costs, controller workload, airline perceptions, predictability (to the pilot) of when an aircraft will actually land, etc.

d. Suppose that, at some particular time, 9 aircraft are queued up waiting to land at this runway. How many possible sequences would have to be examined, in the worst case, under strategies 3 and 4, and how many under CPS (with a maximum shift of one position)? (Think systematically in answering the CPS part.)

2. You may be familiar with the *traveling salesman problem* (TSP), a famous problem in operations research and applied mathematics. One form of the TSP can be stated as follows: Suppose one is given a set of n points on a plane and the Euclidean distances between every pair of these points. The TSP is the problem of finding the shortest-distance tour that begins at one of the points, visits all the other points exactly once, and returns to the starting point. A variation of the TSP is the *Hamiltonian path problem,* which is identical to the TSP as described above, but does not require a return to the starting point (i.e., visit all points exactly once, but end at the nth point visited). Suppose that n airborne aircraft are waiting to land on a runway and that air traffic controllers can sequence them in any way they wish. Suppose also that the sequence will be determined so as to implement strategy 4 of Exercise 1, i.e., so as to land the last aircraft to reach the runway as soon as possible. Argue that this problem is equivalent to solving the Hamiltonian path problem, but with asymmetric distances between the points (i.e., the distance from i to j may be different from the distance from j to i).

3. Consider ground delay programs (GDPs). It is the practice currently to exempt long-range flights from ground holding. For example, a flight between Frankfurt/Main and Boston/Logan, which takes roughly 8 h, will be allowed to take off on time, even if it is likely (but not certain) that, at the time of the flight's arrival, Boston/Logan's capacity (or AAR—see Secs. 13-4 and 13-5) will be low. This policy means that the brunt of ground holding delays is borne by short- and medium-range flights. Does this policy make sense? What is the rationale for it? What are its advantages and disadvantages?

References

Allan, S., Gaddy, S., and Evans, J. (2001) "Delay Causality and Reduction at the New York City Airports Using Terminal Weather Information Systems," Project Report ATC-291, MIT Lincoln Laboratory, Lexington, MA.

Ball, M., Hall, W., Hoffman, R., and Rifkin, R. (1998) *Collaborative Decision Making in Air Traffic Management: A Preliminary Assessment,* National Center of Excellence in Aviation Operations Research (NEXTOR), University of Maryland, College Park, MD.

Ball, M., Hoffman, R., Odoni, A., and Rifkin, R. (2002) "A Stochastic Integer Program with Dual Network Structure and Its Application to the Ground Holding Problem," *Operations Research,* in press.

Barnett, A. (2001) "Air Safety: End of the Golden Age?," *Journal of the Operational Research Society, 52,* pp. 849–854.

Barnett, A., Paull, G., and Iadeluca, J. (2000) "Fatal U.S. Runway Collisions over the Next Decade," *Air Traffic Control Quarterly, 8,* pp. 253–276.

Barnett, A., and Wang, A. (2000) "Passenger Mortality-Risk Estimates Provide Perspectives about Airline Safety," *Flight Safety Digest, 19,* pp. 2–13.

Braff, R., Powell, J., and Dorfler, J. (1994) "Applications of the GPS to Air Traffic Control," in *Global Positioning System: Theory and Applications: Volume II,* B. W. Parkinson and J. J. Spilker, Jr. (eds.), American Institute of Aeronautics and Astronautics, Washington, DC.

Dear, R. (1976) "The Dynamic Scheduling of Aircraft in the Near-Terminal Area," Ph.D. dissertation, Technical Report R76-9, Flight Transportation Laboratory, Massachusetts Institute of Technology, Cambridge, MA.

Erzberger, H. (1995) *Design Principles and Algorithms for Automated Air Traffic Management,* AGARD Lecture Series 200, Brussels (also at http://www.ctas.arc.nasa.gov/).

FAA, Federal Aviation Administration (1989) *Airport Design,* Advisory Circular 150/5300-13, incorporates subsequent Changes 1 thru 5 (Change 5 in 1997), U.S. Government Printing Office, Washington, DC.

FAA, Federal Aviation Administration (1992a) "Designation of Federal Airways, Area Low Routes, Controlled Airspace, Reporting Points, Jet Routes, and Area High Routes," *Federal Aviation Regulations,* Part 71, FAA, Washington, DC.

FAA, Federal Aviation Administration (1992b) "General Operating and Flight Rules," *Federal Aviation Regulations,* Part 91, FAA, Washington, DC.

FAA, Federal Aviation Administration (1999) "Blueprint for NAS Modernization: An Overview of the National Airspace System Architecture Version 4.0," http://www.faa.gov/nasarchitecture/blueprnt/index.htm.

FAA, Federal Aviation Administration (2000) *Airport Design, Change 6,* Advisory Circular 150/5300-13, U.S. Government Printing Office, Washington, DC.

FAA, Federal Aviation Administration (2001) "Operational Evolution Plan," http://www.faa.gov/programs/oep.

Gilbert, G. (1973) "Historical Development of the Air Traffic Control System," *IEEE Transactions on Communications, 21,* pp. 364–375.

Gilbo, E. (1993) "Airport Capacity: Representation, Estimation, Optimization," *IEEE Transactions on Control Systems Technology, 1,* pp. 144–154.

Hall, W. (1999) "Efficient Capacity Allocation in a Collaborative Air Transportation System," Ph.D. dissertation, Operations Research Center, Massachusetts Institute of Technology, MIT, Cambridge, MA.

ICAO, International Civil Aviation Organization (1999) *Aerodromes: Annex 14 to the Convention on International Civil Aviation, Volume I: Aerodrome Design and Operations,* 3d ed., ICAO, Montreal, Canada.

Idris, H. (2001) "Observations and Analysis of Departure Operations at Boston Logan International Airport," Ph.D. dissertation, Department of Aeronautics and Astronautics, Massachusetts Institute of Technology, Cambridge, MA.

Idris, H., Anagnostakis, I., Delcaire, B., Hansman, R., Clarke, Feron, E., and Odoni, A. (1999) "Observations of Departure Processes at Logan Airport to Support the Development of Departure Planning Tools," *Air Traffic Control Quarterly, 7,* pp. 229–257.

Metron, Inc. (2000), "CDM Website," http://www.metsci.com/cdm/.

MIT Lincoln Laboratory (1998) "Air Traffic Control Overview: Kansas City ARTCC," Group 41 Report, MIT Lincoln Laboratory, Lexington, MA.

Mundra, A. (1989) "A Description of Air Traffic Control in the Current Terminal Airspace Environment," Report MTR-88W00167, The MITRE Corporation, McLean, VA.

NASA, National Aeronautics and Space Administration (no date), "The Center TRACON Automation System," http://www.ctas.arc.nasa.gov/.

Neuman, F., and Erzberger, H. (1991) "Analysis of Delay Reducing and Fuel Saving Sequencing and Spacing Algorithms for Arrival Traffic," NASA Report TM-103880, Ames Research Center, Moffett Field, CA.

Nolan, M. (1999) *Fundamentals of Air Traffic Control,* 3d ed., Brooks/Cole Wadsworth, Pacific Grove, CA.

Pujet, N. (1999) "Modeling and Control of the Departure Process at Congested Airports," Ph.D. dissertation, Department of Aeronautics and Astronautics, Massachusetts Institute of Technology, Cambridge, MA.

Richetta, O., and A. Odoni (1993) "Solving Optimally the Static Ground-Holding Policy Problem in Air Traffic Control," *Transportation Science,* 27(3), pp. 228–238.

Volckers, U., Brokof, U., Dippe, D., and Schubert, M. (1993), "Contributions of DLR to Air Traffic Capacity Enhancements within a Terminal Area," in *Proceedings of AGARD Meeting on Machine Intelligence in Air Traffic Management,* Report AGARD-CP-538, Brussels, Belgium.

Part Four
The landside

14

Configuration of passenger buildings

Designers face a crucial design issue when they select the configuration of airport passenger buildings. Several major airports have suffered extensive problems because inappropriate choices hurt their finances and operations. This chapter shows how to evaluate the performance of airport passenger buildings at the aggregate level appropriate to overall planning, and discusses which designs are preferable in which circumstances.

Airport passenger buildings serve the many needs of different types of users. They process check-in, security and customs clearances, and baggage for travelers; provide for their waiting and transfers between flights; cater to their shopping and other activities; and often constitute the port of entry for international traffic. Equally important, they must also perform efficiently and profitably for their distinct stakeholders, including, in addition to passengers, the airlines managing the aircraft, the owners who provide the capital, and the operators of the many services, including the government operating frontier controls.

Airport passenger buildings come in five basic configurations: finger piers; satellites with and without finger piers; midfield, either linear or X-shaped; linear with one airside; and transporter. At large airports, the buildings may be centralized or dispersed. Their performance depends significantly on three characteristics of the traffic: its overall level, its seasonality, and the percentage of transfers. It also depends on their flexibility to expand or accommodate different types of traffic. Which configuration is best depends on these factors and the perspective of the stakeholders.

In general, finger piers are preferable when the level of transfer traffic is low, and linear midfield concourses are best when transfer traffic is high and the airfield permits. Transporter solutions are economical when the seasonal traffic peaks are more than twice that of the low season. The overall configuration at major airports should usually be a hybrid that blends configurations. Hybrid configurations are appropriate both for the distinct needs of the variety of traffic and airline alliances and for flexible expansion in the future.

14-1   Importance of selection

Designers face a crucial issue when they select the configuration of airport passenger buildings. The decisions they make about the shapes of the buildings, their layout on the airport, and their location have important consequences for the performance and profitability of the airport enterprise. Shapes with significant reentrant corners (90° or less*) are inefficient. X-shaped buildings, for example, waste space by making it impossible to park aircraft along significant portions of the building, toward the center where the crosspieces meet. Designs that require aircraft to make numerous turns and stops will be expensive for the airlines. Arrangements that do not feature a central common place, for check-in or other services, are likely to be ineffective in generating commercial revenues. Buildings that park aircraft only along one side will not work well for transfer passengers.

Where designers failed to select appropriate configurations for airport passenger buildings, these bad choices have led to major financial and operational difficulties. The most obvious ones occurred when planners chose buildings that could not process transfer traffic effectively. For example:

- Kansas City built an airport with three separate buildings, each accommodating aircraft on one side and automobiles on the other. This could be an attractive solution if all passengers go directly between the roadway and the aircraft. However, it is a poor configuration for transfers, who have to walk twice as far as they would in buildings designed to serve aircraft on two sides. The Kansas City arrangement is particularly bad for transfers because they also have to walk long distances between separate buildings, in a climate that can be hot and

*Any shape involving buildings that meet at right angles, in the form of a T or an L for example, features reentrant corners. Since aircraft need more than twice their own length for access and parking, that length of building is thus unavailable near the junctions of the buildings. For an X-shaped building serving transcontinental aircraft, this fact effectively implies that no aircraft stand can exist within about 100 m of the geometric center of the structure (allowing for the width of the arms of the building).

rainy. This bad choice of configuration was one of the reasons that, after the economic deregulation in 1978, the major airline that used to have its base of operations in Kansas City moved to St. Louis. Kansas City lost significant traffic, jobs, and economic visibility when TWA shifted its operations. This was a regional economic disaster.

- Frankfurt/Main built a new billion-dollar international terminal, primarily to serve Lufthansa, the German national airline. Unfortunately for the airport, shortly before the opening the airline realized that it could not workably transfer passengers between this terminal and its facilities serving domestic and European cities. Lufthansa therefore declined to move to the new terminal. Smaller airlines, with infrequent schedules and relatively few passengers, thus moved into the new building. As a result, this huge capital investment was for a long time underused and largely wasted, while the national airline remained in the crowded old facilities.

Poor choices of the configuration of the airport passenger buildings are related to many other issues besides transfer passengers. Here are some examples.

- *Decentralized buildings.* Baltimore built an international terminal in the 1990s, largely to serve US Airways. When that airline moved the focus of its international operations to Philadelphia at the end of the decade, however, Baltimore was left with an underutilized building. The airport, meanwhile, had to provide new facilities for Southwest, an expanding domestic airline. Yet because of separation and lack of flexibility, it was unable to make the unused space available to Southwest. Because it had an inappropriate configuration, Baltimore had to spend around $100 million on duplicate facilities. (Baltimore-Washington International Airport, 1998; Little, 1998).

- *Midfield concourses.* The British Airports Authority (BAA) built London/Stansted with only midfield concourses, that is, passenger buildings widely separated from the main building and check-in facilities. While these are efficient for transfer operations, they are inconvenient for passengers using ground transport because they require their own transport system that both adds to flight time and is expensive. Estimating the cost of the underground "people mover" for London/Stansted in the range of £0.50–1.00 (about $1.00) per passenger, the extra cost of this configuration is about $5 million per year— the equivalent of about $50 million in extra capital cost.*

Airport owners and designers frequently make poor choices of the configuration of their passenger buildings. This happens when they do not adequately recognize the diversity and unpredictability of the requirements that passenger buildings should meet. Indeed, designers have not had a general process for carrying out detailed analyses, either of the role of an airport in the systems of regional or metropolitan airports or of the variety of its needs. Furthermore, they have usually based their designs on forecasts fixed on specific numbers (X million annual passengers, or Y thousands of passengers per peak hour in major categories) instead of on ranges. (See, for example, Ashford and Wright, 1992; Horonjeff and McKelvey, 1994; IATA, 1995; ICAO, 1987.) The designers' focus on fixed forecasts leads them to adopt inflexible plans, which causes problems for the airport operators. The next section suggests important factors airport owners need to consider before deciding on the configuration of an airport passenger building. Chapters 3 and 20 discuss how to deal with the risks associated with the unpredictability of airport forecasts.

A root problem is that airport owners normally define the architectural program in terms of numbers of passengers and aircraft operations. They do not refer to the variety of distinct types of traffic, the market segments, or the needs of the several stakeholders. The consequent failures to make good choices of the configuration of the passenger buildings have been due to systemic deficiencies in the master planning process as it has been practiced (Chap. 3). While the failures involve architects, they are not particularly attributable to them. On the contrary, many buildings with severe operational or financial problems were hailed as architectural masterpieces when they opened. The architects generally followed the specifications they got from their airport bosses or the international design manuals.

---

*The translation between capital costs and annual costs is an imprecise art that depends on local circumstances. In the United States, the capital for airport passenger buildings normally comes from municipal bonds exempt of interest to the owner. It thus costs the airport on the order of 5 percent a year, which is about 2–3 percent a year in real terms. Factoring incremental depreciation and maintenance over the cost of the older building, the total additional annual cost of a replacement building is of the order of 10 percent of its capital cost. Private companies typically have a higher cost of capital, but may still be able to raise money through real estate mortgages on their buildings at about 5 percent a year in terms of real cost net of inflation. The factor of 10 used to translate annual savings into justifiable additional capital expenditures is only an order-of-magnitude approximation.

To avoid poor choices of configuration of airport terminal buildings, airport owners need to take a systems approach to the specification of their requirements. They need to:

- Understand the current and possible future role of their airport in relation to other airports nationally and internationally and, as applicable, to other airports in their metropolitan region (Chap. 5)
- Determine the priorities and requirements of the significant stakeholders in the airport
- Provide the designers with guidance on how these distinct priorities should be weighted with respect to each other

In short, responsible effective owners will anticipate the long-term risks and the variety of future needs, and select a flexible configuration suitable for the plausible evolutions of the airport.

14-2   Systems requirements for airport passenger buildings

Airport passenger buildings serve the many needs of different types of users. In addition to passengers, the buildings cater to the airlines managing the aircraft, the owners who provide the capital, and the operators of the many services. The airport buildings will be successful to the extent that they meet the requirements of all these constituencies.

This introductory statement is fundamentally important, although it appears obvious. It is significant because it defines a starting point that is critically different from what designers have done in practice. The reality has been that the design process has typically ignored major stakeholders in the airport. It inherently claims to take into account the needs of the several stakeholders, such as airlines and retail operators, by setting aside space for them according to generalized industry norms. In most cases, however, the design teams responsible for configuring the airport buildings neither bring the stakeholders into the process nor listen to them.

Three examples illustrate how the design process routinely ignores important stakeholders. They concern the airlines, the stores and other commercial operations in the buildings, and the owners. All are crucial to the success of the enterprise.

Airlines have a considerable stake in the choice of the configuration of the buildings. This is because the arrangement of the buildings affects the time it takes an aircraft to maneuver into position, and implies millions of dollars in costs. Yet designers do not involve airlines in this choice (except in the United States, where the airlines typically have a veto on major expenditures and consequently are routinely involved—see Chap. 8). For example, an international design review for the planned T5 passenger building at London/Heathrow pointedly did not invite representatives of British Airways, which was certain to be the major occupant of the facility. The implicit attitude of the designers was, first we design the building for the owner, after that we consult with the airlines about their special needs.

Designers normally set aside areas for commercial space, according to the amounts specified in their design program. Not until several years later, shortly before the airport opens the building, does the airport lease these spaces to commercial activities, which must cope with the available opportunities as best they can. Thus, in Terminal 2 at Tokyo/Narita, the shops are largely hidden on a mezzanine out of the flow of passenger traffic; in the International check-in lobby at San Francisco/International, the stores can hardly be seen. Experienced designers of shopping malls would never tolerate such poor arrangements. To avoid such problems, the master designers should consult in advance with retail experts and thus learn how to lay out profitable commercial spaces. Unfortunately, the authors know of no case in which the architects and designers have undertaken such early consultations. Typically, the expert retailers arrive on site when the building is nearly completed. For example, this was the experience of BAA when it took over retailing at Pittsburgh.

Investors must be concerned with getting good returns on this capital, yet the design process typically operates within a fixed budget limited to the amount available from government grants or bond issues. When inevitable cost increases occur, many items are dropped, even when their extra cost might generate significant returns. Thus, a cost-reduction scheme eliminated a whole floor of commercial space during the design of Kuala Lumpur/International. A superior design process will make sure that long-term profits will not be sacrificed for short-term savings.

In fairness, it is not easy for the traditional design process to determine the needs of the important stakeholders in the airport. Getting useful information from them is difficult. For example, airlines do not normally have airport-planning departments, and are unlikely to provide competent institutional links with the design team. Nobody can speak directly for the operators of retail stores who have not been selected. Government officials running national or city airports generally do not have the authority to approve deviations from prescribed budgets.

A more effective design process needs to make a special effort to determine and take into account the needs of the important stakeholders in the airport. This can be done, although the specific companies and institutions that will be operating at the airport may not be able to specify their concerns themselves. To take a comprehensive approach to the selection of the best configuration of the airport buildings, the master designers can hire experts on the different issues to work with them. Although the airlines may not have airport-planning groups, for example, there are many consultants with airline and airport experience who understand the needs of airlines at airports. Master designers can thus hire appropriately qualified consultants to speak for the interests of the stakeholders in the future facilities. Toronto/Pearson Airport, for example, employed "airline liaison officers," paid for from project funds, to represent the airline interests in the construction of the major new passenger building.*

The rest of this section identifies some major stakeholder issues that a systems approach to the selection of the configuration of airport passenger buildings should address. It focuses on four perspectives: those of the passengers, the airlines, the owners, and the commercial services.

Passenger perspective

The major categories of passengers that deserve special consideration in the design process are:

- Domestic or intercommunity travelers, those who do not require passport or customs controls
- International travelers who require government controls
- Business and commercially important travelers, generally more accustomed to travel, often with less baggage, but requiring special amenities such as luxury lounges
- Vacationers and personal travelers, often with families and much baggage and using charter airlines requiring inexpensive facilities

*Personal communication from Lloyd McCoomb, Vice President, Planning and Development of the Greater Toronto Airport Authority.

- Transfers, travelers who are at the airport simply to transfer from one flight to another

In addition, specific locations may identify other groups that systematically require special treatment. These include charter flights and annual pilgrimages, for example. Except for transfers, passengers require ground transportation, check-in facilities, security and other clearances, easy passage between the landside and the aircraft, waiting lounges, commercial services, and baggage delivery. See Chap. 16 on detailed design of airport passenger buildings.

Transfer passengers deserve special emphasis because their needs are very different from those of the other travelers. The crucial difference is that they require fast, reliable, and easy-to-find connections between aircraft. Their connections should be:

- *Fast,* because the airlines using the airport for hubbing operations need to be competitive with other carriers using other hubs. Thus British Airways, serving Boston to Athens through London/Heathrow, may compete with Lufthansa operating through Frankfurt/Main, Alitalia through Milan/Malpensa, and others. Moreover, it is unproductive for airlines to hold their aircraft on the ground; they need to transfer the passengers and get the aircraft flying.

- *Reliable,* because the cost to the airlines of stranded passengers or delayed bags is very high, due to the direct costs of empty seats and having to deliver bags by taxi to the traveler, and the bad reputation that unreliable service generates.

- *Easy to find,* because complicated routes confuse and delay passengers, and are thus unreliable. Simple, direct routes in a single building are best, as within the United Airlines midfield concourse at Denver/International, or at Amsterdam/Schiphol or within the satellites of Terminal 1 at Tokyo/Narita.

Additionally, transfer passengers obviously do not require check-in facilities, baggage delivery, or easy access to and from ground transportation. Overall, transfer passengers have special needs distinct from those of the other passengers.

Wherever airport operators and airlines have a strategy of promoting a hub in which transfers constitute half or more of the total passengers of an airport, their needs should dominate the choice of the configuration of the airport passenger buildings. This occurs at many hub airports, as Table 14-1 indicates. It also occurs at parts of airports for specific dominant airlines. For example, Munich Airport built a

Table 14-1. Approximate transfer rates at major hub airports in 1999

Airport	Transfer rate (%)		Airport	Transfer rate (%)	
	Airport	Hub airline		Airport	Hub airline
Amsterdam	39	KLM	Frankfurt/Main	48	Lufthansa
Atlanta	60	70, Delta	Houston/Bush	55	65, Continental
Chicago/O'Hare	50	60, American	London/Heathrow	30	50, BA
		60, United			
Cincinnati	70	80, Delta	Minneapolis/St. Paul	55	65, Northwest
Dallas/Fort Worth	60	70, American	Pittsburgh	60	70, USAir
Denver	50	65, United	Salt Lake City	50	65, Delta
Detroit/Metro	50	60, Northwest	Washington/Dulles	30	45, United

Sources: U.S. Department of Transportation, 2000; Salomon, Smith Barney, 2000; Moody's Investor Services, 2000.

new passenger building specifically to cater to the transfer activities of Lufthansa, its dominant carrier.

The needs of transfer passengers deserve special emphasis for two reasons. The first is that designers have often forgotten them, and thus chosen inappropriate configurations for the airport passenger buildings, as the examples cited in the previous section indicate. The second is that the standard references hardly mention the needs of transfers. Indeed, the standard vocabulary used in airport planning tends to ignore their existence. Thus traditional terminology refers to "airport terminals" when at many airports, many and sometimes most of the passengers do not "terminate" at that airport. To avoid the false implications associated with "terminals," this text consistently refers to "airport passenger buildings."*

Airline perspective

Airlines care about the configuration of the airport passenger buildings because it affects their operating costs. Poor designs impose a heavy burden on the airlines. Good designs give them a desirable competitive edge. United Airlines, for example, benefited greatly from the new Denver/International airport: its midfield concourse and completely paved apron dramatically improved the efficiency of its aircraft operation compared to Denver/Stapleton airport (now closed). Because the new design both reliably reduced the average taxi time on each operation (by a few minutes) and delays (by another few minutes), United was able to tighten its schedules by about 15 min on a round trip through Denver. According to their facility manager, the Denver configuration thus enabled United to get an extra round trip a day from short-haul flights, a terrific boost in aircraft productivity. Small savings in time may appear insignificant, but when cumulated over a day they can have a major impact. When these small savings cumulate over hundreds of thousands of operations a year, as they do at major hubs, the savings can be very large indeed.

Airlines and airline alliances that operate transfer hubs benefit from configurations that facilitate these operations. With regard to transfers,

---

*Edwards' (1998) architectural text on the design of airports illustrates the mindset that ignores transfers. The entire book does not seem to mention transfers. Moreover, this quote from the start of the section on the "functional requirements of the terminal building" illustrates the point: "The basic organization of a terminal can be separated into two parallel functional patterns of passenger and baggage movement: departures and arrivals....The terminal consists of two main public spaces: the departures concourse and the arrivals concourse" (p. 118).

the airline perspective is aligned with that of the passengers. Both want fast, reliable, and easy-to-find transfers. Configurations that facilitate these objectives are worth considerable money to the airlines, and they are willing to pay for it. This is why airlines have backed the construction of midfield concourses at Atlanta, Chicago/O'Hare (United Airlines), Denver/International, Detroit/Metro, and Munich.

Airlines are especially sensitive to the costs of maneuvering their aircraft on the ground. They are willing, if given the opportunity, to pay the increased leases or charges associated with configurations of airport passenger buildings that save them time. Significant reductions in ground times can justify hundreds of millions of dollars in new construction. This factor is worth emphasizing, as it is typically omitted from the discussion of the choice of airport terminal configurations.

A simple calculation illustrates the great value to the airlines of easy ground movements. Recognize first that the direct operating cost of a large commercial jet is of the order of about $100 per minute.* Consider next an airport with 100,000 operations per year, which might serve approximately 10 million passengers a year. A configuration that saved just 1 min per operation at this airport would save the airlines around $10 million a year in direct costs alone—the equivalent of about $100 million in extra capital cost.† At airports with configurations that delay aircraft operations significantly, such as the former Denver/Stapleton or London/Heathrow, the savings would be much larger. At London/Heathrow, an efficient configuration of the airport buildings might save the airlines several minutes per flight, for over 300,000 operations a year. The implied possible savings could be over $50 million a year—the equivalent of about $500 million in capital cost. Moreover, this estimate is conservative, in view of the way early-morning delays propagate throughout the day, with a multiplier effect that can double their importance. Such savings provide a strong rationale for tearing down inefficient configurations and starting all over again, as Denver did in closing Stapleton and opening its International Airport.

---

*The U.S. Federal Aviation Administration (FAA) and the International Civil Aviation Organization (ICAO) regularly publish estimates of the direct operating costs of aircraft, including fuel, pilots and crew, insurance, depreciation, and maintenance. However, they neither attempt to measure the multiplier effect that delays have on lower productivity of the aircraft or facilities in the airport passenger buildings, nor do they account for any value of time for the passengers. These estimates are thus conservative estimates of the cost of delays to the airlines and their passengers. Recent figures estimate the per-minute costs of the Boeing 747, Boeing 777, and Airbus 340 at $110, $140, and $70, respectively (FAA, no date; ICAO, 2001).

†See footnote on p. 562.

Owners' perspective

Owners generally want their airports to be both glorious and economically efficient. Airports are major public facilities that can adorn the community. National leaders and governments thus often want their airports to be monumental gateways to their country, as indicated by the chief architect for the Aéroports de Paris:

> *First and foremost, airports are places of great symbolic force...airports are now the place where travelers first come into contact with their destinations: an age-old legacy reminiscent of the "entrance gate" of walled cities....As a gateway in its own right, an airport is almost inevitably destined to be a landmark of great symbolic force, embodying the ambitions of a nation in which it stands. (Andreu, 1997, p. 11)*

To further this natural ambition, airport owners hold international competitions to choose architects who will celebrate a grand vision. Their results are often spectacular. Examples include:

- London/Stansted, a project the British Airports Authority (the name of BAA plc before it was privatized) commissioned. Sir Norman Foster created a translucent glass box unencumbered by air bridges (hence the separate satellites) and designed it right down to details of the shape of the check-in counters
- Kuala Lumpur/International—Dr. Mahathir, President of Malaysia, personally validated the design and opened the airport for the Commonwealth Games of 1998
- Washington/Dulles—the U.S. government built this as a national gateway based on the design by Eero Saarinen (before it devolved this airport and Washington/Reagan to the Metropolitan Washington Airports Authority).

The concept of the airport as a national monument often conflicts with the goal of economic efficiency. Magnificent curved structures (like the Renzo Piano design for Osaka/Kansai) are difficult to construct, expensive to maintain, and nearly impossible to expand compatibly. Unique interior details (such as Sir Norman Foster's for London/Stansted) have to be custom-tailored and correspondingly are both expensive and difficult to maintain. (See Binney, 1999, for details.) This conflict is not a problem so long as the airport owners remain committed to paying the extra costs required to maintain a monumental concept. Once the airport is operational, however, airport operators typically become more interested in costs and profits.

Airport operators generally neither want nor can afford to maintain airports as monuments. Moreover, the owners who actually run the airport are not those who commissioned the passenger buildings. The personalities change. Most important, the institutions change and have new responsibilities. Thus now:

- BAA plc, a privatized stockholder company, operates London/Stansted and requires all projects to meet stringent financial objectives.
- A privatized company runs Kuala Lumpur/International and expects to do so profitably.
- The Metropolitan Washington Airports Authority depends on private capital to construct its expansion and operations, and must operate economically to repay the loans it incurs.

Airport operators thus typically emphasize airport economics. In practice, they generally insist on controlling the costs of airport passenger buildings. Because they normally operate within specified budgets, they focus on overall costs. They routinely scale back the scope of their projects, or cut out various functions (as Kuala Lumpur did when they eliminated the shopping area, as cited in an example at the beginning of this section).

Sophisticated airport operators who expect to be in business for the long term will think about costs over the entire life of the passenger buildings. They consider both the immediate cost of construction and the future costs of expansion. This perspective has two consequences for how they evaluate the configuration of the airport passenger buildings. Airport operators with vision will insist these facilities allow for economical:

- Expansion, for example by allowing space for extending finger piers, as Kuala Lumpur/International did
- Adjustment to different operating conditions, for example, a shift in the level of transfer or international traffic, as occurred at Frankfurt/Main and Baltimore/Washington, respectively

Perceptive airport owners and operators thus insist that the configuration of airport passenger buildings allow for flexible future use. They will correspondingly reject designs that, while great for specific types of traffic, are difficult to reconfigure for alternative airline operations, types of passengers, or even different political situations (that result in changes in government regulations regarding security controls, the separation of arriving and departing passengers, etc.).

Retail perspective

Retail operators want traffic, persons ready to shop, visibility, access, and a coherent environment. This is what they look for when choosing store locations; this is what good developers of shopping centers provide. When planners choose the configuration of passenger buildings, they can significantly affect the ability of the airport to meet these criteria and provide a profitable retail area.

The number of people flowing by a storefront is an immediate measure of the potential attractiveness of a retail location. More people equal more potential customers. Configurations that concentrate traffic thus provide more attractive commercial areas. Any arrangement that features a central area is therefore more attractive to stores than buildings that have many entrances and exits. A good example of this is at Amsterdam/Schiphol, where the airport has created a busy shopping plaza at the main entrance to its building complex, right above the underground railroad station.

The traffic must not only exist, it must also be ready to shop. Passengers in a hurry or anxious to get somewhere else are not good for retail operators. Thus, at Washington/Reagan, the stores in the new building that are between the ground access and the security checks get relatively little business. The travelers typically rush by to get to their gates, to see if they can get on the next hourly flight. People waiting around, looking for things to do and ready to respond to impulsive desires, are best for the stores. Passenger buildings with centralized common waiting areas, such as those at Frankfurt/Main or Singapore, are thus attractive to retailers.

Visibility is also crucial for retail operators. A store not seen is a store not used. The retail area must not only be in some kind of central area, but also be able to announce itself. (Airport operators need to be careful how they allow stores to advertise, however. Too many signs can hide important airline signs from passengers.) Configurations that lead the pedestrian traffic through shopping streets or pass the stores will be more successful. The arrangement at Pittsburgh is a good example of this: a large percentage of the travelers flow through the central space of the X-shaped building that functions as a shopping mall. Conversely, facilities located out of sight, on a floor above or behind the pedestrian flow, will not get many customers.

Retail operators are also concerned with accessibility for their goods. They cannot function economically unless they are able to get their merchandise easily in their stores and to their customers. Access is more a matter of detailed design than of configuration, but poor arrangements can constrain retail operations, and planners need to verify access when they choose the overall design.

The important point is that designers need to incorporate the perspective of retail operators from the beginning. The airport needs a coherent overall plan for how it will organize its commercial space. Good retail operations, producing the best revenues for the airport, result from good locations. The choice of the configuration of the passenger buildings is an important factor determining the success of the commercial operations. Master planners should have knowledgeable retail detail consultants at hand when they are designing the configuration of airport passenger buildings.

Government agencies

Government agencies constitute a particular set of stakeholders that needs to be consulted carefully. They deserve special attention, both for their specific needs, and because their procedures and modes of negotiation are different from those of other stakeholders.

Government agencies have a particularly strong stake in the airport whenever it is an international port of entry, and provides border control and customs services. Border control agencies often impose a broad range of requirements on the area, reaching far beyond their needs in the specific areas they occupy. They typically require tight controls on the flows of the international passengers, and on the mingling of international flights and crews with local employees and services. These requirements greatly complicate the design of the building, and may preclude more efficient arrangements that airport operators can develop for domestic services.

Negotiations with government agencies can be very different than dealing with other stakeholders such as the airlines, travelers, and retailers. Government security groups are typically inflexible. Officials normally do not have the authority or the right to alter established rules, even when the proposed changes might be in everyone's interests. Most especially, they are rarely amenable to arguments based on economic efficiency, in sharp contrast to stakeholders with commercial interests, who are ready to listen to proposals that may save them money or increase their efficiency. Listening to the interests of government stakeholders is thus fundamentally important for developers of international gateways.

Balance

Overall, airport operators need to balance their own strategy and the interests of the several stakeholders. The airport operators need, first, to be clear about their own strategy. Are the operators managing the airport as a public service or as a profit-making venture? Are they catering primarily to local customers, or do they have the vocation to provide a transfer hub? Do they see their primary business as serving travelers and shippers, or developing stores and businesses on and around the airport? (See Chap. 2.) Their strategic decisions along these lines should influence how they balance the many conflicting interests of the stakeholders in the airport.

To preserve their own interests, airport operators must also carefully weigh the demands of the several stakeholders. Major clients, such as a dominant airline at an airport, would gladly advance their own interests to the harm of other users and have the airport arrange the design to suit themselves. Airport operators, however, need to maintain control of their properties, to mediate appropriately between the distinct needs and desires of their stakeholders. It is essential to this process that they listen to these stakeholders—not doing so is the same as treating their needs as insignificant.

Since forecasts are "always wrong" (Chap. 3), the design process should also consider the performance of passenger buildings under multiple scenarios. Modern design seeking the best overall performance over the long term thus needs to take a much broader perspective than has been traditional (Table 14-2).

14-3   Five basic configurations

Designers of airport passenger buildings face a fundamental problem: they need both to concentrate and spread them out. They need to bring passengers into common areas to facilitate check-in procedures, retail opportunities, and access by public transport. They must also spread out the passengers so that they can board their aircraft. The large wingspan of aircraft (see Table 14-3) imposes correspondingly long separations between the gates leading to adjacent aircraft. The sideways distances between gates must be in the range of 50 to 85 m, allowing for clearance between aircraft. All configurations of passenger buildings represent approaches to resolve this fundamental dilemma.

The possibilities for resolving this conflict changed greatly in the last decades of the twentieth century with the development of cost-effective

Table 14-2. Modern design needs to evaluate the performance of passenger buildings with criteria from major stakeholders and under a range of possible scenarios, rather than from a narrow perspective under a single forecast

	Criteria considered	
Forecast	**Single (or few)**	**Multiple**
Single	Traditional Approach: "terminals"	
Broad		Recommended Approach: "airport passenger buildings"

Table 14-3. Representative aircraft wingspans

Aircraft	Wingspan (m)	Aircraft	Wingspan (m)
Airbus A380	80	Boeing 747-400	64
Airbus A300	44	Boeing 777	61
Airbus A310	43	Boeing 767-200	47
McDonnell MD11	52	Boeing 757-200	37

"people movers" (Chap. 17). These devices are small trains or horizontal elevators. They speed people away from a central point, such as a check-in hall, to buildings spread out over the airport. They make it practical to spread passenger buildings over several kilometers, and have led to the widespread implementation of midfield concourses at major airports. This technical innovation has made obsolete many previous conclusions about which configurations of airport passenger buildings are best.

This section describes the basic configurations of airport passenger buildings from a functional point of view. The subsequent sections first show how to analyze some of their essential elements of performance, and then summarize the overall attractiveness of these buildings. The underlying questions throughout are whether, and to what extent, these facilities fulfill the functional requirements of the several stakeholders.*

There are five basic configurations of airport passenger buildings suitable for a major airport. (For minor airports, needing only four gates

---

*This perspective is different from the many architectural descriptions of airport buildings, such as those of Andreu (1997), Blow (1996), Hart (1985), and Powell (1992).

for example, the passenger building can be a simple box.) Designers shape these possibilities in a number of ways. They also combine these forms into hybrid configurations incorporating two or more distinct forms. As Fig. 14-1 illustrates, the basic configurations are:

- Finger piers
- Satellites, with or without finger piers
- Midfield, either linear or X-shaped
- Linear, with only one side devoted to aircraft
- Transporters

At large airports, the buildings may be centralized or dispersed.

Finger piers

Finger pier configurations are simply relatively narrow extensions to a central passenger facility. In plan view as seen from the air, they resemble fingers attached to the palm of a hand—hence the name. Their obvious form places aircraft gates on both sides of the building extending away from the central core. This arrangement has the advantage of placing some aircraft gates close to the central facility, and thus more conveniently for the passengers than the gates at the end of the finger pier.

An alternative arrangement widens the end of the finger pier so that it looks like a T in plan view. This arrangement is also known as a "hammerhead." The end of this pier thus serves a number of aircraft around a small central core (located in the crosspiece of the T). This concentration of passengers in a single space has the advantage of permitting shared use of facilities and thus of decreasing the space required for lounges by 30 percent or more (see Chap. 16). The

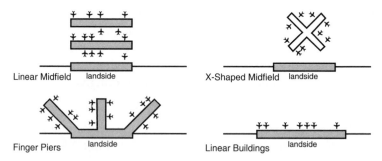

Fig. 14-1. *Sketches of basic configurations of passenger buildings.* (**Source:** de Neufville, et al., 2002.)

number and steady flow of passengers using the range of gates also increases the attractiveness and profitability of retail opportunities. A principal disadvantage of this plan, however, is that it places many aircraft and passengers far from the central part of the main passenger building, and forces passengers to walk farther.

Designers introduced the use of finger piers in the 1950s as the first response to the need to serve dozens of gates from a central check-in hall. For several decades, finger piers constituted the standard configuration. Airports all over the world built passenger buildings with finger piers—for example, New York/LaGuardia, Chicago/O'Hare, San Francisco/International, London/Heathrow, Paris/Orly, Frankfurt/Main, and others.

The difficulty with finger pier configurations is that, at large airports with many gates, they lead to long walking distances for passengers. To avoid long hikes, airport designers no longer propose finger piers as extensively as they formerly did. Instead, they prefer when possible to replace the longer fingers with people movers that serve independent buildings such as satellites or midfield concourses, as discussed below.

Many airports continue to implement finger piers in some fashion. By the end of the twentieth century, however, designers tended to minimize walking distances either by designing short finger piers (as at the new building at Washington/Reagan) or by incorporating a people mover (as at Osaka/Kansai). In this vein, Fig. 14-2 shows the design for the new Nagoya/Chubu airport.

Satellites

Satellites are the logical extension of T-shaped finger piers. They eliminate the gates along the fingers and concentrate gates at the end. Generally, the connection between the satellite and the central check-in area is aboveground. In some designs, designers place the finger underground and it becomes invisible. The satellite is sometimes connected to the central part of the passenger building by a people mover, sometimes not. Table 14-4 and Fig. 14-3 indicate some examples of the possibilities.

Satellites with underground connections to the main building have a singular advantage. They make it possible for aircraft to maneuver freely around the satellite. This facilitates aircraft operations and saves the airlines time and money. This is the arrangement for Terminal 1

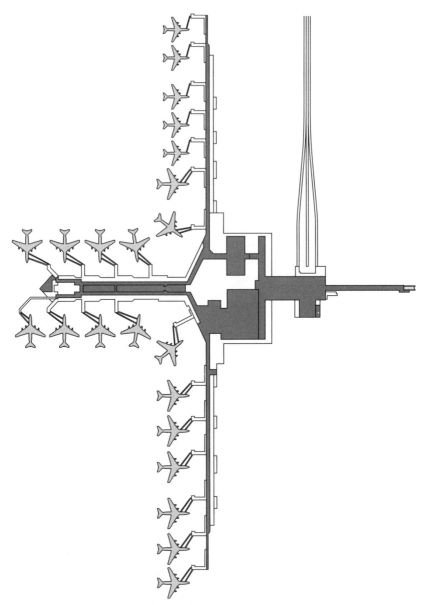

Fig. 14-2. *Finger pier design for proposed Nagoya/Chubu airport.*
(*Source:* Pacific Consultants International.)

at Paris/de Gaulle and at Seattle/Tacoma. The detached facilities at Seattle/Tacoma are connected to the main passenger building by a people mover and are functionally very close to what airport professionals now label as midfield concourses. (Whether the configuration at Seattle/Tacoma consists of satellites or midfield concourses

Table 14-4. Examples of arrangements for satellite passenger buildings

Connection to main building is	People mover	
	No	Yes
Aboveground	Milan/Malpensa	Tampa
	Tokyo/Narita (Terminal 1)	Tokyo/Narita (Terminal 2)
Underground	Paris/de Gaulle (Terminal 1)	Seattle/Tacoma
	Geneva	

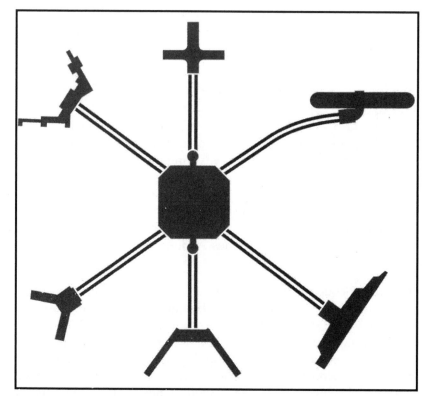

Fig. 14-3. *Layout of satellites at Tampa.*

is debatable. The local authorities have always called their buildings satellites. The Seattle/Tacoma buildings are a bit smaller than the typical midfield facilities that have been constructed around the turn of the century. From a practical perspective, the name does not matter.)

Midfield concourses

Midfield concourses are major independent passenger buildings, often located far from the passenger building, which passengers access from the groundside. They may easily have around 50 gates and be about a kilometer long. The linear Concourse B occupied by United Airlines at Denver/International is 990 m (3300 ft) long and has 46 air-bridge gates at the main building (a ground-level extension to this building serves additional positions for small aircraft). The X-shaped midfield concourse at Pittsburgh serves 75 gates. Midfield concourses are typically between two parallel runways and separated from the other passenger buildings by major taxiways. However, the midfield concourse can be located on the edge of the runways as part of a complex of passenger buildings, as the United Airlines midfield concourse is at Chicago/O'Hare. Generally speaking, midfield concourses differ from satellites in their size and distance from the groundside, but this distinction is not firm.

Because of the distances and number of people involved, passengers usually access midfield concourses by self-propelled people movers.* Indeed, this innovation has been almost indispensable for the development and operation of midfield concourses. (Washington/Dulles is an exception to this rule. In that case, United Airlines developed its midfield concourse around the existing transporter system. As of 2002, however, Washington/Dulles plans to build an underground people mover to serve the midfield concourse.) Reliable, economical people movers have in fact transformed the possibilities for the design of airport passenger buildings and other landside facilities such as parking lots and car rental facilities. Midfield concourses are a major expression of this innovation.

Midfield concourses come in two basic shapes: linear and X-shaped. Linear concourses are simply long buildings with aircraft positions on both sides (see Fig. 14-4). They are frequently wider in the middle section, around the people mover station, to accommodate this facility, provide a central shopping area, and serve larger aircraft and their numerous passengers. They are typically flanked by dual parallel taxiways that allow aircraft to move between their gates and the runways with a minimum of turns and delays. Atlanta and Denver/International have built their entire airport around linear

---

*People movers are either self-propelled or pulled by cable as an elevator. Whereas long cable systems are possible, and even necessary for safety reasons in the tallest skyscrapers, self-propelled vehicles have been more economical over longer distances with complex, multiple routes. Thus, all midfield concourses so far have been accessed by self-propelled people movers. See Chap. 17.

Fig. 14-4. *Midfield passenger buildings at London/Stansted.*
(***Source:*** BAA plc.)

midfield concourses. Chicago/O'Hare and Munich have built single mid-field linear concourses for the transfer operations of United Airlines and Lufthansa, respectively.

X-shaped midfield concourses feature intersecting fingers that give them the X-shape. Normally, the crosspieces are oriented at about 45 and 135° with respect to parallel runways that flank the midfield concourses. This version appears particularly suitable when space is limited. Indeed, Pittsburgh and Hong Kong/Chek Lap Kok airports feature this configuration, and both airports are constrained by difficult terrain that limit the space between their parallel runways. (The former is on leveled hills surrounded by deep valleys; the latter is on a man-made island.) At airports where the distance between the parallel runways is relatively tight, designers effectively have the choice between many short linear mid-concourses perpendicular to the parallel runways, or intersecting slanted midfield concourses that allow them to place more aircraft at a single building. Although parallel diagonal linear concourses should be technically possible, they have not been considered. The X-shaped configurations implemented so far have only used one midfield concourse, in contrast to Atlanta, Denver/International, and London/Stansted, which use several parallel concourses

The cross or +-shaped midfield concourse is a variant on the X-shaped pattern. It is an X rotated 45° so that the crosspieces are perpendicular or parallel to the flanking parallel runways. This is the design implemented at Kuala Lumpur/International. In practice, the arms of the cross-shaped midfield design are relatively short, to minimize the maximum walking distances within the building. As of 2002, the plans to implement cross-shaped midfield concourses at Kuala Lumpur/International and at the Second Bangkok International Airport call for two rows of satellites between the widely spaced parallel runways. The motivation for cross-shaped midfield concourses has not been to make the best use of a tight space, but to try to provide a better environment for passengers.

Some designers have preferred cross-shaped concourses on the grounds that this configuration reduces walking distances. This was the rationale for the design of Kuala Lumpur/International. The argument is that the maximum distance from the center of the building to the farthest end is less for an X-shaped building than a linear concourse, if both have the same number of gates. In fact, however, the X-shaped concourse actually increases the effective walking distances for most passengers. This is because the center of the X-shape cannot accommodate aircraft, which means that:

- Gates cannot be located at the center of the concourse as they can be for linear concourses
- Larger aircraft are positioned at the ends of the X

Both features combine to raise the average walking distance experienced by most passengers. The X-shaped design also complicates airline operations, as compared to linear midfield concourses. This is because the X-shaped configuration involves more turns and delays. Section 14-3 explains these points further.

Linear buildings

A *linear building* is a long, relatively thin structure with one side devoted to aircraft and the other faced by roads and parking lots.* Designers came up with the concept of linear buildings in response to

---

*Blow (1996) uses the term "linear" to refer to buildings whose finger piers, with aircraft gates on both sides, form a straight line with the face of the central block. This configuration is rather like a ⊣ in plan view. An example of this shape is Terminal 4 at London/Heathrow, occupied by British Airways. From the functional perspective, Blow's linear terminals are equivalent to finger piers. Note that when these finger piers can only be accessed from a central block, but have aircraft gates only on one side, they combine some of the worst features of finger piers and linear buildings. They use the building inefficiently (since they need twice as much length for the same number of gates) and they entail long walks (being finger piers).

the great walking distances associated with finger piers. They originally called it the "gate arrival" concept. The idea was that people could drive or be driven right up to their departure gate, park their cars if necessary, and get to their flight by walking through a narrow building.

Several airports, including Dallas/Fort Worth, Kansas City, Paris/de Gaulle (Terminals 2 A–D), and Munich (see Fig. 14-5) built gate arrival buildings in the 1970s and 1980s. Several of these "linear" passenger buildings in fact curved around an interior landside parking area. This curved plan gives architectural interest to a structure that might otherwise appear boring, and has the advantage of providing more frontage on the aircraft side, where it is needed to accommodate aircraft wingspans (see Andreu, 1997). The disadvantage of curved plans in general is that they complicate both the initial construction and subsequent landside traffic flows.

Designers have now generally dropped the notion of "gate arrival" buildings. This is because it is on balance inefficient, unproductive, and impractical to have passengers flow directly from curbside to their aircraft. This scheme is inefficient because it implies that check-in and security facilities have to be duplicated in front of each gate, instead of combined in central services that cater to passengers for

Fig. 14-5. *Linear passenger building at Munich.* (*Source:* Munich International Airport.)

many gates simultaneously. It is unproductive because it virtually eliminates the possibility of significant retail areas, since single gates do not provide enough traffic to justify important stores. It is impractical because passengers cannot count on finding parking spaces or ground transport right in front of their gates, let alone on returning to the same gate to have easy access to their car.

As a practical matter, designers now generally give linear passenger buildings only a few access points. Passengers thus arrive broadly along the front of the building, depending on where space is available, flow to some central area for check-in, security, and shopping, and then proceed out to their gates. The linear concept in this configuration thus amounts to a finger pier. As a result, this linear configuration does not minimize walking distances as much as designers originally imagined. Ironically, the "gate arrival" concept, designed to minimize walking distances, has evolved into a configuration in which these distances are significant.

Transporters

Transporters are the broad category of rubber-tired vehicles that move passengers between passenger buildings and aircraft. Most simply, these are specially designed buses with low platforms and wide aisles for easy access for passengers with bags. These airport buses require passengers to walk up and down the stairs between the airport apron and the aircraft door. Major airports in Europe and Asia commonly use these vehicles, for example, Paris/de Gaulle, Zürich, Berlin (see Fig. 14-6), and Tokyo/Narita.

More complex versions of transporters have a cabin that can be raised and lowered. These are generically known as lift lounges. Mechanically, they are similar to the range of catering vehicles that service aircraft, in which large hydraulically operated scissors or screw jacks move the cabin up or down. Lift lounges are much larger than catering vehicles. They are designed to carry about 80 to 100 passengers. In operation, the lift lounges let passengers board at the normal elevated level associated with departures through airbridges, lower the cabin for travel to the aircraft parked at a remote stand, and then raise the entire passenger compartment to the level of the aircraft door, making it possible for passengers to enter the aircraft horizontally. These devices avoid the many problems associated getting passengers between the aircraft and the ground by means of stairs. They keep the passengers inside at all times, thus avoiding the climatic extremes of

Fig. 14-6. *Transporter in operation in Berlin.* (*Source:* Berlin International Airports.)

heat or cold, snow or rain. They also speed up the loading and un-loading processes (since for many people getting up a long flight of stairs with carry-on baggage is an athletic challenge) and thus increase the productivity of the aircraft and crew. Lift lounges are particularly expensive, however.

Designers developed the transporter configuration as a way of avoiding the long walks and cost of construction of finger piers. The trans-porters take passengers to their aircraft, parked wherever convenient on the apron, directly from a central passenger building housing check-in, baggage claim, stores, and other facilities. This configuration has the obvious advantages of minimizing walking distances, elimi-nating significant construction costs, and freeing aircraft from the difficulties of docking at passenger buildings.

Transporters pose significant operational difficulties, however. Most obviously, to airport operators and airlines, they are expensive. The manufacturers produce these peculiar vehicles in small numbers and cannot obtain the economies of scale associated with using their equipment to make many similar items. Transporters require specially trained drivers who can navigate safely on the airfield and deal with the special requirements of their vehicles. Furthermore, trans-porters in operation must be constantly fueled and maintained (see the discussion on transporter economics in Example 14-2). Finally, transporters present a peculiar risk for airport operators: they place a small group of skilled workers, the drivers of the transporters, in the position of being able to shut down airport operations if their demands are not met.

Transporters also present difficulties for the airlines and their passengers. The use of transporters adds 10–15 min to a flight, because of the time it takes to load and unload these vehicles. These delays are particularly inconvenient on short-haul flights and in transfer operations. Additionally, the less expensive airport busses offer inferior levels of service because they force passengers to go out into the weather and cope with stairs.

All these inconveniences of transporters typically outweigh their advantages except in special situations, which Secs. 14-4 and 14-6 identify. The result is that the pure transporter configuration is practically unique to Washington/Dulles airport.* Even that configuration is vanishing, since the operator of the airport plans to complete its transformation to a midfield concourse serviced by an underground people mover instead of lift lounges. Although the pure form of transporter configuration is practically obsolete, transporters are now used widely and effectively as part of hybrid configurations, in particular for seasonal and low-fare operations, as Sec. 14-5 describes.

Centralized and dispersed

Designers can also choose to centralize or decentralize whatever configuration they choose. A centralized version, such as the design for the new Second Bangkok International Airport, provides a single point of access to the airport and is convenient for rail and other forms of public transport. A dispersed or decentralized concept substitutes smaller buildings, at a more human scale with shorter walking distances, for the single massive structure. This arrangement is sometimes called the "unit terminal" concept. The decentralized configuration works well for airlines or airline alliances that have distinct operations. At New York/Kennedy for example, American, Delta, and United Airlines all have their own complexes serving themselves and their alliance partners. Figure 14-7 illustrates a decentralized configuration of passenger buildings.

*Montreal/Mirabel was also designed to rely on transporters. As of 2002, however, this airport was virtually closed to regularly scheduled passenger flights. All scheduled airlines had relocated to Montreal/Dorval, close to downtown, leaving only charter and cargo flights at Montreal/Mirabel.

Fig. 14-7. *Decentralized passenger buildings at New York/Newark.*
(*Source:* Port Authority of New York and New Jersey.)

The decentralized configuration has several disadvantages, however. Separate buildings

- Complicate transfers between them, as the example of Kansas City cited in Sec. 14-1 indicates
- Inhibit the growth of airlines at an airport, as they find it difficult to operate in distinct buildings, as the experience of American Airlines at Dallas/Fort Worth demonstrates
- Make it difficult to have a central rail station that is convenient for all passengers, as the cases of London/Heathrow and Paris/de Gaulle show

14-4 Evaluation of configurations

Which configuration of airport passenger buildings is best? That is a basic question for airport owners and designers. The discussion and examples so far indicate that there are too many factors, and too many stakeholders with different concerns, as to make it impossible to satisfy all stakeholders completely. There can be no universal answer best for all. Furthermore, the variety of designs being implemented shows that there is no consensus among designers. How should owners and designers proceed?

The basic principle in deciding which configuration is best is to take a comprehensive, analytic view of the issues involved—a systems approach in short. Any single-dimensional approach will leave out too many factors and risk operational disasters (consider, for example, the "gate arrival" design at Kansas City, which focused narrowly on walking distances of local passengers). Acting on intuition alone without analysis will frequently lead to error (for example, assuming that cross-shaped midfield concourses minimize walking distances).

Systems designers recognize that, while it is impossible to define a solution that is best in all cases, it has been possible through systems analysis to determine some fundamental characteristics of good design. Experience shows that the best design:

- Depends on the specific circumstances, the site, the type of traffic, and the needs of the several stakeholders
- Includes features that cater to the specific needs of the variety of clients and stakeholders, and is thus unlikely to be described by a simple concept
- Is flexible, so that it can deal with the changing needs of the clients and stakeholders over the life of the project

Section 14-5 provides a guide of which configurations are best in which circumstances. This section presents four key considerations that support those guidelines. These focus on:

- *Walking distances,* a factor that has motivated designers to search for better configurations—the analysis demonstrates that simplistic intuitions based on geometrical measures of distances are deceptive and frequently wrong.
- *Aircraft taxiing around the buildings,* which involves substantial costs and is fundamentally important to the major clients of an airport and, ultimately, to the competitiveness of an airport compared to other airports.
- *Transporter economics,* to indicate when these vehicles provide a cost-effective complement to passenger buildings
- *Flexibility,* specifically the ability to adapt to different types of traffic as they evolve, such as the development of transfer or international traffic into a significant share of the market at an airport.

Walking distances

Spreadsheet programs such as Excel provide the way to investigate the effect of the configuration of airport passenger buildings on walking distances (de Neufville, de Barros, and Belin, 2002). Using spreadsheets, designers can rapidly analyze the performance of different designs for any distribution of traffic, both between the landside entrances and the gates, and between gates for transfer traffic. Using the "data table" functions embedded in spreadsheets, they can also easily run parametric analyses for a wide range of conditions. These analyses require little time to create and run. Anyone familiar with spreadsheet programs can create their basic elements in a day or so for any specific airport.

The procedure for analyzing walking distances in airport buildings uses two origin-to-destination matrices: *impedance* and *flow matrices*. Each of these matrices of basic data captures different aspects of the traffic within the airport passenger building.

An *impedance matrix* defines the level of difficulty in transiting between any gate (or access point to the passenger building) and any other. It describes the physical aspects of the facility.

Most immediately, it reflects the geometry of the building. Most straightforwardly, it defines the walking distances between these origins and destinations. It could, however, represent travel time or some modified measure of distance that accounts for either the benefits of moving sidewalks, people movers, or other devices, or the inconvenience and delays due to stairways, security checks, or other barriers to movement.

A *flow matrix* defines the volume of passengers moving between each origin and destination represented in the impedance matrix. It embodies operational information about the airport passenger building. It accounts for two elements of the issue that purely geometric analyses ignore. Specifically, it reflects

- *Transfer patterns,* that is, the passenger flows from one aircraft to another
- *Intelligent management of gate assignments,* whereby either airline managers or airport operators place flights with significant transfer traffic at gates close to each other

The impedance and flow matrices conveniently represent all the basic information on the passenger movements within airport passenger buildings. Their distinct functions facilitate analyses of different issues. Architects, for example, can investigate the effect of alternative configurations of the buildings for any airport with specific level of transfers. Airport managers can examine the implications of different operational strategies for assigning gates to aircraft.

It is easy to calculate all the interesting statistics on walking distances using these matrices. Multiplying them gives a *passenger-impedance matrix* in which each cell represents the impedance between each origin and destination, weighted by the number of passengers. For example, if the impedance is measured in meters to be walked, each cell in the passenger-impedance matrix represents the passenger-meters walked by the traffic between the corresponding origin and destination in the building. Summing these results and dividing by the total traffic gives the average walking distance (see Example 14.1). Sorting the cells by distance, and summing the corresponding passenger impedances expressed in terms of percentage of the total, permits the analyst to develop cumulative passenger-impedance diagrams that show the proportion of passengers walking specified distances for any situation, so that the best performance is one farthest to the left.

EXAMPLE 14.1

This example illustrates the spreadsheet method for calculating walking distances. It also demonstrates that the average walking distance, when the airport operators allocate the aircraft to gates intelligently so that connecting, originating, and terminating traffic are close to their gates, is considerably less than a purely geometric analysis would suggest.

Consider a finger pier with 4 gates and a point of access to the main body of the passenger building. It is 18 m (60 ft) wide. Its gates are 60 m (200 ft) apart and laid out for entry on the left hand side of the aircraft (Fig. 14-8). Table 14-5 shows its impedance matrix in meters.

Each gate is occupied by aircraft with 100 passengers; there are transfers, particularly from gate 4 to gate 2, and the airport operator has intelligently placed these aircraft close together. Table 14-6 shows the assumed flow matrix. The summations at the right hand side and the bottom indicate that each of the 4 aircraft arrives and departs with 100 passengers, and that 220 passengers enter and exit through the end of the finger pier. This means that 180 passengers transferred or stayed on board their aircraft.

The resulting passenger-impedance matrix is then as in Table 14-7. The totals on the right-hand side indicate that the total passenger-meters walked in the finger pier is 42,630. This implies an average distance per person of 68.76 m (= 42,630/620). As readers can verify, this is far less than either the average of 74.67 m if there were no transfers or the maximum possible distance of 129 m. Looking carefully at what passengers actually experience, especially with intelligent management of the aircraft stands, substantially improves on the impression that might be derived from a simplistic assessment of the geometry of a facility.

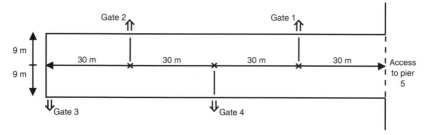

Fig. 14-8. *Sketch of finger pier for Example 14.1.*

Table 14-5. Impedance matrix (meters) for example finger pier

To point	From point				
	1	**2**	**3**	**4**	**5**
1	0	78	108	48	39
2	78	0	48	48	99
3	108	48	0	78	129
4	48	48	78	0	69
5	39	99	129	69	0

Table 14-6. Flow matrix (passengers) for example finger pier

To point	From point					
	1	**2**	**3**	**4**	**5**	**Total**
1	10	0	0	0	90	100
2	0	20	10	40	30	100
3	25	10	20	15	30	100
4	10	0	0	15	70	100
5	50	70	70	30	0	220
Total	100	100	100	100	220	620

Table 14-7. Passenger-impedance matrix (passenger-meters) for example finger pier

To point	From point					
	1	**2**	**3**	**4**	**5**	**Total**
1	0	0	0	0	3,510	3,510
2	0	0	480	1,920	2,970	5,370
3	2,700	480	0	1,170	3,870	8,220
4	720	0	0	0	4,830	5,550
5	1,950	6,930	9,030	2,070	0	19,980
Total						42,630

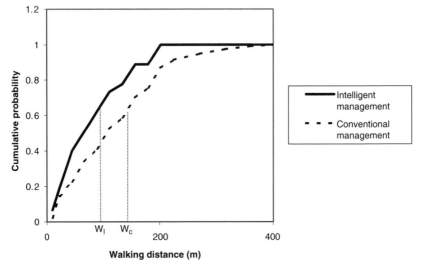

Fig. 14-9. *Passenger walking distances with and without intelligent gate assignment for an example linear midfield concourse (high transfer rate, 60 percent).* (*Source:* de Neufville et al., 2002.)

Designers should note carefully how Fig. 14-9 demonstrates the importance of intelligent management of the gates on walking distances. Managers of gate assignments can dramatically reduce the walking distances passengers experience, by placing connecting flights near each other, by placing larger aircraft near the exits, etc. Thus, although they cannot affect the maximum distance passengers experience, they can sometimes reduce the average by as much as a third.

Table 14-8 summarizes the results of applying this analysis to example buildings (see de Neufville, et al., 2002, for details). Designers should

Table 14-8. Relative performance of airport passenger buildings assuming intelligent management of gates, in terms of walking distances, when serving high and low levels of transfer traffic (data for example 20-gate building)

Configuration overall	Specific form	Average walking distance (m/person)	
		60% transfer rate	No transfers
Midfield concourse	Linear	90	109
	X-shaped	134	136
Finger piers		202	316
Linear building,	3 entrance points	109	98
One airside	1 entrance point	144	157

Source: de Neufville, et al., 2002.

carefully notice that the results depend significantly on the level of transfer traffic. Moreover, the reader should appreciate that these results assume that the airport operators practice intelligent gate assignment; if connecting aircraft are far apart, the results degrade significantly. In any event, the desirability of the configuration depends on the current and anticipated level of transfer traffic. Overall, this analysis suggests the advantages of intelligently managed linear midfield concourses and the disadvantages of finger piers with regard to walking distances.

Linear midfield concourses minimize average walking distances better than X-shaped configurations. Comparing buildings serving the same number of gates, the X-shaped buildings of course reduce the maximum walking distance, since they spread the gates in four directions rather than two. This advantage is lost, however, when it comes to average walking distances. X-shaped buildings have reentrant corners (that is, with about 90° of open area) at the center that not only make it impossible to locate aircraft there, but in fact force managers to place larger aircraft toward the ends of the piers.

The relative performance of linear midfield passenger buildings and finger piers depends on the effort required to transit between the midfield concourse and the landside, and the number of passengers who are not transferring and must cross this distance. For transfer passengers, either building performs well, to the extent that the managers can cluster the aircraft along the pier or can use a shared lounge area at the end of the finger pier. For local passengers, the midfield linear concourse is superior with regard to walking distances within the building itself. This is because it enables managers to position large aircraft at the entrance located conveniently in the middle of the building, and thus to minimize walking between the entrance and the aircraft. Which of the two configurations is better with regard to overall walking distance depends on the importance and discomfort of the travel between the midfield concourse and the landside.

Linear buildings with one airside and one landside perform well for originating passengers but poorly for transfers. In principle, they minimize the walking distance between the curb and the plane. In practice, however, this advantage may be lost because airport managers may limit the number of access points and thus reduce the cost of security checkpoints—as they do at Dallas/Fort Worth, for example. When the number of access points to a linear building is limited, the walking distances can be relatively long for any sizable building.

Transfer passengers in linear buildings with one airside find that, even with intelligent management of the gate assignments, their walking distances are necessarily relatively long. This is because a linear building with one airside is basically twice as long as it would be if gates were on both sides of the building (as they are in a midfield concourse or a finger pier). The excessive walking distances for transfer passengers was one reason TWA transferred its base of operations from Kansas City to Saint Louis when it set up a hub-and-spoke system to serve transfer traffic. However, the walking distances for local passengers can be reduced significantly if decentralized facilities are used, including the provision of several entrance points to the building.

Aircraft delays

Planners can estimate aircraft delays due to terminal configurations in the same way as they can estimate walking delays, provided the area around the passenger building is not very busy and aircraft do not have to wait for each other. Such situations are likely to prevail around new buildings, since designers will be careful to provide sufficient taxiway capacity. Where congestion is likely or already exists, designers should use detailed simulations of airfield traffic (see Chap. 19).

A general result from the analysis of aircraft delays concerns midfield concourses. Equivalent linear midfield concourses incur less delay for aircraft than the +-shaped concourses at Kuala Lumpur/International. Because the linear concourse allows for direct access to the gate from the taxiway, whereas the alternative arrangement may require several turns around the X or +-shape, the linear concourse reduces the average taxi distance around the passenger building by 25 percent and halves the number of turns. This implies savings of about 1 min or at least $50 per operation. Summed over several tens of thousands of operations a year, the advantage of the linear midfield concourse is of the order of millions of dollars a year (Svrcek, 1994).

Transporter economics

Managers of airports with strong seasonal variations in passenger traffic should seriously consider using transporters to provide "gate" capacity in these peak periods. This is because the cost of transporters can be minimized when they are not needed; they can be parked and then do not need drivers, power, cleaning, or maintenance. Transporters contrast in this way with aircraft gates in passenger buildings. Even when unused, the gates in buildings have to have climate control and be cleaned, and their depreciation costs continue relentlessly. This section demonstrates the economic analysis that defines the number of transporters suitable for a particular airport.

A key question for airport operators is: What is the cheapest, most cost-effective way to provide gate capacity during peak periods? In economic terms, the focus should be on the marginal costs per passenger or aircraft (see Example 14.2). The question is, what is the cost of supplying the gates for the peak period, per unit operation? Determination of the most cost-effective solution depends on two factors:

1. *The difference in cost structure* between operating buildings and transporters—which are vastly different between the two solutions

2. *Utilization of the gate over the season.*

The difference in cost structure between the operation of buildings and transporters is the key to the analysis. This is the reason why transporters can be more economical in peak periods. About three-fourths of the costs of transporters come from operating expenses: the drivers' salaries and the running expenses, which can mostly be

EXAMPLE 14.2

Suppose that a gate is needed only 10 days a year, for 5 operations a day. Suppose further that a gate at a significant airport costs $5 million or about $500,000 a year, including depreciation, maintenance, climate control, and so on. The marginal cost of this gate needed only for the peak period is then $10,000 per operation (= $500,000/50). This is the figure to compare to the cost of operating a transporter for these days.

Note that this marginal cost at the peak period is much higher than the average cost per operation for all the gates, which might be about $238 [= $500,000/(365)(6)]. The marginal cost of a gate in a building rises dramatically as its usage falls. This is the fact that can make transporters economical. This average cost of all the gates is not relevant to the analysis of what to do for the peak periods.

eliminated when the vehicles are not needed. In contrast, almost none of the costs of a gate built into an airport building can be turned off when the gate is not needed. The interest and depreciation on the capital invested in a building will continue even if the gate is not used. Moreover, the gate area will have to be climate controlled and cleaned along with the rest of its passenger building. Consequently, the marginal cost per passenger of a gate in a building rises as its use decreases. If the gate is used only half the year, its cost per passenger doubles (see Example 14.2). Most important, the marginal cost of a gate rises much faster than that of transporters. Eventually, for the low utilization rates associated with stands used only in peak periods, transporters can be more economical than gates in buildings. Example 14.3 illustrates the analysis. Note that the costs used in the example are plausible but not definitive. Building costs can easily differ by a factor of 2, depending on the standards adopted. Labor costs can differ by even more, depending on the salary levels, the benefits, and the number of hours worked. Readers should focus on the method and general results.

Airport managers need to do their own analyses to determine when transporters are economical for them. No general rule is possible due to the large variation in transporters, building costs, and labor conditions worldwide. Much depends, for example, on the extent to which the cost of drivers can be eliminated when transporters are not needed; seasonal or part-time employment is acceptable in some circumstances and not in others. Managers also need to consider the current and likely future degree of seasonal use, which differs significantly between airports and is generally highest at airports catering to holiday traffic. As Fig. 14-10 indicates, at airports catering to business travelers such as New York/LaGuardia, gates provided for the peak months will be used most of the year. At an airport catering more to leisure travelers, however, such as New York/Kennedy, gates provided for the two peak months are not needed throughout the rest of the year. Transporters provide an economically attractive solution at such airports and have been used for years at New York/Kennedy.

Prospective users of transporters should note that while these devices may provide for a large fraction of the number of gates needed in

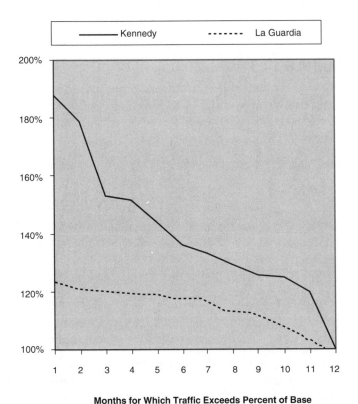

Months for Which Traffic Exceeds Percent of Base

Fig. 14-10. *Seasonal variation in passenger traffic at New York/Kennedy and New York/LaGuardia in 1995.* (*Source:* Port Authority of New York and New Jersey, 1995.)

peak periods, they will be used by relatively few passengers. Consider the implications of the data on New York/Kennedy airport in Fig. 14-10: the extra traffic in the peak two months (over 180 percent of the traffic in the lowest month) is one-sixth or 15 percent higher than that of the next highest month. This extra traffic, which is the sum of the extra traffic over the peak months (mathematically, this is the integral under the curve down to the level of the third month), is only about 2 percent of the total traffic throughout the year (represented by the area under the entire curve plus the 100 percent of the lowest month traffic for the entire year). If an airport operator in a similar situation provided 15 percent of the peak "gate" capacity using transporters, it would only serve only about 2 percent of the total yearly traffic with these devices. Such disparities between the amount of peak capacity and the number of people needing it are common when airports have seasonal peaks. Therefore any inconveniences associated with transporters used during peak periods affect only a small fraction of the traffic.

EXAMPLE 14.3

For the cost of the building, the example assumes that each gate space costs an average of $5 million to build, which translates into $500,000/year (see footnote on p. 562), and that each gate caters to 125 passengers per flight. Then:

$$\text{Building gate \$/passenger} = \$500,000/\text{(annual flights at gate)}$$
$$\text{(passengers/flight)}$$
$$4000/\text{(annual flights at gate)}$$

For the transporter, the example assumes that it costs $30,000/year in capital costs (based on an original cost of $300,000), and $200/h in operating costs when used. If the transporter can effectively serve 50 passengers per flight, and 2 flights per hour, then:

$$\text{Transporter \$/passenger} = 200/(50 \times 2) + 30,000/\text{(annual}$$
$$\text{flights served) (50)}$$
$$= \$2 + (600)/\text{(annual flights served)}$$

If a gate is used regularly throughout the year, building a gate is cheaper. For example, at 2000 flights/year, the example building costs $2/passenger and the transporter $2.30. On the other hand, if a gate is needed only in the high season for 400 flights, the marginal cost of the building gate is $10/passenger whereas that of the transporter is only $3.50/passenger.

Flexibility

Whatever is built today will somehow be inappropriate sometime in the future, perhaps soon. Most obviously, traffic may be expected to grow, and larger facilities are likely to be needed. More subtly, and generally more importantly, the mix of traffic and its demands will somehow be different both from what they are today, and from current expectations. For example:

- The mix of traffic may change to include a much higher or lower level of transfers, as happened in the 1990s at Salt Lake City and Raleigh-Durham, respectively.

- The proportion of passengers requiring international clearances may change, as countries enter into common customs areas such as the European Community, Mercosur, Nafta, or Asean.

- The requirements of the airlines may change as they merge or form alliances, as happened at New York/Kennedy when many foreign carriers joined up operationally with major U.S. airlines and thus no longer needed much of a strictly international passenger building.

The airport configuration will need to evolve. It must also be able to evolve flexibly, to serve different types of customers and needs. Designers are not able to anticipate correctly exactly what will be required. As Chaps. 3 and 20 emphasize, forecasting is not a science, and the forecasts are "always wrong." This means that the configuration of today must be able to adapt to different circumstances other than those that now appear most likely.

In general, centralized complexes of passenger buildings can accommodate change more easily than decentralized facilities. This is because, in a centralized facility, as one airline or type of service grows relative to another, it can move over gradually into other parts of the passenger building (see discussion in Chap. 15 on shared use). This kind of incremental change is difficult in decentralized buildings, as the case of Baltimore/Washington demonstrates. As Sec. 14-1 indicates, this airport could not make use of the international gates vacated by US Airways when Southwest needed extra capacity, in large part because the facilities were decentralized.

A prime way to achieve flexibility is to ensure that space is available for whatever may eventually be required. Having space available is more than having land available somewhere on the airfield. The space should be next to the existing facilities, and the facilities should be designed so that they can grow into this space. The Aérogare 2 complex at Paris/de Gaulle provides a good example of this. In this case the Aéroports de Paris developed their initial phase of buildings along a spine road with lots of space, so that they could extend the facilities almost indefinitely according to need (Andreu, 1997).

Hybrid configurations are also generally more flexible. Kuala Lumpur/ International provides a good example of flexibility in general and in particular due to a mix of configurations. This complex of passenger buildings not only allows plenty of space for various forms of mid-field concourses, but also carefully provides ample space to lengthen the main passenger building, as well as clear zones available for the expansion of baggage facilities and the introduction of rail access. Furthermore, it combines a midfield concourse and finger piers that can flexibly handle varying proportions of international and domestic traffic. This combination allows it to expand easily the specific kind of facilities that may be most appropriate in the future.

14-5   Assessment of configurations

The evaluation of the configurations demonstrates that it is impossible to define a solution that is best in all cases. Table 14-9 provides a subjective summary of the results of the analyses. Others may look at the results of the analyses and other factors and come to different conclusions. In any case, it should be clear that no single design is best overall. Some designs, such as finger piers and transporters, are less attractive, but no design is dominant for all stakeholders.

**Table 14-9. Subjective comparison of configurations
of airport passenger buildings**

| Configuration | Passenger | | Airline | Owner | Retail |
	Local	Transfer			
Finger pier	Fair	Poor	Fair	Fair	Good
Linear, one side aircraft	Fairly good	Poor	Good	Fair	Poor
Transporter	Fair	Poor	Good	Poor	Good
Midfield linear	Fair	Good	Good	Fair	Good
Midfield X-shaped	Fair	Fair	Fair	Good	Good

The transfer passengers may be the single most important factor for designers to consider in the choice of configurations. This is because what is good for them may be poor for local traffic and vice versa. When designers know they are developing a destination airport that will not have much transfer traffic, they may feel freer to develop linear passenger buildings. On the other hand, if they know or anticipate that the airport will be a transfer hub, they should focus on midfield concourses. The shape and indeed the possibility of a midfield concourse depend on the availability of space. When space is particularly tight, designers may find that satellites are a better solution, as they did in developing Milan/Malpensa. In this case, satellites similar to those at Seattle/Tacoma may provide attractive possibilities as they can offer many of the advantages of midfield concourses (see Sec. 14-3).

No consensus exists about centralized or decentralized passenger buildings (see Table 14-10). Airport operators and transfer passengers tend to prefer centralized buildings. The owners appreciate the advantages of concentrating services and retail areas (as at Amsterdam/Schiphol). The transfer passengers find transferring between buildings difficult, especially when the distances are huge as they are between the domestic and international passenger buildings at Sydney, Australia, for example. On the other hand, major airlines, especially those operating a transfer hub, like to control their own space in a self-contained building as they do at New York/Kennedy (American, Delta, and United buildings), London/Heathrow (British), and Tokyo/Narita (Japan Airlines).

The final choice of the configuration of new airport passenger buildings should depend on its specific circumstances, the site, the type of traffic, and the needs of the several stakeholders. Good designers will pay attention to each of these prospective users of the airport, although they are not the immediate clients. They will thus provide the better overall design for their main clients in the long run.

Table 14-10. Subjective comparison of centralized and decentralized airport passenger buildings

	Passenger				
Configuration	Local	Transfer	Airline	Owner	Retail
Centralized	Good	Good	Fair	Good	Good
Dispersed	Good	Poor	Fairly good	Poor	Poor

14-6   Hybrid configurations in practice

Major established airports as a rule do not exhibit a single configuration of passenger buildings. Over the years, they build a variety of facilities designed to serve the particular needs of many users. For example:

- New York/LaGuardia first built a Marine Air Terminal that became a decentralized building exclusively serving Delta Shuttle services to Boston and Washington. Its next major building featured finger piers and provided centralized service to many airlines. Its latest buildings are decentralized unit terminals serving specific airlines. (It functions well with these distinct buildings because it has virtually no transfer traffic.)

- Paris/de Gaulle started with a centralized unit terminal that was going to be the first of five. However, its Aérogare 2 was completely different and was a decentralized linear design (see the discussion under flexibility in Sec. 14-4). To meet the special needs of charter traffic for economical facilities, it then built Terminal 9 as a larger hangar by itself. Finally, at the turn of the century, it built a satellite for American Airlines and Terminal E, designed to function similarly to a linear midfield concourse.

Airports routinely evolve to include features that cater to the specific needs of their variety of clients and stakeholders. The result is thus unlikely to be described by a simple concept. Airports normally end up by adopting a hybrid concept, one that brings together important features of distinct configurations. The hybrid concept ultimately emerges for either of two reasons:

- Needs change, or
- The airport changes its preferences.

Either way, airports implement hybrid configurations as they learn what they really need.

Designers should plan for hybrid configurations since they are the likely final configuration of the airport. At a minimum, they need to make sure that their initial plans can respond flexibly to the changing needs of the airport clients and stakeholders over the life of the project. Better, they should from the start build in the elements that most appropriately serve the various needs.

Exercises

1. Assess the configuration of your local airport or some airport with which you can become familiar. How would you describe the configuration? To what extent does it appear to meet the needs of the principal stakeholders? How flexible does it appear to be, to meet the requirements of plausible future traffic?

2. At an airport you can visit personally, identify by inspection how the airport allocates gates to flights. To what extent does management locate flights intelligently to minimize walking distances for priority classes of traffic (such as heavily traveled domestic flights), for airlines, and for connecting passengers?

3. Describe the movement of aircraft near the passenger buildings at some major airport. To what extent are these patterns direct and free of congestion? Require many turns and involve possible delays as other aircraft block flow on the taxiway? How could these flows be improved (if at all) with an alternative configuration of the passenger buildings?

4. Obtain monthly traffic data from one or more airports. To the extent possible, get data for several years from airports in which you are interested. To what extent do they exhibit seasonal patterns? How many gates might the airport need only for the peak periods? Think about how these peak requirements could be served by transporters.

References

Andreu, P. (1997) *Paul Andreu, The Discovery of Universal Space,* L'Arca Edizioni, Milan.

Ashford, N., and Wright, P. (1992) *Airport Engineering,* 3d ed., Wiley, New York.

Baltimore-Washington International Airport (1998) "Board of Public Works Gives BWI Airports the Green Light to Begin Design Work on Expansion and Renovation," press release, Mar. 4.

Binney, M. (1999) *Airport Builders,* Academy Editions, Chichester, U.K.

Blow, C. (1996) *Airport Terminals,* 2d ed., Butterworth-Heinemann, Oxford, U.K.

de Neufville, R., de Barros, A., and Belin, S. (2002) "Optimal Configuration of Airport Passenger Buildings for Travelers," *Journal of Transportation Engineering,* ASCE, 128(3), pp.211-217.

Edwards, B. (1998) *The Modern Terminal—New Approaches to Architecture,* E & FN Spon, London, U.K.

FAA, Federal Aviation Administration (no date) "Aircraft Operating Costs, Based on BTS Form 41 for 1996," http://www.api.faa.gov/economics.

Hart, W. (1985) *The Airport Passenger Terminal,* Wiley-Interscience, New York.

Horonjeff, R., and McKelvey, F. (1994) *Planning and Design of Airports,* 4th ed., McGraw-Hill, New York.

IATA, International Air Transport Association (1995) *Airport Development Reference Manual,* 8th ed., IATA, Montreal, Canada.

ICAO, International Civil Aviation Organization (1987) *Airport Planning Manual, Part 1, Master Planning,* 2d ed., Doc. 9184-AN/902, ICAO, Montreal, Canada.

ICAO, International Civil Aviation Organization (2001) "Base Line Aircraft Operating Costs," http://www.icao.org/icao/en/allpirg/allpirg4/wp28app.pdf.

Little, R. (1998) "New BWI Pier Is a Dud so Far," *Baltimore Sun,* p. 1D, Nov. 29.

Moody's Investor Services (2000) "All Airport Hubs Are Not Created Equal," Report 57130, June, Moody's New York.

Powell, K. (1992) *Norman Foster and the Architecture of Flight,* Blueprint Monograph, Fourth Estate, London, U.K.

Svrcek, T. (1994) "Planning Level Decision Support for the Selection of Robust Configurations of Airport Passenger Buildings," Flight Transportation Laboratory Report R94-6, Ph.D. dissertation, Department of Aeronautics and Astronautics, Massachusetts Institute of Technology, Cambridge, MA.

15

Overall design of passenger buildings

This chapter presents procedures for defining overall space requirements of passenger buildings. It shows how planners can translate a concept and requirements for a passenger building into an architectural program that can proceed into detailed design. The ideas in this chapter are essential for the detailed design of interior spaces of passenger buildings that the next chapter discusses. This chapter focuses on the analyses needed in the early phases of design, whereas the next treats the issues associated with the later detailed design and management of the airport. The two chapters form a whole. Readers should cover Chap. 15 before Chap. 16.

The emphasis is on process rather than on design standards and specific numbers. Because the design context differs substantially between airports, no single set of standards can be valid for all airports. Indeed, the clients—airport operators, airlines, government inspection agencies, and concessionaires—may insist on different standards. Moreover, the range of traffic patterns at various locations and times permit airport operators to manage passenger and airline traffic more or less efficiently, and thus affect the design standards appropriate in different situations. The text presents illustrative, commonly used standards. These cannot be definitive.

Three ideas are essential for efficient design of a passenger building. These deserve special note because they are not readily available in the general literature on airport planning. Indeed, these ideas are contrary to the widespread notion that projected traffic (for example, X thousand passengers per hour) translates mechanically through formulas or tables into design requirements (such as square meters of space). Knowledgeable designers recognize that other factors strongly affect this translation. These issues are:

- *Sharing of facilities:* To what extent will it be possible to share facilities between different categories of users, and thus to economize on space?

- *Performance objectives:* How does the management of the passenger building want to balance economic efficiency (which leads to smaller spaces) and quality of service (which implies more space)?

- *Management of operations:* How does the operator of the building control the traffic flows and thus shift the loads among the various functions in the facility?

Designing space so that it can be shared among different functions or users is a prime way to reduce overall requirements. Such multifunctional or common space can easily reduce facility requirements substantially compared to the traditional practice of providing each user and each function with its own separate space. In some instances, the reductions in required space can be as high as 50 percent. Additionally, multifunctional space is inherently much more flexible than space dedicated to a single user or function. It thus provides insurance against fluctuations and uncertainties in traffic. Section 15-2 provides guidelines for the design of shared facilities.

To design space appropriately, planners and airport operators must be clear about their economic objectives. They need to be clear about the trade-offs they are prepared to make between quality of service and economy. Indeed, management choices influence the technical analyses. For some facilities, or indeed entire airports, the airport operator will be looking to emphasize economy. Thus, the Aéroports de Paris has created Terminal 9 as an inexpensive space to accommodate "cheap fare" traffic at Paris/de Gaulle, and London/Luton has designed its passenger buildings on the same basis. Singapore, by contrast, emphasizes quality of service and builds generous spaces.

The formulas for translating design levels of traffic into space requirements are arithmetically simple. However, they implicitly make assumptions about what the operator of the passenger building wants to achieve and how he will manage daily operations. Failure to appreciate these assumptions has led to notable design errors. Section 15-3 therefore discusses these assumptions in detail, so that designers will be able to use the formulas properly. Section 15-4 extends the topic to the sizing of passageways such as corridors and stairs. Finally, Section 15-5 covers the issue of space for baggage handling systems. Baggage space is crucial for the effective functioning of the passenger building, and designers should include it in the original overall space estimates for the passenger building. The analysis for baggage rooms is totally different, however, from the estimation of passenger space.

15-1 Specification of traffic loads

The issue

Planners seeking to define the overall dimensions of the various spaces in a passenger building must first face a difficult task. They must define the traffic for each of the functions these spaces serve. To do this, they must translate general forecasts of overall annual traffic into statements concerning the design loads at peak periods, for detailed operations. They must, for instance, specify the design loads for passport control facilities during its peak, which may be 30 min or an hour, depending on the prevailing traffic patterns. Simple formulas cannot solve the problem adequately. Too many local variations exist. For example, international arrivals at some airports, such as Sydney in Australia, cluster around a few peak hours early in the morning, whereas elsewhere, such as Miami, international arrivals are spread throughout the day. The translation between forecast annual international passengers and the loads on the immigration facilities needs to recognize such differences. To define the design loads correctly, analysts need an understanding of local conditions and the evolution of traffic patterns.

Planners need to appreciate that their design loads cannot be accurate. The process for determining these numbers is not scientific, and the results are not conclusive. Different analysts may look at the same overall forecasts and develop quite different conclusions about the design loads. The U.S. Federal Aviation Administration (FAA) has stressed this point in its graphs of the relation between the level of annual and peak hour traffic. It pointedly labels each of its graphs with the caption, in capitals in the original for emphasis (FAA, 1988, pp. 10 and 11):

CAUTION!
NOT TO BE USED FOR DESIGN OR ANALYSIS.
FOR USE IN OBTAINING ORDER-OF-MAGNITUDE
ESTIMATES PRIOR TO IN-DEPTH ANALYSIS

This warning about the inaccuracy of estimates of design loads needs emphasis.

The unavoidable inaccuracy of the estimates of future traffic loads has an important consequence. Planners need to build considerable flexibility into their designs to ensure that their buildings will function properly when the actual loads differ from the design loads. This is a prime motivation for the use of shared, multifunctional facilities.

Peak-hour basis for design

Planners should design airport facilities for peak traffic, but not the absolute maximum traffic. Clearly, they should design for the maximum range of loads rather than average loads. If they did not, then the facilities would be under capacity at all the times when the passengers and airlines needed them the most. On the other hand, if planners design facilities for the single busiest period in the design year, then the facility would be over capacity and oversized for every other hour of that year. In practice, the design load is a compromise between economy and the provision of enough capacity to meet peaks (see Chap. 24).

As a practical matter, it is impossible to define the design load that provides the best compromise between economy and quality of service. To calculate an optimum balance, it would be necessary to know both the local cost of the specific facilities and the value of the quality of service to the particular users. As there is no satisfactory way to measure these quantities, and as these are changeable in any case, no calculation of an optimal balance is ever likely to be defensible. All procedures for defining the design load are thus approximate.

Traditional procedures define the suitable design load in terms of statistics that analysts should be able to obtain from available airport data. In this connection, it should be noted that airport operators typically can have good data on aircraft movements, even by hour of the day. They may have much greater difficulty trying to get data of passenger flows, especially by hour of the day. They generally must rely on airlines to supply data on passengers, and in some locations this may be difficult.

Each definition of a design peak represents a compromise between efficiency and quality of service. No definition is analytically better than the others. Different organizations use different measures. Different practices prevail in different regions. Two widely used definitions of the design load are the *peak-hour activity* in North America, and the *standard busy rate* in Britain.

The usual procedure for defining *peak-hour activity* applies a factor to the "average day of the peak month." The factor is about 9 percent, decreasing steadily as the traffic at the airport builds up, as discussed later. The "average day of the peak month" is simply defined as the traffic in the peak month divided by the number of days in the month (FAA, 1988). (Note that this level of traffic does not necessarily represent any actual day, let alone a median day.) As almost all airports have data on daily traffic, this estimate of the design load is widely applicable.

In Britain, the accepted means of estimating design loads for airports is the *standard busy rate,* defined as the level of traffic during the 30th busiest hour of the year. A disadvantage of this method is that it presumes that the airport operator has reliable statistics on hourly traffic and can thus calculate the number. However, many airport operators worldwide have not collected such detailed statistics, and thus cannot use this approach.

The International Civil Aviation Organization (ICAO, 2000) has proposed a third way of defining peak hour loads. This combines elements of the two previous methods. It also presumes that the airport operator has hourly data, and can thus calculate the historical ratio of traffic in the peak hour to that of the peak day. On the other hand, similarly but slightly differently from the FAA approach, the ICAO approach suggests that the analyst calculate the traffic during the average day of the *two* peak months of the year. The design load is then obtained by multiplying this average peak day by the ratio of peak-hour to peak-day traffic.

For rapidly checking plans, it is possible to estimate peak loads by using rules of thumb. Thus:

$$\text{Average peak-day traffic} \approx \text{annual traffic}/300$$

$$\text{Design peak-hour traffic} \approx \text{annual traffic}/3000 \tag{15.1}$$
$$\text{(airports with } \sim 7 \text{ million total passengers)}$$

$$\approx \text{annual traffic} \times (3/10{,}000)$$
$$\text{(airports with } \sim 25 \text{ million total passengers)}$$

Adjustment for decreasing peaks. An important issue for all approaches to defining peak-hour loads is that the ratio of peak-hour to peak-day traffic usually decreases over time. Uncongested airports typically have sharp peaks in traffic over the day. As traffic builds up, relatively more traffic occurs during the off-hours and the "valleys"

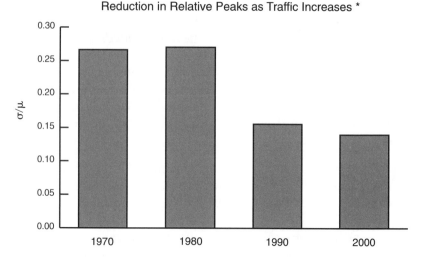

Reduction in Relative Peaks as Traffic Increases *

* Data comes from a typical day of daily operations in October at Lester B. Pearson International Airport in Toronto

Fig. 15-1. *As traffic increases, its relative variability tends to decrease (data for daily airline traffic on typical Wednesday in October for Toronto/Pearson).*

tend to fill up. Saturated airports tend to have steady traffic throughout their periods of operation. Figure 15-1, comparing the airline traffic over 10-year intervals at Toronto/Pearson, illustrates this effect. It shows how the variability of the traffic, measured in terms of the standard deviation as a percentage of the average level over the busy hours, decreased as traffic grew over that period.

Because daily traffic normally evens out as yearly traffic grows, estimates of peak loads that use historical ratios of peak-hour to peak-day traffic ratios overestimate future peak loads. This is the case for the standard busy hour and the ICAO approach.* The FAA approach, however, automatically lowers this factor for larger airports and thus systematically counteracts the overestimate of the alternative approaches.

Nature of loads

Planners must estimate the loads for each distinct activity for which they intend to estimate the space. Some of the major categories to keep in mind are

- Arriving passengers, terminating their travel at the airport
- Transfer passengers going on to other flights

*British data in Ashford and Wright (1992) demonstrate that the "standard busy hour" analysis leads to significantly higher design loads than the FAA approach.

- Originating passengers starting their trip at the airport and needing check-in facilities
- Departing passengers who will need waiting areas
- International and domestic passengers
- Passengers on charter or "cheap fare" airlines who will use special facilities
- Shuttle or commuter passengers needing minimal check-in, lounge, and baggage facilities

Planners should specify the loads according to the crucial periods for the activity under consideration. This may be the peak hour or it may be some other period. For example, the critical period for facilities serving arriving passengers may be the peak 30 min after one or more wide-body or new large aircraft land. This may be the governing period for passport control, baggage rooms, and customs facilities. Likewise, planners should recognize that some forms of traffic have seasonal patterns different from those of the rest of the traffic. Vacation charter flights may be concentrated in a few months, for instance. In short, planners seeking to estimate the overall space required for facilities in passenger buildings need to recognize the distinct local patterns for the traffic they intend to serve.

Note carefully that the total space for an airport passenger building is, as a rule, far less than the sum of the requirements for the distinct groups. Common areas and shared-use facilities reduce overall requirements. This point needs emphasis. Failure to recognize this fundamental fact is a source of major errors in initial planning documents for airport buildings. Airport operators interested in reducing the cost of a construction program may well start by investigating whether the conceptual planners have properly reduced the total size of the program by incorporating the effect of shared use of facilities.

15-2 Shared use reduces design loads

Planners need to recognize that the total space they need to provide can be less than the sum of the space needed for individual activities. Two different activities often can share the same space or facility at different times. This means that this space meets two separate requirements, so the total space that has to be provided is less than the sum of the requirements for each of the different types of activities. For example, when the peaks of international and domestic traffic do not coincide, boarding and waiting areas can serve international passengers at one time of the day, and domestic passengers at another. Such operations have been routine in Atlanta's Concourse E since the early 1990s. By designing suitable facilities to be shared in the passenger building, planners can reduce the total amount of space. This is a crucial fact that designers need to recognize before they proceed to translate the design loads into space requirements.

Economic efficiency is a prime motive for the spread of shared-use, multifunctional facilities in airport buildings. Facilities that can be shared among several uses increase economic performance. They lead to greater rates of utilization and correspondingly lower costs per unit served than facilities designed to serve only one client or function. Moreover, shared-use space provides flexibility to meet unexpected and varying loads. Multifunctional facilities, such as "swing gates" that can serve both international and domestic passengers, have thus become increasingly common internationally (Table 15-1). As of 2001, for example, Toronto/Pearson operated 15 swing gates; management could switch 12 gates between U.S. "transborder" and Canadian domestic flights, and 3 between other international and Canadian flights.*

Drivers for shared use

Two features of airport traffic drive the desirability of shared use:

- *Peaking* of traffic at different times
- *Uncertainty* in the level or type of traffic

Table 15-1. Some airports with international/domestic swing gates

Region	City/Airport	Region	City/Airport
North	Atlanta	North	Orlando/Sanford
America	Calgary	America	Toronto/Pearson
	Dallas/Fort Worth	Austral-Asia	Adelaide
	Denver/International		Wellington
	Edmonton	Asia	Kuala Lumpur/
	Fort Myers		International
	Las Vegas		Osaka/Kansai
	Los Angeles/	Africa	Mombasa
	International	Europe	Birmingham (UK)
	Montreal/Dorval		Athens/Venizelos

Source: de Neufville and Belin, 2002.

*Personal communication from Lloyd McCoomb, Vice-President, Greater Toronto Airports Authority.

**Table 15-2. Primary drivers and time periods
motivating shared use, multifunctional facilities**

Primary driver	Cycle time	Examples
Peaking at different times	Hours	Swing space: sharing gate lounges between flights
	Days	Swing gates for international/ domestic flights
Uncertainty in type of traffic	Days	Additional gates to handle peaks for weather, etc.
	Years	Reserve gates for uncertain future growth

Source de Neufville and Belin, 2002.

The time over which these drivers take place—their *cycle time*—defines the type of sharing that is appropriate (Table 15-2). Moreover, each combination of driver and cycle time requires a distinct form of analysis. This section presents four cases to provide a comprehensive set of tools to define the desirability of shared use.

Peaking of traffic. When distinct parts of the traffic peak at different times, shared use of facilities is economical. This is because the facility required for the peak of traffic A could be used for traffic B as the traffic A drops and traffic B peaks. Shared use reduces the size or number of facilities that the airport needs to provide for a given total traffic, and thus increases productivity and the return on investment.

This driver of shared use has nothing to do with uncertainty. It operates when the flows of traffic are known. This observation is important: it means that the analyses and resulting consequences appropriate to peaking are quite different from those associated with uncertainty.

To make it possible to share facilities between operations that peak at different times, airport operators must be able to "swing" the facility from one use to another. Designers make this possible by building in the features necessary to implement such swings in use. For example, to enable shared lounge space, they have to create *swing space,* that is, joint lounges that can be used for several gates instead of individual gate lounges separated from each other by walls or other barriers. To permit sharing of departure gates, designers have to build corridors that connect these gates with the appropriate users. Specifically, when regulations require these users to be segregated (as is typical for international and domestic operations), designers must provide corridors through a system of doors that can be securely and reliably opened and closed as needed. Atlanta/Hartsfield, for instance, equipped a range of its international gates with electronic switches that immigration authorities use to control the opening of doors and the flow of arriving passengers.

The cycle time between the distinct peaks of traffic influences both the types of analyses and the design of the shared facilities. Two types of intervals are salient with respect to peaking:

- *Hours:* The peaks occur in the range of about an hour, as for example in the case of passengers waiting to board three to four aircraft leaving 10–20 min apart.

- *Days:* The peaks occur at widely separate times of day, as for the peaks of international and domestic traffic at many airports.

When the interval between different peaks is in the range of an hour, two consequences arise. First, the traffic flows associated with the different peaks interact. Second, it is impractical to separate these flows easily. The interaction means that the advantage of sharing the facilities is only a fraction of the total. It also implies that mechanisms requiring substantial operator intervention, such as the opening and closing of secure doors, are difficult to implement.

When the interval between peaks occurs over a much longer period, however, it may be possible to dedicate a facility to an alternative use for a portion of that time. For example, a swing gate can serve international traffic during its peak and domestic traffic at another time. Alternatively, airlines can also "swing" the designation of an aircraft from a domestic arrival to an international departure or vice versa. This could happen when a Star Alliance United Airlines flight arrives domestically at Chicago/O'Hare from St. Louis and proceeds on to Frankfurt, Germany, as a code-shared Lufthansa international flight, or the other way around. Airlines thereby avoid the cost and delay of towing aircraft. Over longer periods, airport operators can also accomplish the tasks necessary to create alternative secure paths to the gate, as required when domestic and international traffic must be separated.

Uncertainty of traffic   Uncertainty in the levels of traffic is the other principal driver motivating shared use. The issues in this case are that:

- Additional facilities are required to "buffer" the system against peaks greater than the scheduled peaks. These extra peaks arise through either short-term delays or uncertainties in long term requirements.

- The efficient size of the buffer depends on either the frequency of the peaks or the range of the uncertainty in the long term needs.

- Thus it is economical to provide these "buffers" jointly for several users, rather than individually for each user.

Shared use of the buffer facilities leads to savings because the peak needs for the entire system are normally considerably less than the sum of the possible peaks for each element of the system. In the shorter term, this is because traffic drops for some users counterbalance the peaks of other users and smooth the variations in the overall traffic. This fact may not reduce the maximum possible peak, such as might occur when all users suffer delays during a major storm, but it does reduce the frequency of peak loads. The economically efficient size of the buffer represents a trade-off between the cost of the buffer and the cost or inconvenience of not having extra space when needed. Therefore, reduced frequency of need reduces the economically efficient size of the buffer space.

In the long term, over many years, the total requirements for a specified level of total traffic are also less than the sum of the anticipated possible maxima for individual uses. In this case, this is because some users do not meet the forecasts. The space reserved for them can then be used by other airlines.

To enable the economies of shared buffer space, designers should physically place this capacity between the core facilities of major users, so they can easily use it when needed. This implies that the airport passenger buildings should somehow be connected rather than independent, as they are in a "unit terminal" configuration. At Singapore, for example, all the buildings are connected, so it is easy in principle for one airline to use additional gates when its neighbor does not require them. Similarly, Toronto/Pearson is developing a single passenger building to replace the three that it had at the end of the twentieth century, which had proved to be inflexible in the era of hub-and-spoke operations and airline alliances.* On the other hand, at Baltimore/Washington it was impractical to allocate gates at the separate international buildings to domestic services. This has caused substantial operational and financial problems. In this case, US Airways moved much of its international traffic to Philadelphia, while Southwest Airlines grew domestic traffic rapidly (Little, 1998).

*Personal communication from Lloyd McCoomb, Vice-President, Greater Toronto Airports Authority.

This led to much lower traffic in the international building, and thus to unused capacity. However, the extra international gates could not be used when the traffic at the international building dropped.

The timing of the uncertainty influences both the types of analyses and the design of the shared facilities, as it does for the peaking factor. Two types of interval appear meaningful with respect to uncertainty:

- *Days:* The uncertainty arises from operational factors such as mechanical and weather delays, and is resolved over one or two days.
- *Years:* The uncertainty is about the level of future operations for different users, and may only be resolved over years.

When the uncertainty is resolved over days, the analysis is conceptually simple. Its essence lies in a trade-off between the cost of the additional facilities and the costs associated with delays and schedule disruptions that arise when facilities are not available when needed. However, airport designers have no credible basis for estimating the costs of disrupted schedules for airlines years in the future, so it is impossible in practice to calculate these trade-offs accurately. Fortunately, exact calculations are unnecessary. Because the base requirements at any airport change constantly with the level of traffic, so does the size of the remaining buffer capacity. In this circumstance, the analyses and rules of thumb cited in the next section are adequate.

The analysis is more complex when the uncertainty covers several years. This is because the cost of constructing flexible buffer facilities now must be traded off against benefits far enough in the future that they should be discounted (see Chap. 21). A "real options" analysis is needed to assess the value of the flexibility provided by the buffer space to meet future demands, as the next section and Chap. 22 discuss.

Analysis methods

Four major cases need to be considered. These are associated with each of the two drivers of shared use and shorter and longer periods, for which the appropriate analyses are different. See Table 15-3.

Peaking, hourly variation: The classic example of this situation is shared use of gate waiting areas. Other facilities, such as check-in counters, baggage handling, and security checks, operate under similar load patterns. In all these cases, the total space required for all users is less than that needed by the sum of several classes of users, because their peaks occur at different times. In practice, a shared-use departure lounge is simply a large room serving several gates (Fig. 15-2).

Table 15-3. Analysis methods recommended for factors motivating use of multifunctional facilities

Primary driver	Cycle time	Analysis methods	References
Peaking at different times	Hours	Simulation (site-specific) or available tables	Paullin, 1966; FAA, 1988; Belin, 2000
	Days	Analysis of Operations (site specific)	Belin, 2000
Uncertainty in type of traffic	Days	General formula or stochastic analysis	de Neufville, 1976; Steuart, 1974
	Years	Real option or decision analysis	de Neufville and Belin, 2002; Belin, 2000

Source: de Neufville and Belin, 2002.

Fig. 15-2. *Shared-use gate lounges at Washington/Reagan.*
(*Source:* Zale Anis and the Volpe National Transportation Systems Center.)

The analysis determines the maximum requirement by calculating the sum of the fluctuating needs of many individual users. It tracks the dynamics of how the traffic builds up and abates for each user, and how these flows add up overall. The procedure divides the period of interest into many smaller intervals. Thus, if the focus were a bank of departures over 2 h, the smaller intervals would be 3, 5, or 10 min long. For each interval, the analysis calculates the total traffic in place, as derived from the cumulative flow patterns appropriate to this activity (see Chap. 23). To calculate the waiting room requirements, for example, the analysis calculates the number of passengers who have arrived for the flights for each small interval and takes away the number who have boarded. These numbers depend on the size of the aircraft, its load factor, the pattern of arrival of the passengers, the time between departures, and the boarding procedures.

Note carefully in this connection that the patterns of arrival of passengers vary by time of day and type of flight. For example, passengers arriving for early-morning flights tend to arrive much closer to departure time than passengers showing up for flights late in the evening. Likewise, passengers for international flights generally arrive much earlier than those for domestic flights. Table 15-4, for Example 15.1, shows a tabular description of a cumulative arrival pattern.

Spreadsheets provide a versatile and cost-effective way to analyze the effect of overlapping peaks. These computer programs easily handle the basic additions and subtractions needed to estimate the number of passengers who need space in the waiting room. Analysts can define entries in the spreadsheet cells parametrically and can thus vary these factors at will. They can also explore all kinds of combinations automatically by means of the "data table" function of spreadsheet programs (de Neufville and Belin, 2002).* Analysts can develop an appropriate spreadsheet model for such situations in about a day. The web site for this book provides an operational example of a spreadsheet that readers may download and use.

Space sharing can cut requirements dramatically under some assumptions, as spreadsheet analyses demonstrate. The space needed depends mostly on the time between departures. This is the factor that permits a preceding flight to empty space and thus make room for passengers for the later flight. If as little as 5–10 min separate the flights, this is enough time to allow many passengers from earlier flights to leave the waiting rooms and provide space for passengers in the later flights. Example 15.1 illustrates the analysis, and Table 15-4 presents typical spreadsheet results for a specific situation. Analyses for different aircraft and different procedures lead to similar results.

It is important to understand that passengers share space over time. Figure 15-3 illustrates the dynamics of the process, in this case for

---

*The spreadsheet process is identical in spirit to the analyses defined by Paullin (1966) and Wirasinghe and Shehata (1988). However, it is far more general because it lets designers examine different circumstances easily.

Example Pattern of Occupancy:
Shared Lounge for 4 Gates

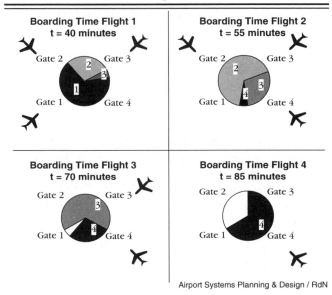

Fig. 15-3. *Schematic evolution of occupancy of departure lounge used by four aircraft.* (***Source:*** de Neufville and Belin, 2002.)

passengers sharing a departure lounge serving four gates. The area starts filling up with passengers for the first flight. The number of travelers for this flight peaks as they start to board their flight. Meanwhile, passengers for the second, third, and eventually fourth flight arrive as the passengers for the first flight all leave, followed by those of the second, and so on. The shared-use space accommodates waves of traffic that peak and recede to leave space for the next wave.

EXAMPLE 15.1

A simple case illustrates both the procedure for calculating the savings due to shared use of lounge space and the results. Consider a departure lounge serving 235 passenger aircraft operating at an 85 percent load factor and thus boarding 200 passengers. Assume that passengers arrive over an hour and begin to board the aircraft about 20 min before departure. Table 15-4 shows the number of passengers who might have arrived at the gate, boarded the aircraft and—by subtraction—be in the waiting lounge. Using the numbers for this case, the maximum number in the lounge for this aircraft is 140 persons.

Table 15-4. Passengers into and out of
a waiting lounge for a single aircraft

Number of passengers	Minutes before departure of aircraft						
	60	**50**	**40**	**30**	**20**	**10**	**0**
Arrived at gate	20	30	50	100	150	180	200
Boarded aircraft					10	140	200
In waiting lounge	10	30	50	100	**140**	40	0

Table 15-5. Passengers into and out of a waiting
lounge for three aircraft leaving at 10-min intervals

Number of passengers	Minutes from start of bank of departures								
	0	**10**	**20**	**30**	**40**	**50**	**60**	**70**	**80**
For first flight	10	30	50	100	140	40	0	0	0
For second flight	0	10	30	50	100	140	40	0	0
For third flight	0	0	10	30	50	100	140	400	0
Total in lounge	10	40	90	180	**290**	280	180	40	0

Suppose now that three identical aircraft, with the same pattern of arrivals and boarding procedures, serve the same departure area. Such a situation might happen when an airline flying a fleet of identical aircraft operates a bank of departures. Assume further that these aircraft leave at 10-min intervals. Table 15-5 shows the number of passengers for each flight in the waiting lounge, and the cumulative number for all three flights. In this case, the maximum number of passengers in the lounge is 290. This total is only 69 percent of the 420 passengers who would have to be provided for in three individual lounges.

Sharing of lounge space can easily reduce the total size needed by 30–50 percent, as extensive analyses along the lines of Example 15.1 demonstrate.* Table 15-6 and Figure 15-4 show consolidated results for specific flight characteristics (size of the aircraft and its load factor) with many combinations of operational characteristics (time available for boarding, time between aircraft departures, and time of gate occupancy).

*Note that recent analyses indicate that the savings are considerably more than those suggested by earlier reports (FAA, 1988). See de Neufville and Belin (2002) for details.

Table 15-6. Shared-use lounge space needed for N flights, as percent of space needed for N separate lounges (200-passenger aircraft, 60 min occupancy at gate; NA represents impossible combinations of hourly frequency and time between departures)

Flights (N)	Time between flight departures (min)					
	0	3	6	9	12	15
2	100	94	87	81	75	70
3	100	87	75	64	56	48
4	100	81	63	51	42	37
5	100	76	54	41	34	NA
6	100	72	48	34	NA	NA
7	100	67	42	NA	NA	NA
8	100	63	37	NA	NA	NA

Source: de Neufville and Belin, 2002.

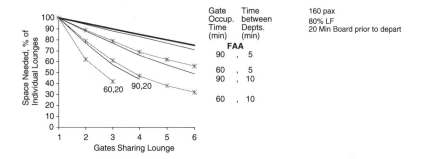

A/C Size	160							
Board time	20		FAA	60/5	60/10	60/20	90/5	90/10	90/20
	1		100	100	100	100	100	100	100
	2		95	89	79	62	94	88	77
	3		90	79	61	42	88	77	57
	4		85	69	47		83	66	44
	5		80	62	38		77	57	
	6		75	56	32		71	49	

Fig. 15-4. *Space required for departure lounges depends importantly on number of gates sharing the space, as well as on the time between departures and the size of aircraft.* (*Source:* de Neufville and Belin, 2002.)

The more flights sharing the lounge space, the greater are the savings. The effect diminishes as the number of gates increases. As Figure 15-5 indicates graphically, combining more than about six gates leads to relatively small additional improvements. Good designers recognize this pattern and usually have four to six gates share lounge space, unless constrained by space or local regulations.

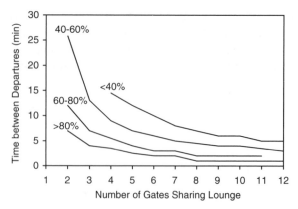

Fig. 15-5. *The percentage of space required for a shared-use departure lounge as a fraction of the amount needed for separate lounges depends on the time between departures and the number of gates sharing the space.* (**Source:** de Neufville and Belin, 2002.)

The above analysis comes with an important warning. Being deterministic, it does not take into account variations in the times between successive flights. Delays in departure may shorten the time between departures, and imply the need for more space in the shared lounge to accommodate passengers who may not be allowed to board on schedule. Table 15-6 indicates the increase in space needed if the interdeparture times decrease. The amount of extra space needed to provide a buffer that accommodates these variations depends on local conditions, such as weather and airline practices, that affect the frequency and distribution of the delays in boarding aircraft.

Peaking, daily (or longer) variation. A prime instance of this situation is the use of gates. Different airlines or services will often exhibit distinct patterns of peaks over a day. Short-haul business traffic, for example, domestically in the United States or internationally between Singapore and Kuala Lumpur, may have traffic peaks in the morning and evening. On the other hand, intercontinental services may have peaks determined by time zones, as when European flights arrive in New York in the early afternoon and leave late in the evening. Distinct peaks in traffic generally occur most significantly between international and domestic services.

When peaks of different users do not overlap, the opportunity for sharing arises because one set of users can use facilities when other users do not need them. Airlines can share not only gates but also all the supporting facilities such as check-in counters and baggage services. At Boston/Logan, for example, American Airlines designed the extension to its passenger building to have swing gates. Their plan is to have two to three gates and Federal Inspection Services serve international traffic, yet be available for domestic flights when there are no international arrivals. In practice, these kinds of shared gates require a carefully conceived system of doors and passageways that can channel the several kinds of passengers appropriately. Figures 15-6a and 15-6b illustrate how Edmonton/International in Canada arranges for shared space among three different types of traffic: domestic, transborder traffic that clears into the United States while still in Canada, and international traffic. Their solution is three-dimensional. In plan view, they have holdrooms separated by movable walls to provide more or lesser space for the international and domestic traffic. In elevation, they use secure corridors to channel arriving and departing passengers.

Spreadsheets provide a versatile and easy way to analyze the possibilities of sharing in these cases as well. A simple way to do this analysis involves creating a table of the requirements for gates by

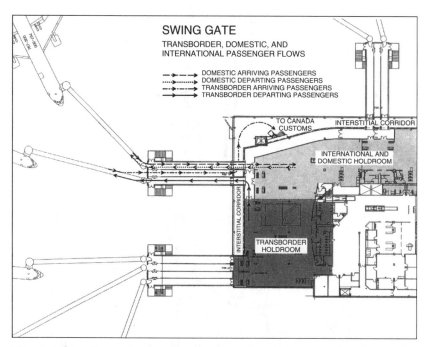

Fig. 15-6a. *Shared-use space used by domestic, transborder and international passengers at Edmonton.*
(**Source:** Lionel Oatway and Edmonton International Airport.)

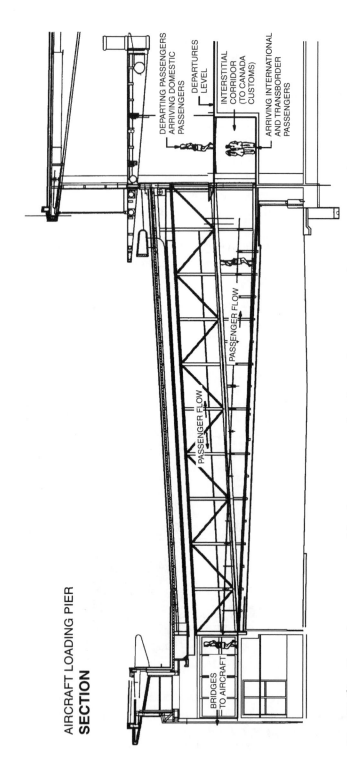

AIRCRAFT LOADING PIER
SECTION

DEPARTING PASSENGERS
ARRIVING DOMESTIC
PASSENGERS

DEPARTURES
LEVEL

INTERSTITIAL
CORRIDOR
(TO CANADA
CUSTOMS)

ARRIVING INTERNATIONAL
AND TRANSBORDER
PASSENGERS

PASSENGER FLOW

PASSENGER FLOW

BRIDGES
TO AIRCRAFT

Fig. 15-6b. *Shared-use space used by domestic, transborder and international passengers at Edmonton.*
(**Source:** Lionel Oatway and Edmonton International Airport.)

time of day and type of use (airline, aircraft, type of service, etc.). Designers can obtain the total requirements of gates or other facilities separately, or for categories merged in various ways. They can thus easily test different forms of sharing. They can display the results in the Gantt charts that airlines use to plan and display their gate assignment schedules, as Fig. 15-7 shows. These can effectively communicate the advantages of sharing to nonexperts.

When peaks of different users are hours apart, it is not practical to give general guidelines for the possible reduction of the size of facilities

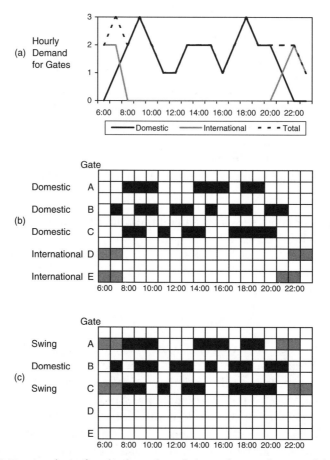

Fig. 15-7. *Analysis for the benefits of shared-use of gates: (a) hypothetical international and domestic operations; (b) Gantt chart showing gates needed for separate international and domestic gates; (c) Gantt chart for the number of shared gates.* (**Source:** de Neufville and Belin, 2002.)

through sharing. This is because the flows of aircraft to various airports are not so universally regular as flows of passengers to aircraft. The savings can certainly be considerable. In Kenya, for instance, tourist flights from Europe arrive in Mombasa at daybreak and leave around midnight, and are almost totally out of phase with the domestic flights from Nairobi that operate during the day. Thus, when the Kenyan government built the new Mombasa passenger building in the 1990s, it was possible to share facilities almost totally. Instead of constructing an international and a domestic building with about four gates each, the designers constructed a single building that would share gates between international and domestic services. Shared use led to about 35 percent savings in this case.

Uncertainty, daily variation. This uncertainty occurs routinely due to weather, traffic, and mechanical delays. These stochastic delays drive the airlines to request additional gates, G^*, and backup facilities beyond their scheduled peak needs, G. They want the flexibility of this buffer space so that they can service their flights when late or delayed aircraft block gates scheduled for other flights. This thinking leads airlines to use rules of thumb such as scheduling gates for only up to 80 percent occupancy during peak hours, knowing that they will need extra facilities to cope with random occurrences (Horonjeff and McKelvey, 1994).

Probabilistic analyses can estimate the total amount of buffer space needed when the schedule requires more gates (see Steuart, 1974; Bandara and Wirasinghe, 1988, 1990; Hassounah and Steuart, 1993). A simple way to estimate the number of gates needed by any airline is (de Neufville, 1976):

$$\text{Total gates required} = G + G^* = G + (G)^{0.5} \qquad (15.2)$$

This formula exploits two facts. One is that the standard deviation of a Poisson random variable equals the square root of its mean. The second is that, for a sufficiently large number of gates ($G > 10$), an upside buffer [$G^* = (G)^{0.5}$], equal to one standard deviation, will cover about 85 percent of the random variations in the Poisson process. The formula thus leads designers to provide a total number of gates that should give immediate access to a gate for about 85 percent of the peak hour flights, assuming that the peak demand for gates during a busy period can be approximated as a Poisson random variable—a reasonably adequate assumption when large numbers of flights (>10–15) are involved.* Notice that the fraction of extra gates needed does decrease as the number of scheduled gates increases. Thus:

$$G^*/G = (G)^{0.5}/G = 1/(G)^{0.5} \qquad (15.3)$$

The major result is that the fraction of extra gates needed in any situation, G^*/G, decreases when the number of scheduled gates is larger. This is because random effects tend to cancel out more when G is larger. The practical consequence is that combining the requirements for individual airlines reduces the total number of gates needed. As airlines share their requirements, they increase G, and thus reduce the relative size of the total buffer space, resulting in savings to all concerned. Example 15.2 illustrates this result.

This thinking provides a rationale for centralized management of all the gates at an airport. Airport operators indeed normally assign gates in most of the world except the United States. (The United States has been an exception for at least two reasons. It features a handful of very

EXAMPLE 15.2

According to Eq. (15.2), if three airlines each have to provide for simultaneous peaks of 10 flights, they would each independently need about 13 (= 10 + $\sqrt{10}$) gates, that is, 39 in all. If the airport as a whole defined their joint requirements, it would establish the overall number of gates at around 36 (= 30 + $\sqrt{30}$), a savings of three gates or around 7 percent.

If the airlines were smaller but collectively had the same demand for scheduled gates, the savings would be greater. Thus, if six airlines each wanted five gates at the peak, the sum of their individual needs would be 45 [= 6(5 + $\sqrt{5}$)] gates. The savings then achieved through joint use would be 9 (= 45 − 36) or 20 percent. Gaudinat (1980) confirmed these results empirically at specific airports.

*Technically, this argument has important hidden assumptions. It assumes that the demands for gates persists at the level of about G gates per hour for a considerable period, so that the variation is not transient. Furthermore, it assumes that there is significant variability in the actual arrival times compared to the scheduled arrival times, so that arrivals in any particular time interval constitute, approximately, a Poisson process. These assumptions would probably not be valid in the case of airlines with a scheduled peak requirement of only five gates. Furthermore, airline requirements are always changing. Equation (15.2) is thus not precisely accurate. However, it is useful as a rule of thumb to demonstrate the underlying trends.

large airlines whose savings through common use of gates or check-in counters would be less than at European or Asian airports, which may feature many airlines with small requirements. Moreover, airlines in the United States have been guarantors of the revenue bonds used to finance the passenger buildings, and therefore have more control over the use of these facilities.)

Applying the same argument to check-in facilities justifies shared use of counter space. This reasoning promotes the installation of CUTE (Common User TErminal) computer systems at both check-in counters and gates. It also promotes the use of a common backbone for information technology in the passenger building that permits decentralized check-in facilities (see Chap. 16).

A convenient design solution may be to place the extra facilities needed for uncertainty delays between major blocks of airlines or airline groups. This arrangement allows airlines to establish their brand at the airport, and enables the airport operator to manage the overall facilities efficiently. For example, Toronto/Pearson allocates a core of gates and check-in facilities to airlines in proportion to their traffic. It places the balance of the facilities in a common pool from which it assigns positions at peak hours according to the varying needs.

The bottom line is that designers can reduce the number of gates by incorporating shared gates among airlines, particularly among smaller airlines. They must balance this opportunity against both the management costs and passenger confusion associated with varying gate assignments.

Uncertainty, long-term variation.   The issue here is that the mix of traffic at an airport varies over the years, so that designers run the risk of getting the proportions wrong. The future proportion of traffic represented by international traffic, by an airline, by a type of aircraft (e.g., narrow or wide body) will almost certainly be neither what it is, nor what it is forecasted to be, at the time of design. Airport owners build major new facilities infrequently, only every decade or so when airport traffic grows at about 5–7 percent annually and doubles every 10 to 15 years. Designers thus normally plan facilities about twice as big as needed immediately. In doing so, they have to make decisions about the proportion of facilities to create for each class of user, at a time when these are highly uncertain.

The uncertainty in the mix of traffic motivates shared use that the airport operator can allocate flexibly to the users who will need it in the future. Shared-use facilities provide some insurance that the appropriate facilities will be available when needed. Airports that fail to provide this kind of insurance may end up with considerable wasted space and resources. The experience of the Baltimore/Washington airport illustrates this phenomenon. In the 1990s, the airport built a major international terminal principally for US Airways. However, US Airways then soon moved much of its international traffic to Philadelphia. Meanwhile, Southwest Airlines was growing rapidly, and required more space. It would have been sensible to let Southwest use the gates vacated by US Airways. Unfortunately, the airport did not have flexible space and could not effect the transfer. It thus could not use its existing facilities to capacity and thus wastefully had to build new gates (Little, 1998; Belin, 2000).

The provision of facilities that can be used in the future for several purposes is known technically as a *real option* (Trigeorgis, 1996; Amran and Kulatilaka, 1999). Technically, an *option* has a precise meaning going beyond the ordinary connotation of "choice." Formally, an option represents the capability of doing something at the owner's discretion, without being under the obligation to do so. In finance, an option is a contract to buy or sell something (a "put" or a "call," for example). In design, a "real" option represents a tangible capability to initiate some action. Flexible space is an option because it provides the ability (but not the obligation) to use a facility in an alternative use sometime in the future. As an option, it has been acquired at some expense, that is, the cost of equipping the facility to serve a multifunctional use. The option is "real" because it consists of physical facilities instead of the contractual arrangements that characterize financial options (see Brealey and Myers, 1996).

Options analysis provides the mechanisms to calculate the value of flexibility, and thus to decide how much to acquire. When applied to projects, it permits designers to calculate the value of flexible physical arrangements (such as shared space) and to determine how much to include in the design. In practice, decision analysis provides a convenient way to calculate the value of real options. It provides the essential rigorous analysis comparing possible levels of investment in shared-use facilities, with the prospective expected value of these investments over the scenario of future outcomes (Faulkner, 1996).

The decision analysis to determine the appropriate amount of shared space to ensure against uncertainty about future traffic proceeds along conventional lines as Chap. 22 describes. The fundamental decision concerns the amount of capacity that will be flexible, that the airport operator will be able to swing from one use to another. The subsequent chance nodes reflect the mix of capacity actually needed at the end of the planning period. The outcomes sum the extra cost of providing the swing capacity (in terms of sterile corridors and other necessary mechanisms) and the cost of meeting any shortfall in the capacity required for any particular use.

The analysis for Phase 1 of the Second Bangkok International Airport illustrates the structure of the procedure (Fig. 15.8). In this case, the

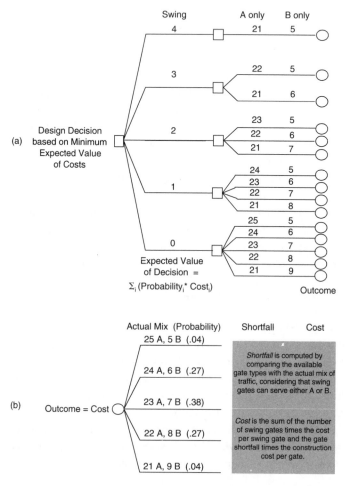

Fig. 15-8. *Decision trees for the analysis of the choice of the number of shared gates: (a) overall decision tree; (b) detail of one outcome node.* (**Source:** de Neufville and Belin, 2002.)

designers originally projected a planned capacity of 30 million annual passengers (MAP), but thought the international traffic (A) might need anywhere from 21 to 25 MAP, and the domestic traffic (B) from 5 to 9 MAP. As can be imagined, the rational design does not provide for the sum of the maximum possible for each use, 34 MAP (= 25 + 9). It should provide for a lesser amount consisting of dedicated and multifunctional gates that can eventually be shared or allocated to one use or the other.

The real options analysis should deal with two scenarios: normal and extraordinary variability in the mix of traffic.

- *Normal variability* refers to the routine variation around historical trends in the mix of traffic that can be observed at an airport. For example, as the proportion of international traffic at Bangkok decreased from 80 to 71 percent in the 1990s, fluctuations occurred around this trend (ICAO, 1990-97).

- *Extraordinary variability* is due to major shifts in the mix of traffic. These occur when operators radically reorient their traffic, as did United Airlines when it unexpectedly built up a hub operation at Washington/Dulles, or US Airways when it pulled much of its international traffic out of Baltimore/Washington. Extraordinary variability implies much greater risk and provides a stronger motivation for shared facilities.

Normal and extraordinary variability have different probability distributions of outcomes.

Analysts can estimate normal variability from historical records. The following illustrates how this can be done. Consider an airport with two types of traffic, *A* and *B*, which could be international and domestic. Airport data over the years will allow the designer to calculate:

- The share of traffic in any year t, $A/(A + B)$
- The trend in that share over time using regression through the n years of data on $A/(A + B)$
- The standard deviation of this share around the trend,

$$s = \left[\sum_n (\text{actual share in a year } t - \text{trend estimate of} \quad (15.4) \right. \\ \left. \text{share for year } t)^2/(n - 1) \right]^{0.5}$$

Assuming a normal distribution of outcomes, these data permit an extrapolation to the end of the planning period of both the expected mix and the likely range defined by 2 standard deviations.

The probability of extraordinary variability is speculative. Analysts can only subjectively estimate the chance of a major change in traffic probability and its maximum shift (in terms of percentage of the mix). Thus, analysts might reasonably assume that there is little likelihood that a congested airport such as Boston/Logan has a future as a transfer hub. However, they might well assign a significant probability to the possibility that an airport with large capacity and convenient runways might become a transfer hub. For instance, there is a possibility that Orlando/International could become a hub at the expense of Miami. Likewise, Miami/International runs the risk of losing transfer traffic. The fact that these risks are subjective and impossible to estimate precisely does not mean that they should be neglected. Good designers have the responsibility to provide some appropriate level of flexibility for dealing with these real risks.

Doing the decision analyses for the normal and extraordinary variability parametrically leads to some useful results (Belin, 2000). The following results derive from analyses for a wide range of:

- Relative extra cost of swing gates (beyond the cost of single function gates)
- Risk of the normal and extraordinary variability
- Range of possible shift in mix for the case of extraordinary variability

To provide insurance against normal variability, the percent of shared use gates should be on the order of the standard deviation around

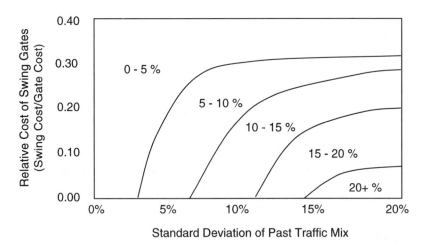

Fig. 15-9. *The proportion of swing gates to provide against normal variability depends on the extra cost of swing gates and the amount of variability.* (***Source:*** de Neufville and Belin, 2002.)

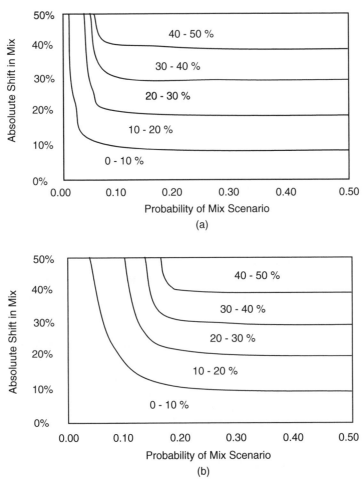

Fig. 15-10. *The proportion of swing gates to provide against extraordinary variability depends on the possible size of the shift and its probability as well as on the relative extra cost of swing gates: (a) lower cost, 5 percent extra; (b) higher cost, 20 percent extra. (**Source:** de Neufville and Belin, 2002.)*

historical trends calculated in Eq. (15.4). This is true so long as the additional relative cost of implementing swing gates is not too high. This makes sense intuitively. One standard deviation covers the bulk of the distribution of the risk, and it is natural that designers should provide less insurance as the risk diminishes. Figure 15-9 is thus a reasonable general guide for conceptual planning.

To provide insurance against extraordinary variability, the desirable percent of shared gates depends both on the absolute size of the possible shift, and the probability this may occur (Fig. 15-10). As for normal variability, the upper limit on the percent of shared gates is about the size of the possible shift. Thus, if the share of the international traffic might jump from 15 to 30 percent (a 15 percent increase) if an airport became an international transfer hub, about 15 percent of the gates might be planned as international to domestic swing gates to provide the flexibility to deal with this extraordinary variability. The desirable percent of shared gates decreases as the probability of the shift is smaller and the cost of the shared gates increases, as the comparison of Figs. 15-10a and 15-10b illustrates.

Overall implications of sharing

Multifunctional, shared facilities are economical because they can significantly reduce the amount of space that must be provided. This is important because space in passenger buildings is expensive—it easily costs around $2000 per square meter to build,* to which must be added annual costs for cleaning, climate control, and maintenance. Responsible planners will thus incorporate facility sharing to the extent possible in their preliminary designs.

Planners should correspondingly incorporate the implications of sharing in their overall estimates of space requirements. This means that they should reduce the total requirements by recognizing that some spaces will be used for several different purposes. They need to do this before they proceed to calculate the overall space requirements. Naturally, they should estimate the cumulative effect of the several different kinds of sharing that may occur. Example 15.3 illustrates how this can be done.

EXAMPLE 15.3

Suppose that, for a specific project, planners have determined that they need to serve 2000 domestic and 1000 international departing passengers per peak hour, corresponding to 16 domestic and 8 international flights. (Using the rule of thumb of Sec. 15-2, these figures correspond roughly to about 10 million departing, and thus about 20 million total passengers per year at the passenger building. They also correspond to about 48

---

*Unit costs of construction vary enormously, depending on the site, the scale of construction, details of the project, the quality of the finish, and other factors. Davis Langdon (2000) provides an indicative detailed analysis of cost. The company reports that airport passenger buildings cost from $1600 to $3500 per square meter as of 1998 in Britain.

operations per hour and 160,000 per year, assuming an average of 125 passengers per operation.)

Waiting room space.   Suppose that, in order to facilitate shared space, the planners cluster 4 aircraft gates around modules using a common waiting room. Each module would have a load of 500 passengers in the design peak period. This means that domestic passengers would require 4 modules in their peak period, and international passengers 2 modules in their peak. Suppose furthermore that planning assumes that, at any module, a flight leaves every 9 min at the peak period. (This implies a domestic departure about every 2 min at the peak period). Referring to Table 15-6, (p.621), one can estimate that the space required for a shared lounge serving 4 gates with aircraft leaving at 9-min intervals is only about 50 percent of the amount required for individual lounges. (In practice, the analysts would have created their own version of Table 15-6, appropriate for the local rate of arrival of passengers at the waiting room and the size of aircraft.) This means that each module could be sized for half the maximum peak load for the 4 gates, that is, for 250 passengers. (However, as mentioned in the text, the size could be increased to provide a buffer to account for variations in departure times.)

Swing gates.   Suppose that at this airport, the peaks for domestic and international traffic do not overlap, so that one of the modules of 4 gates can serve international traffic during the peak for that traffic, and domestic traffic when that peaks. The design of the building could then provide 5 modules of 4 gates, instead of the 6 that would seem to be needed by simple addition of the peak requirements without taking shared facilities into account. Specifically, the design would provide 3 modules exclusively for domestic service, 1 module exclusively for international service, and 1 module that could swing between domestic and international service as required.

The cumulative effect of sharing waiting rooms and gates is most significant. Instead of having to provide waiting room space for 3000 passengers in the peak hour, the design consists of 5 waiting rooms for 250 passengers each, or space 1250 passengers in all. Sharing in this case reduces the space requirement by almost 60 percent. This translates into enormous economies. Since waiting rooms require about 1.9 m² per passenger space and cost about \$2000/m², shared facilities would reduce construction costs by about $(1750)(1.9)(2000) = \$6.65$ million.

15-3   Space requirements for waiting areas

The translation from design loads for an activity to its space requirements is through a table. The standards published by the International Air Transport Association (IATA) are used most widely for determining the areas needed for the several types of waiting areas in passenger buildings. These and any other standards represent subjective judgments about what is desirable. They are not scientific facts. The generally accepted standards change over time. Different organizations may use different standards. Table 15-7 presents the latest version at the time of this writing.

The IATA table defining space requirements properly incorporates two features that reflect on how the airport operator plans to run the passenger building. Most obviously, as the heading in Table 15-7 indicates, it specifies standards according to "level of service." It also subtly refers to operational practices, specifically to how fast the airport operator intends to move passengers through spaces. Specifically, it refers to the passengers' "dwell time" in these spaces. Understanding these two issues is crucial for the correct use of IATA or any other space standards. The next two sections discuss these points in detail.

Importance of level of service

Level of service refers to the quality of the context in which a service takes place.* With regard to passenger buildings, it refers specifically to the amount of space available for the activity. The idea is that the level of service is higher when passengers have more space. For example, passengers waiting for check-in might use larger or smaller spaces, for example 1 or 2 m² per person. In both cases, they could expect to receive their boarding passes. With the greater space, they will be able to move around easily and feel comfortable. With the smaller space, they will be squeezed and uncomfortable.

Traditional practice uses 6 levels of service, from A (best) to F (worst). Their explicit definitions are subjective (Table 15-8). They are also ambiguous, as some people may perceive an environment to be uncomfortably crowded when others may not. Notions of when people are too close for comfort are certainly personal and often cultural.

---

*Historically, highway authorities in the United States developed the notion of level of service. In that context, the level of service refers to the number of cars in a lane of traffic. The U.S. roots of the level of service concept accounts for its grading from A to F. These marks mirror the traditional U.S. way of grading students from best to failure. Most of the terminology associated with the concept of level of service derives directly from highway practice.

Table 15-7. Space to be provided for passengers in different functions (m²/passenger)

Activity	Situation	Level of service standard					
		A	B	C	D	E	F
Waiting and circulating	Moving about freely	2.7	2.3	1.9	1.5	1.0	Less
Bag claim area (outside claim devices)	Moving, with bags	2.0	1.8	1.6	1.4	1.2	Less
Check-in queues	Queued, with bags	1.8	1.6	1.4	1.2	1.0	Less
Hold room; government inspection area	Queued, without bags	1.4	1.2	1.0	0.8	0.6	Less

Source: Adapted from IATA, 1995.

Table 15-8. Definition of level of service standards

Level of service	Description of standard		
	Quality and comfort	Flow condition	Delays
A	Excellent	Free flow	None
B	High	Stable, steady	Very few
C	Good	Stable, steady	Acceptable
D	Adequate	Unstable; stop-and-go	Barely acceptable
E	Inadequate	Unstable, stop-and-go	Unacceptable
F	Unacceptable	Cross flows	Service breakdown

Source: Adapted from IATA, 1995.

Note also that the concept of level of service involves ideas about both the flows and the delays. As Chap. 23 indicates, these characteristics of any service system are tightly related. As the traffic increases toward the level of saturation of the server, both the delays and the variance of the delays increase, leading to increasingly unsteady flow. The descriptions of the several levels of service reflect this fact.

The level of service planners should provide for a facility depends on the performance objectives of the airport operator. More space costs more money for construction and subsequent operation and maintenance. The question is: How does the management of the passenger building want to balance economic efficiency (which leads to smaller spaces) and quality of service (which implies more space)? Airport operators differ significantly in this regard, both for the overall airport and specific facilities. Thus, Singapore Airport has traditionally favored higher levels of service, consistent with their national objectives to make their city-state a premier location for business and tourists. London/Luton focuses on charter traffic and low-fare airlines, and prefers to offer lower levels of service. Aéroports de Paris has a different level of service for its charter facilities than for Air France. Some airport operators may have explicit ideas about the level of service they wish to provide. Thus BAA, the private company that owns the major London airports, has specific targets. However, many airport operators have not defined their space standards and planners then have to use standard assumptions about the desirable levels of service. As the standards are not absolute, planners should feel free to adjust them up or down if the airport operator desires a more luxurious or a less expensive passenger building, respectively.

The standard assumption is that planners should design for level of service C for ordinary use. This implies that, on the worst days, the facility may fall to level of service D. Practitioners generally assume that level D is tolerable for short periods. Note that level of service in any area varies automatically as the number of people in it changes. Any facility therefore actually provides a distribution of service, as Fig. 15-11 suggests. These distributions can be estimated by simulation analyses, as Sec. 15-5 indicates.

Note further that if planners adopt level of service C for the design year, the level of service when the passenger building opens will probably be about A on average. This is because the design year is normally 10 years or more in the future, and anticipates growth in traffic of 30 percent or more. Thus, at opening the space per person will be 30 percent or more greater than at design-year, and thus at level A as Table 15-7 indicates.

It is useful to look carefully at the structure of the space standards shown in Table 15-7. For any level of service, passengers standing in line without large bags take the least space. They require more space, 0.4 m^2 per person according to Table 15-7, when they are standing in line with large bags and perhaps baggage trolleys. They

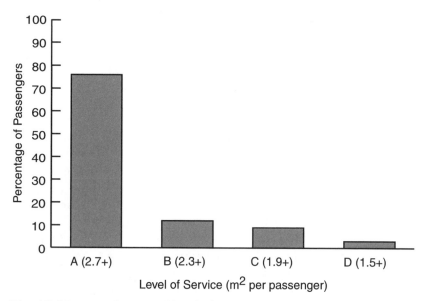

Fig. 15-11. *Distribution of level of service provided by a space over a period.* (*Source:* Greater Toronto International Airport.)

need even more space when they are moving around with their baggage, as in the bag claim area—according to Table 15-7, a further 0.2 m² per person at any level of service. The passengers need the most space when they are actively moving about. In this case, the differential over the preceding case is no longer constant, and may be negative. For example, the space standard for waiting at level of service E is, at 1.0 m² per person, less than the space standard for picking up bags at the baggage claim. Presumably, this is because, in the worst case for waiting, people simply cannot circulate. Using this understanding of the factors that define the space standards, planners can adapt the standards to situations that Table 15-7 does not specifically identify.

Importance of dwell time

Dwell time refers to the typical length of time passengers stay in an area waiting for service. It is a fundamental concept that needs to be carefully understood in order to use the space standards in Table 15-7 correctly. Failure to understand this concept has led to numerous design errors, as Example 15.4 indicates. The fundamental idea is that space is available over time, in units of time-area. Correspondingly, people use space over a specific time, and thus consume some amount of time-area according to the length of time they occupy an area. A given space offering a specific amount of time-area can then cater either to many people each staying a short time—provided their arrival is reasonably spread out—or to fewer people staying longer. This is the "time-space" concept developed by Fruin and Benz (1984) and Benz (1986).

Dwell time is important because it indicates how fast a space can be reused by another batch of passengers. The shorter the dwell time, the sooner a first group will leave the space, and another group can refill it and use the space again. Thus, if 1000 passengers per peak hour have to wait for an hour in a departure lounge, this space should be sized to fit these 1000 people. However, if the airport operator controls the traffic so that passengers wait on average only 30 min in the lounge, then only 500 passengers use the space in the first half-hour, and a second 500 use it in the next half-hour. When the dwell time drops in half—from one hour to half an hour—the amount of space needed also drops in half. The space needed is thus directly proportional to the dwell time.

Dwell time is implicitly incorporated in the space standards. Unfortunately, this fact is not immediately obvious. A superficial glance at the IATA standards seems to indicate that the space required for a design load of 1000 passengers per hour, is simply 1000 times the figure in Table 15-7 appropriate for the activity and the design level of service. Proceeding on this basis is wrong, however, as Example 15.4 illustrates. A dimensional analysis shows why this is so. The design load is in terms of (persons/hour) and the space standard in terms of (area/person). Multiplying these two factors gives a result in terms of (area/hour). In order to obtain the dimensionally correct answer in terms of (area), it is also necessary to factor in the dwell time, expressed in hours. The correct expression for estimating the area required for an activity thus is

Area = (design persons/hour) (space standard m²/person)   (15.5)
     (dwell time in hours)

The space required for an activity in a passenger building thus depends on the management objectives of the airport operator in two ways. The airport operator can influence the space requirements not only by setting standards for level of service, but also by defining the speed of service and the dwell times in specific areas, as Eq. (15.5) and Example 15.4 suggest. Example 15.5 shows how this can be done.

EXAMPLE 15.4

This example, taken from an actual design review, illustrates how designers can misuse the IATA space standards in Table 15-7. It concerns the space to be provided for international arriving passengers to wait for passport control services. The design load was 2000 passengers in the peak hour. Naturally, these passengers do not arrive all at once; they filter into the space as each of their aircraft arrives and allows the passengers to disembark. The airport operator in this case had a design standard specifying that passengers should not wait more than 20 min for passport control. They could achieve such a standard by speeding up the process, most obviously by providing and staffing more passport control booths.

The question was: How large should the waiting space in front of the passport control be? The designer's original answer was to multiply the design load by the space standard for Government Inspection at level of service C, which is 1.0 m^2/person, to obtain 2000 m^2. This answer was wrong because it ignored the dwell time limit of 20 min, or $^1/_3$ h. The correct answer was thus:

$$\text{Area for passport queue} =$$
$$(2000 \text{ persons/hour})(1.0 \text{ m}^2 \text{ standard})(^1/_3 \text{ h dwell time}) = 667 \text{ m}^2$$

The mistake implied construction of 1333 m^2 in unnecessary space. Assuming about $2000 per square meter for passenger buildings, this mistake would easily have cost around $2.5 million!

EXAMPLE 15.5

An airport operator planning the processing of arriving international passengers through passport control and the bag claim normally has to recognize that passengers will somehow have to wait to get their bags. Bags will easily take 15–30 min to reach the claim area at a busy airport.

The airport operator can influence where this wait takes place. By slowing the passport control processes, say by having fewer control booths, the passengers will wait longer at the passport control—and then less in the bag claim area. What should the airport operator do?

Note that the space standard for government inspection is about 40 percent less than for bag claim areas. At level of service C, it is 1.0 m² instead of 1.6 m². Therefore, the airport operator may achieve substantial savings in space if passengers wait at passport control. If the airport operator and the government agencies can manage the traffic cooperatively, passengers can arrive at the claim area at the same time as their bags. This often happens at London/Heathrow.

The ability of management to alter dwell time and thus to reduce space requirements is most important. Management can reduce space by speeding up service and reducing space requirements in many ways. For instance:

- Airlines in the United States almost never weigh baggage on check-in, another U.S. practice that has been different from common European operations (see Chap. 2). This speeds up the check-in process by up to 30 s per passenger, or about 25 percent. The airlines can translate this faster service into either smaller areas in front of the check-in counters, along the lines of Example 15.4, or fewer check-in agents.

- Electronic kiosks for automatically checking in passengers similarly speed up the check-in process, reduce the dwell time, and the requirement for large check-in halls. Since they also can be situated almost anywhere, in parking garages or train stations, for instance, they further reduce the need for check-in space. Huge check-in halls like those in the International Building at San Francisco/International or Terminal 4 at London/Heathrow may become obsolete. Check-in halls may be much smaller in the future.

- Airport operators can reduce the size of waiting rooms for passengers by accelerating the boarding of aircraft. In Singapore, for example, the waiting rooms between the security check and the wide-body aircraft are often small and incapable of holding the passengers destined for the aircraft. This design is successful, however, because the airport operator boards passengers practically as soon as they clear security, and thus cuts the dwell time in the secure waiting room to only a few minutes.

- Government officials can speed up passport control and reduce the space required. Thus, the U.S. Immigration and Naturalization Service now receives electronic data on incoming airline passengers from the airlines. It uses this information to check passports in advance. The result is a faster process on arrival and shorter dwell times (or, alternatively, fewer agents). Similar developments can be expected in other countries.*

The conclusion to this section is that dwell time is an essential factor for determining space requirements. In order to estimate overall space requirements properly, planners need to consult not only with the airport operators, but also the airlines and inspection services about how they intend to manage the dwell time of passengers in the different spaces.

Estimation of areas. Once planners have developed an understanding of the management objectives of the airport operator concerning level of service and the operation of the facilities, the estimation of overall space requirements is easy, as Example 15.3 demonstrates. All analysts have to do is apply Eq. (15.5). It is the preliminary work to determine the level of service and the dwell times that takes effort.

Note that the master Eq. (15.5) can be reversed to provide estimates of the capacity of a space that already exists, or to define its level of service under specific conditions. Example 15.6 shows how this can be done. The relevant formulas are

$$\text{(Capacity persons/hour)} = \qquad (15.6)$$
$$\text{area}/[(\text{space standard, m}^2/\text{person})(\text{dwell time in hours})]$$

$$\text{(Space, m}^2/\text{person)} = \qquad (15.7)$$
$$\text{area}/[(\text{design persons/hour})(\text{dwell time in hours})]$$

15-4 Space requirements for passageways

The analysis for determining the overall size of corridors and stairways is conceptually similar to that for determining the size to provide for various activities in which people wait or spend time. It features level of service and considerations of dwell time. However, the formulas are quite different.

In designing passageways, width is the dimension that must be specified. Width is a prime determinant of the capacity of passageways,

---

*For example, a British government report stated that "there is no reason in principle, subject to…availability of appropriate technology, why there should not be an electronic record of leave to enter or remain rather than persist in every case with a system of stamps in passports which was designed for another age" (United Kingdom Home Office, 1998, p. 27).

EXAMPLE 15.6

A secure hold room for departing passengers dedicated to a single 300-person aircraft is 8 m × 25 m. What is its normal capacity at level of service C, if passengers flow through and have a dwell time of 30 min? What level of service does it afford if aircraft boarding is delayed and the dwell time increases to 1 h?

Under normal conditions, level of service C implies 1.0 m² per person. By Eq. (15.6), the capacity is

$$\text{Capacity, persons/hour} = (200 \text{ m}^2) / [\,(1.0 \text{ m}^2)\,(1/2 \text{ h})\,]$$
$$= 400 \text{ persons/h}$$

The hold room thus has plenty of space under normal conditions. For the extraordinary conditions specified, by Eq. (15.7) the space per person is

$$\text{Space per person} = (200 \text{ m}^2)/[(300 \text{ persons})(1 \text{ h dwell time})]$$
$$= 0.67 \text{ m}^2$$

The hold room would then offer level of service E, which is unacceptable except under very unusual circumstances.

and the factor that the designer can specify. All the formulas for corridors thus focus on this dimension.

The parameter that determines the capacity of a corridor is the rate of flow. This is conveniently specified in terms of the width of the passage. The units are "persons per unit width per unit time," specifically, persons/meter/minute (PMM). As with the space standards for waiting rooms and similar activities, higher levels of service imply less congestion and less capacity for a given design. The space standards can thus be specified in terms of PMM, as in Table 15-9, which represents the authors' recommended standards for airports. They recognize that passengers routinely have baggage with them, either in trolleys or towed behind them. These pedestrians take up much more space than the ones observed by Fruin (1971). To account for this fact, all the figures in Table 15-9 are half those proposed by Fruin for pedestrians without baggage. Experience indicates this adjustment is a reasonable approximation until better values are available.

Table 15-9. Level of service standards for passageways in terms of PMM (passengers/meter of effective width/minute)

Type of passageway	Situation	Level of service standard					
		A	B	C	D	E	F
Corridor	Regular pace	10	12.5	20	28	37	More
Stairs	Slower pace, longer dwell time	8	10	12.5	20	20	More

Source: Modified from Fruin, 1971, to account for airport realities.

The formulas

The capacity of a corridor for any level of service is simply the effective width times the standard. Note carefully that the result should be multiplied by 60 to give the capacity in hours that can be compared to standard design loads specified in terms of passengers/hour. Thus:

$$\text{Corridor capacity per hour} = \qquad (15.8)$$
$$\text{(effective width) (level of service standard) (60)}$$

Conversely, the effective width of corridor needed to carry a design flow is:

$$\text{Effective width needed, meters} = \qquad (15.9)$$
$$\text{(design flow/hour)/[(level of service standard)(60)]}$$

$$= \text{(Design flow/minute)/(level of service standard)}$$

The capacity of corridors is very large. For example, if a corridor has 3 m (10 ft) available for traffic flowing in one direction, its capacity at level of service C is

$$\text{Capacity (persons/hour)} =$$
$$\text{(3 m effective width) (20 persons/meter/min) (60 min)} = 3600$$

The great capacity of passageways needs emphasis. Designers often fail to grasp this point. They commonly plan corridors to be much wider than necessary. This can be very expensive. Corridors in passenger buildings extend over many hundreds of meters, so a couple of unnecessary extra meters of width can imply millions of dollars in wasted construction cost or wasted space that could be better utilized. Hathaway (1999), for example, demonstrated "savings of approximately $10 million through reduction of corridor sizing" at Washington/Reagan. His use of simulation to do this is a good example of the value of carefully targeted simulation studies (see Sec. 16-5).

Because corridors are so frequently oversized for the traffic, airport operators often convert them into unplanned retail areas with push-carts, storage space for miscellaneous facilities, and extensions of stores. All these activities may indeed merit space in the passenger building. Better design would provide for these activities directly rather than letting them randomly take over unused corridor space.

The capacity of corridors is much greater than that of comparably sized waiting rooms and other spaces because people pass through corridors quickly and their dwell times are short. Example 15.7 illustrates this phenomenon. This fact provides the analytic basis for the common-sense observation that the capacity of stairs, up or down, is much less than that of corridors. People move more slowly on stairs, the dwell time is longer, and the capacity is less.

An important design detail results from the difference in capacity between corridors and stairs. A stream of travelers going down a corridor will slow down and form a queue when they reach stairs (see Fig. 15-12). Designers need to anticipate this queue and provide space for it. Importantly, they need to consider carefully situations in which intersecting corridors meet in front of a staircase. The queue backing up in front of the stairs can easily block cross-traffic from the intersecting corridors and severely degrade the performance of that section of the building.

Fig. 15-12. *Jam of pedestrians around a stairway at Dallas/Fort Worth.*
(***Source:*** Harley Moore.)

EXAMPLE 15.7

Consider a space that is 3 m wide and 50 m long. If this area is viewed as a lobby for people to circulate in for an hour, at level of service C that requires 1.9 m^2 per person, its capacity is

$$\text{Capacity for walking/circulating} =$$
$$(150 \text{ m}^2 \text{ area})/(1.9 \text{ m}^2) \text{ (1 h dwell time)} = 79 \text{ persons}$$

Now consider this same space as a corridor through which people walk lengthwise. If they move normally at 3.2 km/h (2 mph), they are going at the rate of 53 m/min. They will traverse the corridor in about 1 min. Their dwell time in this space is then only about 1/60th of an hour. The capacity of this space as a corridor with this assumption for speed is then

$$\text{Capacity as a corridor} =$$
$$(150 \text{ m}^2 \text{ area})/[(1.9 \text{ m}^2) \text{ (1/60 h)}] = 5260 \text{ persons/h}$$
$$\approx 80 \text{ persons/min}$$

Since the effective width of a 5-m corridor is only about 3.5 m because 0.5 m must be deducted from the total width to allow for edge effects (see text) and oncoming traffic, the throughput of this corridor is about 23 PMM. This result is close to the value for level of service C in passageways shown in Table 15-9, and thus justifies these standards.

Effective width

The *effective width* of a passageway is the width that is effectively available to pedestrians. It is critical in determining the capacity of a passageway. The central idea is that, as a practical reality, pedestrians do not use some of the geometric width of the passageway. They avoid the edges and stay away from people coming in the opposite direction. Some distances must thus be subtracted from the geometric width of the passageway to obtain the effective width.

Three elements should be deducted from the geometric width to obtain the effective width of a passageway. These are:

1. *Edge effects.* To reflect the fact that pedestrians shy away from walls, 0.5 m should be deducted for each side of the passageway, that is, 1 m in all for this factor.

2. *Counterflow effect.* Pedestrians also avoid oncoming traffic. Another 0.5 m should be deducted from the geometric width of the passageway to account for this.

3. *Obstacles.* The width of any obstacle intruding into the passageway should also be deducted from the geometric width. These obstacles could include video monitors that attract a cluster of passengers looking, vending machines and shopping stands, etc.

The effective width of a passageway for pedestrians is thus a minimum of 1.5 m less than the geometrical width. To use the standards in Table 15-9 correctly, analysts should be sure to use the actual effective width of the corridor. Example 15.8 shows how they can do these calculations.

Designers should recognize that actual corridors in passenger buildings will normally be wider than the minimum amount obtained from the calculations defined in this section. For esthetic reasons, architects will want wide passageways. For functional reasons, the airport management may also require wider corridors to allow for moving sidewalks or electric vehicles to convey disabled travelers. In short, passageways for passengers in airport buildings will almost invariably be much more spacious than what might be economically preferable (Seneviratne and Wirasinghe, 1989). Planners should

EXAMPLE 15.8

What is the recommended width of a corridor to handle peak design loads of 600 passengers per quarter-hour, in each direction? Note that these peak flows of 1200 persons per quarter-hour correspond to about 4000 persons per hour, assuming that peaks over short intervals do not continue for the full hour. This flow implies about 10–12 million total passengers a year. The design flow of 1200 per quarter-hour equals 80 persons per minute.

For level of service C, the tolerable rate of flow is 20 persons/m/min from Table 15-9. The total required width is thus, using Eq. (15.9):

Required width = effective width + 1.5 m
= [(design flow/minute)/(20)] + 1.5
= (80/20) + 1.5 = 5.5 m

accept this reality, while being aware of the danger of making corridors far wider than they need to be.

Intersecting flows of traffic provide an exception to this discussion, emphasizing the problem of sizing corridors too generously. Managing crossing flows can be difficult. People slow down and stop, to avoid bumping into each other. This increases their dwell time and reduces the capacity of the space. Areas with streams of pedestrians crossing each other should recognize this phenomenon and allow extra space to account for it.

15-5   Areas for baggage handling and mechanical systems

Planners need to provide the right kind of space for baggage handling in the planning stage. Failure to do so has led to extensive cost overruns, difficult workaround solutions, and excessive operating costs. The most glaring example of these difficulties occurred at Denver /International in the 1990s. In that case, a complex combination of planning failures resulted in a 17-month delay in the opening of the new airport, the installation of a second baggage handling system, and numerous unanticipated operational difficulties. The delays alone cost $30 million a month in interest and other payments. All together, inadequate arrangements for the baggage handling system resulted in over half a billion dollars of extra cost and a 15 percent increase in airport costs (de Neufville, 1994; Dempsey, Goetz, and Szyliowicz, 1996).

Planners likewise should provide for the range of mechanical systems that are essential for the operation of a modern airport building. These systems include:

- Security devices for 100 percent screening of bags
- Heating, air conditioning, and ventilation ducts
- Water and sewer pipelines
- Electric substations and transformers
- Telecommunication lines
- Elevators and lifts
- And possibly right-of-ways for people movers, for use immediately or in the future

Planning adequately for these life-support systems of the building will save substantial costs during construction and for many years in the future. One of the better examples of this is the way the Port Authority responsible for New York/Newark planned passenger buildings with an aerial right-of-way for a people mover. When the Port Authority built this system about 25 years later, it was able to do so with a minimum of disruption to ongoing operations (see Sec. 17-5). The airport operator will have to reconfigure these systems many times over the life of the building, just as for the systems for processing passengers. They must be put in place with flexibility in mind.

Space planning for baggage and mechanical systems differs fundamentally from planning for passengers. In planning for humans it is sufficient to think in terms of areas—architects will always be sure that floor-to-ceiling heights will be adequate for people. In planning for machines, however, planners must think in three dimensions. Baggage systems need enough height to cope with layers of pathways crossing each other. The conveyor or tilt trays that carry bags through the building need unobstructed space to bend, rise, and drop as necessary for the proper operation of the baggage system. Planners need to allow both enough area for baggage and other mechanical systems, and sufficient height and pathways for the connecting lines.

Baggage space is frequently most problematic for two reasons. On the one hand, baggage systems normally are in basements, which are confined spaces broken up by large columns supporting the building. On the other hand, the conveyor belts or tilt trays associated with a large system are large and inflexible. Designers attempting to detail the arrangements within an overall plan for a passenger building frequently find themselves trying to squeeze hard objects into an inadequate confined space. This is not easy to do well. Failure to anticipate the needs of the baggage systems can result in convoluted, inefficient arrangements that degrade the performance of the airport passenger building.

Space is also vitally necessary for the maintenance of all the mechanical elements. If there is not reasonable access to the equipment, it will be neglected. This will lead to long-term difficulties, costs, and degradation of service. Unfortunately, this space tends to disappear as the detail designers cope with installing baggage handling tracks and sorters that are needed immediately. In designing these facilities, planners need to consult carefully with all those involved in operating this equipment— the airlines, the baggage handlers, and the maintenance crews.

To compound this basic difficulty, the design of baggage systems is not standard. Each major new airport built in the end of the twentieth century appears to have its own unique system. This lack of standardization has several causes:

- New security regulations introduce X-ray or scanning devices whose design and performance is evolving.
- The technology of handling bags is also changing due to advances in information technology such as laser reader and radio-frequency identification (RFID) systems.
- Each major industrial country has one or more of its own devices.

This variety of systems for screening and moving bags means that there are few general rules about planning baggage systems at the conceptual stage. For preliminary planning purposes, the area of the room for outbound bags should be a minimum of about 0.5 m² (5 ft²) per peak-hour passenger for simple baggage handling systems. It should be up to three times as large for systems using some form of automated sorting (FAA, 1988).* The space needed for outbound bags obviously depends on the number handled. The figures cited assume that people check an average of 1.3 bags per person. This number varies at different locations. It might be half that for business travelers using a shuttle service for short trips; or almost double that for vacation or other flights for which passengers bring a lot of luggage. As the FAA Advisory Circular emphasizes, "caution should be used in applying these [rules]"! Moreover, the size of the bag systems depends on the number and type of screening devices, and the dwell time of the bags in the system. Practice in this regard is changing rapidly, so airport operators and designers will have to pay careful attention to future developments.

The height of the room handling outbound bags should be at least about 2.5 m (8 ft) greater than the height of the first obstruction, with additional space for conduits and utilities—for the simplest systems. Rooms that will accommodate automated baggage systems need about an additional 2 m (6 ft) for each layer of belts or trays that must be accommodated. The total height of a modern automated baggage room might thus be about 9 m (30 ft). It may need to be as high as two levels of ordinary floors!

Because of the lack of general rules and the changing technology, planners should consult with security and baggage handling specialists early. Making sure that there is enough space for the crucial baggage

*FAA nomograph 5-13 suggests a median value of about 5000 ft² per 10 "equivalent-aircraft" carrying up to 100 passengers.

handling function will enhance the operation and efficiency of the airport passenger building.

Exercises

1. For some airport for which you can obtain data on traffic, estimate peak-hour design loads for a specific activity, such as check-in.

2. If a local airport you can visit has shared waiting rooms for aircraft departures, go look at how these perform. To the extent possible, estimate the average number of passengers per departure. Estimate the size of the waiting lounges if individual lounges were provided at each gate. How does this total compare with the space actually available in the shared waiting lounge? Given the typical time between aircraft departures, how does this ratio compare with what would be calculated from data comparable to Table 15-6? In your estimation, is the shared lounge oversized? Undersized? Or just about right?

3. If a local airport you can visit does not have shared waiting rooms, go look at it and estimate how shared waiting rooms might be implemented. To the best of your ability, referring to data comparable to Table 15-6, estimate the savings in space that might result from shared waiting rooms.

4. For an airport you can visit or for which you have sufficiently detailed plans, estimate the capacity of spaces dedicated to specific activities, such as check-in facilities. If you can visit these spaces during peak periods, determine from a visit whether these spaces are in fact adequate to the actual loads. How do the theoretical calculations compare to the reality at the site? What are the discrepancies? What might account for them?

5. Repeat exercise 4, focusing on the adequacy of the corridors and passageways.

References

Amran, M., and Kulatilaka, N. (1999) *Real Options: Managing Strategic Investment in an Uncertain World,* Harvard Business School Press, Boston.

Ashford, N., and Wright, P. (1992) *Airport Engineering,* 3d ed., Wiley, New York.

Bandara, S., and Wirasinghe, S. (1988) "Airport Gate Position Estimation Under Uncertainty," *Transportation Research Record, 1199,* pp. 41–48.

Bandara, S., and Wirasinghe, S. (1990) "Airport Gate Position Estimation for Minimum Total Cost—Approximate Closed Form Solution," *Transportation Research B, 24,* pp. 287–297.

Belin, S. C. (2000) "Designing Flexibility into Airport Passenger Buildings: The Benefits of Multifunctional Space and Facilities," S.M. thesis, Department of Civil and Environmental Engineering, Massachusetts Institute of Technology, Cambridge, MA.

Benz, G. (1986) *Pedestrian Time-Space Concept: A New Approach to the Planning and Design of Pedestrian Facilities,* Parsons Brinckerhoff Quade & Douglas, New York.

Brealey, R., and Myers, S. (1996). *Principles of Corporate Finance,* 5th ed., McGraw-Hill, New York.

Dada, F., and Wirasinghe, S. (1994) "Cost Considerations in Planning Pier Terminals," in *Aviation Crossroads: Challenges in a Changing World, Proceedings of the 23rd International Air Transportation Conference,* pp. 183–192, ASCE, New York.

Davis Langdon & Everest (1999) "Cost Model—Airport Facilities," *Building,* Jan., http://www.davislangdon-uk.com/dle/cost_data/airports.htm.

Dempsey, P., Goetz, A., and Szyliowicz, J. (1996) *Denver International Airport, Lessons Learned,* McGraw-Hill, New York.

de Neufville, R. (1976) *Airport Systems Planning: A Critical Look at the Methods and Experience,* MIT Press, Cambridge, MA, and Macmillan, London, U.K.

de Neufville, R. (1990). *Applied Systems Analysis: Engineering Planning and Technology Management,* McGraw-Hill, New York.

de Neufville, R. (1994) "The Baggage System at Denver: Prospects and Lessons," *Journal of Air Transport Management, 1*(2), Dec., pp. 229–236.

de Neufville, R., and Barber, J. (1991) "Deregulation Induced Volatility of Airport Traffic," *Transportation Planning and Technology, 16*(2), pp. 117–128.

de Neufville, R., and Belin, S. (2002) "Airport Passenger Buildings: Efficiency through Shared Use of Facilities" (with Steven Belin), *ASCE Journal of Transportation Engineering, 127(3), pp. 201-210.*

FAA, Federal Aviation Administration (1988) *Planning and Design Guidelines for Airport Terminal Facilities,* Advisory Circular 150/5360-13, U.S. Government Printing Office, Washington, DC.

Faulkner, T. W. (1996). "Applying Options Thinking to R & D Valuation," *Research Technology Management,* May–June, pp. 50–56.

Feldman, J. (1999). "Controlling the Airport Data Grid," *Air Transport World, 36*(6), p. 34.

Fordham, G. (1995) "Planning Airport Terminals for Flexibility and Change," *The Institution of Engineers, Australia, National Conference Publication,* pp. 45–50.

Fruin, J. (1971) *Pedestrian Planning and Design,* Metropolitan Association of Urban Designers and Environmental Planners, New York.

Fruin, J., and Benz, G. (1984) "Pedestrian Time-Space Concept for Analyzing Corners and Crosswalks," *Transportation Research Record 959*.

Gaudinat, D. (1980) "A Study of Gate Sharing at Airports: Boston as an Example," S.M. thesis, Department of Aeronautics and Astronautics, Massachusetts Institute of Technology, Cambridge, MA.

Hassounah, M., and Steuart, G. (1993) "Demand for Aircraft Gates," *Transportation Research Record*, no. 1423, pp. 26–33.

Hathaway, D. (1999) "Landside Simulations of Washington National Airport," in Mumayiz and Schonfeld (eds.) *Airport Modeling and Simulation*, pp. 253–264, ASCE, Reston, VA.

Horonjeff, R., and McKelvey, F. (1994). *Planning and Design of Airports*, 4th ed., McGraw-Hill, New York.

ICAO (International Civil Aviation Organization) (1990-97) *Digest of Statistics—Airport Traffic*, ICAO, Montreal, Canada.

ICAO, International Civil Aviation Organization (2000) *Aerodrome Design Manual, Part 1, Runways*, 2d ed. 1984, reprinted July 2000 incorporating Amendment 1, Doc 9157, ICAO, Montreal, Canada.

Little, R. (1998) "New BWI Pier Is a Dud so Far," *Baltimore Sun*, Nov. 29, p. 1D.

Paullin, R. (1966) "Passenger Flows at Departure Lounges," Graduate Report, Institute of Transportation and Traffic Engineering, University of California, Berkeley, CA.

Paullin, R., and Horonjeff, R. (1969) "Sizing of Departure Lounges in Airport Buildings," *Transportation Engineering Journal, ASCE*, May, pp. 267–277.

Ralph M. Parsons Co. (1975) *The Apron and Terminal Building Planning Report*, Report FAA-RD-75-191, FAA, U.S. Department of Transportation, Washington, DC.

Reiss, S. (1995). "Down-to-Earth Terminal Design," *Civil Engineering*, 65(2), pp. 48–51.

Seneviratne, P., and Wirasinghe, C. (1989) "On the Optimum Width of Pedestrian Corridors," *Transportation Planning and Technology*, 13(3), pp. 195–203.

Steuart, G. N. (1974) "Gate Position Requirements at Metropolitan Airports," *Transportation Science*, 8(2), pp. 169–189.

Trigeorgis, L. (1996) *Real Options: Managerial Flexibility and Strategy in Resource Allocation*, MIT Press, Cambridge, MA.

United Kingdom Home Office (1998) *Fairer, Faster, Firmer—A Modern Approach to Immigration and Asylum*, HMSO Cm. 4018.

Wirasinghe, S., and Shehata, M. (1988) "Departure Lounge Sizing and Optimal Seating Capacity for a Given Aircraft/Flight Mix—(I) Single Gate, (ii) Several Gates," *Transportation Planning and Technology*, 13(1), pp. 57–71.

16

Detailed design of passenger buildings

This chapter addresses the detailed design of facilities within an overall configuration that either exists or will be built. It concerns design once the outer shell of the building and many of its interior spaces cannot be altered significantly. This assumption is often realistic even for buildings that have not yet been constructed. Country leaders, star architects, and financial backers will often have settled on an overall design, and designers of the details in the building will typically not be able to challenge such agreements effectively. Thus:

- Once the president of Malaysia, Dr. Mahathir, approved the preliminary concept for Kuala Lumpur/International, the designers could not alter the shape of the satellites.

- The engineers for the Osaka/Kansai passenger building could not alter the design by Renzo Piano, winner of the international architectural competition, despite the fact that rapid differential settlements in soft soil made the construction of an enormous glass roof extremely challenging and expensive.

- The linear configuration for the Athens/Venizelos airport was an integral part of the financial package arranged by the build/operate/transfer consortium.

This chapter presents a process for design rather than specific standards or formulas. This is because there is no agreement in the airport industry about standards. These depend on cultural expectations, local practices, and the current objectives of the airport operator. Moreover, airport operators and their clients often want to achieve multiple objectives that conflict, and will therefore compromise their own standards to achieve a suitable balance. It is therefore pointless to look for a unique set of design standards. Section 16-1 addresses this issue.

The chapter emphasizes the need to identify "hot spots" that may seriously degrade the performance of the passenger building. These *hot spots* are concentrations of traffic in time and space that create bottlenecks. These blockages may define the capacity of all or part of the building. Experience indicates that failure to detect and eliminate these obstructions during detailed design has been a primary source of difficulties for the effective and efficient operation of passenger buildings. Section 16-2 illustrates this issue and suggests how hot spots can be identified by careful inspection of the design.

To understand how hot spots arise, it is necessary to examine in detail the varying rate at which people arrive at potential points of congestion such as check-in counters, security points, doorways, and stairs. Section 16-3 shows how this can be done using a graphical analysis of the cumulative arrivals and departures from a service area. The method enables the analysts to see how the hot spots might be eliminated by adding servers, changing operational procedures, and altering design. This approach is effective for rapid analyses of specific bottlenecks.

Detailed simulation of the flow of passengers and bags is a prime way to explore the overall performance of the building and its facilities. It can be especially useful in showing how many different processes interact and sometimes create unexpected hot spots. Section 16-4 illustrates the effective use of simulation in the design of airport passenger buildings.

Finally, Sec. 16-5 provides information for the configuration and sizing of the range of specific facilities in a passenger building. These include check-in counters, security controls, waiting lounges, baggage claim areas, and curbside areas for leaving off and picking up passengers. This section provides ranges of values for general guidance, with the caution that designers need to adapt these to the specific situations they encounter.

16-1   Design standards

Design standards represent judgments about desirable service. They express the values of owner, operator, or users of the airport. They are not scientific facts. Correspondingly, they differ between countries, as Chap. 2 indicates. They may also differ within the same culture. A privatized commercial organization may be concerned primarily with economy and efficiency. A public agency may be motivated by other issues, such as the desire to build impressive gateways to a city or region, and may have quite different standards. Moreover, any given airport operator may apply different standards to different groups. Airports routinely apply higher standards for major airlines of their own country than for foreign carriers, and for regular carriers than for charter airlines.

In short, detailed standards for any particular part of the passenger building are not universal. Table 16-1 illustrates this fact by showing design standards various agencies have used for sizing departure lounges. At a general level, the standards are similar. They each provide seats for about half the passengers and may allow more space for seated passengers. In detail, however, the standards do imply different sizes for departure lounges for identical aircraft.

Detailed standards often refer to time as well as space. Table 16-2 illustrates this fact. Airport operators may specify how long they expect check-in, security and immigration checks, baggage claim, and other procedures to take. In this case, too, national standards may differ tremendously. For example, Mumayiz and Ashford (1986) report that passengers through Birmingham (U.K.) thought that spending 6.5 min in either the security check or customs represented "good/tolerable" service. Travelers in the United States, unaccustomed to standing in queues, would probably not agree with such perceptions. Indeed, the U.S. Federal Aviation Administration (FAA) guidelines indicate that 5 min is the maximum tolerable wait (FAA,/2001).

Note that there is a strong relationship between time and space standards. The faster the procedure, the less time people will be in the process and the less space will be needed. (See the discussion of dwell time in Sec. 15-3.) The exact interaction between space and time standards depends critically on the service rate, i.e., on the average number, μ ("mu"), of people that a server (such as a check-in agent) can process per unit of time. As Chapter 23 indicates, the product of the service rate, μ, and the number of available servers, n, determines the average number of persons served in a specified interval, t,

$$\mu nt = \text{number of persons served} \qquad (16.1)$$

if all n servers are continuously busy during t. Conversely, given the number of persons to be served and a time standard, t, that meets a desired level of service, the number of service positions required is

$$n = (\text{number of persons to be served}) / \mu t \qquad (16.2)$$

To achieve time standards, designers will provide enough positions to make it possible for the service provider to achieve the level of

Table 16-1. Illustrative differences in design standards for departure lounges

Source of standard	Standard for		
	Seated passengers (m²/passenger)	Standing passengers (m²/passenger)	Mix, percent seated
Aéroports de Paris	1.5	1.0	50 to 75
Amsterdam/Schiphol	1.0	1.0	50
BAA	1.0	1.0	60
IATA	1.0 to 1.5	1.0 to 1.2	50

Source: Adapted from Ashford 1988.

Table 16-2. Space and time design standards for immigration facilities

Source of standard	Space standard (m²/passenger)	Time standard	
		National passengers	Others
Aéroports de Paris	0.6	95% < 12 min	95% < 12 min
BAA	0.6	95% < 4 min	95% < 12 min
IATA	0.6	80% < 5 min	95% < 12 min

Source: Adapted from Ashford, 1988.

service that the airport operator desires (see Example 16.1). Note, however, that the actual operators of the service may thwart the intent of the airport owner or designer. Service times may exceed design standards because the operators do not staff the positions. Immigration authorities may not have enough personnel or airport operators may decide to save money by cutting services during weekends or holidays. Alternatively, the service provider may simply have a different set of values than the designers. "Cheap fare" airlines routinely staff a minimum number of counters to save money, and expect passengers to wait.

Experienced airport operators and designers recognize that they should consider other factors besides space and time when evaluating the performance of passenger buildings (see Seneviratne and Martel, 1991). As Chap. 14 emphasizes, walking distances are a major determinant of preferable configurations of passenger buildings. The reliability of the service is also important: How often do long delays occur that result in major consequences such as missed flights and connections? Airport operators may be interested in parameters such as the maximum time people have to wait for a service, or the minimum time it takes transfer passengers or bags to connect from one flight to another. One of the advantages of properly executed simulations of passenger buildings is that they can explore these issues and suggest solutions (see Sec. 16-4).

EXAMPLE 16.1

Consider an airline that has to check in 300 passengers in 50 min. Suppose that its procedures require 1.5 min on average to serve a passenger. The service rate, μ, is thus $2/3$ passengers per minute. (This is representative of situations in which airlines weigh bags and charge for excess weight.) How many check-in counters should be provided? The answer is

$$n = 300/ [(^2/_3) (50)] = 9$$

Observe that if the airline arranged to reduce the average time their check-in agents need to service passengers, they would need fewer agents. They could achieve this by abandoning the practice of weighing and charging for excess baggage, along the lines of current practice in the United States. They could also use electronic check-in so that their agents could dispense with asking customers' names and typing them in. Such practices might double the average service rate per passenger, and halve the number of check-in counters needed.

16-2   Identification of hot spots

Successful operation of a passenger building requires that travelers can flow through many different areas and processes on time and with enough space to meet expected levels of service. All too often, unfortunately, relatively insignificant aspects of the configuration or organization of elements of the airport building cause bottlenecks that jam a significant portion of the operation. Some major artery clogs up and restricts the flows—and capacity—of a major portion of the system. At one major European airport, an obstruction at a crucial point reduced the productivity of a new passenger building by about a third, implying extra capital expenses of about $100 million in current terms (de Neufville and Grillot, 1982). Generically, these small obstructions are "hot spots"—specific points that cause major problems for the efficient operation of the passenger building.

Ironically, the critical bottlenecks typically are most evident when designers have provided enough space overall for some activity. The hot spots arise subtly. They do not result from gross errors in sizing spaces. They usually stem from some seemingly minor architectural detail or lack of understanding of routine operational procedures in the airport building. Because of these details, airport users following their natural instincts concentrate at specific points at specific times. Example 16.2 provides short sketches of cases illustrating how hot spots occur.

In order to identify potential hot spots, it is necessary to focus on the dynamics of processes. It is of course important to review the detailed design to see that it meets overall design standards such as those Chap. 15 describes. However, this kind of review is not sufficient, as the cases in Example 16.2 illustrate. To avoid possible major problems, it is necessary to examine and think through in detail how the airport users will surge through the building over time.

The review for potential hot spots needs to be conducted with persons knowledgeable about airport operations and flows over a broad spectrum of situations. This point deserves emphasis, because the design teams for airports often do not include such persons. Architectural teams engaged in detailed design rarely have employees with long experience in airport operations. Few architectural firms have a steady stream of airport work that could justify this kind of staff. Moreover, many large international firms engaged in airport work use local architecture or engineering firms that have the advantage

EXAMPLE 16.2 Three instances of hot spots

Check-in counters. In a linear passenger building, check-in counters for tickets and bags are arrayed along one side of the circulation passageway that is about 12 m (40 ft) wide. This corridor serves passengers going to many gates farther to the left in Fig. 16-1. The airline occupying the building originally intended to effect 1-h turnaround times at each gate. It thus planned to start check-in for each flight 50 min before departure. Most of its passengers, meanwhile, arrive more than an hour ahead of flight time, and thus well before check-in. These passengers would therefore flock to the check-in the moment it opened. The problem is that although there was plenty of space along the entire length of passageway for all travelers present, the concentration of 200 people in the narrow space in front of the check-in would completely block the passageway and prevent other passengers from reaching their gates. The solution was to increase turnaround times to about 1.5 h, so that fewer people would be in front of the check-in counters. This meant that the productivity of each gate dropped from about 10 to 6 or 7 turns a day—a major and expensive loss of efficiency (de Neufville and Grillot, 1982).

Secure passageways. The design for a major international passenger building features a secure corridor along the airside front. It serves international arriving passengers, who proceed along the length of the corridor and down some stairs to immigration desks at the far end. Departing passengers are supposed to

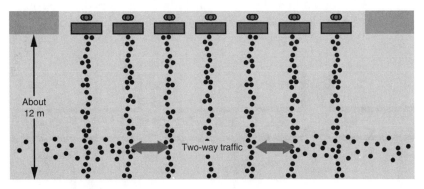

Fig. 16-1. *Check-in counters in a linear passenger building.*

cross the corridor to board their aircraft (see Fig. 16-2). Problems occur whenever several major aircraft land in the same hour. The corridor is narrow, only about 3 m wide. The arriving passengers thus occupy it more or less continually, especially since they back up along the corridor as they reach the stairs. This effectively prevents the passengers in the waiting rooms from departing, since they should not mix with the arriving passengers who have not cleared customs. Even if the arriving and departing passengers were allowed to mix, the intersection of their flows in the narrow space in front of each airbridge would cause major confusion and delays. This situation substantially reduces the maximum number of flights that can use this passenger building. The solution to this problem was to create a second immigration area and to divert some international arrivals to it, thus freeing up space in the corridor.

Underground station.   The overall design of the underground station and platform for the train serving a major airport allowed enough space for a trainload of people. However, the detailed design placed a staircase somewhere in the middle of the platform, as in Fig. 16-3. Travelers descending to the platform naturally clustered at the platform level in front of the stairs (area A). Few passengers would either see the space behind the stairs (area B) or pick up their bags and move around the staircase to get there. The result was that when trains came in, too many people tried to board its cars next to area A, and could not all get on. Their perception was that there was not enough train capacity, although the cars next to area B had lots of space. The solution being considered is to lengthen the platform in area A by extending the tunnel ahead of the stairs. This effort would be extraordinarily expensive, but may be inevitable.

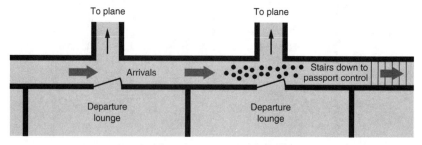

Fig. 16-2. *A secure corridor along the airside front of a passenger building.*

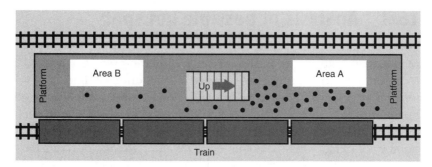

Fig. 16-3. *Underground train platform for a major airport.*

of knowing the local codes but the disadvantage of little experience in airport design. Likewise, employees who have had years of experience working at a particular airport are extremely knowledgeable about that airport, but unlikely to have had the opportunity to learn about how new airports operate in different situations or other countries. In short, appropriate reviewers must be recruited.

The review for points of critical congestion needs to follow passengers through the several processes they encounter. The analysts should carefully consider how the flows of different streams of traffic might interact—for example, the flows of arriving, departing, and transfer passengers. The designers need to focus attention on peak instances when traffic surges over short periods. These may occur over much smaller intervals than the peak hours of traffic. For example, the blockage in front of the check-in counters cited in Example 16.1 was an issue only during the period of time immediately following the opening of the counters for check-in. Designers need to focus on the pattern of cumulative arrivals of passengers at specific areas or processes, as Sec. 16-3 describes.

The review should also consider the psychological behavior of individuals and crowds. People normally, for example, proceed to the first line of service they encounter, rather than turn to the side to look for servers who have shorter lines. Coming down a corridor toward a row of security or border controls, for example, people will tend to cluster around those immediately in front of them, back up into the corridor, and block access to the other servers. Likewise, designers should be careful where they place flight information displays, information booths, and telephones. The intuitive impulse is often to place them where these services are visible to the most people. Unfortunately, the effect may be to create blockages just where traffic is most crowded.

16-3   Analysis of possible hot spots

The cumulative arrival diagram is the basic element for the analysis of potential bottlenecks in the passenger building (see Chap. 23). It represents the total number of arrivals over a period of time. This interval may refer either to a specific time, such as the hours since the beginning of an operation, or to the hours before some event, such as the departure of an aircraft. Thus, one might look at the total number of persons who have entered a waiting lounge during a day. Most usually, the cumulative arrival diagram represents the number of persons who show up for service before some critical moment, such as the departure of a flight. In this case, the time axis is in terms of relative time, of hours before the event (as in Fig. 16-4).

Cumulative arrival diagrams represent an empirical phenomenon. They should be based on observations of actual passenger flows. Designers should expect different diagrams for distinct types of flights and locations. Thus, Fig. 16-4 sketches arrival diagrams for a major European airport, for departures early in the morning and in the middle of the afternoon. People arrive at the airport for the afternoon flights as much as 3 h early, for many reasons. They may have come in on a connecting flight, come to the airport with a

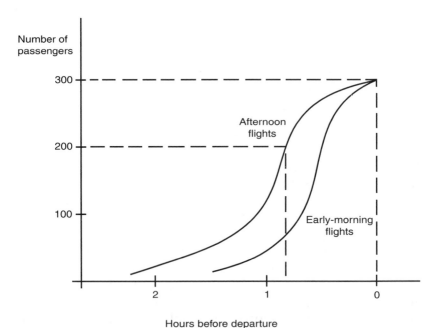

Fig. 16-4. *Arrival diagrams for a major European airport, for early-morning and afternoon departures.*

friend, or simply been anxious about the flight. For flights leaving at 7 a.m., however, people rarely arrive at 4 a.m.; they rise later and get to the airport closer to their flight time. Therefore, at this location, the cumulative arrival diagram for early-morning flights is different from those of later flights. Other reasons for distinct cumulative arrival diagrams include:

- *International flights* requiring various controls on departing passengers tend to have passengers arriving much earlier than domestic flights.

- *Hourly shuttle flights* serving business travelers, as between New York and Washington, or Singapore and Malaysia, usually have most traffic arriving in the half-hour preceding the departure.

Cumulative diagrams also reflect the point at which they are taken. People do not flow directly through the airport from their point of arrival to their exit. They spend time in various airport facilities such as shops, restaurants, and other activities. As a rule, the earlier they arrive, the longer they spend in these ancillary activities. Therefore, the cumulative diagram for arrivals at the boarding gate is typically much more compressed than that for arrivals at the curb (Fig. 16-5).

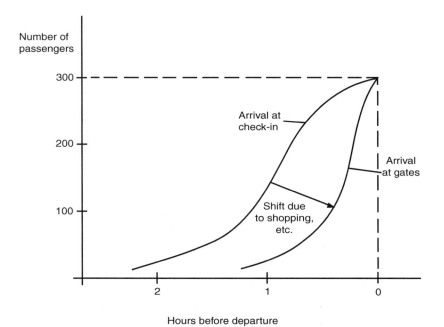

Fig. 16-5. *The cumulative diagram for arrivals at the boarding gate is typically much more compressed than that for arrivals at the curb.*

A cumulative departure diagram represents the number of persons who complete a process over time. As with the arrival diagram, it represents the total number on one axis and time on the other. Departure diagrams generally look different than arrival diagrams because they feature straight lines (Fig. 16-6). These reflect the notion that the number of persons served in any period is a constant multiple of the rate of service per server, μ, and the number of servers, n, [Eq. (16.1)]. This means that the slope of any straight-line segment in the departure diagram simply equals μ times the number of servers then operating.* Furthermore, the departure diagram may reflect a definite starting time. Figure 16-6 thus indicates that the service begins 50 min before departure.

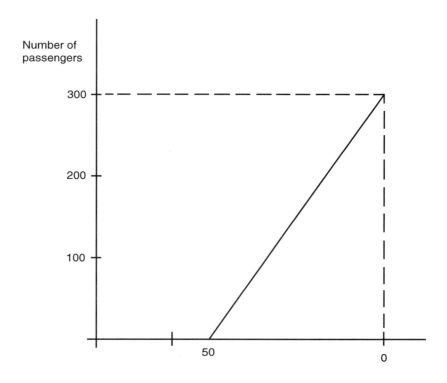

Fig. 16-6. *A cumulative departure diagram represents the number of persons who complete a process over time. It generally features straight lines.*

---

*This statement assumes that the server is continually busy. When there is no queue and the rate of arrivals is less than the possible rate of service, the average service rate equals the arrival rate. The slope of the departure diagram is then defined not by the possible average service rate, but by the average rate of arrivals.

Departure diagrams thus reflect management decisions about when to start service and how fast to run it. They can thus change daily according to short-term management decisions at the airport. In this sense, they differ conceptually from arrival diagrams, which reflect long-term passenger decisions that are difficult for airport managers to affect.

Designers can estimate the length of queues and the waiting time by combining the arrival and the departures diagram (see Fig. 16-7). They simply have to read the differences between these two diagrams. The vertical distance between the two lines is the difference between the number of people who have arrived and those that have left, and therefore is the number in the queue waiting for service. Similarly, the horizontal distance is the average time between when a person arrives and is served, and thus is an indication of average waiting time for a person arriving at the queue at any particular time.* (Chapter 23 reviews cumulative diagrams and provides details on deriving various queuing statistics from them.)

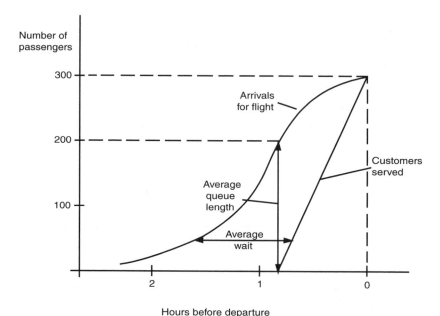

Fig. 16-7. *Combining cumulative arrival and departure diagrams to estimate length of queues and waiting time.*

*In cases in which people may receive service out of sequence (either because there is no snake queue or because the servers offer preferential service to first-class and other customers), the waiting time defined by the diagram might not represent the actual waiting time for the passenger arriving at the moment for which the waiting time is estimated, but gives a good indication of the typical ("average") waiting time to be experienced.

Designers can use this analysis to estimate if a potential hot spot is likely to cause a difficulty. They can then also use it to identify possible solutions to these difficulties by exploring the possibility of alternative forms of operation that change the cumulative departure diagram—and thus the average length of queue and wait time. Example 16.3 shows how this can be done.

EXAMPLE 16.3

Consider the hot spot in front of a check-in counter described in Example 16.2. The issue was that the space from the front of the counter to the opposite wall is about 12 m (40 ft) long and that this space has to serve as a passageway for other passengers. Consequently, only about 10 m (33 ft) are available for the queues. Suppose that Fig. 16-7 represents the cumulative arrival and departure diagrams for this situation if the check-in counter does open 50 min before the flight closes (that is, at Departure T-50 min). As Example 16.1 indicates, this requires a service rate of 6 persons per minute or 9 servers if agents take an average of 1.5 min to process each traveler.

Under the circumstances shown in Fig. 16-7, the airline operator could expect about 200 passengers to be in queue for service when the counters open. (A further 100 would arrive prior to the closure of the flight). This implies that about $200/9 = 23$ persons would be in each queue. Persons in line with trolleys require about 0.6 m (2 ft) per person for queue length (see the discussion for check-in counters in Sec. 16-5). Therefore, the average depth of queue in front of the counters would be about 13.3 m $= (200/9) (0.6)$. This queue length exceeds the space available. It would result in great confusion and cut off all circulation to other gates. Thus the hot spot.

The solution adopted was to open some check-in counters earlier, so as to reduce the maximum length of queue. For example, if 5 counters opened 30 min earlier (at Departure T-80 min) they would expect to serve $(5)(30) (2/3 \text{ persons/min}) = 100$ persons by Departure T-50 min. The queue at that time would then be only 100, and the queue length of about 7 m $= (100/9)(0.6)$ would easily fit into the 10-m space available. Figure 16-8 illustrates this possibility.

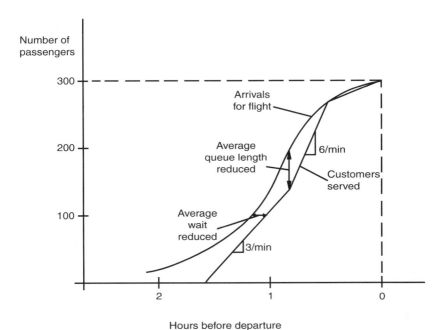

Fig. 16-8. *Opening some check-in counters earlier to reduce the maximum length of queue.*

Analyses using cumulative arrival and departure diagrams provide a simple way to identify major difficulties and ways to resolve them. They have successfully resolved important design issues in many situations. They constitute a powerful tool that any professional designer should be able to employ. However, these analyses have limitations. They are:

- *Deterministic:* they represent typical situations and do not describe the variations that occur in practice.
- *Suitable for one block of processes at a time:* they give no indication of the interaction between processes, as when delays in one part of the terminal affect the patterns of arrivals at subsequent processes.

They are therefore not suitable for all situations. A simulation analysis is necessary to explore the flux of traffic over time throughout the passenger building.

16-4 Simulation of passenger buildings

Good-quality simulation models can be powerful tools in the detailed design of passenger buildings. They can be particularly useful in "fine-tuning" a design by studying the operation of the building *in its entirety*. Some of the tasks that planners can accomplish efficiently in this way, include:

- Checking that the successive processes and services that passengers go through interact smoothly, so that there are no obvious bottlenecks in the building
- Estimating the total processing time for all the services that each class of arriving, departing, or connecting passengers utilizes
- Identifying the correct number of agents, machines, desks, etc., required at each part of the building
- Estimating the level of service at each part of the building both with respect to space per occupant and to waiting times

Example 16.4, based on the authors' experience, is a typical case of effective use of simulation for a passenger building. (Chapter 19 reviews land-side simulation models and their characteristics, as do Mumayiz, 1990, and Mumayiz and Schonfeld, 1999.)

EXAMPLE 16.4

A major European airport used a simulation model to "validate the operational concept" for a very large passenger building scheduled to open about a year later. It first prepared a full, detailed schedule of flights for the "design peak day" (Chap. 24), including the number and characteristics of passengers on each flight, the number of well-wishers or greeters, bags, etc. It then simulated the operation of the building with these loads.

The project served two purposes. First, it helped the airport operator address the thorny issue of allocating stands and check-in counters among the airlines. While all the gates of the building were intended for shared use (Chap. 15), it was understood that each airline's flights would consistently be assigned to the same particular group of gates on a day-to-day basis. The airlines considered the gates in the middle of the long linear building to be much more desirable than those at the satellite to the main building. Similarly, check-in counters in the center of the linear building were believed to be preferable to ones in the far reaches. The operator used the simulation model to experiment with several schemes for allocating flights to gates. It generated data on average walking distances, gate utilization, passenger and well-wisher concentrations, and loading on the various facilities and services. These provided the basis for selecting the eventual allocation scheme.

The second major thrust was concerned with identifying and eliminating potential hot spots. In checking the equipment and personnel requirements and the adequacy of the available space in the various parts of the building, it discovered several potential problems. For example, the simulation soon showed that the number of check-in desks allocated to the national airline during peak hours would result in long queues. These would have been long enough to block circulation in the main departures hall of the building and create a hot spot similar to that of Example 16.2. This finding led to a significant revision of the initial allocation. Similarly, the simulation found that the space available for the queuing of international passengers for passport control was totally insufficient. Consequently, the airport operator modified the configuration of that part of the building and developed a second passport control area elsewhere.

The project lasted about 5 months and cost about $350,000 for model acquisition, programming, validation, and consulting services by the model's provider. These services included development of the simulation of the baggage handling system in the building and the training of two airport staff members in the use of the model. In addition, the airport operator allocated several staff members to the project, including some managers, full and part-time. This group held all-day meetings every few weeks, provided guidance to the programmers, and reviewed the results carefully for reasonableness. Most of their time, however, was spent on the unexciting task of collecting available data for the model or conferring among themselves on the appropriate input values for certain model parameters.

The airport operator concluded that the project was highly successful and cost-effective and adopted the simulation model for in-house use. Its trained staff has operated the model repeatedly to address problems in the passenger building, as they arise.

This example illustrates several of the typical issues related to the use of simulation models in support of the design process. First, all of the analyses performed with the model could have also been done without it. For example, a simple model using cumulative diagrams (Sec. 16-3) would also have shown that the number of check-in counters allocated to the national carrier during the peak period was inadequate. However, the simulation model, once developed, greatly facilitated the performance of many analyses concerning alternative allocation schemes for check-in desks. Moreover, the simulation model made it possible to observe all at once how each of these alternatives would affect other parts of the building.

A second issue concerns the resources required for simulation-based design efforts. Quite often the up-front costs for model acquisition and adaptation to a specific site represent only a part of the total costs. In Example 16.4, the cost of the salaries and benefits of the team that the airport operator assigned to the project probably matched the cost paid to the model provider. The cost of assembling the required data was also substantial. Moreover, significant staff costs continued to be incurred after the completion of the project, to maintain and update the model and to develop and run new simulation scenarios. Nonetheless, given the large capital and operation costs associated with any large passenger building, the successful use of a simulation tool can be very cost-effective.

Unfortunately, not all simulation efforts are successful. In fact, many of them have proved utter failures. Chapter 19 discusses the reasons for such failures and suggests strategies for avoiding them. One point that needs emphasis is that simulation models must be sufficiently advanced to capture the behavior of passengers and well-wishers realistically. In particular, the models need to recognize that passengers and others do not simply flow through the building, passing directly from one service to the next. As Sec. 16-3 indicates, they spend time in various ancillary locations such as shops and restaurants. If the simulation cannot model this behavior adequately (and a number of current models still cannot), the results can be ludicrous. A simulation that does not recognize that passengers spend time in shops and elsewhere will greatly overestimate the number of departing and connecting passengers who arrive early at their departure lounges. Such a simulation will show large numbers of people spending all their time there, while the rest of the building is practically empty!

Finally, Example 16.4 illustrates another important aspect of the simulation of passenger buildings: the use of a detailed schedule of flights. Planners cannot avoid using this kind of schedule: it is a required input to detailed simulation models. It is essential, however, that planners recognize the great uncertainty associated with these hypothetical detailed schedules. They should always perform a large number of sensitivity tests with variations of the schedule. As emphasized in this and the previous chapter, a design that performs well under a wide range of assumptions regarding the future is much preferable to a design that performs "optimally" under a narrow set of conditions.

16-5 Specific facilities

This section offers some guidelines for the design of specific parts of the passenger building. It does not intend to be definitive. Indeed, no statement about these facilities could be definitive, given the variety of standards that exist at different airports, as Sec. 16-1 discusses. Persons needing more detailed guidance should consult the latest version of the International Air Transport Association (IATA) *Airport Development Reference Manual* (IATA, 1995). FAA Advisory Circular 150/5360-13 (FAA, 1988) on design guidelines, the report of the Committee for the Airport Capacity Study (1987), and Panero and Zelnick (1979) contain useful material, even though they refer primarily to the United States and may be out of date. Designers may also find inspiration in Hart (1985). Perhaps unfortunately, there is no rulebook on detailed design of passenger buildings.

Table 16-3 gives an example of design standards for a major international passenger building in the United States. It largely reflects practice in the year 2000. In this respect, it is a state-of-the-art benchmark. However, the design standards do reflect the perspective then prevailing in the United States. Experienced designers will notice that the values and objectives implicit in these standards might not be acceptable elsewhere. For example:

- The allowance for concessions is small compared to the area set aside for circulation and the arrivals hall, and does not reflect the kind of aggressive marketing practiced by some privatized airports.
- Circulation corridors in the ticketing area and at the gates are 40–50 ft (12–15 m) wide and might be considered excessively generous (see Sec. 15-4), especially when this space might be devoted to concessions.
- The amount of seating at the gate lounges is much greater than necessary, and does not reflect any sharing of the facilities (see Sec. 15-3).

Table 16-3. Example planning factors for a major international passenger building in the United States

Item	Amount U.S.	Amount SI	Per	Remarks
Ticket counter area	4.64 ft^2	0.42 m^2	Peak-hour passengers	Based on 3.5 min/passenger for processing
Length	0.39 ft	0.12 m		5.5 ft (1.65 m) per agent position
Queues	9.68 ft^2	0.87 m^2		25 ft (7.5 m) depth for queues
Circulation	15.48 ft^2	1.4 m^2		40 ft (12 m) beyond check-in queues
Public circulation	10,000 ft^2	900 m^2	Million annual passengers	
Security	600 ft^2	54 m^2	Checkpoint	Has two primary and backup devices; Processes 450 passengers/h/checkpoint
Departure lounge	14 ft^2	0.65 m^2	Seat in aircraft using gate	Design for 90% load factor, with 80% of these seating and 20% standing
Concessions	10,000 ft^2	900 m^2	Million annual passengers	Comparable airports
Circulation corridor at gates	45–50 ft	13.5–15 m		This is about 5000 ft^2 (450 m^2) per gate; compare with concessions
Toilets	3.5 ft^2	0.31 m^2	Peak-hour passengers	Comparable airports

Baggage claim, border controls	85 ft²	7.65 m²	Peak-hour passengers	Comparable airports; includes all baggage claim and inspection facilities
Arrivals hall	15 ft²	1.35 m²	Peak-hour passengers	

Source: Adapted from R. L. Brown Associates and HNTB Corporation Joint Venture, 2000.

In practice, designers often have to establish design standards suitable for their project. They may best adopt criteria associated with airports that are comparable to the ambitions of the airport operator or owner. Table 16-3 illustrates this approach; many of its standards reflect what is done at competitive airports in the United States.

Queues

All passenger-handling processes involve lines of people waiting. How much space should designers allow for this? As a rough rule of thumb, the length of a queue is about 0.6 m (2 ft) per person. This number assumes that families and friends normally stand side by side. The actual average separation depends on whether people are using trolleys and on whether many people are in line, which tends to compress the total length of the queue. As indicated in Chap. 15, the level of service for a queue depends on the available space. Figures 16-9 and 16-10 illustrate levels of service C and E.

Snake queues generally use space more efficiently. This arrangement involves a single queue for all persons waiting for several check-in agents or other servers. It is called a snake queue because it normally follows a back-and-forth channel between stanchions. Because it uses all the available floor space, instead of leaving spaces between queues in front of individual servers, snake queues reduce

Fig. 16-9. *Snake line at level of service C.* (*Source:* Zale Anis and Volpe National Transportation Systems Center.)

the amount of space needed for queues in front of counters. (This is a reason the standard for the depth of queues in front of counters in Table 16-3 is relatively short compared to standards elsewhere.) Snake queues have the further advantage of being fairer, since it is impossible for any passengers to be held up for a great length of time by a single customer who has some difficulty. However, snake queues have not been popular outside the United States as of the turn of the century.

Check-in areas

As of 2002, the standards for the design of check-in facilities are undergoing rapid change. The twentieth-century norms presented in available references are unlikely to apply much longer. Security concerns, the internet, and related electronic developments are radically changing the procedures for ticketing and check-in. The exact result of this process is impossible to specify, but may transform the notion of check-in and make conventional check-in halls obsolete.

Electronic ticketing and check-in reduces the need for large check-in halls for two reasons:

- *It reduces processing time at the airport,* thus reducing the number of check-in positions.

Fig. 16-10. *Snake line at level of service E.* (*Source:* Zale Anis and Volpe National Transportation Systems Center.)

- *It distributes many aspects of the process,* implying that there is less need for unique large check-in areas.

Through internet sales or travel agents, as of 2002 it was already common practice that passengers in North America would arrive at the airport with their seat selected. This simple step reduces check-in time by about 10 percent compared to traditional practices. A significant number of passengers do not check bags and thus already bypass the check-in halls, proceeding instead directly to the boarding area, where they confirm their presence with the gate agent. Moreover, it was already common for passengers to check in electronically at special kiosks in which they inserted their credit cards. This eliminates the time-consuming process of spelling out names (try it with "de Neufville") and having agents type them in—less time implies fewer counters. As registration kiosks become more common and appear in parking garages, rental car agencies, and throughout the passenger building, there will be even less need for conventional check-in facilities. The implications of electronic check-in will require careful attention.

Check-in facilities in the United States are likely to experience a further change. Increasingly, U.S. airports are installing Common User TErminal (CUTE) facilities that permit any airline to use any check-in counter.* These devices thus permit sharing of these facilities when peak demands for airlines do not overlap. They thus significantly reduce the total check-in space necessary (see Sec. 15-3). In this respect U.S. practice lags worldwide practice. Most major airports outside the United States have installed CUTE. They need it more because they typically require airlines to share check-in positions—different airlines will use the same position at different times of the day.

Security and border checkpoints

The crucial issue for security screening is the average service rate of the facility (or its inverse, the average time per person). This determines the number of devices needed. This time varies considerably due to different local practices and the nature of the passengers. Domestic and business travelers on short trips take less time; families with several carry-on bags take more. Mandle, LaMagna, and Whitlock (1980) report average times per person of around 30 s. However, note that the standards in Table 16-3 imply an expected average of about 15 s (450 persons/h for a set of two primary screening devices). During

---

*The 2002 version of CUTE is actually CUTE2. The text refers generically to all versions.

periods of special concern, the average time per person will increase, which will increase waits and queues and may necessitate additional screening devices and staff. It is possible that the average screening times per person will be much higher than 15 s/person permanently.

The formulas for calculating loads on security screening, the number of required devices and queue lengths are all permutations on Eqs. (16.1) and (16.2). As proposed by the FAA (2001) the expected load on the security checkpoints can be estimated as

$$L = P(1 - T)(1 + K)R \tag{16.3}$$

The number of checkpoints is then

Number of checkpoints $= L/(\text{service rate}) = L/(S \times F) \tag{16.4}$

In the specific case of equipment needed to screen hand-carried bags this can be written as

$$\text{Number of X-ray stations} = L(B/X \times F) \tag{16.5}$$

where $P =$ peak-hour enplanements

$T =$ percent of transfers who bypass the security checkpoint

$K =$ a factor that accounts for other airport traffic (employees, etc.) as a proportion of P

$R =$ a factor between 1 and 1.5 to provide additional capacity to cope with fluctuations in loads over the peak hour, a higher number for greater variability over the hour

$B =$ number of carry-on bags per passenger

$S \times F =$ the checkpoint nominal service rate, S, adjusted by a utilization factor, F, reflecting breaks by personnel, etc.

$X \times F =$ nominal service rate of the X-ray machine, X, adjusted by a utilization factor

Good design of security checkpoints features secondary devices behind the primary metal detectors. These make it possible to deal separately with persons who set off the alarm when they first walk through the detector. If such devices are not present, each person who has to go back and walk through the metal detector again will delay the entire process considerably while fumbling with forgotten change, taking out a cell phone, or removing a belt. For this reason, the standards in Table 16-3 refer to checkpoints with a set of devices.

Passport control processes, like passenger check-in, are changing in the electronic era. Bar codes in passports now reduce the time to enter data on passengers. For visitors to the United States, airlines forward passenger lists electronically to immigration authorities so that they can effect clearances before passengers arrive. Although immigration authorities worldwide move cautiously, their processes may become more efficient—and require relatively fewer positions—in the years ahead (see, for example, United Kingdom, Home Office, 1998). Political developments, such as the Schengen agreement among many European countries, which abolished passport controls for passengers moving between signatory nations, may also have a tremendous impact in this respect.

Moving walkways

Many airports use moving walkways to reduce the distance people have to walk. These devices extend about 50 m (150 ft). Designers use them over shorter distances, where people movers would not be effective. (Chapter 17 discusses people movers in detail.) For example, in Terminal 3 at London/Heathrow the airport operator placed several moving walkways end to end to cover the distance from the central block to the departures wing. Moving walkways are also useful over longer distances when people need to access many aircraft gates along a passenger building. Many pairs of moving sidewalks are thus an integral part of the design of the midfield linear passenger buildings at Denver/International.

Moving sidewalks do not, on average, speed up pedestrian traffic. Some people will move faster using these devices, to the extent that they can walk freely on them. However, many people will stop and rest on the moving sidewalks, and block others from moving on them. The speed of moving sidewalks is about 40 m/min or 2.4 km/h (1.5 mph). This is half the average pedestrian speed in airports of about 80 m/min or 4.8 km/h (3 mph) based on empirical observations of pedestrian movements in various airport terminal corridors (Young, 1999). On balance, moving sidewalks do not seem to speed up traffic (Young, 1995). They act as a convenience when distances are long.

Waiting lounges

There has been enormous disagreement in the industry about how much space should be devoted to waiting lounges at the gate. First, designers and operators disagree about the space to be provided for each person. Table 16-4 makes the point. The contrast in the standards of United Airlines and the others also suggests a divergence of opinion between airlines that have to pay for the space, and thus want less of it, and airport operators who would like to create higher levels of service (and be happy for users to pay for this).

Airport professionals further disagree about the number of passengers to be provided for in the gate area. The data in Table 16-3 allow for 90 percent of the aircraft capacity. Horonjeff and McKelvey (1994) suggest allowing for 80 percent. Some airlines claim that such a large amount is unnecessary because prospective passengers are in shops, airline clubrooms, or elsewhere, and in any case filter into the boarding area as others begin to board. In Singapore, moreover, the gate rooms are often extremely small, as the airport tends to call flights only when they are ready to board, and thus accommodates very few persons in these lounges. The situation at London/Heathrow is similar.

Finally, the amount to be provided depends greatly on the extent to which gates share common waiting lounges. As Sec. 15-3 indicates, it is possible to achieve up to 50 percent reductions in the total space allocated to waiting areas if these are arranged so that they can be shared (see also de Neufville, de Barros, and Belin, 2002; Horonjeff and McKelvey, 1994).

Economically efficient design will plan on shared waiting lounges and expect that many passengers will spend time outside these areas. It is

**Table 16-4. Some possible space standards
per person for waiting lounges**

	Space standard per			
	Passenger in lounge		Aircraft seat	
Source	**U.S. (ft²)**	**SI (m²)**	**U.S. (ft²)**	**SI (m²)**
Horonjeff and McKelvey (1994)	14.0	1.26	11.2	1.0
IATA, Level of Service C (1995)	11.1	1.00	N/A	N/A
R. L. Brown Associates (2000)	14.0	1.26	12.6	1.14
United Air Lines (Brown, 2001)	N/A	N/A	6 to 9	0.5–0.8

neither attractive for passengers to sit in sterile holding areas, nor profitable for the airport operator, who would prefer to have passengers spend time in concessions. Such design is not yet typical of current practice, however.

Concession space

The financial success of concessions in a passenger building depends on the amount of time people spend in the airport. The longer they have to wait around, the more likely that they will chose to shop [as Hasan and Braaksma (1986) demonstrated in their survey of Canadian airports]. Put another way, commercial success depends noticeably on the inefficiency of the airport from the perspective of passengers who want to get to their plane or destination as quickly as possible.* Moreover, the longer people stay in the airport, the more space the designers will have to provide. The airport operator thus faces a dilemma: commercial success comes at a significant cost in terms of space and possible operational inefficiency. Decisions about how much commercial space to provide should thus reflect management objectives. They cannot be derived from strictly technical considerations.

Planning commercial space is risky. Many shopping developments at airports have been very successful. Overall, commercial activities at airports have become much more significant worldwide as knowledgeable retail operators have replaced governmental airport operators who were not marketing experts (see Chap. 5). The sales volumes at the busiest airports near half a billion dollars a year when 1995 data are brought up to date (Table 16-5). Some companies have obtained their highest sales volumes per unit area at airports (Chesterton, 1993), and a number of airports have achieved remarkable success (Humphreys, 1998). However, many commercial developments in new airport passenger buildings have been significant failures. Airport operators need to be cautious about developing commercial space at airports.

Planners need to place commercial space carefully. The saying that the three most important factors in successful sales are "location, location,

---

*Some people do want to shop at the airport. Travelers may wish to use duty-free stores or take advantage of the convenience of finding many leading brands or souvenir shops in one place. Others use the airport for routine shopping in countries, such as the Germany and the Netherlands, where transportation centers are exempt from restrictive laws limiting when stores can operate. Thus Frankfurt/Main and Amsterdam/Schiphol feature grocery and clothing stores. These situations do not represent many travelers who would be glad to get through the airport rapidly.

Table 16-5. The world's top
duty-free outlets in 1995

Airport	Sales (millions of $)
London/Heathrow	524
Honolulu	419
Hong Kong/Kai Tak	400
Singapore	359
Amsterdam/Schiphol	327

Source: Freathy and O'Connell, 1998.

and location" has a basis in truth. Location is a key factor influencing the success of any commercial space. As a rule, the best locations for commercial space are in the direct line of passenger traffic, after the various barriers between them and the gates. Commercial space needs to be seen. Stores placed in out-of-the way places, on mezzanines people reach by changing floors, or where there is little traffic are at a disadvantage. For example:

- Many stores in the linear buildings at Kansas City and Munich withered for lack of traffic.

- The commercial space on the mezzanine at Tokyo/Narita Terminal 2 is avoided by passengers, who move directly from check-in to their departure gates without making a detour to an upper floor.

- The expensive restaurant and various stores hidden from view behind the check-in counters in the International Terminal in San Francisco/International closed after only a few months of operation.

Commercial space also benefits from being in a place where passengers feel comfortable browsing. In the United States, this means that it works best "after security," after passengers have passed through all the barriers that impede their way to the gate. Thus, in the new passenger building at Washington/Reagan, people walk by the stores "before security," wanting to be sure that they will make it to their gate and will get the next shuttle flight. Once passengers have passed through security and are at their gates, however, they relax, shop, eat, and drink.

Operators of airport stores estimate the area needed per store in terms of the number of persons passing by, the percentage of passengers they can attract (the "penetration rate"), and the time passengers remain ("dwell") in the shop. They also add extra space to account for inventory space. A simple version of this calculus is

$$\text{Passengers in store} = \tag{16.6}$$
$$[(\text{passenger flow rate (passengers/hour)}] \times [\text{target penetration}]$$
$$\times [\text{average dwell time in shop in hours}] \times [\text{peaking factor}]$$

The target penetration might be 0.65 for a general duty-free store, and much lower at gate lounge shops, for example, 0.25. The peaking factor might be 1.2. The resulting estimate of passengers in the store leads to an estimate of the appropriate size of the store. For example:

$$\text{Space/in m}^2 = 4 \text{ [number of passengers in store]} \tag{16.7}$$

The factor of 4 allows 1 m^2 for personal space, 2 m^2 for circulation, and 1 m^2 for cashiers and checkout space (Freathy and O'Connell, 1998).

Baggage claim areas

Baggage claim areas must first provide enough *claim presentation length,* that is, length along the conveyor belt or racetracks, for people to identify and pick up their bags. The IATA (1995) standards recommend about 70 m (230 ft) for wide-body, and about 40 m (130 ft) for narrow-body aircraft being served at the same time. This standard implies about 0.3 m (1 ft) of claim presentation per passenger. In practice, the amount available appears to deviate ±50 percent from this recommendation. The FAA (1988) alternatively defines the length required in terms of the number of aircraft arriving in a peak 20 min, and assumes that passengers check 1.3 bags per person. Either standard leads to approximately the same results. However, these standards should be modified according to local realities such as the average number of bags checked.

The standards most particularly need to account for the intensity of the transfer traffic at the airport. The more travelers connect between flights, the less space is needed in the baggage claim area. The FAA standards include this factor. A major exception to this rule concerns international arrivals in the United States. According to U.S. practice, all passengers entering the United States must claim all their baggage upon arrival and present it for inspection even if they are transferring to another city (they then have to recheck it at a special facility beyond customs). In this case, baggage claim devices need to serve all passengers.

Congestion hot spots can be a major difficulty in baggage areas. This is because passengers will naturally cluster at the places where bags appear on the claim devices. These areas can thus present problems even if the entire baggage area meets the IATA or other space standards (see Sec. 15-3). As a general precaution against such problems, the IATA (1995) recommends that the distance between adjacent baggage claim devices be at least 9 m (30 ft).

Curbside and equivalent areas

Passenger buildings need extensive space for people to get out of and into cars, taxis, and buses. This space typically consists of one or more long stretches of sidewalk along a roadway in front or alongside of the passenger building. Passengers and visitors will either alight from their vehicles on this sidewalk and proceed into the building, or emerge from the building onto this sidewalk and get into their vehicle. Airport operators and designers generally refer to this space as the *curbside area.*

Many airports provide curbside space through short-term parking areas located in front of the passenger buildings. These areas permit drivers to stop their cars to drop off or pick up passengers, and are functionally equivalent to the curbside area. These short-term parking areas are beneficial both to the airport operator and the traveling public (de Neufville, 1982). They allow airports to expand their curbside area relatively easily, in situations where it might be impossible to extend the actual curb in front of the passenger building. Moreover, short-term parking generates revenue for the airport, which curbsides do not, generally at much higher rates than longer-term parking spaces. Short-term parking areas also allow drivers to stay in these drop-off and pick-up areas longer than they could at the curb in front of the building. Boston/Logan and Montreal/Dorval airports have such arrangements (see Sec. 17-4).

Arriving passengers need considerably more curbside space than departing passengers. This fact needs emphasis. Inexperienced designers might assume wrongly that since the numbers of arriving and departing passengers are generally equal, so should be their need for curbside space. This is incorrect, because people have different needs when they are arriving than when they are departing. In the departure process, drivers merely need to drop off the passengers—this can happen quickly. On arrival, however, drivers first need to locate the arriving passengers—this takes time. The dwell

Table 16-6 Example differences in dwell times for enplaning and deplaning passengers

| City | Passenger dwell time (min) | |
	Enplaning	Deplaning
Denver	1.2–2.8	4.8–6.9
Miami	1.6–4.5	2.3–4.5
New York	1.0–1.6	2.1–4.8

Source: Mandle et al., 1980.

time for cars picking up passengers is thus longer, and this translates into more space (see Table 16-6). The exact amount of extra space needed for arriving passengers depends on local circumstances. As a rule of thumb, the curbside area for deplaning passengers should be about 20 percent more than for departing passengers (de Neufville, 1982; Mandle, Whitlock, and LaMagna, 1982).

The length of curb depends on many factors (see Committee for the Airport Capacity Study, 1987; Brimson and Caldwell, 1992). A general formula for estimating the desirable length of "curbside" area appears below (adapted from Mandle et al., 1982). It is the sum of the length provided for private cars and taxis (subscript 1) and larger vehicles such as buses and courtesy vans (subscript 2). Thus:

$$\text{Length of curb space} = C = C_1 + C_2$$

$$C_I = P[(M_I)\,(F_I)\,(D_I/60)\,(L_I)]/V_I \tag{16.8}$$

where P = number of originating and terminating passengers (no transfers) in the peak hour

M_I = fraction of passengers using mode I

F_I = fraction of passengers in mode I using the curb (as opposed to long-term parking)

$D_I/60$ = the average dwell time in hours for vehicles in mode I

V_I = the average number of passengers in vehicles in mode I

L_I = average length of space needed to park vehicles, about 7.5 m (25 ft) for cars and taxis, about 13.5 m (45 ft) for buses.

To put the curbside requirements into perspective, it is useful to express them in terms of length per passenger, C/P. As a very rough approximation, airports need on the order of 5 cm (2 in.) per peak-hour passenger, or about 0.15 m (6 in.) per 1000 annual passengers excluding transfers. (This would be the result if one-fifth of the passengers used the curb, their dwell time was 3 min on average, and the average number of passengers per car was 1.5.) This could work out to about 1.5 km or almost a mile for an airport serving 10 million originating and terminating passengers. Such calculations demonstrate the potential importance of curbside requirements. However, airport operators can reduce the amount of curbside space considerably by careful design and management.

Airport operators can increase the productivity of their curbside primarily by reducing the dwell time. The length needed is directly proportional to the dwell time. Thus, halving the dwell time halves the length needed or, alternatively, doubles the capacity of existing space. Airport managers can reduce dwell time by enforcing regulations on the amount of time cars can take to load or unload passengers. "Hiring policemen" is, in many cases, the simple answer to lack of curbside capacity.

Airport operators can also influence the number of people using the actual curbside along the passenger building by providing short-term parking convenient for the pick-up and drop-off of passengers. This solution is inappropriate for taxis, but can reduce the amount of curbside needed by about a third, depending on the circumstances.

Major airports serving 10 million and more passengers annually must still provide extensive curbside areas, even with clever design and extensive use of public transport that reduce requirements. They do this by:

- *Double-decking the access roads:* placing the departing passengers on the upper level and arriving passengers on the lower level. This design doubles the amount of curbside available for a given frontage of the building.
- *Providing parallel curbsides,* consisting of long pedestrian islands that duplicate the actual curbside along the passenger building. Two or even three separate parallel islands may exist at the busiest or most congested airports, such as Boston/Logan, London/Heathrow, and Atlanta (Fig. 16-11).
- *Establishing taxi pools,* which hold waiting taxis until they are needed and called, which reduces their dwell time at the curb and the amount of curbside that is needed.

In designing parallel curbsides, good practice separates the buses and vans from taxis and cars. This is because larger vehicles need

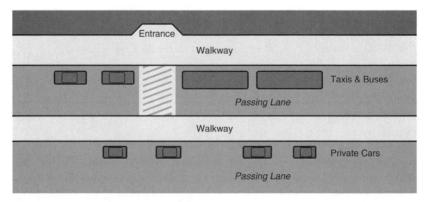

Fig. 16-11. *Typical layouts of curbs with multiple lanes.*

more time to load and unload. If they are mixed in with the smaller vehicles, they could block their flow. Similarly, when the curbside is particularly long, as in front of a linear passenger building, it is important to provide places where traffic can bypass vehicles double-parked somewhere along the curb. This is conveniently done by using parallel curbsides with crossover points. Simulation analyses can usefully explore the exact design of these situations (see Saffarzadeh and Braaksma, 1994).

Exercises

1. Rework Example 16.1 on the assumption that the airline could reduce the average time to check-in passengers by 20 s by eliminating the weighing of bags. How would that affect the number of counters and personnel needed? Using local figures that appear reasonable to you, is it worthwhile to spend the time to weigh bags? Is the potential revenue worth the cost?

2. Explore the potential for electronic check-in. By how much time would it reduce the check-in process if the agent did not have to read your name and destination from a printed form, but got it directly (for example, because the traveler swipes a credit card)? Apply these savings to a major international airport featuring a check-in hall with 100 positions. Using local estimates of the cost of staff, check-in counters, and space, does electronic check-in appear worthwhile?

3. Rework Example 16.3 assuming that there are only 6 counters and agents take an average of 1 min per passenger to check in travelers.

4. Visit an airport and observe the check-in process for one or more flights. (If you choose several, look at different kinds of flights, such as domestic, international, shuttle, or "cheap fare.") How long is the average processing time? How long are the queues? How much space does each person in queue take? What suggestions would you make about how this process could be improved?

5. If possible, observe the security procedures at some airport. On average, how far apart are the persons in the queue who have a minimal amount of baggage? What is the average time per person for this process? Are there secondary facilities for screening persons who cause an alarm on the primary metal detector? In your estimation, how well is this process working?

6. Observe how people use waiting lounges at gates in one or more airports. Considering the size of the aircraft for the flight, what fraction of the passengers is in the lounge at any time? Where else are they? What is the ratio of seats in the lounge to the size of the aircraft? Are there enough seats? What kind of changes would you suggest?

7. Choosing a specific flight, note the time it actually arrives at the gate. Then observe the time the first and last bags appear in the baggage claim area. Also observe, if possible, when the first passengers from the flight arrive. Use these data to evaluate the quality of the baggage delivery service for that flight.

8. Visit an airport and observe the operation of the curbside in a peak period. Does it appear adequate? What is its configuration? How much of it is there? How do these measurements correspond to the general guidelines? Observe 10 vehicles of a similar type (taxis, private cars, or buses) and estimate their dwell times for arrival and departures. How do these match the general guidelines?

References

Ashford, N. (1988) "Level of Service Design Concept for Airport Passenger Terminals—A European View," *Transportation Planning and Technology,* 12(1), pp. 5–21.

Brimson, T., Caldwell, R. (1992) "Kerb Use at Airport Terminals," *Proceedings, Conference of the Australian Road Research Board,* 12(6), pp. 79–89.

Brown, T. (2001) Personal communication on behalf of United Airlines.

Chesterton International Property Consultants (1993) "Airport Retailing: the Growth of a New High Street," London, U.K.

Committee for the Airport Capacity Study (1987) *Measuring Airport Landside Capacity,* Special Report 215, Transportation Research Board, National Research Council, Washington, DC.

de Neufville, R. (1982) "Airport Passenger Parking Design," *ASCE Journal of Urban Transportation, 108,* May, pp. 302–306.

de Neufville, R., de Barros, A., and Belin, S. (2002) "Optimal Configuration of Airport Passenger Buildings for Travelers," *ASCE Journal of Transportation Engineering,* 128 (3), pp. 211-217.

de Neufville, R., and Grillot, M. (1982) "Design of Pedestrian Space in Airport Terminals," *ASCE Transportation Engineering Journal, 108*(TE1), pp. 87–102.

FAA, Federal Aviation Administration (1988) *Planning and Design Guidelines for Airport Terminal Facilities,* Advisory Circular 150/5360-13, U.S. Government Printing Office, Washington, DC.

FAA, Federal Aviation Administration, Office of Policy and Plans (2001), *Recommended Security Guidelines for Airport Planning, Design and Construction,* (draft) Washington, DC.

Freathy, P., and O'Connell, F. (1998) *European Airport Retailing: Growth Strategies for the New Millennium,* Macmillan Business, London, U.K.

Hart, W. (1985) *The Airport Passenger Terminal,* Wiley-Interscience, New York.

Hasan, N., and Braaksma, J. (1986) "Planning of Concessions in Airport Terminals," *ASCE Journal of Transportation Engineering,* 112(2), Mar. pp. 145–162.

Horonjeff, R., and McKelvey, F. (1994) *Planning and Design of Airports,* 4th ed., McGraw-Hill, New York.

Humphreys, I. (1998) "Commercialisation and Privatisation: the Experience of Cardiff Airport," Occasional Paper 49, Department of Maritime Studies and International Transport, University of Wales, Cardiff, U.K.

IATA, International Air Transport Association (1995) *Airport Development Reference Manual,* 8th ed., IATA, Montreal, Canada.

Mandle, P., LaMagna, F., and Whitlock, E. (1980) *Collection of Calibration and Validation Data for an Airport Landside Dynamic Simulation Model,* Report TSC-FAA-80-3, Transportation Systems Center, FAA, Cambridge, MA.

Mandle, P., Whitlock, E., and LaMagna, F. (1982) "Airport Curbside Planning and Design," *Transportation Research Record 840,* pp. 1–6, Transportation Research Board, National Research Council, Washington, DC.

Mumayiz, S. (1990) "Overview of Airport Terminal Simulation Models," *Transportation Research Record 1273,* pp. 11–20, Transportation Research Board, National Research Council, Washington, DC.

Mumayiz, S., and Ashford, N. (1986) "Methodology for Planning and Operations Management or Airport Terminal Facilities," *Transportation Research Record 1094,* pp. 24–35, Transportation Research Board, National Research Council, Washington, DC.

Mumayiz, S., and Schonfeld, P., eds. (1999) *Airport Modeling and Simulation*, ASCE, Reston, VA.

Panero, J., and Zelnick, M. (1979) *Human Dimension and Interior Space: A Source Book of Design Reference Standards,* Whitney Library of Design, New York.

R. L. Brown Associates and HNTB Corporation Joint Venture (2000) *RLB/HNTB Analysis of Atlanta Terminal Plan,* Atlanta, GA.

Saffarzadeh, M., and Braaksma, J. (1994) "Optimum Design of Airport Enplaning Curbside Areas," *Transportation Engineering Journal, ASCE, 120*(4), pp. 536–551. *Note:* The lead author currently refers to himself as Saffarzadeh. In the published version of the paper, however, he is actually listed as Parizi. To find the paper using an author search, use Parizi.

Seneviratne, P., and Martel, N. (1991) "Variables Influencing Performance of Air Terminal Buildings," *Transportation Planning and Technology, 16*(1), pp. 3–28.

United Kingdom Home Office (1998) *Fairer, Faster, Firmer—A Modern Approach to Immigration and Asylum,* HMSO Cm. 4018.

Young, S. (1995) "Analysis of Moving Walkway Use in Airport Terminal Corridors," in *Transportation Research Record 1506, Airport and Air Transportation Issues,* pp. 44–51, National Academy Press, Washington, DC.

Young, S. (1999) "Evaluation of Pedestrian Walking Speeds in Airport Terminals," in *Transportation Research Record 1674,* pp. 20–26, *Pedestrian and Bicycle Research,* National Academy Press, Washington, DC.

17

Ground access and distribution

Good ground access is vital for an airport. It is an indispensable ingredient of good service and maintains the attractiveness of the airport for the users and the value for the airport operators. Providing adequate and appropriate means of getting to and from the airport can be a major challenge for airport planners and operators.

Because most of the users and employees of the airport are widely dispersed over the metropolitan area, it is inevitable that cars and buses provide most airport access at most airports. For particularly busy or remote airports associated with cities that have a substantial rail transit system, rail access will increasingly be an important means of airport access. In general, however, airports need to provide substantial highway access and parking facilities.

People-mover systems can distribute users efficiently around airports. Planners are thus installing them increasingly at major airports. The use of these systems on the airside has been one of the major innovations in airport design of the last generation. It has allowed designers to disperse gate positions widely over the airfield, thus facilitating efficient aircraft ground maneuvers, without making passengers walk excessively.

The distribution of baggage around the airport is a major issue especially as baggage systems are large, cumbersome, and located in basements and other areas that are difficult to expand. Effective sorting systems are crucial to the performance of major airports. These require operational and design procedures that minimize sorting, cost-effective and reliable mechanical devices, and fallback systems that deal with the inevitable sorting failures.

Overall, designers of mechanical distribution systems that serve many points, baggage systems in particular but also people movers, need to recognize that they need to provide about twice as much capacity in their systems as the design load. This margin is needed to guarantee that there are enough empty spaces to serve all the prospective users in good time.

17-1   Introduction

This chapter covers the range of networks for connecting passengers and their baggage to the airport, and for distributing them around the airport passenger buildings. These range from metropolitan systems dealing with airport access to mechanical devices operating within the limits of the airport. The discussion proceeds from the large-scale, regional issues of access to and from the airport, to the smaller-scale issues of people movers and baggage handling systems at the airport.

Specifically, the next two sections first define the overall patterns of flows between the city and the airport and then indicate which access systems best serve these needs. As much of the traffic to and from most airports almost inevitably moves by automobile, Sec. 17-4 discusses strategies for providing adequate parking at appropriate prices. Sections 17-5 and 17-6 then describe the several ways of moving passengers around airports, focusing on the people movers that have been the central innovation in the design and operation of airports in the past generation. Finally, Sec. 17-7 closes with a presentation of the crucial issues of baggage handling systems, without which airports cannot function.

Airport connection and distribution systems are highly complex. The flows on these networks are typically "many-to-many." Passengers and bags originate from many different points and go to many distinct destinations. For example, people come from homes spread around the city, and go to one of several passenger buildings at the airport. Passengers and bags similarly flow from dozens of check-in counters to as many aircraft. In simpler cases, the flows may be considered to be "many-to-one," for example, when passengers coming to the airport go to only a single passenger building. For practical purposes, however, the traffic flows around the airport always have multiple origins, multiple destinations, or both. They essentially are never "one-to-one," serving a single origin and destination.* The interactions

*The few exceptions to this rule are predominately people movers that shuttle passengers between a main passenger building and a single satellite or midfield passenger building. Examples are found at Miami/International, London/Gatwick, and Tokyo/Narita. The people mover at Kuala Lumpur/International operated as a shuttle when the airport opened. However, the designers conceived and built it as part of a network that would eventually connect the main passenger building with several midfield passenger buildings.

among the many intersecting flows increase the complexity of the connections between origins and destinations. Many traffic streams either merge with each other or, worse, merge and then must be sorted according to their several destinations.

Because airport distribution systems are complex, their effective capacity is much less than the maximum flow the system could carry. A complex distribution system involves many sequential queues, as passengers and bags wait for vehicles to take them somewhere, and as these vehicles wait their turn to merge into other flows of traffic. As Secs. 17-5 and 17-7 explain in detail, queuing systems can consistently provide stable service with reasonable delays only when they operate at a fraction of their maximum capacity (see also Chap. 23). Moreover, a second factor worsens this generic problem. In order to provide adequate service to traffic from each of the origins on the network, the system must deliberately feature numerous windows of empty capacity. These empty spaces ensure adequate service to traffic originating down the line. The net effect of the inevitable queues and need for empty capacity is that the practical capacity of complex distribution systems is perhaps only half their mechanical ability to process flows. Failure to appreciate this point has led to numerous expensive and embarrassing mistakes.

17-2   Regional airport access

Getting to the airport can be a challenge. The airport is generally relatively far away, roads may be congested, and traffic patterns at the airport are often confusing. For most travelers, especially the many going to the airport at rush hours, the trip to the airport can be a most annoying part of their journey. Equally, the daily trip to and from the airport can be difficult and expensive for the employees who work at the airport or for the airlines serving the airport. Together, the difficulties of getting to the airport conveniently and reliably constitute the *airport access problem*. What should airport operators and regional authorities do about it?

A popular idea is to build high-speed rail connections between the airport and the center of the city. The concept of modern vehicles speeding travelers between the airport and their local destinations resonates with civic leaders and their constituents, as well as with the local construction industry. This vision has motivated long-term efforts to build such systems worldwide. In this vein, the BAA built the Heathrow Express connecting London/Heathrow with Paddington

Station in London. Malaysia similarly built a rail link between the capital and the new Kuala Lumpur/International airport. Meanwhile, San Francisco authorities have been promoting the extension of the Bay Area Rapid Transit (BART) to San Francisco/Oakland and in fact opened the connection to San Francisco/International in 2002. The Port Authority of New York and New Jersey is also developing a rail connection to New York/Kennedy. At the end of the twentieth century, authorities have built or extended rail systems to Atlanta, Chicago/O'Hare, Copenhagen, Oslo, Paris/de Gaulle, and Tokyo/Narita airports, among others (see Tables 17-3 and 17-4 for details). To what extent is this popular notion a useful and effective approach to the airport access problem?

To answer such questions and to deal appropriately with the airport access problem, it is necessary to understand the nature of the traffic to the airport, the costs and performance of the alternative systems, and the preferences of the users. This section describes the nature of the traffic flows, their distribution, and the preferences of the users. Section 17-3 defines the performance of the technical possibilities, how they match the criteria of actual and potential investors, and recommends preferred choices for different circumstances.

Nature of airport access traffic

Airport traffic has three major components. Planners must appreciate their special characteristics. Each group has its own patterns and needs. These distinct markets include:

- *Originating and terminating travelers,* who have only one access trip per flight, since either their departure or arrival is by air (however, their access may involve a round trip if someone delivers them to the airport, or if the taxi is not allowed to pick up at the airport)
- *Employees,* who commute to and from the airport each day
- *Supply, delivery, and other commercial vehicles* that come and go from the airport

Each of these categories of airport traffic normally accounts for at least about 20 percent of the total trips to the airport. This fraction depends strongly on local conditions, however. If the airport is a transfer hub, it will have relatively few originating and terminating passengers and thus these will account for a smaller proportion of the whole. Similarly, if the airport is a maintenance or training base for an airline, it will have more employees and other commercial traffic. Regardless of these significant variations, each of the categories of traffic is of the same order of magnitude. This observation is important because it corrects the mistaken popular definition of airport access that focuses on the passengers and neglects the other traffic. Airport passengers are only part of the airport access problem.

A misplaced focus on the number of access trips due to passengers is easy to understand. The number of airport passengers may easily be several times the population of the metropolitan area it serves. For example, the annual number of passengers through the Boston and Paris airports is six to eight times the local population. The relative number of airport passengers is large even in areas where air travel is less frequent. Thus, the number of airport passengers through Jakarta and Mexico City about equals the population of these cities. In contrast to the apparent importance of passengers, the number of employees at an airport is relatively low. Although the busiest airports have tens of thousands of workers, they are generally less than one per thousand passengers (Table 17-1). Such figures easily concentrate attention on passengers.

Data on the number of passengers are misleading, however, because they neglect the frequency of travel. Originating or terminating passengers each make one trip to or from the airport. They each account for about one or less vehicle trips (more than one if a driver has to return empty, less than one when they share rides—see Table 17-2). Employees and other commercial traffic, however, make round trips every day of the year—and they may make additional trips to go on various errands. Each employee thus makes about 500 or more access trips a year. This frequency of employee trips compensates for their low number and makes employee traffic the same order of magnitude as passenger traffic (see Example 17.1).

The salience of passengers in the airport access problem is due their particular characteristics, not their number. First, they concentrate at the "main door" to the airport, whereas employee and commercial traffic disperse to locations all around the airport, such as cargo areas and other facilities away from the main passenger buildings. Second, passengers tend to be anxious. They need to make a flight and are often unfamiliar with the airport. Finally, the passengers are important customers. Thus, although they are only part of the airport access problem, they do command attention.

Table 17-1. Sample data on employees at U.S. airports

Airport	Average daily employees	Employees/1000 total annual passengers
Dallas/Fort Worth	48,000	0.80
Chicago/O'Hare	40,000	0.57
Los Angeles/International	40,000	0.65
San Francisco/International	31,000	0.79
Phoenix	23,700	0.76
St. Louis/Lambert	19,000	0.66
Denver/International	17,400	0.47
Boston/Logan	14,500	0.57
Houston/Bush	14,400	0.46
Salt Lake City	13,000	0.65
Seattle/Tacoma	11,400	0.44
San Francisco/Oakland	10,500	1.14
Tampa	8,200	0.59
Las Vegas	7,500	0.37
Portland (Oregon)	5,000	0.38
San Francisco/San Jose	3,500	0.34
San Diego	2,600	0.17
Sacramento	2,300	0.32
Median value		0.57

Source: Leigh Fisher Associates et al., 2000, p. 10.

Table 17-2. Vehicle trips per passenger by access mode

Access mode	Vehicle trips/ passenger
Pick-up or drop-off	1.29
Taxi	1.09
Parking	0.74
Rental car	0.69
Courtesy bus	0.33
Scheduled bus	0.10

Source: Shapiro et al., 1996, from a Massport study.

EXAMPLE 17.1

Consider an airport serving 10 million total passengers a year. The number of passenger trips to and from the airport will be less than 10 million, because transfer passengers do not leave the airport. It might be between 7 and 9 million a year. This implies an average in the range of 20,000–25,000 passenger trips a day to and from the airport.

Suppose that, on average, 5000 people work at this airport every day. This is consistent with a ratio of about 0.5 employees per thousand passengers (see Table 17-1). They include the persons operating the airport, the airline employees, and all the staff working for concessionaires, hotels, freight companies, car rental agencies, and the like. These workers are not all at the airport at the same time, of course, because they work different shifts during the day. If all employees commute both ways, they account for about 10,000 trips a day on average.

Trips by commercial vehicles depend on the situation. They include suppliers of all sorts, deliveries and pick-up of cargo, persons coming to sell products, and other visitors. These easily account for as many trips as employees do.

These conditions imply about 40,000 trips to and from the airport each day. Passengers account for about half of these trips. Employees and commercial traffic each generate about a quarter of the airport access traffic in this case.

Distribution of airport access traffic

The general rule is that only a small fraction of the traffic to and from an airport goes to or comes from any specific destination. Most of the ground traffic for the airport spreads out over a wide area. This fact is crucial for proper understanding of the airport access problem.

The center of the city generates only a small fraction of the trips to and from the airport. Passengers living in the metropolitan area typically start their trips to the airport from their homes, even if they are taking a business trip. Furthermore, their homes are generally located in the suburbs or at least in apartment buildings some distance from the financial or commercial center of the city. Travelers from elsewhere may be going to offices or hotels in the city center. However, if they are visiting friends or relatives, they are likely to go to their homes in the suburbs. Even business travelers may not be destined for the city center, because they are either calling on companies in industrial areas distributed around the metropolitan area, or they are seeking less expensive hotels at some remove from the city center. According to a 1996 survey, only about 8 percent of the passengers originating or terminating at Boston/Logan came from or went to the whole city of Boston, spread out over some 20 square miles (Leigh Fisher Associates et al., 2000, p. 43). Only a fraction of these trips was associated with the center of Boston. In general, the percentage of passengers coming from or going to the city center is quite small.

Employee and commercial traffic goes primarily to the edges of the city. This traffic links the airport to the parts of the metropolitan area that are less expensive for housing and industry. Only exceptionally is it connected directly to the center of the city.

Overall, only about a tenth of the airport access traffic wants to come from or go to the city center. Example 17.2 shows how this works out. The exact amount depends, of course, both on how planners define "city center" and the local situation. In any case, the fraction of airport access traffic connected with the city center is small. Most of the airport access traffic is spread widely over the metropolitan area. For the Boston/Logan, San Francisco/Oakland, and San Francisco/International airports, for example, about half the passengers traveled more than 15 mi (25 km) to get to the airport (Leigh Fisher Associates et al., 2000, p. 47).

The fraction of traffic destined for the city center tends to decrease for the largest airports. This is because the airports with the most traffic, such as Atlanta, Chicago/O'Hare, London/Heathrow, and Tokyo/Narita, generally are transfer hubs. They thus have relatively fewer originating or terminating passengers and therefore relatively fewer trips to the center of the city.

EXAMPLE 17.2

The 10-million-passenger airport of Example 17.1 generated about 40,000 access trips a day. Suppose that 20 percent of the passengers connect with the city center. Suppose also that essentially none of the employees or commercial visitors goes there. This means that about 4000–5000 passengers a day go to or come from the city center. This is roughly 10 percent of the total.

Preferences of the users

Individually, a prime concern for passengers is getting to the airport on time. They tend to be most concerned about the reliability of their travel time to the airport. Missing a flight can have severe consequences in terms of:

- Delays until the next flight hours or a day later
- Costs due to the need to replace or upgrade tickets
- Missed connections or appointments

Reliable connections are much more important than speed. To deal with unreliable access and to get to their flights on schedule, passengers routinely allow substantial extra time for their trip to the airport. This provides a buffer to ensure that they will not miss their departures. Accepting added time allowed to compensate for unreliable connections is equivalent to accepting a much slower average speed for the trip. Such actions demonstrate the passengers' willingness to sacrifice speed in favor of reliability.

Collectively, passengers also want access systems that can distribute them to their destinations spread widely over the metropolitan area. Employees likewise share this requirement for a system that can take them between the airport and their homes scattered over the suburbs and other bedroom communities. Systems that serve only a few points and do not connect conveniently with a wide network of transportation do not adequately address the requirements of most of the airport access traffic.

Price can be a significant consideration. It is generally a secondary concern for passengers, compared to reliability and accessibility. Business passengers paying hundreds or thousands of dollars for an airfare may be prepared to pay reasonable amounts to get to and from the airport. Families and others may be more careful. Moreover, the price of the airport access trip is salient for employees, who have to pay for daily round trips. In short, whereas some passengers may be willing to pay for premium service, many passengers and most employees cannot afford this kind of service.

Price considerations tilt passengers and employees toward the use of automotive forms of access. When they have access to privately owned vehicles, these will generally provide cheap (and convenient) access. Passengers may also favor public taxis because these vehicles normally cost the same for as many people as can fit inside—contrary to other forms of access, for which each traveler has to pay an additional fare. This makes the average cost per trip per person much lower for the many passengers who travel in groups, as with family, friends, or associates. When people consider the total cost of getting to the airport, automobile access can appear much more economical.

Overall, the users of airport access systems need reliable systems that can distribute them broadly throughout the metropolitan area. Most of the market is also unlikely to pay premium fares for individual trips so long as they have some form of inexpensive automotive transport available.

Needs of airport operators

Airport operators often feel obliged to promote rail and other forms of public transport. Some major airports face an inability to build highway capacity for the airport, and must either develop rail access or face gridlock. Toronto/Pearson, for example, foresees that as the city grows it will not be able to count on the local highways to deliver traffic reliably. It has therefore acquired and is reserving right-of-way for a connection to the local mass transit system. Likewise:

- New York/Kennedy is being connected to the metropolitan rail and mass transit system.
- Japan has connected Tokyo/Narita to two rail lines.
- BAA has built the Heathrow Express connection between London/Heathrow and the Central London rail and mass transit network,
- The major island airports, Hong Kong/Chek Lap Kok and Osaka/Kansai, have rail links to the mainland.

Rail links are sometimes not a choice but a necessity to ensure maximum use of the airport.

Many airport operators face powerful pressures to restrict automotive access to the airport. Because major airports often are the largest single destination and economic activity in a metropolitan area, environmental interests see them both as good targets for environmental improvements and as "deep pockets" that can afford to subsidize public transport. Thus Massport, the operator of Boston/Logan, is paying for regional bus service and water taxis to induce a few people not to come by car. Meanwhile, however, its economic interest is to attract cars, since it derives about 30 percent of its total revenues (about $90 million of $325 million in 2000) from parking and fees from rental cars (see Sec. 17-4).

Over the last generation, operators of many major airports have in-stalled tracked forms of airport access. In the United States, they have focused on metropolitan rail systems (Table 17-3). In Europe and Japan, where people use long-distance railroads regularly, many airport railroads connect to the national intercity and even bullet trains (Table 17-4). Correspondingly, the Asian and European sys-tems tap into existing markets and serve a higher fraction of the air-port traffic than the U.S. airport rail systems (Table 17-5).

17-3 Cost-effective solutions

The issue

Highways provide the dominant mode of airport access. Automobiles, taxis, vanpools, and buses dominate the traffic. They are the people's choice. For most people going to the airport, automotive vehicles provide the best value for money. Passengers and employees appre-ciate that this form of airport access fulfills the essential function of distributing traffic conveniently throughout the metropolitan area. It also caters to their varied needs by providing a range of more or less convenient and luxurious service, more or less expensively.

Airport owners appreciate that, for all but the largest airports in the most congested areas served by an extensive rail network, high-ways are relatively easy and inexpensive to build—compared to rail

Table 17-3. Airports served by metropolitan rail systems in the United States in 2002 (recent developments in boldface)

City	Airport	Status
Atlanta	Hartsfield	**System built**
Baltimore	Baltimore/Washington	**Light rail**
Boston	Logan	Traditional transit
Chicago	O'Hare	**Transit extended**
Chicago	Midway	Traditional transit
Cleveland		Transit extended
New York	Kennedy	**Under construction**
Philadelphia		**Transit extended**
San Francisco	International	**System extended**
St. Louis	Lambert	**Light rail**
Washington	Reagan	**System built**

Table 17-4. Airports served by rail systems in Europe, Asia, and Australia in 2002 (recent developments in boldface or under construction, U.C.)

Region	Country	City	Airport	Metro-politan transit	Intercity railroad	Bullet train
Europe	Belgium	Brussels	Zaventam	Yes	Yes	Yes
	Denmark	Copenhagen	Kastrup	**Yes**	**Yes**	
	France	Lyon			**Yes**	
		Paris	de Gaulle	**Yes**	**Yes**	**Yes**
		Orly		**Yes**		
	Germany	Berlin	Schönefeld	Yes		
		Dresden		**U.C.**		
		Düsseldorf		**Yes**	**Yes**	**Yes**
		Frankfurt	Main	**Yes**	Yes	**Yes**
		Hamburg		**U.C.**		
		Hannover		Yes	**U.C.**	
		Köln-Bonn		Yes		**U.C.**
		Leipzig-Halle		**U.C.**	**U.C.**	**U.C.**
		Munich		**Yes**		
	Italy	Milan	Malpensa		**U.C.**	
	Netherlands	Amsterdam	Schiphol	Yes	Yes	

	Norway	Oslo			**Yes**
	Sweden	Stockholm	Arlanda	**Yes**	Yes
	Switzerland	Geneva			Yes
		Zürich			**Yes**
	United Kingdom	Birmingham			**Yes**
		London	Gatwick		Yes
			Heathrow	**Yes**	**Yes**
			Stansted		**Yes**
Asia and Australia		Manchester		**Yes**	
	Australia	Sydney		**Yes**	
	China	Hong Kong	Chek Lap Kok	**Yes**	**Yes**
	Japan	Osaka	Kansai		**Yes**
		Sapporo	Shin Chitose		**Yes**
		Tokyo	Haneda	Yes	
			Narita	**Yes**	**Yes**
	Korea	Seoul	Incheon		**Yes**
			Gimpo		U.C.
	Malaysia	Kuala Lumpur	International	**Yes**	U.C.
	Philippines	Manila		U.C.	
	Singapore	Singapore	Changi	U.C.	
	Thailand	Bangkok	Don Muang		Yes

Table 17-5. Market share of passengers served by rail systems in the United States compared to those in Europe and Asia

United States		Europe and Asia	
Airport	Market share	Airport	Market share
Washington/Reagan	14	Oslo	43
Atlanta/Hartsfield	8	Tokyo/Narita	36
Chicago/Midway	8	Geneva	35
Boston/Logan	6	Zürich	34
San Francisco/Oakland	4	Munich	31
Chicago/O'Hare	4	Frankfurt/Main	27
St. Louis/Lambert	3	London/Stansted	27
Cleveland	3	Amsterdam	25
Philadelphia	2	London/Heathrow	25
Miami/International	1	Hong Kong/Chek Lap Kok	24
Washington/Baltimore	1	London/Gatwick	20
Los Angeles/International	1	Paris/de Gaulle	20
		Brussels/Zavendam	11
		Paris /Orly	6

Source: Leigh Fisher Associates et al., 2000, pp. 4 and 19.

connections. Moreover, highways create profitable demand for parking, which for many airports is a major source of revenue (see Sec. 17-4). Compared to tracked forms of airport access, such as railroads, highways generally require less investment. Highways are everywhere, and the airport operator often needs only to build a short piece of highway to connect to this existing system.

The issue for planners is this: What kinds of complementary airport access systems should airport operators develop, under what circumstances? New rail projects—either intercity railroads or metropolitan transit systems—can be very expensive. The development of the special rail service to London/Heathrow from Paddington Station in London reportedly cost £450 million ($675 million) (Harris, 2001). (See Fig. 17-1.) The Arlanda Express to Stockholm reportedly cost $600 million for a distance of 25 mi—$24 million per mile, on average. As of 2001, the AirTrain connection being built between New York/Kennedy and the Long Island Railroad and the New York Metropolitan Transit Authority had a budget of over $1.2 billion (Jane's Airport Review,

Fig. 17-1. *Heathrow Express connecting London/Heathrow and Paddington Station, London.* (***Source:*** BAA plc.)

2001). Rail projects, being so expensive, can also be highly controversial and difficult to implement. The project for New York/Kennedy was under discussion for a quarter-century before construction began. Likewise, the railroad system to Tokyo/Narita opened over 20 years after the airport operator built the airport and the railroad station. Rail systems to the airport can be highly visible and impressive, but when and where do they make sense?

Door-to-door analysis

When considering a new airport access system, planners first need to estimate its prospective number of users. If only a few people use the system, it will not be serving its intended function. Moreover, if traffic is low, the service is likely to be infrequent, and this discourages prospective riders. Under such circumstances, fares may not cover the operating costs, let alone repay the initial investment, and the system could represent an enormous financial loss. Adequate traffic is essential for the operational and economic success of any airport access system.

It is generally difficult for alternative airport access systems to compete with highways. They typically enter the market at a disadvantage, after the highway systems are well established. By the time a new service opens, people have their cars, and fleets of rental cars, taxis, and buses exist. When it begins service, it may have few passengers and thus may find it difficult to afford to offer the frequent services that will make it competitive with automobiles that operate on demand. To make a rail or other mass transit system successful, planners need to define carefully how this complementary system will be competitive.

A *door-to-door analysis* of trips by competitive transportation modes establishes their relative advantages. It focuses on the total time, cost, and convenience of each mode for the entire trip, door to door, between the home or office and the airport. The need to consider the door-to-door experience is a most important point, often neglected. Travelers look at the entire experience of a trip, not just a single segment. Popular presentations of new rail systems, however, frequently focus on their speed from station to station, and lose sight of the fact that people generally have to wait for departures and spend time and money getting to and from the station. For example, publicity about the new rail connection to New York/Kennedy has boasted that it "will cut the travel time to Manhattan to 45 min" (Jane's Airport Review, 2001). This is true if the traveler enters the station just before the train departs, makes the connection at Jamaica station with no delay, and wants to go to Pennsylvania Station. Normally, however, travelers walk to the station, wait for the train, and then have to find their way from the train stop to wherever they want to go—which may take them another 20 min and cost $10 or more. All these connections add considerable time and cost to the trip, as well as the inconvenience of having to get in and out of several vehicles. Considering the total door-to-door experience, the trip by the fast train may be, for many people, slower and more expensive than a taxi directly between home or office and the airport.

To understand the competitive position of a proposed new airport access system, analysts need to estimate carefully the total door-to-door time and cost of the competitive modes. In addition to the costs associated specifically with the mode (the taxi fare, the rail ticket, or the parking charges), they need to add the time and effort it takes to get to these modes. They may have to include the cost of taking a taxi from a hotel to the railroad, for example, and then the wait to buy tickets and for the train to begin its trip. Example 17.3 illustrates how this can be done.

Analysts should also consider the trip costs for the owner of a rail or alternative system. Indeed, the fare the passenger pays for a service

EXAMPLE 17.3

Consider the excellent Heathrow Express service that BAA, the operator of the major London airports, provides between London/Heathrow and Paddington Station in Central London. As of 2001, it offers trains every 15 min during most of the day and delivers a direct ride to the airport. It is one of the best services of its kind.

Table 17-6 shows one estimate of the door-to-door cost for the traveler of this BAA service from a central London hotel to an airport check-in counter, for peak and off-peak traffic.* Table 17-7 shows a comparable estimate for a taxi journey. These numbers are approximate, of course. Fares change and sometimes feature special discounts. Travel times by taxi vary considerably.

As the comparison of the two tables indicates, the Heathrow Express does offer competitive service. Although the train moves quickly, getting to London/Heathrow using this service is not particularly fast on average. It is reliable. It is cheaper than a taxi for people traveling alone, but not necessarily for groups. It is generally less convenient than a taxi because it requires repeated loading and unloading of bags. Placing this into context, it is faster and more expensive than the London Underground (metropolitan transit system), which also goes directly to the airport. As of 2001, the Heathrow Express carried over 4 million passengers a year. This was equal to about 10 percent of the approximately 50 million passengers originating and terminating at London/Heathrow.

often does not represent its true cost. In the case of the Heathrow Express, for example, the BAA ought to net over £45 million a year over their operating costs to cover depreciation and return on their £450 million investment. Yet their revenues from passengers in 2001 appear to have been not much more than this amount (4 million passengers times £12 full fare less discounts). At least in the initial years, this service operated at a loss. For the airport operator, the actual cost of the service per rider was substantially greater than the fare. A public agency may consider this part of normal subsidy of

---

*As of 2001, passengers could check their bags at Paddington Station, and thus bypass this process and its delays at the airport. However, they had to arrive early at Paddington to take advantage of this service.

Table 17-6. Example door-to-door times and cost for a passenger to London/Heathrow from a central London hotel, on BAA's Heathrow Express

Journey element	Cost (£) (1£ ≈ 1.5 US$)	Time (min)	
		Peak	Off-peak
Taxi to station	10	15	10
Get to train		5	5
Buy ticket	12	2.5	2.5
Wait for train		7.5	7.5
Train ride		15–23	15–23
Walk to check-in		5	5
Total	**22**	**50–58**	**45–53**

Table 17-7. Example door-to-door time and cost for a passenger to London/Heathrow from a central London hotel by taxi, for a single person and per person for a family of three

Journey element	Cost (£; 1£ ≈ 1.5 US$)		Time (min)	
	Total	Per person	Peak	Off-peak
Taxi to airport	60	20	75	40
Get to train				
Buy ticket				
Wait for train				
Train ride				
Walk to check-in			2	2
Total	**60**	**20**	**77**	**42**

metropolitan transit. However, a private operator such as the BAA has to be particularly concerned with covering costs.

Rail solutions

Comparative analyses of door-to-door travel times and costs generally demonstrate that rail systems of airport access can be competitive with highway modes in favorable circumstances. Among the factors that favor competitive rail service are:

- *Airport size,* to generate enough passengers to cover costs and to sustain frequent service that lessens the time people have to wait for trains

- *Existing local rail service,* which lowers the cost of the airport connection (this is the basis for the rail service at Amsterdam/ Schiphol, Frankfurt/Main, Geneva, London/Gatwick, Lyon, and Zürich)

- *Easy connections to a wide metropolitan transit system,* as exists at London/Heathrow, Paris/de Gaulle, and Washington/Reagan

- *Difficulty of automobile access to the airport,* a factor for airports on man-made islands, such as Hong Kong/Chek Lap Kok and Osaka/Kansai, and for distant airports such as Oslo or Kuala Lumpur/International.

Conversely, the same analyses demonstrate that rail access systems are difficult to justify for most airports, even in big cities. They cost too much. There are no easy connections to an efficient transit system. They serve too few people. They do not provide enough value for the money spent. They are not cost-effective. As Table 17-5 indicates, about half the rail connections to airports in the United States contribute only marginally to the solution of the airport access problem.

Highway solutions

For most airports, rubber-tired modes of transport provide the most cost-effective forms of airport access (see de Neufville and Mierzejewski, 1972). These systems can flexibly serve the entire metropolitan area without enormous investment. Operators can adjust the vehicles and routes of these systems to meet the needs of the range of travelers and employees. The vehicles can range from private automobiles and taxis to various forms of collective transport such as vans, luxury motor coaches, and ordinary busses.

Collective transport using rubber-tired vehicles is not glamorous and does not get the attention it deserves. Precisely because access systems using buses and vans distribute people effectively over the metropolitan area, they are small and spread out compared to rail systems that are big and concentrated. Public transportation based on buses and vans has proven, however, to be successful at many locations. For example:

- Fleets of courtesy buses connect Atlanta/Hartsfield to a wide array of local destinations.

- Airport shuttle vans connect San Francisco/International with the Bay Area.

- London/Heathrow, New York/Kennedy, and Tokyo/Narita connect to their metropolitan areas with fleets of large buses.

Coping with rubber-tired traffic requires substantial effort. A 10-million-passenger airport may receive 40,000 vehicles a day, or about 4000 vehicles in the peak hour (see Example 17.1). Three lanes of traffic each way are necessary to serve this level of traffic. These figures give an impression of the importance of the effort that has to be made to provide suitable airport access. Readers should note, however, that the numbers cited are only illustrative. The actual amount of highway required depends on local conditions. The number of lanes needed is relatively less for transfer hubs that cater to many passengers who do not need airport access. It is also less in regions in which people do not use their own cars to get to the airport, because people either own relatively few cars or are accustomed to using public transport. Correspondingly, the access highways to airports in North America tend to be larger than they are elsewhere. These differences are only a matter of degree, however. Major airports inevitably require extensive connections to the regional highway network.

Major airports need to develop strategies to reduce the number of vehicles coming to their airports. Although cars provide parking revenues, they cause large costs in terms of the miles of curb frontage, the cost of elevated roads and garages, and their use of valuable space around crowded passenger buildings. Moreover, as Chap. 6 stresses, private cars cause much air pollution. Airport operators thus need to encourage the use of various forms of high-occupancy vehicles. Rail access can be one good approach in some circumstances. Often, however, a range of access modes provides the best overall service. Analysts and planners need to consider each of the possibilities and try to optimize the overall service that the mix of services can provide to the variety of prospective customers.

17-4 Parking

Modern airports need a lot of parking. Worldwide surveys indicate that major airports typically provide between 200 and 1200 parking spaces per million total passengers a year, plus 250–500 spaces per thousand employees (de Neufville and Rojas Guzmán, 1998; Ashford and Wright, 1992). The largest airports in the United States provide between 200 and 1700 spaces per million total passengers, with a slight tendency for larger airports to offer fewer spaces. The total number is thus large: Dallas/Fort Worth provides about 30,000 spaces and Chicago/O'Hare and Los Angeles/International each around

20,000 (ACI-NA, 2001). The actual number needed or desirable depends on local circumstances, such as the level of automobile ownership in the population, the availability of public transport, and current environmental policy. In any case, operators of major airports need to consider thousands of parking spaces.

Airport parking generates substantial revenues. In the United States, it accounts for over 20 percent of the revenues at some major airports such as Boston/Logan and Tampa. The total revenue from parking at the 30 largest airports was nearly $1.2 billion in 2000, thus averaging about $40 million per airport. Chicago/O'Hare obtained over $ 91 million (see Table 17-8). Their annual revenues per space ranged from $1000 to $4000 (ACI-NA, 2001; FAA, 2001).

Airport operators need to recognize that the market for parking has at least five distinct segments. The best designs will provide the facilities appropriate to each. The categories are:

1. Short-term parking for a few hours, used primarily to pick up arriving passengers

Table 17-8. Parking revenues at 30 largest U.S. airports in 2000

Airport	$, Millions	Airport	$, Millions
Chicago/O'Hare	91.3	Miami/International	33.6
Denver/International	77.3	Houston/Bush	33.8
Boston/Logan	71.1	Tampa	33.6
Dallas/Fort Worth	70.9	Orlando/International	32.8
San Francisco/International	65.8	Washington/Dulles	31.0
Atlanta/Hartsfield	65.1	Washington/Reagan	23.4
New York/Newark	62.9	Philadelphia	21.9
Los Angeles/International	59.4	Pittsburgh	19.5
Baltimore/Washington	50.1	Salt Lake City	18.0
Seattle/Tacoma	47.1	Charlotte	17.7
Minneapolis/St. Paul	43.0	Cincinnati	16.2
Phoenix	40.3	Las Vegas	15.4
New York/LaGuardia	35.5	San Diego	14.0
Detroit/Metro	35.5	St. Louis/Lambert	12.4
New York/Kennedy	33.9	Honolulu	10.8

Source: FAA, 2001.

2. Structured parking close to passenger buildings, mostly serving persons on short trips or business travelers who can afford this expensive facility

3. Long-term, remote, and less expensive parking (often provided privately near the airport)

4. Rental car parking

5. Employee parking

Hourly parking

Hourly parking serves drivers meeting arriving passengers and, sometimes, delivering passengers. These people want spaces close to the passenger buildings, so that they will not have to carry bags a long way. If this is not available, they will tend to wait at the curb in front of the building, or circle around the airport. Short-term parking thus both meets a demand and helps relieve the congestion on the curbs in front of the passenger building (see Sec. 16-6 and de Neufville, 1982).

To provide hourly parking space, the airport operator needs to set aside some convenient area, generally a parking garage, segregated from daily parking. This can be done by installing separate entrances and exits. Toronto, for example, plans to dedicate to short-term parking the floors of its parking garage that are closest to the entrance of the passenger building. The airport operator ensures that spaces are available for hourly parkers by charging high hourly rates that are affordable for persons staying only a hour or so, but prohibitive to anyone leaving a car for one or more days.

The total number of spaces dedicated to hourly parking can be relatively small. Because people park in these spaces for only a short time, a single space is used many times a day. On average in the United States, a short-term parking space serves around 1150 cars a year, whereas a long-term space serves only about 220. Because of this high turnover, short-term parking serves the bulk of airport visitors. It constitutes only about a quarter of the spaces on average, yet it serves about two-thirds of the visitors (ACI-NA, 2001). See Example 17.4.

Hourly parking provides airport operators with two significant benefits. It relieves crowding at the curbs in front of the passenger buildings. It is also profitable, since the revenues per space are high. By differentiating the market for parking into those who do and do not want to pay for

EXAMPLE 17.4

The number of spaces required to serve a group of clients depends on the turnover. If parkers come and go every few hours, as in hourly parking, relatively few spaces will meet the demand. If they stay for several days, the number of spaces needed increases by a factor of 10 or more. It increases directly with the length of time users stay in the space, their *dwell time* (see Sec. 15-3). Table 17-9 illustrates this phenomenon.

Table 17-9. Number of parking spaces needed to serve 1000 clients a day

Situation	Length of stay	Spaces needed
Hourly parking	~2 h	~200
Daily parking	2 days	2000
Long-term parking	1 week	7000

convenience, the airport operator can charge premium prices for the most desirable parking spaces. Some airports are even introducing valet parking.

The introduction of hourly premium parking is a recent innovation. Traditionally, airport operators charged uniform prices for all spaces in a given location. At the beginning of the twenty-first century, airport operators have retrofitted most hourly parking facilities into parking garages built originally for a single range of prices. This was the situation at Montreal/Dorval and Boston/Logan, for example. Future designs should make hourly parking an integral part of garages from the start.

Structured parking

Multilevel garages are generally a practical necessity at busy airports. The demand for convenient space next to the passenger buildings is high. The only way to make efficient use of this valuable property is to build it up. Thus, the busiest airports tend to feature four- or five-level garages with spaces for thousands of cars.

Multilevel garages are expensive. Airport garages require about 35 m^2 (350 ft^2) per parking space, including ramps, stairways, and columns.

At about $350/m^2$ including all mechanical equipment, they easily cost about $12,000 per space—and they can be much more expensive (see Table 17-10).* These figures translate into high costs to the users. The actual daily charges for parking depend on the financial objectives of the airport operator, but are easily up to $20 a day and more.

Structured parking thus normally serves persons on short trips or business travelers who can afford this expensive facility. It is too expensive for long-term parkers who might want to stay a week or more. This reality motivates the development of cheaper parking for these users.

Long-term parking

At-grade lots provide less expensive parking suitable for long-term stays. At busy airports these are necessarily some distance away, typically in old industrial sites (as around Los Angeles/International) or open spaces (as around Pittsburgh). These facilities often develop independently, in competition with the airport operator. When this is the case, they constrain the airport operator's profits from parking.

Rental car parking

Rental car agencies often require substantial parking for their fleets. They typically operate out of individual lots in remote areas tucked into corners of the airport, or even off the airport. These are often confusing to the renters, who often do not know the area and have to find their way to and from these lots. They also require a constant

Table 17-10. Approximate construction cost of parking at largest airports in the United States in 2000

Type of facility	Cost, $ thousands		
	Minimum	Average	Maximum
Structured garage	7.2	23.3	85.7
Surface, close in	1.1	2.4	6.9
Surface, remote	0.4	1.3	3.2

Source: ACI-NA, 2001.

*Airport garages normally require more space than other residential garages, since they do not feature parking on ramps. Additionally, structures procured through a public process typically cost more than those negotiated privately. For example, Orlando/International paid about $11,000 per parking space, whereas Orlando/Sanford, which is run by a private company, built its garage for about $7000 per space. [Information supplied by Kelsey Construction (http://www.kelseygccm.com) and Finfrock Design-Manufacturing-Construction, specialists in parking garages in Orlando, Florida.]

flow of courtesy vans or buses to ferry clients to and from the passenger buildings. These require special parking spaces and generally congest the curb areas.

A *consolidated rental car facility* is a multilevel garage that houses several companies. This innovation simplifies airport access at congested airports. By having a single major facility, the airport operator can direct all rental cars uniformly, thus reducing confusion, and can implement a single transfer system between the passenger buildings and the rental car building. This can either be a bus service, as at Washington/Reagan, or a people mover, as at New York/Newark and San Francisco/International. Consolidated rental car facilities are likely to become prevalent at major airports.

Employee parking

Employees require hundreds, even thousands of spaces at the busiest airports. Many of these spaces will be dispersed around the edge of the airport, close to the cargo centers, maintenance bases, and other facilities distant from the congested passenger buildings. The airport operator, however, does have to establish some facilities for persons who work in and around the passenger buildings. The airport operator must likewise establish some links between the passenger complex and the parking or public transport locations the employees use. Bus services usually provide these connections. A few airports, such as Dallas/Fort Worth and San Francisco/International, serve employee parking lots with people movers.

17-5   On-airport access

Major airports typically feature a network of bus routes. Courtesy buses serve remote parking, local hotels, and move airline and airport employees around the airport as needed. Numerous shuttle services may link the airport with homes and businesses throughout the metropolitan area. Major bus companies may also serve a variety of downtown and suburban destinations. This traffic can be confusing to the passengers and congest the curb space in front of the passenger buildings. Airport operators need to organize these networks coherently.

To deal with the confusion of access and distribution systems around the airport, a number of operators have invested in major automated systems to connect principal points of access to the airport. These are known generically as *people movers,* as discussed in Sec. 17-6.

New York/Newark has a monorail that passes through its three main passenger buildings, the rental car and parking facilities, and connects with the regional transit and rail system. San Francisco/International has built a short tracked system that connects the consolidated rental car facility, the metropolitan transit system (the Bay Area Rapid Transit, BART), and the passenger buildings. Paris/de Gaulle has committed to install some system to connect its several terminal and hotel complexes.

The desirability of automated systems of airport access needs to be considered carefully. They are expensive. Their large extra cost over the alternative of bus routes between the passenger buildings and other facilities can be difficult to justify. Boston/Logan, for example, decided it could not afford to pay over $300 million to eliminate its shuttle bus service. In many circumstances, however, people movers may be essential to the smooth functioning of the airport. At some point, the buses necessary to move people around the airport will not be able to handle the loads, and the buses themselves will over-crowd the roadways. Thus Dallas/Fort Worth depends on its auto-mated people mover for airport access and distribution, and is committed to building a completely new people-mover system by 2005. This will be nearly 5 mi (8 km) long and the biggest in the world (Nichols et al., 2000).

For many airports, the use of people movers for airport access might be justified in the future if not immediately. In this connection, it is interesting to note that the airport operator of New York/Newark built the passenger buildings with aerial rights-of-way to enable a people mover to be installed, as was done some 20 years later. In con-cept, this is an excellent example of a "real option" that gives the air-port operator the flexibility to implement new features when needed. (See Chap. 22 for details on using "real options" to give airport management the flexibility to respond effectively to future develop-ments and opportunities.) Unfortunately, in this particular case, the space set aside was inadequate, so the provision was not as effective as it should have been.

17-6   Within-airport people movers

People movers is the generic name for the broad range of automated vehicles for moving people horizontally over relatively short distances, up to 1–2 mi for the most part. Sometimes, these devices are called *horizontal elevators,* a name that evokes their function. Specialists generally refer to them as APMs (automated people movers).

The use of people movers for moving people around the airport is perhaps the single most important innovation in the design of airport layouts in the last generation. It has enabled airport designers to disperse aircraft across the airport, thus facilitating their movement, while still providing passengers with short walking distances and travel times within the passenger buildings. Specifically, this development has led to implementation of midfield concourses as a standard feature of most new major airport complexes (see Elliott and Norton, 1999.)

A crucial question for airport operators is whether a people-mover system is worthwhile. People movers can be very expensive. The budget for the new Dallas/Fort Worth APM, proposed to be the largest on any airport to date, was over $845 million in 2001 (In Motion, 2001).

Technologies

Airport people movers share some basic characteristics. They are

- *Automated,* operating without drivers, much like modern elevators
- *Designed for people* and, to some extent, their baggage
- *Confined to special-purpose guideways* reserved for their use
- *Generally run as trains* of 2 or 3 vehicles
- *Operated horizontally* (exceptionally, the system at Kuala Lumpur/International drops two levels from above-grade boarding areas into a tunnel underneath the taxiways).

Beyond these common features, people movers come in a wide array of technologies and models. Many U.S., French, Japanese, and other manufacturers anticipated a broad market for people movers beyond airports—in shopping areas, amusement parks, and central cities—and have installed experimental models at many locations. They have different suspension systems (rubber tires, steel wheels, and levitation) and operate on various guideway structures. Most are self-propelled, but others are pulled by cables, exactly like elevators. Moreover, people movers come in a wide range of sizes (see Moore and Foschi, 1993). In recent years, this complexity has been decreasing, as about five manufacturers are emerging as the industry leaders.

The most common design installed at airports is a rubber-tired self-propelled system currently sold by Bombardier (http://www.bombardier.com/). Westinghouse developed the original design, which AEG, ADtranz, and now Bombardier have taken over and transformed (see Fig. 17-2). Versions are in place at Atlanta/Hartsfield, Denver/International, Frankfurt/Main, London/Gatwick, London/Stansted, Singapore/Changi, and other locations. A newer version is scheduled for Dallas/Fort Worth (Fig. 17-3). Self-propelled vehicles such as these are essential for any system longer than about 600 m (2000 ft).

Fig. 17-2. *Bombardier Adtranz CX-100 automated people mover operating land-side at Frankfurt/Main.* (*Source:* Harley Moore, Lea1Elliot, and Bombardier.)

Fig. 17-3. *Bombardier Adtranz Innovia scheduled for introduction at Dallas/Fort Worth.* (*Source:* Harley Moore, Lea+Elliot, and Bombardier.)

For short distances between two points, cable-driven systems can work well. These systems have numerous advantages. Their motors are in the structure at the end of the track, rather than on-board, so the vehicles and their supporting structure are lighter and less expensive than for self-propelled APMs. In addition, the drive mechanisms may be easier to maintain. However, these systems are not competitive for large systems with long hauls and complicated routes. The Otis systems installed at Tokyo/Narita, Minneapolis/St. Paul, and Detroit are excellent examples of this technology (Figs. 17-4 and 17-5).

Fig. 17-4. *Otis cable-driven automated people mover operating air-side at Tokyo/Narita.* (*Source:* Harley Moore, Lea+Elliot, and Otis.)

Fig. 17-5. *Poma Otis indoor automated people mover operating at Detroit/Metro.* (*Source:* Harley Moore, Lea + Elliot, and Otis.)

Location

Airport designers can locate people movers either landside or airside (see Table 17-11). They serve different functions in either application.

Landside people movers   *Landside people movers* tie together airport access systems and the distinct passenger buildings spread around access roads. They are "landside" because they are used by the general public, not just the passengers, crews, and employees who have been cleared through security. They typically serve older-style configurations of airport passenger buildings.

Landside people-mover systems substitute for bus transport systems. Although bus systems have the great advantage of being inexpensive to build, easy to implement, and flexible (see Shen and Zhao, 1997.), they have many disadvantages. Their level of service is relatively low. They require many specialized drivers and can thus be expensive to operate and maintain. If they operate on conventional fuels, they emit a great deal of pollution as they idle at stop. Moreover, busses are neither modern nor glamorous. The development of landside people-mover systems thus often depends on the objectives of the airport operator. Is it willing to spend large sums for the sake of higher levels of service and cleaner air? In some cases, it may not have a choice.

Airside people movers   *Airside people movers* serve passengers and others who are moving through the airport beyond the security checks. They are a prime characteristic of the designs of many of the newer airport configurations. They typically connect a central passenger building with one or more satellites or midfield buildings. Sometimes, as at Detroit/Metro or Osaka/Kansai, they serve long passenger buildings.

An airside people-mover system enables the development of midfield passengers buildings that offer significant operational advantages to the airlines, in terms of reduced aircraft taxi times (see Sec. 14-3). These savings are particularly valuable to hub airports serving many banks of transfers. Correspondingly, landside people-mover systems are an integral part of the new airport facilities at the hub airports of Atlanta/Hartsfield, Denver/International, Hong Kong/Chek Lap Kok, and Pittsburgh.

The size of the airports serving the largest aircraft also motivates the use of people movers. The 79.6-m (265-ft) wingspan of the A-380 means that the distances between gates approaches 90 m (300 ft). Moving sidewalks can deal adequately with distances of up to perhaps 500 m (1500 ft). Beyond that distance, people movers with current technology appear to perform better. So, for airports needing to move people great distances, people movers are the attractive solution.

Table 17-11. Examples of landside and airside people mover systems

Landside	Airside	
Chicago/O'Hare	Atlanta/Hartsfield	Minneapolis/St. Paul
Dallas/Fort Worth	Dallas/Fort Worth	Orlando/International
Frankfurt/Main	Denver/International	Osaka/Kansai
Houston/Bush	Detroit/Metro	Pittsburgh
New York/Newark	Hong Kong/	Seattle/Tacoma
Paris/Orly	Chek Lap Kok	Singapore/Changi
San Francisco/	Kuala Lumpur/	Tampa
International	International	Tokyo/Narita
	London/Gatwick	
	London/Stansted	
	Miami/International	

Capacity of network

The practical capacity of simple shuttle systems, carrying passengers back and forth between only two points as at Tampa, is easy to determine. The maximum capacity is simply the capacity of each vehicle or train of vehicles times the number of departures per hour. As for any system that must serve varying loads, the practical capacity over a sustained period is somewhat less than this maximum (see Chap. 23).

Beyond the simple case of a shuttle, the effective capacity of a people-mover system can be difficult to determine. It is much less than its ability to carry a number of passengers per hour. The system must also deliver reasonable service, with tolerable delays, to all points on the network. These requirements mean that the system should operate with substantial excess capacity so that users along the line will be able to board without excessive delays. This subtle issue is easy to overlook.

The essential problem is the likelihood of unequal service that can make the system a functional failure for many users. The problem can occur on any network with many stops, which is the typical arrangement. In multistop systems, the users at the beginning of the line have the opportunity to take all the places. When they do this, users farther down the line may not be aboard to get on to the system. They may have to wait a long time for service, indeed until all the users at the start of the line have been served. This situation means that some users—those at the beginning of the system—may get good service with minimum delays. Other users—those trying to use the system down the line—may face intolerable delays (see Example 17.5). Considering the overall average delay to all users, the service might not appear to be bad. In detail, however, an important section of the public might see that the system was a failure.

If a system with many stations is operating near capacity, it is impractical to overcome this unequal service for passenger service. It is possible to imagine complicated operations that prevent passengers from boarding at the beginning of the service or involve trains skipping stops. These do not appear feasible when dealing with people, however. (With baggage systems, such arrangements are somewhat possible because bags do not complain and themselves resist delays. However, see the discussion in Sec. 17-7.)

The solution to providing adequate service to all users is to provide sufficient excess capacity. When this is available, users at the beginning

EXAMPLE 17.5

A people mover serves a central passenger building (T) and three midfield concourses (A, B, and C), an arrangement similar to Atlanta/Hartsfield or Denver/International. An APM with a capacity of 120 persons leaves C destined for T every 5 min. Suppose that demand surges over 10-min: 100 travelers arrive at A, B, and C each 5 min, wanting to go to T. To simplify calculations, imagine there is then no further traffic after this peak.

Table 17-12 traces how the APM carries this peak and shows the delays at the stations down the line. Passengers at the beginning of the run (at C) see an empty train and get immediate service. They leave a few spaces open for the next stop (B), but most of those passengers have to wait. Passengers at the third stop (A) face full trains until all passengers at the earlier stations have obtained service. The result is that passengers at C get immediate service; those at B have to wait for about 2 trains; and passengers at A have to wait for 3 to 4 trains and will perceive that the APM is totally inadequate. This demonstrates the potential problem caused on a network when certain users can seize priority. The distribution of delays is unequal, and can make a system an operational failure for many users.

Table 17-12. Unequal distribution of delays occurs when some users can seize priority

Time from start (min)	Persons in state	Midfield concourse		
		C	B	A
5	Waiting	100	100	100
	Served	100	20	0
10	Waiting	100	180	200
	Served	100	20	0
15	Waiting	0	160	200
	Served	0	120	0
20	Waiting	0	40	200
	Served	0	40	80
25	Waiting	0	0	120
	Served	0	0	120

of the line will not crowd out prospective users farther down the line. The right level of excess capacity is not easy to define, however. It depends strongly on the distribution of passengers along the line and their destinations (Daskin, 1978). Thus, the right amount of excess capacity needed for an airport people mover can change as airlines alter their schedules or move operations from one midfield concourse to another, as happens regularly. Good design requires careful, expert examination of this issue, most likely with some form of simulation. In any event, the practical capacity of a people-mover system featuring several stops can be as low as half its maximum capacity.

The capacity needed to ensure good service might not have to be provided at the start of the operations. Airport operators can size the stations to maximum capacity to give themselves the flexibility to expand the number and length of their people-mover trains when the need arises. In this case, paying for large stations in advance of need would give the airport operator the flexibility to add people-mover capacity as needed. This approach in effect provides insurance in a tangible form, that is, a "real option" (see Chap. 22).

17-7 Within-airport distribution of checked bags

Networks for handling bags can be very large. Large systems involve miles of conveyor belts or thousands of baggage vehicles. For example, the design for the baggage handling system for the new Toronto/Pearson passenger building featured over 11 km (7 mi) of conveyors (Table 17-13). The baggage handling system just for the international passenger building at San Francisco/International has 12,000 m

Table 17-13. Design characteristics of baggage system for new Toronto/Pearson passenger building

Feature	Extent
Conveyors	11,600 m (7.25 mi)
Belts for check-in counters	168
Sorting pushers	177
Levels of conveyors	5
Screening system	100% of bags

Source: Jane's Airport Review, 2001b.

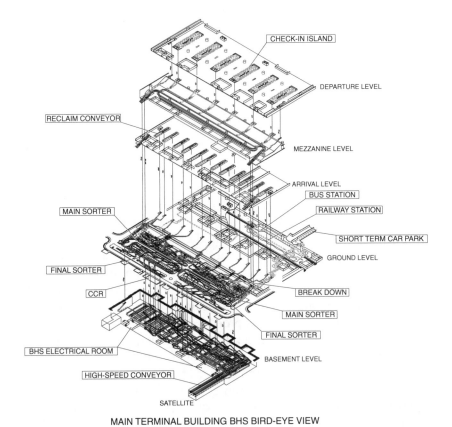

MAIN TERMINAL BUILDING BHS BIRD-EYE VIEW

Fig. 17-6. *Schematic drawing of baggage handling system at Kuala Lumpur/International.* (***Source:*** Pacific Consultants International.)

(40,000 ft) of conveyors (Jane's Airport Review, 2001a). Modern baggage systems are major projects by themselves.

Furthermore, baggage handling networks can be enormously complex. They may assemble bags from well over 100 check-in points, merge them, and then sort them out to at least as many destinations. Along the way, they might pass the bags through up to five levels of screening for security. The streams of bags thus merge and separate and merge again. The layout of the system might remind one of noodles on a plate (see Fig. 17-6 and refer to IATA, 1995). Some baggage systems can be simple, of course. At smaller airports, straightforward arrangements of carts and perhaps containers cope with the makeup of bags for flights and their subsequent distribution to baggage claim devices, as described in the FAA design guidelines (FAA, 1988). For significant airports, however, the baggage systems are inevitably complex.* Designers need to pay special attention to these networks.

Security systems

Security systems involve two parallel processes:

1. *Screening of the bags* by X-ray machines and other devices
2. *Baggage reconciliation,* the process of making sure that each bag placed on an aircraft belongs to someone who is definitely on board

Baggage reconciliation is a way to prevent people from placing explosives on board an aircraft. It is a process that must be performed on all international flights to and from the United States, the member countries of the European Union, and many other nations. It is also required by national law for domestic flights within many countries. [Barnett et al. (2001) provide a detailed description of the process and of related issues, along with many statistics applicable to U.S. airlines and airports.] To implement this process efficiently, an airline must be able to identify rapidly, through electronic processing of boarding passes or by other means, all the passengers who have boarded a flight. The airline must also know the owner and location of each of the bags placed on board so that it can effect the baggage reconciliation and remove bags quickly if there is a discrepancy. Some airlines and airports are better prepared to accomplish this task than others, due to the different approaches they use to sort bags and place them on the aircraft. [People may miss flights for all kinds of harmless reasons: they fall asleep in a waiting room, get lost trying to make a transfer, stay too long in a shop or restaurant, etc., as Spake (1998) describes.] The next subsection describes the issues involved in creating information systems for bags.

Complete, 100 percent screening is the standard to be met. It is now accepted that all checked bags should be examined by X-ray and other devices. In the United States, the Airport Security Federalization Act of 2001 called for 100 percent screening of checked bags by early 2003. The countries belonging to the European Civil Aviation Conference have also committed themselves to 100 percent screening of

---

*Partial exceptions to this rule are possible. At Paris/de Gaulle, many check-in counters for Air France are positioned right in front of the gate, so that bags flow directly to the makeup area. This design eliminates baggage sorting on departure. However, once Air France decided to develop a transfer hub at this airport, it had to deal with bags from arriving aircraft. Air France then needed to retrofit procedures for sorting and distributing bags—which cannot have been easy since the designers had not allowed space for such processes.

hold-room baggage by 2003. Although airports in Europe and the United States may miss these schedules, concerns about terrorism have imposed this norm.

The trade-off between speed and accuracy governs the design of screening processes. A complete search of any bag requires either a highly sophisticated (and expensive) device or a personal inspection of the contents. This is too long and too expensive to apply to every bag. On the other hand, machines that are faster and cheaper often provoke a high number of false alarms. Airport operators cannot rely on any one type of machine for baggage screening. They must consider a range of machines: fast ones for most of the bags and slower, more sensitive machines for the more suspicious pieces.

The design of screening systems thus consists of cascades of screening devices. All bags go through at least a preliminary screen. Some are cleared at this point, and some are diverted to a more intensive examination. Some of those clear this inspection, but some are set aside for even more investigation. Depending on the design, this process may stop there or go further. The baggage system for the new passenger building at Munich anticipates five levels of screening, the last of which involves a bomb squad (see Example 17.6).

Information systems

Airports deal with complex baggage systems by automating the process. This involves two distinct types of technologies:

1. Information systems, to identify the bags and their destinations
2. Mechanical sorting systems, to distribute each bag through the screening process and to its makeup area, where handlers place it in appropriate containers. (The next subsection discusses these systems.)

Most baggage handling systems rely on bar-code labels to identify the bags. These have the significant advantage of being inexpensive. As of 2001, they cost around 7 cents each (Pilling, 2001a). The baggage system reads these labels and uses the information to direct the bags down appropriate chutes. The basic reading device consists of laser scanners mounted on an apparatus over the baggage conveyors. As the reader may have experienced at a grocery store, lasers do not always read on the first try, so they require substantial redundancy. A typical arrangement for bags has 6–8 laser readers mounted circularly around the path of the bags, and each laser attempts to read several times.

EXAMPLE 17.6

The design for 100 percent screening of 6000 bags in the peak hour for the new passenger building at Munich has five levels of screening. According to Pilling (2001b) these are:

- Level 1, which consists of automatic X-ray detection devices, each having a rated capacity of 1200 bags/h (3 s/bag). However, the designers expect that these fast machines will clear only half the bags—the rest, or about 3000 bags/h, will be diverted to level 2.

- Level 2 consists of another kind of automatic X-ray devices that can create detailed images of suspicious items in a bag. These machines have a rated capacity of only 240 bags/h (15 s/bag). The designers expect that around 4 percent of the bags (or 120 bags/h) will trigger alarms at this stage.

- Level 3 involves more detailed automatic detection. As of 2001, the Munich airport was still testing these devices to determine the ones that would provide the right combination of reliability and accuracy.

- Level 4 requires inspectors to conduct a detailed inspection of the contents of the bag. This should be totally reliable—and take several minutes per bag.

- Level 5 is the S-Kammer, "a chamber that simulates the time, pressure, etc., of an actual flight." Only the most suspicious bags would go through this process or call for a bomb squad.

Overall, laser scanners work generally well—provided the tag is not hidden underneath the bag, dirty or torn, and that the lasers are cleaned regularly. As a practical matter this means that they work best when bags are first checked in, and less well at transfer points, after mechanical devices and baggage handlers have pushed the bags around and thrown them in and out of containers. In a major baggage system that is performing well, about 3 percent of the bags may not be read correctly on their first circuit through the system. This error rate may be much greater after the bags have been handled many times, specifically at transfer points. British Airways thus reported that 10 percent of its bags at London/Heathrow could not be read automatically and had to be read by hand (Barker, 1998). Therefore, baggage systems must have substantial systems for sorting these misread labels. Despite these issues, bar-code systems have been more cost-effective than the alternative.

The other means to identify bags is through *radio frequency identification* (RFID). This technology uses radio antennas to read individual chips that airlines attach to each bag. This process has its own operational difficulties. Obstacles may block or distort the communications, for example, or the chip may fall off the bag. Cost is the major problem with RFID systems. The reading equipment may be less expensive than that for a bar-code system, since it consists of antennas rather than an array of lasers. As of 2001, however, an RFID chip cost around 40 cents, that is, five to eight times more than a bar-code label. If this cost decreases substantially, then RFID systems may replace the existing bar-code system (Pilling, 2001a). Otherwise, they will be limited to special applications. For example, San Francisco/International installed an RFID system to identify the small percentage of bags that check-in agents mark for thorough inspection (level 2 or 3). The BAA has also tested a premium service that delivers bags to hotels, which uses RFID tags to track bags from London/Heathrow, through the Heathrow Express train, onto a courier service and the hotel.

Management and design decisions can "encode" information and thus significantly enhance the performance of baggage handling systems. For example, arrangements that sort prospective bags into containers destined for their ultimate destination capture this information physically. They make it possible to transfer these bins of bags directly from one aircraft to another, thus bypassing the sorting system at the transfer hub. In a similar vein, designs that require passengers to bring their bags directly to the gate for check-in eliminate the need for a local sorting of outgoing bags. This feature determined the configuration of the first passenger buildings built for Air France at Paris/de Gaulle. Airport managers should always be on the lookout for similar arrangements that might make it easier to solve their baggage handling problems.

Mechanical systems

Baggage systems need to separate bags from each other by about 1.5 m (5 ft), so that the sorting mechanism can distribute them correctly. The bags cannot be right next to each other, as they often are at a bag claim device. Most commonly, the bag system simply spaces the bags out along conveyor belts. This arrangement allows mechanical pushers to enter behind one bag, divert the next one to its destination, and retract before the next bag comes down the line. Alternatively, the system consists of individual conveyances. These can be tilt-trays linked together that tip the bag into a chute or *destination-coded vehicles* (DCVs) that operate independently (Figs. 17-7 and 17-8). The International Air Transport Association (IATA, 1995) gives detailed general descriptions of the various technologies, but designers need to consult baggage handling experts for current details of this rapidly progressing field.

The requirement for individual spaces between the each bag has two important implications for the design and operation of an automated bag system. Most obviously, its mechanical devices require a lot of space. The separation standard limits the capacity of each baggage line, and increases the total size of the system. More important, however, the need for individual spaces means that the baggage system is a highly complex system of queues.

Fig. 17-7. *Destination-coded vehicle (DCV) for handling baggage.*
(*Source:* BAE Automated Systems, Inc.)

Fig. 17-8. *Tilt-tray baggage handling device.* (**Source:** Siemans Deimatic Corporation.)

The need to place bags in physical or spatial slots slows processing and can lead to operationally disastrous distributions of delay. It slows the check-in process because agents cannot feed another bag into the system until there is an empty space, for which they often have to wait. This kind of delay occurs every time baggage lines merge—when bags from check-in agents merge with the line behind a bank of check-in counters, when these lines merge with those of other banks, etc. Furthermore, unless the system has substantial excess capacity, certain users will have to wait excessively because others use up all the available spaces (see Example 17.5). Theoretically, the operation of the system might be able to allocate the spaces throughout the system very carefully and avoid excessive delays. In practice, this has not been possible because the pattern of flows changes unexpectedly as flights are delayed. This phenomenon was a principal reason the DCV system at Denver/International could not deliver bags within the time constraints associated with transfer operations (de Neufville, 1994).

Sorting bags at transfer hubs is a very challenging task. Bags must be taken off arriving aircraft, sorted, loaded onto potentially several connecting flights—all within 30–60 min. To do this for all bags individually is not practical. The number of bags to be sorted within a limited time would require enormous capacity. It is also difficult to avoid excessive delays for many connections. Airlines solve this problem by presorting bags at the airports delivering passengers to the transfer hubs. For example, an airline might sort bags in Boston into containers destined, via connections in Miami, for Bogota, Caracas, São Paolo, etc. In Miami, the airline can then ship each of these containers directly from the Boston flight to the appropriate connection, in a "tail-to-tail" operation that bypasses the sorting system at Miami. [See Robusté and DaGanzo (1992) for an analysis of this situation.]

Delivery of bags to passengers is, by contrast, simple. Normally, bags from any flight go directly to conveyor belts that feed claim devices. Passengers identify and pick up their own bags, as shown in Fig. 17-9.

Fig. 17-9. *A typical baggage claim device.* (*Source:* Zale Anis and Volpe National Transportation Systems Center.)

Capacity

The practical capacity of a baggage handling system is much lower than the theoretical capacity. This is true even for the simplest arrangements. FAA guidelines emphatically stress that the practical capacity is one-third less than the theoretical amount (FAA, 1988). For complex automated systems, the practical capacity is even less due to the need to wait for slots for each bag, and to provide reliable service to and from all points in the system. At Denver/International, for example, the automated baggage system apparently at best reached only about 40 percent of the theoretical amount.*

Exercises

1. Investigate the traffic at a local airport or one in which you have contacts. As best you can, estimate the daily flow of passengers, employees, and commercial vehicles. The airport should have data on either these flows or their contributing factors. Compare these estimates with the ranges in the text and discuss the factors that account for differences.

2. Ask 10 or more persons about where they started and ended their last trips to airports. What percentage went to a downtown area? Used public transport? Taxis? Their own or someone else's private vehicle? To what extent do you think this informal survey is representative of passenger traffic?

3. Ask 10 or more persons (perhaps those selected for Exercise 2, perhaps at the same time) about the factors that matter most to them in the airport access trip. Cost? Speed? Reliability? Convenience? What do these responses indicate about the kinds of public transport that might be suitable for airport access?

4. Estimate the total door-to-door time and cost of your several modes of airport access. Which mode is most attractive to you? Do a similar estimate for your possible modes of travel from your last destination airport. What accounts for any differences in your choice? What factors influence these choices?

5. Go to some facility that has short-term parking—at an airport if possible, otherwise at some store such as a supermarket. Observe a set of about 20 spaces close to the main entrance of the passenger building or store for half an hour. How many cars parked in these spaces in that time? Estimate how long

---

*Personal communication from Gary Lantner, manager of facilities for United Airlines.

people stay on average and the capacity of these spaces, that is, the number of cars that might park in these spaces in a day.

6. Explore the distribution of delays on a network similar to that in Example 17.5. You can do this conveniently using a spreadsheet model organized similarly to Table 17-12. Look at short and longer surges of traffic, different patterns of arrivals and destinations, for various levels of line-haul capacity. Calculate how limits on delays at any station set the practical capacity of the network below its theoretical maximum.

References

ACI-NA, Airports Council International, North America (2001) *Parking Survey Database—2000,* ACI-NA, Washington, DC.

Ashford, N., and Wright, P. (1992) *Airport Engineering,* 3d ed., Wiley, New York.

Barker, J. (1998) "Beware Tunnel Vision—A New Inter-terminal Tunnel Promises Relief for Baggage Delays at London Heathrow," *Jane's Airport Review,* May, pp. 44–45.

Barnett, A., Shumsky, R., Hansen, M., Odoni, A., and Gosling, G. (2001) "Safe at Home? An Experiment in Domestic Airline Security," *Operations Research,* 49(2), Mar.–Apr., pp. 181–195.

Daskin, M. (1978) "The Effect of the Origin-Destination Matrix on the Performance of Loop Transportation Systems," Ph.D. dissertation, Department of Civil Engineering, Massachusetts Institute of Technology, Cambridge, MA.

de Neufville, R. (1982) "Airport Passenger Parking Design," *ASCE Journal of Urban Transportation, 108,* May, pp. 302–306.

de Neufville, R. (1994) "The Baggage System at Denver: Prospects and Lessons," *Journal of Air Transport Management, 1*(4), Dec., pp. 229–236.

de Neufville, R., and Meirzejewski, E. (1972) "Airport Access—A Cost-Effective Analysis," *ASCE Transportation Engineering Journal, 98*(3), Aug. pp. 663–678.

de Neufville, R., and Rojas Guzmán, J. (1998) "Benchmarking Major Airports Worldwide," *ASCE Journal of Transportation Engineering,* TN 11639, *124*(4), July/Aug. pp. 391–396.

Elliott, D., and Norton, J. (1999) "Introduction to Airport APM Systems," *Journal of Advanced Transportation, 33*(1), pp. 35–50.

FAA, Federal Aviation Administration (1988) *Planning and Design Guidelines for Airport Terminal Facilities,* Advisory Circular 150/5360-13, U.S. Government Printing Office, Washington, DC.

FAA, Federal Aviation Administration (2001) *Database on Form 5100-127.*

Harris, K. (2001) "On the Rails," *Jane's Airport Review,* July/Aug., pp. 14–15.

IATA, International Air Transport Association (1995) *Airport Development Reference Manual,* 8th ed., IATA, Montreal, Canada.

In Motion (2001) "DFW Plans World's Largest Airport People Mover," *In Motion, 9*(1), pp. 1–2.

Jane's (2002) "World Market for Automated People Movers," Special Report, ISBN 07106 1939 1, http://catalogue.janes.com/wmapm.shtml.

Jane's Airport Review (2001a) "More Records at San Francisco," *Jane's Airport Review,* May, p. 25.

Jane's Airport Review (2001b) "New Baggage Handling in Toronto," *Jane's Airport Review,* May, p. 4.

Jane's Airport Review (2001c) "Change at New York," *Jane's Airport Review,* pp. 4–5.

Leigh Fisher Associates, Coogan, M., and MarketSense (2000) *Improving Public Transportation Access to Large Airports,* Transit Cooperative Research Program Report 62, Transportation Research Board, National Academy Press, Washington, DC.

Moore, H., and Foschi, M. (1993) "Leonardo da Vinci International Airport APM Systems," *Proceedings of the 4th International Conference on Automated People Movers,* pp. 504–513, ASCE, New York.

Nichols, D., Solis, P., and Brown, D. (2000) "Airport on the Move," *Civil Engineering,* Sept., pp. 34–39.

Pilling, M. (2001a) "RFID Waits to Tune in," *Airport World, 6*(1), Feb.–Mar., pp. 44–46.

Pilling, M. (2001b) "Munich Seeks Another Level for Bag Screening," *Airport World, 6*(1), Feb.–Mar., pp. 44–46.

Robusté, F. and Daganzo, C. (1992) "Analysis of Baggage Sorting Schemes for Containerized Aircraft," *Transportation Research A, 26A*(1), pp. 75–92.

Shapiro, P., Katzman, M., Hughes, W., McGee, J., Coogan, M., Wagner, E., and Mandle, P. (1996) *Intermodal Ground Access to Airports: A Planning Guide,* Prepared by US Federal Highway Administration and Federal Aviation Administration, Bellomo-McGee, Vienna, VA.

Shen, L., and Zhao, F. (1997) "Airport Automated People Movers: Costs and Benefits," Creative Access for Major Activity Centers Proceedings of the International Conference on Automated People Movers, pp. 353–362, ASCE, Reston, VA.

Spake, J. (1998) *Jeremy's Airport,* BBC Worldwide, London, U.K.

Part Five
Reference material

18

Data validation

Data are often not what they seem to be. Any single series may contain different definitions, errors, and limitations that distort its meaning. Different series, from different contexts and sources, are unlikely to be fully comparable.

Analysts using historical data need to validate this information. To make forecasts, to plan facilities correctly, they need consistent data that they understand. Consequently, they need to go through a process of examining their data to understand its definitions, to ensure that these definitions are consistent and to eliminate errors.

18-1 The issue

The data available for airport planning and design are often not what they seem to be. Because of particular local definitions or needs, the data may not mean what the analyst thinks they imply. They may mean different things at different times, as definitions or methods of collection change what officials measure and report.

Errors

The data may simply contain significant errors. A few examples illustrate some of the possibilities.

Different, changing definitions The government of Mexico employed a team to review long-term forecasts of passengers for Mexico City prepared for the World Bank. As part of its careful examination, the team found that the official data implied that the number of passengers per flight in domestic traffic had dropped significantly over a 6-year period. This was surprising, since airlines normally adjust their schedules to maintain consistent loads. What accounted for this pattern? Why did it persist for 6 years?

741

Discussions with the airlines indicated that from their perspective nothing special had happened during that period. Aviation officials then remembered that the Mexican government had changed the definition of international passengers during a specific presidential administration (a term of 6 years in Mexico). Originally, passengers leaving Mexico City for a final destination in the United States were counted as "international" from the moment they boarded the aircraft, even if it landed at a midway point such as Guadalajara. During this one administration, however, they were considered "domestic" passengers when they left Mexico City for the intermediate destination inside Mexico. This second definition is perfectly reasonable. However, the inconsistency of the definition over time led to confusing statistics. (It also provoked some imaginative explanations of the factors influencing traffic from the economists working for the World Bank.) The review team fixed the problem by adopting a consistent definition of the data. (For details, see de Neufville et al., 1980)

Distorting methods of collection   In designing the new passenger building for Mombasa, Kenya, it was important to know the number of international passengers, since these would determine the size of the international departure lounges, the number of passport control booths, etc. The Kenyan government supplied data derived from their in-house aviation experts, the nationally owned Kenya Airways. After working in Kenya for a while, designers found that the actual international traffic was much less than shown in the official figures. What was the story?

It turned out that Kenya Airways reported data from its International Division, that is, from the crews and aircraft normally engaged in international flights. Many of these aircraft actually flew regularly on domestic routes. Thus, an aircraft arriving in Nairobi from England early in the morning would make one or two round trips from Nairobi to Mombasa during the day, before it returned to England late at night. The traffic on these domestic flights was quite properly ascribed to the work of the International Division. Somewhere in the statistical reporting process, this distinction between the International Division and actual international passengers was lost. The designers worked with Kenya Airways to solve this problem and, with the assistance of data from the passport control officials, derived a correct estimate of the international passengers.

Simple errors   A master planning process for Los Angeles/Burbank airport used official forecasts from the U.S. Federal Aviation Administration (FAA). A careful reviewer of this plan looked at previous data and noticed about a 25 percent jump in traffic in a single year—not only extraordinary in itself, but also very different from the normal low steady growth. What was going on here?

The reviewer went back to the airport officials and the original airline data they used to collect the annual statistics on enplanements. It turned out that, somewhere between California and the FAA headquarters in Washington, there had been a simple transposition of two similar numbers, a 3 switched for an 8 in an important column. The actual 2.3 million passengers had unintentionally grown to 2.8 million.

Incompleteness

In addition to statistical inconsistencies, some data series are inherently incomplete. A couple of examples illustrate this problem.

Systematic undercounting   One example of incomplete data concerns the premier source of data for the international air transport industry, the World Air Transport Statistics published by the International Air Transport Association (IATA, Annual). This publication presents information from airlines and air carriers that have chosen to pay to be members of the IATA. It is thus systematically incomplete, since many airlines, including some of the largest, have not been members of IATA. Furthermore, the membership normally changes from year to year, so that the series is not consistent over time.

Methodological blindness   The statistics on air cargo for Canada present another example of incompleteness. In that case, the responsible government agency, Transport Canada, has developed data on air cargo from waybills on individual shipments. This procedure fails to capture data on all-cargo aircraft carrying freight for a single shipper, such as UPS or FedEx. Thus, the official statistics have shown virtually no cargo going through Toronto/Hamilton airport, when it is actually one of the Canadian hubs for integrated cargo carriers.

18-2   The resolution

As the examples suggest, airport planners need to validate carefully the data they will use for design. However, no single procedure for checking statistics will resolve all the issues. The available statistics may be misleading for all kinds of reasons. Analysts should be looking constantly for potential problems.

Analysts should adopt two basic approaches to validating their data. They should

- Understand what the data mean
- Double-check the data using alternative sources.

These guidelines are in fact the basis of all good investigations.

Airport planners should first be clear about what the data mean. They should understand both the formal and practical definition of the statistics. Although the formal definition may seem obvious, it often is not. For instance, professionals might assume that there could be no confusion about what it means to be an "international passenger." However, this term represents different concepts in various situations. Thus, in the European Community many "international passengers" are indistinguishable from what planners elsewhere consider domestic passengers, in that they require no customs clearance or passport checks (this applies to travelers between the countries party to the Schengen agreement).* Conversely, passengers arriving in the United States from Canada are generally not treated as international passengers, since they formally entered the country at special facilities in the Canadian airports. Planners often make significant errors when they assume they know what the categories of data mean and should routinely check the local definitions.

Analysts should also check the practical definition of the data. How the statisticians collect data gives this information an operational definition that can significantly modify the formal definition. The examples concerning the practical definition of international passengers in Kenya and Mexico illustrate how local practices give different meanings to equivalent formal concepts. Planners should find out how the local authorities collect data.

Analysts should also double-check their data to the extent possible. They need to obtain independent confirmation of the accuracy and completeness of the data. Any data series may contain biases and errors for many reasons, such as those cited in the introductory examples. It is easiest to corroborate data with alternative series that supposedly mean the same thing. Thus, in the examples of Los Angeles/Burbank

---

*This agreement is between members of the European community that have accepted that passengers between them do not have to go through passport control and customs. The number of countries participating in this treaty has been changing, as additional countries sign up. France and Germany have participated from the start, but the United Kingdom did not. Members of the European Community established this compact during a meeting in the city of Schengen.

and Mexico City, the analysts were able to identify airline statistics that should have agreed with the original information. The discrepancies between the parallel series of data helped identify what was really happening. Analysts should seek out complementary data series to verify their sources of data.

When independent data sources are not available, analysts may have to find more creative ways to validate their data. They can look at operational patterns, for instance, to see if the behavior implied by the official statistics agrees with what would be good practice. The original clue that the data on international passengers was inconsistent in the Mexico City example was that the implied ratios of passengers per flight were abnormal. Sometimes, the analysts will have to go beyond statistics and rely on interviews with local operators and observations in the field. Canadian planners may not find formal alternatives to the Transport Canada data on air cargo through Toronto/Hamilton because the shippers may not wish to release these figures. They can, however, verify that integrated cargo carriers do have frequent flights to this airport and thus determine that the shipments are important. They can also estimate the tonnage shipped, knowing the number and size of the cargo aircraft. In short, airport analysts should also think "out of the box" when validating their data.

In summary, airport planners and designers should validate their data in four ways. They should

- Check the local meanings of the data, to establish the formal definitions
- Find out how the local authorities collect data, to determine the practical definitions
- Seek out complementary data series to verify their sources
- Think "out of the box" when validating their data

Exercises

1. Consider any series of airport data (for example, passengers, aircraft operations, or cargo). Determine its definition. Then attempt to find out how the relevant authorities collect the data. Ask yourself what elements the data might leave out. Do the authorities count crew members flying on passes as passengers, for example? How do they count military or training operations? And so on.

2. For the same series or some other, think about how you could double-check these data in practice. What alternative independent sources could confirm their accuracy? What field observations or interviews could help you validate the formal series?

References

de Neufville, R., Zuniga, S., Kanafani, A., and Olivera, A. (1980) "Forecasting Airport Traffic: Mexico City as a Case Study," *Transportation Research Record, 732,* pp. 24–29.

IATA, International Air Transport Statistics (Annual) *World Air Transport Statistics,* IATA, Montreal, Canada.

19

Models of airport operations

Computer-based models of airside and landside operations have become indispensable tools for airport planners and designers. The capabilities of these models have improved greatly in recent years, but poor model selection and underestimation of resource requirements are persistent problems.

Models can be classified according to level of detail, methodology, and coverage or scope. Highly detailed (or microscopic) models are usually simulations. Less detailed (or macroscopic) models can be mathematical constructs, as well as simulations.

On the airside, several good-quality macroscopic models supplement the available highly detailed simulation packages. Although some of the latter are better known internationally, the macroscopic models are often more appropriate for answering many of the questions that typically arise in airport planning. On the landside, most of the models in use are microscopic simulations. No single model has attained the status of an "international standard." Many of the products offered provide programming environments for the development of models that are adapted to specific local conditions.

Airport operators and managers undertaking model-supported projects must endeavor to select the most appropriate models for the types of issues to be addressed and to create the proper contractual, organizational, and staffing arrangements for the effective use of the models.

19-1 Background

A large number of models of airport operations have been developed over the years as tools for assisting planning, design, and operations management, on both the airside and the landside. Indeed, this is one of the most "mature" areas of computer-based transportation modeling, with the first significant efforts dating back to the late

1950s. By now, several quite advanced modeling capabilities exist. Properly selected and used, such tools can be extremely cost-effective and valuable to airport planners and managers. Unfortunately, practical results to date have not quite lived up to this potential. The main reasons are poor model selection, underestimation of resource requirements, and improper use of the models.

Any major airside or passenger terminal project should be supported by studies that utilize one or more carefully selected computer-based models. Airport planners and managers should therefore endeavor to become familiar with the range of capabilities that are available and with the strengths and weaknesses of some of the most widely used models. This chapter provides a brief review along these lines and offers some suggestions to prospective model users based on the authors' extensive experience.

19-2   Classification of models

It is useful to classify models of airport operations with respect to three aspects: level of detail, methodology, and coverage.

Level of detail

Models can be classified as macroscopic or microscopic, corresponding respectively to low and high levels of modeling detail. The objective of *macroscopic models* is to provide approximate answers to planning (primarily) and some design issues, with emphasis on assessing the relative performance of a wide range of alternatives. To this purpose, macroscopic models omit a great deal of detail. For example, air traffic demand may be described simply by the hourly rate of arrivals at an airport and a probabilistic description of how these arrivals occur over time (e.g., Poisson arrivals). The cumulative diagrams described in Chaps. 16 and 23 are also macroscopic in nature. In these models one is interested in predicting passenger or aircraft accumulation while waiting to be processed at an airport facility as well as the aggregate flow of these entities over time, without being overly concerned about the exact way the flows are handled or about what happens to any individual passenger or aircraft. Macroscopic models are useful primarily in policy analysis, strategy development, cost–benefit evaluation, and approximate analysis of traffic flows. Ideally, they should be fast, in terms of both input preparation and execution time, so they can be used to explore a large number of alternative scenarios.

Microscopic models are designed to deal with more tactical issues and aim at a highly faithful representation of the various processes that take place at an airport. Typically, such models represent aircraft, passengers, or bags on an individual basis (or as groups that consist of only a few members). The model moves these entities or "objects" through the airside or landside elements under study, taking into consideration each aircraft's performance characteristics or each passenger's or bag's attributes. Operational details such as taxiway and gate selection, pushback maneuvering, queuing discipline, etc., are generally included only in microscopic models.

Methodology

It is important to draw a distinction between *analytical* and *simulation* models. The former are abstract, necessarily simplified mathematical representations of airport operations, typically consisting of sets of (simple or complex) equations. The simple runway capacity model described in Chap. 10 is an example. By solving or processing such sets of equations (either in closed form or numerically), analytical models derive estimates of quantities of interest—most often of measures of performance, such as capacity or delays. In contrast, simulation models create objects (aircraft, people, bags, etc.) that move through those parts of the airport that the model describes. By observing the flows of such objects past specific locations (e.g., the threshold of a runway or a check-in desk) and the amount of time it takes to move between such points, simulation models compute appropriate measures of performance and of level of service. There is a strong correlation in practice between model methodology and level of detail. Specifically, analytical models tend to be mostly macroscopic. By contrast, practically all the microscopic models are simulations.

Models, whether analytical or simulations, can be further classified with respect to methodology, according to whether they are (1) dynamic or static and (2) stochastic or deterministic. *Dynamic models* accept input parameters that are time-dependent and capture the fluctuations of traffic and of level-of-service over time; *static models* assume that the model's parameters remain constant over time. *Stochastic models* accept input parameters that are probabilistic quantities (i.e., random variables) and capture the impacts of uncertainty on the chosen performance metrics. Stochastic simulation models are often referred to as *Monte Carlo simulations*.

Coverage

Models can be limited in scope or comprehensive. For example, an airside model may be concerned only with apron operations, while another may represent the entire range of airfield and terminal airspace operations.

19-3　Airside models and issues in model selection

A large number of airside models have been developed over the years and are available to airport operators, planners, and managers—some at little or no cost. Because the features and capabilities of such models tend to evolve rapidly over time and because new models are being developed, any comparisons among them can be valid only for a limited period. For this reason, the discussion in this section is intended primarily as an indication of the range of choices available. The principal goal is to suggest how to think about model selection in general terms, not to recommend any specific models.

Principal existing airside models

Table 19-1 lists a number of models available in 2001, classified according to level of detail and coverage. Analytical models are indicated with an asterisk; the others are simulations. The table includes some simulation models (SIMMOD, TAAM, The Airport Machine, and RAMS) that are well known internationally.

As Table 19-1 suggests, macroscopic models concentrate on modeling operations at runways and final approaches, i.e., those parts of the airport airside that constitute the principal bottlenecks of traffic flow in the great majority of cases. Thus, despite their limited coverage/scope, these macroscopic models can be highly useful when computing the cost of aircraft delays at an airport—a standard issue in most policy and cost–benefit studies—or in estimating when in the future (or at what number of annual movements) the runway system will be saturated and additional capacity will be needed. The unavailability of general-purpose macroscopic models of taxiway/apron and of terminal airspace operations is due primarily to the fact that analyses of traffic performance at these parts of the airside must be very location-specific, taking account of the local geometry and other airport and air traffic management system characteristics (Chaps. 10 and 13).

Of the macroscopic models listed in Table 19-1, the first refers to the simple analytical model for single-runway arrivals due to Blumstein (1959), and its extensions to single runways used either for departures only or for both arrivals and departures, all of which are discussed in Chap. 10. These simple models can be easily programmed on a computer and provide good approximations to single-runway capacities for any particular set of circumstances. The FAA Airfield Capacity Model extends the same fundamental approach to 14 different common runway configurations, ranging from one to four simultaneously active runways. It was initially developed in the late 1970s by an FAA contractor and further modified for the FAA by the MITRE Corporation (Swedish, 1981). DELAYS is a model developed at the Massachusetts Institute of Technology. It is concerned solely with computing runway-related delays under dynamic and stochastic conditions. DELAYS views the runway complex of an airport as a queuing system whose

Table 19-1. Classification of analytical and fast-time simulation models of airside operations

Level of detail (type of study)	Model coverage/scope		
	Aprons and taxiways	Runways and final approach	Terminal area airspace
Macroscopic (policy analysis, cost-benefit studies, approximate traffic flow analysis)		Blumstein model and extensions* (see Chap. 10) FAA Airfield Capacity Model* DELAYS* LMI Capacity and Delays Model*	
	MACAD†	MACAD†	
Microscopic (detailed traffic flow analysis, preliminary and detailed design)	SIMMOD TAAM The Airport Machine HERMES	SIMMOD TAAM The Airport Machine HERMES	SIMMOD TAAM RAMS

*Indicates an analytical model.
†MACAD is an analytical model, except for its apron model, which is a simulation.

"customers" are aircraft demanding to land or take off and whose capacity is equal to the arrival, departure, or total capacity of the runway system, depending on whether one is interested in delays to arrivals, to departures, or to the "average operation," respectively. The model is based on a fast approximation scheme, refined by several researchers over the years, for solving the differential equations that describe quite general dynamic queuing systems (Koopman, 1972; Kivestu, 1976; Malone, 1995).

The Logistics Management Institute's (LMI) Runway Capacity and Delays Model (Lee et al., 1997) attempts to combine approaches similar to those of the previous three models into a single package and includes several improvements and generalizations of the logic of the Blumstein and FAA models for computing runway capacity. For a single runway and for any given aircraft mix and set of separation requirements, it computes (1) the "all-departures" capacity of a runway (i.e., the capacity when the runway is used for departures only), (2) the all-arrivals capacity, (3) the number of "free" departures that can be performed without reducing the all-arrivals capacity, and (4) the capacity of the runway if a departure is always inserted between two arrivals, so that arrivals alternate with departures on the runway (see Chap. 10). The capacity of the runway for any other mix of arrivals and departures and any other sequencing of arrivals and departures can then be computed approximately by utilizing the four estimates above. For configurations involving the simultaneous use of more than one runway, the model has to be extended in an ad hoc way for each airport of interest. The LMI model also includes a delay-estimation model—similar in concept, but less general than the one in DELAYS.

MACAD (for Mantea Capacity and Delay Model) builds on these earlier models and includes many of their best features (Stamatopoulos, 2000; Stamatopoulos et al., 2002). It consists of an enhanced version of the LMI model's single-runway capacity analysis; an extension of these models to pairs of (dependent or independent) parallel runways and pairs of intersecting runways; the DELAYS model for delay-estimation purposes; and a simple simulation model of remote and contact stand occupancy that extends coverage to encompass apron operations as well. MACAD needs only a fraction of a second on a reasonably fast personal computer to run through a 24-h period of operations at some of the busiest airports in Europe.

Most of the microscopic models in Table 19-1 are well known. SIMMOD, TAAM, and The Airport Machine have been used in numerous airspace and/or airport studies in many parts of the world. The first is a model developed in various stages beginning in the late 1970s, with support from the FAA. It is available at nominal cost from a small number of vendors. The other two models are proprietary and require a license fee, which, as of 2001, was very substantial in the case of TAAM. SIMMOD and TAAM cover both airspace and airport operations, while The Airport Machine is limited to airport and final approach operations only. RAMS is an airspace operations modeler, developed by EUROCONTROL in the early 1990s. EUROCONTROL is also the organization that controls access to RAMS by qualified users. The least-known model, HERMES, has been developed by the Civil Aviation Authority (CAA/NATS) in the United Kingdom. Up to 2001, its use had been limited to simulating in detail operations at London's Heathrow and Gatwick Airports. HERMES is a good example of a highly detailed simulation model adapted to a very specific set of local conditions. Detailed descriptions of these and other microscopic simulation models, including their computational requirements and characteristics can be found in Odoni et al. (1997), while a review of some specific applications and related issues is provided in Transportation Research Board (2001).

An important distinction in the case of microscopic models is between node-link and three-dimensional (3D) airport representations. *Node-link models* represent airports and airspace as networks consisting of nodes and links. Aircraft move from node to node along paths consisting of sequences of nodes and links. Conflicts occur when more than one aircraft tries to occupy the same node simultaneously. These conflicts are resolved by delaying all but one of the aircraft vying for a node at any particular time. By recording the amount of delay incurred at each node by each aircraft, the model compiles the requisite aggregate and distributive delay statistics for every location on the airport or in airspace and for every aircraft processed. SIMMOD and The Airport Machine are node-link microscopic models. By contrast, *three-dimensional (3D) models* allow aircraft to fly arbitrary three-dimensional routes,* according to specified equations of motion. RAMS and TAAM are microscopic 3D models.

Selection criteria

Table 19-1 shows only a subset of available airside models. Given the broad range of choices, it is useful to consider the criteria that should

---

*When simulating airport surface traffic operations, these routes are, of course, reduced to two dimensions.

be used for selecting a model for any particular project. One should start with an appreciation of the fact that there is no such thing as a "best" model. Different models are "best" for different kinds of questions. Thus, the first consideration in selecting a model should be to match the model's level of detail, flexibility and coverage/scope to the questions at hand. Experience suggests that by far the most common and costly mistake in airside model selection is the use of microscopic models (e.g., SIMMOD or TAAM or The Airport Machine) to address questions that can be answered better, much more quickly— and at a small fraction of the cost—by a macroscopic model. It makes little sense to spend a great deal of time and resources developing a detailed simulation of an airfield, just to compute the ultimate capacity of a system of runways or the approximate level of airside delays associated with some hypothetical future number of annual movements. Yet this happens all the time! Part of the reason is that much of the user community is not familiar with the capabilities of macroscopic models, such as those listed in Table 19-1. Another, somewhat paradoxical reason may be the simplicity and low cost of these models ("If it's that simple to use, it cannot be good").

A second obvious selection criterion is cost, both in economic terms and in terms of time. The amount of time and effort needed to develop a high-quality, detailed simulation model of an airfield is almost invariably underestimated at the outset of new projects. Sophisticated simulation models such as SIMMOD, TAAM, and RAMS are associated with steep "learning curves" and require expert, well-trained programmers. Such models can be very cost-effective in support of detailed airfield design projects, but only if adequate time and human resources are invested in adapting and using them in any particular context. The initial model acquisition cost also may not be a good indicator of eventual project cost. For instance, a licensed copy of SIMMOD can be obtained for a few thousand dollars, but any nontrivial application of the model will typically cost more than $100,000—possibly far more—when all expenses are taken into account. The development of a good SIMMOD model will also usually require expert support from a consultant or in-house staff with prior experience with the model.

An important related issue—particularly when selecting among complex microscopic models—concerns the anticipated frequency of use. At one end, a model may be adopted for a one-time (or occasional, e.g., every few years) use—for instance, as a tool in addressing a specific design issue or in support of an airport expansion project. At the other extreme, it may be anticipated that the model will be used repeatedly to address various airfield-related problems as they arise. In the former case, the airport operator will be better off hiring a consultant to answer the question at hand. Selecting a model appropriate to the task and adapting it to local conditions should be left up to that consultant, but early "milestones" and escape clauses should be written into the contract to make sure the selected model meets expectations. The supposedly less expensive approach of "let's buy a license for simulation model X and program the model in house" is often begging for trouble. The airport operator will usually discover that "programming the model" is far more difficult and time-consuming than expected and will eventually end up hiring a consultant anyway.

By contrast, adopting a model as an in-house analysis tool will be the most effective approach in the long run if it is anticipated that a microscopic simulation tool will be used repeatedly, at frequent intervals. In such cases, the best approach will be to contract with a model vendor for provision of a package of services that will include: (1) providing a license for a simulation model and its future enhancements; (2) training of the airport operator's staff in model programming and use; (3) adapting the model to the airport in question by developing a basic representation of the airfield and its traffic; and (4) committing to provide occasional support in the future, e.g., by answering technical questions about the model, as they may arise. Experience suggests that (3) is particularly important, as it provides a starting point for future projects that would presumably rely primarily on in-house expertise. If a basic model of the airfield is already available, potential alternative configurations can be studied by modifying that basic model.

An airport operator who decides to develop an in-house modeling capability for airside studies should be aware of two critical points. First, such a decision implies the need for permanent in-house staff of at least two or three specialists responsible for maintaining, upgrading, and programming the models as needed. The need for a multiperson staff stems from the fact that turnover rates in positions of this type tend to be very high. Model specialists, as a rule, are smart, well trained, and in high demand. They often move to other organizations—or are promoted to other jobs within the airport authority. Their departure often deprives their employer of the kind of institutional memory necessary to sustain an in-house modeling capability. Having more than one such individual on hand provides insurance against this type of disruption. Second, it is a good idea not to rely on a single model, but to develop instead a "toolkit" of two or three models. Such a toolkit should include at least one macroscopic model, to support policy analysis, strategic planning, and cost–benefit studies, in addition to a microscopic simulation model. For example, a combination of MACAD and TAAM or SIMMOD will enable an airport operator to address efficiently a far broader set of questions than if the operation relies on TAAM or SIMMOD alone.

Model flexibility and adaptability is another very important criterion. For example, 3D microscopic models (such as TAAM and RAMS) hold an inherent advantage over node-link models (such as SIMMOD), with respect to flexibility for studies that consider issues relevant to terminal area airspace (or en-route airspace). This is especially true when it comes to evaluating concepts, such as "free flight," which give airspace users the freedom to select their own optimized flight paths (Chap. 13). It is extremely difficult to adapt node-link models to such an environment. However, when it comes to airfield simulations, node-link models can be just as powerful as 3D ones.

Two additional criteria for model selection should also be mentioned briefly. One is user-model interfacing and ease of model use. Input preparation and scenario development can be a major part of the task of adapting a model local conditions. Some models facilitate this task through appropriate interfaces, dialogue boxes, default input data, etc.; others do not. The second consideration concerns a model's graphics and animation capabilities. Significant technical advantages can obviously be derived from high-quality graphics and animation in airport models. Less obviously, important public relations benefits can be obtained. Airport projects everywhere have high visibility and typically undergo considerable scrutiny by the public and the media. Project adoption often hinges on approval by elected officials or political appointees with limited technical background. Graphics and animation can be extremely valuable in demonstrating the benefits of airport projects in this decision-making environment.

19-4 Models of passenger building operations

The second major category of airport operations models concerns the processing of people and bags in passenger buildings. The growing size and enormous cost of these buildings, the large number of passengers processed, and especially the complex interactions that take place between successive phases of processing (Chaps. 15 and 16) have stimulated demand for such models. A comparison between airside models and models of passenger building operations reveals many similarities, but many important differences as well. The following discussion identifies some of the principal characteristics of models of passenger buildings and highlights some of their similarities and differences with airfield models.

Model availability

Table 19-1 indicates the existence of several good-quality, computer-based macroscopic models of airside operations, which utilize an analytical approach (as opposed to simulation). Only very few analogous, low-level-of-detail models are available for passenger buildings, primarily due to the complexity of the processes and interactions that take place in these buildings. Moreover, the few models that exist are very recent, and little practical experience has been obtained with their use. A typical example is SLAM (Simple Landside Airport Model), which is based on the notion of cumulative diagrams (Chaps. 16 and 23), but extends that methodology to cover the entire sequence of service processes taking place at an airport (Brunetta et al., 1999). Given a flight schedule and probability distributions for the time (prior to the flight) when departing passengers show up at the passenger building (Chap. 16), SLAM computes (1) the approximate waiting times at each processing facility in the building and (2) the level of service (A–F, see Chap. 15) in each part of the building, based on available space per passenger, using the IATA space standards.

As can be inferred from the previous paragraph, studies of passenger buildings to date have relied primarily on either the approximate formulas presented in Chaps. 15 and 16 or on highly detailed simulation models. In this last respect, it is interesting to note that, in airfield modeling, a few simulation models (SIMMOD, TAAM, The Airport Machine) dominate the field internationally, whereas there are no dominant models of *passenger building* operations. Part of the reason is that the development of landside models, in general, has been mostly left up to private initiative, in contrast to airside modeling for which there has been significant direct (SIMMOD) or indirect (TAAM) government funding. Therefore, the market for passenger building models is more competitive and dynamic and, in recent years, more innovative from the technical point of view than the market for airside models.

A second reason for the absence of dominant simulation models is that passenger buildings and associated processes differ in many respects from one location to another. It is thus very difficult to develop a "standard" simulation model that can be adapted to any given local conditions in a simple way. During the 1980s there were, indeed, a number of attempts to develop detailed simulation models that could be easily adapted to any particular passenger building (Mandle et al., 1980; Mumayiz and Ashford, 1986; FAA, 1988; Mumayiz, 1990). These attempts were not particularly successful in practice, and model adaptation to local conditions proved cumbersome. As a result, the entire modeling "philosophy" in this area changed. The approach most commonly adopted today is to develop each simulation of a passenger building more or less from scratch. Increasingly, the "models" of passenger buildings that vendors and consultants offer are, in fact, only *simulation environments,* i.e., software that greatly facilitates the modeling of the service processes and interactions between successive services that characterize a passenger building. The simulation environments used are typically those tailored to simulating either queuing networks (e.g., EXTEND©) or manufacturing systems (e.g., ARENA© or WITNESS©). It is easy to see that the sequence of service processes that take place in a passenger building can, in fact, be abstracted as a network of queuing systems (where the "customers" are passengers and bags that move from one queuing system to the next) or as a manufacturing plant (where the units "produced" are processed passengers and bags that must be handled by a series of "machines," i.e., the various services and facilities in the building).

The selection criteria for landside models have therefore shifted considerably. Choice is now based largely equally on the technical features, flexibility, and adaptability of the simulation environment being offered, and on the record, reputation and previous experience of the vendor or consultant offering the simulation model.

Data requirements

A striking feature of models of passenger building operations is their requirement for enormous quantities of often difficult-to-obtain input data. Compared to the data requirements of models of passenger building operations, the data requirements of airside models are modest. Detailed information must be supplied to the model on such items as characteristics of arriving, departing, and connecting passengers; characteristics of persons accompanying passengers; features of the baggage handling system; flight schedules and number and classes of passengers on each flight; paths through the building that the different types of passengers will follow; number of servers (staff or equipment) at the various service facilities (see, e.g., Raymond, 1999); processing times at each facility; etc.

It is highly unlikely that, in any practical application, all the data needed for a detailed simulation of a building will be available. It is usually necessary either to "borrow" some of the input data from other airports—under the often-questionable assumption of similar prevailing conditions—or simply to guess at the data. Moreover, there is no guarantee that any existing data from current airport operations will be valid at the future time to which the simulation refers. Typical examples of types of data that are extremely difficult to obtain include:

- The fractions of passengers, on a flight-by-flight basis, who choose any one of several alternative routings and select various alternative sequences of services through the building

- The range of characteristics of *connecting* passengers (routing through the building, percent using various services, etc.)

- The ways departing and connecting passengers will choose to allocate any "slack" (or "free" or "idle") time among such activities as shopping, visiting a food concession, or simply sitting at the departure gate waiting to board an airplane

To appreciate the difficulty of obtaining reliable data for these examples, simply note that, in all three cases, the amount of time left until flight departure critically affects passenger behavior.

Repeated versus one-time model use

Unlike the case of airside models, it makes little sense in most instances to develop a detailed simulation model of a passenger building for one-time use only. The reason is that in any practical application there is a virtually endless set of issues that the model will be asked to address. It is impossible to know in advance what questions will arise in the course of studying the building, what alternative solutions will be examined, and what future scenarios will emerge. The following "real-life" example illustrates this point.

EXAMPLE 19.1

An airport operator contracted with a consulting firm for the development of a detailed simulation model of a large new passenger building. The contract did not call for the transferring of the model, once completed, to the airport operator. Consequently, it did not provide for training of the airport operator's staff in the use of the model. The consultant was also required under the contract to examine the performance of the building for two future demand scenarios and to report on the results. Study of any additional scenarios would require an additional negotiated fee per scenario, depending on the amount of effort involved.

The simulation project was then carried out. The contractor spent several days on the airport operator's premises gathering data and interviewing key individuals to understand better the building's future operating procedures. Several weeks later they presented their results. By that point, the managers of the project on the airport operator's side had been told by their top management that the two demand scenarios that the simulation examined were no longer "valid" and that several alternative scenarios had emerged that required a significant revision to the model. The cost of revising the model and of studying the new scenarios with the revised version was deemed prohibitive and the entire project was abandoned, ending up as a total waste of effort, time, and money.

The prudent course in the case of detailed passenger simulations is to go through the model-acquisition steps outlined in Sec. 19-3. The airport operator should enter into a contract with a reputable vendor for a license to use the simulation model, training of the airport operator's staff, development of a working prototype simulation, and future occasional technical support services.

Model development process

The process followed in developing a detailed simulation of a passenger building is critical. Experience suggests that an airport operator can benefit as much (or more) from the model development process as from the results obtained subsequently from operating the model. The typical model development process requires the close cooperation between the vendor/consultant supplying the model and the airport operator over a period of several months. During that time, the vendor gathers input data with the assistance of the airport operator and discusses with the operator the service processes at the airport, desirable features in the model to be prepared, types of output reports to be generated (see below), etc. This provides an excellent opportunity for astute airport operators to use the simulation development process as the focal point of an internal review of the building's design and operations. For example, a model development team can be formed that includes representatives of the various organizational units involved in the planning and operation of the building (engineering, operations, commercial, security, airline services, passenger services, public information, etc.). The task of providing input data and descriptions of operating procedures for the model forces such a team to consider every aspect of the building's operations in a way and at a level of detail not previously done. Very often, this detailed process brings to the fore issues that had never been considered (e.g., what is the best route through the building of a particular class of connecting passengers, or how to get departing passengers of a flight with a long delay to a restaurant). It also makes it possible for the various organizational units to find out exactly what the plans of the other units are and brings any discrepancies or problems in this respect to the attention of senior management.

Communicating the results

Somewhat paradoxically, one of the serious practical problems associated with highly detailed simulation models of passenger buildings is the processing of the results. The models often simulate the processing of many hundreds of flights and tens of thousands of passengers per day and generate massive raw output on every transaction that takes place at every part of the building every second. Summarizing this information into meaningful reports is not an easy task. Good summary reports should capture not only average behavior, but also extreme occurrences (e.g., maximum queue lengths observed) of which planners and managers should be aware. Moreover, because of the uncertainty inherent in many of the inputs to these simulations, ways have to be found to present appropriate sensitivity analyses to decision makers without overwhelming them with information. Good graphics and animation capabilities can be particularly useful in this context.

Exercises

1. Consider a passenger building used for international flights at an airport with which you are familiar. Try to identify a building that processes arriving, departing, and connecting passengers. Prepare a flow chart that shows the sequence of processes through which each type of passenger will go from curbside to boarding gate and vice versa. Show the points where connecting passengers merge with departing passengers. (Different airports use different procedures in this respect.)

2. The behavior of departing passengers in airport passenger buildings is hard to simulate. The reason is that many of these passengers have a lot of "slack" time on their hands and allocate the time they spend at the various parts of the building accordingly. Consider, for example, a passenger who has just checked in for a flight and still has 90 min to go until aircraft boarding time. This passenger is obviously much more likely to take advantage of the commercial services available in the building than a passenger who has only 15 min left until boarding time and must rush to his or her gate. Describe in a few paragraphs what general approach you would use in a simulation to capture this type of behavior. What data would you need for your simulation?

3. The efficiency of use of apron stands is always a matter of interest to airport operators. Several simulation models are available for this type of analysis. In this exercise, you are asked to develop a flow chart for a macroscopic (low level of detail) simulation of stand occupancy at an apron area that consists of approximately 50 stands, of which about 30 are at contact gates and the remaining are remote. Let the stands be subdivided into two categories: those that can hold all types of aircraft; and those that can hold aircraft with a wingspan up to and including that of FAA Group III or ICAO Aerodrome code letter C (see Chap. 9). Assume that (a) all airlines have access to all the stands and (b) airlines have a preference for contact stands, so that when such a stand (of appropriate size) is available, the next arriving aircraft will be directed to occupy that stand. Assume for convenience that, at time t = 0, all the stands are empty and that a schedule of aircraft arrivals and departures to/from the apron area has been provided. What additional data would you need to implement this simulation?

References

Blumstein, A. (1959) "The Landing Capacity of a Runway," *Operations Research, 7,* pp. 752–763.

Brunetta, L., Righi, L., and Andreatta, G. (1999) "An Operations Research Model for the Evaluation of an Airport Terminal: SLAM (Simple Landside Aggregate Model)," *Journal of Air Transport Management, 5,* pp. 161–175.

FAA, Federal Aviation Administration (1988) *Planning and Design Guidelines for Airport Terminal Facilities,* Advisory Circular 150/5360-13, U.S. Government Printing Office, Washington, DC.

Kivestu, P. (1976) "Alternative Methods of Investigating the Time-Dependent M/G/K Queue," S.M. thesis, Department of Aeronautics and Astronautics, Massachusetts Institute of Technology, Cambridge, MA.

Koopman, B. (1972) "Air Terminal Queues under Time-Dependent Conditions," *Operations Research, 20,* pp. 1089–1114.

Lee, D., Kostiuk, P., Hemm, R., Wingrove, W., and Shapiro, G. (1997). *Estimating the Effects of the Terminal Area Productivity Program,* Report NS301R3, Logistics Management Institute, McLean, VA.

Malone, K. (1995) "Dynamic Queuing Systems: Behavior and Approximations for Individual Queues and Networks," Ph.D. thesis, Operations Research Center, Massachusetts Institute of Technology, Cambridge, MA.

Mandle, P., LaMagna, F., and Whitlock, E. (1980) *Collection of Calibration and Validation Data for an Airport Landside Dynamic Simulation Model,* Report TSC-FAA-80-3, Transportation Systems Center, FAA, Cambridge, MA.

Mumayiz, S. (1990) "Overview of Airport Terminal Simulation Models," *Transportation Research Record 1273,* pp. 11–20, Transportation Research Board, National Research Council, Washington, DC.

Mumayiz, S., and Ashford, N. (1986) "Methodology for Planning and Operations Management or Airport Terminal Facilities," *Transportation Research Record 1094,* pp. 24–35, Transportation Research Board, National Research Council, Washington, DC.

Odoni, A., Deyst, J., Feron, E., Hansman, R., Khan, K., Kuchar, J., and Simpson R., (1997) "Existing and Required Modeling Capabilities for Evaluating ATM Systems and Concepts," International Center for Air Transportation, MIT, http://web.mit.edu/aeroastro/www/labs/AATT/aatt.html.

Raymond W.T. (1999) "Intelligent Resource Simulation for an Airport Check-in Counter Allocation System," *IEEE Transactions on Systems, Man and Cybernetics, Part C: Applications and Reviews, 29*(3), pp. 325–335.

Stamatopoulos, M. (2000) "A Decision Support System for Airport Strategic Planning," Ph.D. thesis, Department of Management Science and Marketing, Athens University of Economics and Business, Athens, Greece.

Stamatopoulos, M., Zografos, K., and Odoni, A. (2002) "A Decision Support System for Airport Strategic Planning," *Transportation Research C,* in press.

Swedish, W. (1981) *Upgraded FAA Airfield Capacity Model Supplemental User's Guide,* Report MTR-81W16 and FAA-EM-81-1, The MITRE Corporation, McLean, VA.

Transportation Research Board (2001) *Airport-Airspace Simulations for Capacity Evaluation,* J. Rakas and S. Mumayiz (eds.), Transportation Research Circular E-C035, Washington, DC, http://trb.org/trb/publications/circulars/ec035/ec035.pdf.

20

Forecasting

Forecasting is an art, not a science. Any forecast of phenomena involving people is inherently unreliable and likely to be wrong. All forecasts of social activities involve many arbitrary assumptions and opinions. Users of forecasts need to understand and appreciate this reality.

The basic concept of forecasting is to estimate past trends and project them forward. The idea is simple in concept, questionable in execution. To estimate past trends, analysts have to make many assumptions. They have to select a span of data (over time or different circumstances), principal drivers of the phenomenon being forecast, the form of these factors and of the mathematical model, and future values of their drivers. Different analysts choose these factors differently, and obtain quite different results.

The mathematics involved in forecasting can look frightening, but all procedures call upon a simple idea. They attempt to create a formula that correlates well with past experience. Analysts can generally do this easily, whatever assumptions they make. They do this by adjusting the parameters of the formulas, the form of these formulas, and the variables included. Mathematics is thus not the decisive factor.

Ultimately, all forecasts about airport activities reflect judgment and opinion. In the short run, existing rates of change are likely to persist due to inertia. In the longer run, airport forecasts are based on the fallible opinions of experts.

This chapter first focuses on the selections analysts have to make to create a formula for forecasting, using data from forecasts prepared for a Master Plan of Los Angeles Airports. It thereby underlines the important role of judgment in the creation of forecasts. It next proceeds to describe the fundamental mathematics of forecasting, principally regression analysis. This discussion shows how analysts can easily develop many models that fit past trends well, demonstrating that mathematics alone cannot validate a forecast. The discussion then indicates how experts can use their knowledge directly to develop scenarios of future developments. Finally, this section presents an overall approach to airport forecasting that blends analysis and expert judgment, giving more weight to trend extrapolation in the short run and judgment in the long run.

20-1 Forecasting assumptions

Forecasters use the past to describe the future. They create some model of how things happened to project what may occur. They look at some factor in which they are interested (such as the number of passengers) and attempt to describe how some other secondary factors (such as airline fares and the level of local economic activity) made it change.* When they are satisfied with their model, they estimate the future level of the secondary factors and use them to project the future values of the activity in which they are interested.

Alert readers will notice that this process is immediately questionable. Analysts start with a desire to forecast one factor, and end up having to forecast several factors in order to do so. If these secondary factors can be known in advance with some assurance, the process may be fine. If they cannot, as is often the case, the process actually makes a hard problem harder—instead of forecasting one factor, the analysts have to forecast several.

When analysts develop forecasts mathematically, they create formulas to describe the evolution of the factor in terms of other factors. For example, the consultants on the Los Angeles Master Plan wished to

---

*The standard literature refers to the secondary factors as the "independent" or "explanatory" variables that "explain" the "dependent" variable that the analyst wishes to predict. These terms are not used here because they are misleading. Only in laboratory situations can the researcher control the situation and definitely know which variables cause a phenomenon (for example, weights on the beam cause it to bend). In dealing with social situations, such as travel, we cannot assume which factors cause which events. In fact, many factors interact with each other and the direction of causality is ambiguous. For example, lower prices stimulate demand for a product, and higher demand can lead to lower costs and prices. Labeling one set of variables as explanatory or independent presumes that one knows what is happening, when actually the analyst is trying to find out.

forecast the number of passengers for Los Angeles. Formally, they identified the variable they wish to forecast, Y, and many other factors, X_i, and tried to develop formulas linking the two. These were of the form

$$Y = f(X_1, X_2, \ldots, X_n) \tag{20.1}$$

To describe a model numerically, analysts have to make four kinds of assumptions. These concern:

- The *span of the data,* in terms of periods or different situations
- The *variables* to be included in the formulas
- The *form of these variables*
- The *form of the equation* itself

Each of these choices affects the calculations of the forecasts, yet analysts have no firm scientific basis for making these choices. The forecasts thus inherently reflect opinion.

Analysts choose the span of their data. For the Los Angeles Master Plan, for example, the consultants considered the data in Table 20-1. Their choice of 30 years was reasonable, yet it was not the only possible choice. An alternative analysis might have dropped some of the earlier years. It might do this on the grounds that the eras before the first oil crisis (1973), or before the economic deregulation of the airlines in the United States and the second oil crisis (1978), were so different from the more recent situation that they represented different trends. Either of these alternatives is defensible; either gives different models.

Aviation analysts typically work with time series, as in Table 20-1, and have to decide on the number of periods. However, they may also work with data on several different situations for a common period, using what is known as a cross-sectional approach. For example, in trying to model the number of passengers, researchers could look at the data for many cities in some year, and would have had to choose the number of cities. Using either time series or cross-sectional data, analysts have to choose the span of the data.

Analysts next have to choose the factors they believe are the most important drivers of the factor they wish to forecast. For Los Angeles, the consultants focused on population, employment, per-capita personal income for the region, and national average yield for the airlines. This choice is reasonable, but not the only one that they could

Forecasting

Table 20-1. Total annual passengers at Los Angeles International Airport

Year	Domestic	International	Total
1965	12,133,915	444,994	12,578,909
1966	14,690,615	560,657	15,251,272
1967	17,422,725	702,427	18,125,152
1968	19,449,393	896,618	20,346,011
1969	20,112,355	1,197,713	21,310,068
1970	19,388,269	1,392,449	20,780,718
1971	18,808,943	1,538,104	20,347,047
1972	20,195,606	1,881,983	22,077,589
1973	21,335,527	2,166,170	23,501,697
1974	21,240,809	2,343,780	23,584,589
1975	21,228,961	2,490,067	23,719,028
1976	22,996,544	2,986,535	25,983,079
1977	25,069,588	3,292,275	28,361,863
1978	28,746,053	4,155,308	32,901,361
1979	29,925,964	4,997,241	34,923,205
1980	27,386,068	5,652,091	33,038,159
1981	27,280,810	5,441,724	32,722,534
1982	27,646,646	4,736,459	32,383,105
1983	28,516,910	4,909,821	33,426,731
1984	28,978,424	5,383,291	34,361,715
1985	31,758,581	5,889,402	37,647,983
1986	34,968,077	6,449,790	41,417,867
1987	37,408,359	7,464,754	44,873,113
1988	36,339,707	8,058,904	44,398,611
1989	35,824,232	9,142,989	44,967,221
1990	35,969,037	9,841,184	45,810,221
1991	35,284,399	10,383,805	45,668,204
1992	35,508,568	11,455,987	46,964,555
1993	35,899,762	11,945,032	47,844,794
1994	38,371,410	12,678,865	51,050,275

Source: Landrum and Brown, 1996,

have made. For a similar study of Miami, for example, other analysts considered the regional share of national traffic. They could also have looked at the national level of income, the economic growth of Latin America, etc. There is no definite set of variables on which all analysts agree.

Analysts also have to choose how they will represent the data. Normally, most traffic forecasts include some measure of cost (such as fares) as one of the factors to include in the analysis. Basic economics has taught us that prices strongly influence demand. Lower prices increase traffic, for example. However, there are many ways to represent this factor. Analysts can use figures on average price (as the consultants did for Los Angeles), on comparative price (with respect to road or rail, for example), on differential price compared to competition, etc. For each of the basic concepts that might be involved, different analysts can and do justify different versions. These choices will affect the forecasts.

Finally, the analysts have to choose the form of the model. Most simply, it could be a linear model, such as

$$\text{Passengers} = \hspace{2cm} (20.2)$$
$$(\text{population})[A_0 + A_1(\text{income}) + A_2(\text{employment}) + A_3(\text{yield})]$$

Alternatively, the model could be a power relationship, which economic theory indicates more accurately represents the relationship between price and demand:

$$\text{Passengers} = B_0[\text{yield}]^{B_1} \hspace{2cm} (20.3)$$

It could also be exponential in time, a form that conveniently represents constant rate of growth C_1 over each period of time T:

$$\text{Passengers} = C_0(e)^{C_1 T} \hspace{2cm} (20.4)$$

For Los Angeles, the consultants in fact tried all these three and others. Even with these, they had not exhausted the possibilities. Analysts can create all kinds of complicated models.

The point the reader should retain is that analysts have to choose between the models without any clear indication of which is best. This and the other subjective judgments are at the root of all the analysis, and influence the mathematical results. The forecasts thus have no firm scientific basis. All forecasts rest on opinions.

20-2 Fundamental mathematics

To obtain a specific model of the behavior of the factor of interest, such as the number of passengers, analysts have to "fit" the model to the historical data. The standard approach is known as *linear regression analysis* (see, for example, Lewis-Beck, 1980; Kanafani, 1983; ICAO, 1985; Makridakis and Wheelwright, 1989; Pindyck and Rubinfeld, 1998). Formally, it applies to linear models such as that in Eq. (20.2). In fact, it applies as well to exponential and power relationships [such as Eqs. (20.3) and (20.4)], since these can be made into linear equations once they are expressed in terms of logarithms. For example,

$$\text{Passengers} = C_0(e)^{C_1 T} \qquad (20.5)$$
$$=> \log(\text{passengers}) = \log(C_0) + C_1 T(\log e)$$

Linear regression analysis fits the model to the data by minimizing the sum of the squared differences between the actual observations of the data at different times, Y_t, and the values indicated by the model, $\mathbf{Y}_t$:

Difference to be minimized for best fit $= \sum_t (Y_t - \mathbf{Y}_t)^2 \qquad (20.6)$

The squared term has a double advantage:

- It ensures that positive and negative differences are taken in absolute terms and do not cancel out.

- It penalizes large deviations in favor of small ones so that the distribution of differences resembles a Normal distribution of random errors.

Regression analysis determines the constants of the equation that best fits the actual data. It minimizes the sum of squared differences by differentiating Eq. (20.6), with the model expressed in terms of the X_i replacing the $\mathbf{Y}_t$, with respect to the coefficients of the X_i. Setting these expressions equal to zero and solving gives the values of the coefficients that define the best-fit model. Forecasters can ignore these details, since spreadsheet programs such as Excel do linear regression analyses automatically upon request.

Spreadsheet programs also provide measures of how well the model fits the data. The most usual measure is R^2. Formally, R^2 represents the amount of variance in the Y_t accounted for by the model. $R^2 = 1.0$ indicates perfect fit. Any R^2 above 0.9 indicates good fit.

In practice, it is easy to get good fit to time-series data. This is because most statistics associated with humans change fairly constantly over time (for example, population grows or shrinks at a few percent a year, as does employment, income, travel, and so on). These statistics can thus be expressed reasonably well in a form similar to Eq. (20.4). Therefore, any two time series (say, C_1 and D_1) can in general be easily expressed in terms of each other, regardless of their relative rates of growth; for example,

$$C_0(e)^{C_1 T} = [C_0/D_0 \ (e)^{C_1/D_1 T}] \ D_0 \ (e)^{D_1 T} \qquad (20.7)$$

This means that it is easy to obtain good correlation, with high R^2, between factors that have nothing to do with each other. Past efforts have shown good correlation between airport traffic and such diverse factors as the egg production of New Zealand or the prison population of the state of Texas. Correlation is not causality, however: good fit does not necessarily imply that a model is meaningful.

20-3 Forecasts

Once the analysts have developed a model to fit the past data, they can use it to forecast future levels of traffic. Thus, one of the models that the consultants fitted to the Los Angeles data was similar to Eq. (20.2):

$$\text{Domestic passengers} = \qquad (20.8)$$
$$(\text{population})[-3074.4917 + 0.1951 \ (\text{income}) + 8713.6280 \ (\text{yield})]$$

They used this to project possible future levels of traffic for Los Angeles, by inserting forecasts of the future values of the population, per-capita income, and airline yield for the following 20 years.

As indicated right at the start, the standard forecasting process makes a simple problem more complex. To get the forecast of passengers, the process for Los Angeles had to predict three other variables. This is not easy, especially as the experts in those areas are not agreed. Thus, four reputable planning agencies had four different estimates of future levels of per-capita income. Moreover, they were not even agreed on past levels, as Table 20-2 demonstrates.

The bottom line is that an honest forecasting process leads to a wide range of results. This is because it is possible to fit several models [such as Eqs. (20.2)—(20.4)] to any specific span of data, and because there are many forecasts of the secondary factors. For example, Table 20-3 provides some of the possible forecasts for domestic passengers between 1995 and 2015. The spread of forecasts in this case is between 89.2 and 52.7 million. The range is 36.5 million, or between 41 and 70 percent of the value. This is comparable to what one expects in practice, based on retrospective analyses of the accuracy of forecasts, as Chap. 3 indicates. To put the matter simply, the analysis does not provide confidence in the first digit, let alone the second or third decimal.

Table 20-2. Per capita personal income (PCPI) for Los Angeles as reported and projected by different planning agencies (National Planning Associates, Woods and Poole Economics, Regional Economic Models, Inc., and the Southern California Association of Governments) (SCAG data for 1995 and 2005 interpolated from original data)

	LA Region PCPI in Constant 1987 Dollars			
Year	NPA	W&P	REMI	SCAG
1975	$14,234	$14,231	$14,828	$13,832
1980	$16,528	$16,528	$16,563	$15,200
1985	$17,669	$17,669	$17,870	$15,870
1990	$17,984	$17,984	$17,658	$16,540
1995	$18,175	$17,641	$17,764	$18,865
2000	$19,362	$19,276	$19,206	$21,190
2005	$19,985	$21,113	$20,411	$23,750
2010	$20,988	$23,082	$21,180	$26,310
2015	$21,873	$25,174	$21,723	$26,550

Source: Landrum and Brown, 1996.

The only way to develop a single forecast is to apply judgment. Mathematics or analysis by themselves cannot resolve the differences between the outcomes of the formulas. Judgment and opinion are essential parts of the process, both to select the mathematical forms and to reconcile the discrepancies between possible results. Thus, for Los Angeles and elsewhere, the selection of a single number as the forecast for 20 years ahead represents some effort to pick an acceptable value, a middle value, or an allegedly most likely value. It is an artful, not a scientific decision.

The case of Miami/International illustrates how art rather than analysis ultimately drives forecasts in practice. The consulting team prepared a range of 30-year forecasts for domestic passengers (Table 20-4). They generated these forecasts by assuming

- Different spans of data (either Dade County alone, Dade augmented by Broward County, time series for Miami/International, and time series for the entire United States)
- Different variables (population, yields, and per-capita personal income)

Table 20-3. Forecasts for domestic passengers for Los Angeles generated by several models associated with the master planning effort

Year	Actual data	Master plan	Plan without employee factor	Yield to power	Exponential
1975	19,370,427				
1980	29,545,572				
1985	34,295,720				
1990	35,756,429				
1995	38,023,561	38,023,561	38,023,561	38,023,561	38,023,561
2000		50,875,069	49,372,075	45,076,610	48,513,473
2005		63,987,762	61,770,188	47,385,522	56,588,900
2010		78,543,601	75,327,580	49,921,458	66,008,543
2015		89,157,117	85,408,668	52,718,434	76,996,154

Source: Landrum and Brown, 1996.

Table 20-4. Preferred forecast selected from range of forecasts, for domestic passengers at Miami/International

Forecast method and variant		Forecast	Actual	
Method	**Data used (form)**	**2020**	**1990**	**2000**
Population	Dade County	13.96		
	Dade and Broward	15.35		
	Dade and Broward (nonlinear)	17.74		
Yield and per-capita personal income	Dade County	19.87		
	Dade and Broward	19.69		
	Dade and Broward (nonlinear)	19.13		
Time series	Dade County	17.41		
	Dade and Broward	18.67		
	Dade and Broward (nonlinear)	40.05	**9.92**	**17.4**
Per-capita personal income	Dade County	26.58		
	Dade and Broward	24.34		
	Dade and Broward (nonlinear)	42.40		
Share of U.S.		23.48		
	Maximum	42.40		
	Average	22.97		
	Medium	19.69		
	Minimum	13.96		
	Preferred	**15.35**		

Source: Landrum and Brown, 1992; Miami Airport, 2001.

- Different forms of the variables (linear and nonlinear)
- Different forms of the equations (the five different ones listed)

Once they had these 13 different forecasts, the forecasters decided to select one. In this case, their judgment led them to a "preferred" forecast for the year 2020 (a linear correlation with the regional population). Having to select a single forecast from the range of possibilities is not a scientific process.

In the case of Miami/International, note that the actual traffic in the year 2000 already surpassed the level of the preferred forecast for the year 2020. As stressed in Chap. 3, individual forecasts are "always wrong."

20-4 Scenarios

The mathematical approach to forecasting presented in the previous sections is not appropriate to long-term forecasts. Regression and other statistical analyses presume that past trends continue. In the short run, over a few years, the inertia in the system is certainly likely to continue existing trends. In the longer run, however, over 10–20 years, the assumption that trends continue indefinitely is likely to be false. Newer trends eventually displace older ones. For longer-run forecasts, it is thus important to use judgment explicitly to develop forecasts.

A *scenario,* in terms of forecasting, is a concept of what might happen at some future time. Analysts develop scenarios to create a context that defines a long-term forecast. Scenarios can deal with all the factors relevant to airport development. For example, they might focus on the following:

- Macroeconomics, which might suggest the growth or fall of local industry
- Demographics, such as the stagnation and aging of the population and of the travelers
- Competitive airports, whose congestion might lead to a shift in traffic
- Industry structure, where consolidation of airlines might change the location of traffic hubs
- Environmental and resource constraints, which might inhibit growth
- Saturation of demand, in line with experience elsewhere

Analysts develop scenarios in consultation with experts in the several fields of interest. The scenarios thus represent judgments about the future.

Forecasters use scenarios to provide a rationale for bending trends into new directions. They might argue, for example, that a new airport such as Kuala Lumpur/International or Athens/Venizelos would attract new traffic to a region because it provides more competitive, less congested facilities that will attract airlines. Alternatively, they might suggest that recent rapid rates of growth would become lower, due to saturation of the market for air travel. Scenarios provide the means to override the continuation of past trends that seem obsolete.

20-5 Integrated procedure

A responsible forecasting procedure recognizes that

- Careful analysis of data is important
- Judgments are an integral part of the exercise

It will thus not rely exclusively on either a mathematical analysis, or pure judgment and opinion. It will combine both appropriately.

Good forecasting will also acknowledge that current trends are likely to dominate during the near term. Over the short term, trends persist due to the inertia in the air transportation system. Passengers and shippers maintain their habits and relations. Airlines change their operations and fleets slowly. In the longer run, however, new trends will emerge. These new patterns can only be guessed at, since they do not constitute a major portion of the existing trends. Analysis should thus dominate short-term forecasts and judgment the long-term estimates.

The recommended procedure balances analysis and judgment (de Neufville et al., 1980). It has the following five elements:

1. *Obtain and verify data on past traffic and relevant factors.* These observations should be carefully examined for consistency and correctness, along the lines indicated in Sec. 18-1.

2. *Do regression analyses to develop a model of traffic.* Examine several forms of models [such as Eqs. (20.2)–(20.4)] over different spans of time. Use judgment to decide which models are most suitable for local conditions.

3. *Project statistical models over the short term (5–10 years).* It is a good idea to look at the range of forecasts implied by the most relevant models developed in step 2; the analyst can then estimate a middle forecast, and a range of possible outcomes.

4. *Develop scenarios of future conditions suitable to local region and situation.*

5. *Estimate long-term (10–20 years) forecasts with wide ranges.* Do this by using the scenarios of step 4 to modify the short-term trends. In recognition of the unavoidable uncertainty in

guesses at the future, be sure to associate wide ranges with the median long-term forecasts. Based on past experience, suitable ranges on 20-year forecasts are about plus or minus 30 percent (see Chap. 3).

Exercises

1. Obtain traffic data for an airport of interest. Use a spreadsheet program to establish a trend line using one or all of Eqs. (20.2)–(20.4). Project these trend lines forward—how do they differ?

2. Repeat Exercise 1, using a different span of data (for example, 10 years back instead of 15 or 20). How do the results differ from those obtained in Exercise 1? Think about which is the appropriate span of data to consider, considering the relevant local circumstances.

3. Think about how you would obtain data on future prices, income levels, and other factors that might be relevant to a model of air traffic. How would you obtain reliable, credible forecasts of these factors?

4. What scenarios are relevant to the future levels of traffic for some airport of interest? How might these affect the forecasts? How would you develop consensus that these scenarios are appropriate for your airport and region?

References

de Neufville, R., Zuniga, S., Kanafani, A., and Olivera, A. (1980) "Forecasting Airport Traffic: Mexico City as a Case Study," *Transportation Research Record, 732,* pp. 24–29.

ICAO, International Civil Aviation Organization (1985) *Manual on Air Traffic Forecasting,* 2d ed., Doc. 8991-AT/722/2, ICAO, Montreal, Canada.

Kanafani, A. (1983) *Transportation Demand Analysis,* McGraw-Hill, New York.

Landrum and Brown (1992) *Draft Master Plan for Miami International Airport,* Miami, FL.

Landrum and Brown (1996) *LAX Master Plan—Chapter III, Forecasts of Aviation Demand,* prepared for Los Angeles Department of Airports, Los Angeles, CA.

Lewis-Beck, M. (1980) *Applied Regression: An Introduction,* Sage, Beverly Hills, CA.

Makridakis, S., and Wheelwright, S. (1989) *Forecasting Methods for Management,* 5th ed., Wiley, New York.

Miami Airport (2001) "Traffic Statistics," http://www.miami-airport.com/.

Pindyck, R., and Rubinfeld, D. (1998) *Econometric Models and Economic Forecasts,* 4th ed., McGraw-Hill, New York.

21

Cash flow analysis

Capital investments into airport facilities, services, and equipment generate cash flows that occur over many years. *Cash flow analysis* refers to the set of techniques used to compare the economic performance of alternative investments.

The concept of the *discount rate* is the key to performing these comparisons. It converts *present dollars* into *future dollars* and vice versa. The discount rate used to evaluate any project should reflect the opportunity cost associated with that project. The present value and the annual value of any stream of costs or benefits are computed by means of discount factors, which are easily derived from the fundamental relationship of discounting.

The three most commonly used measures of economic performance of a project are the net present value, the benefit–cost ratio, and the internal rate of return. Each of these measures has its strengths and weaknesses, depending on the case at hand. Simpler measures, such as the payback period and cost-effectiveness ratios, are also used occasionally in practice. When comparing among a set of alternative projects, the recommended criterion is to select that set of projects which maximizes net present value, subject to satisfying any economic and technical constraints present.

21-1 Introduction

Airport operators and planners must often compare the economic performance of alternative capital investments into airport facilities and equipment. These investments typically generate *cash flows* (or *streams*) of revenues/benefits and of expenditures/costs that occur over many years. For example, in planning a new runway, airport operators should determine whether the estimated future flow of revenues from landing fees over the economic lifetime of the runway, typically 20–50 years, will be sufficient to cover the flow of costs associated with building, maintaining, and operating the runway, including any interest paid on borrowed capital. Another way to make this assessment is to compute the landing fee rate (e.g., the charge per 1000 kg of takeoff weight) required to pay back the investment in the new runway plus its variable costs. If this computed rate is deemed reasonable, then the runway will be considered economically viable. If the rate is excessive, compared to what has been charged in the past or to what the airport's competitors charge, then the project may be rejected.

No matter the precise terms in which such questions are posed, one common theme runs through them. They require intertemporal comparisons between amounts of money disbursed or received at different times. Such comparisons are the subject of cash flow analysis, one of the principal topics of cost–benefit analysis, of engineering economy, and of finance theory. This chapter reviews briefly this subject at an introductory level. Far more extensive presentations are provided in many textbooks, for example, Brealey and Myers (1993), de Neufville (1990), Steiner (1993), Sullivan et al. (1999), and White et al. (1998).

21-2   Discounting and the discount rate

Discounting is the fundamental concept used in making intertemporal comparisons between flows of costs and benefits occurring at different times. Discounting makes these flows "commensurable" and permits comparisons among them. The discount rate is the instrument used to convert present dollars into future dollars and vice versa. Typically denoted as i, the *discount rate* is defined as the marginal rate of substitution between one dollar at $t = j$ and one dollar at $t = j + 1$, for any integer value of j. Specifically, given a discount rate i, it is assumed that decision makers are indifferent between \$1 at $t = j + 1$ and \$ $1/(1 + i)$ at $t = j$ [or, conversely, between \$1 at $t = j$ and \$$(1 + i)$ at $t = j + 1$].

This definition leads, by induction, to the following fundamental relationship:

$$\text{\$1 at } t = n \text{ is equivalent to } \$\frac{1}{(1 + i)^n} \text{ at } t = 0 \qquad (21.1a)$$

or, more generally, $\$Q$ at $t = n$ has a *present value, PW (Q)*, given by

$$PW(Q) = \$\frac{Q}{(1 + i)^n} \qquad (21.1b)$$

The unit of time in the above is most often defined as 1 year and, in that context, i is the *annual discount rate*. However, any other unit of time may also be specified, with the concepts and results being exactly the same. Note, as well, that the discount rate, i, may often be a function of time; i.e., different values of i may apply at different times t.

In Eq. (21.1), and in similar formulas given below, the discount rate i is expressed as a decimal fraction—e.g., an 8 percent discount rate means $i = 0.08$. Thus, at a 10 percent annual discount rate, an amount of $Q = \$1$ million at $t = 6$ years from now has present value $PW(Q)$ $= \$(1\ \mathrm{M})/(1.1)^6 = \$564,474$, i.e., is equivalent to $\$564,474$ today (at $t = 0$).

Clearly, the discount rate can be interpreted as an "interest rate" of a more general nature. The mathematics of discounting is, in fact, identical to the traditional mathematics of financial transactions involving compound interest, as suggested by Eq. (21.1). However, the context of the analysis may be far broader. In fact, depending on the application area, the discount rate (or "marginal rate of substitution") may refer to:

- An applicable *interest rate,* as in the case of a family that takes a bank loan to purchase a home
- The *opportunity cost of capital,* i.e., the rate of return that capital invested in a program or project might otherwise have generated
- A *reasonable rate of return* that, e.g., the shareholders of a company would find acceptable (also referred to as the *minimum acceptable rate of return)*
- The *social discount rate,* which is applied by government organizations in evaluating public projects/programs.*

The notion underlying all these interpretations is that the discount rate should reflect *opportunity cost,* i.e., the rate of return associated with opportunities that are foregone in order to undertake a particular

---

*The social discount rate is usually computed, at least in principle, by estimating the rate of return that resources expended by the government would otherwise have fetched if utilized by the private sector.

project, program, or activity. The discount rate thus serves to limit the alternatives under consideration in a cash flow analysis to the set of any projects proposed (e.g., a new passenger building of size X versus one of size Y) plus the "do nothing" alternative. The "do nothing" (or *null*) alternative is then implicitly assigned a rate of return equal to the discount rate in the cash flow analysis.* This important point will be discussed further in Section 21-5.

21-3 Present and annual value of monetary flows

An important notion in cash flow analysis is that of the present value of a flow (or stream) of costs or benefits. Let $X(T) = \{X_0, X_1, X_2,...,X_T\}$ be a stream of $T + 1$ monetary amounts over T consecutive years, with X_j denoting the amount associated with the end of year j. Note that X_0 is the amount associated with $t = 0$, i.e., with the beginning of the first year. Let i be the discount rate. Then the *present value* of the stream, $X(T)$, denoted as $PW[X(T)]$, is the sum of the present values, $PW(X_j)$, of each of the amounts X_j that compose $X(T)$. Stated differently, $PW[X(T)]$ is equal to the amount of money that, if deposited in a bank at $t = 0$ at an interest rate equal to the discount rate i, would generate a stream of payments identical to the stream $X(T)$. Formally:

$$PW[X(T)] = X_0 + \frac{X_1}{(1 + i)} + \frac{X_2}{(1 + i)^2} + \cdots + \frac{X_T}{(1 + i)^T} \qquad (21.2)$$

$$= \sum_{j=1}^{T} \frac{X_j}{(1 + i)^j}$$

In the context of the economic evaluation of engineering projects, if the amount X_j ($j = 0, 1,...,T$) is positive, it is usually referred to as a "benefit," and if it is negative it is a "cost." T is called the *lifetime* (or *economic lifetime*) of the project.

Another very common operation in cash flow analysis involves the computation of *annual value*. If R is a monetary amount at time $t = 0$, its annual value, $AW(R)$, for a specified period of n years and a discount rate of i is defined by the equation

*This use of the discount rate is particularly important in the case of large organizations that wish to decentralize decision making. The "social discount rate" serves this purpose for government.

$$R = \frac{AW(R)}{(1 + i)} + \frac{AW(R)}{(1 + i)^2} + \cdots + \frac{AW(R)}{(1 + i)^n} \qquad (21.3a)$$

$$= AW(R)\left[\frac{1}{(1 + i)} + \frac{1}{(1 + i)^2} + \cdots + \frac{1}{(1 + i)^n}\right]$$

$$= AW(R)\left[\sum_{k=1}^{n}\left(\frac{1}{(1 + i)^k}\right)\right]$$

Intuitively, if n annual payments, each equal to $AW(R)$, are made at an interest rate equal to i, and with the first payment at $t = 1$, the second at $t = 2$, and the last at $t = n$, then the present value of these payments will be equal to R. Note that, by convention, it is assumed that the first payment occurs at $t = 1$ (*not* at $t = 0$) for a total of n equal annual payments, as shown in Fig. 21-1. Each payment is equal to $AW(R)$. Such a series of n equal annual payments is often referred to as an *annuity*.

After some algebraic manipulation, the term in brackets on the right-hand side of Eq. (21.3a) can be rewritten in a form that is particularly easy for computation:

$$R = AW(R)\left[\sum_{k=1}^{n}\left(\frac{1}{(1 + i)^k}\right)\right] = AW(R)\left[\frac{(1 + i)^T - 1}{i(1 + i)^T}\right] \qquad (21.3b)$$

It is now natural to extend the notion of annual value to *flows of costs or benefits*. As before, let $X(T) = \{X_0, X_1, X_2, \ldots, X_T\}$ be a stream of $T + 1$ monetary amounts over T consecutive years, with X_j denoting the amount associated with the end of year j. Let i be the discount rate. Then, the annual value of the flow, $X(T)$, denoted as $AW[X(T)]$, can be computed by simply multiplying $PW[X(T)]$, the present value of

Fig. 21-1. *An annuity consisting of a stream of* n *equal payments that begins at* t = 1 *and ends at* t = n. *Note that there is no payment at* t = 0.

$X(T)$ as computed in Eq. (21.2), by the inverse of the term in brackets on the right-hand side of Eq. (21.3b):

$$AW[X(T)] = PW[X(T)] \left[\frac{i(1 + i)^T}{(1 + i)^T - 1} \right] \quad (21.4)$$

The application of Eqs. (21.2)–(21.4) is illustrated in Example 21.1.

EXAMPLE 21.1

Assume an airport project with an economic lifetime of 10 years requires that three payments of $5, $10, and $15 million be made at the end of years 3, 6, and 10, respectively, as shown in the second column of Table 21-1.

If the discount rate is 10 percent, the present value of this schedule of payments is found from Eq. (21.2) to be equal to approximately $15.18 million (at $t = 0$, i.e., on the first day of the first year of the project). The annual value of the schedule of payments, from Eq. (21.4), is $2.47 million at the end of years 1–10. Thus, the three streams of payments shown in columns 2–4 of Table 21-4 are economically equivalent at the 10 percent annual discount rate. This would indeed be true in an environment in which the airport operator undertaking this particular project can borrow, without restrictions, at a 10 percent annual interest rate.

Table 21-1. Three streams of payments, which are equivalent at a 10% discount rate

Year	Original schedule ($ million)	Present value ($ million)	Annual value ($ million)
0	0	15.18	0
1	0	0	2.47
2	0	0	2.47
3	5	0	2.47
4	0	0	2.47
5	0	0	2.47
6	10	0	2.47
7	0	0	2.47
8	0	0	2.47
9	0	0	2.47
10	15	0	2.47

Expressions (21.2)-(21.4) are essential in estimating the costs of borrowing, through the bond market or from financial institutions, in order to finance airport projects. This is illustrated in Example 21.2.

21-4   Notes on computing

The following comments on performing cash flow computations will be useful at this point.

EXAMPLE 21.2

At $t = 0$ the operator, AO, of a large airport borrowed $100 million through a short-term loan from a consortium of banks. The interest rate on the loan is 10 percent per year and the loan must be paid back in five equal annual installments, the first of which takes place at the end of year 1 ($t = 1$) and the last at the end of year 5.

Applying Eq. (21.3b) with $AW(R)$ as the unknown and $R = \$100$ million,

$$AW(R) = R\left[\frac{i(1 + i)^T}{(1 + i)^T - 1}\right] = (\$100 \text{ M})\left[\frac{(0.1)(1.1)^5}{(1.1)^5 - 1}\right]$$

$$\approx \$26.38 \text{ million}$$

Thus, the loan will be paid back in five equal annual payments of $26.38 million each, beginning at $t = 1$. These payments include both interest and paying back of the principal of $100 million.

Table 21-2 shows how much of each installment goes toward paying interest on the loan and how much is used to pay back the principal on a year-to-year basis. At the end of the first year $10 million is due as interest [= ($100 M)(0.10)]. This means that of the $26.38 M of the first installment, $10 million will be allocated to interest and the remaining $16.38 million to retirement of the outstanding principal of $100 million. This will leave an outstanding principal of $83.62 million (= $100 − 16.38$) at the end of year 1. As a result, the interest due at the end of the second year is $8.36 million [= ($83.62 M)(0.10)]. The remainder of Table 21-2 is self-explanatory. Note that the part of each installment that goes toward payment of interest diminishes from year to year and that AO will pay a total of $31.9 million in interest during the 5-year period.

Table 21-2. Schedule and allocation of installment payments for AO's $100 million loan (all amounts in millions of dollars)

	Year 1	Year 2	Year 3	Year 4	Year 5	Total of row
Installment amount	$26.38	$26.38	$26.38	$26.38	$26.38	$131.9
Interest	10.0	8.36	6.56	4.58	2.4	31.9
Payback of principal	16.38	18.02	19.8	21.80	23.98	100.0
Principal remaining	83.62	65.60	45.78	23.98	0	

1. Cash flow analysis can be performed very efficiently through computer spreadsheets. Expressions like

$$\frac{1}{(1 + i)^n} \quad or \quad \frac{(1 + i)^n - 1}{i(1 + i)^n}$$

often referred to as "discounting formulas," are programmed as standard functions in Excel, Lotus 1-2-3, etc., as well as on many hand-held calculators. Even better, for any specified flow of costs and benefits, spreadsheet users can call on standard functions to compute easily the flow's present and annual value [see Eqs. (21.2) and (21.4), respectively] as well as such measures of effectiveness as net present value, cost–benefit ratio, and internal rate of return (see Sec. 21-5).

2. The terms "annuity" and "annual value" should not be interpreted to mean that the operations performed by Eqs. (21.3) and (21.4) can only refer to time units of years. Exactly, the same expressions can be used to compute any sequence of n equal payments at the end of each of n consecutive equal intervals, 1 through n, whose present value is equal to a specified amount. The length of each of these n equal intervals can be a month, a quarter, a day, etc.

3. In deriving all the discounting formulas so far, it has been assumed that *discrete compounding* is used. In other words, discounting takes place at the end of specified discrete, evenly spaced time intervals of 1 year, 1 month, 1 day, etc. In the limiting case—i.e., when the time intervals become infinitesimally small—compounding becomes *continuous*. For every formula referring to discrete compounding, an analogous formula assuming continuous compounding exists. All these formulas are based on the observation that, if i is the *nominal* discount rate per time period (e.g., per year) and if compounding takes place at m evenly spaced intervals during that period, each time at the rate of i/m, \$1 at $t = 0$ is equivalent to \$ $[1 + (i/m)]^m$ at $t = 1$. In other words, the *effective* discount rate is given by

$$i_{\text{eff}} = \left(1 + \frac{i}{m}\right)^m - 1 \qquad (21.5)$$

When m becomes infinite (i.e., when continuous compounding is in effect), a well-known limit from calculus is useful:

$$\lim_{m \to \infty} \left(1 + \frac{i}{m}\right)^m = e^i \qquad (21.6)$$

Thus, \$1 at $t = 0$ is equivalent to \$$e^i$ at $t = 1$ under continuous compounding, and the effective discount rate is given by

$$i_{\text{eff}} = e^i - 1 \qquad (21.7)$$

Example 21.3 illustrates Eqs. (21.5)–(21.7).

Using Eq. (21.6), expressions for discounting under continuous compounding can be easily derived. For example, given an amount

EXAMPLE 21.3

Assume that the nominal annual discount rate is 9 percent. If compounding takes place once a year, \$1 at $t = 0$ (i.e., at the beginning of the year) is equivalent to \$1.09 at the end of the year ($t = 1$). The nominal *and* the effective annual rate are equal to 9 percent in this case.

If compounding takes place on a monthly basis, there are 12 compounding points during the year at which a (monthly) discount rate of 0.75% ($= 9/12$) is applied. This means that \$1 at $t = 0$ will be equivalent to \$1.0938 ($= 1.0075^{12}$). The nominal annual discount rate is 9 percent, but the effective annual rate is 9.38 percent.

With continuous compounding, \$1 at $t = 0$ will be equivalent to \$1.0942 ($= e^{0.09}$) at the end of the year. The effective annual rate is 9.42 percent.

of $\$Q$ at $t = n$ years with an annual nominal discount rate of i and continuous compounding, one obtains for the present value, $PW(Q)$,

$$PW(Q) = \$\frac{Q}{e^{i \cdot n}} \tag{21.8}$$

Equation (21.8) is the continuous compounding equivalent of Eq. (21.1b). Similarly, given an amount of $\$R$ at $t = 0$, its annual value, $AW(R)$, over a period of n years with continuous compounding [compare with Eq. (21.4)] is obtained from

$$AW(R) = R \cdot \frac{e^{i \cdot n}(e^i - 1)}{(e^{i \cdot n} - 1)} \tag{21.9}$$

21-5   Measures of project effectiveness

Several alternative measures exist for summarizing and comparing the economic effectiveness of projects or programs that involve flows of costs and benefits over time. The three most commonly used of these measures will now be defined and their relative merits will be discussed. A couple of more informal measures are also described briefly.

Consider an airport capital improvement project (e.g., the construction of a new passenger building) with an economic lifetime of T years. The project is forecast to generate flows of costs, $C(T) = \{C_0, C_1,...,C_T\}$, and of benefits, $B(T) = \{B_0, B_1,...,B_T\}$ during its lifetime—with the subscript in each case indicating the year when each cost or benefit will occur. (Usually $B_0 = 0$, i.e., benefits do not commence until the end of the first year of the project; B_T may include the *salvage value* of the project, i.e., it may reflect whatever economic value is left at the end of the project's economic lifetime.)

Let the discount rate be equal to i and assume discrete yearly compounding. The most commonly used measures of the economic effectiveness of projects are the following.

- The *net present value, NPV,* defined as the difference between the present value of the benefits, $PW[B(T)]$, and the present value of the costs, $PW[C(T)]$:

$$NPV = PW[B(T)] - PW[C(T)] = \sum_{j=0}^{T} \frac{B_j - C_j}{(1 + i)^j} \tag{21.10}$$

- The *benefit–cost ratio, B/C,* defined as the ratio of the present value of the benefits, *PW[B(T)],* divided by the present value of the costs, *PW[C(T)]:*

$$B/C = \frac{PW[B(T)]}{PW[C(T)]} = \frac{\sum_{j=0}^{T} \dfrac{B_j}{(1 + i)^j}}{\sum_{j=0}^{T} \dfrac{C_j}{(1 + i)^j}} \qquad (21.11)$$

- The *internal rate of return, IRR,* defined as the discount rate at which the present value of the benefits is equal to the present value of the costs, i.e., the discount rate at which

$$PW[B(T)] = PW[C(T)] \qquad (21.12)$$

Equation (21.12) is equivalent to setting *IRR* equal to the value of the discount rate, x, that solves the equation

$$\sum_{j=0}^{T} \frac{(B_j - C_j)}{(1 + x)^j} = 0 \qquad (21.13)$$

Solving Eq. (21.13) is, of course, equivalent to finding the discount rate at which *NPV* = 0. Note that the unknown in Eq. (21.12) or (21.13) is the discount rate. In other words, the value of *IRR* is determined solely by the flow of benefits, *B(T),* and costs, *C(T),* and is thus independent of the prevailing discount rate i. The term "internal rate of return" stems, in fact, from this observation, namely, that *IRR* is determined by the "internal" economic characteristics of each project.

Intuitively, the net present value (*NPV*) corresponds roughly to the notion of "profit." It estimates the amount by which benefits/revenues associated with a project will exceed (or fall short of) costs/expenditures, when both costs and benefits are discounted to $t = 0$ at a given discount rate i. Recall (Sec. 21-2) that the discount rate typically should reflect the opportunity cost associated with the capital and other resources needed to carry out a project. Thus, if a proposed project A has an *NPV* value of 0, this means that A earns an average rate of return equal to i during its economic lifetime. It is thus more accurate to say that *NPV* indicates how much "profit" a project will make *over and above* what is needed to cover its opportunity costs. It follows that the *critical value* of *NPV* is $0: if *NPV* is greater than $0, this means that the project earns a rate of return that exceeds the rate of return that the resources it utilizes would earn elsewhere. The converse is true if *NPV* is less than $0.

For the same reason, the critical value of the benefit–cost *(B/C)* ratio, which essentially measures the "economic cost-effectiveness" of investing in any particular project, is 1.0. The *B/C* ratio measures how many dollars one gets back for every dollar invested in the project *after* discounting at the applicable discount rate, i.e., after taking opportunity cost into consideration.

Finally, in the case of the internal rate of return *(IRR),* the critical value is the applicable discount rate *i*. If *IRR* is greater than *i,* the project under consideration has a higher rate of return than other opportunities elsewhere and is therefore desirable.

If, for a given project, any one of *NPV, B/C,* or *IRR* exceeds its critical value ($0, 1.0, or *i,* respectively), then the other two measures will exceed their critical values as well. This can be seen from the definitions of the measures. However, when it comes to *comparing* among alternative projects, the three measures may give contradictory recommendations, as Example 21.4 shows.

EXAMPLE 21.4

Consider three proposed projects, A, B, and C. The three are mutually exclusive: if one of them is accepted, the other two must be rejected. Each project has an economic life of 4 years and flows of costs and benefits as shown in Table 21-3. The discount rate to be used is 6 percent. Which one, if any, of the projects should be chosen?

NPV, B/C, and *IRR* can be computed from Eqs. (21.10), (21.11), and (21.12), respectively, for each of the three alternatives. For instance, for alternative A:

$$PW(\text{benefits of A}) = \frac{40}{(1.06)} + \frac{40}{(1.06)^2} + \frac{40}{(1.06)^3} + \frac{300}{(1.06)^4}$$
$$\approx \$345,000$$
$$NPV(A) = PW(\text{benefits of A}) - PW(\text{costs of A})$$
$$\approx 345,000 - 200,000 = \$145,000$$
$$B/C(A) \approx 345/200 = 1.725$$

**Table 21-3. Flow of costs and benefits
for three mutually exclusive alternatives
A, B, and C (in thousands of dollars)**

	A	B	C
Initial cost ($t = 0$)	$-200	$-220	$-310
Year 1	40	100	90
Year 2	40	100	90
Year 3	40	100	90
Year 4	300	100	350

**Table 21-4. Summary of measures of
effectiveness for the three alternatives**

Alternative	PW (benefits)	PW (costs)	B/C	IRR	NPV
A	$345,000	$200,000	1.725	~25%	$145,000
B	347,000	220,000	1.58	~29%	127,000
C	518,000	310,000	1.67	~27%	208,000

To find *IRR*(A), the equation

$$\frac{40}{(1 + x)} + \frac{40}{(1 + x)^2} + \frac{40}{(1 + x)^3} + \frac{300}{(1 + x)^4} = 200$$

is solved for x to obtain $x \approx 0.25$, i.e., *IRR*(A)≈25 percent.

Table 21-4 summarizes the measures of effectiveness for all three alternatives. Note that these three measures provide three different recommendations, as A has the highest *B/C* ratio, B the highest *IRR*, and C the largest *NPV*.

What happens in such cases? The answer—subject to some qualifications to be discussed below—is that *NPV should be the dominant decision-making criterion in most instances.* For instance, in Table 21-4, alternative A has the highest benefit–cost ratio of 1.725. However, it can be seen that C has an additional present value of costs equal to $110,000 (= $310,000 − $200,000) compared to A, but an additional present value of benefits equal to $173,000 (= $518,000 − $345,000). While it is indeed true that A offers the highest *B/C* ratio, it is also clear that, as long as it

is possible to spend $310,000 or more at $t = 0$, it is worth
spending the additional $110,000 to obtain benefits worth
$173,000. In fact, the "marginal" B/C ratio associated with aban-
doning A in favor of C is equal to 1.57 (= $173,000/$110,000);
i.e., every additional dollar spent provides $1.57 in benefits—
certainly a most worthwhile proposition. Remember that the
discount rate of 6 percent represents, in principle, the rate of
return that can be obtained by investing resources in projects
other than A, B, or C. An investment that has exactly a 6 percent rate of return will have a B/C ratio equal to 1.0. The B/C
ratio of 1.57 indicates that the additional ("marginal") resources
needed to undertake C rather than A earn more than the 6 percent return they would have earned elsewhere. An entirely
similar argument can be used to justify why alternative C, with
the largest NPV, should be preferred over alternative B, which
has a higher IRR.

In general, when faced with the task of choosing among a number
of alternative projects, one should select that subset of projects that,
at the given discount rate, maximizes total NPV, subject to satisfying
any existing economic or technical constraints. An upper limit on the
total amount that can be invested (i.e., a "budget") is an example of an
economic constraint: obviously, the total cost of the subset of projects
selected should not exceed that limit. Examples of technical constraints
may be the mutual exclusivity of two projects (i.e., one project cannot
be included in the selected subset, if the other has been included) or
a technical relationship between two projects that would make one
project a prerequisite to the other (e.g., "B cannot be selected unless
A is selected"). It can be shown that selecting the alternative (or set
of alternatives) that maximizes total NPV is equivalent to following
the "marginal B/C ratio" approach, described in Example 21.4, to select
among alternative investments. But, absent a budget constraint, this
last approach is equivalent to investing up to the last dollar that
earns more than one dollar back in revenues, after discounting. It is
a fundamental result of economics that this is also the level of investment that maximizes economic welfare. In the presence of a budget
constraint, the "marginal B/C ratio" approach provides an optimal
stopping rule for maximizing economic welfare. Thus, the criterion
of maximizing total NPV, subject to satisfying any existing economic
or technical constraints is consistent with maximizing economic welfare under any set of conditions.

Several cautionary notes should now be made regarding the above decision-making criterion. First, from the practical point of view, the "maximize NPV" criterion should be applied only in cases where the economic sizes of the alternatives under consideration are roughly similar. It really makes little sense to compare the "profit" (after discounting) associated with investing a few thousand dollars versus many millions.

Second, in order to compute the NPV (as well as the B/C ratio) of a project, the discount rate must be specified. But, in some cases, the discount rate may not be known. The IRR can be useful in such cases since, as mentioned earlier, it is independent of the discount rate. It may be safe to select under such circumstances projects whose IRR is very high. This will make it highly probable that the NPV of the selected project or projects will be positive, no matter what the (yet unknown) discount rate actually turns out to be. Readers should also be aware, however, that Eq. (21.13), which must be solved to determine IRR, may in some cases* yield more than one value for IRR. When this is the case, it is essentially meaningless to talk about the IRR of a project. A statement such as "the project has an internal rate of return of either 5 percent or 12 percent" has little practical value.

Use of the B/C ratio as a selection criterion should also be avoided. The B/C ratio suffers from two obvious disadvantages. It favors projects with small costs, since small denominators produce high ratios; and it can be easily manipulated in how benefits and costs are recorded to produce higher or lower B/C ratios.[†]

A third, more important, caveat is that NPV is an aggregate measure that simply compares the present value of the flow of project costs against the present value of project benefits. NPV is not concerned with how these flows are distributed over time. For example, it is very possible that a positive NPV will be computed for an airport project that incurs high costs for several years at the beginning of its economic lifetime and generates even higher benefits toward the end of its lifetime. Despite having a positive NPV, a project with this type of distribution of costs and benefits over time may not be practically feasible. Operators of small airports, for instance, usually cannot afford the financial exposure and risk implied by strongly negative

---

[*]Equation (21.13) is a polynomial equation in x (the IRR). The necessary conditions under which Eq. (21.13) is solved by more than one positive values of x are well known (see, for example, Steiner, 1993).
[†]By contrast, NPV is a more "robust" measure in this respect.

cash flows accumulating over several years in a row.* Example 21.5, reflecting a situation that often arises in practice, illustrates this point.

Finally, one should not forget that forecasts of future economic performance of airport projects are fraught with uncertainty (see Chaps. 3 and 20). The computation of *NPV* and of the other measures of effectiveness introduced in this chapter has treated the forecasts of the various cash flows, as well as the discount rate as given constants when, in fact, they are usually subject to significant variability.†

EXAMPLE 21.5

An airport operator, AO, is considering construction of a new passenger building for which the construction cost is estimated to be $100 million. To raise this capital, AO will have to borrow money (in the bond market or from banks) at an effective interest rate of 8 percent per year (discrete annual compounding). Assume that the economic lifetime of the new building is 20 years and that the $100 million will have to be repaid in 20 equal annual payments ("annuity") due at the end of years 1, 2,…,20 after the opening of the building. (The payments will, of course, include interest on the borrowed capital.)

It is estimated that net annual operating revenue (= gross annual revenue − annual operating costs) will be $8 million in the first year of operations and will increase at 6 percent per year (in current prices) thereafter. (Operating costs include maintenance, operations, and administrative costs, but not capital costs.)

Table 21-5 lists the projected annual capital costs and net operating revenues for each year in the building's economic lifetime. Net annual cash flows are shown in the rightmost column. If AO is a state- or city-owned airport authority, as is typically the case in the United States, it is reasonable to assume that the discount rate it uses is approximately equal to the interest rate that it pays for borrowing funds in capital markets. A discount rate

---

*In principle, the undesirability of having a negative cash flow over a period of several consecutive years can be seen as another economic constraint that must be satisfied while maximizing *NPV*.

†In advanced analyses, *NPV* is often treated as a random variable (as are the other measures of effectiveness). The goal of such analyses is then to estimate the expected value and the variance (and, if possible, the probability distribution) of *NPV*, in order to make decisions that take uncertainty into consideration [see, e.g., Brealey and Myers (1993) or Luenberger (1998)].

**Table 21-5. Schedule of annual capital costs
and net operating revenues for Example 21-5**

Year	(1) Capital costs*	(2) Net operating revenue	(2) – (1) Net annual cash flow
1	$10,185,221	$8,000,000	($2,185,221)
2	10,185,221	8,480,000	(1,705,221)
3	10,185,221	8,988,800	(1,196,421)
4	10,185,221	9,528,128	(657,093)
5	10,185,221	10,099,816	(85,405)
6	10,185,221	10,705,805	520,584
7	10,185,221	11,348,153	1,162,932
8	10,185,221	12,029,042	1,843,821
9	10,185,221	12,750,785	2,565,564
10	10,185,221	13,515,832	3,330,611
11	10,185,221	14,326,782	4,141,561
12	10,185,221	15,186,388	5,001,167
13	10,185,221	16,097,572	5,912,351
14	10,185,221	17,063,426	6,878,205
15	10,185,221	18,087,232	7,902,011
16	10,185,221	19,172,466	8,987,245
17	10,185,221	20,322,813	10,137,592
18	10,185,221	21,542,182	11,356,961
19	10,185,221	22,834,713	12,649,492
20	10,185,221	24,204,796	14,019,575

*Capital costs include interest and repayment of principal.

equal to the interest rate of 8 percent will therefore be used to discount the net operating revenues from the building to time $t = 0$. One then obtains:

$PW(NR)$ = (present value of the stream of net revenues)
$= (\$8\ M)/(1.08) + (\$8.48\ M)/(1.08)^2 + \cdots$
$+ (\$24.204796\ M)/(1.08)^{20} \approx \$132.25$ million

The NPV of the building is therefore equal to ($132.25 M − $100 M =) $32.25 million. This can be interpreted as the present value of the "profit" that AO will make from the new building over the building's lifetime at a discount rate of 8 percent.

(Note that the present value of the annual capital costs in the second column of Table 21-5 is $100 million, by definition.)

By solving for x in the equation

$$(\$8\ M)/(1 + x) + (\$8.48\ M)/(1 + x)^2 + \cdots$$
$$+ (\$24.204796\ M)/(1 + x)^{20} \simeq \$100\ M$$

the internal rate of return, *IRR*, of the investment in the building can be computed. It is equal to approximately 11.1 percent. This means that the project has an annual average rate of return during its lifetime of about 11.1 percent, significantly higher than the discount rate (and cost of capital) of 8 percent per year. Note that for any discount rate less than 11.1 percent the project will have a positive *NPV*.

Although both its *NPV* and *IRR* are favorable, it should be noted that the project has significant risks. For example, note that during the first 5 years the project has a negative cash flow that adds up to a total of $-5,829,361 by the end of the fifth year. This amount has to be somehow "covered" by revenues from other sources or through additional borrowing. It can also be estimated that, at the 8 percent discount rate, the project does not break even until roughly 14.5 years into its 20-year lifetime; i.e., the present value of the annual net revenues becomes equal to $100 million only in the middle of the fifteenth year of the project. It is therefore conceivable that, given the usual uncertainties about the actual size of forecast revenues, some airport operators might decide against going forward with construction of the building or might postpone the project for a few years.

The difficulties that the strictly economic measures of project effectiveness, including *NPV*, run into—as illustrated by Example 21-5—have motivated the development of additional, less formal measures that are occasionally used in practice. One such measure is the *payback period*. This is defined as the amount of time it takes for the cumulative *undiscounted* net inflows of revenues into the project to equal the initial investment. In Example 21-5 the payback period is about 10 years, determined by simply adding the amounts in the net operating revenues column, beginning at $t = 1$ year until the total exceeds $100 million. The payback period is clearly a simplistic measure. Nevertheless, it is widely used by companies of all sizes in practice. One of its perceived advantages is that it eliminates from consideration projects with speculative benefits that occur far in the future.

A second, more substantive family of measures consists of various types of *cost-effectiveness ratios* that can be applied to the context at hand. For example, in the airport environment, alternative projects aimed at increasing airside capacity can be compared on the basis of a ratio such as "additional maximum throughput capacity per dollar spent" (Chap. 10) or, conversely, "dollars spent per unit of maximum throughput capacity gained." Analogous cost-effectiveness ratios can be devised for projects aimed at reducing noise impacts, increasing airport safety, increasing passenger handling capacity, etc. The main advantage of these ratios is that they do not require quantification in economic terms of performance indicators such as noise, capacity, safety, etc. This type of quantification is difficult, at best. The principal disadvantages are two. First, only projects aimed at the same objective (e.g., reducing noise) can be compared to one another through such ratios. The second is that cost-effectiveness ratios are of limited use in assessing overall economic performance of an airport and in making a case for project financing.

Exercises

1. The following potential projects and alternative projects are being considered by airport operator AO.

Project	Initial cost ($000,000)	Annual net benefits ($000,000)
A1	30	14.5
A2	40	19.0
B1	50	18.5
B2	60	21.5
C1	5	1.8
C2	10	3.9
D1	10	4.0

Each project has a 10-year life and no salvage value. The discount rate in use by AO is 15 percent. Projects whose code name begins with the same letter are mutually exclusive.

a. Which projects should be selected by AO with an unlimited budget?

b. Which projects should be selected with an initial budget of $100 million?

2. On January 1, 2005, the Big Aircraft Company (BAC) must make a decision on whether to launch a new large airplane, the BA-1, with 600 seats. The major development costs of this airplane (in millions of dollars) are estimated to be as follows:

Year	Design/development costs
2005	$800
2006	$2500
2007	$2500
2008	$2500
2009	$2500
2010	$2000

The first BA-1 will be ready for delivery in January 2011. Manufacturing costs per airplane are expected to decrease gradually due to learning-curve effects and economies of scale.

Year	Manufacturing cost per airplane	No. of airplanes produced (estimate)
2011	$200 million	40
2012	$160 million	50
2013	$140 million	70
2014–2015	$130 million	80
2016–2020	$120 million	100

BAC plans to sell all BA-1's for $200 million and uses a discount rate (opportunity cost of capital) of 10 percent. For convenience, you may also assume that all costs and benefits occur on the last day of each year (e.g., the $800 million cost occurs on December 31, 2005).

BAC uses *NPV* as its criterion for decision-making.

a. Assuming all the estimates given above are correct, should BAC launch the BA-1 program on January 1, 2005? (Take your time horizon to be the 16 years until the end of 2020.)

b. If your answer was "no," find the approximate price of the BA-1 at which BAC would be indifferent between launching and not launching the program on January 1, 2005? If your answer was "yes," at the end of what year will the BA-1 program become "acceptable" for the first time to BAC? (In other words, what is the minimum time horizon needed for the BA-1 to generate a positive *NPV* for BAC?)

c. What is the approximate price of the BA-1 at which the *IRR* of the program would be 12 percent?

3. Consider three 6-year projects, A, B, and C, with time streams of costs and benefits (in $000) as shown below (a "+" indicates a net benefit):

t	0	1	2	3	4	5	6
A	−6000	0	+x	+x	+x	+x	+x
B	−4000	+1000	+1000	+1000	+1000	+1000	−1300
C	−8000	+2000	+2000	+2000	+2000	+2000	0

a. Given that A and C have *NPV* = 0 at exactly the same discount rate, find the value of x.

b. For the value of x found in part *a*, compute the internal rates of return of A, B, and C.

c. Compare the *NPV*s of A, B, and C at a discount rate of 5 percent.

4. The government of country A is examining the construction of a new airport for one of its major cities. Construction would begin on January 1, 2005, and would last exactly 5 years, so that the airport would open on January 1, 2010. A 20-year time horizon has been adopted for the economic evaluation of the project and a table of costs and benefits has been prepared (see below).

(i) All the figures in the table are in constant (January 1, 2005) millions of dollars, except for the concession fees (column 4), which are in current (inflated) dollars, because the government has signed a contract with a concessionaire under which the latter would pay $100 million dollars (in current dollars) each year in exchange for exclusive rights to all the concessions (restaurants, duty-free shops, etc.) at the new airport.

(ii) Maintenance and operations costs (column 2), are expected to increase, from $50 million, at a rate of 3 percent per year (in constant prices) after year 6 (see *).

(iii) Revenues from landing fees (column 3) will increase, from $100 million, at a rate of 5 percent per year (in constant prices) after year 6 (see †), due to traffic growth.

(iv) Time savings to airline passengers (column 5) are estimated at $100 million in year 6 (in constant, 1/1/2000 prices). However, it is anticipated that the relative value of time of airline passengers will increase at a rate of 2 percent per year in constant prices after year 6 (see ‡); in addition to this, there will be a 5 percent growth in the amount of time saved per year, due to the aforementioned traffic growth at the airport.

Time	Construc-tion costs (1)	M+O costs (2)	Landing fees (3)	Conces-sion fees (4)	Time savings (5)
0	200				
1	200				
2	200				
3	200				
4	200				
5	200				
6		50	100	100	100
7		*	†	100	‡
8		*	†	100	‡
9		*	†	100	‡
10		*	†	100	‡
11		*	†	100	‡
12		*	†	100	‡
13		*	†	100	‡
14		*	†	100	‡
15		*	†	100	‡
16		*	†	100	‡
17		*	†	100	‡
18		*	†	100	‡
19		*	†	100	‡
20		*	†	100	‡

Country A uses a 7 percent discount rate in constant prices. Economists estimate that the inflation rate will be 4 percent per year between years 2005 and 2025.

What is the *NPV* of this airport project (as of 1/1/2005)? Feel free to make appropriate approximations; great accuracy is not the issue here.

References

Brealey, R., and Myers, S. (1993) *Principles of Corporate Finance,* 4th ed., McGraw-Hill, New York.

de Neufville, R. (1990) *Applied Systems Analysis: Engineering Planning and Technology Management,* McGraw-Hill, New York.

Luenberger, D. (1998) *Investment Science,* Oxford University Press, New York.

Steiner, H. (1993) *Engineering Economic Principles,* McGraw-Hill, New York.

Sullivan, W., Bontadelli, J., and Wicks, E. (1999) *Engineering Economy,* 11th ed., Prentice Hall, Upper Saddle River, NJ.

White, J., Case, K., Pratt, D., and Agee, M. (1998) *Principles of Engineering Economic Analysis,* 4th ed., Wiley, New York.

22

Decision and options analysis

The future is always uncertain, as Chap. 3 emphasizes. Planners, managers, and designers need to recognize the wide range of situations that may occur, examine the implications of these scenarios, and develop strategies that enable them to seize opportunities and protect them from risks. This chapter presents decision analysis and options analysis as two of the most effective ways to understand and develop effective strategies for dealing with risks.

Decision analysis is a standard method for determining the best sequences of decisions or developments from among the hundreds that might be available. Several user-friendly computer programs exist to carry out the necessary analyses. The advantage of this method over others is that it considers the whole range of uncertainties that exist over time, and enables planners and managers to define flexible strategies that are responsive to the way the situation develops.

Options analysis defines the value of flexibility in a physical or management plan. It is thus an essential part of effective contingency planning. It enables analysts to determine which forms of flexibility are desirable and should be included in the development strategy. Specifically, it identifies forms of insurance that are worthwhile, and opportunities to incorporate physical flexibility into designs.

22-1 The issue

The recognition that the future is uncertain complicates any analysis. This reality means that the analyst needs to apply special techniques to sort through the possibilities. This issue motivates the use of decision and options analyses.

Most obviously, uncertainty multiplies the number of possibilities to be reviewed and complicates the analysis considerably. It means that responsible professionals should examine how their projects will

803

perform in a variety of circumstances. If the design process focuses on a single forecast, the analysis needs to look at only one set of future conditions. However, when planners recognize that the immediate growth of traffic might be low, medium, or high, this implies at least three scenarios. Moreover, the traffic growth over the life of a major project is unlikely to be constant. It may be low in the first years of the project and high in the next period. It could be the reverse. An analysis might easily consider three possible scenarios in each of two periods—this would still be a simple description of the major possibilities. It would then have to consider a total of nine scenarios. The complexity increases exponentially with the possibilities in each period and with the number of periods.

An unknown future is risky. This fact can further complicate the analysis of major endeavors that develop over time. In situations that feature fluctuating risks, a fully correct analysis recognizes that higher risk demands a higher discount rate. (Fluctuating risks are a feature of commodity markets, for example, for foreign exchange and fuel.) Moreover, as the level of risk changes, so should the discount rate. This complication is often an essential element of the analysis of options, which is described in Sec. 22-4.[*]

Decision and options analyses provide the means to deal with these two related issues associated with uncertainty. They are closely related approaches with different applications. As indicated in Sec. 22-3, decision analysis provides a straightforward way to examine complicated sets of scenarios and outcomes, and to determine a preferred strategy. Options analysis provides a framework for evaluating actions or design features in an uncertain environment. In particular, it enables analysts to determine the value of specific features of a design or plan that provide flexible. Options analysis thus helps planners and managers determine which contingency plans they need to incorporate.

22-2   Decision analysis concept

The process of planning for a possible Second Sydney Airport in Australia motivates the use of decision analysis and illustrates the basic concepts. It also provides a context for the subsequent presentations in Secs. 22-3 and 22-4. The general issue was whether to develop a second major commercial airport for the Sydney metropolitan area.

---

[*]A theoretically correct analysis of options accounts for the fact that discount rates should vary according to the amount of risk. This is not easy to do. The developers of the mathematical methods for doing the basic cases won a Nobel Prize in economics for their work.

The expansion of the existing airport is constrained by a densely inhabited urban area and the environmentally sensitive Botany Bay. Moreover, there are not many sites suitable for an alternative or complementary commercial airport around Sydney. Mountains ring the area and limit the available sites. These sites, furthermore, are disappearing as the city expands.

In the 1970s and early 1980s, two major studies examined the question of the Second Sydney Airport in the traditional way. Each of these efforts prepared a long-term forecast and then evaluated the two alternatives of building or not building a new airport. One report concluded that a new airport should be built, but the other did not. The political process and public sentiment disregarded both reports equally, however. As discussed in detail in Chap. 3, the forecast is "always wrong." In the case of Sydney, world developments rapidly moved the levels of actual traffic in Australia away from each trend that formed the basis for the planning exercises. These developments discredited the entire process—if the transportation experts could not forecast accurately two to three years ahead, why should anyone believe their 20-year forecasts, let alone any conclusions they might draw from them?

A third planning process achieved success by using decision analysis. This approach first recognized that the future was uncertain, and that the essential task of the airport planners was to prepare a plan that would make sense whatever happened. Second, it recognized that the development of a major project is a long-term process that does not require all decisions to be made right away. At the start, managers only have to decide whether to take a first step. They can decide later, depending on the circumstances, whether to take subsequent steps. In this case, the crucial first decision concerning a possible Second Sydney Airport was whether to secure a site. This step would take at least 5 years and give managers the opportunity to observe the development of traffic and then determine if it would be desirable to proceed further. In short, the planning process based on decision analysis reformulated the planning question:

- *From:* Does the single forecast justify the construction of a major new airport?
- *To:* Does the range of possible scenarios justify preparation, in the form of site reservation, for a possible new airport?

The decision analysis of the Second Sydney Airport looked at the question of whether to acquire land for a possible major airport under three scenarios of traffic growth. Table 22-1 summarizes the results.

If the planners could know that future traffic would be high, then the Second Airport could be justified. If, on the other hand, they could know that future traffic would be low, then the Second Airport would not be justified. In fact, since planners cannot tell what the future will bring, they should look at the range of possibilities and their consequences. In this light, Table 22-1 makes it clear that the better decision is to anticipate the possibility that the airport will be needed. If the government did not acquire a site, it might have a bad outcome. Buying the site provides insurance that the overall performance will be satisfactory.

The recommendation of the decision analysis of the Second Sydney Airport was thus that the government should:

- Acquire a site for a new major airport, as insurance against possible need, but
- Not make a decision about whether to build a site until later, when they knew more about the evolution of traffic.

The government accepted this recommendation and proceeded to buy a site (Australia, 1985; de Neufville, 1990b, 1991). Subsequently, aviation in Australia changed completely, as later governments privatized airlines and airports. So far, at least, the site has not been needed.

The decision analysis process views airport development as a series of decisions taken over years. Decisions taken now prepare the way to the future, but do not commit the airport operators to a specific sequence. The process considers that the airport operators will have the opportunity, in the future, to take the decisions that are appropriate to the circumstances that then prevail. It identifies good first decisions that are part of a strategy for dealing with the range of events that unfold.

22-3   Decision analysis method

Decision analysis is an organized way to structure and analyze the complex combinations of possibilities and probabilities of occurrence. It is now a standard analytic procedure (see, for example, de Neufville, 1990a; Keeney and Raiffa, 1993). Once the analyst has provided the information in the correct form, user-friendly computer programs will process these data and determine the most desirable strategy for these circumstances. The analyst can then validate this recommendation by repeating the analysis with alternative estimates of the risk, thus determining if the strategy is robust over a range of assumptions.

Table 22-1. Possible outcomes for initial two alternatives for the Second Sydney Airport

Alternative for first stage	Traffic scenario		
	Low growth	Medium growth	High growth
Acquire a site	**OK result.** Site not needed, but government owns valuable property it can sell.	**OK result.** Site may or may not be needed. Government can wait and see.	**Good Result.** Site is needed and government is prepared.
Do not acquire a site	**Good result.** Site not needed. Neither money nor effort spent on getting site.	**Questionable result.** Site may be needed and, with growth of city, is more difficult to acquire.	**Poor result.** Site needed and is not available.

Source: Adapted from Australia Department of Aviation, 1985.

The essential task in doing a decision analysis is to structure the problem. The planner, manager or designer needs to identify:

- The possible decisions or choices at each stage
- The important risks that will affect the outcomes of each possible decision

Additionally, the analyst also needs to quantify the value of the possible outcomes and probabilities of occurrence. These data determine which strategy is most desirable.

To define the decisions or choices at each stage, it is necessary to focus on which choices absolutely have to be made at the beginning, and which can be deferred until later. Thus, in the case of the consideration of the Second Sydney Airport, the important decision at the first stage concerned the site. Decisions about the size of the airport, the configuration of its facilities, its financial structure, and its role in a multi-airport system (see Chap. 5) could all be left for later, closer to the time when construction and operations might begin. In particular, the decision about whether to build runways could be left to the second stage.

Likewise, it is necessary to focus on the important risks that will affect the consequences of each choice. The airport operator needs to determine the one or two factors that will most critically shape the value of the first choices and subsequent choices. These will normally be a combination of chance events, such as the timing of booms and recessions, and deliberate actions by major participants in the industry such as airline mergers, choices of aircraft, and so on. Thus, in the case of Sydney, the essential uncertainty associated with the development of a Second Sydney Airport was the growth rate of the traffic. This factor would define whether the existing airport had sufficient capacity or required complementary facilities. Other factors, such as the composition of the airline fleets and the number of international passengers, would be important in the design of an airport, but would not be critical to decisions about whether to build additional capacity.

Decision analysis conventionally represents the structure of the series of choices and possible outcomes associated with each phase of development as a "tree." The metaphor is that at the beginning (the "root") the decision makers have several choices to make in the first stage (these are the first "branches"). Each first-stage decision can have several outcomes, depending on how the industry develops (these are subsequent "branches"). This sequence continues as long as desired. The whole structure constitutes the "decision tree." This imagery accounts for the names of many computer programs for decision analysis (Treeplan, TreeAge, etc.).

The series of choices for the decision analysis of the Second Sydney Airport is as follows. At the first stage, the decision was whether to select a site or not, as Fig. 22-1 shows. Correspondingly, the significant risk or chance events concerned the growth rate of traffic over the next years. Using the same range of possibilities shown in Table 22-1, this leads to the branches associated with the growth rate to be as in Fig. 22-2. The decision tree corresponding to the first stage of the decision analysis for the Second Sydney Airport combines the above two elements. It appears in Fig. 22-3.

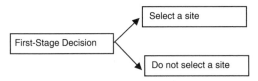

Fig. 22-1. *First-stage decisions for the Second Sydney Airport.*

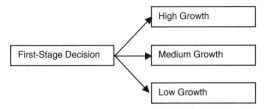

Fig. 22-2. *Possible circumstances determining the outcome of the first-stage decisions for the Second Sydney Airport.*

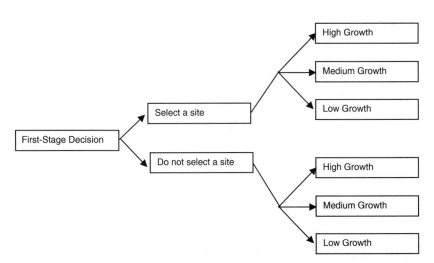

Fig. 22-3. *Decision tree for first stage of Second Sydney Airport example.*

Decision trees are useful because they inform the analysis. In practice, a full picture of the decision tree is much too complicated to represent in detail. Imagine Fig. 22-3 with five initial choices and five possible outcomes in the first stage. If the choices and outcomes were as numerous in the second stage, the tree would have 625 branches at the end. This is not useful to draw. Fortunately, it is not necessary to draw decision trees to do the analysis.

The decision analysis cannot be carried out conceptually along the lines presented in Sec. 22-2. This version was good for presentation to the political process, editorial writers, and general public discussion. Airport operators making significant investments require a quantitative analysis, however.

A full decision analysis thus requires explicit evaluation of the costs and benefits of each outcome. These can be obtained through financial, environmental, or other models that are relevant to the situation. Since these models should all be based in computers, estimating the value of the consequences for many scenarios is not especially difficult. Of course, this effort needs to be tailored to the importance of the decision. A decision concerning an investment of $100 million in a new building deserves more effort than some $1 million expense. In practice, leading consultants routinely have many analysts performing sensitivity analyses on all major planning, management, and design decisions.

A full decision analysis also needs best estimates of the likelihood that different outcomes will occur. Since the future is unclear, estimation of the probabilities of possible events is not a precise science. The proper approach to this question is first to develop best estimates from experts and, second, to vary these estimates to test the sensitivity of the results of the analysis to the assumptions. If the decision analysis recommends the same strategy for the range of possible probability distributions, the recommendation is robust. Thus it was for the analysis of the Second Sydney Airport. In that case, as Table 22-1 illustrates, possible consequences of not having a site were never better than those associated with having a site available. Thus, no reasonable distribution of the probability of low, medium, or high growth rates would change the recommendation.

The numerical decision analysis determines the preferred solution by calculating the expected values of each series of decisions. In this case, the expected value for a decision J is the sum of the values of the outcomes weighted by their probability of occurrence:

$$E(V)_J = \sum_I \text{(probability of event } I\text{)(value of} \quad (22.1)$$
$$\text{resulting outcome } I)$$

Example 22.1 illustrates how this can be done. For examples of more complicated situations involving many stages, readers should consult standard references.

EXAMPLE 22.1

Suppose an airport operator faces a decision about the size of the passenger building to build. The forecasters estimate that traffic growth will require enough capacity to serve 6 million additional passengers. Suppose further that the airport operator is evaluating three designs:

- A large passenger building, good for 6 million passengers.

- A small passenger building, good for 3 million passengers initially but extendable to 6 million. It would avoid costs if no expansion were needed, but if expanded to the large size, would cost considerably more than the construction of a large building as one project.

- A small passenger building with facilities for transporters, good for 6 million if the vehicles are fully used. It would cost more than the small passenger building due to the transporters. However, these vehicles could serve 3 million additional passengers as needed relatively inexpensively.

Suppose that the costs for these designs in the first period are, respectively, in millions of dollars, [100, 60, 70]. Suppose that the costs of upgrading the designs to serve 6 million passengers in the second period, if necessary, are [0, 70, 40] millions of dollars. Finally, suppose that experts indicate that the forecast of 6 million passengers is most likely, with a probability of 60 percent. However, there is also a 30 percent probability of only 3 million passengers, and a 10 percent probability that traffic will stagnate due to poor economic and other conditions.

Figure 22- 4 shows the decision tree for this case created by the TreeAge (1999) program. Reading from the left-hand side, the diagram shows

- Possible decisions, represented by the 3 designs

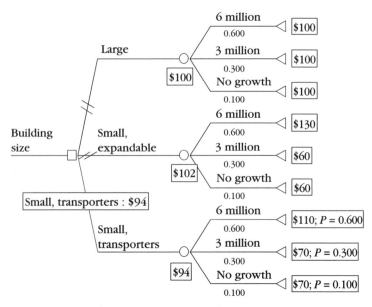

Fig. 22-4. *Decision tree for example analysis of optimal size of passenger building (executed in TreeAge).*

- Possible levels of traffic, each with the estimated probability of occurrence stated as a decimal

- Outcomes for each combination of design and traffic, expressed in millions of dollars

The boxes indicate the value of each design and define the optimal solution.

The optimal choice in this case is a small building with transporters. It is less expensive on average ($94 million versus 100 million). This solution is relatively cheap if the large building is not needed, and yet not especially expensive otherwise. In this example, the transporters provide a more economical, more flexible way to acquire the option to expand if and when needed.

22-4   Options analysis concept

An *option* has a special specific meaning in the world of financial planning and investment analysis. The word connotes something quite different from what it means in ordinary public speech. In the common language, an "option" is a synonym for a "choice" or an "alternative." This is not the meaning relevant to the evaluation of projects. Readers should note the following definition carefully.

An *option* is the *right, but not the obligation* to take a course of action. For example, contracts to lease cars usually give the person leasing the car the option to buy it at the end of the lease for some stated amount that depends on the condition of the car. People holding such contracts can decide at the end of the lease whether they are interested in actually owning the car for the contract price. They are not obliged to buy the car, but they can do so.

Options provide flexibility. This is their central feature for planners, managers and designers.* They give the owners of the option the capability to respond to new situations if it is in their interests to do so. This flexibility is valuable. Options provide insurance that the systems operators will be able to meet future conditions well. They may "save the day," and this is what gives them value to an airport operator.

The point of options analysis, from the perspective of the airport operator and developer, is to define the value of the flexibility. The planner or manager can then compare the value of the option with the cost of acquiring it. If the option provides significantly more value than its cost, it should be incorporated into the airport plan, its design, or management strategy.

There are two kinds of options, financial and "real" options. *Financial options* are defined by contracts. The lease on a car giving the option to buy is a typical financial contract. The whole of the option is contained in the agreements about the price and other conditions of exercise. Financial options are common and concern all kinds of assets such as stocks, bonds, and commodities. Financial options on commodities can provide system managers with an excellent way to deal with fluctuating prices for the range of assets they use in the course of business, such as aviation gasoline, electric power, and foreign exchange. *"Real" options* are defined by physical facilities or devices that enable the system operator to react to changes. Thus, the installation in a power plant of burners that can use either natural gas or diesel fuel provide a "real" option to the owner. These

---

*This discussion takes the perspective of the owner and operator of the system, whose prime concern is the optimal development of that facility over time. Speculators who trade options are more interested in the fact that options offer the potential for spectacular gain. Options provide high leverage. Small investments in options can lead to 10-fold returns— or to a total loss.

dual-fuel burners give the plant operator the option of switching from one fuel to another if their relative prices change and it is advantageous to do so. "Real" options can be extremely important for airport operators.

Financial options

The basic financial options are "calls" and "puts." A discussion of these is sufficient to indicate why airport operators should be interested in them, and how they might use them in practice. For more complete descriptions, see a standard text such as Brealey and Myers (1996).

A *call* is the right to buy an asset at a fixed price, no matter what the market price is otherwise. For example, an airline might have a "call" on buying 1 million euros at the price of $1 per euro. They might want to have this contract because they have revenues in dollars but have expenses to pay in euros, and they would like to protect their ability to pay their debts at a known price, rather than be subject to fluctuations in the rate of exchange. A *put* is the right to sell an asset at a fixed price, no matter what the market price is otherwise. For example, a supplier might have the right to sell jet fuel to an airline at a fixed price. It might want to have this option so that it could protect itself against losses it might suffer if the price collapsed. A "put" is essentially the inverse of a "call."

To understand the value of options, it is important to appreciate how their value changes asymmetrically according to fluctuations in the markets. Consider the call on euros mentioned above. If the price of euros in the foreign exchange market is $1 or less, the call has no value—the company will find it cheaper to buy the euros in the market rather than exercise its option and pay $1 per euro. Whenever the price of the euro in the foreign exchange market is greater than $1, the call will be valuable. Specifically it is worth $10,000 for every penny the price of the euro exceeds $1 (= $1 million × 0.01). Figure 22-5 illustrates the phenomenon. The important feature of this behavior is that the value of the option is not symmetric around a specific price (this is the *strike price*, the amount beyond which it may be valuable to exercise the option). In one direction it changes, in the other it is flat. In one way it is all profit, in the other it has no loss—beyond the price paid to acquire the option in the first place. This asymmetry gives the option its value.

Options are particularly remarkable because their value increases with risk. They differ in this from other kinds of assets. Normally, the

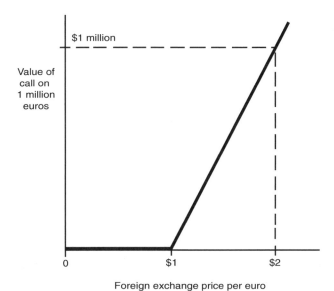

Fig. 22-5. *Payoff value of the call on $1 million euros at $1 per euro.*

riskier the project, the less it should be worth, all else being equal. Options on the contrary become more valuable with risk. This is because of the asymmetry in return. For example, if the dollar/euro exchange rate were stable and fluctuated only between $0.95 and $1.05, the maximum value of the call on a million euros would be $50,000. However, if the exchange rate were more volatile, fluctuating between $0.50 and $1.50, the maximum value of the same call would be $500,000 or 10 times as much. Options are particularly valuable in risky situations.

To the degree that the airline/airport industry is constantly changing, airport operators face major risks in many areas. This is particularly true in a deregulated, privatized environment, as Chap. 4 indicates. The airport traffic and thus revenues may be volatile and the airport clients may change. In such an uncertain environment, options are particularly valuable.

"Real" options

Conceptually, "real" options are entirely similar to financial options [for a detailed discussion, see Trigeorgis, (1996)]. What distinguishes them is the fact that they are "real" in the sense of being physical. Financial options are also real, in a different sense of the term, in that they are not imaginary and have real consequences. Nonetheless, the term "real" options refers to physical possibilities. "Real" options are embodied in design. They are thus permanent, whereas financial options are contracts typically limited to short periods such as a year or less.

Any part of a design that has flexibility to adapt to new conditions represents a "real" option. Properly understood, "real" options exist throughout any design. For example:

- Reserve land provides a "real" option on the construction of new facilities, as for Sydney.
- Preserving a right-of-way for public transport provides a "real" option on the eventual development of such a system, as at New York/Newark or Paris/de Gaulle.
- Facilities designed for shared use provide "real" options on changing the space allocated to different airlines, as at Las Vegas.
- Glass or other non-load-bearing walls dividing international and domestic baggage areas constitute a "real" option on expanding either area, as at Athens/Venizelos.

Understanding the design features that provide flexibility as "real" options is important. Once they are understood to be options, it is possible to calculate their value as options. When the flexible features are not viewed as options, this important value is neglected and the airport operator will not maximize the value of the facilities. Understanding flexible features as options enables the designers to appreciate their full value, and will lead to the design of more flexible facilities, more capable of responding to the changing demands of the airline/airport industry.

22-5   Options analysis method

The theoretically correct determination of the value of an option requires considerable information. It needs basic information about the terms of the contract, such as its length and the strike price under various conditions, as for a car lease with option to buy. Additionally, the determination of the value of an option requires extensive statistical data. These are needed to determine the volatility of the asset under option. This information, in turn, is necessary in order to calculate the expected value of the option, as alluded to in the previous section. If all these data are available, it is possible to construct computer programs to provide quick estimates of the value of standard options. Financial analysts use such programs routinely.

Planners, managers, and designers of technical systems such as airports generally cannot use standard programs to calculate the value of their options. This is for several reasons. First, the situations they face are unique to airport operation, and standard programs for these operations are very rare, if they exist at all. Second, the uncertainties they face, such as the effects of a New Large Aircraft such as the A-380, may be unprecedented, so that historical data do not provide a good indication of future volatility. They thus will probably have to do with approximate methods. Many approaches are currently being developed, generally using either simulation or decision analysis (for example, de Neufville and Neely, 2001; Longstaff and Schwartz, 2001).

Even when airport operators cannot calculate an exact value for an option, they can benefit from "options thinking" (see Faulkner, 1996; Amran and Kulatilaka, 1999). Recognizing that flexibility is a great asset in a constantly changing world, and associating even an approximate value to it can lead them to design more effective, and thus more valuable, airport facilities.

Exercises

1. Consider the design of a major new airport passenger building at an airport of your choice. What major decisions have to be taken first? (For example, these could concern the configuration and the size of the building.) Which uncertainties might most significantly affect the performance of this facility? (For example, these might concern the level of traffic or airline decisions about establishing hub operations.) First, structure a decision tree that sets out the first phase of the decision analysis. Next, think about the important decisions for the second phase, and the relevant uncertainties.

2. Consider the same or a different airport. Think about how it might cope in 20 to 25 years with traffic three to five times as large as today. What aspects could be built into the current designs to provide "real" options for the range of possible developments?

References

Australia Department of Aviation (1985) *Second Sydney Airport: Site Selection Programme,* Australia Department of Aviation, Canberra.

Amran, M., and Kulatilaka, N. (1999) *Real Options; Managing Strategic Investment in an Uncertain World*, Harvard Business School Press, Boston, MA.

Brealey, R., and Myers, S. (1996) *Principles of Corporate Finance*, 5th ed., McGraw-Hill, New York.

de Neufville, R. (1990a) *Applied Systems Analysis: Engineering Planning and Technology Management*, McGraw-Hill, New York.

de Neufville, R. (1990b) "Successful Siting of Airports; The Sydney Example," *ASCE Journal of Transportation Engineering, 116*(1), Feb., pp. 37–48.

de Neufville, R. (1991) "Strategic Planning for Airport Capacity: An Appreciation of Australia's Process for Sydney," *Australian Planner, 29*(4), Dec., pp. 174–180.

de Neufville, R., and Neely, J. (2001) "Hybrid Real Options Valuation of Risky Product Development Projects," *International Journal of Technology, Policy and Management, 1*(1), Jan., pp. 29–46.

Faulkner, T. W. (1996) "Applying Options Thinking to R & D Valuation," *Research Technology Management,* May–June, pp. 50–56.

Keeney, R., and Raiffa, H. (1993) *Decisions with Multiple Objectives: Preferences and Value Tradeoffs,* Cambridge University Press, Cambridge, U.K.

Longstaff, F., and Schwartz, E. (2001) "Valuing American Options by Simulation: A Simple Least-Squares Approach," *The Review of Financial Studies, 14*(1), pp. 113–147.

TreeAge Software (1999) *Data 3.5 User's Manual,* TreeAge, Williamstown, MA.

Trigeorgis, L. (1996) *Real Options: Managerial Flexibility and Strategy in Resource Allocation,* MIT Press, Cambridge, MA.

23

Flows and queues at airports

Practically every airside and landside facility and service can be viewed as a queuing system. Prospective users (aircraft, passengers, bags, or other entities) form queues at these facilities and services and wait for their turn to be served. Flow analysis and queuing theory provide important tools for studying and optimizing these processes.

Queuing systems consist of three fundamental elements: a user source, a queue, and a service facility that contains one or more identical servers in parallel. Users arrive at the queuing system at instants described by the probability distribution of the demand interarrival times. Demand rates can be constant over time, but at airports they usually vary with the time of the day, the day of the week, and the season. Service is described by the service rate and by the probability distribution of the length of service times.

Many measures of performance and of level of service at airport queuing systems are of interest. Some of the principal ones include the utilization of the facility or service, the expected number of users in queue and the expected waiting time, the variability of queuing time, the reliability and predictability of the system, and the extent to which users perceive the system to be orderly and "fair."

Overloads exist whenever the demand rate exceeds the service rate. During overloads, the average delay per facility user increases linearly with the length of the overload period. Cumulative flow diagrams provide a convenient way of visualizing and analyzing what happens at a queuing system under overload conditions. The mathematical analysis of cumulative diagrams is simple and intuitive.

Delays and congestion may also be present during periods when the demand rate is less than the service rate. Such delays are due to the probabilistic fluctuations of demand interarrival times and of service times. They are called "stochastic delays" to distinguish them from overload delays. When the demand rate is lower than but close to the service rate, stochastic delays can be very significant. Queuing theory provides a number of important closed-form expressions for estimating stochastic delays under certain conditions.

A queuing system cannot be operated with a demand rate that exceeds the service rate in the long run, as this results in unacceptable delays. Delays will increase nonlinearly as the demand rate approaches the service rate. Small changes in the demand rate or the service rate can then have a large impact on the magnitude of delays and the length of queues. Moreover, the variability of delays and of queue lengths increases as the demand rate approaches the service rate.

23-1   Introduction

Queuing theory explores the relationship between demand on a service system and the delays suffered by the users of that system. Since all airports—in their entirety or broken down into their individual elements—can be viewed as networks of queuing systems, queuing theory often plays a central role in the analysis of airport operations and in the planning and design of airport facilities and services.

Queuing theory is the mathematical study of congestion. Those who wish to apply its results should appreciate the kinds of questions that the theory can answer. They should also understand the nature of the assumptions behind the answers.

In working with queuing theory one must take the particular airport facility of interest, study this facility, and specify a mathematical model to represent it. The analyst can either create this model or simply choose from the list of models in queuing theory. Through the model, one then computes the statistics that describe the behavior of the facility under the postulated conditions. Inherent to the process of creating and working with a mathematical model are the notions of *simplification* and *approximation*.

To make the analysis tractable, many details about the facility are necessarily disregarded as superfluous (or of minor importance) to the central points of interest. The details to be omitted from the model must be chosen carefully if the model is to resemble reality adequately. Data about the airport facility will also often be incomplete necessitating further assumptions and "educated guesses." Under the circumstances, the estimates of the quantities of interest obtained through a queuing analysis are, in most applications, only approximate indicators of the magnitude of these quantities in the real world. Consequently, the application of queuing theory is most useful in helping identify the inadequacies of existing facilities and services. It indicates the general directions in which to proceed for improving these facilities and services. It can also suggest the approximate values that some of the controllable variables must have if the queuing system is to achieve a satisfactory level of performance.

This chapter presents a general introduction to the application of flow analysis and queuing theory in the airport environment. The emphasis is on the fundamental concepts, on describing the behavior of queuing systems from a short- and a long-term perspective, and on the implications of this behavior for airport planners and managers. Many operations research textbooks (see, for example, Hillier and Lieberman, 1995, or Larson and Odoni, 1981) offer more detailed introductions, while specialized books (e.g., Gross and Harris, 1998, or Wolff, 1989) provide advanced treatment.

23-2   Describing an airport queuing system

The generic model of a *queuing system* (Fig. 23-1) consists of three elements: a user source, a queue, and a service facility that contains one or more *identical servers in parallel*. Each user of a queuing system is "generated" by the user source, passes through the queue where (s)he may remain for a period of time (including possibly zero time), and is then processed by one of the parallel servers.

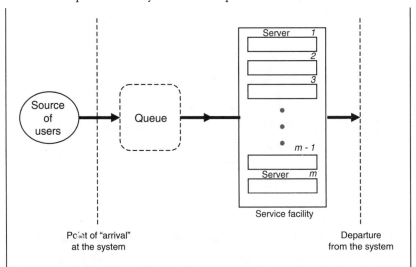

Fig. 23-1. *A generic queuing system.*

A *queuing network* is a set of interconnected queuing systems (Fig. 23-2). In a queuing network, the user sources for some of the queuing systems may be other queuing systems in the network. For example, the output of a check-in desk may be the source for the queues in front of a security checkpoint. As noted above, any airport or its elements can be viewed as a queuing network.

To specify fully a queuing system, information must be supplied about its three generic elements, the user generating process, the queuing process, and the service process. The following discusses briefly these processes in the specific case of airports.

The user generation process

Flows of prospective users to airport facilities are described primarily by

- The *rate* at which they occur over time (known as the *demand rate*), i.e., the expected ("average") number of demands per unit of time.

- The *probability distribution* of the time intervals between successive demands. These time intervals are often referred to as *demand interarrival times* or simply as *interarrival times*.

The demand rate is typically denoted by the Greek letter λ ("lambda").* For example:

$\lambda = 40$ aircraft departures per hour between 07:00 and 10:00, or

$\lambda = 2000$ arriving passengers per hour from 09:00 to 13:00 at a busy passenger building.

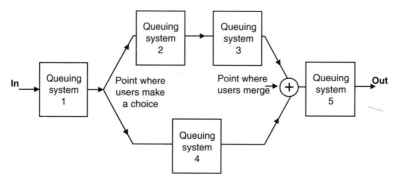

Fig. 23-2. *A queuing network consisting of five queuing systems.*

*The use of such symbols in this chapter is consistent with the standard notation used in queuing theory.

If the demand rate is dynamic, i.e., varies as a function of time, this may be made explicit by using the notation $\lambda(t)$.

For any particular demand rate, the probability distribution for the length of the time intervals between successive demands is very important in determining the performance of a queuing system. Example 23.1 further explains this statement and introduces two fundamental types of distributions.

The probabilistic behavior of user demands at airport facilities typically falls somewhere between perfectly deterministic and perfectly random (Poisson) behavior, i.e., between the opposite poles represented by the constant and by the negative exponential demand interarrival times, respectively. Random events (weather, unforeseen schedule disruptions and operational problems, flight delays due to problems elsewhere, ground traffic delays, etc.) are the rule rather than the exception in the airport environment. Thus, it is not surprising that the Poisson process often turns out to be a good approximate model for the generation of user demands at many airport facilities. At busy airports, for example, the instants when arriving airplanes come within a 100-km radius from the airport can often be approximated statistically as random events generated according to a Poisson process with a dynamic demand rate $\lambda(t)$, which varies according to the time of day. It is true that these instants are related to a preset flight schedule. However, daily deviations from this schedule are typically sufficiently large to make plausible the use of a (time-varying) Poisson demand model.

In addition to the variability of interarrival times, a further complication is that demands for airport facilities and services are often generated in groups (*batch demands*) rather than individually. For example, departing passengers often show up at check-in desks in family groups of two or more people. More important, batch demands usually dominate when it comes to services and facilities for arriving passengers. These passengers typically come into the passenger building within a short interval of time, often in groups of 100 or more, following the arrival of an airplane at a gate. Batch demands are a major consideration in the analysis and planning of operations on the landside of airports, whereas they do not play a role on the airside.

The service process

Entirely analogous ideas apply to the description of the service process. The *service rate* (or *capacity*), i.e., the expected number of

EXAMPLE 23.1

Suppose the demand rate for some airport facility is equal to 60 per hour. At one extreme, demand requests at the queuing system could occur at intervals of exactly 1 min. This is the case of *constant* (or *deterministic*) demand interarrival times at the rate of $\lambda = 60$ per hour.

Another extreme is the case in which demands, while *on average* occurring at the rate of $\lambda = 60$ per hour, are *completely randomly distributed* over time. This naturally implies that some 1-h intervals will have more than 60 demands, while others will have fewer. Moreover, within any interval (1 h, 93 min, or whatever), the instants when the demands occur are distributed completely randomly in time and independently. For instance, suppose it is known that there were 86 demands over a 93-min interval. Then the instants when these demands occur could be "simulated" in the statistical sense, by taking a 93-min timeline and throwing randomly, as if blindfolded, 86 darts on it. Each dart would have equal probability of "landing" anywhere in the interval between $t = 0$ and $t = 93$ min, no matter where the other darts have landed. The process just described has a special mathematical meaning and is known as the *Poisson process*.

The Poisson process is described by demand interarrival times with the *negative exponential* probability density function shown in Fig. 23-3. Observe that, in qualitative terms, short demand interarrival times occur with high probability, while some very long interarrival times are also possible but with low probability. The expected (or "average") length of a demand interarrival interval is equal to $1/\lambda$, the inverse of the number of demands per unit of time. When the occurrence of demands at an airport facility is approximately Poisson, there is a significant probability of observing "bunches" of demands within relatively short intervals of time, interspersed between periods with low demand activity. Due to the potential bunching of demands, users of queuing systems where the occurrence of demands can be approximated by the Poisson process are much more prone to experiencing delays than users of systems with approximately constant demand interarrival times.

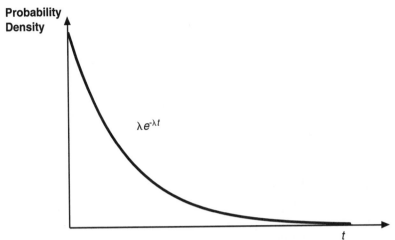

Fig. 23-3. *The negative exponential probability density function.*

users that can be served per unit of time, can be constant or vary dynamically over time. Service rates are usually denoted by μ ("mu")—or $\mu(t)$ if time-varying.

The service rate of a runway system usually depends strongly on weather conditions (see Chap. 10). Given a weather forecast at the beginning of a day, one can specify a function $\mu(t)$ for the capacity expected to be available per runway during, e.g., the next 24 h. When a queuing system contains s identical parallel servers and the service rate for each server is μ, the total service rate for the queuing facility equals $s\mu$.

The probability distribution that best describes the duration of *service times* changes as one moves from one airport facility to another. In some cases, service times may be more or less equal for all users—and can sometimes even be considered as approximately constant. In other cases there may be more variability, requiring probability distributions, such as the negative exponential or others, whose characteristics include a wide range of possible values and a large variance. For example, service times at well-operated security ("X-ray") checkpoints can be fairly constant. However, check-in counters often display wide variability, with average service times in the order of 1.5–2 min per passenger at many airports, but with some passengers taking as little as 1 min and others as long as 5 min or more.

There are practical instances in which the service rate of a queuing system increases or decreases as a function of queue length. Several such examples can be found at airports. For instance, officers performing passport controls often speed up service when a long line forms in front of their desks—sometimes limiting "control" to a nod to the passenger to move on. The reverse may also be true. Long queues at check-in sometimes create so much confusion and frustration that check-in agents at some airports have responded by slowing down service.

The queuing process

To describe the queuing process one must provide details on how prospective users line up for access to system service, as well as on their behavior while in queue. Numerous variations of queuing processes exist in practice and, interestingly, most of them can be encountered at airports.

The most obvious issue about a queue concerns the *priority discipline*. Most queues at airports operate on a first-come, first-served basis (FCFS). Under this regime, prospective users (passengers or aircraft) line up for service according to the order of their arrival at the relevant queue. However, at some airports, conditions often become sufficiently chaotic that the next user to be served is chosen more or less randomly (service in random order—SIRO). The priority discipline for baggage retrieval for incoming flights is also nearly SIRO. In most cases, the order in which deplaning passengers arrive at carousels and other retrieval devices has no bearing on the order in which their bags arrive.[*]

Many queuing systems divide users into classes. These classes then receive different priorities for access to the system's servers. The typical example at airports is the subdivision of passengers into first, business, and economy classes. First- and business-class passengers, as a rule, have priority for check-in and possibly other services as well. They may either go to the head of queues (bypassing economy-class passengers) or, more often, have some service counters reserved for them. These counters may also serve economy-class passengers when no higher-priority passengers are present. Similarly, when a runway serves both landings and takeoffs, arrivals generally receive priority over departures, although the details of this practice vary considerably according to country and location.

[*]Some airlines try, not always successfully, to make sure that the bags of their most prized customers are the ones to appear first at the carousel.

Another important design parameter at airports is whether prospective users line up in a single queue or parallel queues. Check-in is again the prototypical example. At most airports, when several adjacent check-in desks serve a flight,* a separate queue forms in front of each desk. If there are s parallel counters, there are also s queues. However, at a growing number of airports, airlines utilize a *single* queue for such a group of counters. This is sometimes called the *snake queue* because of the twisting shape that it is usually forced to take. One important advantage of the snake queue is that it saves space in crowded terminal buildings, as the contours of the waiting line are clearly designated and more passengers are forced into a smaller area. Another advantage is that it gives passengers a sense of fairness, as everyone is served in a FCFS way, eliminating the possibility that the queue that one chooses to join will prove to be a "slow" one. On the negative side, snake queues sometimes grow to lengths of literally hundreds of people, causing anxiety and occasionally anger to those that must join it. Note that the overall service rate equals $s\mu$ in either case.

A related issue when parallel queues are allowed to form in front of servers is whether to allow those joining a particular queue to switch to another queue if their own queue moves slowly. The issue arises in passenger processing, for example, at check-in, passport control, or security control. Some airlines and/or airport operators try to prevent such switching by placing barriers between queues. The rationale is that this makes for a more orderly process. On the negative side, preventing queue switching may lead to underutilization of some service capacity. Some servers may sometimes be idle while persons queued in front of other servers are unable to take advantage of the presence of idle servers.

Airlines and airports can manage queues by monitoring queue lengths and adjusting the number of active servers accordingly. Whenever the number of passengers waiting for service exceeds a certain limit (exactly or approximately specified), an airline or the airport operator can activate one or more additional desks/counters/servers. When the number waiting falls back below the same or some other limit, the number of active servers can be reduced as well. This practice naturally requires the availability of idle counters and related equipment, as well as of standby employees. Typically, these employees can be primarily engaged in some other activity, but are available to staff a service position (e.g., a check-in counter) as necessary.

*Each of the desks may, in fact, provide check-in service for more than one flight—and possibly all of the flights of an airline.

A crucial parameter in describing and designing airport queuing systems is *queue capacity*. This is the maximum number of prospective facility users that the waiting line can accommodate at any single time. At airports, space limitations typically determine this capacity. Examples are the length of taxiway that departing aircraft can use to line up for takeoff, the area available for waiting passengers in front of check-in counters, and the volume of terminal airspace available for "stacking" arriving aircraft waiting to land. As these examples suggest, it is usually somewhat difficult to pinpoint exactly the capacity of airport queues. After all, many people can be crammed into any particular part of a passenger terminal for a short period of time. Queue size limitations nevertheless do exist at airports and often present a real problem. For example, they occur when departing aircraft have to be held at their apron stands, because the taxiway system is saturated with other airplanes waiting for takeoff (Chap. 10), or when passengers waiting to check in block circulation in a departures concourse (Chap. 16). It is important to recognize these queue capacity limitations explicitly and to consider their effects in analyzing and designing airport facilities and services.

23-3 Typical measures of performance and level of service (LOS)

Many measures describe the performance of a queuing system and the resulting level of service (LOS). This section reviews those most relevant to airport facilities and services.

Utilization ratio

The *utilization ratio* is quite possibly the most fundamental measure of LOS. It "drives" all other measures of a queuing system's performance, as indicated below. It is denoted as ρ ("rho"). For a single-server queuing system with demand rate λ and service rate (i.e., capacity) μ, it is given by

$$\rho = \lambda/\mu \qquad (23.1)$$

When the queuing system has s parallel and identical servers,

$$\rho = \lambda/s \cdot \mu \qquad (23.2)$$

Intuitively, ρ indicates the "intensity" of utilization of the queuing system. It is often referred to simply as the *demand-to-capacity ratio.* A queuing system with ρ greater than 1 is called "saturated," for obvious reasons. Values of ρ close to but less than 1 are desirable if one's objective is to make maximum use of the productive capacity of a facility or resource. As shown later, however, this may also entail important inefficiencies, specifically long waiting times for access to the facility or resource.

Expected waiting time and expected number in queue

It is convenient to define two quantities central to the description of the performance of a queuing system. Consider a queuing system over a particular interval of time. Define

> W_q = the waiting time in queue experienced by a user selected randomly among all those who visited the system during that interval of time

> N_q = the number of users waiting in the queue at a randomly selected instant during that interval

Both W_q and N_q are *random variables.* This is because both the demand interarrival times and the service times at the queuing system are, in general, probabilistic quantities as discussed in Sec. 23-2. Therefore, both the waiting time experienced by a system user and the length of the queue at the system will also vary probabilistically.

The characteristics of W_q and N_q are of great importance in describing the performance of any queuing system. The most obvious and by far most commonly used of these characteristics are their expected ("average") values, denoted $E[W_q]$ and $E[N_q]$, respectively. The waiting time that a random user of the queuing system will experience, on average, at an airport service or facility is of vital interest. So is the average length of the queue of passengers or aircraft waiting for service. Many of the airport airside delay statistics that are often cited by airlines, government agencies, and the media involve $E[W_q]$, the expected waiting time.

Variability

Expected values tell only one part of the story, however. Almost equally important to passengers and airlines is the *variability* of W_q and N_q. For example, it is one thing for departing passengers to know that the total delay while being processed at an airport (check-in, security control, etc.) is 20 min on average (i.e., total $E[W_q]$ = 20 min) with a typical range of 10–35 min. It is quite another to know that total $E[W_q]$ is equal to 20 min, but with a range of 5–90 min. Experience shows that departing passengers will behave differently in the two cases. They will get to the airport considerably earlier, relative to their scheduled flight departure time, in the second case than in the first. Similarly, high variability of delay from day to day means that airlines have to construct the daily itineraries of aircraft, cockpit crews, and cabin crews with turnaround times on the ground that include considerable "slack time" between successive flights.* If they do not, they will find it difficult to execute their schedule of flights reliably, as discussed below. The most common measure of variability of delay is the *variance* of W_q, denoted here as $\sigma^2(W_q)$, or the variance's square root, the *standard deviation* $\sigma(W_q)$. A large variance or standard deviation indicates high variability of delay.

Reliability

Reliability and variability are closely interrelated. The more variable the behavior of a queuing system, the more difficult and costly it will be to ensure that it operates reliably. Airports are a prime example of this relationship. Airlines and airports measure reliability, especially when it comes to airside operations, primarily through statistics regarding the frequency with which long delays occur. In effect, they measure the probability that W_q will exceed certain threshold values (or "tail of the distribution values") that are considered critical to maintaining a reliable schedule of flight operations. For example, both the U.S. Federal Aviation Administration (FAA) and EUROCONTROL regularly collect and report statistics on the percent of flights arriving more than 15 min behind schedule at each of the major airports in the United States and in Europe. The 15-min value has been chosen because it is considered critical to both the passengers and the airlines.

Typically, a delay of less than 15 min on arrival will not prevent the aircraft involved from departing on time for its next flight. Turn-around times on the ground between flights—as scheduled by the airlines—usually include sufficient "slack" to absorb a 15-min delay on arrival without affecting the scheduled time of departure. However, arrival delays of more than 15–20 min usually "propagate" to the subsequent departure and thus have a more disruptive effect. If the probability of long delays of this type is high, then the on-time execution of an airline's overall schedule of flights is problematic.

*Schedules of aircraft, flight crews, etc., are said to have a lot of "slack" in such cases.

In trying to design reliable flight schedules, the scheduling departments of major airlines examine carefully the probabilities of long airside delays at each airport they serve. This can all be put in quantitative terms. For instance, suppose that the estimated probability that W_q for a particular arriving flight will exceed 15 min is 10 percent during the course of a scheduling season. Then a ground turnaround time that includes a 15-min slack will ensure 90 percent reliability on the departure time of the next flight to be performed by the same aircraft.[*] Analogously, a number of airports specify LOS standards for the design and operation of their passenger terminals (see Chap. 15) partly in terms of the probability of extreme delays. For example, they may specify that "80 percent of passengers checking in should experience a waiting time of less than 12 min."

Maximum queue length

In much the same spirit, planners and designers of passenger building facilities and services are interested in estimates of the *maximum queue length*. This measure of performance is not particularly well defined. In theory, the length of any queue at a busy airport may, with very low probability, become extremely long under certain combinations of events for short periods of time. What planners and designers truly want to know is the amount of space they should provide at each terminal facility or service to run only a small risk that this space will prove inadequate. To answer this question planners take two approaches. The first, and more correct one, is to compute a value of N_q that will be exceeded only with a small prespecified probability, e.g., 5 percent, and define this value of N_q to be the maximum queue length. This differs little conceptually from the approach described in the previous paragraph for estimating reliability. The second approach is to perform a detailed simulation of airport operations on the design peak day (DPD) or design peak hour (DPH)—see Chap. 24—and use for planning and design purposes the maximum queue length observed at the facility or facilities of interest.

The psychology of queues

When measuring and evaluating the performance of queuing systems, one should not underestimate the importance of *psychological factors*. This is a subject of growing interest among queuing specialists (Larson, 1988). The basic point is that psychological factors play a central role

---

[*]This assumes that the delay caused by airport congestion, W_q, is the only source of delay. Unfortunately, this is usually untrue.

in user assessments of the severity of delays and congestion at any facility or service. Thus, airport operators and airlines can, to some extent, ease the unavoidably negative reactions of air travelers to airport delays by taking appropriate steps to influence *perceptions* of the situation.

At the most obvious level, the *physical environment* is a central influence on perceptions about the severity of delays. The more comfortable the environment (area per occupant, availability of seating, ventilation and temperature, ambience of space, etc.), the more tolerant airport users are of delays. Ashford (1988) has presented results from interviews showing that passengers react less negatively to delays as the number of square meters per passenger in the area used for waiting increases.

The availability of *information* is also crucial in shaping perceptions. Airport users generally react less severely to delays if given reliable information on the reasons for the delays and/or some advance estimate of how long a delay will be. A number of airports, for example, now display electronically the estimated time until the bags of passengers on each individual incoming flight will start arriving on the bag retrieval carousels. In recent years, airlines have also been displaying increasingly detailed information on the reasons for flight delays.

A third important aspect of user perceptions regarding delays has to do with the notion of *fairness.* Larson (1987) suggests that perceptions of fairness ("social justice") in a queuing system are strongly related to the number of "slips" and "skips" that take place per unit of time. A *slip* is an event in which a user receives service before another user who arrived at the queuing system (joined the queue) before him or her. A *skip* is the opposite. The higher the number of slips and skips, the more "unfair" the system is perceived to be—leading to increasingly negative user reactions to any delays they experience. The snake queue (Sec. 23-2) is one way in which slips and skips can be controlled. Some airlines and airport operators place a high priority on preserving an image of orderliness and fairness at their facilities and services.

Perceptions about the severity of delay usually increase nonlinearly with its length. Passengers typically perceive a 20-min wait for check-in as being more than twice as "bad" as a 10-min wait. Thus the *avoidance of extremely long waiting times* at individual facilities and services not only increases reliability but also contributes psychologically to a more positive assessment of LOS by airport users.

23-4   Short-term behavior of queuing systems

This section and the next two briefly discuss the behavior of queuing systems. They give a general description of how delays and queue lengths grow over time and as a function of the utilization ratio, ρ, the "intensity" of use of the queuing system. The presentation is mostly qualitative and conceptual. The detailed behavior of queuing systems within a specific short period of time (short-term behavior) is described first. Section 23-6 takes a more macroscopic point of view (long-term behavior). In the airport context, "short-term" usually means periods ranging from 1 h to a full day of operations, while "long-term" typically involves entire seasons or years.

Consider an airport queuing system where, over a 24-h period, the demand rate λ and the service rate (or capacity) μ undergo a number of changes. Suppose also that the demand interarrival times and the service times are random variables, i.e., take on different values according to their respective probability distributions.[*] Three general statements can be made about the behavior of W_q and N_q during this 24-h period.

1. *Overload delays* will certainly occur (and queues will form) during any time when the demand rate exceeds the service rate. This is because prospective users will arrive on average at the queuing system at a rate greater than the capacity of the system to serve them. Colloquially, "demand exceeds capacity" during such periods. In general, when there is a time interval during which the demand rate exceeds the service rate, then both the expected queue length, $E[N_q]$, and the expected waiting time, $E[W_q]$, for users arriving at the queuing system during that interval will grow. The growth will be in direct proportion to the length of the interval, T, as the next section shows.

2. *Stochastic delays* may also occur when the demand rate is *less* than the service rate. As Example 23.1 noted, this is due to the probabilistic fluctuations in the demand interarrival times and/or the service times, i.e., to the likely presence of time intervals with "clusters" of short interarrival times of demands and/or of long service times. This

---

[*]At a more technical level we shall also assume that the lengths of successive demand interarrival times are (statistically) independent of each other and so is the duration of successive service times.

type of delay is often called *stochastic* delay* to distinguish it from *overload delay*. If the utilization ratio, ρ, is close to but less than 1 during a fairly long time period, then stochastic delays can be very significant, i.e., both $E[N_q]$ and $E[W_q]$ may become large for users arriving at the queuing system during such periods. In general, the higher the variability (as measured by the variance) of the demand interarrival times and of the service times, the higher the stochastic delays will be. Section 23-6 returns to these points.

3. *The dynamic behavior of queues is complex.* The complex behavior of airside delay is described in Chap. 11. The same complexity can also be observed at landside facilities. The waiting times and queue lengths experienced during any particular time interval depend strongly on the waiting times and queue lengths during previous intervals. Consider, for example, two different hours of the day at an airport, one early in the morning (e.g., 06:00–06:59) and the other in late afternoon (e.g., 18:00–18:59). Assume that conditions within the two hours are identical: the demand rates, λ, are equal in the two hours—and so are the service rates, μ. However, the first hour is preceded by a period of little demand and the second by a period of high demand. The delays and queue lengths will then be far greater in the afternoon hour than in the morning. Moreover, the exact magnitude of these delays and queue lengths will depend on the time history of the queuing system before the two hours of interest, the values of λ and μ, and the probability distributions of the demand interarrival times and of the service times.

An interesting aspect of the dynamic behavior of queues is that a lag may exist between the times when demand peaks and when delay (and queue length) peaks. This phenomenon is sometimes called *hysteresis*. It is entirely analogous to a daily experience of people who drive home from work on urban road networks. Those leaving work during the peak demand hour (e.g., between 16:00 and 17:00) will usually experience less delay than drivers going home one or two hours later—when the number of those starting their commuting trip is smaller. The reason is that by 17:00 or 18:00, traffic congestion has already built up, so that those entering the traffic join "the end of the queue" and experience longer delays. In effect, they suffer the consequences of following on the heels of the earlier overload. Airside traffic and passenger traffic at busy airports are liable to experiencing exactly the same sequence of events.

---

*"Stochastic" is synonymous with "probabilistic"; we use the term here because it is also widely used in queuing theory.

23-5   Cumulative diagrams

Cumulative flow diagrams (or simply *cumulative diagrams*) provide a convenient way of visualizing and analyzing what happens at a queuing system under overload conditions. Cumulative diagrams can be very useful in obtaining approximate estimates of delays at both landside and airside facilities of busy airports. Their simplicity and intuitive appeal make it possible to present this graph-based approach in considerable detail in this section. The presentation is based on the analysis of a generic situation. The "users" of the facility described below can be passengers, aircraft, bags, or other relevant entities.

Consider Fig. 23-4, in which the functions $\lambda(t)$ and $\mu(t)$ denote, respectively, the demand rate and the service rate over time at a service facility in an airport. Note that the service rate indicates the "maximum throughput capacity" of the facility in the terminology of Chap. 10. The units for both the demand rate and the capacity are "users per unit of time."

Figure 23-4 approximates a situation that arises frequently at airports. The time axis shows the busy part of a day at the airport, e.g., the origin may correspond to 07:00 local time. To simplify the analysis, it is assumed the demand rate is constant throughout the busy part of the day, i.e., $\lambda(t) = n$ users per hour for all t. The normal service rate b (for "high") is greater than n. However, for the time interval between $t = a$ and $t = b$, the service capacity is reduced to l (for "low") users per unit of time. It will be convenient later on to denote with $T = b - a$ the interval of time during which the capacity is low.

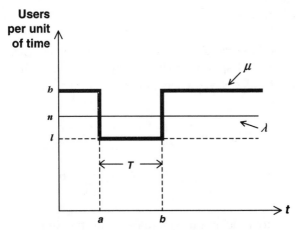

Fig. 23-4. *Demand and service rates at a queuing system.*

The objective is to explore quantitatively the implications of the temporary overload on delay levels at this airport facility.

An important simplification in the analysis is achieved through the assumption that demands and service completions occur at evenly spaced intervals. For example, if $n = 60$, it is assumed that a user arrives at the facility to obtain service exactly every 60 s. Similarly, for any value of the service rate, h or l, the service times are assumed constant. If, for instance, $l = 30$ and the service facility is continually busy, a service is completed exactly every 2 min. This assumption, which is tantamount to adopting an entirely deterministic model, is used in practically all applications of the cumulative diagrams technique.

Clearly, the situation shown in Fig. 23-4 will result in some delays to users during the time period between $t = a$ and $t = b$, at the very least. In fact, it is very easy to plot, as a function of time, the number of users waiting in the queue for access to the service facility. This is done in Fig. 23-5. There is no queue until $t = a$ because the facility's capacity is greater than the demand rate. Beginning at $t = a$ the queue builds up at a rate of $n - l$ users per unit of time, so that, by the time b, the queue will build up to a length of $(n - l) \cdot (b - a) = (n - l) \cdot T$ users. After time b, when the facility's capacity is "high" again, the queue length will decrease at the rate of $h - n$ users per unit of time, i.e., at the rate at which users receive service minus the rate at which new users join the queue. Thus, it will take an amount of time equal to $(n - l) \cdot T/(h - n)$ for the queue to dissipate and get back to zero. Therefore,

$$\begin{matrix} \text{Total amount of time} \\ \text{with a queue present} \end{matrix} = T + \frac{(n - l) \cdot T}{(h - n)} = T \cdot \frac{(h - l)}{(h - n)} \quad (23.3)$$

The area under the triangle in Fig. 23-5 represents the total amount of time that users will spend waiting in the queue, i.e., the total delay time suffered by all delayed users due to the reduction in capacity from $t = a$ to $t = b$:

$$\text{Total delay time} = \frac{1}{2} \cdot (n - l) \cdot T \cdot \frac{(h - l)}{(h - n)} T \quad (23.4)$$

$$= \frac{1}{2} \cdot T^2 \cdot \frac{(n - l)(h - l)}{(h - n)}$$

Note that the total delay time increases with the *square* of the duration of the low-capacity interval. Equation (23.3) indicates that the number of users demanding service during the period when a queue

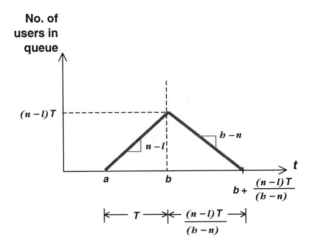

Fig. 23-5. *The number of users in the queue.*

exists is equal to $n \cdot [(b - l)/(b - n)] \cdot T$, the demand rate multiplied by the duration of the queue. This is the number of users that will suffer some delay. It is then possible to compute another quantity of interest by dividing the total delay time, given by Eq. (23.4), by the number of delayed users to obtain

$$\begin{matrix} \text{Expected delay per} \\ \text{delayed facility user} \end{matrix} = \frac{1}{2} \cdot T \cdot \frac{(n - l)}{n} = \frac{1}{2} \cdot T \cdot \left(1 - \frac{l}{n}\right) \quad (23.5)$$

Note the meaning of Eq. (23.5): given that a user was delayed, this is the delay that the user suffers "on average." It is remarkable that this expected delay is a function only of the "low" capacity and of the demand rate and independent of the "high" capacity, b.

An informative way to display the behavior of this queue is through the *cumulative flow diagrams* shown in Fig. 23-6. The cumulative flow diagrams typically show (*a*) the total ("cumulative") number of user requests for service that have been made between $t = 0$ and the current time t, and (*b*) the total number of these requests that have been admitted for service up to the current time t. The former is the *cumulative demand diagram** and the latter is the *cumulative admissions-to-service diagram*. Obviously, the difference between the cumulative demands and the cumulative admissions to service at any time t is the number of users *queued for admission* to the service facility. Figure 23-6 shows these two cumulative diagrams for the case at hand. Note that the vertical axis of Fig. 23-6 maintains a count of the

*In fact, the cumulative demand diagram simply plots the value of $\int_0^t \lambda(x) \cdot dx$ for all values of t.

cumulative number of demands and of admissions-to-service as they occur. For this example, the demand and admissions-to-service cumulative flow diagrams coincide up to $t = a$, because demands are admitted as soon as they show up at the service facility, due to the high capacity h. At $t = a$, however, the number admitted begins lagging behind the demand, as it increases only with a slope of l users per unit of time. This lasts until $t = b$, when the number admitted starts increasing at the rate of h per unit of time, until it eventually "catches up" with the demand curve at time $(n - l) \cdot T/(h - n)$ later. Thereafter the two cumulative flow diagrams coincide again and increase at the rate of n per unit of time, because capacity is higher than the demand rate.

Note that the cumulative flow of admitted-for-service users can never exceed, by definition, the cumulative flow of the demands. At best, it can always be equal to the cumulative flow of demands, in cases where no delays occur due to the capacity being equal to or greater than the demand rate for all t. The (positive) vertical distance and the (positive) horizontal distance between the cumulative flow diagrams in Fig. 23-6 both have important physical interpretations. The *vertical distance* at any time t gives the *number of users in queue* at that time. It shows the difference between the number of users that have demanded service up to time t and the number admitted for service. Plotting that vertical distance as a function of time for Fig. 23-6 leads to Fig. 23-5. Similarly, the *horizontal distance* gives the *delay suffered by the i-th demand*, i.e., the time that elapses between the instant when the i-th demand requests service and the time when that demand is admitted for service, assuming that demands are admitted

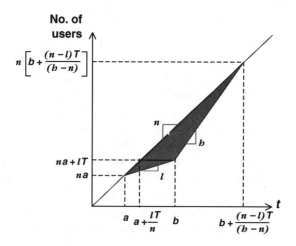

Fig. 23-6. *Cumulative demand and service diagrams.*

for service in first-come, first-served (FCFS) order.[*] Note that the total "area" between the two cumulative flow diagrams is equal to the total delay time in units of user time (e.g., passenger-hours). Thus, this area is equal to the quantity shown in Eq. (23.4) (see Exercise 1).

The user that demands service at the instant when the horizontal distance between the cumulative flow diagrams is greatest, assuming a FCFS priority discipline, will suffer the longest delay of all. From Fig. 23-6 it is clear that this is the user that will be *admitted* for service at $t = b$, because the horizontal distance between the two cumulative flow diagrams begins decreasing immediately thereafter. It can be seen that this is the $(n \cdot a + l \cdot T)$th user to demand service (and to be admitted for service). This user demands service at the time t that satisfies $n \cdot t = n \cdot a + l \cdot T$, that is, at $t = a + (l \cdot T)/n$. Since it is already known that this user will be admitted for service at exactly $t = b$, the delay the user will suffer is given by

$$\text{Longest delay suffered by any user} = b - \left(a + \frac{l \cdot T}{n}\right) = T \cdot \left(1 - \frac{l}{n}\right) \quad (23.6)$$

From Fig. 23-6, taking advantage of the observations in the last paragraph, one can now prepare Fig. 23-7. This shows, for all values of t, the amount of delay that a demand *requesting service* at time t will suffer, under the FCFS assumption.

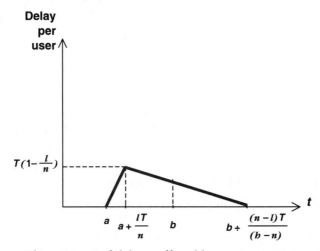

Fig. 23-7. *The amount of delay suffered by a user requesting service at time* t.

---

[*]If a FCFS priority discipline is not in effect, then the horizontal distance is the time that elapses between the i-th demand by a user for service and the i-th admission for service of a *possibly different user.*

EXAMPLE 23.2

It is instructive to assign some realistic values to the various parameters of Fig. 23-4 and look at the implications. Assume that
the situation examined is the queuing of aircraft for use of the
runway system at a major European airport. This example uses
values that may be typical of such an airport.

Let the time units be hours, and set $t = 0$ to correspond to
07:00 local time, $a = 1.0$ and $b = 4.0$, so that the duration, T,
of the low-capacity period is 3 h. The time a may correspond
to the beginning of a period of fog lasting from 08:00 to 11:00.
Let also $n = 60$ aircraft movements per hour, $h = 70$ aircraft
movements per hour, and $l = 35$ aircraft movements per hour.
The poor weather conditions reduce the capacity to one-half its
normal value—not an unusual phenomenon in practice.

Equation (23.3) indicates that a queue will be present for a period
of 10.5 h beginning at 08:00 local time. The after-effects of the
poor weather thus persist for 7.5 h after the weather event ends
at 11:00! With 60 aircraft movements scheduled per hour, a total
of 630 movements will suffer some delay. The peak length of
the queue is 75 movements (Fig. 23-5 or 23-6) and occurs at
11:00 local time, but the peak delay time is suffered by the movement scheduled for $t = 2.75$ or for 09:45 local time. This
movement will suffer a delay equal to 1.25 h, or 75 min [from
Eq. (23.6)] and will actually reach the runway system at 11:00,
exactly the instant when the weather event ends. The total
amount of delay time incurred during the day, from Eq. (23.4),
is equal to 393.75 aircraft-hours!

The economic cost of this delay depends primarily on the mix of
aircraft at this airport and on whether the delays will be absorbed
while the aircraft are airborne or on the ground (Chap. 13). Just
to indicate the order of magnitude of the cost, assume $2000 as
the direct operating cost to airlines of one aircraft-hour—a typical
amount for aircraft using major airports. This gives approximately
$800,000 as the total cost of delay due to the 3-h weather event,
not including the cost of delay time to the passengers! From Eq.
(23.5), the average delay per movement for the 630 movements
delayed is 0.625 h or 37.5 min, for a cost of approximately
$1250 per aircraft at the assumed $2000 per aircraft-hour.

The simple case examined in this section demonstrates, among other things, how cumulative flow diagrams should be drawn and interpreted. It is straightforward to extend this approach to the more general case in which (1) both the demand rate and the capacity vary over time and (2) there are several, not just one, instances during the day when the demand is higher than the capacity. Figure 23-8 shows a typical example of a cumulative flow diagram for such a more general case. The only differences from Fig. 23-5 are that (1) the "demand" and the "admitted for service" cumulative flow diagrams undergo several slope changes and (2) there may be several instances during the day when there is a backlog of demands waiting to be served, i.e., when delays will occur. The cumulative diagrams need not be piecewise linear functions, as in Fig. 23-8. They can have other shapes, depending on the functional forms used to describe the demand rate, $\lambda(t)$, and capacity, $\mu(t)$, during the time interval of interest.

The deterministic model and approach described in this section have a fundamental deficiency: they capture *only* overload delay. For instance, the deterministic model in Example 23.2 states that there would be *no delay* at the runway system if the airport could operate all day at the high-capacity level of 70 movements per hour, which exceeds the demand of 60 movements per hour. In practice, however, one might see some very significant delays under these conditions, if the demands occurred sufficiently randomly in time to create periods of traffic surges and/or if the service times on the runway system exhibited significant variability. *Stochastic* queuing models and analysis can capture delays of this type.

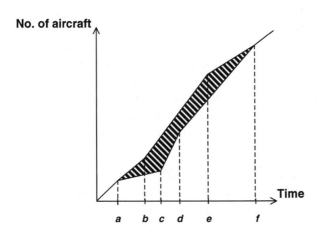

Fig. 23-8. *Cumulative demand and service diagrams for a more general case.*

23-6 Long-term behavior of queuing systems

This section discusses the behavior of queuing systems from a *long-term perspective*. It looks at the typical characteristics of a queuing system under *equilibrium conditions,* i.e., over a span of time that is sufficiently long to allow the system to "settle down" to a statistically repetitive type of behavior. This does *not* mean that there will be no variability in W_q, N_q, and other related quantities over time, but that this variability involves probabilistic fluctuations around certain long-term average characteristics. These long-term average characteristics are precisely those one wishes to observe by adopting the macroscopic viewpoint. Stated more mathematically, the discussion focuses on the statistical description of the queuing system as the period, T, during which it operates tends to infinity.

A fundamental condition needs to be satisfied if a queuing system is to reach long-term equilibrium behavior, or *steady state* in the terminology of queuing theory. That condition illustrates in a simple way the meaning of "long-term equilibrium." The last section examined the behavior of a queue when, for a specific interval of time, T, the demand rate, λ, exceeds the service rate, μ, i.e., the utilization ratio ρ is greater than 1. From the "short-term" point of view, operating a queuing system in overload conditions ($\rho > 1$) is perfectly normal: queue length and delays will increase during the interval T and then will presumably decrease when the period of overload is finished. However, in the long term, one cannot operate a queuing system in overload conditions and reach any type of equilibrium. If ρ is greater than 1 *on average* over time, for the infinite period of observation, T, the length of the queue and the waiting time at the queuing system will grow without bound over time. This would happen even if demand exceeded capacity by only a minuscule amount per unit of time. The server(s) would on average "fall behind" demand during every unit of time, so that more and more prospective users would accumulate in the queue. It follows from this argument that a queuing system can reach long-term equilibrium conditions only if $\rho < 1$, i.e., only if the long-term demand rate is strictly less than the long-term service rate.[*]

Two fundamental observations from queuing theory can now be presented: Little's law and a description in quantitative terms of the nonlinear behavior of queuing systems.

---

[*]We have not discussed what happens when $\rho = 1$. It can be proved that queuing systems do not reach equilibrium (i.e., queues and waiting time go to infinity) if $\rho = 1$; the only exception is the very special case when both the demand interarrival times and the service times are constant with $\lambda = \mu$.

Little's law

In addition to W_q and N_q, queuing theory is interested in the characteristics of two other random variables:

W = total amount of time spent by a user in a queuing system

N = total number of users in the queuing system

Note that W is simply equal to the sum of the amount of time, W_q, that a user spends waiting to be admitted to service and the time the user spends in service. Similarly, N is the sum of the number, N_q, of users waiting in queue and the number of users being served. (In a single-server queuing system the number of users being served is at most one.) When a queuing system is in steady state, the expected values of the four important random variables, W, W_q, N, and N_q, satisfy the following three relationships:

$$E[N] = \lambda E[W] \tag{23.7}$$

$$E[N_q] = \lambda E[W_q] \tag{23.8}$$

$$E[W] = E[W_q] + 1/\mu \tag{23.9}$$

Equation (23.9) follows directly from the definition of W. If a server processes μ users per unit of time on average (the service rate), then the expected service time is equal to $1/\mu$. The expected value of the total time in the system, $E[W]$, is then given by Eq. (23.9).

Equations (23.7) and (23.8) are both statements of Little's law (Little, 1961). The following argument provides an intuitive explanation (but not proof) of Little's law, as expressed by Eq. (23.7). In the steady state, i.e., with the system in equilibrium, the average number of users that a random user finds at the queuing system upon arrival should be equal to the average number (s)he leaves behind upon departure, with both of these numbers equal to $E[N]$. But the average number of users left behind is simply the demand rate per unit of time, λ, times the average amount of time, $E[W]$, that a random user stays in the system!

Relationship between congestion and utilization

Queuing theory has led to the discovery of the following relationship between congestion at a queuing system and the intensity with which such a system is utilized:

Under steady-state conditions, $E[W]$, $E[W_q]$, $E[N]$, and $E[N_q]$, at any queuing system, increase nonlinearly with ρ, in proportion to the quantity $1/(1 - \rho)$.

A graph of $E[W_q]$ for a particular queuing system is shown in Fig. 23-9, where the horizontal axis is the long-term demand rate as a fraction of the long-term service rate. This graph is typical of the behavior described by the above relationship. As ρ approaches 1, or as the demand rate approaches the service rate (or, colloquially, as "demand approaches capacity"), $E[W_q]$ grows nonlinearly. It reaches infinity at $\rho = 1$, as explained in the discussion on the necessary condition for equilibrium ($\rho < 1$).

The mathematical expressions for $E[W]$, $E[W_q]$, $E[N]$, and $E[N_q]$ depend on the specifics of the queuing system under consideration. For example, consider a single-server system at which demands arrive at entirely random times according to a Poisson process. Interarrival times are thus described by the negative exponential probability distribution with parameter λ, the demand rate per unit of time. Suppose that the service rate is equal to μ and that the service time, S, has a variance equal to $\sigma^2(S)$. Finally, let this system have infinite queue capacity. The queuing system just described is known in queuing theory as an M/G/1 system.[*] It is important because of its many applications. For the M/G/1 system, it can be shown that

$$E[W_q] = \frac{\lambda \cdot [(1/\mu)^2 + \sigma^2(S)]}{2 \cdot (1 - \rho)} = \frac{\rho^2 + \lambda^2 \cdot \sigma^2 (S)}{2\lambda \cdot (1 - \rho)} \qquad (23.10)$$

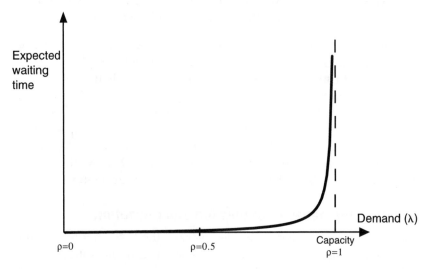

Fig. 23-9. *Expected time in queue as demand increases.*

[*]The code "M/G/1" means that the demand is Poisson (indicated by the letter "M" for "memoryless"), the service times can have any probability distribution (indicated by the letter "G" for "general"), and the system has one server.

From Eq. (23.10), one can also compute $E[W]$, $E[N]$, and $E[N_q]$, using Eqs. (23.9), (23.7), and (23.8), respectively. Note that all one needs to know to use Eq. (23.10) are the values of λ, μ, and the variance of the service time $\sigma^2(S)$. The equation is valid only as long as $\lambda < \mu$, or equivalently, $(\lambda/\mu) = \rho < 1$. Example 23.3 illustrates through Eq. (23.10) the sensitivity of delay to even small changes in the demand rate when ρ is close to 1.

Figure 23-10 presents another example that compares the values of $E[N_q]$ for two different M/G/1 queuing systems, A and B. Both systems have $\mu = 60$ per hour, so that the expected service time $1/\mu$ is equal to 1 min. However, system A has deterministic service times, meaning that $\sigma^2(S)$ is equal to zero, while system B exhibits significant variability of service times with a standard deviation equal to 0.9 min [or $\sigma^2(S) = 0.81$ min^2]. While the overall shape of the two curves for $E[N_q]$ in Fig. 23-10 is dominated by the $1/(1 - \rho)$ term, the values that the two expressions take as a function of ρ differ significantly, with $E[N_q]$ for the high-variability system B increasing faster. In general, the higher the variability (as measured by their variance) of the demand interarrival times and of the service times, the faster $E[W_q]$ and $E[N_q]$ increase as λ and ρ increase or as μ decreases.

Equation (23.10) is valid only if the demand rate and the service rate are constant over time. Queuing theory does not provide any closed-form expressions, such as Eq. (23.10) for cases in which λ and μ vary with time, as usually happens at airports due to demand peaks, weather changes, etc. For these time-varying systems, $E[W]$, $E[W_q]$,

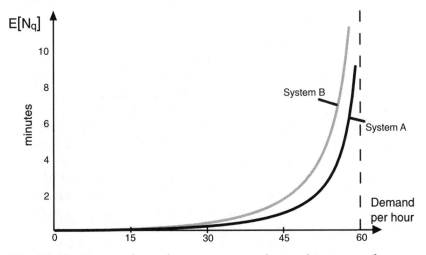

Fig. 23-10. *Expected number in queue as demand increases for two different queuing systems.*

EXAMPLE 23.3

Consider a single-server airport facility with a capacity of approximately 48 per hour (μ=48 per hour). (This, for instance, could be a runway used for both arrivals and departures, where it takes about 75 s on average between successive operations.) Assume that the service times at this facility can be approximated by a random variable whose expected value is 75 s and whose standard deviation is 25 s. Demands at this facility occur at a steady rate throughout the busy hours of the day (e.g., for 15 or 16 h). Demand occurrences can be reasonably approximated as Poisson, so that demand interarrival times have an approximately negative exponential probability distribution.

These assumptions are consistent with the description of the M/G/1 queuing system above. Therefore Eq. (23.10) can be used to compute $E[W_q]$. Consider the case where $\lambda = 36$ demands per hour. Applying Eq. (23.10) with* $\lambda = 36/3600 = 0.01$ demands per second, $1/\mu$ 75 s, $\rho = 36/48 = 0.75$, and $\sigma^2(S) = (25)^2$ gives $E[W_q] = 125$ s. One then obtains $E[W] = 125 + 75 = 200$ s [from Eq. (23.9)]; $E[N_q] = (0.01)(125) = 1.25$ aircraft in queue [from Eq. (23.8)]; and $E[N] = (0.01)(200) = 2$ aircraft in the system [(from Eq. (23.7)].

Table 23-1 shows indicative approximate values for the expected length of the queue and the waiting times, for several values of the demand rate. It also shows the changes resulting from a 1 percent increase in the demand rate, λ, at the 0.625, 0.75, 0.875, and 0.9375 levels of utilization.

This model may approximate reality only roughly. Table 23-1 nonetheless underscores several points. Note, for example, that a 1 percent increase in demand at the 87.5 percent level of system utilization results in an almost 10 percent increase in expected queue length and 8.6 percent increase in delay. These percentages jump to roughly the 20 percent level when the system operates at about 94 percent of its capacity in the long run. Table 23-1 also suggests that the expected delay reaches the 4-min level when the demand rate, λ, is equal to roughly 41 per hour. If this system were a runway, its practical hourly capacity, PHCAP, would be about 41 aircraft per hour and it is reached at about 85 percent utilization (= 41/48)—see also Chap. 10.

---

*Note the importance of using consistent units of time for all the parameters; in this case "seconds" is the time unit used.

Table 23-1. The indicative approximate values for the expected length of the queue and the waiting times, as derived from queuing theory

Arrival rate λ (per hour)	Service rate ρ	Expected number in queue		Expected waiting time	
		$E[N_q]$	$E[N_q]$ (% change)	$E[W_q]$ (seconds)	$E[W_q]$ (% change)
30	0.625	0.58		69	
30.3	0.63125	0.60	3.4%	71	2.9%
36	0.75	1.25		125	
36.36	0.7575	1.31	4.8%	130	4%
42	0.875	3.40		292	
42.42	0.88375	3.73	9.7%	317	8.6%
45	0.9375	7.81		625	
45.45	0.946875	9.38	20.1%	743	18.9%

$E[N]$, and $E[N_q]$ can only be estimated through numerical techniques or through computer-based simulation (Chap. 19). However, as long as the queuing system reaches a long-term equilibrium, the observation that the long-term expected values $E[W]$, $E[W_q]$, $E[N]$, and $E[N_q]$ are proportional to $1/(1 - \rho)$ holds true, even if the demand rate and the service rate are functions of time. In such cases, one should be careful to interpret λ, μ, and ρ as the long-term averages of the demand rate,[*] the long-term average of the service rate (capacity), and the long-term average utilization of the system.

23-7   Policy implications

The observations of the last section and the "generic shape" of $E[W]$, $E[W_q]$, $E[N]$, and $E[N_q]$ shown in Figs. 23-9 and 23-10 have important implications for airports at the *policy level*. First, they are a warning to airport operators, airlines, and civil aviation managers not to operate airport facilities and services at levels of utilization which are close to 1 over an extended period of time. Doing so risks having long delays, long waiting lines, and a poor LOS.

Moreover, queuing theory has also shown that not only do the expected values of W, W_q, N, and N_q increase in proportion to $1/(1 - \rho)$,

[*]More formally, we compute as $\left(\int_0^t \lambda(x) \cdot dx\right)/T$ as $T \to \infty$.

but so also do their standard deviations, $\sigma(W)$, $\sigma(W_q)$, $\sigma(N)$, and $\sigma(N_q)$. This means that, when ρ is close to 1, a queuing system not only experiences serious congestion, but is also subject to great variability. Under the same set of *a priori* conditions (e.g., for the same λ, μ, and probability distributions for demand interarrival times and service times), delays on a particular day may be modest and tolerable and, on the following day, extremely long and unacceptable. This is a phenomenon observed very often at the busiest airports throughout the world.

It is difficult to make any precise general statements about the utilization ratio at which an airport facility or service should ideally be operated. The most appropriate value depends both on the particular operating characteristics of the system (probability distribution of service times and of demand interarrival times, variability over time of the demand rate and of the service rate, number of servers, etc.), and on the measures of performance considered most important (economic and other perceived costs of delay times, cost of the queuing system when idle, emphasis on avoiding extreme delays, etc.).

In most cases, it is fair to say that any facility or service operating at the range of 80 to 95 percent of its capacity for the duration of the consecutive active traffic hours of the day ("long-term" ρ between 0.8 and 0.95) is near the "danger zone," or already in it, as far as serious delays are concerned. A long-term utilization ratio of more than 0.9 (when $1/(1 - \rho)$ is 10 or greater) usually means long delays, low LOS on many days, and unstable conditions. These reflect the large expected value and standard deviation of waiting times and queue lengths. The reference to *active* traffic hours in the above should be noted. The consecutive hours of truly active traffic in a day are considerably less than 24 at the great majority of airports. Moreover, certain facilities or services may be utilized for only some of those hours.

The second major point at the policy level is that when a queuing system operates at high levels of utilization, small changes in demand or capacity can cause large changes in delays and queue lengths. This simple practical observation is a direct consequence of the proportionality of both the expected value and the standard deviation of W_q and N_q to $1/(1 - \rho)$. It motivates much that is being done at major airports around the world. Many initiatives are aimed at either managing/controlling demand (see Chap. 12) or at increasing the airside and landside capacity of these airports (Chaps. 13 and 16). Airport operators generally recognize that most of these initiatives will only produce small changes in demand or capacity. However, because many facilities and services at busy airports operate at very high utilization ratios, airport operators expect that these small changes will produce significant reductions in delay. The reductions may be sufficient to maintain acceptable levels of service for a few additional years until more dramatic improvements in capacity might be achieved.

Exercises

1. Show that the area between the cumulative diagrams in Fig. 23-6 is equal to the area of the triangle in Fig. 23-5. Both areas are given by Eq. (23-4).

2. Demand for the runway system at an airport is 90 movements per hour throughout the busy hours of the day, except for the period 10:00–12:00 when it is 70 movements per hour. Suppose that, on a given day, the airport capacity was 100 movements per hour until 7 a.m. However, due to a weather front, the capacity was only 60 movements per hour between 7 a.m. and 9 a.m. From 9 a.m. to 11 a.m. the capacity increased to 80 movements per hour and, finally, at 11 a.m., the capacity went back to 100 movements per hour, where it stayed for the rest of the day.

   Under the usual assumptions described in Sec. 23-5, draw carefully the cumulative diagram for the number of demands as a function of time and the number of aircraft "admitted" for service at the runway system as a function of time. Begin your picture at 6 a.m. What is the longest delay suffered by any aircraft during this day? What is the total amount of delay suffered by all aircraft during that day?

3. Consider an airport with a runway used exclusively for landings during peak traffic hours. Under such peak conditions, the arrivals of airplanes at the vicinity of the airport can be assumed to be approximately Poisson with a rate $\lambda = 55$ aircraft per hour. Of these airplanes, 40 on average are commercial jets and 15 are small general aviation and commuter airplanes. The probability density function for the duration of the service time, S, to a random aircraft landing on the runway is uniformly distributed between 48 and 72 s. Peak traffic conditions occur during 1000 hours per year, and the average cost of 1 min airborne waiting time (i.e., of time spent in the air while

waiting to land) is $40 for commercial jets. (This accounts for additional fuel burn, extra flight crew time, and other variable operating costs.) Estimate the yearly costs to the airlines of peak traffic conditions. Assume that Eq. (23.10) for estimating waiting time is valid for this case.

References

Ashford, N. (1988) "Level of Service Design Concept for Airport Passenger Terminals—A European View," *Transportation Planning and Technology, 12*(1), pp. 5–21.

Gross, D., and Harris, C. (1998) *Fundamentals of Queuing Theory,* Wiley, New York.

Hillier, F., and Lieberman, G. (1995) *Introduction to Operations Research* 6th ed., McGraw-Hill, New York.

Larson, R. (1987) "Perspectives on Queues: Social Justice and the Psychology of Queuing," *Operations Research,* 35, pp. 885–905.

Larson, R. (1988) "There Is More to a Line than Its Wait," *Technology Review,* pp. 60–67.

Larson, R., and Odoni, A. (1981) *Urban Operations Research,* Prentice-Hall, Englewood Cliffs, NJ.

Little, J. D. C. (1961) "A Proof of the Queuing Formula $L = \lambda W$," *Operations Research, 9,* pp. 383–387.

Wolff, R. (1989) *Stochastic Modeling and the Theory of Queues,* Prentice-Hall, Englewood Cliffs, NJ.

24

Peak-hour analysis

Airport facilities are typically designed to accommodate loads during a design peak hour. Many alternative definitions of the design peak hour have been proposed and are used in practice. All these definitions share a common characteristic: they specify a level of traffic that is exceeded only during a small number of hours, e.g., 30 or 40, in the target year. The intention is to ensure that airport facilities have adequate capacity to handle demand at a desired level of service practically throughout the year, while not being overdesigned just to handle a few instances when extreme peaks may occur.

To estimate design peak-hour loads planners must review carefully historical data to understand seasonal, monthly, daily, and hourly peaking patterns at the airport under consideration. Judgment must also be used in assessing how these patterns will change in the future. Demand peaking at airports usually becomes less acute as traffic grows. Additional care must be exercised in distinguishing between the peaking characteristics of passengers versus those of air traffic, as well as of arriving passengers versus departing passengers.

Annual demand forecasts can be converted into design peak-hour traffic estimates through the application of conversion factors. These factors are based on experience with peaking patterns at airports of various sizes. They usually provide good first-order approximations, but should not be used uncritically.

24-1   Introduction

Much of airport planning and design revolves around the notion of the *design peak day* (DPD) and, especially, of the *design peak hour* (DPH)—also referred to as the *typical peak hour* (TPH) or sometimes simply as the *design hour* or the *peak hour*.

Airport facilities are designed to accommodate loads that are typically specified both on an *annual* and an *hourly* or a *daily* basis. Thinking in terms of annual totals is useful for "macroscopic" planning purposes, because forecasts of traffic are, as a rule, given in annual terms. One can readily compare, for instance, the capacity of a passenger building designed to handle 20 million passengers per year with a forecast of annual passenger demand in a particular year in order to determine whether the building will be adequate to handle traffic that year. The same is true when it comes to planning for other parts of the airport, such as cargo facilities or the airfield—in which case the relevant demand forecast is the annual number of aircraft movements.

However, hourly figures and daily figures are typically far more important for the purpose of detailed design. It is short-term loads that determine the required size of a facility, the number of servers in each of its constituent parts, etc. For example, in designing an airfield, one would like to know the expected number of design peak-hour movements. If, for instance, that number is 40, a single runway will probably suffice; but if it is 65, a two-runway system will be necessary (see Chap. 10). Similarly, as indicated in Chap. 15, space requirements in each part of a passenger building are determined with reference to the number of simultaneous occupants during the peak *hours* of the year. The number of required processing units (check-in desks, passport control desks, security-check machines, etc.) is also determined by the flows in the building during these hours. Likewise, when wishing to determine whether a group of contact and remote stands at an airport has adequate capacity, one typically works with a scenario involving a *daily* schedule of arriving and departing aircraft and must necessarily select for examination a day when the number of movements is high.

This chapter reviews the estimation of flows during design peak hours and thus has a bearing on several other topics in this book. It first discusses the many existing alternative definitions of the DPH along with a common condition that all these definitions should satisfy. It then describes a simple approximate process for converting annual forecasts of traffic into DPH and DPD forecasts. The final two sections address two specific questions that often come up in practice. The first concerns estimating the number of DPH aircraft movements, while the second deals with the number of DPH arriving passengers and of DPH departing passengers.

24-2 Definition of the design peak hour

Many alternative definitions of DPH are in use. The following is a nonexhaustive list of possibilities:

1. The 20th, 30th, or 40th busiest hour of the year

2. The peak hour of the average day of the peak month of the year

3. The peak hour of the average day of the two peak months of the year

4. The peak hour of the 95th percentile busy day of the year, i.e., the peak hour of, roughly, the 18th busiest day of the year

5. The peak hour of the 7th or 15th busiest day of the year

6. The peak hour of the 2nd busiest day during the average week in a peak month

7. The "5 percent busy hour," i.e., an hour selected so that all the hours of the year that are busier handle a cumulative total of 5 percent of annual traffic.

All of these definitions have been used at times or been recommended by various organizations. For example, in the United Kingdom it is standard practice to use definition 1, specifically, the 30th busiest hour of the year. The corresponding level of traffic is called the *standard busy rate.* In the United States, definition 2 is often used (FAA, 1988), while the International Civil Aviation Organization (ICAO, 2000) has recommended the third.

For practical purposes, it makes little difference which definition is used as long as it fulfills the following condition: the DPH should not be the hour of the year with the highest traffic demand, but one with a demand that is exceeded only during a reasonably small number of days of the year. "Reasonably few" may mean something like 10 to 30 days, depending on the context and on the intensity of demand peaking at the airport of interest. This condition is intended to ensure that airport facilities have adequate capacity to handle demand practically throughout the year, while not being overdesigned just to handle a few instances when extreme peaks may occur. Such extreme peaks may be associated with a few days each year when traffic is exceptionally heavy (e.g., a holiday period or an annual religious pilgrimage) or with certain special events (e.g., the annual Super Bowl game in the United States or the Olympic Games somewhere in the world every four years). The practice of selecting the DPH in this way implicitly recognizes that some deterioration of the level of service during a few hours or days of extremely high demand should be tolerated in the interest of reducing capital and operating costs.

It can be seen that all of the definitions listed at the beginning of this section satisfy this condition. Moreover, the differences among the estimates of DPH demand that are obtained from these alternative definitions are typically insignificant compared to the uncertainty stemming from: (1) the likely errors in the forecast (Chap. 20), especially when the target date for which the facility is being designed lies 5–10 years in the future, as is often the case; and (2) the use of many other simplifying assumptions and approximations in all design methodologies.

Airport planners and designers should feel free (unless restricted to adhere to local practice, as is the case with the use of the "standard busy hour" in the United Kingdom) to select any one of the definitions, as appropriate to the case at hand and the available data. Note that definition 2 is often the least demanding, in terms of data, as it requires detailed hour-by-hour information only for the peak month of the year.*

24-3   Conversion of annual forecasts into DPH forecasts

It is usually the case that planners obtain DPH forecasts—whether for passengers or aircraft movements or some other measure of demand for an airport facility—in a "top-down" fashion from annual forecasts. In other words, one typically begins from a given annual forecast and obtains a DPH forecast by applying appropriate "conversion coefficients," based on historical data which are often adjusted judgmentally. The reason for starting with annual forecasts is that these are in most cases the only forecasts available. Airport forecasting methodologies (Chap. 20) are, by nature, strongly oriented toward predicting annual figures of demand. The unadjusted conversion coefficients can be estimated rather easily, once the DPH has been defined (e.g., according to one of the alternatives 1–7 in the previous section). Finally, in making the judgmental adjustments to the conversion coefficients, planners should recall that peaking patterns at an airport are usually quite stable from year to year, but peaking becomes less acute as traffic increases. This empirical observation is discussed further, after Example 24.1.

*If using definition 2, it is best to exclude from consideration days on which demand is consistently much lower than on typical days; for example, at many airports in the United States, Saturdays have much lower demand than all other days of the week.

EXAMPLE 24.1

Suppose an airport that today handles 12 million passengers per year is forecast to have a demand of 18 million 10 years from now, a 50 percent increase. Suppose also that local practice defines the 30th busiest hour of the year as the DPH and, based on the current year's traffic data, the number of arriving and departing passengers during that hour is 4500. Thus, the fraction of annual passenger traffic processed during the DPH is now equal to 0.000375 (= 4500/12,000,000) or 0.0375%. This would suggest a projected 6750 DPH passengers 10 years from now [= (0.000375) × (18 million) = (1.5) × (4,500)]. Planners, however, should probably adjust this estimate. For example, by looking at the historical evolution of peaking patterns at the airport and speculating on the most likely ways these might change, one may decide to reduce the conversion coefficient 0.000375 to account for a reduction in peaking as the annual traffic increases. A value such as 0.00033 may be deemed more appropriate for use with the 18-million-passenger level, resulting in an estimate of about 6000 DPH passengers. Plausibility checks should also be performed. Will the runway system, for example, be capable of handling aircraft operations consistent with a rate of 6750 (or 6000) passengers per hour, given the projected mix of aircraft types at the airport?

The stability of *monthly* peaking patterns is illustrated in Fig. 24-1, showing the total passenger traffic at the three airports that serve the New York City area in 1990, 1993, and 1997. Note that while the volume of traffic changed considerably during these years, the peaking pattern remains remarkably consistent. Figure 24-2 illustrates a similar stability for the monthly passenger demand at New York/Newark. Again, despite a very significant increase in the number of passengers, the monthly pattern is very stable. In a similar fashion, the *hourly* peaking patterns in February, the lowest month of the year, and in August, the highest month, closely follow each other (Fig. 24-3), despite the fact that the peak demand in August is over 30 percent greater than in February. However, the peaking at any given airport may become less acute as its traffic increases over the years, because the distribution of the number of aircraft movements, as well as of the number of passengers, across both the months of the year and the hours of the day often tends to become progressively "flatter" as the annual numbers grow. Figure 24-4 illustrates this for the case

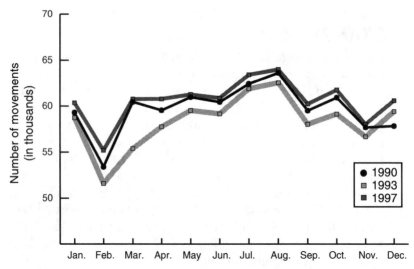

Fig. 24-1. *Total monthly runway traffic at the three New York City airports in 1990, 1993, and 1997; the downturn in December 1990 reflects the onset of the Gulf War.*

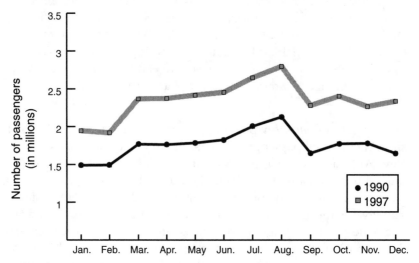

Fig. 24-2. *Monthly passenger traffic at New York/Newark in 1990 and in 1997.*

of the average weekday profile of aircraft movements at Boston/Logan in 1993 and 2000. Note how operations have spread more evenly by 2000.

The U.S. Federal Aviation Administration (FAA) long ago recommended a set of conversion coefficients, shown in Table 24-1, for estimating

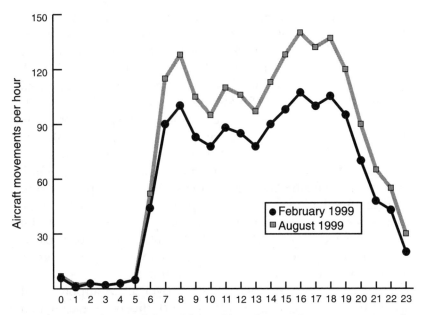

Fig. 24-3. *Similarity of the 24-h profiles of aircraft movements on an average weekday in February and in August 1999 at Boston/Logan.*

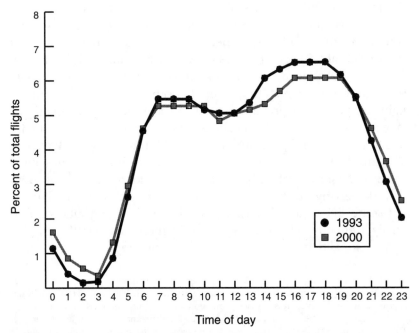

Fig. 24-4. *Hourly peaking patterns at Boston/Logan in 1993 and in 2000; the vertical axis indicates the percent of total daily movements taking place in each hour.*

approximately the number of DPH passengers* from annual forecasts of demand (FAA, 1969). Experience indicates that these coefficients work very well in practice, *as first-order approximations*. Their use is recommended when a rough "figure of merit" is needed quickly or in the absence of detailed historical data on an airport's peaking patterns. Table 24-2 offers further insight into the peaking characteristics of airports. It summarizes the statistics on the peak-month traffic obtained from an Airports Council International (ACI, 1998) survey of 80 airports worldwide.[†] For example, for the 23 airports in the survey that had more than 20 million annual passengers, the *monthly peaking ratio*, i.e., the "average number of passengers per day during the peak month of the year" divided by the "average number of passengers per day during the entire year," ranged between 1.09 and 1.43, with an average value of 1.18. Stated differently, average daily passenger traffic during the peak month was from 9 to 43 percent greater than the average daily passenger traffic during the entire year[‡]—and was 18 percent greater when averaged over all 23 airports. In the case of 6 of these 23 airports (26 percent), the average daily traffic during the peak month was more than 20 percent greater than the average during the entire year.

The third and fifth columns of Table 24-2 suggest that what is true for individual airports ("monthly peaking tends to become less acute as traffic increases") is also generally true across airports. However, the considerable range of values in the fourth column underlines the importance of utilizing, whenever possible, data which are specific to the airport of interest. In general, airports with primarily domestic flights and those serving large numbers of business passengers will have relatively low monthly peaking ratios, in contrast to those serving mostly vacation and pleasure travelers. The three New York airports offer an excellent case in point. LaGuardia, with a large amount of domestic business traffic, has a monthly peaking ratio of only 1.10. Newark, serving mostly domestic—including many low-fare flights— and some international traffic, has a higher peaking ratio of 1.20. Finally, Kennedy, which serves primarily international traffic, exhibits high seasonality, with a 1.31 monthly peaking ratio.

---

*The FAA uses the term "typical peak hour passengers" (TPHP).
[†]This is the sample of airports that have provided seemingly reliable answers to the relevant question in the ACI survey. It is not clear how representative this sample is, as it includes 38 European, 30 North American, 8 Asian, 2 African, 1 Australian, and 1 South American airports.
[‡]The six airports in the sample with more than 35 million passengers (Atlanta, Denver, Frankfurt, London/Heathrow, Los Angeles/International, and San Francisco/International) all had very similar monthly peaking ratios (1.18, 1.19, 1.19, 1.14, 1.19, and 1.22, respectively).

Table 24-1. Conversion coefficients for estimating the number of DPH passengers from annual figures

Total annual passengers	DPH passengers as a percentage of total annual passengers
More than 20 million	0.03%
10 million–20 million	0.035%
1 million–10 million	0.04%
500,000–1 million	0.05%

Source   FAA, 1969.

Table 24-2. Monthly peaking characteristics of the 80 airports in the ACI survey

Total annual passengers	Sample size	Average monthly peaking ratio*	Range of monthly peaking ratios	Monthly peaking ratio greater than 1.2
More than 20 million	23	1.18	1.09–1.43	6 of 23 (26%)
10–20 million	13	1.25	1.08–1.55	9 of 13 (69%)
1–10 million	44	1.35	1.11–1.89	34 of 44 (77%)

*Monthly peaking ratio = (average number of passengers per day during peak month)/ (average number of passengers per day during entire year).

Table 24-3. Hourly peaking characteristics of some U.S. airports

Total annual passengers	Peak-hour available seats as percentage of total daily number of available seats
More than 20 million	7–10%
10–20 million	8–12%
1–10 million	9–20%

In an analogous way, Table 24-3 shows ranges of values for the *hourly* peaking of air traffic movements for a sample of U.S. airports. For instance, at airports with more than 20 million passengers a year, the number of (arriving or departing) aircraft seats during the peak hour of a typical day constitutes anywhere between 7 and 10 percent of all aircraft seats available at the airport during that day.

Tables 24-2 and 24-3 explain why the conversion coefficients shown in Table 24-1 provide reasonable DPH estimates. Consider, for example, an airport with more than 20 million annual passengers. Selecting the values 1.18 from the third column of Table 24-2 and 9 percent, from the range of 7–10 percent in the second column of Table 24-3, gives:

(Conversion coefficient for airports with more than 20 million

$$\text{annual passengers}) = \frac{1}{365} \cdot (1.18) \cdot (0.09) \approx 0.000291 \text{ or } 0.0291\%$$

This is within 4 percent of the 0.03 percent value suggested by Table 24-3. Similarly, for an airport with 10–20 million annual passengers, one obtains

$$\frac{1}{365} \cdot (1.25) \cdot (0.10) \approx 0.00034 \text{ (or } 0.034\%)$$

which is within just 3 percent of the 0.035 percent in Table 24-1. Finally, for 1–10 million annual passengers, a typical estimate might be

$$\frac{1}{365} \cdot (1.35) \cdot (0.12) \approx 0.000444$$

or within about 10 percent of 0.04 percent.

These observations should not be interpreted as absolving planners from the responsibility of seeking, whenever possible, peaking data specific to the airport they are concerned with. As Tables 24-2 and 24-3 indicate, local monthly and hourly peaking characteristics span a considerable range of values. The resulting conversion coefficients may thus differ significantly from one airport to the next.

24-4 DPH estimates of aircraft movements

Peaking characteristics for aircraft movements are, in general, quite similar to those for passengers, but with some noteworthy differences. For one, whereas the peak month for aircraft movements almost always coincides with the peak month for passengers, this does not necessarily hold true for the peak hours.

More important, the peaking of aircraft movements is usually somewhat less acute than the peaking for passengers. The reason is that passenger load factors are generally higher during the peak season of the year and during the peak traffic hours of the day. In most cases the differences are not large and the magnitude and ranges of the conversion coefficients in Tables 24-1, and of the peaking characteristics in Tables 24-2 and 24-3 are, for practical purposes, nearly as reasonable for estimating DPH aircraft movements as DPH passengers. Airport planners, however, are advised to check carefully local peaking data, when it comes to aircraft movements, especially at those airports whose runway systems impose serious capacity constraints. In fact, some major airports that operate at the limits of their airside capacity for most of the day (e.g., New York/LaGuardia, Chicago/O'Hare, London/Heathrow) are characterized by essentially flat profiles of aircraft movements for 10–16 h of the day, as the number of movements is "capped" by the runway system's capacity. At such airports, the number of aircraft movements during the peak hours of the day is of the order of only 7 percent of the total movements in the day, i.e., at the lowest end of the ranges shown in Table 24-3. For such runway capacity-constrained airports, the simple FAA conversion coefficients of Table 24-1 may result in estimates of DPH aircraft movements that are considerably higher than the available runway capacity.

24-5   DPH estimates of flows of arriving passengers and of departing passengers

A second issue concerns DPH estimates specific to the number of peak-hour *departing* passengers and peak-hour *arriving* passengers. Such estimates are necessary for the planning and design of passenger buildings and other landside areas, where many facilities and spaces are dedicated to either serving arrivals only or departures only. Two observations are useful in this respect.

First, the peak hour for departing passengers, the peak hour for arriving passengers and the peak hour for total passengers need not coincide. In fact, it is not unusual for them to occur at three different hours of the day. As well, peaking will probably be more acute for the arriving or the departing flows alone than for total passengers. For example, the scenario of Table 24-4 is entirely plausible. It is therefore necessary to collect data specific to the *hourly* peaking pattern of arrival passengers only and of departing passengers only.*

*Note, however, that the *monthly* peaking ratios for arriving and for departing passengers— see, e.g., Table 24-2—will be virtually identical with the monthly peaking ratio for total passengers.

Table 24-4. A plausible hourly peaking pattern at a major airport

	Peak hour of day	Peak-hour flow as percentage of total daily flow
Total passenger flow	16:00–16:59	8%
Departing passenger flow	07:00–07:59	10%
Arriving passenger flow	18:00–18:59	9%

Second, the hourly peaking of arriving and departing passengers depends critically on the details of airline schedules. For this reason, arrival and departure peaking characteristics, considered separately, are less stable over the years than the peaking characteristics of overall traffic. Thus, predictions of DPH flows of departing passengers alone or of arriving passengers alone are less reliable, in general, than predictions of total DPH flows, especially when they concern target dates 5 or 10 years into the future. This is particularly true at hub airports, where the hourly peaking of arriving and of departing passengers can be expected to be very acute and may change rapidly as the hubbing airlines change their schedules and strategies.

Exercises

1. Explain why demand peaking at an airport usually becomes less acute as traffic grows. Consider seasonal peaking, day-of-the-week peaking, and hour-of-the-day peaking. How do airlines schedule additional flights on individual markets as traffic on these markets grows? Are there differences between short-haul and long-distance markets? What is the influence of airport capacity? What additional factors play a role in the gradual "de-peaking" of traffic?

2. For an airport with which you are familiar, prepare a set of graphs or tables showing the evolution of traffic peaks on a monthly, day-of-the-week, and hour-of-the-day basis over a recent 10-year period. Select, for example, the first, fifth, and tenth years of the period.

References

ACI, Airports Council International, 1998, *Airport Capacity Demand Profiles: 1998 Edition*, ACI, Geneva, Switzerland.

FAA, Federal Aviation Administration (1969) *Aviation Demand and Airport Facility Requirement Forecasts for Medium Air Transportation Hubs through 1980,* FAA, Washington, DC.

FAA, Federal Aviation Administration (1988) *Planning and Design Guidelines for Airport Terminal Facilities,* Advisory Circular 150/5360-13, U.S. Government Printing Office, Washington, DC.

ICAO, International Civil Aviation Organization (2000) *Aerodrome Design Manual, Part 1, Runways,* 2d ed., 1984, reprinted July 2000 incorporating Amendment 1, Doc 9157, ICAO, Montreal, Canada.

Index

Acapulco, 68
Accelerate stop-distance available (ASDA), 330, 332
Access (*see* Airport access)
Access control (*see* Demand management)
Accidents
  aircraft and vehicles, 36
  rates, 12
Accounting practices, 276–277
ACI (*see* Airports Council International)
Adelaide, 612
ADP (*see* Aéroports de Paris)
Adtranz, 719, 720
Advanced Surface Management and Control Systems (A-SMGCS), 550
Advani, A., 100, 124
Advisory circulars (*see* FAA)
AEA, 290
AEG, 719
AENA, 221, 222, 273
Aer Lingus, 19
Aer Rianta, 221, 222, 273
Aérogare 2, Paris, 600, 602
Aerolineas Argentinas, 173
Aeronautical revenues (*see* Revenues)
Aeroplane reference field length (*see* Airplane reference field length)
Aeroporti di Roma, 222
Aéroports de Montréal, 68, 108, 124, 138n
Aéroports de Paris, 45, 69, 138, 161, 205, 219, 222, 224, 570, 600, 606, 638, 658
Aeropuertos Argentina 2000, 223
Aeropuertos y Servicios Auxiliares, 96
Africa, 155–156, 279, 502, 612
  North, 131
  West, 97
African Development Bank, 224
Agee, M., 801
AIR-21 legislation, 243, 466, 490
Air bridge, 33–34, 570, 584
Air Canada, 7, 17, 19, 23, 97
Air conditioning (*see* HVAC)
Air Florida, 206
Air France, 7, 17, 19, 23, 49, 97, 145, 638, 728n, 731
Air Inter, 23, 97, 113
Air navigation fee, 262
  en route, 266
  terminal area, 262
Air New Zealand, 17, 19
Air pollution (*see* Pollution, air)
Air quality (*see* Pollution, air)
Air route surveillance radar (ARSR), 502, 515
Air Route Traffic Control Centers (ARTCC), 506
Air traffic control (ATC), 370–430
  (*See also* Air traffic management)
Air Traffic Control System Command Center (ATCSCC), 527

Air traffic control radar beacon system (ATCRBS), 503
Air traffic controllers, 371–398, 501
Air traffic flow management, 463, 513, 525–534
Air traffic management (ATM), xxix, 325, 370, 499–556
  critical components of, 501
  generations, 502–503
  longitudinal separation requirements, 378
  separation requirements, 377–388
  system, 368, 397–398
Air Transport Association, 44
Air transport industry (*see* Airport/airline industry)
Airborne delays, 542
Airbus Industrie, 30, 304, 364
  aircraft, 306, 308, 569, 575
  A380, 80, 183, 304, 575, 723, 817
Aircraft
  size of, 156, 618, 635
  characteristics, 303, 305–310
Aircraft approach category, 301
Aircraft boarding, 642
Aircraft certification (*see* Emissions certification; Noise certification)
Aircraft costs (*see* Seat-mile costs)
Aircraft delays, xxviii, 568, 594–595, 616
  stochastic, 626
Aircraft engines (*see* Emissions certification)
Aircraft maneuvers, 564, 693
Aircraft mix (*see* Mix of aircraft)
Aircraft operations, 577, 608
  arrivals, 122
  departures, 122
Aircraft parking charge, 262–263
Aircraft situation display (ASD), 529
Aircraft stands
  contact, 33–35, 352–354, 424
  exclusive use, 424
  positioning time, 425
  remote, 33, 352, 424, 584
  scheduled occupancy time, 425
  shared use, 424
  stand blocking time, 427
Aircraft towing (*see* Towing, aircraft)
Aircraft types
  short range, 111
Aircraft wingspan, 301, 303, 306, 308, 310, 574–575
Airfield capacity, 367–434
  (*See also* Airside capacity)
Airfield delays, 435–459
  (*See also* Delays)
Airfield design, 295–365
Airline
  alliances (*see* Airline alliances)

Airlines
  as stakeholders, 40, 563–564
  charter, 657
  cheap fare, 79, 99, 163, 659
  integrated cargo, 99, 155, 162, 163
  largest, 5, 7
  no-frills, 66, 154
  specialized, 154
Airline accidents, 10, 36, 93, 95
Airline alliances, 1, 38, 59, 71, 80, 114, 568, 599,
    615
  Oneworld, 8, 19
  Star, 8, 19, 614
Airline industry, 97–98
  size of, 4
  structure of, 113
Airline liaison officers, 565
Airline mergers, 71, 80
Airline operations centers (AOC), 509, 536
Airplane design groups, 301, 389
Airplane reference field length, 301, 302
Airport acceptance rate (AAR), 531
Airport access, xxix, 104, 139, 142, 159, 693–725
  by aircraft, 199
  by commercial vehicles, 210, 696,
  by employees, 696, 703
  control of, 209–210
  cost-effective solutions, 703–712
  distribution of, 699–700
  ground access, 572, 693–737
  nature of, 696–699
  needs of operators, 702–703
  problem, defined, 695
  regional, 695–703
Airport access, door-to-door, 52, 707
  analysis of, 707–710
Airport access, modes
  airport shuttles, 711, 717
  bullet train, 703
  busses, 685–687, 693, 703, 708, 717
  cars, 87, 685–687, 693, 703
  courtesy vans, 210, 686–687, 717
  high-occupancy vehicles, 712
  highway access, 693
  mass transit, 693, 702
  rental cars, 698, 702, 707, 709
  taxis, 52, 209, 682, 685, 686–687, 696, 698,
    701, 703, 707
  trains, 52, 87, 209, 586, 600, 702–703, 710
  vanpools, 703
  vehicles, low-emission (see Low-emission
    vehicles)
  water taxis, 702
Airport access, user preferences
  price, 701
  reliability, 701, 709
  speed, 701
Airport/airline industry, 3–5, 18, 93, 814
Airport alliances, 18, 59, 71, 80
Airport and Airways Trust Fund, 68, 243, 275
Airport Authorities, 15, 40, 69, 95, 108
  (See also: British Airports Authority; Calgary
    Airport Authority; Port Authority of
    New York and New Jersey; Massport;
    Ottawa Airport Authority; Toronto
    Airport Authority; Vancouver Airport
    Authority; Metropolitan Washington
    Airports Authority)
Airport City, 231
Airport classification codes, 300–312
Airport closings, 141–142
Airport companies, 15, 70
  (See also Aéroports de Paris; Amsterdam; BAA;
    Copenhagen; Frankfurt; Milan; Naples;
    Rome; South African Airports; TBI;
    Vancouver; Vienna; Zürich)

Airport competition, 124–125, 129
  (See also Multi-airport systems)
Airport Improvement Program, 243, 275
Airport islands (see Islands, man-made)
Airport layouts, 314–328
Airport Machine, The, 750–754, 757
Airport operators, 217–249
  (See also Organizational structures;
    Ownership)
Airport organizations, 225–234
  (See also Organizational structures)
Airport reference code, 300–304
Airport Security Federalization Act, 728
Airport services, 129
Airport sites, 140, 805–810
  securing of, 805
Airport surface detection equipment (ASDE), 515
Airport surveillance radar, 515
Airport terminals, 568
Airport traffic control tower, 506–507, 509–510
Airport Trust Fund, 68
Airport types
  commercial, xxii
  heliports, xxii
  maintenance base, 696
  military, 160
  primary, reliever, secondary (see Multi-airport
    systems)
  seaplane, xxii
  STOL, xxii
  transfer (see Transfer hubs)
Airport usability factor, 313
Airports, busiest, 5, 7, 318
Airports Council International (ACI), 26, 44,
    133n, 304, 318, 364, 736, 858, 862
Airside, xxix
  roadways, 36
Airside capacity, xxix, 122, 435–457
  annual, 375, 450–455
  declared, 373, 453
  maximum throughput, 367, 371, 414
  practical annual, 375
  practical hourly, 371, 453
  saturation, 371
  sustained, 372, 373, 453
Airside delays, 437–444
  (See also Delays)
Airside models, 750–756
Airspace structure, 504–506
Airtrain, 706
Alitalia, 17, 142, 566
Allan, S., 554
America, 1, 32
  Latin, 97, 768
  North, 14, 53, 80, 93, 108,117,120, 207, 209,
    280, 317, 369, 502–503, 514, 608, 612,
    678, 712
  South, 124, 155–156, 502
American Airlines, 7, 19, 23, 79, 112–113,
    116–117, 120, 122, 124, 155, 394n, 567,
    586, 601–602, 623
Amortization, 255
Amran, M., 629, 652, 817, 818
Amsterdam, 67, 162, 222, 567, 706
  /Schiphol, 6, 7, 48, 70, 94, 96, 100, 116,
    196–197, 199, 201, 231, 250, 283, 316,
    318, 328, 351, 376, 378, 394n, 400, 566,
    572, 601, 658, 683, 704, 711
  airport company, 20, 48, 232, 234
ANA, 19, 97, 273
Anagnostakis, I., 205, 212, 555
Anchorage, 24
Ancillary locations, 672
Andreatta, G., 418, 432, 763
Andrews AFB (see Washington)
Andreu, P., 570, 575, 583, 600, 603

Andrus, K., xxv
Anis, Z., xxv
Annual capacity, 375, 450–455
Annual value, 782, 786
Annuity, 783, 786
APM (*see* Automated people mover)
Approach control facilities, 506
Approach feeder fix, 511
Apron, aircraft, 584
    capacity of, 424–430
    dynamic capacity, 425, 429
    static capacity, 425
Apron taxiways, 345
Aprons, 351–355
Architects, 105, 570, 589, 648, 650, 655
Architectural program, 562
Area control center (ACC), 527
Area navigation (RNAV),524
ARENA ©, 758
Argentina, 16, 20, 42n, 101, 108, 234
Arlanda Express, 706
    (*See also* Stockholm)
Armstrong, J., 89
Armstrong, M., 108,124
Arrival fixes, 511
Arthur, H., 108,124
Arup/NAPA, xxv
Ascher, W., 71, 89
Asean, 599
Ashford, N., 9, 26, 180, 212, 562, 603, 610, 652,
    657–658, 689–690, 712, 736, 758, 763, 850
Ashraf, J., 198n
Asia, 5, 71, 74, 275, 317, 395, 502–503, 584, 612,
    705
ATA, 9, 27
ATFM (*see* Air traffic flow management)
ATM (*see* Air traffic management)
Athens, 218,504, 566
    /Hellenikon, 161, 323, 407,
    /Venizelos, 6, 16, 68, 69n, 100, 160, 221, 244,
        247, 316, 323, 356, 612, 655, 755, 816
Atlanta, 50, 120, 124, 135, 204, 242, 311, 315–316,
    318–319, 326–327, 354, 401, 421, 436n, 567,
    569, 580, 612, 687, 696, 700, 703, 858
    /Hartsfield, 6–7, 116–117, 123, 377, 614, 703,
        706, 711, 713, 720, 723, 725
Auckland, 223–224
Austin, 160
Austral-asia, 612
Australia , 15–17, 41, 43, 62, 68, 80, 98, 100–101,
    144, 233–234, 601, 607, 705, 804–806
    Department of Housing and Construction 9,
        27, 96n, 124
    Federal Department of Aviation, 77, 89, 96,
        159, 164, 817
    Ministry of Aviation, 77
Australian Airlines, 17,19
Austria, 16, 101, 277
Authority (see Airport Authorities)
Automated people movers, 8, 70, 85, 210, 575,
    580, 586, 589, 649, 693, 694, 717, 718–726
    airside, 723
    cable-driven, 719
    landside, 722
    monorail, 718
    self-propelled, 719
Automated Radar Terminal System (ARTS), 514
Automatic dependent surveillance (ADS), 515, 547
Automation, 501
Auxiliary power units, 196
Average-cost pricing, 283–288, 476
Aviation Trust Fund, 265
A-weighted sound, 172

BAA plc, 16, 20, 42n, 69, 70, 100n, 140n, 224,
    564, 571, 638, 658

BAA (*see* British Airports Authority)
Baggage, carry-on (*see* Carry-on bags)
Baggage distribution, 693, 726
Baggage facilities, 53, 559, 566, 600, 611, 616,
    649, 675
    claim area, 585, 637, 639, 656, 657, 675,
        684–685, 734
    claim presentation length, 684
Baggage handlers, 105
Baggage screening, 649, 729
    metal detectors, 679
    scanners, 651, 729
    x-ray devices, 651, 679, 728, 730
    (*See also* Security systems, baggage)
Baggage sorting, 693, 734
    bar codes, 680, 729
        (*See also* RFID)
    information systems, 728, 729
    mechanical, 731–734
    pre-sorting, 734
    pushers, 731
    tail-to-tail, 734
Baggage space, 651, 732
    height for, 651
Baggage systems, 70, 88, 566, 600, 606, 694
    automated, 651, 735
    belts, 52
    capacity of, 735
    distribution, 726–735
    handling, 37, 161, 727
    maintenance of, 650
    trolleys, 52, 638, 644
    vehicles, 726
    (*See also* Conveyor belts, DCV)
Bailey, E., 98, 124
Ball, M., 543, 554
Baltimore/Washington, 79–81, 89, 113, 116, 145,
    149, 162, 401, 436n, 561, 571, 599, 603,
    615, 629, 630
Bandara, S., 626, 652, 653
Bangkok, 6, 139, 194, 318, 631, 705
    /Don Muang, 705
Bankruptcy, 114, 115, 117
Banks, 96, 105
Banks of traffic (*see* Waves of traffic)
Barber, J., 79, 89, 112, 124, 653
Bar codes, 680
    (*See also* Baggage sorting)
Barker, J., 730, 736
Barnett, A., xxv, 434, 501, 554, 728, 736
Barrett, C., 483, 496
Barrett, S., 155, 163
Barriers to entry, 111
BART (*see* Bay Area Rapid Transit)
Base leg, 512
Basel, 149
Basements, 650
Batch demands, 823
Bay Area Rapid Transit, 696, 718
Beatty, S., 108, 110, 124
Beauvais (*see* Paris)
Beijing, 219, 224, 241, 323
Belfast, 136, 137
Belgium, 704
Belin, S., xxv, 589, 603, 612, 613, 617–619, 625,
    629, 630, 632, 653, 690
Benchmarking, 48, 93, 673
Benedict, R., 38n, 55
Benefit-cost ratio, 789
    (*See also* Cash flows)
Benz, G., 640, 653, 654
Benz, H., 72, 89
Berlin, 42, 67, 133, 136, 137, 222, 584, 585, 681,
    704
    /Schönefeld, 6, 704
Berlin Wall, 133

Bermuda 2 Agreement, 258
Best practice, 48
Bijker, W., 31n, 55
Binney, M., 570, 603
Birds, 211
Bird strikes, 211, 212
Birmingham (UK), 41, 612, 657, 705
Block, J., 41, 55
Blow, C., 575, 582n, 603
Blumstein, A., 408, 432, 751, 763
Boarding areas (*see* Departure gates)
Boarding passes, electronic, 728
Boeing, 30
    aircraft, 212, 306, 308, 569, 575
      757, 378
Boeing Field (*see* Seattle)
Bogota, 734
Bombardier, 719, 720
Bombay, 218
Bonds, 40, 96, 220
    coverage, 246
    general obligation, 245
    revenue, 245, 285, 628
Bonn (*see* Düsseldorf)
Bontadelli, J., 801
Boston, 44, 135, 137, 145, 148, 152, 155, 204, 566,
      602, 623, 697, 703, 734
    /Hanscom Field, 220
    /Logan, 16, 40, 44, 53, 67, 95, 104, 114–118,
      133, 159, 168, 191, 196–198, 202, 205,
      209, 219–220, 242, 248, 271, 280, 311,
      316, 318, 354–355, 373, 376–377, 390,
      392–394, 398, 401–402, 404–407, 423,
      436n, 437, 450, 452–454, 483, 484, 513,
      623, 632, 685, 687, 698, 700, 702–703,
      706, 713, 715, 718, 856, 857
    /Manchester, 67, 133, 147, 148, 155
    Massport, 40, 44, 95, 159, 191, 219–220, 229,
      230, 248, 271, 398, 483, 702
    /Providence, 66–67, 131, 133, 155
    /Worcester, 147, 148, 159
    runways at, 40–41
BOT, 246, 269, 655
Botany Bay, 805
Bottlenecks (*see* Hot spots)
Braaksma, J., 682, 688, 690–691
Braff, R., 501, 504, 546, 554
Braniff, 113
Brealey, R., 629, 653, 780, 794n, 801
Brimson, T., 686, 689
Brisbane, 20, 233
Britain, 8, 16, 17, 38, 97, 104, 144, 608, 634, 742
British Airports Authority, 41, 42n, 46, 69, 70,
      94, 107, 133, 140, 144, 219, 224, 236–237,
      240, 268n, 273, 277, 283, 485, 488, 561,
      570, 658, 695, 702, 709
British Airways, 7, 17, 19, 23, 24, 42, 97, 111,
      223, 564, 566, 567, 582, 601, 730
British Caledonian, 97, 113
B-Cal (*see* British Caledonian)
British Midland, 19
Brokof, U., 556
Broward County (*see* Miami)
Brown, D., 737
Brown, T., xxv, 681, 689
Browne, T., xxv
Brunetta, L., 432, 757, 763
Brussels, 136, 137, 155, 202, 222, 315, 326, 487,
      527, 704
    /Charleroi, 155
    /Liège, 155
    /Zavendam, 704, 706
Budgets, 564, 571
Buenos Aires, 136–137, 139, 143,
    /Aeroparque, 143, 157
    /Ezeiza, 146, 156, 316

Buffalo, NY, 99
Buffer space, 614, 616
Buffer time, 413, 426
Build/operate/transfer (*see* BOT)
Building codes, 191
Burbank (*see* Los Angeles)
Burke, E., 43, 55
Business, participation of, 29
Business travel, 651
Buses (*see* Transporters)
Bussolari, S., xxv
Buy-and-sell, 475, 489

CAA, 46, 48, 56, 105, 107, 124, 237, 240, 249, 753
CAB, 16, 97, 105
Cabotage, 18
Cairo, 218
Caldwell, R., 686, 689
Calgary, 142, 280, 612
Calgary Airport Authority 108, 124
California, 67, 98, 743
Calls, 629, 814
Cambridge University, 45
Canada, 16–17, 41, 61–62, 68, 99, 101, 108, 623,
      743, 744
    (*See also* Transport Canada)
Canadian Airlines, 23
Cancellations, 537
Cancun, 68
Cao, J.-M., 480, 496
Capacity, xxix, 438
    expansion, 160
    passenger, 96,117
    (*See also* Airside capacity; Apron, aircraft;
      Runways; Taxiways)
Capacity coverage chart, 368, 401–408, 450
Capacity regulation, 97
Capital costs, 794
Capital investments, 243–247, 259
Capacity standards (*see* Standards, capacity)
Caracas, 734
Cargo, 22–23, 70, 743, 745
    area, 697
    belly, 153
    integrated, 8, 99, 120, 153
Cargo buildings, 115
Cargo handling, 266
Cargo service charge, 265
Caribbean, 124, 131
Carlin, A., 288, 291, 477
Carry-on bags, 585, 678, 679
Cars, 688
    parking charges, 269
    rentals charges, 269
    (*See also* Airport access)
Cartels, 97, 111
Case, K., 801
Cash flows, 779–801
    analysis, 791
Catchment areas, 139, 149
Category I (approach), 389, 518, 520, 546
Category II (approach), 389, 390, 518, 520, 546
Category III (approach), 389, 390, 518, 520, 546
Catering, 105
    vehicles, 584
Cathay Pacific, 19, 24
Causality, 766, 771
Caves, R., xxv, 64, 89
Ceiling (*see* Weather conditions)
Census Bureau, 132n, 166
Center TRACON Automation System (CTAS),
      549
Central Flow Management Unit (CFMU), 527
Centralized management, 627
Certification (*see* Emissions certification; Noise
      certification)

CFR, 212
Chapter 1, 2, 3 aircraft, 182, 200, 263, 436n
Characteristics of commercial jet airplanes (*see* Aircraft, characteristics),
Charlotte, 120, 204, 401, 436n, 713
Charter flights, 98, 111, 566, 602, 611
Check-in, 53, 559, 566, 574, 584, 585, 657
    agents, 32, 37, 733
    counters, 616, 628, 656, 661, 669–671, 694, 728
    electronic, 21, 659, 677
    electronic kiosks for, 678
    facilities, 32–33, 48, 636, 677
    hall, 577, 642, 677
    weighing of bags, 642
    workstation, 37
Checkpoints (*see* Security)
Chess, playing of, 83, 86
Chesterton International Property Consultants 682–689
Chevallier, J.-M., xxv
Chicago, 95, 122, 135, 137, 145, 204, 703
    /Midway, 146–147, 162, 703, 706
    /O'Hare, 7, 24, 79, 95, 112–113, 146, 198, 219, 280, 316, 318, 401, 436n, 454, 474, 530, 567, 569, 577, 580, 581, 614, 696, 698, 700, 703, 706, 712–713, 723, 861
Chicago Convention, 258, 295
Chile, 5
China, 71, 705
Ciattoni, J.-P., 212
Cincinnati, 79, 112, 120, 125, 242, 401–402, 436n, 567, 713
Citizen Advisory Groups, 190
Civil Aeronautics Board (*see* CAB)
Civil Aviation Authority (*see* CAA)
Clarke, J.-P., xxv, 196, 212, 555
Clean Air Act, 210
Clear zones, 159
    obstructions, 159
Clearance delivery, 509
Cleary, E., 210, 212
Cleveland, 703, 706
Climate, global, 202, 203
Clubrooms, airline, 681
Cockpit display of traffic information (CDTI), 547
Cogeneration, 4
Cohas, F., 149, 164
Cohen, S., 38n, 55, 60, 89
Collaborative decision making (CDM) (*see* Decision making)
Collaborative routing, 544
Commercialization, 3, 9, 14–18
Commercial activities, 269
Common User Terminal Equipment (*see* CUTE)
Commonwealth Games, 570
Communications, 501, 503, 548
Community noise equivalent level, 180
Community relations, 190–191
COMPAS, 548
Compensatory system, 238, 241–242, 284
Competition, 13, 59, 97, 98, 112–113, 129, 131
    dynamics of, 129, 150, 156
    (*See also* Airport competition)
Composite noise rating, 180
Compression, 540
Computer models, xxiii, xxix, 588, 747–764, 803
    airside, 750–756
    analytical, 749
    classification, 748–750
    deterministic, 85, 749
    dynamic, 749
    frequency of use, 754
    in-house modeling, 755
    macroscopic, 748–749, 753

Computer models, (*Cont.*)
    mathematical, 395, 408, 418
    microscopic, 749
    Monte-Carlo, 749
    node-link, 753
    passenger building operations, 757
    probabilistic, 85
    simulation, 749
    spreadsheet, 85, 590, 618, 623, 770, 786
    static, 749
    stochastic, 749
    3-dimensional, 753,
    user-model interfacing, 756
Concentration of traffic (*see* Traffic concentration)
Concessions, 101, 103, 256, 646, 672, 673, 674, 682–684
    location of, 682
Concourses (*see* Passenger buildings)
Confederation of Independent States, 38
    (*See also* Russia)
Configuration (*see* Passenger buildings)
Congestion, 436
Congestion pricing, 475–486
    theory, 476–480
    practice, 480–486
Congress, U.S., 41, 100, 102, 128, 214
Constrained position shifting (CPS), 550
Construction costs (*see* Costs)
Construction management, 53
Consultants, 96, 105
Continental Airlines, 7, 23, 120, 124–125, 147, 567
Contours (*see* Noise contours)
Controlled airspace, 504–505
Controlled time of arrival (CTA), 532
Controlled time of departure (CTD), 532, 543
Controller-pilot data link communications, 548
Converging runways (*see* Runways)
Conversion coefficients (*see* Peak hour)
Conveyor belts, baggage, 32, 650, 684, 733
    (*See also* Baggage)
Coogan, M., 52, 55, 210, 212, 737
Coordinated airport, 469
Copenhagen, 125, 205, 196, 222, 224, 236, 273, 283, 696, 704
    airport company, 16, 223
    /Kastrup, 201, 704
Cork, 221
Corps des
    Ponts et Chaussées, 45
    Travaux Publics de l'Etat, 45
Correlation, statistical, 771
Cost center, 253–257, 283
Cost, current, 288–290
Cost estimates, 72–74
Cost recovery, 253
Cost structure, 595
Cost-benefit analysis, 780
Cost-effectiveness, 25, 85, 574, 595, 618, 790
    ratios, 797
Costs
    annual, 562
    average, 596
    capital, 103, 117, 259, 562
    construction, 36, 72, 159, 571, 585, 611, 634, 649
    maintenance, 36, 259
    operating, 36, 569, 649, 794
    seat-mile (*see* Seat-mile costs)
Costs, of air travel, 10, 11
Costs, marginal
    per aircraft operation, 569, 595, 596
    per passenger, 595, 596
Cowan, S., 124
Crayston, J., 203, 212
Crews, flight, 573

Critical aeroplane, 302
Critical aircraft, 334
Critical engine-failure speed (*see* Decision speed)
Crosswind component (*see* Winds)
Crosswind coverage (*see* Winds)
Cultural context, 29, 655
Cumulative diagrams, 439n, 748, 835–841
  arrival, 664, 667–668, 672
  departure, 664, 667–668, 672
  limitations of, 669
Cumulative flow diagrams (*see* Cumulative
  diagrams)
Curb area, 656, 685–688, 714, 717
  parallel curbs, 687
Curbside equivalent, 685
Curfews, 199
Customs, 559, 573, 611, 684, 744
CUTE, 116, 628, 678
Cycle time, 613

Dada, F., 653
Dade County (*see* Miami)
Daellenbach, H., 350, 364
Daganzo, C., 734, 737
Daily patterns (*see* Peaking patterns)
Dallas, 155
  /Fort Worth, 6–7, 24, 69n, 79, 112, 122–125,
    135, 137, 156, 158, 242, 280, 316–318,
    327n, 328, 329, 377, 394n, 400–401,
    436n, 549, 567, 583, 587, 594, 612, 646,
    698, 712–713, 718–720, 723
  /Love Field, 155–156, 158
Daniel, J., 477, 496
Daskin, M., 726, 736
Data
  accuracy, 741–743, 744
  completeness of, 744
  cross-sectional, 767
  time-series, 767, 770
  validation, 741–746
Data links, 503
Data, span of, 772
Data table function, 589, 618
Davis, Langdon & Everest, 634, 653
Day-night average sound level, 175, 176
Day-night noise level, 399
Dayton, 24
DCV, 732
Dear, R., 415, 433, 550, 554
De Barros, A., xxv, 589, 603, 681, 690
Decentralization (of ATM), 503, 534
Decision analysis, xxix, 59, 617, 629,
    803–812
  branches, 808
  trees, 630, 808, 812
Decision making criterion (*see* Cash flows,
    Analysis and Decision analysis)
Decision making process, 49
  centralized, 39, 52–53
  collaborative, 525, 536
  pluralistic, 39–41, 53
Decision speed, 333
Decision support systems, 530, 535, 548–551
Declared capacity, 383, 448, 469, 471
Declared distances (*see* Runways),
De Gaulle (*see* Paris)
Deicing, methods of, 206–208
  fluids, glycol, 206
  forced hot air, 207
  infrared heaters, 207
Delay (*see* Aircraft delay)
  delays, 626, 637
  overload, 438, 833–834
  stochastic, 440, 833–834
  sensitivity of, 435, 442, 845
  (*See also* Queues, Queuing systems)

DELAYS model, 441, 751, 752
Delcaire, B., 212, 555
Delta Airlines, 7, 23, 49, 79, 112, 120, 124, 125,
    152, 567, 586, 601
  Delta shuttle, 602
Demand management, xxix, 461–497
  access control, 464
  administrative approaches, 467–475
  economic approaches, 475–486
  hybrid approaches, 476, 486–492
Demand interarrival times, 822
  constant, 824
  deterministic, 824
  negative exponential, 824
  Poisson, 824
Demand patterns, 446, 450
  daily, 453
  seasonal, 453
Demand rate, 438, 822
Dempsey, P., 649, 653
de Neufville, R., 15, 27, 48, 55, 61, 71, 74, 79,
    89, 106, 112, 124, 139, 144, 150, 153,
    164–165, 168, 212, 477, 496, 589, 592, 603,
    612–613, 617–619, 625–626, 630, 649, 653,
    660, 681, 685–686, 690, 711–712, 714, 736,
    742, 746, 776–777, 780, 801, 806,
    817–818
Denmark, 16, 704
Denver, 95, 125, 141, 157, 567, 686, 219, 436n,
    549, 858
  costs of, 45
  /International, 7, 69n, 82, 95, 122–124, 141,
    208, 248, 280, 315–316, 318, 327n, 328,
    401, 549, 566, 568–569, 577, 580, 612,
    649, 680, 698, 720, 723, 725, 735
  /Stapleton, 141, 248, 568–569
Departure control, 513
Departure gates, 49, 52, 88, 115, 584, 611, 621,
    662, 672, 674, 678, 680
  international, 162, 614, 713, 742
  shared (*see* Shared use)
  swing gates, 612, 623, 632, 633, 635
  time of occupancy, 619–620
Depreciation, 255, 284, 562, 569, 596
  (*See also* Costs)
Deregulation of airlines, xxix, 8, 13, 17, 59, 72,
    80, 93–128, 147, 561, 767
  economic, 79
  implications for airports, 111–125, 147
  "open skies", 8, 17, 80
Descent advisor (DA), 549
Design, 103
  criteria, 18, 25, 29, 44
  detailed, xxix
  overall, xxix
Design loads, 611, 851–852
Design peak day, 670, 851
Design peak hour, 851, 853–854
Destination-coded vehicles (*see* DCV)
Detroit,
  /City, 147–148
  /Metro, 7, 115, 124, 280, 318, 401, 436n, 567,
    569, 713, 721–723
Deutsche Bank, 216, 249, 273, 280, 291
Development
  incremental, 160–162
  staging of, 160
Development banks, 244
Deyst, J., 459, 763
D/FW (*see* Dallas/Fort Worth)
DHL, 155
Differential GPS (*see* GPS),
Dippe, D., 556
Direct operating costs (*see* Costs, operating)
Disabled passengers (*see* Passengers,
    disabled)

Discounting, 780–782
  annual value, 782, 786
  annuity, 783, 786
  compound interest, 781
  continuous compounding, 787
  discrete compounding, 786
  formulas, 786
  present value, 781
Discount rates, 780,
  effect of risk on, 804
  effective, 787
  nominal, 787
  social, 781
Displaced runway threshold (*see* Runways)
Distance measuring equipment (DME), 524
Distribution (*see* Baggage)
Distribution, Normal (*see* Normal distribution)
Diverging runways (*see* Runways)
Doganis, R., xxv, 291
Dokken, D., 214
Dolbeer, R., 211--212
Dominican Republic, 20
Door-to-door access (*see* Airport access)
Dorfler, J., 554
Dorval (*see* Montreal)
Double-deck access roads, 687
Doucette, R., xxv
Downwind leg, 512
Drazen, M., 496
Dresden, 704
Dual till, 237–241
Dubai, 218
Dublin, 221–222, 273
Dulles (*see* Washington)
Düsseldorf/Bonn, 42, 136–137, 273, 704
  Köln/Bonn, 704
Dwell time, 636, 639–643, 646, 651
  ability to manage, 642
  of cars, 686
Dynamic strategic planning, xxii, xxix, 9, 59–91, 65
  concepts, 81–83
  process and methods, 83–88
  (*See also* Planning)

Eastern Airlines, 99, 113, 114, 117
East River, 140
Easyjet, 155, 163
Ecole Nationale des Ponts et Chaussées, 97, 127
Ecole Polytechnique, 45
Economic efficiency, 68, 106, 254, 570, 573, 608, 612, 638
Economic incentives, 201
Economic lifetime, 782
Economic performance, 44, 46
Economic surpluses, 244
Economic trends, 80
Economies of scale, 585
Edinburgh, 146
Edmonton, 69n, 139, 142, 158, 612
  /International, 158, 623–624
  /Municipal, 158
Edwards, P., xxv
Edwards, B., 604
Effective jet operations, 398
Effective perceived noise level, 178
Effective width (*see* Passageways, effective width)
Efficiency (*see* Economic efficiency)
EIS, 76, 204, 210
Electric power, 649
Electronic
  check-in (*see* Check-in)
  commerce, 3, 21–25
  data, 642
  ticketing, 3, 8, 21–23, 677

Elevators, 649
Elliott, D., 719, 736
Emissions, total, 170, 204
Emissions certification of aircraft engines, 204
Engineered Performance Standards, 372n
Engineering economy, 780
Engineering services, 270
Engineers, 105
England (*see* Britain)
Enhanced Traffic Management System (ETMS), 529
Enriquez, R., 240, 242, 249
En-route control center, 506, 507, 522–525
Environment, 104
Environmental impacts, 167–214
  opposition to airports, 167, 169
  restrictions, 167
  zoning, 184, 189, 191
Environmental Impact Statement (*see* EIS)
Environmental Protection Agency, 203, 210, 212–213
Equivalent noise level, 175–176
Equivalent sound level, 175–176
Erzberger, H., 549, 550, 555, 556
Essential Air Service, 489
Estimated time of arrival (ETA), 532
EUROCONTROL, 266n, 267, 436, 463, 475–476, 527–528, 544n, 753, 830
Europe, 5, 14, 32, 53, 66, 80, 118, 120, 124, 154, 209, 270, 272, 275, 317, 369, 378, 395, 491, 502–504, 514, 522, 527, 584, 612, 626, 664, 703–704, 729, 830
European Civil Aviation Conference 527, 728
European Commission, 262, 281, 474, 550
European Common Market (*see* European Union)
European Community, 17, 53, 68, 80, 599, 744
European Investment Bank 244, 247
European Union, 113, 118, 247, 254, 260n, 262, 280, 317, 473, 545n, 728
Evans, J., 554
Excel, 588, 770, 786
  (*See also* Computer models)
Excess baggage, weighing of, 659
Exit taxiways, 396
  (*See also* Runways, high-speed exits)
Expected departure clearance time (EDCT), 532
Expected value, of a decision, 810
Expedite Departure Path, 550
EXTEND©, 758
External cost, 478
External delay cost, 477, 479

FAA, 8n, 9,10n, 27, 30, 40, 44, 45n, 56, 61–63, 67, 69, 74, 82, 89, 95, 118, 124, 127, 145, 183, 186, 196, 204, 213, 244, 264, 271–272, 291, 295–296, 300, 302–304, 328, 341, 345, 348–350, 356, 361, 364, 370–371, 378, 384, 421, 433, 436, 454, 456, 459, 463, 466–467, 496, 500, 505, 514, 519, 522, 546, 549, 569, 604, 607, 610, 651, 653, 673, 679, 684, 690, 727, 735, 736, 742, 751, 763, 830, 853, 856, 858, 863
  advisory circulars, 44, 297
  Federal Aviation Regulations (FAR), 182, 186, 297
FAA Airfield Capacity Model, 751
Faburel, G., xxvi, 41, 55
Failures, 130, 560, 588, 611, 649
Falcons, 211
Fan, T., 443, 459, 467, 472, 497
Fares, 97, 766
  long-term decrease in 12
  regulation of, 97
Faulkner, T., 629, 653, 817, 818

Federal Aviation Administration (*see* FAA)
Federal Inspection Services , 264, 623
FedEx, 7, 14, 23, 24, 66, 70, 99, 111, 115, 153, 486, 743
Fees
  landing, 103, 205
  passenger, 103
Feldman, J., 653
Feron, E., 212, 459, 555, 763
Final approach fix, 518
Final approach path, 409
Final Approach Spacing Tool (FAST), 549
Final approach speed, 415
Final control position, 512
Finance theory, 780
Financing, xxix, 102, 103, 215–250
Finfrock Design-Manufacturing-Construction, 716
Finger piers, 50, 117, 352, 354, 559, 576–577, 584, 591, 594
Finnair, 19
Fitch ICBA, 247, 248, 250
FitzGerald, G. P., 109, 127
Fixed-base operators, 279
Flexibility, 59, 81, 85, 87, 104, 588, 599–600, 608, 612, 650
  value of, 616, 629, 803
Flexible design, 115–117, 157, 161–162
Flexible space as real option, 629
Flight management system (FMS), 508
Flight operations, 195–197
Flight plan, 508
Flight Schedule Monitor (FSM), 536
Flight strip, 508
Florence, 225
Florida, 124
Flow (*see* Matrix)
Flow management positions, 527
Flow patterns, 618, 694
  many-to-many, 694
  many-to-one, 694
  one-to-one, 694
Flow rates, 644
Flows, 618, 637, 819–830
  crossing, 649
Fog (*see* Weather conditions)
Fordham, G., 653
Fortner, B., 319, 364
Forecast variables, 767
  "explanatory", 766n
  "independent", 766n
Forecasting, xxix, 3, 765–778
  (*See also* Volatility),
Forecasts, 10, 741
  aggregate, 74–76
  "always wrong", 59, 70–80, 574, 599, 805
  errors in, 74–75, 80
  fixed, 562
  standard deviation of, 73
  (*See also* Peak hour)
Forecasts, assumptions needed, 765
  equations, 767
  long-term, 775, 805
  models, 765
  short-term, 775
  span of data, 765, 767
Forecasts, drivers of, 765, 767
Forecasts, models for
  exponential, 770
  linear, 769
Fort Lauderdale (*see* Miami)
Fort Myers, 612
Formulas for
  space requirements, 606
Foschi, M., 719, 737
Foster, N., 570

France, 17, 38n, 41, 43, 47, 60, 62, 69, 97, 267, 277, 704, 744n
Frankfurt/Main, 7, 20, 42, 53, 67, 221n, 222, 224–225, 273, 282–283, 316, 318, 321–322, 378, 395, 471, 472, 549, 561, 566–567, 571–572, 577, 614, 704, 706, 711, 720, 723, 858
  airport company, 20, 222, 224–225
Fraport (*see* Frankfurt/Main, Airport company)
Freathy, P., 683, 684, 690
Free departures, 395, 419
Free flight, 545, 756
Free float (of shares), 218
Frequency,
  of flights, 97
  matching of, 153
Frequency share, 152
Frequent flyer programs, 99
Fron, X., 437, 459
Frontier Airlines, 113
Fruhan, W., 150, 165
Fruin, J., 640, 644–645, 653–654
Fuel prices, 13
  (*See also* Petroleum prices)
Fully coordinated airport, 469

Gaddy, S., 554
Galileo, 545n
Game theory, 131n
Games, competitive, 131n
Gantt charts, 625
Garages (*see* Parking)
Gate arrival concept, 583–584, 588
Gate assignments
  management of, 574, 589
Gate utilization, 671
Gates (*see* Departure gates)
Gateways, 570
Gatwick (*see* London)
Gaudinat, D., 654
Gelerman, W., 150, 153, 165
General aviation, 149
General Aviation and Manufacturers' Association, 44
Geneva, 42, 206, 264, 320, 577, 705, 706, 711
Geographic Positioning System (*see* GPS)
Geometric characteristics (of airports), 319–328
  (*See also* Airport layouts; Passenger building, configurations)
German Aerospace Research Establishment (DLR), 549
Germany, 8, 16, 17, 42, 43, 61, 66, 67, 97n, 277, 614, 704, 774n
Gifford, F., 205, 213
Gilbert, G., 502, 555
Gilbo, E., 418, 433, 544, 555
Gimpo (*see* Seoul)
Glide slope, 517
Global Navigation Satellite System (GLONASS), 545n
Global Navigation Satellite System (GNSS), 545
Global Positioning System (*see* GPS)
Globalization, 3, 9, 18–21
Goetz, A., 649
Gold, R., xxv
Gomez-Ibanez, J., 108, 127
Gosling, G., 64, 89, 736
Government,
  local, 95
  national, 94
Government funding, 274–275
Government grants, 242
Government subsidies, 260
GPS, 8, 21, 525, 545–547
  differential, 546
GPS-based precision approaches, 503, 522

Graham, D., 124
Greece, 16, 100, 101, 277, 504
Greensboro, 79
Griggs, D., 214
Grillot, M., 660, 690
Gross, D., 821, 850
Ground access (*see* Airport access)
Ground controller, 509
Ground delay programs (GDP), 530
Ground handling, 266, 277–283
Ground holding, 526, 528
Ground transport (*see* Airport access)
Ground vehicles, 34, 210
Growth
  doubling of, 9
  long-term, 3, 9–14
Guadalajara, 742
Gulf War, 74
Gusts (*see* Weather conditions)
Guzmán, J., 48, 55
Hall, W., 544–555
Hamburg, 241, 704
Hamiltonian path, 553
Hammerhead, 576
  (*See also* Finger piers)
Haneda (*see* Tokyo)
Hangars, 160
  charge, 262
Hani, H., 89
Hannover, 42, 704
Hansman, R., xxv, 212, 459, 763
Hansen, M., 736
Harris, C., 821, 850
Harris, K., 737
Harris, R., 418, 433
Harrisburg, 108, 219
Hart, W., 575, 604, 673, 690
Hasan, N., 682, 690
Hassounah, M., 626, 654
Hathaway, D., 645, 654
Hax, A., 64, 65, 90
Heathrow (*see* London)
Heathrow Express, 695, 702, 706, 709–710, 731
Heating (*see* HVAC)
Heimlich, J., xxv
Helsinki, 218
Hemm, R., 763
Hennessy, P., 38n, 46, 55
HERMES model, 751, 753
Hibon, H., 71, 90
High Density Rule, 466, 474, 489
High-occupancy vehicles, 712
  (*See also* Airport access)
High-risk flights, 266
High-speed runway exits (*see* Runways, high-speed exits)
High-speed rail, 52, 695
  (*See also* Airport access, modes)
Hillier, F., 821, 850
Hiroshima, 61, 68, 168
Historical cost, 288–290
Historical precedent, 470
Hobart, 20
Hobby (*see* Houston)
Hochtief, 221
Hockaday, S., 418, 433
Hoffman, R., 554
Hoffman, W., 496
Hogarth, R., 71, 90
Hold rooms, secure, 644
Honduras, 108
Hong Kong, 135, 137, 155–156, 323, 705
  /Chep Lap Kok, 156, 318, 581, 702, 705–706, 711, 723
  /Kai Tak, 683
  /Macao, 155

Hong Kong, (*Cont.*)
  /Shenzhen, 155–156
Honolulu, 219, 683, 713
Horizontal elevators (*see* Automated people-movers)
Horonjeff, R., 9, 27, 85, 90, 213, 348, 350, 364, 562, 604, 626, 654, 681, 690
Hot points (*see* Hot spots)
Hot spots, 369, 429, 424, 526, 656, 660, 685
Hotteling, H., 477, 497
Houston, 69n, 136, 137, 204, 401
  /Bush, 7, 318, 436n, 567, 698, 713, 723
  /Hobby, 163
Howard, G., 17
Hub-and-spoke networks, 118, 594, 615
Hubbing operations, 566
Hubs, by volume of traffic, 67, 87
Hubs, transfer (*see* Transfer hubs)
Hughes, T., 31n, 55
Hughes, W., 737
Humphreys, I., xxv, 682, 690
Hupe, J., 203, 212
HVAC, 649
Hybrid concourses, 352, 559, 586, 600, 602
Hye, H., 213
Hysteresis, 198, 834

Iadeluca, J., 554
IATA, 9, 27, 62, 82, 90, 318, 356, 364, 461, 469, 472–473, 497, 562, 604, 636, 640–641, 658, 673, 681, 684, 685, 690, 727, 732, 737, 743, 746, 757, 853, 863
Iberia, 17, 19
ICAO, 9, 27, 30, 56, 62, 82, 90, 182–183, 203, 213, 234, 241, 250, 252, 255, 259–261, 267n, 277, 291, 295–297, 300–304, 328, 330, 341, 345, 348–350, 356–360, 365, 387, 433, 488, 522, 555, 562, 569, 604, 609, 610, 654, 770, 777
Ice (*see* Weather conditions; Deicing, methods of),
Ice cream sellers, parable of, 150, 151
Icing (*see* Weather conditions; Deicing, methods of)
Idris, H., xxv, 212, 405, 423, 550, 555
Imaginary surfaces, 356, 361
  (*See also* Obstacle limitation surfaces)
Immigration service, 642, 657, 659, 662
Impedance (*see* Matrix)
In Motion, 737
Incheon (*see* Seoul)
Independent runways (*see* Runways)
Indianapolis, 20, 24, 219, 436n
Inertial navigation, 525
Inflatable structures, 117
Inflation, 562n
Information technology, 628
Inner marker, 528
Innovation, 21
Instrument flight rules (IFR), 377, 504
Instrument landing system (ILS), 390, 517
Instrument meteorological conditions (IMC), 389, 504
Instrument runway, 311
Insurance, 158, 629, 806, 813
  (*See also* Real options)
Integrated Noise Model, 180–182
Integrated Terminal Weather System (ITWS), 548
Inter-American Bank, 244
Interarrival times, 822–824, 844
  (*See also* Demand interarrival times)
Interest, 255, 596,
  (*See also* Discounting)
Internal delay cost (*see* Private delay cost)

Internal rate of return, 789
International Air Transport Association (*see* IATA)
International Civil Aviation Organization (*see* ICAO)
International Convention on Civil Aviation (*see* Chicago Convention)
International terminals, 601
Internet, 677
  (*See also* Electronic)
Intersecting runways (*see* Runways)
Inventory, of current conditions, 86
Ireland, 221
Islands, man-made, 141, 581, 702
Italy, 16, 42–43, 279, 382, 704
Itami (*see* Osaka)

Jakarta, 697
Jamaica, NY, 708
Jan, A., xxv
Jane's Airport Review, 706, 726, 737
Jansson, M., 480, 497
Japan, 8, 14, 38n, 43, 47, 53, 60, 62, 66–68, 75–76, 78, 97, 145, 168, 207, 277, 703, 705
Japan Airlines, 17, 24, 97, 601
Jeddah, 218
Jet routes, 524
JICA, 96, 127
Johnson, C., 60, 90
Jordan, W., 97, 127

Kahn, A., 16, 98, 127
Kanafani, A., 418, 433, 480, 496, 746, 770, 777
Kansai (*see* Osaka)
Kansas, 149
Kansas City, 69n, 82, 158, 316, 560, 583, 587, 594, 683
Kansas City ARTCC, 523, 524
Kaplan, D., 124
Kapur, A., 127
Kasper, D., xxv
Katzman, M., 737
Keeney, R., 806, 818
Kelsey Construction, 716n
Kennedy (*see* New York)
Kenya, 84, 626, 742, 744
Kenya Airways, 742
Kiernan, L., xxv
King, C., xxv
King, D., 74, 89
Kiosks, electronic (*see* Check-in)
Kirshner, S., xxvi
Kitakyushu, 168
Kivestu, P., 752, 763
KLM, 19, 394n, 567
Knudsen, T., 73, 90
Kobe, 168
Köln/Bonn (*see* Düsseldorf)
Koopman, B., 752, 763
Korea, 17
Korean Airlines, 24
Kostiuk, P., 433, 763
Kuala Lumpur, 138, 139, 158, 570, 571, 612, 705
  /International, 50, 51, 69n, 86, 87, 142, 158, 323, 564, 571, 582, 595, 600, 622, 655, 694n, 696, 705, 711, 719, 723, 727, 775
  /Subang, 142, 158
Kuchar, J., 459, 763
Kulatilaka, N., 629, 652, 817, 818

LaGuardia (*see* New York)
LaMagna, F., 678, 686, 690
Lan Chile, 19
Land acquisition (*see* Landbanking)
Land and hold short, 388n
Land area requirements, 315–319

Landbanking, 14, 157, 160, 816
  (*See also* Airport sites)
Landing distance available (LDA), 330, 332
Landing fees, 201, 238, 261, 283–288, 464
Landrum and Brown, 768, 772–774, 777
Landside, xxix, 14, 559–603
Land-use policies, 191–194
Lantner, G., 735
Larson, R., 821, 832, 850
Las Vegas, 7, 135, 318, 401, 436n, 612, 698, 713, 816
Lauda, 19
Lea + Elliott, xxv
Leases
  by airlines, 96
  long-term, 96, 101, 104, 117
  short-term, 116–118
  30-day, 117
Le Bourget (*see* Paris)
Lee, D., 418, 433, 752, 763
Leigh Fisher, 698, 700, 706, 737
Leipzig-Halle, 704
Leo, F., xxv
Lesage, Y., 89
Level of service, 371, 372, 450, 636–639, 643, 676, 828–832
  definition of, 636, 637
Level of service standards (*see* Standards)
Levine, M., 477, 497
Lewis, S., 496
Lewis-Beck, M., 770, 777
Liberalization, 93
  (*See also* Deregulation)
Licenses, 97
Lieberman, G., 821, 850
Life, economic, 115
Lift-lounges, 584
  (*See also* Transporters)
Lifts (*see* Elevators)
Lightning (*see* Weather conditions)
Lima, 20
Linate (*see* Milan)
Linear concourses, 352, 354, 580, 582–584, 594, 661, 670
Lipson, W., 110
Lisbon, 118, 139, 143, 159, 160
  /Ota, 159
Lister, E., 214
Little, I. M. D., 288, 291, 485, 497
Little, J. D. C., 843, 850
Little, R., 79, 90, 561, 604, 615, 629, 654
Load factors, 153, 607, 618, 619
Loans, 245
Local area augmentation system (LAAS), 546
Local controller, 509
Localizer (of ILS), 517
Location theory, 150
Lodge, D., 52, 56
Logistics Management Institute, 751, 752
London, 42, 46, 66, 129, 130, 135, 137–139, 145, 155, 504, 510, 638, 705, 709
  /City, 111, 133, 155, 163
  /Gatwick, 20, 97, 133, 140, 144, 155–156, 237, 241, 277, 316, 318–320, 378, 485, 488, 694n, 705–706, 711, 720, 723, 753
  /Heathrow, 7, 20, 42, 52, 97, 104, 107, 133, 140, 141, 144, 170, 199n, 201, 237, 240, 241, 277, 316–318, 323, 326, 373, 378, 471–472, 485, 488, 564, 566–567, 569, 577, 582n, 587, 642, 680, 681, 683, 695, 700, 702, 705–707, 709–711, 730, 731, 753, 858, 861
  /Luton, 82, 131, 133, 155, 163, 606, 638
  /Stansted, 66, 69, 133, 139, 140, 155, 160, 162, 237, 485, 488, 561, 570, 581, 705–706, 720, 723

Long Island Railroad, 706
Long-range (civilian) navigation system
    (LORAN-C), 525
Longitudinal separation requirements (*see* Air
    traffic management)
Longstaff, F., 817, 818
Los Angeles, 67, 95, 102, 132, 135, 137, 139,
    148, 160, 204, 316, 318, 327, 436n,
    765–766, 769, 771–772
    /Burbank, 94n, 742, 744
    /El Toro, 160
    /International, 7, 24, 25, 93, 100, 115, 198,
        213, 219, 242, 280, 401, 454, 612, 698,
        706, 712–713, 716, 768, 858
    /John Wayne, 199
    /Long Beach, 147–148
    /Ontario, 66, 163, 219
    /Palmdale, 219
    /Van Nuys, 132, 219
Lösch, A., 150, 165
Lottery (of slots), 466, 468
Louisville, 24–25, 66
Love Field (*see* Dallas)
Low-emission vehicles, 210
Lounges (*see* Departure gates)
Low IFR, 466, 468
Luenberger, D., 794n, 801
Lufthansa, 7, 17, 23–24, 42, 97, 145, 561,
    566–568, 581, 614
Luton (*see* London)
Lyon, 52, 704, 711

MACAD model, 751–752, 756
Macao (*see* Hong Kong)
McCann, L., xxvi
McCoomb, L., xxv, 565, 612n, 615n
McDonnell aircraft, 575
McFarland, M., 214
McGee, J., 737
McGraw, P., xxv
McKelvey, F., 9, 218, 348, 364, 562, 604, 626,
    654, 681, 690
McLeod, K., 288, 291, 485, 497
Madison Avenue, 149
Madrid, 139, 221–222, 318
    /Barajas, 6
Mahathir, M., 570, 655
Maintenance (*see* Costs)
Maintenance facilities, 117, 160
Majluf, N., 64, 90
Majority in interest agreements, 102
Makridakis, S., 71, 90, 770, 778
Malaysia, 16, 87, 142, 570, 665, 696, 705
    Airports Holdings Berhad, 225
Malaysian Air System, 87, 142
Maldonaldo, J., 78, 90
Malone, K., 19
Malpensa (*see* Milan)
Managed arrival reservoir (MAR), 543
Manchester (UK), 41, 97n, 205, 273, 705
Mandle, P., 678, 686, 690, 737, 758, 763
Manhattan, 140, 708
Manila, 705
MAP, 631
Marginal-cost pricing, 287
    (*See also* Pricing)
Marginal VFR, 389
Marine Air Terminal, 602
Marketing, of secondary airports, 162
Market share, 112, 150, 152
Markets, 129, 142, 154
    dynamics of, 129, 132, 145
    (*See also* Airport services)
Marchi, R., xxv
Martel, N., 659, 691
Massport (*see* Boston/Logan)

Master Plans (*see* Plans, master)
Mathematical models (*see* Computer models)
Matrix
    flow, 589–590
    impedance, 589–590
    origin-destination, 589
Maximum certificated takeoff weight, 302
    (*See also* Maximum takeoff weight)
Maximum landing weight, 261, 306, 308, 310
Maximum queue length, 831
Maximum sound level, 173
Maximum structural takeoff weight, 378
    (*See also* Maximum takeoff weight)
Maximum takeoff weight, 261, 306, 308, 310,
    378, 464n, 485
Maximum throughput capacity (*see* Airside
    capacity)
Measures of effectiveness, 788–797, 828–832
Mechanical systems, 649, 650
Meeters and greeters, 670
Melbourne, 20
Memphis, 24, 25, 66, 401, 436n, 486
Mendoza, 143
Mercosur, 599
Metering, 529
Metron, Inc., 555
Metropolitan areas, 132
    consolidated, 132n
Metropolitan Washington Airports Authority, 69,
    95, 103, 571
Meyer, J., 16, 27, 98, 108, 127
Mexican Southeast Airport Group, 223–224
Mexico, 16, 41, 43, 62, 68, 236, 241, 744
Mexico City, 113, 139, 143, 194, 697, 741, 745
Miami, 95, 120, 135, 137, 607, 632, 686, 734,
    772, 778
    /Fort Lauderdale, 66
    /International, 6–7, 24–25, 35, 95, 124, 219,
        316, 318–319, 326, 436n, 694, 706, 713,
        723, 772
    /Palm Beach, 191
Microbursts (*see* Weather conditions)
Microwave landing system (MLS), 521
MidAmerica (*see* St. Louis)
Middle East, 277
Middle marker, 518
Midfield concourses, 8, 36, 50, 122, 354, 561,
    568, 577, 586, 595, 601, 719, 723, 726
    linear, 559, 593
    x-shaped, 559, 580–581, 588, 593
Midway Airlines, 146–147, 162
Midway (*see* Chicago)
Midwest, 123
Mierzejewski, E., 711
Milan, 42, 69n, 136, 137, 139, 142, 222, 273, 383,
    510, 704
    airport company (SEA), 20, 139, 142, 222
    /Linate, 139, 142, 163, 168, 382, 384
    /Malpensa, 139, 142, 168, 316, 321, 382, 566,
        577, 601, 704
Miller, J., xxv
Minimum acceptable rate of return, 781
Minimum landing fee, 482
Minneapolis/St. Paul, 7, 242, 318–319, 401–402,
    436n, 454, 567, 713, 721, 723
Mintzberg, H., 65, 90
Mira, L., 477, 496
Mirabel (*see* Montreal)
Missed approach, 355
MIT, 184, 213
    Lincoln Laboratory, 500, 522–523, 556
MITRE, 751
Mix of aircraft, 353, 391–394
Mix of movements, 394–396
Mode C transponder, 516
Mode S transponder, 516

Models (*see* Computer models)
Mombasa, 84, 612, 626, 742
Monopolies, 17, 46, 93, 100–101, 104, 110
  regulation of, 46
Monopolies and Mergers Commission, 237
Monorail (*see* Automated people movers)
Montreal, 136–137, 139, 143, 158
  /Dorval, 138, 143, 208, 280, 586n, 612, 685, 715
  /Mirabel, 68, 136, 143, 158, 160, 586n
Moody's Investor Services , 45, 56, 70, 90, 96,
  127, 246–250, 567, 604
Moore, H., xxv, 719, 737
Morita, S., xxv
Morrison, S., 477, 480, 496
Moscow, 135–137, 218
  /Sheremetyevo, 322
Mosley, P., xxvi
Moving sidewalks, 52, 589, 680, 723
Multi-airport systems, xxix, 66, 69, 129–166, 562,
  808
  definition of, 132
  development of, 132, 138
  factors favoring, 154–157
  political considerations, 131, 135, 143
  primary airports, 135, 152
  secondary airports, 69, 129, 135, 137, 145
  technical factors, 131, 135, 143, 156
  worldwide, 133
  (*See also* Boston; Buenos Aires; Chicago;
    Dallas/Fort Worth; Hong Kong;
    London; Los Angeles; Miami; Milan;
    Montreal; New York; Orlando; Paris;
    Rio de Janeiro; São Paulo; Tokyo;
    Washington)
Multi-functional facilities (*see* Shared use)
Mumayiz, S., 657, 670, 690–691, 758, 763
Mundra, A., 500, 511, 517, 519, 556
Munich, 42, 67, 69n, 157, 202, 207, 222, 282,
  283, 316, 323–324, 566, 569, 581, 583, 683,
  704, 706, 729
Municipal bonds, 562
Murphy, R., 496
Myers, S., 629, 653, 780, 794n, 801, 814, 818

Nafta, 599
Nagoya/Chubu, 6, 3, 61, 68, 168, 577
Nairobi, 84, 626, 742
Naples (Italy), 20, 42
Narita (*see* Tokyo)
NASA, 549, 550, 556
NASA Space Center, 163
National Airlines, 99, 113
National Association of State Aviation Officials,
  44
National Plan of Integrated Airport Systems (*see*
  NPIAS)
National Route Program, 525
National Transportation Safety Board (*see* NTSB)
National Weather Service, 514
Nationalization, 94
Navigation, 31, 501, 503, 517–522
Neely, J., 817, 818
Negotiation, 573
Nelkin, D., 213
Net present value, 169, 788
Net value, 284
Netherlands, 8, 38n, 67, 233, 277, 704
Networks
  cargo, 66
  hub-and-spoke (*see* Hub-and-spoke)
  international, 66
  national, 66
  regional, 66
Networks, on-airport, 717–726
  capacity of, 724–726
  multi-stop systems, 724

Neuman, F., 550, 556
New entrants, 470
New Hampshire, 133
New Jersey, 140
New large aircraft (*see* NLA; Airbus A380)
New Orleans, 145
New South Wales, 473
New York, 129, 135, 137–138, 148–149, 204,
  219, 510, 513, 548, 622, 686, 703, 855, 856,
  858
  /Islip, 148
  /Kennedy, 6, 7, 20, 24–25, 48, 109, 115, 140,
    196, 203, 233, 242, 280, 316, 318, 394,
    401, 436n, 454, 474, 513, 586, 597, 599,
    601, 696, 702–703, 706–708, 711, 713,
    858
  /LaGuardia, 114, 140, 203, 242, 315–316,
    325–326, 401, 421, 474, 491, 436n, 454,
    513, 530, 577, 713, 858, 861
  /Newark, 7, 24, 69, 79, 125, 139–141, 160,
    162, 210, 316, 318, 321, 401, 436n, 454,
    466, 468, 474, 530, 536, 587, 650, 713,
    717–718, 723, 816, 855–856, 858
  /Stewart, 148
  /White Plains, 147–148
  (*See also* Port Authority of New York and
    New Jersey)
New York Air, 114–115
New York, Metropolitan Transit Authority, 706
New Zealand, 771
Next generation air/ground communications
  (NEXCOM), 548
Newark (*see* New York)
Nichols, D., 718, 737
Nishimura, T., 75, 90
NLA (New Large Aircraft), 21, 71, 80, 817
  (*See also* Airbus A380)
Noise, 165, 202
  abatement procedures, 195–196
  budgets, 200
  certification of aircraft, 180, 182, 200
  charge, 263
  contours, 178, 180
  cumulative measures, 175–177
  effects on neighbors, 178
  goals, 169
  human response to, 178
  impact (on capacity), 368–400, 405
  measurement, 170–186
  mitigation, 169, 186–189
  monitoring systems, 186–190
  physics of, 169
  single-event measures, 172–175, 184
Noise and number index, 180
Noise exposure forecast, 180
Nolan, M., 500, 545n, 546, 556
Nonaeronautical revenue, 268–270, 238
Non-airport revenue, 270
Noncoordinated airport, 469
Normal distribution, 631, 770
Northeast Airlines, 113
Northwest Airlines, 7, 19, 23, 24, 66, 115, 120,
  124, 567
Norton, J., 719, 736
Norway, 705
NPIAS, 61, 63, 67

Oakland, 144
  (*See also* San Francisco)
Objectives, 655
  economic, 606
  management, 643
  performance, 606, 637
Obstacle-free zone, 338, 339
Obstacle limitation surfaces, 356–361
  approach surface, 356, 359, 361

Obstacle limitation surfaces, (*Cont.*)
  balked landing surface, 356, 360–361
  conical surface, 356, 359, 361
  inner approach surface, 356, 360–361
  inner horizontal surface, 356–357, 361
  inner transitional surfaces, 356, 360–361
  takeoff climb surface, 356, 360
  transitional surfaces, 356, 360
Obstacles, physical, 355–361
Obstructions (*see* Clear zones)
O'Connell, F., 683, 684, 690
Odoni, A., 6, 212, 417, 418, 432–434, 448, 456,
  459, 467, 472, 480, 496–497, 543, 554–556,
  736, 753, 763–764, 821, 850
Office of Technology Assessment, 71, 74, 90
Official Airline Guide, 145
O'Hare (*see* Chicago)
Olivera, A., 746, 777
Olympic Airlines, 113, 160
Oneworld (*see* Airline alliances)
Ontario (*see* Los Angeles)
Ontario, Province, 109, 127
Open skies (*see* Deregulation)
Operating revenues, 794
Operational Evolution Plan, 546
Operations, management of, 606
Opportunities (*see* SWOT analysis)
Opportunity cost, 781
  of capital, 781
Opposition to airports (*see* Environmental
  impacts)
Options, 812–817,
  financial, 629
  as insurance, 813
  (*See also* Real options)
Options analysis (*see* Real options analysis)
Organization, 215–250
Organizational structures, 225–231
  flat, 226–227
  line units, 226–227
  operating units, 228
  pyramidal, 227–228
  staff units, 226
  supporting units, 226
Origin-destination matrix (*see* Matrix)
Originating traffic, 134, 157
  minimum levels of, 134
  thresholds, 134–135, 142, 156
Orlando, 136, 137, 148
  /International, 63, 163, 315, 318, 327–328,
    401, 436n, 632, 713, 716n, 723
  /Sanford, 63, 111, 144, 147–148, 163, 612,
    716n
Orly (*see* Paris)
Osaka, 61, 68, 69n, 135, 153, 139, 141, 155, 158,
  705
  /Itami, 141, 145, 155–156, 158
  /Kansai, 61, 141, 145, 158, 168, 316, 320, 471,
    570, 577, 612, 655, 702, 705, 711, 723
Oslo, 218, 696, 705–706, 711
  /Tore, 66
Oster, C., 17, 27, 98, 127
Otis, 721
Ottawa Airport Authority, 108, 127
Oum, T. H., 480, 497
Outer main gear wheel span, 301
Outer marker, 518
Ownership, 217–225
Ownership rights, 101
  management control, 101
  residual income, 101–103, 106–107, 217
Oxford University, 45

Pacific Consultants International, xxv
Pacific Rim, 229, 369, 503, 514
Paddington Station, 695, 706–707, 709

Pagliari, R., 474, 497
Palm Beach (*see* Miami)
Palmdale (*see* Los Angeles)
Pan American Airlines, 99, 113
Panero, J., 673, 691
Parallel runways (*see* Runways)
Paris, 113, 129, 132, 135, 137, 155, 222, 273,
  510, 697, 704
  /Beauvais, 133, 155
  /de Gaulle, 6, 7, 36, 41, 52, 66, 69, 100, 123,
    138, 145, 161, 316, 318, 327, 402, 578,
    583, 584, 587, 600, 602, 606, 696, 704,
    706, 711, 718, 728, 731, 816
  /Le Bourget, 132
  /Orly, 41, 66, 69, 131, 138, 145, 155–156, 577,
    704, 706, 723
  (*See also* Aéroports de Paris)
Park, R., 288, 291, 477, 496
Parking, 50, 105, 582, 693, 694, 702, 706,
  712–717
  consolidated rental car facility, 717, 718
  employee, 714
  garages, 21, 163, 642, 678, 714
  long-term, 714, 716
  rental car, 678, 714, 716–717
  restrictions on, 209
  short-term, 210, 685, 687, 713–714
  structured, 714, 715–716
  valet, 715
Parliament, 43
Partnerships,
  public-private, 93, 100, 107
Passageways, 606, 661, 668
  capacity of, 645
  crossing, 646
  secure, 623, 662
  space requirements for, 643–649
Passageways, effective width, 647–649
  counterflow effects, 648
  edge effects, 647–648
Passenger buildings, xxix, 22, 36, 70, 87, 96,
  104, 161, 559–604, 628, 650, 694, 714, 718
  centralized, 559, 586, 599
  configuration of, xxix, 88, 122, 352–354, 559,
    600
  connected, 116
  consequences of, 560
  decentralized, 559, 561, 586, 587, 599
  definition of, 568
  design of, 605–654, 612
  detailed design, 655–691
  evaluation of, 587–600
  international, 113, 115, 116, 598
  (*See also* Finger piers; Hybrid concourses;
    Linear concourses; Midfield
    concourses; Satellite concourses;
    Transporter concourses)
Passenger buildings, perspectives on
  airlines, 568–569
  government services, 573
  owners, 570–572
  passengers, 565–568
  retail, 572–573
Passenger buildings, requirements for, 563–574
Passenger Facility Charge (*see* PFC)
Passenger handling, 266
Passenger service charge, 264–265
Passengers, 97, 703
  arriving, 610, 685
  business, 565, 651, 678
  charter, 611
  departing, 611, 644, 672, 685
  disabled, 648
  domestic, 565, 611, 618, 678, 771
  international, 565, 573, 599, 611, 618, 642,
    742, 744–745

Passengers, (*Cont.*)
  originating, 156, 611, 696
    (*See also* Originating traffic)
  shuttle, 611, 651
  terminating, 696
  transfers, 566, 600, 610, 672
    (*See also* Traffic, transfer)
  vacationers, 565, 597
Passengers, arrival patterns of, 618
Passengers/flight, 741, 745
Passport control, 52, 607, 611, 641, 680, 742, 744
Paullin, R., 618n, 654
Paull, G., 554
Pavaux, J., 97, 127
Payback period, 796
Peak-hour, xxix, 608–609, 851
  aircraft movements, 860
  analysis, xxix, 608, 851–863
  arriving passengers, 861, 862
  basis for design, 608–609
  conversion coefficients, 854, 856
  departing passengers, 861, 862
  forecasts, 854
  passengers, 686
  pricing, 46
  traffic, 117
  typical, 851
Peak-hour/peak-day ratio, 610
Peak loads, 609–611
Peak period pricing, 476
Peaking patterns, 854–860
  daily, 375
  hourly, 855
  monthly, 855
  peaking ratio, 858
  seasonal, 375
  stability of, 855
Peaks, decreasing, 609–610
Peaks, drivers of shared use (*see* Shared use,
  drivers of)
Pearson, L., 496
Pedestrians, 572
Penner, J., 214
Pennsylvania Station, 708
PEOPLExpress, 79, 113–115, 162
People movers (*see* Automated people
  movers)
Performance criteria (*see* Design criteria)
Performance, measures of, 48
  (*See also* Benchmarking)
Peru, 20, 101
Petroleum prices, 73–74
PFC, 8n, 244, 248–249, 264
Philadelphia, 24, 79, 81, 113, 116, 204, 321, 401,
  436n, 454, 561, 615, 629, 703, 706, 713
Philippines, 101, 705
Phoenix, 204, 317, 318, 401, 436n, 698,
  713
Physical obstacles (*see* Obstacles, physical)
Piano, R., 570, 655
Pilgrimage, 566
Pilling, M., 730, 731, 737
Pilot reporting, 514
Pinch, T., 31n, 55
Pindyck, R., 770, 778
Pittsburgh, 120, 280, 401, 436n, 564, 567, 572,
  580–581, 713, 716, 723
Plans
  master, 60–64, 78, 742, 765
Planning, 103
  bottom up, 61, 67
  contingency, 141, 803, 804
  directive, 68
  master, 59, 81, 562
  regional, 59, 107,
  strategic, 60, 64–70, 81

Planning, (*Cont.*)
  top down, 60, 80
  (*See also* Dynamic strategic planning)
Planning factors, 674
PMM, 644, 645
Point-to-point service, 118, 119
Poisson process, 436n, 626, 823–824, 844,
  848
Political power, central, role of, 29
Pollution, air, 167, 202–206, 209
  dispersion models for, 205
  effect on health, 202
  smog, 203
Pollution, air, constituents of
  carbon monoxide, 204, 210
  hydrocarbons, 203–204
  nitrogen oxides, 203–204
  ozone, 210
  sulfur oxides, 203
  volatile organic compounds, 203
Pollution, air, sources of
  aircraft operations, 203
  ground service equipment, 204
  fuel storage, 204
  motor vehicles, 204
Pollution concentration, ambient, 204
Pollution, groundwater, 206–209
  discharge permits for, 209
  fuel leaks, 206, 208
  storm water, 206, 208–209
Pollution, mitigation measures, 205–206
  delay reduction, 205
  monitoring, 205
  towing of aircraft, 205
Pollution, noise (*see* Noise)
Pooling agreements, 15, 97
Port Authority of New York and New Jersey, 69,
  95, 109, 140n, 466, 696
Port of entry, 573
Port of New York Authority, 140
Porter, M., 64, 91, 149, 165
Portland, OR, 698
Portugal, 159, 273, 277
Positive controlled airspace, 504
Powell, J., 554
Powell, K., 575, 604
Practical annual capacity (*see* Airside capacity)
Practical hourly capacity, 371, 448, 449
  (*See also* Airside capacity)
Prague, 118
Pratt, D., 801
Precipitation (*see* Weather conditions)
Precision approach runway, 312
Precision instrument approaches, 517–522
Preferential runway systems, 196, 197
Present value, 782
Prices, 94
Pricing, 104
  flexible, 98
  (*See also* Monopolies; Marginal cost pricing)
Primary radar (*see* Primary surveillance radar)
Primary surveillance radar, 502, 514
Private delay cost, 477
Privatization, xxix, 9, 13, 15–17, 68–69, 93–128,
  217
  airlines, 17, 98, 99
  airports, 16, 98, 100
Probability, estimation of, 810
Productivity, 235, 568, 613
Profits, 101
Property acquisition, 198
Psaraftis, H., 415, 433
Public inquiry, 42
Public interest, 100, 106
Public participation programs, 190–191
Public service, 574

Public transport, 209, 574, 586, 816
  (*See also* Airport access)
Pujet, N., 212, 550, 556
Pushcarts, 646
Puts, 629, 814

Qantas, 17, 19
Quality of service, 608
Queue discipline, 393
  (*See also* Queuing systems, priority
    discipline; Sequencing)
Queue length, 667, 669
Queues, xxix, 9, 676, 819–850
  dynamic behavior of, 834
  fairness, 832
  psychology of, 831–833
  snake queues, 676, 677, 827
Queuing networks, 822
Queuing systems, 446, 695, 821–822
  equilibrium conditions, 445, 842
  expected values, 443, 843
  level of service, 828–832
  Little's law, 843
  long-term behavior, 842–847
  priority discipline, 826
  queue capacity, 828
  reliability, 830–831
  short-term behavior, 833–834
  standard deviations, 443, 848
  steady state, 842
  variability, 443, 829
  variance, 443, 830
Queuing theory, 820–821, 842–847

Radar, 502
Radio frequency identification (*see* RFID)
Raiffa, H., 806, 818
Rail (*see* Airport access; High-speed rail)
Railroad stations, 572, 642, 662
  (*See also* Airport access)
Raleigh-Durham, 113, 117, 122, 156
Ralph M. Parsons Co., 654
Ramp handling, 266
RAMS model, 750, 753, 754
Random variables, 370, 626
Range of airfield capacities, 400–408
Rationing by schedule (RBS), 532
Raymond, W.-T., 759, 763
Real options, 144, 718, 726
  definition of, 629, 813
  maintaining of, 159
Real options analysis, xxix, 60, 616, 629, 803,
    812–817
Reasonable rate of return, 258, 781
Recycling (*see* Deicing fluids)
Reentrant corners, 593
Regional Air Navigation Plan, 267
Regional development, 104
Regional planning (*see* Planning)
Regression analysis, 769, 776
Regulation, 46, 93, 233–242
  price, 105
  price caps, 235–237
  quality of service, 105
  (*See also* Monopolies)
Reiss, S., 654
Reliability, 120, 659, 701
Remaining value, 284
Remote stands (*see* Aircraft stands, remote)
Rental cars, consolidated parking facility (*see*
    Parking)
Rentals, 269–270
Rerouting, 528
Residual system, 238, 241–242, 284
Residual income (*see* Ownership rights)
Restaurants (*see* Concessions)

Return on assets, 259
Return on investment, 613
Revenue bonds (*see* Bonds)
Revenue center, 253–257, 283
Revenues, 96, 103, 238–239, 242n, 560
  "diversion of", 102
  operating, 794
RFID, 651, 731
Rhode Island, 133
Richetta, O., 543, 556
Rifkin, R., 554
Righi, L., 432, 763
Rights (*see* Ownership rights)
Right-of-way, 649, 718, 816
Rio de Janeiro, 136, 137
  /Galeao, 146, 157
Risks, 19, 129
  analysis of, 59, 85
  investment, 112
  sharing, 103, 116
Rivas, V., 248, 250
R. L. Brown Associates, 681, 691
Robusté, F., 734, 737
Rojas-Guzman, J., 712
Rome, 16, 42, 163, 222, 273, 282–283, 382–383
Rose, R., 56
R-squared measure, 770, 771
Rubinfeld, D., 770, 777
Ruijgrok, G., 214
Rules of thumb for
  gate requirements, 626
  ground handling, 282
  daily traffic, 609
  peak-hour traffic, 609, 632
  runway movements, 429
Runway capacity envelope, 418
Runway configurations, 377, 455
Runway occupancy time, 396
Runway visual range (RVR), 518
Runways, 49, 104, 115, 116, 295, 370–400
  active, 122, 377
  blast pads, 336
  capacity (*see* Airside capacity)
  classification, 311–312
  clearway, 330–331
  close parallel, 320
  close-spaced, 386
  converging, 387
  designation, 311–312
  declared distances for, 330–332
  dependence of operations, 377
  diverging, 387
  displaced threshold, 194, 330
  exits, 347, 349, 396–397
  geometric layouts, 376–377
  geometry, 336–343
  high-speed exits, 194, 349, 397
  holdlines, 340
  intersecting, 312, 325, 326, 387
  length, 328–335
  medium-spaced, 321
  noninstrument , 311
  nonprecision, 311
  object-free area, 338
  parallel, 123, 321, 384, 581
  precision object-free area, 339
  protection zone, 339, 340
  resurfacing, 733
  safety area, 336, 345
  separation from other parts of airfield,
    340–341
  shoulders, 336
  staggered, 49, 324, 386
  stopway, 330, 332
  threshold, 332, 381
  usability, 332–334

Runways, (*Cont.*)
  vertical profile, 341–343
  (*See also* Clear zones)
Russia, 71, 527
Ryanair, 66, 79, 111, 118, 154–155, 163

Saarinen, E., 570
Sabena, 16, 99, 113
Sabre, 8
Sacramento, 99, 118–119, 210, 698
Safety (*see* Airline accidents)
Saffarzadeh, M., 688, 691
St. Louis, 114, 125, 136, 138, 242, 319, 401,
  436n, 454, 524, 561, 594, 614, 703
  /Lambert, 318, 698, 703, 706, 713
  /Mid-America, 136
Salomon, Smith Barney, 567
Salt Lake City, 125, 401, 436n, 567, 698, 713
Salvage value, 788
San Diego, 320, 401, 698, 713
San Francisco, 67, 95, 99, 122, 129–130, 132,
  135, 137, 139, 149, 155, 205, 219, 280, 696,
  703
  /International, 6, 7, 24, 95, 104, 116, 133,
    144, 168, 210, 219, 317–318, 328, 401,
    436n, 530, 536, 564, 577, 642, 683, 698,
    700, 703, 711, 713, 717–718, 723, 726,
    730, 858
  /Oakland, 24, 132–133, 139, 144, 155, 162,
    696, 698, 700, 705
  /San Jose, 132–133, 155, 698
San Jose (*see* San Francisco)
Santiago, 143
São Paulo, 136–138, 734
  /Garulhos, 157
Sapporo, 705
SAS, 17, 125
Satellite concourses, 87, 352, 354, 559, 577–579,
  601
Saturation capacity (*see* Airside capacity)
Scandinavia, 125
Scenarios, 82, 85, 574, 775–776, 803, 805
  extraordinary, 631
  normal, 631
Schedule coordination, 428, 469–475
Schedule, of flights, 672
Schengen, 680, 744
Schonfeld, P., 670, 691
Schubert, M., 556
Schwartz, E., 817–818
Scotland, 38
SEA (*see* Milan, airport company)
Seasonal patterns (*see* Peaking patterns)
Seasonal variations, 595
Seat-mile costs, 118
Seattle, 132
  /Bellingham, 99
  /Boeing Field, 132
  /Tacoma, 242, 280, 318–319, 321, 401, 436n,
    454, 578, 579, 601, 698, 713, 723
Second Bangkok Airport, 87, 582, 586, 630
  (*See also* Bangkok)
Second Sydney Airport, 76–77, 144, 159, 803,
  806–809
  (*See also* Sydney)
Secondary surveillance radar, 503, 514, 516
Sectors, 522
Secretary of Transportation, 91
Security, 13, 22, 559, 566, 571, 584, 651, 657,
  674, 677
  charge, 265–266
  check points, 80, 49, 52, 572, 589, 616, 642,
    656, 657, 678–680
Security systems, baggage, 728–729
  reconciliation, 728
  screening, 649, 728

Self-handling, 278
Sendai, 67
Seneviratne, P., 648, 654, 659, 691
Sensitivity analysis, 810
Seoul, 129, 135
  /Gimpo, 7, 318, 705
  /Incheon, 6, 323, 471, 705
Separation requirements (*see* Air traffic
  management)
Sequencing, 389
  of aircraft, 415
  of movements, 395
  strategy, 396
Service rate, 657, 666, 668, 823
Service times, 825
  constant, 825
  negative exponential, 825
Settling ponds, 209
Shanghai, 136–137
Shannon, 221
Shapiro, G., 433, 763
Shapiro, P., 737
Shared knowledge base, 535
Shared use, 84, 116, 576, 599, 606, 608, 611,
  612, 624, 816
  as insurance, 629
  of gates, 163, 576, 611, 613, 625, 670
  of lounges, 576, 593, 613, 618, 681
Shared use, drivers of
  peaking, 612, 613–626
  uncertainty, 612, 626–634
Shared use, savings due to, 626, 634
  shares (*see* Frequency share;Market
    share)
Shehata, M., 618n, 654
Shen, L., 722, 737
Shenzhen (*see* Macao)
Shi, W., xxv
Shock absorbers, 112
Shopping, 559, 571, 584, 672, 681
Shopping malls, 564
Shumsky, R., 736
Shuttle service, 8, 114, 152, 156, 683
Sidewalks, moving (*see* Moving sidewalks)
Sigafoos, R., 99, 127
Signatory airlines, 242
SIMMOD model, 457, 750–754, 756, 757
Simpson, R., 763
Simulation, 617, 645, 656, 659, 669, 749
  limitations of, 672
  (*See also* Computer models)
Singapore, 116, 122, 218, 316, 318, 323, 572,
  606, 615, 622, 638, 642, 681, 683, 705
Singapore Airlines, 24
Singapore/Changi, 6, 82, 87, 224, 705, 720,
  723
Single till, 237–241, 284
Sites (*see* Airport sites)
SLAM, 757
Slots, 428, 487, 489–492
  allocation of, 53
  auctions, 491–492
  exemptions, 466
Snake queues (*see* Queues)
Social discount rate, 781
Soil settlements, 655
Solis, P., 737
Sorting systems, 17
  (*See also* Baggage systems)
Sound, 6
  barriers, 194
  frequency, 172
  insulation, 191, 198
  pitch, 172
Sound exposure level, 173
South Africa, 16, 241

South African Airports Company, 16, 42n, 222, 235n
Southwest Airlines, 7, 23, 66, 79, 99, 111, 116, 118, 131, 154, 155, 158, 162, 163, 561, 599, 615, 629
Soviet Union, 38, 71, 135
  (*See also* Russia)
Space, capacity of, 643
Space requirements
  estimates of, 634, 642–643
  waiting areas, 636–643
Space requirements, formulas for (*see* Formulas)
Spain, 221
Spake, J., 728, 737
Spine road, 600
Split operations, 116
Spreadsheets (*see* Computer models)
S-shaped relationship, 150, 152–153
  (*See also* Frequency share; Market share)
Stage 1, 2, 3 aircraft, 182, 200, 263, 436n
Staggered runways (*see* Runways)
Stairways, 589, 606, 643, 646
Stakeholders, 39, 49, 559, 563, 574, 587, 600
  airlines, 559, 563
  owners, 559, 563
  passengers, 559
  stores, 563
Stamatopoulos, M., 418, 420, 432–433, 752, 763
Standard and Poor's, 247–248
Standard atmospheric conditions, 302
Standard busy rate, 608, 610n, 853
Standard deviations, 78, 626, 631, 633
  (*See also* Forecasts)
Standard Terminal Automation Replacement System (STARS), 514
Standards,
  capacity, 96
  for design, 300–312, 605, 636, 640, 641, 655, 657
  international, xxi, 30
  national, xxi
Stands (*see* Aircraft stands)
Stansted (*see* London)
Star alliance (*see* Airline alliances)
Statistics, discrepancies in, 146
Steiner, H., 780, 801
Steuart, G., 617, 626, 654
Stochastic analysis, 617
Stockholm, 218, 264, 705–706
  /Arlanda, 705
Storm-cell hazards (*see* Weather conditions)
Strategic planning, (*see* Dynamic strategic planning; Planning, strategic)
Strategy, 115, 803
Strike price, 814
Subsidies, 68
Substitutions, 537
Suleiman, E., 38n, 45, 56
Sullivan, W. G., 780, 801
Surcharges (on landing fees), 481
Surface operations, 195–197
Surface radar, 515
Surface Management System (SMS), 550
Surveillance, 501, 503, 514–517
Sustained capacity (*see* Airside capacity)
Svrcek, T., 595, 604
Swan, W., 319, 365
Sweden, 705
Swedish, W., xxv, 418, 420, 434, 751, 764
Swing gates (*see* Departure gates)
  (*See also* Shared use)
Swing space, 613
Swissair, 99, 113
Switzerland, 16, 42, 705
SWOT analysis, 65, 86

Sydney, 76, 91, 144, 241, 316, 321, 471, 473, 491, 601, 607, 704, 805, 808, 816
  (*See also* Second Sydney Airport)
Sylvan, S., 191, 198, 199n, 205, 206, 214
Symons, T., xxv
Systems approach, 9, 26
Systems planning, xxix, 65–70
Szyliowicz, J., 649

TAAM model, 457, 750–754, 756, 757
Tail-to-tail (*see* Baggage sorting)
Tailwinds (*see* Winds)
Taipei, 136, 137, 143
  /Chiang Kai Shek, 35, 135, 143, 157
  /Sung Shan, 135
Taiwan, 135
Takeoff distance available (TODA), 330, 332
Takeoff run available (TORA), 330, 332
Tampa, 326, 401, 579, 698, 713, 723, 724
Taneja, N., 150, 166
Taxes, 96, 102, 220, 250, 286
Taxis (*see* Airport access, modes)
Taxiways, 116, 343–351, 422
  aircraft stand taxilanes 345
  apron, 345
  capacity of, 422, 594
  curved segments, 347
  design standards, 345–346
  dual, 351, 355
  exit taxiways (*see* Runways)
  full-length, 422
  high-speed exit taxiways (*see* Runways)
  holding bays, 347, 351
  intersecting, 347
  on bridges, 347–348
  parallel, 580
  taxilanes, 345
TBI, 63, 223, 224
Technical change, 171
Technical experts, importance of, 29
Technology, social construction of, 29
Telecommunications, 649
Terminal area navigation fee, 262
Terminal airspace control center, 506, 507, 510–514
Terminal control airspace, 505
Terminal Doppler weather radar (TDWR), 548
Terminal One, New York/Kennedy, 49
Terminal 4, London/Heathrow, 52
Terminal 5, London/Heathrow, 42
Terminal Radar Approach Control (TRACON), 506
Terminals (*see* Passenger buildings)
Texas, 98, 160, 771
TGV, 52
  (*See also* High-speed rail)
Thai Airlines, 19
Thailand, 139, 705
Third-party handling, 279
Thorpe, K., 42, 56, 104, 128
Threats (*see* SWOT analysis)
Thresholds (*see* Originating traffic)
Thunderstorms (*see* Weather conditions)
Tilt trays (*see* Conveyor belts)
Time-space concept, 640
Time above, 180
Times between
  arrivals (*see* Interarrival times)
  departures, 618
TNT, 155
TRL (Transportation Research Laboratory), 222, 250, 291
Toilets, 674
Tokyo, 129, 130, 132, 135, 137, 145, 155, 160, 705
  /Haneda, 7, 131, 145, 168, 316–318, 705

Tokyo, (*Cont.*)
   /Narita, 145, 155, 156, 320, 471–473, 564,
      566, 579, 584, 601, 683, 694, 696, 700,
      702, 705–707, 711, 721, 723
   runways, 43
   /Yokota, 132, 160
Topeka, 524
Toronto, 99, 136, 137, 143, 159
   Airport Authority, 108, 124
   /Hamilton, 159, 163, 743, 745
   /Pearson, 16, 116, 201, 208, 280, 318, 402,
      565, 610, 612, 615, 628, 714, 726
   /Pickering, 159
Tourism, 14
Towing, of aircraft, 88
Trade-offs, 606, 615, 616
Traffic, 607–611
   characteristics of, 559
   domestic, 84, 600
   doubling of (*see* Growth)
   international, 84, 600, 665
   originating and terminating, 696
   pedestrian (*see* Pedestrian traffic)
   seasonality, 559
   transborder, 612–613
   transfer, 82–83, 161, 559, 590, 599, 632, 684
   waves or banks (*see* Waves of traffic)
Traffic alert and collision avoidance system
      (TCAS), 516
Traffic allocation, 130
   impracticality of, 139
Traffic concentration, 129
   at airports, 149
   at primary airports, 152–154
   on routes, 150, 152
Traffic handling, 266
Traffic management, 9
Traffic management advisor (TMA), 549
Traffic management unit (TMU), 527
Traffic mix, 394, 599, 626
   homogeneous, 393
   international/domestic proportions, 600
Traffic, originating (*see* Originating traffic)
Traffic peaks, 560
Traffic surges, 663
Traffic volatility (*see* Volatility)
Train stations (*see* Railroad stations)
Transborder traffic (*see* Traffic)
Transfer hubs, 8, 19, 82, 94, 98, 100, 112,
      118–124, 138, 568, 574, 696, 723, 730, 734
   criteria for, 122–124
Transfer operations, 49, 161, 559, 586
Transfer pattern, 589
Transfer traffic (*see* Traffic)
Transponders, 516
Transport Canada, 96, 108, 128, 745
Transportation Research Board, 753, 764
Transporter concourses, 352–353, 559
Transporters, 117, 584–586
   as insurance, 811
   economics of, 586, 588, 595–598
Travaux Publics de l'Etat (*see* Corps)
Traveling salesman problem (TSP), 553
Treeage, 808, 811–812, 818
Treeplan, 808
Trends, 93
   (*See also* Forecasts)
Trigeorgis, L., 629, 654, 815, 818
Trolleys (*see* Baggage trolleys)
Trump Airlines, 114
Tupolev, 30n
Turkey, 527
Turnaround times, of aircraft, 425, 661, 830
TWA, 17, 23, 113, 114, 125, 561, 594
Tyrolean, 19

Uncertainty, driver of shared use (*see* Shared
      use, drivers of)
Uncontrolled airspace, 505
Underground, London, 709
Unit charges, 256
United Airlines, 7, 8, 23, 45n, 109, 120–121, 124,
      131, 137, 145–146, 566–569, 581, 586, 601,
      614, 631, 681, 735
United Kingdom, 46, 48, 180, 258, 504, 654, 680,
      691, 705, 744, 753, 853
   (*See also* Britain)
United Nations, 30, 214
United States, 4, 5, 7–8, 11, 40, 43, 47, 49, 53,
      61, 66, 68, 69n, 74, 94, 97, 99, 105, 109,
      113, 118, 124, 154, 184–185, 196, 204, 209,
      218–219, 242, 245n, 258, 265, 269, 286,
      295, 311n, 372, 378, 396, 400, 413, 436,
      451, 461, 474, 489, 503, 505, 513, 562, 564,
      570, 622–623, 626, 642, 657, 673, 678, 684,
      703, 712, 728–729, 744, 767, 772, 853, 830
United States Department of Transportation,
      436, 459, 475, 489–490, 497
United States Federal Aviation Authority (*see*
      FAA)
Unit terminal, 586, 615
UPS, 7, 14, 23, 24, 66, 70, 99, 155, 163, 743
US Airways, 7, 23, 79, 109, 113–114, 116, 120,
      124, 152, 162, 561, 567, 599, 615, 629, 631
User charges, xxix, 233, 251–291
   aeronautical, 261
   non-aeronautical, 268–270
User-preferred routes, 524
Utilities, 270
Utilization rates, 595, 612
Utilization ratio, 444–445, 828

Valet parking (*see* Parking)
Van de Horst, H., 38n, 56
Van Nuys (*see* Los Angeles)
Van Wolferen, K., 38, 56
Vancouver, 99, 280
   airport company, 20
   International Airport Authority, 108, 127
Varig, 19
Vehicles, apron, 37
Vehicle-trips/passenger, 698
Venkatakrishnan, C., 415, 434
VHF omnidirectional range (VOR), 524
Vickers, J., 124, 128
Vickrey, W., 477, 497
Victor airways, 524
Video conferencing, 13
Vienna, 16, 101, 222, 224, 236, 238, 273
   airport company, 20, 224
Visibility (*see* Weather conditions)
Visual flight rules (VFR), 321, 389, 504
Visual meteorological conditions, 389, 504
Visual runway, 311
Voice radio, 548
Volatility, 80, 112–118, 146
   of customers, 130
   of traffic, 94, 112, 115–117, 124, 131, 145
Volckers, U., 549, 556
Volpe National Transportation Systems Center,
      536

Wagner, E., 737
Waiting areas, 611, 618, 635, 644, 656, 661
   (*See also* Departure gates)
Waitz, I., xxv
Wake turbulence, 379
Wake vortex, 379
Wales, 38
Walking distances, 582–585, 588–594, 659, 671,
      719

Waiting time, 446, 667
Walking time, 280
Walls, moveable, 623
Wang, A., 501, 554
Warburg Dillon Read-UBS, 240, 250, 291
Washington, 130, 132, 135, 137–138, 152, 160,
    204, 602, 743
    /Andrews AFB, 132,160
    /Baltimore (*see* Baltimore/Washington)
    /Dulles, 6, 36, 69, 95, 120, 121, 124, 131, 137,
        145, 160, 401, 436n, 567, 570, 580, 586,
        630, 703, 713
    /Reagan, 95, 137, 144, 196, 206, 316, 401,
        436n, 474, 570, 572, 577, 617, 645, 683,
        703, 706, 711, 713, 717
    (*See also* Metropolitan Washington)
Water quality (*see* Pollution, groundwater)
Waves of traffic, 122, 123, 394n, 617, 618
Weather conditions,
    ceiling, 388–391
    fog, 530
    gusts, 548
    icing, 390
    lightning, 548
    microbursts, 548
    precipitation, 388–391
    storm-cell hazards, 548
    thunderstorms, 530
    windshear, 548
        (*See also* Winds)
    visibility, 388–391, 518
Weather data, 548
Weather delays (*see* Aircraft delays)
Weather systems, 503
Web site, xxiii
Weber, A., 140, 166
Weinberg, R., xxv
Wellington, 612
Wellwishers (*see* Meeters and greeters)
Westinghouse, 17
Weston, R., xxv
Wheelwright, S., 71, 90, 770
White, J. A., 780, 801
Whitlock, E., 678, 686, 690, 763

WHO (World Health Organization) 178, 214
Wichita, 149
Wildlife management, 211
Wiley, J., 228–229, 250
Williams, R., xxv
Wind coverage, 312–314
Wind rose, 313–314
Windshear (*see* Weather conditions)
Winds, 312, 391
    crosswinds, 312–313, 391
    tailwinds, 312n, 391
Wingrove, W., 433, 763
Wingspan (*see* Aircraft wingspan)
Wirasinghe, S., xxvi, 618, 626, 648, 652–654
WITNESS®, 758
Wolff, R., 821, 850
Woodhead, W., xxv
Workload unit, 133
Work rules, 48
Workers' rights, 29
World Bank, 244, 741
Wright, P., 9, 26, 180, 212, 562, 603, 610, 652,
    712, 736
WTRG Economics, 74, 91

Xiamen, 225
X-ray (*see* Baggage screening)

Yarrow, G., 128
Yield Management System, 8, 13
Yokota (*see* Tokyo)
Young, S., 680, 691

Zelnick, M., 673, 691
Zhang, Y., 480, 497
Zhao, F., 722, 737
Zografos, K., 432, 764
Zoldos, R., xxv
Zoning, 159, 191
Zuniga, S., 746, 777
Zürich, 16, 36, 42, 196, 205, 264, 316, 328, 584,
    705–706, 711
    airport company, 16, 225

About the Authors

Dr. Richard de Neufville is Professor of Engineering Systems and of Civil and Environmental Engineering, and Founding Chairman of the Technology and Policy Program at MIT. He has worked extensively for Boston, Dallas/Fort Worth, London, Mexico City, Miami, Paris, Sydney, Kuala Lumpur, Bangkok and many other airports and Civil Aviation Authorities worldwide. His expertise is in forecasting, risk management, competition between airports, and the configuration and design of passenger buildings. He received the FAA Award for Excellence in Aviation Education (with Prof. Odoni), the MIT Award for the Most Significant Contribution to Education, the French Chevalier des Palmes Académiques, and an honorary doctorate from the Technical University of Delft. He has also had White House, Guggenheim, and US-Japan Leadership Fellowships.

Dr. Amedeo R. Odoni is the T. Wilson Professor of Aeronautics and Astronautics and of Civil and Environmental Engineering and Co-Director of the Global Airline Industry Center at MIT. He specializes in the use of operations research and other quantitative methods in planning, designing, operating and evaluating airport and air traffic management systems. Over the years, he has consulted at Amsterdam, Athens, Boston, Milan, Munich, New York, Sydney, Stockholm and many other airports, as well as at several Civil Aviation Authorities. He has received the Robert Herman Lifetime Achievement Award of INFORMS for major contributions to Transportation Science and several teaching awards at MIT. He has served as Co-Director of MIT's Operations Research Center, Editor-in-Chief of Transportation Science and Co-Director of the National Center of Excellence in Aviation Operations Research, established by the FAA in 1996.